Lecture Notes in Computer Science   7030

Commenced Publication in 1973
Founding and Former Series Editors:
Gerhard Goos, Juris Hartmanis, and Jan van Leeuwen

Editorial Board

David Hutchison
Lancaster University, UK

Takeo Kanade
Carnegie Mellon University, Pittsburgh, PA, USA

Josef Kittler
University of Surrey, Guildford, UK

Jon M. Kleinberg
Cornell University, Ithaca, NY, USA

Alfred Kobsa
University of California, Irvine, CA, USA

Friedemann Mattern
ETH Zurich, Switzerland

John C. Mitchell
Stanford University, CA, USA

Moni Naor
Weizmann Institute of Science, Rehovot, Israel

Oscar Nierstrasz
University of Bern, Switzerland

C. Pandu Rangan
Indian Institute of Technology, Madras, India

Bernhard Steffen
TU Dortmund University, Germany

Madhu Sudan
Microsoft Research, Cambridge, MA, USA

Demetri Terzopoulos
University of California, Los Angeles, CA, USA

Doug Tygar
University of California, Berkeley, CA, USA

Gerhard Weikum
Max Planck Institute for Informatics, Saarbruecken, Germany

Baoxiang Liu   Chunlai Chai (Eds.)

Information Computing and Applications

Second International Conference, ICICA 2011
Qinhuangdao, China, October 28-31, 2011
Proceedings

 Springer

Volume Editors

Baoxiang Liu
Hebei United University, College of Sciences
Xinhua West Road 46, Tangshan 063000, Hebei, China
E-mail: liubx5888@126.com

Chunlai Chai
Zhejiang Gongshang University
School of Computer Science and Information Engineering
Xuezheng Street 18, Hangzhou 310018, China
E-mail: ccl@mail.zjgsu.edu.cn

ISSN 0302-9743   e-ISSN 1611-3349
ISBN 978-3-642-25254-9   ISBN 978-3-642-25255-6 (eBook)
DOI 10.1007/978-3-642-25255-6
Springer Heidelberg Dordrecht London New York

Library of Congress Control Number: 2011940799

CR Subject Classification (1998): C.2, D.2, C.2.4, I.2.11, C.1.4, D.2.7

LNCS Sublibrary: SL 3 – Information Systems and Application, incl. Internet/Web and HCI

Typesetting: Camera-ready by author, data conversion by Scientific Publishing Services, Chennai, India

Printed on acid-free paper

Springer is part of Springer Science+Business Media (www.springer.com)

Preface

Welcome to the proceedings of the International Conference on Information Computing and Applications (ICICA 2011), which was held in Qinhuangdao, China, October 28-30, 2011.

As future-generation information technology, information computing and applications become specialized, information computing and applications including hardware, software, communications and networks are growing with ever increasing scale and heterogeneity, and becoming overly complex. The complexity is getting more critical along with the growing applications. To cope with the growing and computing complexity, information computing and applications focus on intelligent, self-manageable, scalable computing systems and applications to the maximum extent possible without human intervention or guidance.

With the rapid development of information science and technology, information computing has become the third approach for scientific research. Information computing and applications is the field of study concerned with constructing intelligent computing, mathematical models, numerical solution techniques and using computers to analyze and solve natural scientific, social scientific and engineering problems. In practical use, it is typically the application of computer simulation, intelligent computing, Internet computing, pervasive computing, scalable computing, trusted computing, autonomy-oriented computing, evolutionary computing, mobile computing, applications and other forms of computation to problems in various scientific disciplines and engineering. Information computing and applications is an important underpinning for techniques used in information and computational science and there are many unresolved problems worth studying.

The ICICA 2011 conference provided a forum for engineers and scientists in academia, industry, and government to address the most innovative research and development issues including technical challenges and social, legal, political, and economic issues, and to present and discuss their ideas, results, work in progress and experience on all aspects of information computing and applications.

There was a very large number of paper submissions (865), representing six countries and regions. All submissions were reviewed by at least three Program or Technical Committee members or external reviewers. It was extremely difficult to select the presentations for the conference because there were so many excellent and interesting submissions. In order to allocate as many papers as possible and keep the high quality of the conference, we finally decided to accept 289 papers for presentation, reflecting a 33.2% acceptance rate. And 97 papers are included in this volume. We believe that all of these papers and topics not only provided novel ideas, new results, work in progress and state-of-the-art techniques in this field, but also stimulated future research activities in the area of information computing and applications.

The exciting program of this conference was the result of the hard and excellent work of many individuals, such as the Program and Technical Committee members, external reviewers and Publication Chairs, all working under a very tight schedule. We are also grateful to the members of the local Organizing Committee for supporting us in handling so many organizational tasks, and to the keynote speakers for accepting to come to the conference with enthusiasm. Last but not least, we hope you enjoy the conference proceedings.

October 2011   Baoxiang Liu
                                                                  Chunlai Chai

Organization

ICICA2011 was organized by the Hebei Applied Statistical Society (HASS), College of Science of Hebei United University, and sponsored by the National Natural Science Foundation of China, Northeastern University at Qinhuangdao, Yanshan University and Nanyang Technological University. It was held in cooperation with *Lecture Notes in Computer Science* (LNCS) of Springer.

Executive Committee

Honorary Chair
Qun Lin, Chinese Academy of Sciences, China

General Chairs
Yanchun Zhang, Victoria University, Australia
Baoxiang Liu, Hebei United University, China
Yiming Chen, Yanshan University, China

Program Chairs
Chunfeng Liu, Hebei United University, China
Leizhen Wang, Northeastern University at Qinhuangdao, China
Chunlai Chai, Zhejiang Gongshang University, China

Local Arrangements Chairs
Jincai Chang, Hebei United University, China
Aimin Yang, Hebei United University, China

Steering Committee
Qun Lin, Chinese Academy of Sciences, China
Yuhang Yang, Shanghai Jiao Tong University, China
MaodeMa, Nanyang Technological University, Singapore
Nadia Nedjah, State University of Rio de Janeiro, Brazil
Lorna Uden, Staffordshire University, UK
Xingjie Hui, Northeastern University at Qinhuangdao, China
Xiaoqi Li, Northeastern University at Qinhuangdao, China
Xiaomin Wang, Northeastern University at Qinhuangdao, China
Yiming Chen, Yanshan University, China
Maohui Xia, Yanshan University, China
Chunxiao Yu, Yanshan University, China

	Yajuan Hao, Yanshan University, China
	Dianxuan Gong, Hebei United University, China
	Yajun Zhang, Northeastern University, China
Publicity Chairs	Aimin Yang, Hebei United University, China
	Chunlai Chai, Zhejiang Gongshang University, China
Publication Chairs	Yuhang Yang, Shanghai Jiao Tong University, China
Financial Chair	Chunfeng Liu, Hebei United University, China
	Jincai Chang, Hebei United University, China
Local Arrangements Committee	Li Feng, Hebei United University, China
	Songzhu Zhang, Northeastern University at Qinhuangdao, China
	Jiao Gao, Northeastern University at Qinhuangdao, China
	Yamian Peng, Hebei United University, China
	Lichao Feng, Hebei United University, China
	Dianxuan Gong, Hebei United University, China
	Yuhuan Cui, Hebei United University, China
Secretaries	Jingguo Qu, Hebei United University, China
	Huancheng Zhang, Hebei United University, China
	Yafeng Yang, Hebei United University, China

Program/Technical Committee

Yuan Lin	Norwegian University of Science and Technology, Norway
Yajun Li	Shanghai Jiao Tong University, China
Yanliang Jin	Shanghai University, China
Mingyi Gao	National Institute of AIST, Japan
Yajun Guo	Huazhong Normal University, China
Haibing Yin	Peking University, China
Jianxin Chen	University of Vigo, Spain
Miche Rossi	University of Padova, Italy
Ven Prasad	Delft University of Technology, The Netherlands
Mina Gui	Texas State University, USA

Nils Asc	University of Bonn, Germany
Ragip Kur	Nokia Research, USA
On Altintas	Toyota InfoTechnology Center, Japan
Suresh Subra	George Washington University, USA
Xiyin Wang	Hebei United University, China
Dianxuan Gong	Hebei United University, China
Chunxiao Yu	Yanshan University, China
Yanbin Sun	Beijing University of Posts and Telecommunications, China
Guofu Gui	CMC Corporation, China
Haiyong Bao	NTT Co., Ltd., Japan
Xiwen Hu	Wuhan University of Technology, China
Mengze Liao	Cisco China R&D Center, China
Yangwen Zou	Apple China Co., Ltd., China
Liang Zhou	ENSTA-ParisTech, France
Zhanguo Wei	Beijing Forestry University, China
Hao Chen	Hu'nan University, China
Lilei Wang	Beijing University of Posts and Telecommunications, China
Xilong Qu	Hunan Institute of Engineering, China
Duolin Liu	ShenYang Ligong University, China
Xiaozhu Liu	Wuhan University, China
Yanbing Sun	Beijing University of Posts and Telecommunications, China
Yiming Chen	Yanshan University, China
Hui Wang	University of Evry, France
Shuang Cong	University of Science and Technology of China, China
Haining Wang	College of William and Mary, USA
Zengqiang Chen	Nankai University, China
Dumisa Wellington Ngwenya	Illinois State University, USA
Hu Changhua	Xi'an Research Insti. of Hi-Tech, China
Juntao Fei	Hohai University, China
Zhao-Hui Jiang	Hiroshima Institute of Technology, Japan
Michael Watts	Lincoln University, New Zealand
Tai-hon Kim	Defense Security Command, Korea
Muhammad Khan	Southwest Jiaotong University, China
Seong Kong	The University of Tennessee, USA
Worap Kreesuradej	King Mongkutus Institute of Technology Ladkrabang, Thailand
Uwe Kuger	Queen's University Belfast, UK
Xiao Li	CINVESTAV-IPN, Mexico
Stefa Lindstaedt	Division Manager, Knowledge Management, Austria

Paolo Li	Polytechnic of Bari, Italy
Tashi Kuremoto	Yamaguchi University, Japan
Chun Lee	Howon University, Korea
Zheng Liu	Nagasaki Institute of Applied Science, Japan
Michiharu Kurume	National College of Technology, Japan
Sean McLoo	National University of Ireland, Ireland
R. McMenemy	Queen's University Belfast, UK
Xiang Mei	The University of Leeds, UK
Cheol Moon	Gwangju University, Korea
Veli Mumcu	Technical University of Yildiz, Turkey
Nin Pang	Auckland University of Technology, New Zealand
Jian-Xin Peng	Queen's University of Belfast, UK
Lui Piroddi	Technical University of Milan, Italy
Girij Prasad	University of Ulster, UK
Cent Leung	Victoria University of Technology, Australia
Jams Li	University of Birmingham, UK
Liang Li	University of Sheffield, UK
Hai Qi	University of Tennessee, USA
Wi Richert	University of Paderborn, Germany
Meh shafiei	Dalhousie University, Canada
Sa Sharma	University of Plymouth, UK
Dong Yue	Huazhong University of Science and Technology, China
YongSheng Ding	Donghua University, China
Yuezhi Zhou	Tsinghua University, China
Yongning Tang	Illinois State University, USA
Jun Cai	University of Manitoba, Canada
Sunil Maharaj Sentech	University of Pretoria, South Africa
Mei Yu	Simula Research Laboratory, Norway
Gui-Rong Xue	Shanghai Jiao Tong University, China
Zhichun Li	Northwestern University, China
Lisong Xu	University of Nebraska-Lincoln, USA
Wang Bin	Chinese Academy of Sciences, China
Yan Zhang	Simula Research Laboratory and University of Oslo, Norway
Ruichun Tang	Ocean University of China, China
Wenbin Jiang	Huazhong University of Science and Technology, China
Xingang Zhang	Nanyang Normal University, China
Qishi Wu	University of Memphis, USA
Jalel Ben-Othman	University of Versailles, France

Table of Contents

Computational Economics and Finance

Computational Statistics

Mobile Computing and Applications

Social Networking and Computing

Intelligent Computing and Applications

Internet and Web Computing

Parallel and Distributed Computing

System Simulation and Computing

Implementation of Visualization Analysis Software for Evaluating the Changes of Permafrost on Qinghai-Tibet Plateau

Jiuyuan Huo[1,2] and Yaonan Zhang[1,*]

[1] Cold and Arid Regions Environmental and Engineering Research Institute,
Chinese Academy of Sciences,
Lanzhou 730000, China
[2] Information Center, Lanzhou Jiaotong University,
Lanzhou 730070, China
yaonan@lzb.ac.cn

Abstract. To understand heat transfer process and climatic response process in the Qinghai-Tibet highway subgrade, lots of long-term ground temperature observation systems have been deployed to obtain the periodic temperature values. Contour maps were used to analysis these data for study permafrost changes. This paper is to fully utilize the powerful drawing software Surfer through the embedded programming of Visual Basic 6.0. A visualization and analysis software of permafrost change process has been quickly realized by this hybrid programming method. Compared with the original method, it saves time of developing application program to draw contour maps, but also ensure accuracy and efficiency. It also helps researchers to find characteristics and regular patterns of permafrost change process in Qinghai-Tibet highway easily.

Keywords: Permafrost, Contour Map, Visualization, Surfer Software.

1 Introduction

Qinghai-Tibet Highway, about 1,937 km, is north from Xining, Qinghai, and southern to Lhasa, Tibet. It is mainly located on the permafrost areas, which about 4,000m to 5,000m altitude in the Qinghai-Tibet Plateau. As the presence of permafrost and thick ground ice in permafrost areas, it makes subgrade stability of the Qinghai-Tibet highway have great uncertainty. With the changes in the external environment, the melting of ice in permafrost will lead to rapid and extreme weaken of road foundation, and make a significant impact on the stability of the highway [1], [2], [3].

To understand heat transfer process and climatic response process in the Qinghai-Tibet highway subgrade, a number of long-term ground temperature observation systems have been deployed to obtain the periodic temperature changes of the observed road section [4],[5]. Through these long-term accesses to critical data, it could be possible make accurate grasp of trends and judgments to the stability of the roadbed, and take action to early prevention and diagnosis of engineering diseases.

---

* Corresponding author.

B. Liu and C. Chai (Eds.): ICICA 2011, LNCS 7030, pp. 1–8, 2011.
© Springer-Verlag Berlin Heidelberg 2011

Contour map is a general name of a series of equivalent lines distributed on a map. It is a powerful tool to analysis the spatial characteristics of physical elements, and could grasp the overall features of the spatial variation to promote the analysis and judgments [6]. Especially in geography, large amounts of data are needed to be drawn into the contours map to be used for analysis and research. Thus we took visual analysis of contour map to the collected data of temperature change and facilitate the researchers to discover scientific laws of the change process of frozen soil in a more intuitive manner.

For the powerful capabilities of drawing contour map, Surfer software becomes more and more popular [7]. As it handled with large amounts of data, the way of operating the menu item to processing data and drawing map will result low efficiency. Therefore, drawing contour maps in a timely and efficient manner has been an urgent problem.

In this paper, we want to fully utilize the powerful drawing software Surfer, and call Surfer's drawing functions through the embedded programming of VB6.0. In order to improve the efficiency of plotting of contour maps and provide the researchers an analysis environment, a visualization and analysis system of temperature contour maps of permafrost change process is developed by this hybrid programming of VB and Surfer software.

2   Related Work

The drawing software Surfer has been widely used at domestic and international to plot contour map. It involves in many research fields such as environment, geography, meteorology and others, and has achieved certain results.

Bodhankar adopted the Survey of India toposheet number 64 G/12 as a source of information pertaining to contours, urban features, drainage and lakes and surface water harvesting structures [8]. Surfer software was used in this study to interpolate these data to obtain intermediate contour lines. Using the contour data, a wireframe map was generated to visualize the topography and ground slope variation in three-dimension. It is a very fast and reliable method to evaluate the utility of the proposed structure to be used for surface water harvesting before actual construction.

Vertical electrical resistivity soundings were conducted in order to delineate groundwater potential aquifers in Peddavanka watershed. Contour maps were prepared and interpreted in terms of resistivity and thickness of various sub-surface layers by using computer software Surfer. This study reveals that the weathered and fractured portions in shale and limestone that occur in the southernmost and central portions of the watershed area constitute the productive water-bearing zones categorized as good groundwater potential aquifers [9].

A quantitative method to evaluate the amounts of heavy metals in river sediments is established. Using a section of BT Drainage River as an example, the total amounts of the main pollution indicators and those of their harmful forms are estimated by the Surfer software, which simulates the pollution status within the downstream sediments of the outfall at this section [10].

Zheng et al. achieved the automatic plotting of contour maps by realizing of compatibility of VB6.0 software with Surfer 8.0, a powerful plotting function provided by Surfer software. And a batch processing visualization software for contour maps has been developed. It can improve the efficiency of the post-processing of finite element analysis significantly [11].

3 System Design

Golden Software Surfer (Surfer) is popular software of drawing three-dimensional (3D) graph and contour map. Conventional way of drawing contour maps by Surfer for large quantities of data wastes a lot of labour and time. For convenience to users, Surfer software added automation technology to support the secondary development of other visual and rapid developing program language such as Visual Basic. In this developing environment, Surfer server was running in the background, client application developed by VB can detach from the main control interface of Surfer and could quickly achieve contour mapping. With this development approach, visualization and analysis applications could be achieved quickly and give full play to the ability of Surfer software.

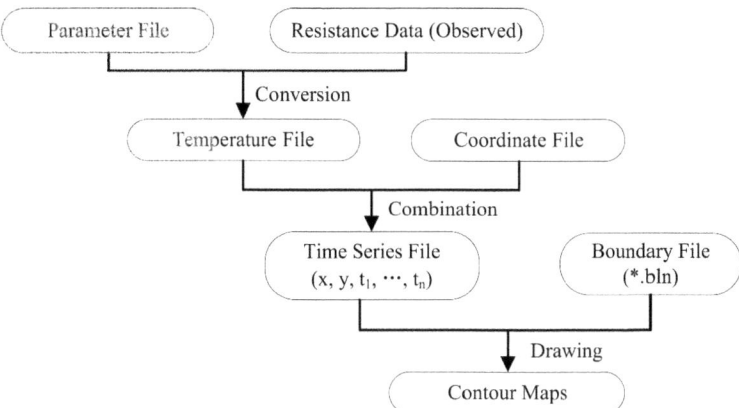

Fig. 1. Flowchart of visual analysis system of permafrost changes process in Qinghai-Tibet Highway

As shown in Fig.1, the flowchart of visual analysis system of permafrost changes process in Qinghai-Tibet Highway mainly has the following three steps:

Step1: According to the conversion formula of resistance to temperature and the sensor's parameters calibrated in factory, convert resistance values to temperature values.

Step2: By examining the position of sensor's groups and separation distance between the sensors, visually defines the coordinate location of each deployed sensor and generate a standard coordinate file. Combine the temperature data file with the coordinate file of temperature sensors, and produce the data file which conformed to the form $(x, y, t_1, t_2, ..., t_n)$.

Step3: In the range of coordinates the limited by blank file, get the data of coordinate *X* column, coordinate *Y* column, and the *i*th column to form (x, y, t_i) data, then call the Surfer process to generate a temperature contour map. At last, a set of temperature contour maps of time series were generated.

4   System Functions

In this section, we will discussed the system functions which provided by this visual analysis system.

4.1   Description of Monitored Data

For the purpose of understanding heat transfer process and climatic response process of subgrade of the Qinghai-Tibet highway deeply, about 43 different forms of long-term ground temperature observation systems have been deployed along the highway to obtain the periodic temperature changes of the observed road section. Many sections have been chosen to deploy the temperature sensors in the subgrade and under the road at distance in a certain interval. We adopted the thermal resistance as the temperature sensor and indoor accuracy of calibration is ±0.05°C.

4.2   Data File Conversion

Temperature sensors could monitor the temperature changes of deployment location in form of resistance value. According to the conversion formula of resistance to temperature and the sensor's parameters which calibrated in factory, resistance values should be converted to temperature values. A sample monitoring resistance data file was shown in Table 1. The first column is monitoring time series of the nodes, and the second column to the last column is resistance values collected by each sensor at the timestamp in the time series.

Table 1. A sample monitoring resistance data file of temperature sensor nodes

TIMESTAMP	S_1	S_2	S_3	S_4
2009-9-14 16:00	1738.314	1885.548	2132.538	1997.72
2009-9-14 18:00	1741.028	1887.019	2134.2	1998.295
2009-9-14 20:00	1742.062	1888.09	2135.588	1998.999
2009-9-14 22:00	1745.681	1880.973	2126.614	1995.986
2009-9-15 0:00	1742.632	1882.845	2123.814	1996.55
2009-9-15 2:00	1741.005	1885.807	2125.805	1997.333
2009-9-15 4:00	1740.37	1888.25	2128.597	1997.684

The sample parameter file of temperature sensor nodes was shown in Table 2, the first column is the name of deployed temperature sensors, and the second column to the last column is conversion parameters of each sensor which calibrated at the factory. There are 7 parameters. The conversion formula of resistance to temperature was shown in Formula.1, where *Pa* is Parameter *A*, *Pb* is Parameter *B*, *Pxa* is Parameter *XA*, *Pyb* is Parameter *YB*, *Pzc* is Parameter *ZC*, *L* is Length of wire, *Wr* is the Resistance of Wire and *Rr* is the resistance measured by sensor.

$$T = Pxa * ((Pb / (LN (Rr - Wr) - Pa) - 273.16)^2) + Pyb * (Pb / (LN (Rr - Wr) - Pa) - 273.16) + Pzc \qquad (1)$$

Table 2. The sample parameters file of temperature sensor nodes

Sensor_No	Pa	Pb	Pxa	Pyb	Pzc	L	Wr
S_1	-3.00753	2943.1336	-0.00051	1.00003	0.45938	3	0.58
S_2	-2.82699	2914.6231	-0.00042	1.00002	0.38427	3.5	0.64
S_3	-2.96090	2944.4563	-0.00047	1.00003	0.42627	4	0.69
S_4	-3.00010	2928.7743	-0.00047	1.00003	0.42628	4.5	0.73

The sample temperature data file converted from resistance data file was shown in Table 3, the first column is the name of deployed temperature sensors, and the second column to the last column is the temperature data which converted from the observed resistance data.

Table 3. The sample temperature data file converted from resistance data file

TIME STAMP	2009-9-14 16:00	2009-9-14 18:00	2009-9-14 20:00	2009-9-14 22:00	2009-9-15 0:00	2009-9-15 2:00
S_1	8.43	8.38	8.37	8.31	8.36	8.38
S_2	8.3	8.28	8.26	8.36	8.34	8.29
S_3	4.37	4.35	4.33	4.44	4.47	4.45
S_4	3.57	3.57	3.56	3.6	3.59	3.58

4.3   Generate the Coordinates of Sensors

To accomplish the visual analysis of contour map of observed temperature data, it is needed to establish a plane coordinate system for each section in the monitoring field, and define the coordinate location for each deployed sensor. According to the position of sensor's groups and separation distance between the sensors, visualization system could determine coordinates of each sensor through visual method, and generate a standard coordinate file. The sample coordinate data file of deployed temperature sensor nodes was shown in Table 4, the first column is the name of deployed temperature sensors, and the second column to the third columns are the coordinates of the sensor coordinate X and coordinate Y.

Table 4. The sample coordinate data file of deployed temperature sensor nodes

Sensor_No	X	Y
S_1	-11	-0.2
S_2	-11	-0.7
S_3	-11	-1.2
S_4	-11	-1.7

4.4   Combination of Data and Coordinates

According to the name of each deployed temperature sensor, the temperature data file which converted from the monitored resistance data file in the first step should be combined with the coordinate file of temperature sensors which generated in the second step to produce a temperature data file of sensor nodes. And it is the final data file which could be used by Surfer software to draw contour maps. The sample drawing data file which combined from the temperature data and the coordinate file of sensor nodes was shown in Table 5, the first column is the name of deployed temperature sensors, the second column to the third columns are the coordinates of the sensor coordinate X and coordinate Y, and the fourth column to the last column is the temperature value which converted from resistance value collected by each sensor at the timestamp in the time series.

Table 5. The sample drawing data file which combined from the temperature data and the coordinate file of sensor nodes

TIME STAMP	X	Y	2009-9-14 16:00	2009-9-14 18:00	2009-9-14 20:00	2009-9-14 22:00	2009-9-15 0:00	2009-9-15 2:00
S_1	-11	-0.2	8.43	8.38	8.37	8.31	8.36	8.38
S_2	-11	-0.7	8.3	8.28	8.26	8.36	8.34	8.29
S_3	-11	-1.2	4.37	4.35	4.33	4.44	4.47	4.45
S_4	-11	-1.7	3.57	3.57	3.56	3.6	3.59	3.58

4.5   Drawing Contour Maps

After completion of temperature data processing, client application could call the Surfer mapping software through the embedded programming method. In accordance with the time series, a temperature contour map was generated at each timestamp. Finally, a set of time series of temperature contour maps were generated. Fig.2 shows a distribution contour map of roadbed temperature of a section of a temperature observation field in Qinghai-Tibet Highway at 10:00 on September 3, 2009.

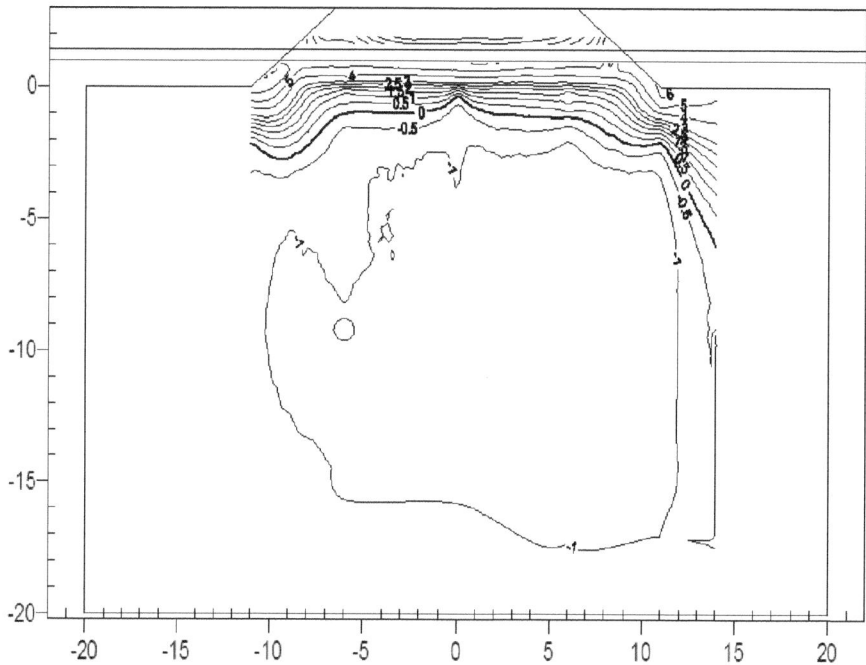

Fig. 2. A distribution contour map of roadbed temperature of a section of a temperature observation field in Qinghai-Tibet Highway (at 10:00 on September 3, 2009)

5   Conclusion and Future Work

A seamless connection could be realized to integrate the programming language VB and drawing map software Surfer. It utilized the powerful graphical user interface (GUI) of VB and the perfect graphics capabilities of Surfer software. Compared with original evaluation method, it saves time of developing application program to draw contour maps, ensure accuracy of mapping and improve efficiency. It helps researchers to find characteristics and regular patterns of permafrost change process in subgrade of Qinghai-Tibet highway, and make the scientific prevention and diagnosis of early engineering diseases. In this paper, we developed contour map visual analysis software, but the contour map is only one of the analysis methods. Our future work in this area is to do further research, carry out other visual analysis such as 3D map and do some comparative analysis for a better understanding.

Acknowledgments. This work is supported by Informationization Foundation of Chinese Academy of Sciences (CAS), "The E-Science Environment for Ecological and Hydrological Model Research in Heihe River Basin" (Grant number: 29O920C61); Project for Incubation of Specialists in Glaciology and Geocryology of National Natural Science Foundation of China (Grant number: J0930003/ J0109); and Second Phase of the CAS Action-Plan for West Development (Grant number: KZCX2-XB2-09-03).

References

1. Feng, G.L.: The Frozen Ground Roadbed Researches in China: Past, Present and Future. Journal of Glaciology and Geocryology 31(1), 139–147 (2008)
2. Wang, S.J., Chen, J.B., Li, X.H.: The Highway Construction Technology in Permafrost Regions: Research and Engineering Practice. Journal of Glaciology and Geocryology 31(2), 384–392 (2009)
3. Pang, Q.Q., Zhao, L., Li, S.X.: Influences of Local Factors on Ground Temperatures in Permafrost Regions along the Qinghai-Tibet Highway. Journal of Glaciology and Geocryology 33(2), 349–355 (2011)
4. Yu, Q.H., Qian, J.: Designing Research of the Qinghai-Tibet High-Grade Testing Highway. Journal of Glaciology and Geocryology 31(5), 907–914 (2009)
5. Yu, H., Wu, Q.B., Liu, Y.Z.: The Long-Term Monitoring System on Permafrost Regions along the Qinghai-Tibet Railway. Journal of Glaciology and Geocryology 30(3), 475–481 (2008)
6. Davis, J.C.: Statistics and data analysis in geology, 3rd edn. John Wiley & Sons (2002)
7. Golden Software Surfer Guide, http://www.goldensoftware.com/Surfer10Guide.pdf
8. Bodhankar, N.: Application of vectors for suitability of landforms in siting surface water harvesting structures. Environmental Geology 44(2), 176 (2003)
9. Gowd, S.S.: Electrical resistivity surveys to delineate groundwater potential aquifers in Peddavanka watershed, Anantapur District, Andhra Pradesh, India. Environmental Geology 46(1), 118 (2004)
10. Wang, C.C., Niu, Z.G., Li, Y., Sun, J., Wang, F.: Study on heavy metal concentrations in river sediments through the total amount evaluation method. Journal of Zhejiang University - Science A 12(5), 399–404 (2011)
11. Zheng, S., Li, S.Y., Huang, L.Z., Chen, Y.L.: Visualization programming for batch processing of contour maps based on VB and Surfer software. Advances in Engineering Software 41, 962–965 (2010)

Statistical Analysis of Number of Applicants for Entrance Examination for MA Degree in China Based on Grey Prediction Model

Gang Li, Xing Wang, and Yala Tong

School of Science,
Hubei University of Technology,
Wuhan, 430068, China
87475853@qq.com

Abstract. As the number of college students and graduate students increasing, taking part in the entrance exams for postgraduate schools has become more and more popular for college students' choice. The increasing number of taking part in the entrance exams for postgraduate schools has the vital significance to our economic construction and education development. By application of grey system theory and its methods, based on the change in the number of taking part in the entrance exams for postgraduate schools in recent years in China, this article constructed grey model GM(1,1) and grey metabolic prediction model, compared two models, and took some relative optimum processing. At last, using grey metabolic prediction model to forecast the number of students who take part in the entrance exams for postgraduate schools in next five years, provided a reference for the relevant departments.

Keywords: graduate applicants, statistical analysis, grey prediction model, MA degree.

1   Introduction

With the college enrollment and graduate enrollment increased, taking an entrance examination for MA degree has become more and more students' choice. The number of applying for it increases every year, and it is inseparable from a number of factors, such as fierce competition for jobs in society, employers in high demand high standards for graduates, and the students' craving for more high-tech jobs. For university four year study can not reach the requirement, it forces more and more students take a longer time to gain more expertise to deal with the community at any time screening of high-quality personnel [1]. Increase in number of graduate students is significant to educational development of Chinese economy. Taking parting in the entrance for MA degree turns more popular year after year, which attracts a number of scholars to study the behind motive and predict the trends of the number in the future, and provides a reference for the relevant departments[2][3].But many scholar only focus using the linear regression to fit the number of applying for postgraduate. For example, in [4], the author select the number of students, the unemployment rate as a variable factor to construct a linear regression model.

B. Liu and C. Chai (Eds.): ICICA 2011, LNCS 7030, pp. 9–16, 2011.

However, the number of candidates usually can be affected by many uncertain factors, and in particular period it may have unconventional changes, such as in 2008 the reform of charging system and other factors make the number of application for entrance examination reduced for the first time in nearly a decade. Linear regression model is too simple, and many factors can not be determined. And regression analysis is the application of statistical methods, is a large amount of observational data collation, analysis and research, so as to come within the laws reflect some of the conclusions of things.

The recent number of application for entrance examination is small sample. This paper argues that, based on the consideration of practical significance, and the characteristics of "small sample", "poor information" of grey system theory[5-7], we can establish the grey prediction model[8-11] for the number of taking parting in the entrance examination for MA degree, and reasonably forecast the future development of the number in the future.

2   Trend Analysis of Number of Candidates

2.1   Comparison of GM(1,1) and Grey Linear Regression Combined Model

This paper selected the original data of applying entrance examination for MA degree from 2002 to 2011 to be analyzed. For choosing the appropriate dimension for the screening of the prediction model, considering that the grey constructed data is usually less than five, we respectively choose 10-dimensional data, observe the data changes, and constructed the GM(1,1) and grey linear regression combined model[6] to forecast.Source: China Education Online

```
http://kaoyan.eol.cn/kaoyan_news_3989/20090111/t20090111_
354041.shtml
```

Table 1. The comparison of GM(1,1) and grey linear regression combined model

sequence	year	Original data	GM(1,1)			Grey linear regression combined model		
			fitted value	Residuals	Relative error (%)	fitted value	Residuals	Relative error (%)
1	2002	62. 4	62. 4	0	0	54. 94	7. 46	11. 9
2	2003	79. 7	94. 42	−14. 72	−18. 46	91. 72	−12. 02	−15. 08
3	2004	94. 5	100. 06	−5. 56	−5. 89	101. 14	−6. 64	−7. 03
4	2005	117. 2	106. 03	11. 17	9. 53	109. 48	7. 72	6. 59
5	2006	127. 12	112. 37	14. 75	11. 6	116. 84	10. 33	8. 13
6	2007	128. 2	119. 08	9. 12	7. 11	123. 36	4. 84	3. 77
7	2008	120	126. 19	−6. 19	−5. 16	129. 12	−9. 12	−7. 6
8	2009	124. 6	133. 72	−9. 12	−7. 32	134. 22	−9. 62	−7. 72
9	2010	140. 6	141. 71	−1. 11	−0. 79	138. 72	1. 88	1. 33
10	2011	151. 1	150. 17	0. 93	0. 62	142. 71	8. 39	5. 55

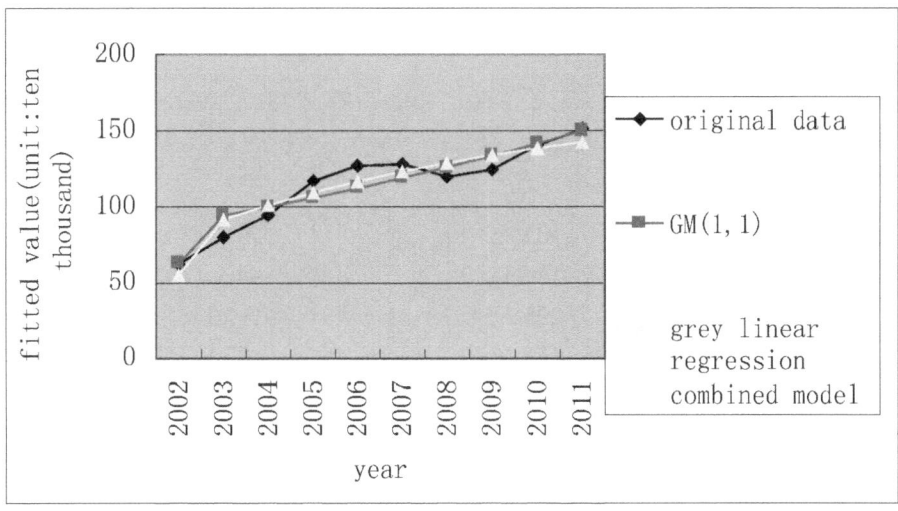

Fig. 1. The comparison of the fitted value of GM(1,1) and grey linear regression combined model

We all know the shortage that the linear regression model lacks the exponential growth and the GM(1,1) lacks the linear trend, so the combined model is a sequence both having exponential growth and linear factor.

Known from the Table two, the mean relative precision $\bar{q}_1$ in GM (1,1) is 93.645%, while the grey linear regression combination model has the better precision. Seen from the picture two, the prediction of applying entrance examination for MA degree after 2010, grey linear regression combination prediction model becomes linear, and the GM (1,1) model more accurately predict the trend of the number.

Prediction accuracy of GM (1,1) model for smooth monotone sequences of non-negative is high. But in this case, for the large abnormal changes in the data of 07,08 year, the model fitted values obtained is not particularly desirable. We should process the related data of the model to enhance the smoothness of the original data. We can choose the data in 07,08 year to amend the original value, taking the median value of the data in 07,08 year to replace the original data, which obtained (128.2 +120) / 2 = 124.1. Select the data after correction to re-establish model.

Processed raw data sequence is:

$$X^{(0)} = [\ 62.4, 79.7, 94.5, 117.2, 127.12, 124.1, 124.6, 140.6, 151.1] \tag{1}$$

Step 1, make $1-$AGO operator;

$$X^{(1)} = [x^{(1)}(1), x^{(1)}(2), x^{(1)}(3), x^{(1)}(4), x^{(1)}(5), x^{(1)}(6), x^{(1)}(7), x^{(1)}(8), x^{(1)}(9)] \tag{2}$$
$$= [\ 62.4, 142.1, 233.6, 353.8, 480.92, 605.02, 729.72, 870.22, 1021.32\ ]$$

Step 2, find the close mean generated sequence $Z^{(1)}$, and construct matrix B and vector Y;

$$Z^{(1)} = [z^{(1)}(2), z^{(1)}(3), z^{(1)}(4), z^{(1)}(5), z^{(1)}(6), z^{(1)}(7), z^{(1)}(8), z^{(1)}(9)] \quad (3)$$

$$= [102.25, 189.35, 295.2, 417.361, 542.97, 667.32, 799.92, 945.77]$$

$$B = \begin{bmatrix} -z^{(1)}(2) & 1 \\ -z^{(1)}(3) & 1 \\ -z^{(1)}(4) & 1 \\ -z^{(1)}(5) & 1 \\ -z^{(1)}(6) & 1 \\ -z^{(1)}(7) & 1 \\ -z^{(1)}(8) & 1 \\ -z^{(1)}(9) & 1 \end{bmatrix} = \begin{bmatrix} -102.25 & 1 \\ -189.35 & 1 \\ -295.2 & 1 \\ -417.36 & 1 \\ -542.97 & 1 \\ -667.32 & 1 \\ -799.92 & 1 \\ -945.77 & 1 \end{bmatrix}, Y = \begin{bmatrix} x^{(0)}(2) \\ x^{(0)}(3) \\ x^{(0)}(4) \\ x^{(0)}(5) \\ x^{(0)}(6) \\ x^{(0)}(7) \\ x^{(0)}(8) \\ x^{(0)}(9) \end{bmatrix} = \begin{bmatrix} 79.7 \\ 94.5 \\ 117.2 \\ 127.12 \\ 124.1 \\ 124.6 \\ 140.6 \\ 151.1 \end{bmatrix} \quad (4)$$

Step 3, the least square estimation of parameters $\hat{a} = \begin{bmatrix} a & b \end{bmatrix}^T$:

$$\hat{a} = (B^T B)^{-1} B^T Y = \begin{bmatrix} -0.0722 & 84.1047 \end{bmatrix}^T \quad (5)$$

The time response equation of 9-demensional GM(1,1) built is:

$$\hat{x}^{(1)}(k+1) = (x^{(0)}(1) - \frac{b}{a})e^{-ak} + \frac{b}{a} = 1227.285e^{0.0722k} - 1164.885 \quad (6)$$

Table 2. The modified GM(1,1) and its fitted value

sequence	year	Original data	Modified GM(1,1)		
			fitted value	Residuals	Relative error (%)
1	2002	62.4	62.4	0	0
2	2003	79.7	91.89	-12.19	-15.29
3	2004	94.5	98.78	-4.28	-4.53
4	2005	117.2	106.18	11.02	9.4
5	2006	127.12	114.13	12.99	10.22
6	07, 08	124.1	122.68	1.42	1.14
7	2009	124.6	131.87	-7.27	-5.83
8	2010	140.6	141.75	-1.15	-0.82
9	2011	151.1	152.37	-1.27	-0.84

Posterior difference test:

Firstly, calculate the mean variance S_0 of original sequence $\bar{x}^{(0)}$:

$$\bar{x}^{(0)} = 102.132; S_0 = 31.338; \quad (7)$$

Secondly, calculate the mean variance S_1 of the residual sequence $\varepsilon^{(0)} = x^{(0)}(i) - \hat{x}^{(0)}(i)$:

$$\overline{\varepsilon}^{(0)} = \frac{1}{n}\sum_{i=1}^{n}\varepsilon^{(0)}(i) = -0.073; S_1 = \sqrt{\frac{1}{n}\sum_{i=1}^{n}[\varepsilon^{(0)}(i) - \overline{\varepsilon}^{(0)}(i)]^2} = 8.02 \tag{8}$$

Calculation mean variance ratio:

$$C = \frac{S_1}{S_0} = 0.256\varepsilon \tag{9}$$

Small error probability:

$$p = \{|\varepsilon^{(0)}(i) - \overline{\varepsilon}^{(0)}| < 0.6475 \cdot S_0\} \tag{10}$$

$$\left|\varepsilon^{(0)}(1) - \overline{\varepsilon}^{(0)}\right| = 0.073 < 20.291372; \left|\varepsilon^{(0)}(2) - \overline{\varepsilon}^{(0)}\right| = 12.117 < 20.291372 \tag{11}$$

$$\left|\varepsilon^{(0)}(3) - \overline{\varepsilon}^{(0)}\right| = 4.207 < 20.291372; \left|\varepsilon^{(0)}(4) - \overline{\varepsilon}^{(0)}\right| = 11.093 < 20.291372 \tag{12}$$

$$\left|\varepsilon^{(0)}(5) - \overline{\varepsilon}^{(0)}\right| = 13.063 < 20.291372; \left|\varepsilon^{(0)}(6) - \overline{\varepsilon}^{(0)}\right| = 1.493 < 20.291372 \tag{13}$$

$$\left|\varepsilon^{(0)}(7) - \overline{\varepsilon}^{(0)}\right| = 1.797 < 20.291372; \left|\varepsilon^{(0)}(8) - \overline{\varepsilon}^{(0)}\right| = 1.077 < 20.291372 \tag{14}$$

$$\left|\varepsilon^{(0)}(9) - \overline{\varepsilon}^{(0)}\right| = 1.197 < 20.291372; \tag{15}$$

Posterior difference test: $c_3 = 0.256; p_3 = 1$
The average relative accuracy: $\overline{q}_3 = 95.193\%$;
Prediction accuracy level is good (first-class).

2.2   The Comparison of the Fitted Value of Model of the Raw Data and the Processed Dada

Table 3. Font sizes of headings. Table captions should always be positioned *above* the tables

model	Small probability of error p value	MSE ratio c	Average accuracy
GM(1,1) established by the original data	1	0.342	93.65%
GM(1,1) established by the modified data	1	0.256	95.19%

From the table, compares GM(1,1) with the original sequence and the that established with the modified sequence, we can see that, in the original sequence established GM (1,1) model and the modified sequence set GM (1,1) model, the average relative precision was 93.645% and 95.193%, standard deviation ratio of 0.342 and 0.256, respectively, found that the modified sequence from established GM (1,1) model has higher prediction accuracy.

2.3   Establishment of Metabolic GM (1,1) Group

In the Conventional GM (1,1) modeling, several close data have the high precision , but as the extension of the prediction, the accuracy of the model is also reduced. The metabolic GM (1,1) model in forecasting process, eliminate aging information dynamically, adding new information. Modeling system is now better reflect the sequence characteristics of the system, can better reveal the trend, usually obtain higher prediction accuracy.

Using the gotten data by GM(1,1) to predict the candidates of application for MA degree in 2012, replace the old data with the new data in the original sequence, remove the old 2002 data, in order to maintain the same number of dimensions. Metabolic modeling data is updated once:

$$X^{(0)} = [\ 79.7, 94.5, 117.2, 127.12, 124.1, 124.6, 140.6, 151.1, 163.79] \qquad (16)$$

Step one: get the 1-AGO sequence of $X^{(0)}$:

$$X^{(1)} = [x^{(1)}(1), x^{(1)}(2), x^{(1)}(3), x^{(1)}(4), x^{(1)}(5), x^{(1)}(6), x^{(1)}(7), x^{(1)}(8), x^{(1)}(9)] \qquad (17)$$
$$= [\ 79.7,\ 174.2,\ 291.4,\ 418.52,\ 542.62,\ 667.22,\ 807.82,\ 958.92,\ 1122.71\ ]$$

Step two: Find the close mean generated sequence $Z^{(1)}$ of $X^{(1)}$, and construct matrix B and data vector Y :

$$Z^{(1)} = [z^{(1)}(2), z^{(1)}(3), z^{(1)}(4), z^{(1)}(5), z^{(1)}(6), z^{(1)}(7), z^{(1)}(8), z^{(1)}(9)] \qquad (18)$$
$$= [126.95,\ 232.8,\ 355.01,\ 480.62,\ 604.92,\ 737.52,\ 883.42,\ 1040.815\ ]$$

$$B = \begin{bmatrix} -z^{(1)}(2) & 1 \\ -z^{(1)}(3) & 1 \\ -z^{(1)}(4) & 1 \\ -z^{(1)}(5) & 1 \\ -z^{(1)}(6) & 1 \\ -z^{(1)}(7) & 1 \\ -z^{(1)}(8) & 1 \\ -z^{(1)}(9) & 1 \end{bmatrix} = \begin{bmatrix} -126.95 & 1 \\ -232.8 & 1 \\ -355.01 & 1 \\ -480.62 & 1 \\ -604.92 & 1 \\ -737.52 & 1 \\ -883.42 & 1 \\ -1040.815 & 1 \end{bmatrix}, \qquad (19)$$

$$Y = \begin{bmatrix} x^{(0)}(2) \\ x^{(0)}(3) \\ x^{(0)}(4) \\ x^{(0)}(5) \\ x^{(0)}(6) \\ x^{(0)}(7) \\ x^{(0)}(8) \\ x^{(0)}(9) \end{bmatrix} = \begin{bmatrix} 94.5 \\ 117.2 \\ 127.12 \\ 124.1 \\ 124.6 \\ 140.6 \\ 151.1 \\ 163.79 \end{bmatrix} \qquad (20)$$

Step three: by the least square estimation for parameters $\hat{a} = \begin{bmatrix} a & b \end{bmatrix}^T$, we can obtain:

$$\hat{a} = (B^T B)^{-1} B^T Y = [-0.0637 \quad 94.8716]^T \tag{21}$$

Then, time response equation of 9 Victoria built by GM (1,1) model is:

$$\hat{x}^{(1)}(k+1) = (x^{(0)}(1) - \frac{b}{a})e^{-ak} + \frac{b}{a} = 1569.05e^{0.0637k} - 1489.35 \tag{22}$$

Put k=5 into(5) the formula above, and we can have the predictive value of 171.72 million in 2013, then add it to the data sequence of metabolic modeling, and remove the data in 2003, predict the new value by the secondary metabolism, similarly, and gradually cycle, obtained the required forecasting value   follow by the order. Forecast results are listed in Table.

Table 4. The predictive value of candidates in the near future

year	Number of candidates
2012	163.78
2013	171.72
2014	179.78
2015	191.78
2016	206.63

2.4   Analysis of Predicted Results

Originally the development of graduate education is an educational progress, but also social progress. Predicted by the above data, we found the number of applicants for entrance examination for MA degree will become more and more, and candidates are expected to exceed 200 million mark around 2016. The increase in the number means that the craze for graduate school will continue.Although the recent emergence of the increased quota of graduate enrollment while decline in the quality on graduate education, graduate employment situation is grim and so on, causing the government, experts, media and civil society's attention, with the economic development and social improvement in the level of civilization, the need to improve quality of highly educated professionals in society, is almost all agree with all the levels of community, which is also consistent to the trend predicted in this paper.

3   Conclusions

This paper mainly analyzed the number of applicants for the entrance examination for MA degree in China based on grey prediction model. Firstly, constructed the GM(1,1)and grey linear regression combined model. The former is mainly used for single index sequence while the latter is more appropriate for the sequence with both linear trends and exponential growth trend. By comparison, the model with the conventional GM (1,1) model can be better fitted values. Secondly, when the data changes gently, using conventional GM(1,1) to fit the sequence can get a better effect; and when the data changes rapidly, the predicted effect poor. After having the data processed by the Close mean generation algorithm , and replacing dramatic changes value with a certain treatment, we can found that the data set in GM (1,1) can be fitted better than that. Finally, since the metabolic GM (1,1) model than the conventional GM (1,1) model in predicting the time to get better accuracy, we use the processed data to establish metabolic model groups to predict the number of applicants for the entrance examination for MA degree in next five years, and got a better predictive value, and then draw the relevant conclusions.

Acknowledgments. We are very grateful to the anonymous referees for their insightful and constructive suggestions, which have led to an improved version of this paper. This work was supported by Hubei Provincial Department of Education Science and Technology Research Project (Grant No. Q20111408), the Programs for Science and Technology development of Wuhan (Grant No. 201010621218), and Hubei Province Natural Science Foundation projects(Grant No. 2009CDB312).

References

1. Yan, S.: Campus Fashion helpless. Graduate Employment in China (1), 30–31 (2011)
2. Zhang, M.: "Graduate Rush" sudden drop in temperature. Zhongguancun (3), 76–79 (2007)
3. Qin, Z., Cai, J.: "Graduate Rush" Phenomenon Exploration and Analysis. Modern Education Science (5), 56–57 (2008)
4. Zhao, Y.: On the number of its influencing factors about applying for entrance examination for MA degree. Decision and Information (8), 100 (2010)
5. Deng, J.: The Basic of the Grey System. Huazhong University of Technology Press, Wuhan (2002)
6. Liu, S., Dang, Y., Fang, Z.: Grey system theory and its application, 3rd edn. Science Press (November 2004)
7. Liu, S., Xie, N.: Grey system theory and its application, 4th edn. Science Press (December 2008)
8. Jiang, H., Liang, Q.-h.: Application Research of the Grey Forecast in the Logistics Demand Forecast. In: Proceedings of 2009 First International Workshop on Education Technology and Computer Science, pp. 361–363. IEEE Computer Society (2009)
9. Wang, Y., Pohl, E.A., Dang, Y.: Reliability Prediction Using an Unequal Interval Grey Model. In: RAMS 2010 Proceedings-Annual, pp. 1. IEEE (2010)
10. Li, J., Shi, X., Qin, Y., Zhang, G.: DSRV Recovery Control Using Grey Prediction Model. In: Proceedings of the 2010 IEEE International Conference on Information and Automation, Harbin, China, June 20-23, pp. 294–300 (2010)
11. Xu, J., Gu, M.: Realtime Calibration of Pulse Oximetry Based on Grey Model. In: 2010 OSA-IEEE-COS Advances in Optoelectronics and Micro/Nano-Optics, AOM 2010. IEEE Computer Society (2010)

A Case-Based Reasoning Approach for Evaluating and Selecting Human Resource Management Information Systems Projects

Santoso Wibowo

Faculty of Business Informatics, CQUniversity Melbourne 3000, Victoria, Australia
wibowos@mel.cqu.edu.au

Abstract. This paper presents a case-based reasoning (CBR) approach for evaluating and selecting human resource management information systems (IS) projects in an organization. The concept on case-based distance is introduced for measuring the degree of similarity between each case in the case base and the new case. To avoid the inconsistency of the decision maker's subjective assessment, an induction technique is applied to help assign the importance of the criteria in the similarity measure. A human resource management IS project evaluation and selection problem is presented to demonstrate the effectiveness of the approach.

Keywords: Case-Based Reasoning, Evaluation and Selection, Multicriteria, Human Resource Management IS Project.

1 Introduction

Organizations are facing tremendous pressure to compete against their competitors in order for them to survive in this dynamic environment. To effectively deal with this competitive nature of the business environment, organizations are focusing their attention on improving their human resource management function as part of their business strategy in order to achieve competitive advantage [1-2].

Human resource management emphasizes on the development of the organization's capacity to respond to the external environment through a better deployment of human resources. There are increasing numbers of empirical studies on the effect of human resource management in increasing organizational performance by matching unique internal processes with environmental opportunities and needs [2]. As a result, human resource management is now seen as an important factor in improving organizational performance and effectiveness.

In order for organizations to support their human resource management effectively and efficiently, the use of information systems (IS) to support human resource management in organizations has increased dramatically in the past few years [1-2]. Human resource management IS facilitates the provision of quality information to management for informed decision making [3] and supports the provision of executive reports and summaries for senior management [1].

The development and implementation of human resource management IS projects has become a crucial activity for organizations nowadays. These IS projects have

B. Liu and C. Chai (Eds.): ICICA 2011, LNCS 7030, pp. 17–24, 2011.

been considered to be the solution for organizations to improve productivity, increase employee satisfaction, and reduce costs [3]. By adopting the appropriate human resource management IS project, modern organizations can gain competitive advantages that are vital to the organization's future growth [4].

To be able to identify which IS project is likely to improve the organizational performance helps in preventing the decision maker (DM) to make incorrect decisions. This evaluation and selection process is however challenging due to the lack of empirical results [4]. It is therefore more feasible to evaluate and select IS projects by using old experiences to solve current problems.

Case-based reasoning (CBR) is a process for solving a new problem case by referring to the solutions of similar past cases [5]. Similar past cases are first retrieved and their associated solutions are then used to aid the DM in developing solutions for the new case. CBR is efficient in simplifying knowledge retrieved from old cases, improving solution to the new situation, as well as in accumulating knowledge for further reasoning [6]. CBR is useful for evaluating and selecting IS projects as this approach is capable of providing further explanation for its results based on reasoning from knowledge retrieved or accumulated from old cases.

To effectively address the problem as above, this paper presents a CBR approach for evaluating and selecting human resource management IS projects. The approach integrates the concept on case-based distance and the induction technique. The concept on case-based distance is introduced for measuring the degree of similarity between each case in the case base and new case. To avoid the inconsistency of the DM's subjective assessment, an induction technique is applied to help assign the importance of the criteria in the similarity measure. An example is presented to demonstrate the applicability of the proposed approach for dealing with the human resource management IS project evaluation and selection problem.

In what follows, a review on the existing approaches for dealing with IS projects is provided. This is followed by the development of the CBR approach for evaluating and selecting the human resource management IS projects. An IS project evaluation and selection problem is exemplified to demonstrate the applicability of the proposed approach.

2   A Review of Existing Approaches

Much research has been done on the development of various approaches for dealing with IS projects evaluation and selection problem. These approaches can be classified into statistical approaches and artificial intelligence approaches [7-9]. Commonly used statistical approaches for dealing with IS projects evaluation and selection include discriminant analysis, logit and probit analyses and linear probability. These approaches generally produce the result as a numerical value or a probability of failure based on a set of explanatory variables for each IS project under investigation. The result is then compared to a predetermined cut-off score in order to determine whether the IS project is likely to fail [4]. Statistical approaches are ideal for handling numerical data. These approaches, however, are inherently limited in their ability to (a) evaluate qualitative data and (b) explain the results of the decision outcome [5].

Artificial intelligence approaches are capable of learning directly from data, coping with imprecise and incomplete data and handling numerical data [7]. The representatives of artificial intelligence approaches include neural networks [9], genetic algorithms [10] and CBR [4-6]. A neural network approach is an adaptive model containing a set of interconnected units that perform computing tasks. Associated with each connection is a weight to be adjusted through learning processes until the network performs well on training data [9]. The weakness of neural network approach is that it cannot explain the results.

A genetic algorithm is a search heuristic that mimics the process of natural evolution. It applies principles of evolution to the challenges of problem solving by creating an initial set of guesses as potential solutions to a problem and then repeatedly rearrange these guesses in the order of how well they solve the problem through selection, cross-over and mutation until a good solution is found [10]. The strength of genetic algorithms is that they only require one to recognize a good solution. However, there is no guarantee of an optimal solution.

CBR uses old experiences to solve current problems. It retrieves similar cases stored in a case base and adapts the solutions of the retrieved cases to solve a new problem [4]. In CBR, most knowledge is represented in cases. A case is comprised of a problem state, its corresponding solution and/or outcome. CBR is particularly useful in dealing with problems that are impossible or difficult to understand completely. It is capable of dealing with various data types and can justify its solution. The weakness of this approach is that the approach is less appropriate to use when cases are difficult or impossible to obtain [5].

From the discussion above, it can be seen that no one approach is more superior over the others. This paper proposes a CBR approach for dealing with IS projects selection problem. The application of CBR helps to (a) increase efficiency in solving new problems due to the reuse of relevant reasoning and (b) improve quality of solutions as past cases can guide the DM towards successful alternatives.

3   The Case-Based Reasoning Approach

Evaluating and selecting IS projects usually involves in the selection or ranking of one or more alternatives from a set of n available alternatives $A = (A_1, A_2, ..., A_i, ..., A_n)$, with respect to each q selection criteria $C_1, C_2, ..., C_q$, where the values of criteria $C_1, C_2, ..., C_q$ are expressed as C_{ij} $(j = 1, 2, ..., q)$.

The proposed approach consists of four phases, including (a) information gathering from the new problem, (b) case retrieval, (c) case adaptation and revision, and (d) case validation as shown in Figure 1.

The first phase starts with the gathering of relevant information required to solve the IS project evaluation and selection problem. This includes (a) the performance of human resource management IS project alternatives with respect to each criterion, (b) criteria used for the evaluation and selection process, and (c) the number of alternatives.

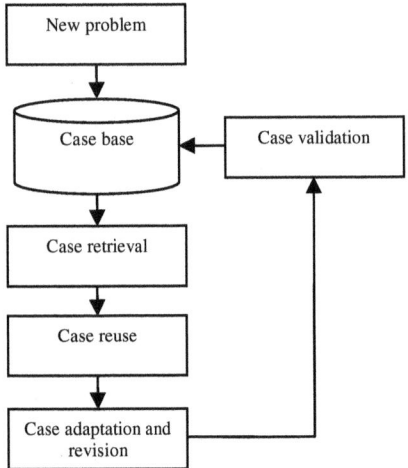

Fig. 1. The Case-Based Reasoning Framework for IS Projects Evaluation

The second stage is the case retrieval phase. This is performed by calculating the degree of similarity between each case in candidate case set and new case, and to sort them in the order of similarity degree. The case with higher similarity is obtained for reuse.

To measure the degree of similarity of the DM's assessments with respect to each criterion, the concept on distance-based [11] is introduced. This measure is used to compare the degree of similarity between each case in the case base and new case. To understand the degree of similarity between each case in the case base and new case, a similarity threshold index is established. This index is pre-determined by the DM before the evaluation process.

Suppose that a DM specifies t selected cases as $T = (Z_1, Z_2, ..., Z_t)$ in the evaluation problem. Let $Z = (Z_1, Z_2, ..., Z_m)$ denote the m acceptable cases and T-$Z = (Z_{m+1}, Z_{m+2}, ..., Z_t)$ denote the t-m unacceptable cases. Denote Z_* as a fictitious alternative at the centre of Z, where

$$C_{*j} = \frac{1}{m} \sum_{r=1}^{m} c_{rj} \tag{1}$$

The case distance between $Z_r \in Z$ and the centre Z_* with respect to criterion j can be obtained as

$$d_j(Z_r, Z_*) = \frac{(C_{rj} - C_{*j})^2}{d_j^{\max}} \tag{2}$$

where $d_j^{\max}$ is the maximum of $(C_{rj} - C_{*j})^2$ and $r = (1, 2, ..., m)$ is used as a normalization factor for criterion j.

Based on (2), the distance between a case Z_r and the target case Z_* with respect to each alternative can be defined as

$$D(Z_r, Z_*) = \sum_{j=1}^{q} w_j \times d_j(Z_r, Z_*) \qquad (3)$$

where w_j represents the weight of criterion j.

To reduce the cognitive demanding on the DM, IF-THEN rules are generated using IF-THEN statements for evaluating the alternatives. IF $D(Z_r, Z_*)$ > threshold index THEN Decision = "Remove the alternative". Otherwise, the alternative is retained. This process helps to reduce the number of alternatives that do not meet the requirements of the DM in an efficient and effective manner.

The process of determining the set of weights that can improve the overall accuracy is an important and challenging task. It is common that the weights are assigned by the DM subjectively or using the trial and error approach which leads to inconsistency in the decision outcome [12]. In order to solve this problem, this paper introduces an induction technique to help assign the importance of the criteria in the similarity measure.

Induction generates a set of weight through information gain [13]. Information gain is a heuristic of iterative dichotomizer three (ID3) decision making tree algorithm [14]. It is used for comparing potential splits in finding the most discriminating criterion for dividing the cases. It calculates the difference between entropy of a case base and its partitions built from a criterion and can be assigned as a criterion weight [13]. The entropy characterizes the impurity of a set of cases T with respect to the target criterion that has n outcomes. It is defined as

$$E(T) = \sum_{i=1}^{n} -p_i \times \log_2 p_i \qquad (4)$$

where p_i is the proportion of T belonging to outcome i.

If T is partitioned on criterion C with q values, the expected value of the entropy is the weighted sum over the subsets given by

$$E_c(T) = \sum_{j=1}^{q} \frac{|T_j|}{|T|} \times E(T_j) \qquad (5)$$

where T_j is the subset of T for which criterion C has value j. The information gain by branching on C to partition T is measured by

$$gain(C) = E(T) - E_c(T) \qquad (6)$$

Information gain is calculated for every criterion and is used as the criterion weight in the similarity measure assessment. The relative importance of the criteria is taken into consideration by using induction to automatically generate a set of weights in order to avoid the inconsistency of the DM's subjective assessments in the decision making process.

The next stage is case adaptation and revision. This is where the new problem may not match the old case problem entirely, and therefore the case should be modified to fit the new problem. The last stage is case validation which verifies the solution of the new problem and stores the problem as well as its solution into the case base for future use.

4   An Example

To demonstrate the applicability of the proposed approach above, a problem of evaluating and selecting suitable IS projects for an organization is presented. Based on a thorough investigation by the management of the company, ten potential human resource management IS project alternatives are identified.

A Delphi approach is used to determine the evaluation criteria which would be appropriate for the evaluation process. Three experts helped prioritize the criteria and reached a consensus about the important criteria for evaluating and selecting the human resource management IS projects. Based on their thorough discussion, five selection criteria are identified including the External criterion (C_1), the Internal criterion (C_2), the Risk criterion (C_3), the Cost criterion (C_4) and the Benefits criterion (C_5) [15].

The External criterion (C_1) concern with the DM's assessment on the expectation of the management of an organization on the use of IS to react to the external environment. This is measured by the ability of IS to ally with partner, its commitment to government requirements, its commitment to societal needs, and its ability to compete with other competitions.

The Internal criterion (C_2) refer to the DM's assessment in regards to the expected contribution of the IS project alternative towards the internal environment of the organization. This is measured by the improvement on organizational learning; the capability of meeting user's requirements, the compatibility with the existing IS portfolio, and the ability to restructure the organization.

The Risk criterion (C_3) involve the DM's assessment on the potential negative impact of the IS project including the failure in the development and the implementation of the IS project [3]. This is often assessed from the technical risk, the development risk, risk of cost overruns, and the size risk of individual projects.

The Cost criterion (C_4) concern with DM's assessment on the economical and financial feasibility of the IS project with respect to the resource limitation of an organization and its business strategy. This is measured by hardware costs, software costs, implementation costs, and maintenance costs involved.

The Benefits criterion (C_5) reflect the perception of the DM on how individual IS projects serve the business strategy and organizational objectives in the long term [15]. Issues such as the contribution to organizational goals, the importance to the organizational competitiveness, the aid to improve information quality, and the relevancy to critical success factors are taken into account.

The evaluation and selection process starts with the DM assigning the performance ratings for the human resource management IS project alternatives. Numerical values between 1 and 5 are used for assessing the performance of the human resource management IS project alternatives where 1 indicates very poor performance and 5 indicates the very good performance. Table 1 shows the results.

Table 1. Performance Ratings of IS Project Alternatives

Alternatives	Criteria				
	C_1	C_2	C_3	C_4	C_5
A_1	3	5	3	3	5
A_2	5	3	5	5	3
A_3	3	4	3	3	4
A_4	2	5	5	2	5
A_5	1	2	3	1	2
A_6	3	2	3	3	2
A_7	5	3	5	5	3
A_8	3	3	3	3	3
A_9	1	2	3	1	2
A_{10}	3	2	3	3	2

In this case, the DM has assigned the similarity threshold index to be 0.60. Using (1)-(6) as well as the values obtained from Table 1 and Table 2, the importance of the criteria weights for IS project alternatives and the degree of similarity of the DM's assessments with respect to each criterion can be obtained.

Table 2. An Example of IS Project Data Set in the Case Base

Criteria					Result
C_1	C_2	C_3	C_4	C_5	
3	2	4	3	2	Success
3	3	3	5	3	Success
5	4	5	3	5	Success
3	3	3	4	3	Success
2	2	2	3	3	Failure
1	2	1	2	1	Failure
4	3	3	4	3	Success
2	1	1	3	2	Failure

Table 3 shows the criteria weights for all IS project alternatives. Based on these criteria weights, the top three alternatives are retrieved. Alternatives A_1, A_2, and A_7 are found to be the top three ranked alternatives.

Table 3. Criteria Weights of IS Project Alternatives

Alternatives	Criteria				
	C_1	C_2	C_3	C_4	C_5
A_1	0.383	0.386	0.405	0.623	0.331
A_2	0.757	0.810	0.433	0.534	0.512
A_3	0.588	0.389	0.204	0.960	0.569
A_4	0.247	0.322	0.217	0.259	0.085
A_5	0.678	0.834	0.526	0.669	0.464
A_6	0.716	0.290	0.504	0.718	0.474
A_7	0.342	0.259	0.295	0.169	0.304
A_8	0.672	0.558	0.651	0.477	0.539
A_9	0.669	0.526	0.569	0.259	0.678
A_{10}	0.322	0.217	0.259	0.389	0.085

5   Conclusion

This paper has presented a CBR approach for evaluating and selecting human resource management IS projects. The approach integrates the concept on case-based distance and the induction technique. The concept on case-based distance is used for measuring the degree of similarity between each case in the case base and new case. An induction technique is applied to help assign the importance of the criteria in the similarity measure. With its simplicity in concept and computation, the approach is effective for solving the human resource management IS project evaluation and selection problem.

References

1. Stone, D.L., Stone-Romero, E.F., Lukaszewski, K.: Factors affecting the acceptance and effectiveness of electronic human resource systems. Hum. Resour. Manage. Rev. 16, 229–244 (2006)
2. Hussain, Z., Wallace, J., Cornelius, N.E.: The use and impact of human resource information systems on human resource management professionals. Inform. Manage. 44, 74–89 (2007)
3. Deng, H., Wibowo, S.: Fuzzy Approach to Selecting Information Systems Projects. In: 5th ACIS International Conference on Software Engineering, Artificial Intelligence, Networks and Parallel/Distributed Computing, Beijing, China, June 30-July 2 (2004)
4. Hsu, C., Chiu, C., Hsu, P.L.: Predicting information systems outsourcing success using a hierarchical design of case-based reasoning. Exp. Syst. Appl. 26, 435–441 (2004)
5. Tung, Y.H., Tseng, S.S., Weng, J.F., Lee, T.P., Liao, Y.H., Tsai, W.N.: A rule-based CBR approach for expert finding and problem diagnosis. Exp. Syst. Appl. 37, 2427–2438 (2010)
6. Du, Y., Wen, W., Cao, F., Ji, M.: A case-based reasoning approach for land use change prediction. Exp. Syst. Appl. 37, 5745–5750 (2010)
7. Zhang, G., Keil, M., Rai, A., Mann, J.: Predicting information technology project escalation: A neural network approach. Eur. J. Oper. Res. 146, 115–129 (2003)
8. Lim, S.H., Nam, K.: Artificial Neural Network Modeling in Forecasting Successful Implementation of ERP Systems. Int. J. Comput. Intell. Res. 2, 115–119 (2006)
9. Tian, L., Noore, A.: Evolutionary neural network modeling for software cumulative failure time prediction. Reliab. Eng. Syst. Saf. 87, 45–51 (2005)
10. Narayana Naik, G., Gopalakrishnan, S., Ganguli, R.: Design optimization of composites using genetic algorithms and failure mechanism based failure criterion. Compos. Struct. 83, 354–367 (2008)
11. Chen, Y., Kilgour, M., Hipel, K.: Case-based distance method for screening in multiple-criteria decision aid. Omega 36, 373–383 (2008)
12. Zhao, K., Yu, X.: A case based reasoning approach on supplier selection in petroleum enterprises. Exp. Syst. Appl. 38, 6839–6847 (2011)
13. Quinlan, J.R.: Induction of Decision Trees. Mach. Learn. 1, 81–106 (1986)
14. Wettschereck, D., Aha, D.W., Mohri, T.: A review and empirical comparison of feature weighting methods for a class of lazy learning algorithms. Artif. Intell. Rev. 11, 273–314 (1997)
15. Chou, T.Y., Chou, S.T., Tzeng, G.H.: Evaluating IT/IS investments: A fuzzy multi-criteria decision model approach. Eur. J. Oper. Res. 173, 1026–1046 (2006)

Data Mining for Seasonal Influences in Broiler Breeding Based on Observational Study

Peijie Huang[1], Piyuan Lin[1,*], Shangwei Yan[1,2], and Meiyan Xiao[1]

[1] College of Informatics,
South China Agricultural University,
Guangzhou 510642, Guangdong, China
[2] Information Centre,
Guangdong Wens Food Group Limited Company,
XinXing 527400, Guangdong, China
pyuanlin@163.com

Abstract. For the modern poultry breeding companies, it is worthwhile to extract valuable knowledge from the massive historical data to help future production and management. However, data analysis and mining of poultry raising dataset is a challenge due to the complexity and uncertainty bring by the influence of environmental and physiological factors. In this paper, data mining based on observational study is proposed for the research of seasonal influences in broiler breeding. Systematic observational study with the statistical analysis and data mining technology is adopted including macro analysis, exploratory data analysis, and modeling and prediction. Case study using the broiler growth dataset of the most famous poultry raising company in China shows the effectiveness of our approach.

Keywords: Observational study, Data mining, Seasonal influences, Broiler breeding.

1 Introduction

Modern poultry breeding companies hope to extract valuable knowledge from the massive historical data to help future production and management. However, because of the complexity and uncertainty bring by the influence of environmental and physiological factors, data analysis and mining of poultry raising dataset is a challenge. In animal production genotype and environment are two main factors that affect output. Ample research has demonstrated that meteorological factor is one of the ambient environmental factors that play important part in broiler production [1-5]. In broiler breeding, growth performance parameters have obvious seasonal variation.

Much existing literature of broiler growth performance deals with the observation or fitting under preselected ambient factor levels [1-3,5-8]. These works have a certain practical significance for broiler breeding. However, for the farmers who carry on the large scale breeding all the year round, it is impractical to adjust all the ambient

---

* Corresponding author.

B. Liu and C. Chai (Eds.): ICICA 2011, LNCS 7030, pp. 25–32, 2011.

factors to suitable levels systematically like in the laboratory. The levels of ambient environmental factors cannot be controlled and thus, much of the time it involves a situation in which the data structure can only be a monitoring of the data from the poultry house across time. In this case it may require an observational study. Based on the observational historical data, we try to uncover the seasonal growth rule hidden among the broiler data, which is regulated by both the natural law of seasons and the biological law of poultry upgrowth. Given a certain uncontrollable environmental condition, the knowledge about seasonal influences in broiler breeding gained from our approach will provide decision support for broiler farmers.

The rest of this paper is organized as follows. In the next section, we briefly discuss the distinctness between observational study and designed experiment, and present the technical framework of the proposed observational study. Detail introduction of data mining based on observational study for broiler breeding is given in Section 3. Finally, Section 4 lists some conclusions and discussion.

2   Observational Study: Method and Technology

2.1   Observational Study vs. Designed Experiment

The goal of our research topic is to determine if and how seasonal influences affect the broiler growth performance. Broiler growth performance parameters are the response variables. The meteorological factors are set as the explanatory variables. There are two methods for collecting and analyzing data, observational study and designed experiment [9]. In designed experiment, researcher assigns the individuals in a study to a certain group, intentionally changes the value of the explanatory variable, and then records the value of the response variable for each group. In contrast, an observational study measures the value of the response variable without attempting to manipulate or influence the value of either the response of explanatory variables. That is, in an observational study, the researcher has to find that group or the thing he wants to study occurring naturally.

The prominent characteristic of observational study is that factor levels could not be preselected. But for large scale poultry raising, it is not possible to conduct an well designed experiment, thus it should confront such challenge. When compared to designed experiments, the disadvantage in observational study is that unlike the former, observational studies are at the mercy of nature, environmental or other uncontrolled circumstances that impact the ranges of factors of interest [10]. Also, another potentially problem in observational study is that differences found in the fundamental response may be due to other lurking variables that are not considered in a study [9]. To tackle these problems, it is much in need of systematic analysis method and powerful technology.

2.2   Technical Framework of the Proposed Observational Study

We perform a systematic observational study based data mining for seasonal influences in broiler breeding, from macro analysis, exploratory data analysis, to modeling and prediction. The technical framework is shown in Fig. 1.

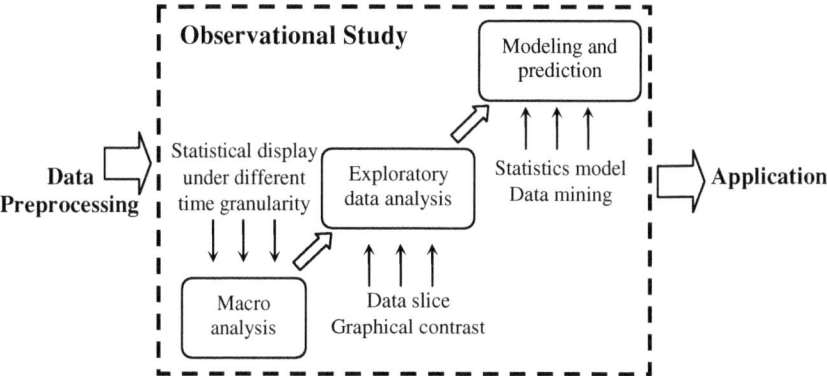

Fig. 1. Technical framework of the proposed observational study

Traditional statistical analyses are used in macro analysis and exploratory data analysis. Often a summary of a collection of data via data slice and graphical display can provide insight regarding the system from which the data were taken. Data mining technology is combined with traditional statistical model for modeling and prediction. It is worthwhile to note that, data preprocessing is an important issue for effective data mining, as real-world data tend to be incomplete, noisy, and inconsistent [11]. The detail about data preprocessing of the broiler production data can be found in our former work [12]. Base on it, this paper foucses on data analysis and mining, the knowledge gained can be applied to large scale broiler breeding.

3   Data Mining Based on Observational Study for Broiler Breeding

3.1   Materials

The broiler growth dataset is provided by Guangdong Wens Food Group Limited Company, which is the most famous poultry raising company in China. The dataset of *short-feet buff B* of the breeding area of Guangdong province of China, which is the breed with the largest numbers, is taken as case study to show the effectiveness of our approach, in which we use the hen growth data during 2004 to 2007 for macro analysis and the hen growth data of 2007, which consists of 5714 data, remaining 4209 data after data preprocessing is selected for deeper studies of both exploratory data analysis and modeling. Each datum corresponds to the broilers raised by a certain farmer in one time period with the amount varying from several thousand to several tens of thousand. In modeling and prediction, we select 70% samples randomly for training, and the rest for testing. The dataset of meteorological factors is provided by Guangdong Provincial Climate and Agrometeorological Center.

3.2   Macro Analysis

At the first stage, the universality of seasonal influences in broiler breeding is validated by macro analysis. Broiler growth performance parameters including rate for sale, mean weight, feed conversion ratio, and mean drug cost of *short-feet buff B* of Guangdong province of China during 2004 to 2007 are shown in Fig.2.

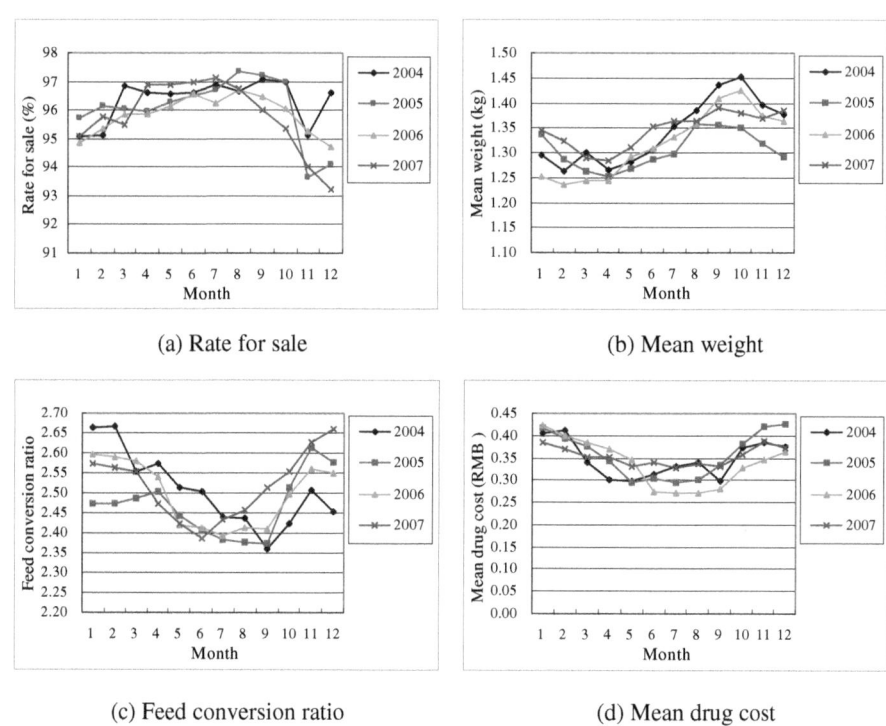

(a) Rate for sale   (b) Mean weight

(c) Feed conversion ratio   (d) Mean drug cost

Fig. 2. Growth performance parameters of short-feet buff B of Guangdong province of China

The data shown in Fig.2 is the mean of the broilers adopted in the same month. As we can see from Fig.2, all the observed growth performance parameters take on entirely seasonal fluctuations. The rate for sale increased steadily after January, reaches its peak at July or August, and then decreases gradually, with worst-case at November or December for possible influence of disease and low temperature environment. It can be observed that feed conversion ratio, and mean drug cost of broilers reach highest and lowest in summer and in winter respectively. That may be because high environmental temperatures depress food intake and body weight and also cause deterioration in the food conversion ratio. Mean weight found to be best in autumn, and worst in summer.

3.3   Exploratory Data Analysis

After the macro analysis, a deeper study is done by exploratory data analysis, in which data slice and graphical contrast are adopted. In general, most people are accustomed to use monthly display, just as we take in the macro analysis above. However, the month time granularity is too large to catch the variety of meteorological factors. In contrast, the short time granularity such as day can reflect the daily change, but it often causes over fitting and leads to unreliable prediction results. As a compromise, we choose moderate time granularity such as ten-day period to make further study. One month is divided into three ten-day periods, i.e. early, middle and late. Some interesting phenomenon is found in the deep analysis according to the observational data. The most representative one is shown in Fig.3.

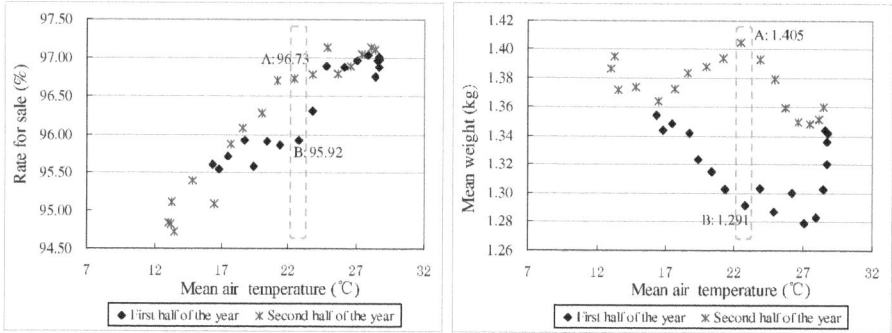

Fig. 3. Growth performance comparison of first and second half of year (sample data: short-feet buff B of Guangdong province of China, 2007)

The data shown in Fig.3 is the mean of broilers with the mid date of breeding period in the same ten-day period. As we can see in Fig. 3, two set of observational data, A (22.48°C) and B (22.81°C), in second and first half of year respectively, which have approximating mean air temperature in the entire growth period, have unbelievable different rate for sale and mean weight. The growth performance of second half of year is better than first half of year significantly, involving an extended span of time. Such phenomenon can confirm some summarization of the experience in poultry science. In poultry science viewpoint, the full growing stage of broiler can be divided into chickling stage (the first 4 weeks) and adult chicken stage, and broiler of these two stages have different seasonal adaptability, that is "chickling fear of cold and adult chicken hot afraid". Replacing the mean air temperature in the entire growth period with the combination in chickling stage and adult chicken stage, we get that A (26.82°C, 20.17°C) and B(18.79°C, 24.80°C). The air temperature combination of datum A is more suit for both chickling and adult chicken, apparently, and thus gets better growth performance. So we use the meteorological factors of both chickling stage and adult chicken stage as explanatory variables when modeling and prediction.

3.4   Modeling and Prediction

Modeling is an effective mean to provide prediction for better decision support. Because seasonal factors only have partial linear influence on broiler growth performance, we adopt multiple linear regression (MLR) [13] and neural network [14] models to catch both linear and nonlinear correlations. Among all the meteorological factors, air temperature is the most relative one to broiler growth performance. In this paper, we take the influence of the air temperature to the rate for sale for example to introduce the modeling. In both MLR and neural network, we use the ten-day mean air temperature in chickling stage and adult chicken stage as inputs, and set the rate for sale as output. For each broiler datum, we compute the ten-day periods of both mid dates of chickling and adult chicken stages. Then, the broiler data are classified by the computed breeding ten-day period combination. The broiler data with the same breeding ten-day period combination are then aggregated into one input, and only those inputs with more than 20 samples are adopted in training and more than 5 samples are used in testing.

The following MLR equation is fit for the training data:

$$y = 0.000711x_1 + 0.000618x_2 + 0.93422 \tag{1}$$

where y is the rate for sale, and x_1 and x_2 are the ten-day mean air temperature in chickling stage and adult chicken stage respectively.

Fig. 4 shows the real observed values and predicted rate for sale for both MLR and neural network (labeled as "NN" in Fig. 4) models, using the testing data.

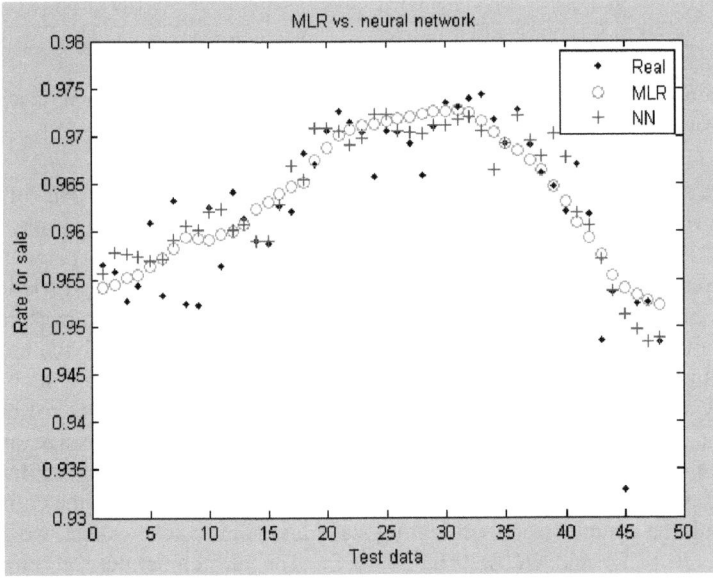

Fig. 4. Comparison of MLR and neural network in prediction

The goodness of fits for the obtained MLR and neural network model was calculated by mean percentage error (MPE) and mean square error (MSE).

The MPE and MSE are computed as

$$MPE = \frac{1}{n}\sum_{t=1}^{n}\frac{y_t - \hat{y}_t}{y_t} \tag{2}$$

$$MSE = \frac{\sum_{t=1}^{n}(y_t - \hat{y}_t)^2}{n} \tag{3}$$

where y_t equals the observed value at time t, $\hat{y}_t$ equals the estimated value, and n equals the number of observations.

Table 1 shows the statistics for the MLR and neural network models for predicting broiler rate for sale.

Table 1. Model statistics for MLR and neural network for predicting rate for sale

Model	Statistic	
	MPE	MSE
MLR	0.10%	2.039E-05
NN	0.11%	2.034E-05

As we can see from Table 1, both neural network model and MLR model get good peformance in prediction, which are of comparable ability.

4   Conclusions and Discussion

This paper presents data mining for seasonal influences in broiler breeding based on systematic observational study, from macro analysis, exploratory data analysis, to modeling and prediction. Case study using the broiler growth dataset of the most famous poultry raising company in China shows the effectiveness of the proposed method.

Our approach validates some summarization of the experience from broiler farmers and industry experts. For example, the phenomenon found in exploratory data analysis as mentioned in Subsection 3.3 gives the validation to the industry knowledge of "chickling fear of cold and adult chicken hot afraid". In addition, from the gained MLR model in Subsection 3.4, we can found that the ten-day mean air temperature in chickling stage has greater influence to rate for sale than that of adult chicken stage, which validates that "higher mortality occur in chickling stage than adult chicken stage". Moreover, compare to the rough evaluation of growth performance based on "high" and "low" ambient temperature, the prediction models gained from systematic observational study can give better guiding for broiler farmers all the year round.

Regarding the prediction models, due to only partial linear correlation between seasonal factors and broiler growth performance, we adopt both linear regression and neural network model for nonlinear fitting. Neural network model and MLR model get comparable ability in prediction. So one can consider the combined prediction in actual application. Future work includes reseach on more prediction technologies, such as Support Vector Machine (SVM) which is a learning technique based on the structural risk minimization principle, and making use of the advantage of fuzzy logical and rough set theory in dealing with uncertainty.

Due to the limited objective conditions, our research needs further improvement. Along with the development of Internet of Things (IOT), automatic monitoring condition of poultry breeding will get rapid promotion. Based on multiview and multilevel granularity observational data, the analysis and mining using systematic observational study proposed in this paper is expected to provide better decision support for large scale broiler breeding.

Acknowledgments. This work is supported by Science and Technology Planning Project of Guangdong Province, China under Grant No. 2007A020300010 and 2010B020315024, the Foundation for Distinguished Young Talents in Higher Education of Guangdong, China under Grant No. LYM09034.

References

1. Donkoh, A.: Ambient temperature: A Factor Affecting Performance and Physiological Response of Broiler Chickens. International Journal of Biometeorology 33, 259–265 (1989)
2. Cahaner, A., Leenstra, F.: Effects of High Temperature on Growth and Efficiency of Male and Female Broilers from Lines Selected for High Weight Gain, Favorable Feed Conversion, and High or Low Fat Content. Poult. Sci. 71, 1237–1250 (1992)
3. Bonnet, S., Greraert, P.A., Lessire, M., et al.: Effect of High Ambient Temperature on Feed Digestibility in Broiler. Poult. Sci. 76, 857–863 (1997)
4. Akyuz, A.: Effects of Some Climates Parameters of Environmentally Uncontrollable Broiler Houses on Broiler Performance. J. Anim. Vet. Adv. 8, 2608–2612 (2009)
5. Olanrewaju, H.A., Purswell, J.L., Collier, S.D., et al.: Effect of Ambient Temperature and Light Intensity on Physiological Reactions of Heavy Broiler Chickens. Poult. Sci. 89, 2668–2677 (2010)
6. Aggrey, S.E.: Comparison of Three Nonlinear and Spline Regression Models for Describing Chicken Growth Curves. Poult. Sci. 81, 1782–1788 (2002)
7. Roush, W.B., Dozier III, W.A., Branton, S.L.: Comparision of Gompertz and Neural Networks Models of Broiler Growth. Poult. Sci. 85, 794–797 (2006)
8. Blahová, J., Dobšíková, R., Straková, E., et al.: Effect of Low Environmental Temperature on Performance and Blood System in Broiler Chickens (Gallus domesticus). Acta Vet. Brno 76, S17–S23 (2007)
9. Sullivan, M.: Statistics: Informed Decisions Using Data, 3rd edn. Pearson Prentice Hall (2008)
10. Walpole, R.E., Myers, R.H., Myers, S.L., et al.: Probability & Statistics for Engineers & Scientists, 9th edn. Pearson Prentice Hall (2011)
11. Han, J., Kamber, M.: Data Mining: Concepts and Techniques, 2nd edn. Morgan Kaufmann (2006)
12. Xiao, M.Y., Lin, P.Y., Yan, S.W., et al.: Data Preprocessing of Poultry Breeding Production Data. Journal of Anhui Agricultural Sciences 38, 20707–20709 (2010)
13. Weisberg, S.: Applied Linear Regression, 3rd edn. Wiley, New York (2005)
14. Haykin, S.S.: Neural Networks and Learning Machines, 3rd edn. Pearson Prentice Hall (2009)

Application of Eco-service Value Theory to Environmental Impact Assessment of Land Use Planning

Ai-Qing Sun[1], Ke-Ning Wu[1], and Duan-Hai Cao[2]

[1] School of Land Science and Technology,
China University of Geosciences,
Beijing, 100083, China
[2] Chinese Academy of Land & Resources Economics,
Beijing, 101149, China
sunaiqing@gmail.com,
knwu@sohu.com,
cdhjyy@hotmail.com

Abstract. The purpose of this paper is to evaluate the environmental impact of the implementation of General Plan for Land Use in Taiyuan (2006-2020). The research is carried out based on the eco-service value theory. Firstly, area change of each land use ecosystem before and after the implementation of the Plan is comparatively analyzed; and then on the basis of former researches, the unit eco-service value of each land use ecosystem is modified combining with the local situation of Taiyuan; finally, the total eco-service values both in 2005 and 2020 are calculated and sensitivity analysis is made to test the accuracy of the eco-service value coefficient and confirm the estimated results. The studies indicate that the eco-service value of Taiyuan will increase by 17.08%, mainly due to the significant area growth of garden plot and woodland, which have relatively high eco-service value. Thus it can be concluded that General Plan for Land Use in Taiyuan (2006-2020) is practicable, and it will contribute to the local sustainable development.

Keywords: land use planning, environmental impact assessment, eco-service value, Taiyuan.

1   Introduction

In china, land use planning acts as the direct guideline for rational land utilization. Now with the constant deterioration of ecological environment, to carry out environmental impact assessment has reached an unprecedented strategic position, and it becomes a necessary scientific basis for decision making of land use planning. The eco-service value theory explains the various ecosystem service functions and their values. Ecosystem service function is the natural condition and utility of ecosystem that we humans depend upon for survival and eco-service value is the quantitative description with monetary form of ecosystem service function [1]. Land is the carrier of all terrestrial ecosystems, so land use changes directly affect the type

B. Liu and C. Chai (Eds.): ICICA 2011, LNCS 7030, pp. 33–40, 2011.

and intensity of ecosystem services. We can learn the environmental impact of land use change through calculating the eco-service value. In recent years, by accounting eco-service value, researchers home and abroad have focused on the study of the environmental impact of land use change in different regions, such as Texas [2], Pinggu district of Beijing[3], Baoan District of Shenzhen[4], Shanghai[5], etc. Implementation of land use planning leads directly to the variation of land use quantity and structure, so it is undoubtedly feasible and practical to evaluate the environmental impact applying the eco-service value theory. To prove this point, an empirical study is carried out in this paper, taking Taiyuan city as an example. Taiyuan city is ecologically sensitive and fragile, which limits land utilization to a certain extent, thus this study is of great theoretical significance and application value. The research raw data and material adopted in this paper are from General Plan for Land Use of Taiyuan (2006-2020). In this plan, the base year is 2005, and the planning period is from2006 to 2020. The procedure may be as follows: firstly, area changes of all land use types should be analyzed; second, the unit service value of land use ecosystem should be estimated, combing with the local conditions; finally, based on the studies above, the change of the total eco-service value can be quantified, from which we can draw a conclusion that whether the land use planning scheme is reasonable from the perspective of environmental protection.

2   Methodology

2.1   Area Changes of Land Use Ecosystems

According to the research needs, we divided the land use ecosystem of Taiyuan into 8 types: cultivated land, garden plot, woodland, rangeland, residential and industrial area, traffic land, water, and other lands, where compared with the usage classification, the residential and industrial area include towns, village land, land for industrial and mining use; water stands for water area and water conservancy facilities; other lands include other farmland and natural reserved area. After processing the raw data, we get the area of the each land use ecosystem of Taiyuan in 2005 and 2020(as shown in Table 1).

Table 1 shows that besides cultivated land, rangeland and other lands, area of most land use ecosystem trend to increase from 2006-2020. Area of woodland increases significantly, as its proportion to the total land area rises from 31.81% to 41.22%, which mainly benefits from the *Grain for Green* project and afforestation. Area of residential and industrial area increases by 2.09%; area of traffic land increases by 0.47%; area of garden plot increases by 0.97%; area of water increases by 0.09%. These data suggest that while accelerating economic development, much attention is paid to environment protection and ecological improvement in Taiyuan during the planning period. Among cultivated land, rangeland and other lands, cultivated land decreases the least, because land comprehensive regulation is greatly promoted in Taiyuan, which is the key to achieve the goal of dynamic equilibrium of the total amount of cultivated land.

Table 1. Area Changes of Land Use Ecosystems from 2005 to 2020 unit: hm2

Ecosystem type	Cultivated land	Garden plot	Wood land	Range land	Residential and industrial area	Traffic land	Water	Other land
Area in 2005	131873	19673	219823	39750	54619	4214	13051	207993
Ratio in 2005/%	19.08	2.85	31.81	5.75	7.90	0.61	1.89	30.10
Area in 2020	121453	26373	284823	17750	69041	7437	13706	150413
Ratio in 2020/%	17.58	3.82	41.22	2.57	9.99	1.08	1.98	21.77
Area change	-10420	6700	65000	-22000	14422	3223	655	-57580
Ratio change/%	-1.50	0.97	9.41	-3.18	2.09	0.47	0.09	-8.33
change rate/%	-7.90	34.06	29.57	-55.35	26.40	76.48	5.02	-27.68

2.2   Unit Eco-service Value of Land Use Ecosystem

On the basis of precedent achievements related, eco-service functions in Taiyuan city may cover 9 items: atmospheric regulation, climatic regulation, water conservation, soil formation and protection, waste disposal, biodiversity maintenance, food production, raw materials production and cultural entertainment [6]. Then the next thing to do is to quantify the functions above, in other words, we should estimate the value of the eco-service functions. We will do it separately, according to different types of ecosystem.

While estimating the unit eco-service value of cultivated land ecosystem, we make a reference to the research result obtained by Xie Gao-Di(2005). In his study, he put forward "the equivalent of ecosystem services evaluation by croplands in China" and "the biomass factors of different provinces in China" [7], based on which we adjust the unit eco-service value of Taiyuan city. The calculation Formula is:

$$si = (f / F) \times Si \tag{1}$$

s_i acts as the unit value of a certain eco-service function of research region, f acts as the biomass factor of research region, F acts as the national average biomass factor, S_i acts as the national average unit value of a certain eco-service function. Take the eco-service function of food production for example. According to the research result of Xie Gao-Di, the biomass factor of Taiyuan city is 0.46, that is f=0.46; the national average biomass factor is 1.00, that means F=1.00; and the national average unit value of food production is 884.9yuan/hm².a, that is S_i=884.9. So we can use the above-mentioned formula to calculate the unit service value of food production of cultivated land in Taiyuan city, which is s_i=0.46/1×884.9=407.05 (Yuan/hm².a). In the same say, we calculate the unit value of other 8 eco-service functions of cultivated land.

The unit eco-service value of woodland results from the modification of correlated research outcomes [8]-[9]. As for rangeland, previous researchers have already identified its unit eco-service value in terms of different regions [10], which can be directly used in this paper. As for the unit eco-service value of garden plot, we take the average value of woodland and rangeland. The unit eco-service value of other lands is the average value of desert and rangeland [11]. As is widely accepted, the residential and industrial area, and traffic land have no eco-service value. Thus, according to methods above, the unit eco-service values of all land ecosystems in Taiyuan are calculated (as shown in Table 2).

Table 2. Unit Eco-service Value of Each Land Use Ecosystem   unit: Yuan/hm2.a

Ecosystem type	Cultivated land	Garden plot	Wood land	Range land	Water	Other land
Atmospheric regulation	203.50	1249.98	2411.35	88.60	0.00	44.30
Climatic regulation	362.25	1039.80	1860.17	219.42	407.00	109.71
Water conservation	244.21	1212.44	2204.63	220.25	18033.20	123.38
Soil formation and protection	594.27	628.06	902.56	353.56	16086.60	181.18
Waste disposal	667.55	1957.41	2686.89	1227.92	8.80	622.81
Biodiversity maintenance	288.97	1413.20	2245.97	580.43	2203.30	440.62
Food production	407.05	435.62	68.91	802.33	88.50	405.57
Raw materials production	40.71	898.12	1791.26	4.97	8.80	2.49
Cultural entertainment	4.05	487.71	881.85	93.56	3840.20	51.18
Total	2812.58	9322.32	15053.59	3591.04	40676.40	1981.22

2.3   The Total Eco-service Value of Land Use Ecosystems

With the data shown in Table.1 and Table.2, the total eco-service value of land use ecosystems and its variation caused by land use planning can be calculated. The calculating formula is as follows:

$$ESV = \sum_{j=1}^{n} s_j \times A_j \qquad (2)$$

In this formula, ESV stands for the total eco-service value of the research region. Sj stands for the unit eco-service value of a certain land use ecosystem. Aj stands for the area of the certain land use ecosystem. n stands for the number of land use ecosystems. The total eco-service values of 2005 and 2020 in Taiyuan city are listed in the table below (Table 3).

Table 3 shows that in 2005 the eco-service value of Taiyuan city is 4949million Yuan, while in 2020 it reaches 5794million Yuan, which means the added value is 845million, up by 17.08%. The eco-service value of garden plot increases the most, varying from 183 million in 2005 to 246 million in 2020, up by 34.06%. The eco-service value of woodland increases the second fastest, rising from 3309 million to 4288 million, up by 29.57%. The eco-service value of rangeland decreases significantly, varying from 143 million to 64 million, down by 55.35%. This is mainly caused by the dramatic reduction in the area of rangeland during the planning period, due to the occupation for non-agricultural construction purposes. Although rangeland drops considerably in area, it has no big influence on the upward trend of the total eco-service value of the whole region, because the proportion of rangeland to total land area is not very high, and the unit eco-service value is relatively low.

Table 3. The Total Eco-service Value of Taiyuan from 2005 to 2020   unit: 108yuan/a

Ecosystem	Cultivated land	Garden plot	Wood land	rangeland	Water	Other land	Total
Total value in 2005	3.71	1.83	33.09	1.43	5.31	4.12	49.49
Value ratio in 2005/%	7.49	3.71	66.86	2.88	10.73	8.33	100.00
Total value in 2020	3.42	2.46	42.88	0.64	5.58	2.98	57.94
Value ratio in 2020/%	5.90	4.24%	74.0	1.10	9.62	5.14	100.00
Total value change	-0.29	0.62	9.78	-0.79	0.27	-1.14	8.45
Value change rate/%	-7.90	34.06	29.57	-55.35	5.02	-27.68	17.08

2.4   Sensitivity Analysis

In order to make the results more cogent and precise, the accuracy of the unit eco-service value (value coefficient) is tested by introducing a concept of "the coefficient of sensitivity (CS)". "The coefficient of sensitivity" reflects the sensitivity of the eco-service value to value coefficient, and it can be calculated through the elastic coefficient method, which is commonly used in economics [12]. The calculation formula is as follows:

$$CS = \left| \frac{(ESV_a - ESV_b) / ESV_b}{(VC_{aj} - VC_{bj}) / VC_{bj}} \right| \quad (3)$$

In the formula, VCaj stands for the adjusted unit eco-service value of a certain land use ecosystem; while VCbj stands for the unadjusted unit eco-service value. ESVa stands for the total eco-service value after the adjustment of the unit eco-service value; while ESVb stands for the unadjusted total eco-service value. If the calculation result of CS is less than 1, then the ESV is lack of elasticity; if CS is beyond 1, then the ESV is elastic to the value coefficient, and the larger CS is, the more important is the accuracy of the unit eco-service value.

In this paper, we both up-regulate and down-regulate the value coefficient by 50%, and the total eco-service values after the adjustment are separately calculated. Then by using formula (3) the coefficient of sensitivity is estimated (as shown in Table 4), according to which we can make the sensitivity analysis.

Table 4. The Adjusted Total Value and the Coefficient of Sensitivity

Ecosystem type	The adjusted ESV/10^8yuan		Influence of the adjustment			
			2005		2020	
	2005	2020	ESV change rate	CS	ESV change rate	CS
Cultivated land VC+50%	51.35	59.65	3.75%		2.95%	
Cultivated land	47.64	56.24	-3.75%	0.0749	-2.95%	0.0590
Garden plot VC+50%	50.41	59.17	1.85%		2.12%	
Garden plot VC-50%	48.57	56.71	-1.85%	0.0371	-2.12%	0.0424
Woodland VC+50%	66.04	79.38	33.43%		37.00%	
Woodland VC-50%	32.95	36.51	-33.43%	0.6686	-37.00%	0.7400
rangeland VC+50%	50.20	58.26	1.44%		0.55%	
rangeland VC-50%	48.78	57.62	-1.44%	0.0288	-0.55%	0.0110
Water VC+50%	52.15	60.73	5.36%		4.81%	
Water VC-50%	46.84	55.16	-5.36%	0.1073	-4.81%	0.0962
Other land VC+50%	51.55	59.43	4.16%		2.57%	
Other land VC-50%	47.43	56.45	-4.16%	0.0833	-2.57%	0.0514

Table 4 shows that the sensitive degree of the total eco-service value to the unit eco-service value of rangeland is the lowest, as the coefficient of sensitivity is only 0.0110~0.0288, which means that when the value coefficient of rangeland increases by 1%, the total eco-service value only increases by 0.0110 to 0.0288 points of percentage. The sensitive degree of the total eco-service value to the unit eco-service value of woodland is the highest, as the coefficient of sensitivity is 0.6686~0.7400, which means that when the value coefficient of woodland increases by 1%, the total eco-service value increases by 0.6686 to 0.7400 points of percentage. Moreover, apart from other land use ecosystems, the garden plot coefficient sensitivity and the woodland coefficient sensitivity increase from 2005 to 2020, which means the variation of their value coefficient has growing influence on the total service value. However, overall, the total eco-service value is lack of elasticity to value coefficient, with the coefficient of sensitivity is lower than1 both after up-regulation and down-regulation. Thus the unit eco-service value of each land use ecosystem adopted in this paper is basically reasonable and the estimated result of the total eco-service value is trustworthy.

3   Conclusion

The main contents and conclusions of this paper are as the follows:

(1) After analyzing the land use change due to General Plan for Land Use of Taiyuan (2006-2020), it is found out that while the area of cultivated land, rangeland and other land decreases, mainly due to occupation for construction use, the area of woodland, construction land, garden plot and water trend to increase, especially woodland, its area increases 65,000hm2. The data reflects that the government attaches greater importance to the coordination of economic development and environment protection.

(2) According to the existing research achievements, the unit eco-service value of each land use ecosystem is modified according to the local situation of Taiyuan city, then the total eco-service value is estimated, and the concept of coefficient of sensitivity is introduced to test the accuracy of the value coefficient and the reliability of the research results.

(3) The results show that the regional eco-service value will increase largely by 17.08%, due to the implementation of the land use planning, so the planning scheme is undoubtedly feasible from the angle of environment protection and sustainable development. However, what needs to be stressed is that it is very important to carry out the Grain for Green project and afforestation during the planning period so as to achieve the ecological goal.

(4) From this study we conclude that the method of applying the theory of eco-service value to environmental impact assessment of land use planning does have great advantages, for it is simple and practical, and it realizes the combination of qualitative analysis and quantitative evaluation. However, it still need to be improved and perfected, especially there is no solution accepted universally to the determination of some parameter, like the unit eco-service value of land for construction, such as residential land, industrial land ,traffic land, etc. With the in-depth study of the theory of eco-service value, its application will surely obtain fast development in the near future.

References

1. Huang, Y.-C., Chen, S.-L., Dai, F.: The Ecosystem Services Value Response to LUCC in Zhangzhou. Journal of Fujian Normal University (Natural Science Edition) 25(3), 114–118 (2009)

2. Kreuter, U.P., Harris, H.G., Matlock, M.D.: Change in ecosystem service values in the San Antonio area. Texas. Ecological Economics 39, 333–346 (2001)
3. Bai, X.-F., Chen, H.-W.: The Value of Ecosystem Service of Land Use–a Case Study in Pinggu. Journal of Beijing Agriculture College 18(2), 111–119 (2003)
4. Ran, S.-H., Lu, C.-H., Jia, K.-J., Qi, Y.-H.: Environmental Impact of Land Use Change of Baoan District in Shenzhen. China Population Resources and Environment 16(5), 72–77 (2006)
5. Cheng, J., Yang, K., Zhao, J., Wu, J.-P.: Impact Assessment of Land Use Change in Center District of Shanghai based on Ecosystem Services Value. China Environmental Science 29(1), 95–100 (2009)
6. Wu, K.-N., Zhao, K., Zhao, J.-S.: The Environmental Impact Assessment of Land Use Planning Based on the Theory of Ecosystem Services Value: Taking Anyang as an Example. China Land Resources 22(3), 23–28 (2008)
7. Xie, G.-D., Xiao, Y., Zhen, L., Lu, C.-X.: Study on Ecosystem Services Value of Food Production in China. China Journal of Eco-agriculture 13(3), 10–13 (2005)
8. Zhao, T.-Q., Ou Yang, Z.-Y., Zhen, H.: Forest Ecosystem Services and Their Valuation in China. Journal of Natural Resources 19(4), 480–491 (2004)
9. Wang, B., Ren, X.-X., Hu, W.: Assessment of Forest Ecosystem Services Value in China. Scientia Silvae Sinicae 47(2), 145–153 (2011)
10. Xie, G.-D., Zhang, Y.-L., Lu, C.-X., Zheng, D., Cheng, S.-K.: Study on Valuation of Rangeland Ecosystem Services of China. Journal of Natural Resources 16(1), 47–52 (2001)
11. Xie, G.-D., Lu, C.-X., Leng, Y.-F., Zheng, D., Li, S.-C.: Ecological Assets Valuation of the Tibetan Plateau. Journal of Natural Resources 18(2), 189–196 (2003)
12. Pang, S., Liu, K., Ji, W.-H.: Effects of Land-used Change on Ecosystem Service Value in Yan'an City. Ground Water 33(1), 154–157 (2011)

Numerical Analysis of Western Hills Highway Tunnel Excavation and Support

Deq-qing Gan, Hong-jian Lu, Xiao-na Lu, and Zhong-jian Yang

Hebei United University, Tangshan 063009, China
luhongj2006@sina.com

Abstract. This paper applied the finite element analysis software MIDAS/ GTS, taking use of ground structure method analysis stability in the process of tunnel excavation and support the western hills. The results showed that the different sections during the excavation the maximum principal stress within the rock focused on steel feet, it is timely lock pin bolt when steel erection, taking use of small catheters grouting and other methods reinforcement of the rock arch foot when necessary. From the perspective of lining deformation, the maximum deformation at the end of the foot and the location of the vault subsidence, so we should pay attention to the monitoring of settlement of vault, the gradual excavation of core soil, ensure the construction and structural safety, and should be based on engineering facilities in time to control the chamber deformation.

Keywords: highway tunnel,excavation, supporting, numerical analysis.

1   Project Overview

Western hills tunnel is located in foothills of Yan Shan, The type of landscape is structure of ablation hilly area, the more complex topography, hilly gully development. Tunnel surrounding rock is mainly to Sandy mudstone interceded conglomerate, Jurassic conglomerates in the system after the city group, Jurassic sandstone system after the city group. The tunnel was designed by the standard of two-way six-lane highway, the length of left line was 810m, the right one was 737m, net width of the main tunnel for the tunnel construction clearance was 14.50m, and net height is 5.0m. The surrounding rock of tunnel entrance with low quality and there is bias, the stability of the excavation and support has relatively strong effect. This paper applied finite element analysis software MIDAS / GTS made in-depth analysis to the stability during the process of construction.

2   Establish of Tunnel Excavation Numerical Simulation Model

2.1   Calculation Methods and Underlying Assumptions

The calculation method selected stratigraphic structure method. It was analyzed by two-dimensional plane strain, the rock material is assumed to be isotropic material in

B. Liu and C. Chai (Eds.): ICICA 2011, LNCS 7030, pp. 41–47, 2011.
© Springer-Verlag Berlin Heidelberg 2011

finite element analysis. Rock material criterion use Druker-Prager yield criterion, and selected the associated flow rule to calculate, assuming that material more than tensile stress, which cannot withstand tensile stress. In addition, the concrete lining and shotcrete support, structural materials are assumed to be elastic.

2.2   Finite Element Model and Material Parameters

Select the left line IV rock typical cross section ZK8+790 to simulate.
   Calculation of material parameters was shown as table 1-2.

Table 1. Solid element calculation parameter table

Material name	Modulus/MPa	Poisson's ratio	Bulk density/KN/m³	Cohesion/KPa	Friction angle/°
Weathered	500	0.45	17	50	20
Strong weathering	1200	0.40	19	80	22
Breezed	3500	0.32	22	400	30

Table 2. Beam, bar unit calculates parameter table

Material name	Modulus/MPa	Bulk density /KN/m³	Cross-sectional area /m²	Moment of inertia I/ m^4
Anchor	210000	78.5	0.000380	50
Initial lining	23000	25	0.26	80
Second lining	29500	25	0.45	400

3   Numerical Simulation Results and Analysis

In this paper applied structural design finite element analysis software-MIDAS/GTS, numerical simulation according to the construction steps, the simulation results shown in Figure 1 ~ 8.

3.1   The First Step Results and Analysis

The section after excavation, the maximum horizontal displacement occurs in the cutting face, arch around the waist position was maximum 1.16mm, vertical displacement occurred in the vault, the top was sinking 7.0mm, bottom uplift 6.2mm; Displacement there is a greater value, and the corresponding place with large compressive stress, shear stress is large.

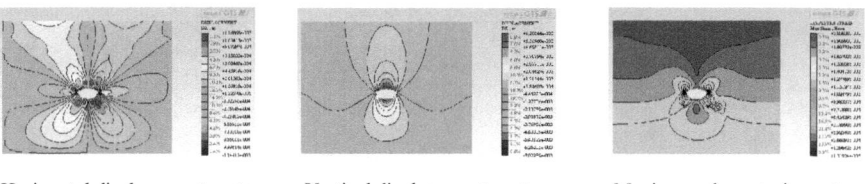

Horizontal displacement contour   Vertical displacement contour   Maximum shear strain contour

Fig. 1. The first step the simulation results

3.2   The Second Step Results and Analysis

The displacement and stress have been effectively controlled after support, Maximum bolt axial force was took place in the vault and the arch lumbar, the maximum was 0.44KN; Axial force, shear force, bending moment which lining bolt in the lock at the foot was greater, respectively, 13.8KN, 51KN, 1.1KN / M.

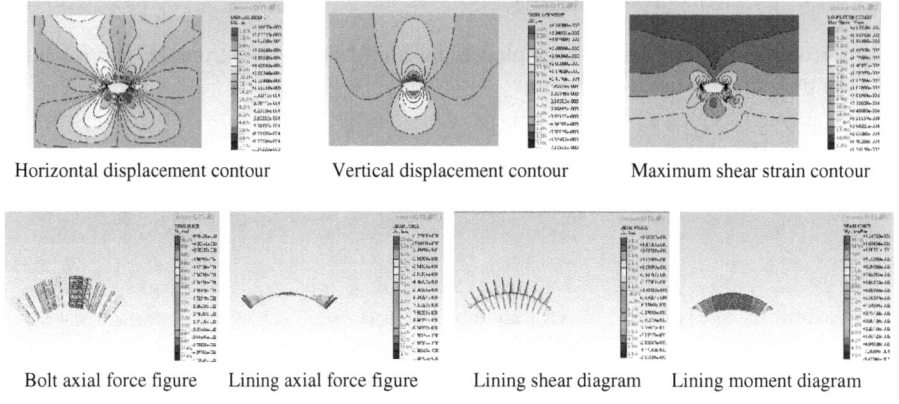

Horizontal displacement contour   Vertical displacement contour   Maximum shear strain contour

Bolt axial force figure   Lining axial force figure   Lining shear diagram   Lining moment diagram

Fig. 2. The second step the simulation results

3.3   The Third Step Results and Analysis

Section on the left after excavation, initial support in the end is controlled under the displacement of small change; horizontal displacement occurred hence excavation face is 2.51mm, because volley surface increases after the left section excavation is completed, vertical displacement of 8.9mm in position at the tunnel crown. The lining on the face and the bolt support force are all into growth trends, the maximum bolt axial force occurred in the position of vault and the arch lumber, the maximum is 13.6KN; lining locks the foot in the axial force of bolt, shear, bending moment greater, respectively, 1037KN, 414KN, 10KN / M.

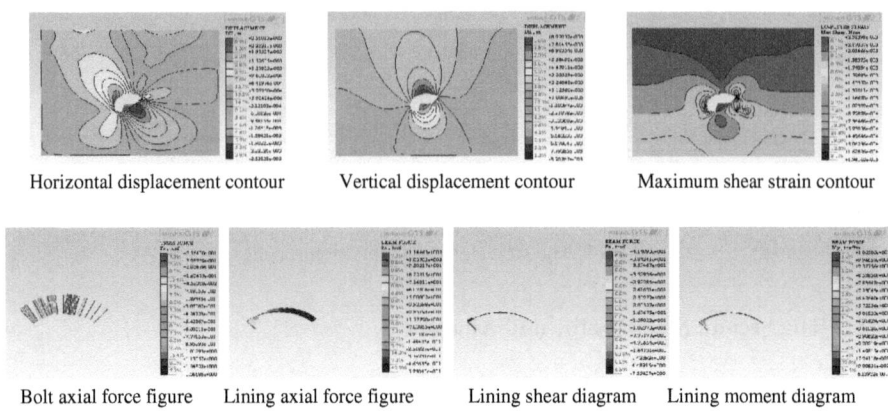

Horizontal displacement contour   Vertical displacement contour   Maximum shear strain contour

Bolt axial force figure   Lining axial force figure   Lining shear diagram   Lining moment diagram

Fig. 3. The third step the simulation results

3.4   The Fourth Step Results and Analysis

Support section on the left, the displacement and the support structure, the resulting change of control value was smaller.

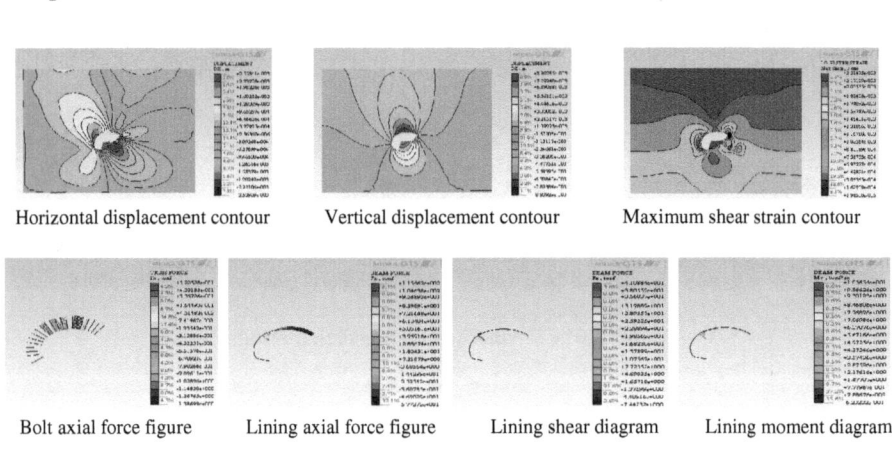

Horizontal displacement contour   Vertical displacement contour   Maximum shear strain contour

Bolt axial force figure   Lining axial force figure   Lining shear diagram   Lining moment diagram

Fig. 4. The fourth step simulation results

3.5   The Results and Analysis of the Fifth Step

Right section after excavation, as volley surface after section excavation was increased, the increase in displacement. The force on the left section support structure was increased, Vertical displacement occurred on the position of tunnel crown 10.6mm. the maximum bolt axial force occurred in the position of vault and the arch lumber, the maximum is 52KN, lining locks the foot in the axial force of bolt, shear, bending moment greater, respectively 1013KN, 410KN, 10.7KN/M.

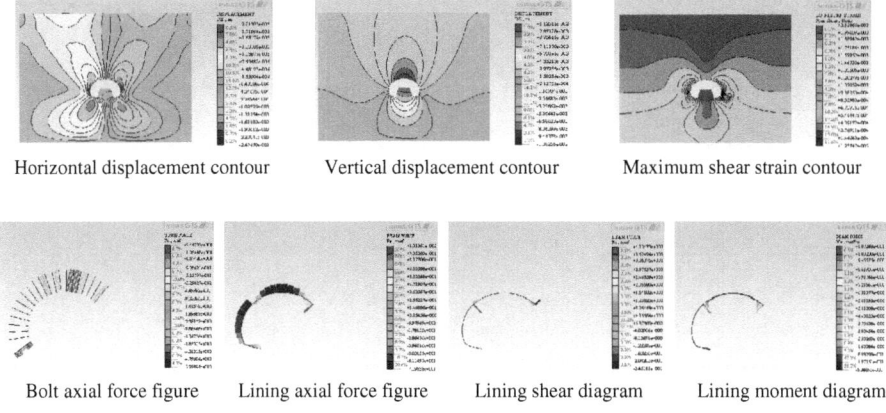

Horizontal displacement contour   Vertical displacement contour   Maximum shear strain contour

Bolt axial force figure   Lining axial force figure   Lining shear diagram   Lining moment diagram

Fig. 5. The fifth step simulation results

3.6   The Results and Analysis of the Sixth Step

Supporting the right cross section, the displacement and force support structure have been effectively controlled, the change is smaller.

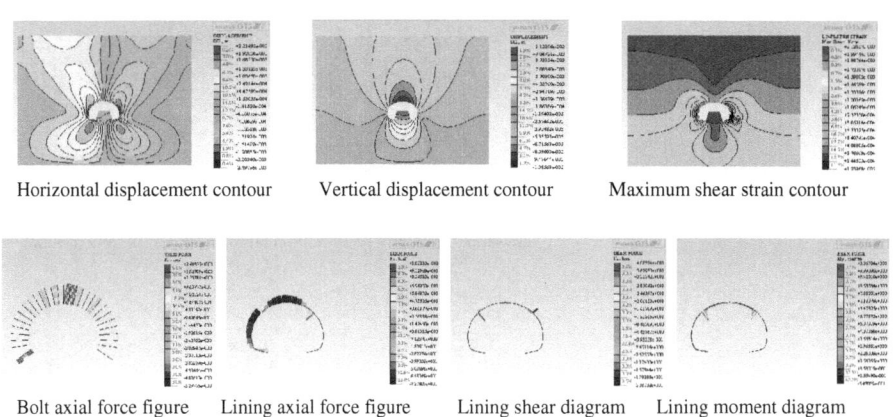

Horizontal displacement contour   Vertical displacement contour   Maximum shear strain contour

Bolt axial force figure   Lining axial force figure   Lining shear diagram   Lining moment diagram

Fig. 6. The sixth step simulation results

3.7   The Results and Analysis of the Seventh Step

Core soil under the section after excavation and support, the force on the displacement and support structure are small changes.

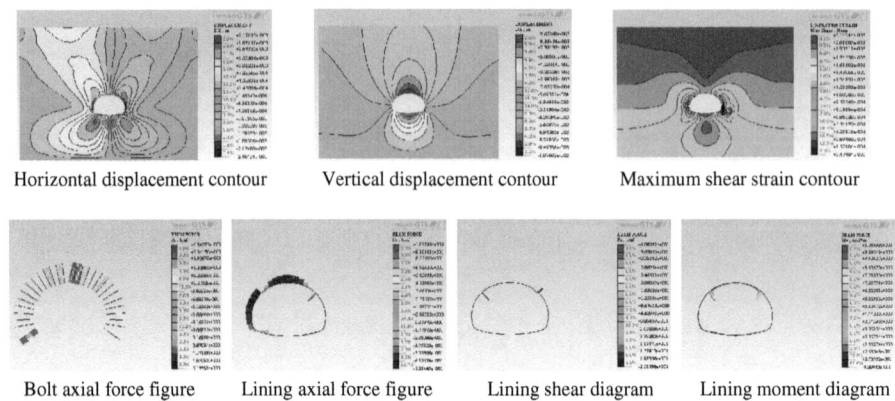

Horizontal displacement contour   Vertical displacement contour   Maximum shear strain contour

Bolt axial force figure   Lining axial force figure   Lining shear diagram   Lining moment diagram

Fig. 7. The seventh step simulation results

3.8   The Results and Analysis of the Eighth Step

After pouring the second lining, two lining displacement and stress values is small, tunnel surrounding rock displacement and structural changes in the value of supporting small, which play a primary support better control effect.

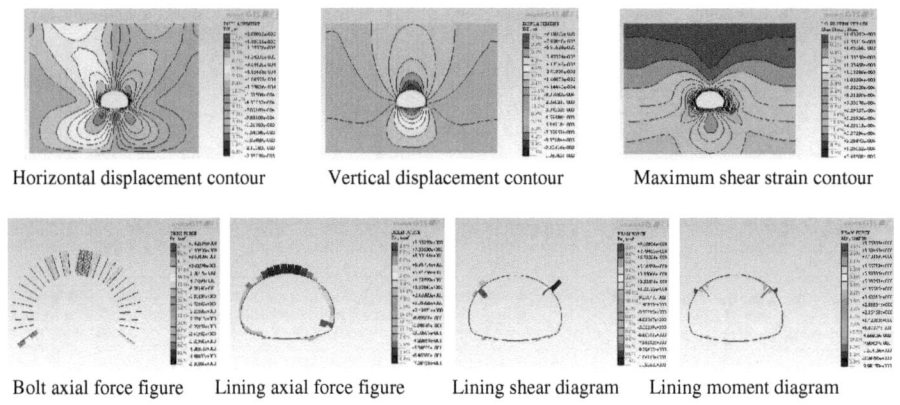

Horizontal displacement contour   Vertical displacement contour   Maximum shear strain contour

Bolt axial force figure   Lining axial force figure   Lining shear diagram   Lining moment diagram

Fig. 8. The eighth step simulation results

4   Conclusions

1) Different sections during the excavation, the maximum principal stress within the surrounding rock concentrated on the steel foot, when steel erection you should lock the foot bolt and construction on time, taking use of small catheters grouting and other methods reinforcement of the rock arch foot when necessary.

2) From the perspective of lining deformation, the maximum deformation at the end of the foot and the location of the vault subsidence, Thus, we should pay attention to the monitoring of settlement of vault, the gradual excavation of core soil, Ensure the construction and structural safety, and should be based on engineering facilities in time to control the chamber deformation.

3) Surrounding the most unfavorable position, appears in the vault and invert both sides, it should be focus on strengthen position.

4) According to the simulation final results: Maximum vault displacement of 10.8mm, the bottom of the uplift of 9.1mm, the maximum lateral wall displacement is 2.38mm; maximum bolt axial force is 51KN, lining axial force is 1144KN, shear is 119KN, moment is 87KN / M, therefore, the stress of surrounding rock displacement and institutions are in a safe range, indicating a reasonable tunnel construction, structural design to meet the strength requirements.

References

1. Tan, R., Wang, C., Yang, Q.: Tunnel engineering. Chongqing University Press, Chong Qing (2001)
2. Liu, T., Lin, T.: Soft rock engineering design theory and construction practice. China Building Industry Press, Bei Jing (2001)
3. Weng, Q.-n., Yuan, Y., Du, G., et al.: Three-dimensional numerical analysis of integrity state of double-arch tunnel. Underground Space and Engineering 2(1), 96–100 (2006)
4. Yu, L.: Soft rock tunnel excavation and support numerical analysis (Master thesis). Dalian University of Technology, Da Lian (2003)
5. He, M., Li, C., Wang, S.: Kenton room large section of soft rock excavation numerical simulation of nonlinear mechanical properties. Public Process of Rock 4, 483–485 (2002)
6. Wang, Z., Li, L.: Analysis of excavation support of tunnel simulation. Shanxi Traffic Technology 5(194), 60–63 (2008)
7. Sanavia, L.: Numerical modelling of a slope stability test by means of porous media mechanics. Engineering Computations 26(3), 245–266 (2009)
8. Park, K.H., Tontavanich, B., Lee, J.G.: A simple procedure for ground response curve of circular tunnel in elastic-strain softening rock masses. Tunneling and Underground Space Technology 23(2), 151–159 (2008)
9. Seung, H.K., Fulvio, T.: Face stability and required support pressure for TBM driven tunnels with ideal face membrane – Drained case. Tunnelling and Underground Space Technology 25(5), 526–542 (2010)
10. Shin, J.-H., Moon, H.-G., Chae, S.-E.: Effect of blast-induced vibration on existing tunnels in soft rocks. Tunnelling and Underground Space Technology 26(1), 51–61 (2011)

Study on the Positioning of Forestry Insect Pest Based on DEM and Digital Monitoring Technique[*]

Feifei Zhao and Yanyou Qiao

Institute of Remote Sensing and Applications,
Chinese Academy of Sciences, Beijing, 100101 China
zff8699@163.com, yyqiao@irsa.ac.cn

Abstract. Forests are extremely important natural resources of terrestrial eco-system. However, forestry pests and diseases always cause prodigious loss. To protect forestry resources effectively and find diseases and pests as soon as possible, advanced digital PTZ (Pan/Tile/Zoom) and video camera were adopted. According to the azimuth angle and pitch angle returned from the digital PTZ in real time, with the classical Bresenham algorithm extended to three-dimensional space, a radial in this direction was obtained. And then, combining the geo-graphic spatial information supplied by DEM (Digital Elevation Model), the forestry diseases and pests could be positioned through searching along this di-rection. This method can position the area of forestry disasters effectively and rapidly, which is beneficial to realize the sustainable development of forest resources.

Keywords: Digital Monitoring Technique, DEM, Forestry Diseases and Pests, Single-point Positioning, Bresenham Algorithm.

1 Introduction

Forests are important parts of terrestrial ecosystem, with the maximum area, most widely distribution, most complicated composition structure and the most abundant material resources on land. Forests are also resource pools with the most impeccable function, which play an irreplaceable role in improving ecological environment and keeping ecological balance [1]. However, insect pests are one of the major disasters to damage forest resources, not only affecting the normal growth and development of trees, but also leading to a certain degree of environmental damage. China is one of the world's countries with more serious forest pests and diseases. There are more than 8000 kinds of various forest pests and rodents, of which over 200 kinds causing serious damage [2].

How to quickly detect and accurately position forestry diseases and insect pests in order to achieve timely and effective monitoring has become an urgent task in forestry management. Currently, some areas of the forestry sector have introduced remote monitoring system of forestry for the daily monitoring of forests, such as a remote

[*] Agriculture Science and Technology Achievements Conversion Fund Project "Forestry Pests Monitoring System based on GIS/GPS", No. YOQ0150040.

B. Liu and C. Chai (Eds.): ICICA 2011, LNCS 7030, pp. 48–56, 2011.
© Springer-Verlag Berlin Heidelberg 2011

forest disaster early-warning monitoring system of Huzhou City [3] and Emei Mountain forest pests remote monitoring system [4]. Among all these forestry disaster remote monitoring systems, accurate and effective positioning of disasters is the core of the monitoring system.

With the rapid development of electronic information technology, digital PTZ (Pan/Tilt/Zoom) and video surveillance technology are relatively mature. What is more, most of the forests have also been equipped with video surveillance system, but mainly for the identification and monitoring of forest fires, which causes a serious waste of hardware resources. On this basis, combined with Digital Elevation Model (DEM) and digital monitoring technique, a remote monitoring and positioning system of forest pests and diseases has been researched and designed. Mainly using the azimuth angle and pitch angle returned from the digital PTZ in real time, combined with the relevant geographic information provided by DEM, accurate and effective positioning of the pests can be realized. Meanwhile, the relevant information is archived, providing important information basis for forest management. Thus, the efficiency of forest pests control is improved effectively, accordingly achieving the sustainable development of forest resources.

2   System Design

2.1   System Architecture

The positioning method proposed in this system is realized through linking two-dimensional view, three-dimensional view and video display module together.

The system framework is described in Fig.1.

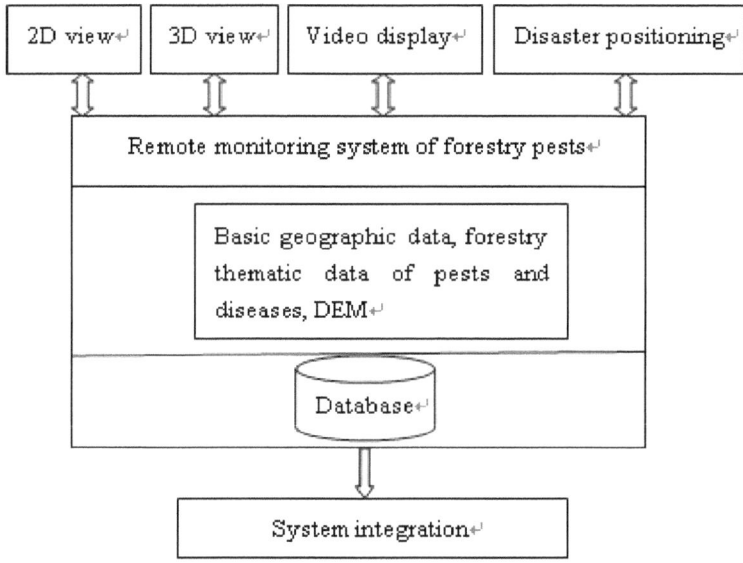

Fig. 1. System framework

Two-dimensional view is mainly to provide friendly search function of monitoring points for users, and meanwhile to update the forestry disasters thematic data in real time, so as to help users to position disasters. Three-dimensional display module is principally based on DEM and DOM (Digital Orthophoto Map) to construct a virtual scene for navigating and roaming. Disaster positioning module is such an important module, through analyzing the video from the camera, on the basis of DEM, realizing the core function of this system. And then record the geographical position, degree and area of disasters into the disaster thematic database. During the process of roaming, complementary with basic geographic information or forestry thematic information, the availability of roaming can be greatly enhanced. Under the assistance of three-dimensional module, users can pan, tilt and zoom (PTZ) the video, in order to further search disasters.

The relationships among the three modules are shown in Fig.2. Two-dimensional view provides the global view, which is conductive to the navigation and positioning and can also skim through disaster thematic maps. Three-dimensional view assists to control video and positioning disasters. Video display view is responsible for capturing disaster pictures and positioning disasters.

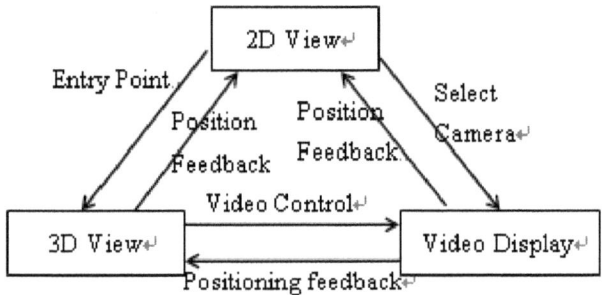

Fig. 2. Relationships among the three views

2.2   Relevant Parameters

Azimuth angle: Counting from due north direction, the camera along clockwise direction moving to the target direction, the horizontal angle between the two directions is called azimuth angle. It ranges from 0 to 360 degree.

Pitch angle: The swing angle of the camera in the vertical direction. It ranges from -90 to 90 degree, 0 degree at horizontal position, up positive and down negative.

Monitoring point's coordinates: The latitude and longitude of the camera at monitoring points.

Monitoring point's elevation: The altitude of the location of lookout towers with surveillance cameras on top.

Tower height: The height in meters from the ground to the top of the lookout tower with monitoring camera placed.

Camera height: The actual altitude of the camera, the value of which is the sum of the monitoring point's elevation and the tower height.

3   Positioning Methods

Generally speaking, positioning methods can be divided into three kinds: single-point positioning, double-point positioning and multi-point positioning. Single-point positioning refers to adopting a single monitory point to position the forestry disasters within the scope of surveillance. Double-point positioning is to use two monitory points, capturing the forestry disasters in overlapping area at the same time, and then obtains the geographical coordinates of the disasters by means of intersection method. Multi-point positioning, based on double-point positioning, uses three or more monitory points to capture the forestry disasters in common shrouded area, and then calculates the geographical position information of the disasters according to basic equations of adjustment and condition equations of angle measurement network.

3.1   Double-Point Positioning

Double-point positioning method is the easiest method to implement and does not require the support of DEM. According to the coordinates of two monitoring points and the returned angles by digital PTZ in real time, the coordinates of forestry disasters can be computed through two-point intersection method. The principal of double-point positioning is shown in Fig. 3.

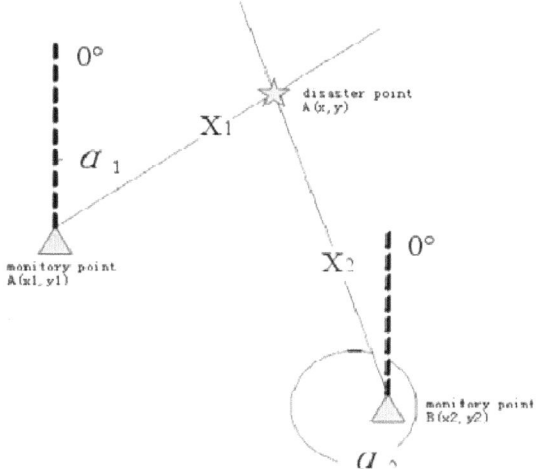

Fig. 3. Schematic of double-point positioning

When the cameras at two monitoring points capture pests and diseases together, in accordance with azimuths α1 and α2 returned by digital PTZ in real time, starting from monitoring points A and B, two rays X1 and X2 are drawn. In this paper, the first intersection of these two rays is considered to be the disaster point. Hereinto, geographical relative coordinates of monitoring points need to transform from space actual coordinates. According to the relevant information obtained, we can derive the formulas to compute the coordinates of the disaster point through two-point intersection method.

$$\begin{cases} x = \dfrac{x_2 \times \tan(\alpha_1) - x_1 \times \tan(\alpha_2) + (y_1 - y_2) \times \tan(\alpha_1) \times \tan(\alpha_2)}{\tan(\alpha_1) - \tan(\alpha_2)} \\ y = \dfrac{x_2 - x_1 + y_1 \times \tan(\alpha_1) - y_2 \times \tan(\alpha_2)}{\tan(\alpha_1) - \tan(\alpha_2)} \end{cases} \qquad (1)$$

Here, (x1, y1) and (x2, y2) are the coordinates of two monitory points; α1 and α2 are azimuth angles.

3.2   Multi-point Positioning

Multi-point positioning is to obtain common coverage of three or more monitoring points, and then, on the basis of adjustment principal, computes high precision geographic coordinates of the disasters. Multi-point observation can be used to rectify the coordinates of forestry pests and diseases, but this is bound to be at the cost of the efficiency of positioning. During the process of positioning forestry pests and diseases, for the area with large amount of data and complicated data structure, only need to get the appropriate range of disasters, then through human interaction to achieve accurate positioning, of which efficiency is a key aspect. Therefore, details of this method are no longer discussed here.

3.3   Single-Point Positioning

In consideration of the actual situation of hilly forest areas, multiple cameras with a common scope of monitoring is unlikely, and human interaction is used in positioning method, so double-point or multi-point positioning can affect the efficiency of positioning. Therefore, this paper adopts single-point positioning method. Utilize a single monitory point to scan uninterruptedly, so as to search and position forestry pests and diseases. Monitoring points are generally set at areas with a higher elevation, thus they can cover a larger scope. Fig.4 [5] shows the schematic of single-point positioning.

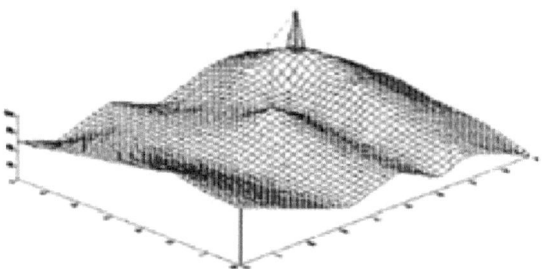

Fig. 4. Schematic of single-point positioning

DEM is the basic data source of single-point positioning. After obtaining vector geographical map and DEM data, according to the geographical coordinate information of the monitoring point and the returned azimuth angle α, take the monitoring camera as the origin to draw a ray on the electronic map along the direction, as is shown in Fig.5 [6].

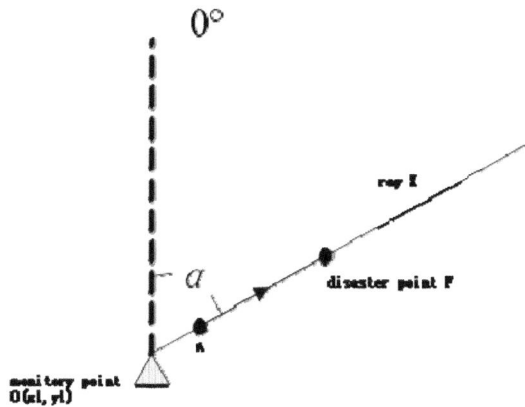

Fig. 5. Single-point positioning plane sketch

According to the ray X, a profile along this direction can be cut out on the digital elevation map. This paper considers the first intersection point of ray X and DEM as the disaster point. Fig.6 shows the schematic diagram of single-point positioning under the circumstance of angle of depression.

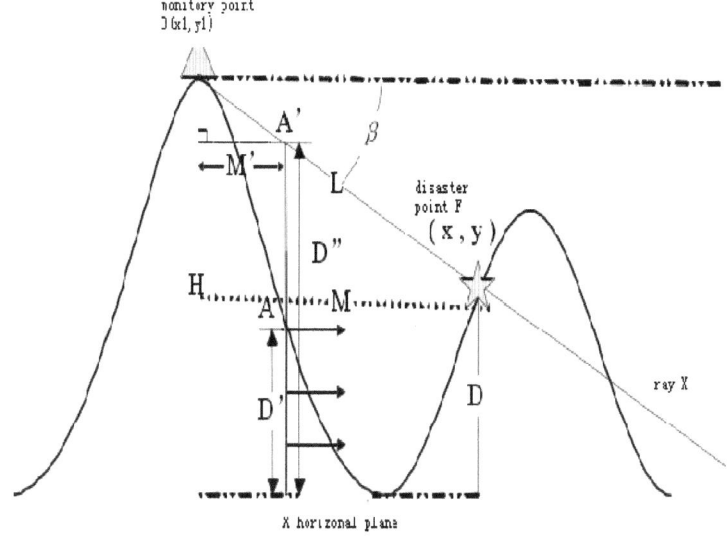

Fig. 6. Schematic profile of single-point positioning of angle of depression

As is shown in Fig.6, point A is on the DEM profile and A' is on the ray X, both have the same geographical plane coordinates. The key of this algorithm is to judge whether point A and A' is fitting. Through finding such a point which has the minimum elevation difference between the points on DEM and ray X with the same plane coordinates [7], this point is considered to be the disaster point. But when considering some certain forestry disaster points, such as the summit and valley of mountain and so on, the convergence criteria of this algorithm are to be refined [8].

During this process, the D' on the DEM profile, which is the actual elevation of point A, can be obtained from DEM; the D'' on the ray X can be computed through the following formula [9].

$$D''=H+M'*\tan(\beta) \tag{2}$$

Hereinto, H is the camera's elevation, being equal to the sum of monitory point's elevation and tower height; β represents the pitch angle, which is negative under angle of depression and positive under angle of elevation ; M' is the horizontal distance between point A and the monitoring point, which can be computed easily.

4   Bresenham Algorithm in Three-Dimensional Space

During the process of positioning forestry pests and diseases, it is one of the critical issues to obtain ray X in high efficiency. As is known to all, line is such an important element in computer graphics [9]. And Bresenham algorithm [10, 11] is among the most classical and effective algorithms. Its basic principle is: firstly, construct a set of virtual grid lines over the pixel center of all each row and column; and then calculate the intersection between the line and the vertical grid lines along the line from the beginning to the end; finally, determine the pixel in the column nearest to the intersection. The trick of this algorithm is that it uses incremental computation. Therefore, for each column, as long as the sign of an error item is checked, the pixel value of this column can be determined [12]. Because Bresenham algorithm does not involve the operation of real numbers, so it can generate a line very rapidly and effectively, which can meet the demand of positioning pests and diseases during the process of real-time scanning.

However, Bresenham algorithm is based on two-dimensional space, and the positioning of forestry pests and diseases is carried out in three-dimensional space. Therefore, in order to better utilize Bresenham algorithm, this paper improves and extends it to three-dimensional space. In this way, it can draw lines in any direction and any slope. Thus, when positioning pests and diseases, we not only get the coordinates information of disasters, but also obtain the elevation values of disaster points, which have great help to validate the results of positioning.

The steps of generating a line by Bresenham algorithm are as follows.

1) According to the elevation of the monitoring point, tower height, azimuth angle and pitch angle, take the monitory point $(x1,y1,z1)$ as the origin, to calculate the other intersecting end $(x2, y2, z2)$ between the line and DEM profile. Hereinto, z1 and z2 represent the elevation values of corresponding geographical coordinates.

2) Set the initial values of pixel coordinates: x=x1,y=y1,z=z1.

3) Calculate respectively: dx=abs(x2-x1), dy=abs (y2-y1), dz=abs (z2-z1).

4) Calculate respectively: sx=sign(x2-x1), sy=sign (y2-y1), sz=sign (z2-z1).

5) Initialize error discriminants: p1=2*dy-dx, p2=2*dz-dx.

6) Compare dx, dy, dz and select the coordinates with the biggest change direction to step one pixel. Meanwhile, the coordinates of the other two directions are determined whether to step one pixel through the error discriminant.

There are three branches. Here, this paper just shows the first branch, and the other two are symmetrical.

If dx>dy and dx>dz, then step one pixel in x direction: x=x+sx. Under this case, if p1>=0,then y=y+sy, p1=p1-2*dx; If p2>=0, then z=z+sz, p2=p2-2*dx.

7) Through looping, generate a line.

The following two schematic diagrams respectively show rays in two cases: slope less than 1 and greater than 1, both are based on Bresenham algorithm.

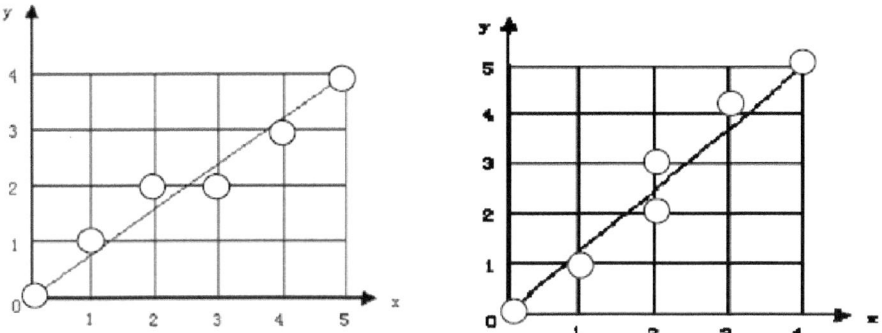

Fig. 7. Two cases of line generation

After generating the ray, traverse the points of the ray and compare the elevation values on the ray with the corresponding elevation value on DEM, to find the points meeting the requirements. And then, input disaster information into disaster thematic database, so as to provide convenience for forest managers to search and analyze forest pests and diseases.

5   Conclusions

To protect forestry resources from pests and diseases, this paper outlines the system architecture of remote monitoring and positioning system of forest pests and diseases, and presents three positioning methods, focusing on studying single-point positioning method to position forestry diseases and pests. Through introducing the classical Bresenham algorithm, being extended to three-dimensional space, which does not involve the computations of real numbers, it can generate a line so rapidly and effectively. Thus, the efficiency of positioning is greatly improved, achieving the core function of remote monitoring system of forestry diseases and pests.

Whereas, many forest regions are located in rugged mountainous areas, therefore, the positioning error from the external environmental conditions and the mechanical jitter of the instrument itself cannot be ignored and avoided. Hence, the positioning method also needs further study and improvement. What's more, the error needed to be corrected so as to provide more accurate positioning and improve the efficiency of diseases and pests control. On the other hand, the principle of deciding the disasters can take the concept of minimum bounding box into consideration, whereby the positioning of forestry pests and diseases is more accurate and complicated.

References

1. Li, Y.: Studying about the key technology used in remote monitoring of forest insect pest. Institute of Remote Sensing and Applications, China Academy of Sciences, Beijing (2009)
2. Ministry of Forestry, Republic of China. On further Strengthening the Work of Forest Pest Management (April 14, 1997)
3. Information on, http://anji.kx108.com/news/2543.html
4. Information on, http://www.scly.gov.cn/article342515.html
5. Wang, Y., Zhang, J., Han, N.: GIS-based positioning methods in video monitoring of forest fires. Forestry Machinery & Woodworking Equipment 36(5), 24–26 (2008)
6. Zhang, H., Wang, L.: Surveillance camera technology to achieve analysis and applications of forest fire positioning in DEM. Forest Fire Prevention (3) (2006)
7. Wang, Y.: Research on automatic positioning techniques for forest fire video monitoring system Based on GIS. Beijing Forestry University, Beijing (2008)
8. Zhang, J.: Research on technology and methods for forestry fire video monitoring. Beijing Forestry University, Beijing (2009)
9. Haque, A.-u., Rahman, M.S., Bakht, M., Kaykobad, M.: Drawing lines by uniform packing. Computers & Graphics 30, 207–212 (2006)
10. Bresenham, J.E.: Algorithm for computer control of a digital plotter. IBM Systems Journal 4(1), 25–30 (1965)
11. Bresenham, J.E.: A Linear algorithm for incremental digital display of circular arcs. CACM 20(2), 100–106 (1977)
12. Information on, http://course.cug.edu.cn/cugThird/CGOL_NET/CLASS/course/2-1-2-a.htm

Assessment of Soil Erosion in Water Source Area of the Danjiangkou Reservoir Using USLE and GIS

Xuan Song[1], Liping Du[1,*], Changlin Kou[2], and Yongli Ma[1]

[1] School of Water Conservancy and Environment,
Zhengzhou University, Zhengzhou 450001, China
[2] Institute of Plant Nutrition, Resources and Environment,
Henan Academy of Agricultural Sciences, Zhengzhou 450002, China
{songxuan,dulp}@zzu.edu.cn,
koucl@126.com, 190161725@qq.com

Abstract. The study is aimed at the evaluation of soil erosion in water source area of the Danjiangkou Reservoir in Henan province, China, using the Universal Soil Loss Equation (USLE) and geographical information system (GIS). The R-factor (rainfall erosivity) was determined by interploation using meteorological data from five stations. The K-factor (soil erodibility) was obtained by using soil survey data. The LS-factors (slope length and steepness) were determined from the digital elevation model (DEM) of the study area, while the C-factor (crop and management) was determined from remote sensing imageries. The P-factor (conservation practice) was estimated from field experiments. The spatial distribution maps of the soil erosion were estimated in study area from 1991 to 2007. The results showed that soil erosion decreased from 44.8 t $\cdot$ hm^{-2} $\cdot$a^{-1} in 1991 to 16.9 t hm^{-2} $\cdot$ a^{-1} in 2007. The total soil loss fell by 9 350 000 t in 2000 and by 13 580 000 in 2007 compared to it in 1991. Soil nutrients loss decreased, such as organic matter loss decreased to 108 397 t$\cdot$ a^{-1} in 2007 from 226 918 t$\cdot$a^{-1} in 1991, total nitrogen and total phosphorus decreased by 7236 t$\cdot$a^{-1} and 5640 t$\cdot$a^{-1} from 1991 to 2007.

Keywords: USLE, GIS, Soil Loss, Nutrients Loss.

1 Introduction

As soil erosion influencing water quality and increasing risk of water eutrophication is a typical non-point source pollution, it has long been recognized as severe problems for environment protection and human sustainability. The Danjiangkou Reservoir is water resource of the middle route project of the South-to-North Water Diversion in China, it is necessary to evaluate the risk of soil erosion for better protecting the water quality of the reservoir

Using theoretical and empirical models to estimate the spatial distribution of soil loss is the most commonly used method, it is difficult to obtain the parameters of these models on the large. Rapid development of remote sensing (RS), geographic

---

* Corresponding author.

B. Liu and C. Chai (Eds.): ICICA 2011, LNCS 7030, pp. 57–64, 2011.

information system (GIS) and global position system (GPS) has brought the new opportunities for application of the soil erosion models. RS can help us quickly access the ground state data, GIS can facilitate the use of remote sensing information, visualize the calculated results in the form of graphical output and describe spatial distribution of soil erosion, it is a trend of soil erosion model with a perfect combination of RS and GIS[1]. Many soil erosion models are put forward towards the practical application stage. The universal soil loss equation (USLE) was widely used because of considering more comprehensive factors influencing soil erosion[2-6]. Kinnell summarized the main methods for estimating these factors influencing soil erosion [7].

Soil erosion and corresponding soil nutrients loss were estimated by use of RS, GIS, and USLE in the water resource of Danjiangkou Reservoir in Henan province,China, it provides the theoretical basis and data support for protection of ecological environment around the Danjiangkou reservoir area.

2   Description of the Study Area

The 7200 km^2 study areas is situated in a junction of Henan, Hubei and Shanxi provinces. It is located between 110 ° 47 'to 111°50' E longitude and 32°35 'to 34 ° 00' N latitude (Fig.1). Annual mean precipitation is about 804mm, summer average rainfall is 560 mm, accounting for 70% of annual precipitation.

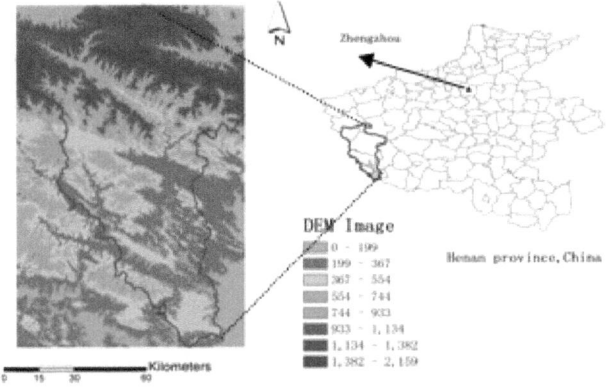

Fig. 1. Location and DEM of the study area

3   Methodology

The Universal Soil Loss Equation (USLE) has been applied broadly for predicting the average annual soil loss, the USLE is expressed as follows [8]

$$A = R \times K \times LS \times C \times P \tag{1}$$

Where A is annual soil loss (t·hm^{-2}·a^{-1}), R is rainfall erosivity factor(MJ·mm·hm^{-2}·h^{-1}·a^{-1}), K is the soil erodibility factor (t·hm^2·h·hm^{-2}·MJ^{-1}·mm^{-1}), L is slope length factor (unitless), S is slope steepness factor (unitless), C is crop and management factor (unitless), and P is conservation supporting practices factor (unitless).

3.1 Rainfall Erosivity (R) Factor

The rainfall erosivity indicator is a critical parameter to evaluate the soil erosion due to rainfall. As there is no record of rainfall intensity in the study area, monthly rainfall data were used to calculate R-factor using the following relationship[8]:

$$R = \sum_{i=1}^{12} 1.735 \times 10^{\left\{ 1.5 \cdot \lg \frac{p_{mi}^2}{P_m} - 0.8188 \right\}} \tag{2}$$

Where R is rainfall erosivity factor, P_{mi} is monthly rainfall in mm and P_m is annual rainfall in mm.

R-factor was determined by using monthly rainfall data from five meteorological stations (Xichuan, Xixia, Lushi, Neixiang, Luanchuan) from 1991 to 2007. To obtain R-factors for the whole study area, the Kriging interpolation was used in ArcGIS 9.2 producing erosivity maps of the study area in 1991, 2000 and 2007.

3.2 Soil Erodibility (K) Factor

The K factor is related to the integrated effects of rainfall, runoff and infiltration on soil loss, accounting for the influences of soil properties on soil loss during storm events on upland areas [9]. The K value was calculated by using EPIC (Erosion-Productivity Impact Calculator) formula [10].

$$K = \left\{ 0.2 + 0.3\exp\left[-0.0256SAN\left(1 - \frac{SIL}{100} \right) \right] \right\} \cdot \left(\frac{SIL}{SLA + SIL} \right)^{0.3} \cdot$$
$$[1.0 - \frac{0.25C}{C + \exp(3.72 - 2.95C)}] \cdot [1.0 - \frac{0.7SN_1}{SN_1 + \exp(-5.51 + 2.95SN_1)}] \tag{3}$$

Where SAN is sand content (%), SIL is silt content (%), SLA is clay content (%), C is organic matter content (%), SN_1 is 1- SAN/100.

The soil data used in the study was drawn during China' second national soil survey in early 1980's, these data included that soil classification map, soil physical and chemical data. The K factor for each soil type was associated with a K value assuming that the same soil type has the same K value throughout the study area. The K factor distribution is between 0.2 and 0.5.

3.3 Topographic Factor (LS)

The LS factor accounts for the effect of topography on soil erosion in USLE, the soil loss increases as the slope length and steepness increases. The slope length factor reflects the effect of the slope length on soil erosion, and the slope steepness represents the influence of the slope gradient on soil erosion. The LS factors are calculated by these following equations [9,11].

The slope length factor (L) is calculated as follows.

$$L = (\lambda/22.13)^m \tag{4}$$

where λ is slope length in meter, m is calculated using the following formula

$$\begin{cases} m = \beta(1+\beta) \\ \beta = 1/2[\sin\theta/0.0896)/ [3.0(\sin\theta)^{0.8} +0.56] \end{cases} \tag{5}$$

where θ is slope.

The S factor was calculated based on the following relationship.

$$\begin{cases} S = 10.8 \sin\theta + 0.03 \quad \theta < 9\% \\ S = 16.8 \sin\theta - 0.05 \quad \theta \geq 9\% \end{cases} \tag{6}$$

where θ is slope (°).

Digital topographic data with 30m resolution used in the study was provided by International Scientific & Technical Data Mirror Site, Computer Network Information Center, Chinese Academy of Sciences (http://datamirror.csdb.cn) . The LS factors were calculated by use of the ArcGIS 9.2.

3.4   Crop and Management Factor (C)

The crop management factor depends on vegetation cover, which dissipates the kinetic energy of the raindrops before impacting soil surface. Therefore, the vegetation cover and crop system have a large influence on runoff and erosion rates. Crop management factor was calculated using the following formula [12].

$$\begin{cases} 1 \dotfill c = 0 \\ C = 0.6508 - 0.3436 \lg c \quad 0 < c \leq 78.3\% \\ 0 \dotfill c > 78.3\% \end{cases} \tag{7}$$

Where C is the crop management factor, c is the vegetation cover, it is calculated by use of the following formula [13].

$$c = \frac{NDVI - NDVI_{min}}{NDVI_{max} - NDVI_{min}} \tag{8}$$

Where $NDVI$ is normalized difference vegetation index, $NDVI_{max}$ is the maximum of NDVI, $NDVI_{min}$ is the minimum of $NDVI$.

Using radiance in NIR and R wavebands, the NDVI was calculated by using the following relationship.

$$NDVI = (L_{NIR} - L_R)/(L_{NIR} + L_R) \tag{9}$$

Where L_{NIR} is the radiance of NIR waveband, L_R is the radiance of R wavebands.

These NDVI maps in three periods were gotten from Landsat 5 Thematic Mapper (TM), Enhanced Thematic Mapper Plus (ETM+) and China-Brazil Earth Resources

Satellite, these data consisting of three remote sensing images acquired in September 1991, April 2000 and May 2007 respectively. The geometric correction was performed using topographic map to establish the transformation equation between the image and map coordinates, the accuracy of geometric correction was ranged within a half pixel.

3.5 Conservation Practice Factor (P)

The conservation practice factor (P) is the ratio of soil loss with specific support practice to the corresponding loss with standard conditions cultivation. In the study area, woodland, grassland, water area, residential land, cultivated land and bare land of P values are 1.0, 1.0, 1.0, 0.8, 0.3, 0.2 respectively by field experiments.

4 Results and Analysis

After the R, K, LS, C and P factors images were developed, they were overlaid to get the soil erosion risk image using ArcGIS9.2. The spatial distribution maps of soil erosion modulus were obtained in water source area of the Danjiangkou Reservoir in 1991, 2000 and 2007.

4.1 Analysis of Soil Erosion Change

According to national standards of soil erosion classification of China, the modulus maps of soil erosion were classified into 6 levels, namely the erosion of very slightly, slightly, moderate, strong, severe and extreme erosion in three periods. The classification maps of soil erosion were gotten (Fig. 2) in 1991, 2000 and 2007. The statistical results of different levels were showed in Table 1.

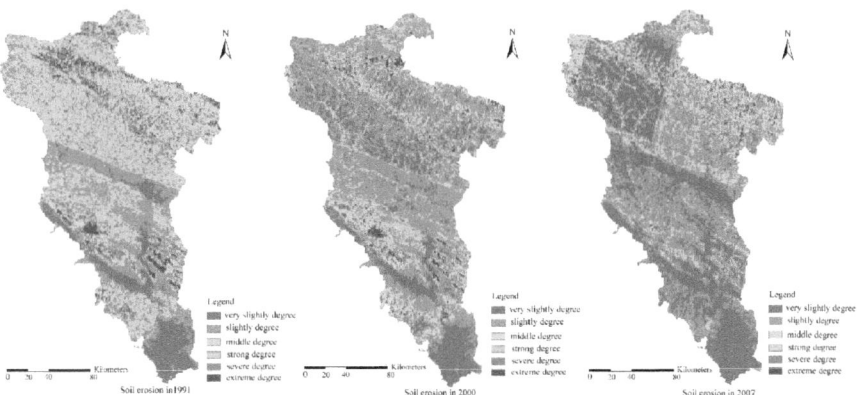

Fig. 2. Soil erosion in 1991, 2000 and 2007

Table 1. Different levels of soil erosion in 1991, 2000 and 2007

Time	Soil Erosion Risk (%)					
	Very slightly	Slightly	Moderate	Strong	Severe	Extreme
1991	28.54	28.72	16.64	10.21	9.31	6.58
2000	19.06	53.01	14.70	6.97	4.35	1.91
2007	34.94	43.25	12.41	5.49	3.13	0.78

The average erosion modulus was 44.8 $t \cdot hm^{-2} \cdot a^{-1}$, about 57 % study area was below the slight erosion, and 43 % was above moderate erosion in 1991. The average erosion modulus in 2000 was less than that in 1991, that was 23.9 $t \cdot hm^{-2} \cdot a^{-1}$, the slight erosion land area was 72% of the research area. The soil erosion didn't become serious and had a tendency of decrease in 2007, that 22% area was more moderate erosion.

From the overall analysis, soil erosion presented a trend of alleviation from 1991 to 2007 in the study area. The area of moderate, strong, severe, extreme erosion showed the decline, they was 12.41%, 5.49%, 3.13%, 0.78% respectively of all study area in 2007.The area of very slight soil erosion continued to increase with relatively large rate, but slight erosion area had a downward trend on the whole, the main reason was the conversion among the grades of soil erosion.

4.2   The Calculation of Soil Loss

When the soil erosion intensity is less than a certain limit, soil erosion has no effect on land productivity, this value is known as soil erosion permit. According to Ministry of Water Resources of China issuing the Standards of Classification of Soil Erosion in 1997, soil erosion permit is set as 200 t $\cdot km^{-2} \cdot a^{-1}$ in study area.

The total soil erosion is calculated by use of soil erosion areas and its corresponding soil loss in different levels, the empirical formula is expressed as follows,

$$S = A2 \times 1150 + A3 \times 3550 + A4 \times 6300 + A5 \times 11300 + A6 \times 15000 \qquad (10)$$

Where, S is total soil erosion in t, A_2 is area of slight erosion, A_3 is area of moderate erosion, A_4 is area of strong erosion, A_5 is area of severe erosion, A_6 is area of severe erosion. The numbers of the formula are difference value between the middle value of soil erosion modulus and the value of the soil erosion permit.

The soil loss were calculated in 1991, 2000 and 2007, the results were showed in Table 2.

Table 2. Soil loss amount change in different periods (10 000t $\cdot a^{-1}$)

Time	Soil erosion amount					
	Slightly	Moderate	Strong	Severe	Extreme	Total
1991	238	426	464	759	712	2600
2000	439	374	312	346	195	1665
2007	359	315	244	252	72	1242

The soil loss showed a trend of falling from 1991 to 2007, the maximum of value was 26 000 000 t per year at 1991, mainly consisting of the severe and extreme soil erosion. Compared to 1991, the soil loss fell by 9 350 000 t in 2000 and by 13 580 000 in 2007.

From the proportion of total soil erosion, the severe and extreme soil erosion gradually decreased 57% in 1991, 32% in 2000, and 26% in 2007 respectively.

With soil loss, the nutrients also lost, the nutrient loss model was expressed as follows [14],

$$N=0.001 \times A \times C \times Er \times SDR \tag{11}$$

Where N is the total nutrient content, 0.001 is conversion coefficient (from kg to t), A is soil erosion amount ($t \cdot km^{-2} \cdot a^{-1}$), C is soil nutrient content, Er is ratio of soil nutrient content in sediment and surface soil by erosion (nutrient enrichment ratio), SDR is sediment delivery ratio.

The soil nutrient content (C) was obtained by analyzing the soil sample collected, nutrient distribution maps were drawn by assignment its value to the diagram in the study area. The nutrient enrichment ratio was obtained by field simulation, SDR was taken as 0.3 by referring the study in similar areas [14]. The soil nutrient loss were calculated by using the above formulas in 1991, 2000 and 2007 (Table 3).

Table 3. Soil nutrient loss in different periods

Time	Organic matter($t \cdot a^{-1}$)	Total nitrogen ($t \cdot a^{-1}$)	Total phosphorus ($t \cdot a^{-1}$)
1991	226918	13854	10798
2000	145315	8872	6915
2007	108397	6618	5158

Soil nutrients loss decreased from 1991 to 2007, the organic matter loss decreased to 108 397 $t \cdot a^{-1}$ in 2007 from 226 918 $t \cdot a^{-1}$ in 1991, total nitrogen and phosphorus decrease by 7236 $t \cdot a^{-1}$ and 5640 $t \cdot a^{-1}$ from 1991 to 2007.

The part of these lost nutrients deposited on the spot, some part got into Danjiangkou Reservoir with river, which would undoubtedly brought about some negative impact to water quality of Danjiangkou Reservoir. The study found that the spatial distribution maps of soil erosion, it provided a scientific basis for the construction and protection of ecological environment in Danjiangkou reservoir area.

5   Conclusion and Discussion

The change of soil erosion is affected by the natural factors and anthropogenic factors, the natural factors are mainly the weak soil resistance to erosion and the concentrated rainfall, the anthropogenic factors affect soil erosion intensity, speed and scale.

Early Danjiangkou reservoir construction, forest resources were not protected scientifically, and the utilization of land didn't combine closely soil conservation in the migration process, these factors led to severe soil erosion in 1991.

Since the State Council approving the National Soil and Water Conservation Plan in 1993, the upper reaches of the Hanjiang River was considered as the key region of soil erosion control. Key soil and water conservation projects around the reservoir area were carried out by Changjiang Water Resources Commission in 1994. So soil erosion in study area showed great change, soil loss reduced by 9 350 000 t.

Since 2000, the soil and water conservation were acted as the main responsibility of local governments, and as the key point for developing rural economy and increasing rural income. Xixia and Luanchuan counties were listed as national ecological restoration counties by Ministry of Water Resources in 2002. The effective measures made the soil erosion decrease to 12.42 million t in 2007, decreasing as much as 13.58 million t compared to 1991 in study area. However the results showed that it was important for us improve the ecological environment of the reservoir area, reduce non-point source pollution and ensure water quality.

Acknowledgments. This research work was supported by the natural science foundation of China (No.40971128) and key project of national scientific and technological support of China (No.2007BAD87B09).

References

1. Zheng, F.L., Wang, Z.L., Yang, Q.K.: Status, Challenge and Tasks of Water Erosion Prediction Model Research in China. Science of Soil and Water Conservation 3(1), 7–14 (2005)
2. Song, X., Fan, R.Q., Wang, B.F., Li, K.: Study on Soil Erosion Using GIS and USLE in LuShan County. Yellow River 32(10), 108–110 (2010) (in Chinese)
3. Meusburger, K., Konz, N., Schaub, M., Alewell, C.: Soil Erosion Modelled with USLE and PESERA Using Quickbird Derived Vegetation Parameters in An Alpine Catchment. International Journal of Applied Earth Observation and Geoinformation 12, 208–215 (2010)
4. Beskow, S., Mello, C.R., Norton, L.D., Curi, N., Viola, M.R., Avanzi, J.C.: Soil Erosion Prediction in the Grande River Basin, Brazil Using Distributed Modeling. Catena 79, 49–59 (2009)
5. Bagarello, V., Stefano, C.D., Ferro, V., Pampalone, V.: Using Plot Soil Loss Distribution for Soil Conservation Design. Catena 86, 172–177 (2011)
6. Liu, H.H., Fohrer, N., Hörmann, G., Kiesel, J.: Suitability of S Factor Algorithms for Soil Loss Estimation at Gently Sloped Landscapes. Catena 77, 248–255 (2009)
7. Kinnell, P.I.A.: Event Soil Loss, Runoff and the Universal Soil Loss Equation Family of Models: A Review. Journal of Hydrology 385, 384–397 (2010)
8. Wischmeier, W.H., Smith, D.D.: Predicting Rainfall Erosion Losses A Guide To Conservation Planning, USDA, Agriculture Handbook (1978)
9. Williams, J.R., Renard, K.G., Dyke, P.T.: EPIC-A New Method for Assessing Erosion's Effect on Soil Productivity. Journal of Soil and Water Conservation 38(5), 381–383 (1983)
10. Renard, K.G., Foster, G.R., Weesies, G.A., McCool, D.K., Yoder, D.C.: Predicting Soil Erosion by Water: A Guide to Conservation Planning with the Revised Universal Soil Loss Equation (RUSLE). In: Agricultural Handbook 703. U.S. Department of Agriculture, Washington, DC (1997)
11. Foster, G.R.: Comments on Length-Slope Factor for the Revised Universal Soil Loss Equation: Simplified Method of Estimation. Journal of Soil and Water Conservation 49, 173–177 (1994)
12. Ma, C.F., Ma, J.W., Buhe, A.: Quantitative Assessment of Vegetation Coverage Factor in USLE Model Using Remote Sensing Data. Bulletin of Soil and Water Conservation 21(4), 6–9 (2001) (in Chinese)
13. Zhao, Y.S.: The Principle and Method of Analysis of Remote Sensing Application. Science Press in China (2003) (in Chinese)
14. Yang, L.: Research of Nutrient Loss From Heimiaogou Watershed in Danjiang Reservoir Area Based on GIS. Master Thesis of Huazhong Agricultural University (2007) (in Chinese)

Influence of Robe River Limonite on the Sinter Quality

Lili Wang[1], Haibin Ke[2], Hao Zhang[1],
Hongyan Zhang[1], and Jiangyan Zheng[1]

[1] Qinggong College, Hebei United University, TangShan HeBei, China
[2] Tangshan Iron and Steel Corp., TangShan HeBei, China
kehaibin928@126.com

Abstract. To study the influence of robe river limonite on the Sinter quality, do some sintering cup experiments which added different proportions of robe river limonite. Through the sintering cup experiments obtained the variation regular of Sinter performance. Results show that: When the addition of robe river limonite reached 10%, the drum strength and sintering velocity and low temperature reduction degradation and beginning soften temperature of agglomerate reached maximum. The appropriate proportion was identified as 10%.

Keywords: Robe River Limonite, Sintering Characteristics, Microstructure.

1 Introduction

Robe River limonite produced from Western Australia, the iron content is lower and the silicon content is higher, the iron content is only 56% and the silica content is 5.7% and the crystal water content is 8.8%, the price is very cheap but China's iron and steel companies which use the Robe River limonite are very few in recent years. Powder ore resources increasingly reducing and material cost increasingly rising, in order to reduce the sinter cost and seek new powder ore resources, do some sintering cup experiments which added different proportions of robe river limonite according to existing material conditions of a China's iron and steel company. Thereby concluded the Influence of Robe river limonite on the Sinter quality and determined the appropriate proportion, for sintering production technology provided guidance. In blast furnace iron burden, sinter proportion generally reaches around 80% in China, and whatever its quality is good or bad plays a vital role.

2 Sintering Experiments

The process of sintering comprises a high-temperature treatment (above 1000°C)of iron ore fines on a moving grate, blended with flexed and coke breeze (finely divided coke) to form hard lumps of iron-rich material suitable for use as blast furnace feed.

2.1 The Materials and Fuel for Doing Experiments

The materials and fuel provided by a China's iron and steel company. The table 1 shows the Chemical composition of the materials and fuel for doing experiments. The particle size of Robe River limonite is very wee, the Content of less than 0.5mm is up to 45%. The fuel is coke powder.

B. Liu and C. Chai (Eds.): ICICA 2011, LNCS 7030, pp. 65–73, 2011.
© Springer-Verlag Berlin Heidelberg 2011

Table 1. Chemical constitution of materials and fuel /%

Iron materials	H_2O	Fe	CaO	SiO_2	MgO	Al_2O_3
1st Indian iron ore powder	10.00	58.00	0.50	6.50	0.23	4.50
2nd Indian iron ore powder	8.00	61.00	0.50	5.50	0.23	4.00
Robe Rive iron ore powder	8.80	56.00	0.86	5.70	0.60	2.70
Brazil iron ore powder	6.50	65.20	0.61	4.40	0.30	1.00
Steel scoria	8.80	19.20	41.00	12.00	9.35	4.00
Red mud	18.00	48.00	9.31	7.44	3.40	2.50
Dust ash	6.50	45.39	7.44	7.00	1.80	1.50
Return mines	1.00	55.80	9.50	5.40	2.39	2.30
White limestone powder	0.00	0.00	75.00	6.50	7.90	0.00
Limestone powder	2.20	0.00	46.50	3.50	6.00	0.00
Dolomite fines	1.50	0.00	29.50	1.50	21.00	0.20
Coke powder	13.50	0.00	0.80	8.00	1.00	3.00

2.2   The Testing Program

To calculate the mixture ratio of experiments, according to the Iron content (T%) 54.5±0.05% and the alkalinity (CaO/SiO₂) 1.75. The mixture ratio of experiments respectively is 5%, 8%,10%,15%,20%. The table 2 shows the testing program.

Table 2. The testing program /%

Iron materials	1st program	2nd	3rd	4th	5th
1st Indian iron ore powder	17.00	16.00	15.00	11.00	10.00
2nd Indian iron ore powder	18.00	16.00	15.00	14.00	10.00
Robe Rive iron ore powder	5.00	8.00	10.00	15.00	20.00
Brazil iron ore powder	26.00	26.00	26.00	26.00	26.00
Steel scoria	2.00	2.00	2.00	2.00	2.00
Red mud	2.00	2.00	2.00	2.00	2.00
Dust ash	2.00	2.00	2.00	2.00	2.00
Return mines	10.30	10.30	10.50	10.70	10.70
White limestone powder	4.60	4.60	4.60	4.50	4.50
Limestone powder	4.00	4.00	4.00	4.00	4.00
Dolomite fines	4.30	4.30	4.10	4.00	4.00
Coke powder	4.80	4.80	4.80	4.80	4.80

2.3   The Method of Experiment

The sintering experiment equipment is done in sintering cup with 230 mm×370 mm. The weight of the total materials is 30kg and the moisture of the materials is controlled in 8.5% or so.   The materials are first blended in the balling disc for three minutes and continue to pelletize in the granulation drum for five minutes for second blending. The water addition time is three minutes in the drum. The ignition pressure is subtractive 3.92 kPa and the ignition temperature is 1150°C or so and the ignition time is one minute and the ignition fuel is liquefied gas. The air draft pressure is subtractive 7.84 kPa. The hearth layer for sinter is 10-25 mm agglomerate with 1kg. When sintering waste gas temperature began to fall down, the sintering process is over .Sintering devices shown below:

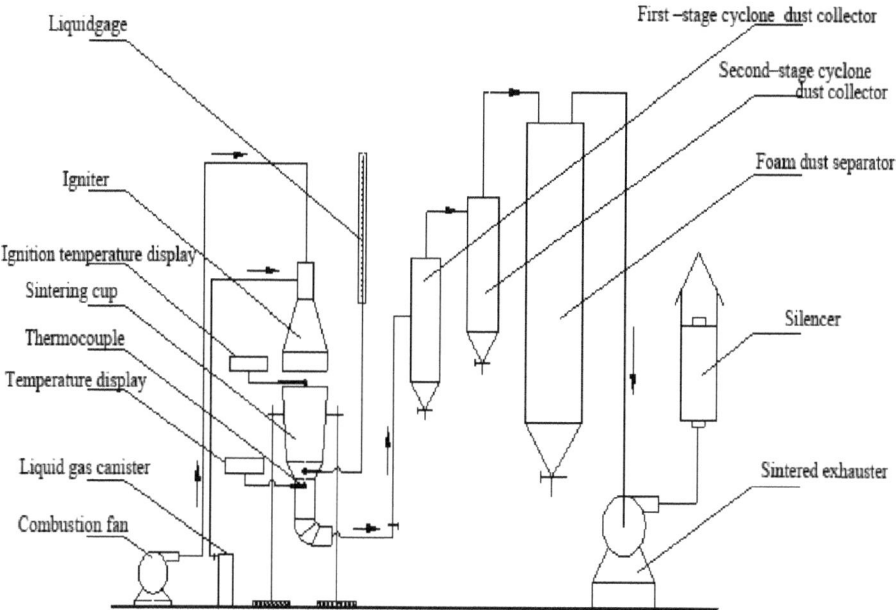

Fig. 1. Sintering devices

To measure the sinter cold strength, according to the standard (GB8029- 87) of China. The sinters are poured out from sintering cup ,and do three times experiments after cooled with falling from 2 meters device. The sinters are shattered and sieved in order to measure its rate of finished products. Sampling > 10 mm sinters are measured drum index with 1/5*ISO* standard rotating drum.

According to the standard (GB/T13242-91) of china measure the sinter Low temperature reduction degradation performance : To make static constant temperature reduction by the constitution of CO and N_2 reduction gas, in the fixed bed, under 500 °C temperature, the diameter of reaction tube is Φ 75mm . After 1h reduction, pull out the sample from the furnace, under the protection of nitrogen cooled the sample to the room-temperature, turn the small rotating drum 300r, separated that by the square hole screen of 6.3mm, 3.15mm and 0.5mm, use the quantity percent of corresponding part to show the property of reduced powder, show that respectively by the sign of $RDI_{+6.3}$, $RDI_{+3.15}$, $RDI_{-0.5}$. The sign of $RDI_{+3.15}$ is evaluation index, and $RDI_{+6.3}$, $RDI_{-0.5}$ is reference index.

According to the standard (GB/T13241-91) of china measure the sinter reduction performance: To make constant temperature reduction by the constitution of CO and N_2 in, grain size is 10.0mm~12.5mm, the sample weight is 500g, the reducing temperature is 900°C, the reducing gas composition is 30% CO,70% N_2,and the reducing gas discharge is 15L/min about 3h,the main qualify for reducing index $(RI)$is the reducing degree after 3h iron ore reducing ,according to the ferric iron (assume the iron of iron ore exists in Fe_2O_3 on the whole, and regard the oxygen of iron ore as 100%). Use mass percent to express.

Measuring the Load softening properties with 1 kg/cm^2 load, the grain size of proof sample is 2-3mm, the height of stock column is 40mm. In the process of heating-up, when the beginning temperature of softening is the temperature that is the height of stock column is contracted to 10%, and the ending temperature is contracted to 40%, the refractoriness under load space interval is $\Delta T=T_{40\%}-T_{10\%}$.

3   The Analysis of Experiment

3.1   The Sinter Cold Strength and Test Index

The table 3 shows the sinter drum strength and test index.

Table 3. The results of the sintering test

Testing program	Strength index /%		Sintering velocity /mm·min^{-1}	Rate of finished products /%	Fire waste /%
	Drum strength	Abrieb index			
1st program	58.23	7.33	23.03	79.40	14.66
2nd program	59.17	7.00	25.21	78.15	15.31
3rd program	61.33	6.33	25.85	77.62	15.33
4th program	58.63	7.00	24.35	77.11	15.61
5th program	58.27	7.00	23.73	76.86	16.32

Drum strength:

The sinter strength is evaluation of sinter quality stand or fall of one of the important indexes. From the table 3,When the addition of robe river limonite is less than 10%, the drum strength increased gradually. The drum strength reaches its maximum when the addition of robe river limonite gets 10%. The main reason is that the robe river limonite is an iron ore powder which contains lots of SiO_2 and Al_2O_3, it is benefit to increasing the silicate bonded phase and improving the sinter strength when properly improve the amount of SiO_2[1].And it is benefit to forming the needle calcium carbonate and improving the sinter strength when properly improve the amount of Al_2O_3[2].The reason why the drum strength gradually decreases when the addition is larger than 10% is that robe river limonite including crystal water, when sintering the crystal water evaporates which easily lead to forming the thin-wall and macroporous structure, which makes the sinter brittle and reduces the strength[3].

Sintering velocity:

Sintering velocity rises at first, then decreases, match the maximum amount for 10%. The main reason is that the robe river powder is thinner and its balling performance is well, and the pelletization of sinter mixture improves, the air permeability improves, sintering velocity quickens when allocated the robe river powder; when the addition is larger than 10% the sintering velocity gradually decreases, the main reason is that the increasing crystal water evaporating leads to sinter bed more shrinking and wet layer more thickening, so that the permeability worsening decreases the sintering velocity[4].

Rate of finished products:

With the addition of robe river limonite growing, the rate of finished products gradually decreases. The main reason is that the decomposition of limonite crystal water in sintered charge leads to the shrinkage of sinter bed increasing, which improves the heat stress of sinter ore and breaks the sinter ore[5]. Another possible reason is that the evaporation of crystal water makes the sinter ore cellular, about 1200°C, which forms the closed pore space of bed of material that is mainly of liquid phase of calcium ferrite system. Thus makes the air permeability of sinter bed non-uniform and aggravates the transverse direction non-uniform formed by the coke combustion of sinter bed [6].

3.2   The Index of Metallurgical Properties

The table 4 shows the index of metallurgical properties.

Table 4. The Index of metallurgical properties of the sinter

Testing program	Degradation index /%			Reducing index		Softening index /°C		
	$RDI_{+6.3}$	$RDI_{+3.15}$	$RDI_{-0.5}$	RI /%	RVI /%·min^{-1}	$T_{10\%}$	$T_{40\%}$	ΔT
1st program	32.6	70.1	8.2	0.47	77	1173	1285	112
2nd program	38.4	71.4	8.4	0.48	78	1178	1295	117
3rd program	43.6	73.1	7.7	0.49	79	1191	1306	115
4th program	40.5	72.2	8.0	0.50	81	1180	1298	118
5th program	38.8	71.6	8.8	0.51	83	1164	1282	118

Low temperature reduction degradation:

The low temperature reduction degradation is one of the important index of metallurgical properties directly affects the air permeability of blast furnace lump belt, and affects the normal operation of blast furnace. From the table 4,the index $RDI_{+3.15}$

with the addition of robe river limonite rises at first, then decreases, match the maximum amount for 10%,and then gradually decreases. This phenomenon is in concordance with the rules of sinter drum strength. The main reason is that when the addition is less than 10%, the increasing of silicate bonded phase and needle calcium ferrite makes $RDI_{+3.15}$ greater; but as the robe river power is a kind of hematite including crystal water, when the robe river powder is addition greatly, the Fe_2O_3 is increasing, while the reducing expansion of Fe_2O_3 is the basic reason of producing the reduction degradation, as a result, the index of low temperature reduction degradation decreases[7]. In general, if the $RDI_{+3.15}$ value of varieties of sinter ore is larger than 70%, all these $RDI_{+3.15}$ can satisfy the requests of blast furnace production.

Reducing performance:

After allocating the roe river powder, the reducing index RI and the reducing rate RVI gradually increased. The main reason is that the crystal water evaporating makes the rate of pore space of sinter ore increasing, which improves the reducing performance[8]. The RI maximum reaches 0.51%.

Softening performance:

Load softening properties decides on the permeability of middle blast furnace. During the blast furnace smelting, as the load softening properties of iron mineral is bad so that often can cause hang-ups and the bosh thick. The blast furnace generally requires that sinter beginning soften temperature is higher than 1100°C and the softening range is about 150°C [9]. Therefore, it is important to explore and improve the load softening properties of iron mineral for the blast furnace smelting.

The beginning temperature of sinter softening is higher after allocating the robe river powder, five styles are higher than 1160°C , when the addition is 10%,the temperature reaches its maximum 1191°C; and the softening range is thinner, about 115°C. Foreign studies show that sinter softening temperature depends on its mineral composition and stomatal structure strength, the changes of the beginning soften temperature is often the outcome that stomatal structure strength plays a leading role[10]. This shows that the stomatal structure strength gradually increases after allocating the robe river powder, when addition into the maximum amount for 10%, then that the evaporation of crystal water easily form thin-walled macroporous structure makes stomatal structure strength reduced, causes the beginning soften temperature lower, but still meet the requirements of the blast furnace smelting.

3.3   Sinter Microstructure

Make microscopic structure analysis from the program 1, 3, 5 sinter specimen. Three of bonded phase is mainly of calcium ferrite the stomatal in the size is different, uneven distribution and irregular formation. Blue-gray is calcium ferrite, black is stomatal, dark grey is kirschsteinite.

Program1 sinter ore sample microstructure from figure 2. Mainly for bonded phase is calcium ferrite, board columnar, partly needle and arborization, some arborization calcium ferrite concentrate local areas. Stomatal size and shape irregular, and the porosity is from 20% to 25%. Stomatal edge is board columnar calcium ferrite weaving structure.

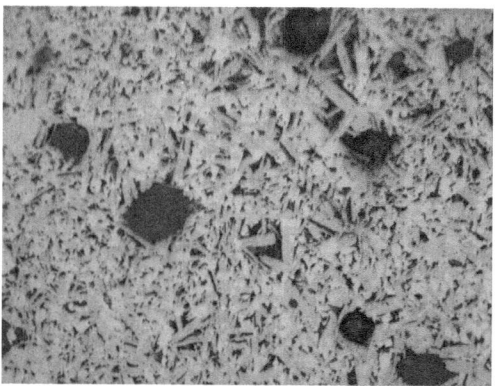

Fig. 2. The microstructure of program1

Program 3 sinter ore sample microstructure from figure 3. Mainly for bonded phase is calcium ferrite, board columnar, partly needle and arborization, some arborization calcium ferrite concentrate local areas. Stomatal in the size is different, distribution is uneven, macro pore is increasing, formation is irregular and porosity is increasing from 30% to 35%. As stomatal edge is needle calcium ferrite weaving structure. So the stomatal structure strength is higher and the sinter drum strength, the index of reduction degradation, and the beginning soften temperature reach their maximum.

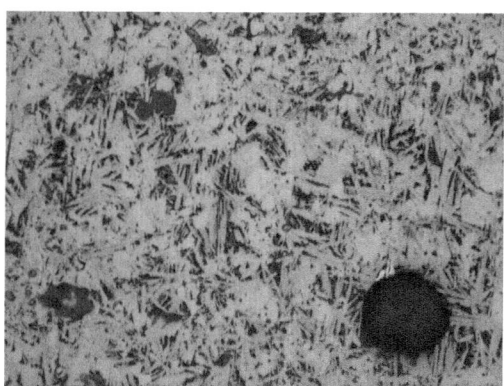

Fig. 3. The microstructure of program3

Program 5 sinter ore sample microstructure from figure 4, This sample bonded phase is still mainly calcium ferrite, anhedron, board columnar, partly ancicular, crystal size relatively coarse, some needle calcium ferrite concentrate local areas. But obviously porosity becomes higher, about 45%-50%, stomatal sizes different and big stomatal more, stomatal wall is thinner, more connectivity, fracture crack development, stomatal edge is kirschsteinite cementation solution structure. So the stomatal structure strength is lower and the sinter drum strength, the index of reduction degradation, and the beginning soften temperature reach their minimum. The increasing porosity makes the sinter reductive strength gradually raised, the style of reductive strength reaches its maximum.

Fig. 4. The microstructure of program5

4   Conclusions

According to the above analysis, some pieces of conclusions gained:

The robe river powder is a kind of limonite and granularity is smaller, after allocating the drum index, the sintering velocity rises at first, then decreases, when reached 10% reached its maximum. Yield gradually reduces, but all can meet requirements of blast furnace smelting.

Low temperature reduction degradation index and beginning soften temperature rises at first, then decreases, when reached 10% reached maximum. Reductive strength is increasing gradually.

After the allocating of robe river powder, the sinter porosity gradually raises, the main bonded phase is calcium ferrite. But when the addition is 10%, the main bonded phase is needle calcium ferrite, stomatal edge is needle calcium ferrite weaving structure, so the stomatal structure strength is higher. Lead to sinter cold strength and low temperature reduction degradation index and beginning soften temperature reached maximum. The increasing porosity makes the sinter reductive strength gradually raised.

From the macro sintering process and determination of sintering indicators, allocate the 5% ~ 20% of rob river powder is feasible. Relatively, it can be the best to allocate 10% of robe river powder.

Suggest increasing sinter alkalinity.

References

1. Wang, W., Lu, Q., et al.: Influence of Australian ore fines ratio on sintering process andmetallurgical properties of vanadium-bearing sinter. Research on Iron & Steel 36, 3–6 (2008)

2. Hong, Y.: Study on Fundamental Properties of Limonite in Sinter Process. Journal of Iron and Steel Research 22, 9–11 (2010)
3. Xu, M., Zhang, Z., et al.: Improve sinter plant load softening properties. Sintering and Pelletizing 24, 3–7 (1999)
4. Kowalski, W., Ersting, K., Werner, P.: The Influence of Sinter Composition on Sintering Rate and phusical Quality of Sinter Ironmaking. In: Conference Proceedings, pp. 415–425 (1997)
5. Dawson, P.R.: Recent Developments in Iron Ore Sintering New Development for Sintering. Ironmaking & Steelmaking, 135–136 (1993)
6. Jin, J., Wu, Y., Jiang, J.-h.: Study on Sintering Technique for High Ratio of Limonite. Iron and Steel 44, 15–19 (2009)
7. Wu, C.-b., Yin, G.-l.: Research on Improving the Low Temperature Reduction Degradation of Low Silicon Sinter. Iron and Steel 45, 16–20 (2010)
8. Hong, Y.-c.: Study on Fundamental Properties of Limonite in Sinter Process. Journal of Iron and Steel Research 22, 9–12 (2010)
9. Hong, Y.-c.: Study on Reasonable Sintering Process Parameters of Limonite. China Metallurgy 20, 4–8 (2010)
10. Wang, S., Gao, W., Kong, L.: Formation Mechanism of Silicoferrite of Calcium and Aluminumin Sintering Process. Ironmaking and Steelmaking, 296–230 (1998)

The Design and Realization
of Power System Load Forecasting Software

Yan Yan[1], Baoxiang Liu[1] , Hongcan Yan[1], and Rong Zhang[2]

[1] College of Sciences, Hebei United University
[2] College of Metallurgy and Energy,
Hebei United University, Tangshan, Hebei, China
yanjxky@126.com

Abstract. Based on CC-2000 system with good openness, portability and scalability, we explain the structure and overall design of the load forecasting software, discuss how to apply meteorological information to load forecasting software steps and used instructions of weather software in detail, and finally point out the limitations of meteorological correction and make relevant recommendations.

Keywords: power system, load forecasting software, meteorological factors.

1   Introduction

Power industry has a history of more than 100 years from the birth to the present, which is the fastest growing period of human society and play a major role in promoting the development of human society. Power industry has completely changed the production of human society and the way of life, and has become an important pillar of modern society. Energy will occupy a greater proportion in the social structure of energy consumption for it is fast, convenient and no pollution, so the pace of the   development of the power industry will not only not slowing down, but will be faster to meet the future of human society. As the core part of the power industry, the function of electric power system is that it economically provides reliable and matching the standards power for various types of users as possible , for contenting each kind of customer's request at any time. Saying with the technical term of electric power system is contented load request[1]. In order to improve the quality of service to users, the power system must know in advance the user side of the load size and characteristics. Although energy is a clean and pollution-free benefits, the energy has the following characteristics: energy production, transmission and consumption are almost running at the same time, and the power can not be stored easily(now some new energy storage technology is only a temporary electrical energy into other Forms of energy, until the time of need and then transformed into electrical energy, such as the pumped storage power station), so in the process of energy production and consumption, we need to know the side of the power load changes for users during the next short period time (such as 1 day or1 week)[2] .In short, the power industry development, production and operation are inseparable from the power load forecast for the future. According to differences between the forecast

B. Liu and C. Chai (Eds.): ICICA 2011, LNCS 7030, pp. 74–80, 2011.
© Springer-Verlag Berlin Heidelberg 2011

period and the uses, load forecasting can be divided into ultra-short term load forecasting, short-term load forecasting, medium-term load forecasting and long-term load forecasting. According to the differences of forecast range, Short term load forecasting can be divided into the system load forecasting and the bus load forecasting. The system load forecasting is to forecast total system load and the bus load forecasting is to forecast every bus's load. Under normal circumstances the system load forecast made by the total system load value at a time, and then assigned to each bus, the bus load distribution coefficient is estimated by the state to maintain. As computing technology and the rapid development of information technology in the new computing and information technology, promotion, load forecasting technology will also towards distributed and network direction.[3]

Power system load forecasting is very important for power system control, operation and planning, and improving its accuracy not only can enhance the safety of electric power system operation, but also can improve the economics of power system operation. Electric power load forecasting is the premise of real-time control, operation plan and development planning, which is to say only when you master to do load forecasting can you achieve the precondition of power production initiative. Assuming power load forecasting software based on existing CC2000 support platform, operating system for Unix 4.0 f, commercial database version for SYBASE 11.0, C/C + + programming language. We have the meteorological information payload to CC2000 system, consider meteorological factors in load forecasting software, which is helpful to improve the accuracy of load forecast software, and can further improve practical level of EMS application software. At the same time, this software comprehensively make use of the advanced software engineering s' design ideas and methods at present, which makes software running speed and precision degree is greatly increased.

2   The Basic Principle of Power System Load Forecasting

According to the load conditions of the present and past time, power system load forecasting estimates the size of the load in the future. So its object is the uncertainty of events and random events. However, power system load forecasting should predict the development trend of loads and conditions possibly. Following we describe some load forecasting basic principles, which will be used to direct the load forecast.[4]

2.1   Knowability Principles

That is to say people can know the forecast object's the development laws, trends and status in the future, which is the fundamental basis for forecasting activities.

2.2   Possibility Principle

Because of the development and change of the things are carried out under the interaction of the internal and external causes, and the change of the internal and the external effects are different, so the development of the things will create lot of possibility. For the forecast of a specific index, we often predict multiple solutions according to its various possibilities of the development and change.

2.3   Continuity Principle

It is also known as the inertia principle. Continuity principle means that the development of the forecasted object is a continuous uniform process, and its development is the continuation of this process in the future .The principles considered the things will remain and last some original features during the process of development and change. The development and change of the power system load also have the inertia, which is the main basis for load forecasting.[5]

2.4   Similarity Principle

In many cases, as a thing of the forecast object, the development process and its present development situation are similar to a certain stage of the past, so it can be based on the known development of conditions to forecast the procedure and conditions of the forecasted objects in the future, which is the similarity principle. At present, in the prediction technology we use analogical method, which is a forecasting method based on the principle.

2.5   Feedback Principle

Feedback is that we use the output to returns to the input, it adjusts the output results again. In the predicting practice activities People found that when there are differences between the predicted results and the actual after a period of practice, we can make use of the gap in the feedback adjustment in order to improve the prediction accuracy. When we carry out the feedback adjustment, the essence is that we combine the predicted theory with the practical. First of all serious value we analyze the gap between the forecasted numbers and the actual numbers as well as the reasons that cause the gap .Then according to reason, we appropriately change input data and parameters and do some feedback adjustments to make further improve the quality prediction.

2.6   Systematic Principle

Systematic principle thinks that the forecasted object is a complete system, which has its own inner system, and it is contacted with the outside things, so it forms its external systems. The future development of the forecasted object is the dynamic development of the system as a whole. However, among the whole system's dynamic development, its component part and the influent factors there are closely mutual effect and influence. Systematic principle stresses that the whole is the best, only the best prediction of the system as a whole is the high quality forecast, and be able to provide the best prediction solutions for policymakers.

3   The Overall Design Thought

To run the power system load forecasting software we must rely on a lot of data about electric network state, in order to improve the reliability of the software running and save the cost of developing the software, it needs some relational functions of the load forecasting software. Load forecasting software includes the following main functions: " area selection ","ultra-short term load forecasting", "short-term load forecasting, ""ultra-short error analysis", "short-term error analysis, ""historical load query, ""historical load modified "," helpful screens ", every functional module obtains the original data according to the historical database, and according to the real-time database to store intermediate data and the predicted results data. Therefore, the first step to run the load forecasting software should be connect the load forecasting software with the local meteorological station in network, and take the meteorological information into the CC2000 system[6] ,then we can consider the effect of the meteorological factors in the load forecasting software. The second step is that we use the load forecasting software to read a real-time data every one minute, and the real-time data is collected from ANALOG table of the real-time library PSBOB, then we find out the appropriate records based on keywords, writing it in the history ORACLE database, and history library is named loadfore. Daily load data form a record form, whose name sets with the corresponding date of access, such as the May 20, 1998 load data, and the stored form named "t19980520", which supports the SQL language to retrieve records, so it will be very convenient to search something. That is the relationship between the historical database, the real-time database and the load forecasting.

4   The Concrete Realization of the Project

4.1   The Exploitation Steps of Meteorological Software

① The data flow table

Meteorological data will be written in SYBASE database of CC2000 system at a fixed time every day, node: his1, service name: H4000A, database name: loadfore. Meteorological information should include the temperature, weather conditions, humidity, rainfall, wind direction, wind and other data during 24 or 96 hours. If you get the meteorological data is difficult during 24 or 96 hours, you can also adopt the traditional daily maximum temperature, minimum temperature, weather conditions, humidity, rainfall, wind direction, wind and other data. Meteorological of provides history weather data in the past. Forecasters provide power load forecasting curve and the relevant information on meteorological factors. Load forecasting software is according to the weather information to amend meteorological factors, and related software needs to amend adaptively.[7]

② The format of meteorological data

Meteorological Station's real-time weather data is sent to the MIS system, and by forwarding process of the regular running we will write the meteorological data in the server his1 the SYABSE database. We proposed real-time weather data in the computer side of the MIS system are stored in text files, daily forming a file, and the file name is "yyyymmdd.txt", for example, "20021208.txt".

We establish a connection of historical server SYBASE database ODBC, the connection name: ems_ts, node: his1, service name: H4000A, user: sa, password: none. Database Name: loadfore, table name: weather.

Executable file name: weather.exe, which can create a desktop shortcut.

Node name of meteorological information data's forwarding process: QX, the operating system is Microsoft Windows 98, the operating environment is the Microsoft Visual Basic 6.0 Enterprise Edition, the history server operating system is DIGITAL UNIX V4.0E, and the remote database is SYBASE 11.5.

4.2   Instructions of Meteorological Receiving Software[6][8]

① Meteorological Station automatically have the weather forecast next day and the live weather data the day before yesterday transmitted to the automated computer side by the dial-up means, by the timing or manual means transmitable programme writes meteorological data in SYBASE database of CC2000 system daily. Load forecasting software is according to the weather information to amend meteorological factors, and related software needs to amend adaptively.

② If we click desktop "meteorological information transmission" icon, it will pop-up forwarding software screen .there are "forecast respectively date" and "the actual date" text box in the screen ,which is used to input the date of the files sent by people ,and the right side of the text boxes is correspond to the "send" button; in bottom of the screen there is a "automatic data transmission of information, " text box which is used to display the results information of automatic transmission of meteorological data.

③ Automatic transmission: after the pop-up screen, the software in the period 16:00:00-18:00:00 a day automatically transmit the weather forecast data next day and live data the day before yesterday every the whole 5 minutes until it is successful. If successful, the "automatic data transfer information " in the text box displays "Meteorological data transfer successfully ", or display failure reasons, which is usually that the data files were transferred to the computer lately or SYBASE database users is temporary surplus and other reasons.

④ Artificial transmission: under normal conditions automatic transmission can be met with demand, without manual transmission. Only when microcomputer fails or meteorological station is failed to timely transmit data, which leads that meteorological data transmission is interrupted in several days, can we need for manual transmission. Operation method: Enter the date of transfer file, you can click on the appropriate button. If successful, the pop-up "Meteorological data transmits successfully" message box, or display failure.

⑤ Load forecast software is amended by considering meteorological information ,the operation method on the screen: Click the "Control Parameters Screen "menu, select the "weather factor into account "check box, the daily load forecast can be.

⑥ Maintenance:

Meteorological File Name:

Weather forecast data files: fyyyymmdd.txt, such as the June 20, 2003 forecast meteorological data, f20030620.txt.

Live weather data files: tyyyymmdd.txt, such as the June 20, 2003 live weather data, t20030620.txt.

Meteorological file format: time, temperature, weather conditions, humidity, rainfall, wind direction, wind- force.

5   Limitations of Meteorological Amendment

If we are able to have the electric network load forecasting software and the meteorological station networked, meteorological station will be automatically have the weather forecast next day and the live weather data the day before yesterday transmitted to the automated computer side by the dial-up means, by the timing or manual means transmitable programme writes meteorological data in SYBASE database of CC2000 system daily .Thus load forecasting software is according to the weather information to amend meteorological factors. Meteorological amendment has an important effect on improving load forecasting accuracy, but as follows the factors still restrict load forecasting accuracy at present.[9]

① The accuracy of meteorological forecast needs to further improve: Summer load is very sensitive to the changes of temperature and humidity .That temperature rise per 1°C may drive load rise100MW, besides , a ding-on may cause the load to fall off 200-400MW, so the temperature and humidity, the rainfall preliminary time and past time, and the rainfall accurate forecast are very important.

② We should improve the accuracy of the meteorological information text which is transmitted daily by the meteorological station, otherwise the conclusion of the load forecasting software will exist warp.

③ The interval of meteorological forecast should be shortened.

④ The transmission time of the meteorological forecast should be advanced.

⑤ Generally, load forecasting researchers are well-known about the relationship between the load and the weather, and their experience has an important effect on improving load forecasting accuracy .Currently the arithmetic of the load forecasting software is universal .It is that according to historical data we calculate the relevant coefficients of the temperature and load, and that we according to the predicted temperature data to predict the load curve. If it can combine with the experience of the load forecasting researchers, it will further improve the accuracy of load forecast software.[10]

Aiming at the above questions, Suggestions are as follows:

① In permitting conditions, we should further improve the accuracy of the weather forecasting, shorten the interval of meteorological forecast, and advance the time of the transmited meteorological forecast.

② Every day in the existing load forecasting the basis of automatically predicting load, strengthen the manual intervention, correction, and combine software with the experience of people. On the basis that the existing load forecasting automatically predicts load daily, we should strengthen the manual intervention, amendment, and combine software with the experience of people.

6   Conclusion

This paper designs and develops the load forecast of meteorological information system, and has a useful discussion on the design ideology development steps and the instructions for the use of the software. After the actual test of the software, some limitations were also overcome during software applications. The software proved successful in the technical and practical application.

References

1. Liu, C.: The Theory and Method of Power System Load Forecasting. Harbin Institute of Technology Press (1987)
2. He, R., Zeng, G., Yao, J.: Application of Weather Sensitivity Neural Networks Model in Short-term Load Forecasting on Area. Automation of Electric Power Systems 25(17), 32–35 (2001)
3. Duan, X.: MATLAB function converts to VB available DLL. Microcomputer and Application (5), 9–11 (2000)
4. Aly, Sheta, W.M., Abdelaziz, A.F.: Development of EvolutionaryModels for Long-Term Load of Power Plant Systems. In: Proceedings on IEEE International Conference of Computer Systems and Applications, pp. 117–120 (2003)
5. Du, X., Zhang, L., Li, Q., Liu, X.: Research on the VB-Matlab interface method in load forecasting software of power system. Power System Protection and Control (19) (2010)
6. Chen, X.: Research of Power Load Forecasting for Power Based on Neural Network and Software Developing. Cenetral South University Master Degree Paper (2010)
7. Bi, Z.: Research of the Medium and Long Term Load Forecasting Algorithms and System Realization. Nan Chang University Master Degree Paper (2010)
8. Liu, J.: Research on Peak Load Forecasting Using RPROP Wavelet Neural Network. Har Bin Institute of Techinology Master Degree Paper (2009)
9. Liu, D.: Technique of Power System Load Forecasting. Heilongjiang Science and Technology Information (31) (2010)
10. Niu, P., Kang, J., Li, A., Li, L.: New operation form of power network started by smart grid. Power System Protection and Control (19) (2010)

The System Framework of Jinan Digital Municipality

Yong-li Zhou[1,3], Shu-jie Ma[2], and Wen Tan[2]

[1] Shandong University Postdoctoral Research Station of Mechanical Engineering,
Jinan city, China, 250101
[2] Shandong Telchina Co., Ltd., Jinan city, China, 250101
[3] Postdoctoral research station of Jinan Hightech Industrial Development Zone,
Jinan city, China, 250101
Zhouyl@telchina.net, 24077161@qq.com

Abstract. This paper elaborated the current situation of facility management and existent question of Jinan city urban municipal services, and surrounded "safety, service, supervision" three subjects of municipal services , proposed the construction object of "Digital Municipality" and discussed "Digital Municipality" design thought, system framework and design content based on hierarchical architecture.

Keywords: Digital Municipality, Internet of Things, GIS, pipenetworks.

1   Preface

Along with the high speed of urbanization progress, municipal services infrastructure develops quickly, the facility management and the safety circulating also become complicated day by day. Supporting safety run, and intensifying a municipal services supervision, have become an important task of manager in the city. Since opening a port a hundred years ago, Jinan city have already formed large-scale and complex underground pipe network and overground municipal infrastructure. For lack of intact and authoritative information-based data of municipal infrastructure, and useful safety early-warning monitor machining, the management of these municipal services facility mainly depended on experience, this city often have occurrence while exploding a tube and digging to break safe accident of etc., that inconvenience people's life, result in serious personnel and resource losing. At the same time for some sudden events, the administration section is hard to make a science, accurate criterion and treatment, cause fast reaction can not be performed to city safe operation, bring hazard to city safety run-time.

Therefore, make use of information-based way settle municipal services facility safety run-time and take charge of a problem, have already more and more received the high respect of each segment of all levels. For this reason, Jinan city decided to finish top-grade digital municipal services system of whole country within 2-3 years since March, 2010, builds "guarantee line" of people living and "Life Line" of city function.

B. Liu and C. Chai (Eds.): ICICA 2011, LNCS 7030, pp. 81–88, 2011.

2   Construction Object

The total object of the Jinan Digital Municipality system is around safety, service and management three aspect: Set up safe early-warning system, raise to meet an emergency command and quickly treat capability, guarantee safety in the city run-time; Create service quality monitor system, raise service quality, guarantee service supply to sustain steady; Propeling city informationization, the digital management and raise industry level of management. Surrounding safety, service and management three aspects, the concrete objects are:

Constructs a network of run-time safety supervising and controlling and early-warning;
Constructs a network of sound service quality monitoring;
Constructs a perfect consumer service system;
Constructs a innovative industry managing system.

3   The System Architecture Designs

3.1   Design Thought

"All-in-one" System design, is also called the system integration design. The system which was designed under the system integration thought instruction except can meet outside the user personalization need in front end, most importantly can effectively realize the backstage integration management, causes the system standardization degree to be high.

Create a reasonable, open and standard-based municipal services resource foundation data management platform, which mainly includes: The foundation geography database, municipal services facility database and special subject database etc. Combine to provide cooperative work service, application integration, information exchange service and information share service etc., gradually perfect and overlay the applied system of all business field [1], consumer can select and assemble correspond business system according to business demand, and construct own integral "all-in-one" network application system on the united platform. Thereby achieve the effect that include background integral management, personalized front-end handling, Single Sign On, whole net transiting , Single exiting.

3.2   Total Frame

According to analysis on the Jinan Municipal Public Utilities Bureau's service present situation, proceeding from the Jinan Municipal Public Utilities Bureau's reality, the overall frame of Jinan Digital Municipality following chart shows:

Jinan Digital Municipality uses "municipal public facility integration" design concept and "from top to bottom" design method, realizes digitized and intellectualized management of the underground and overground municipal infrastructure, joins each different industry's (water, gas, heating and so on) infrastructure, the safety production, the monitor information, the command and adjustment, the customer service closely through the digitized method to a unification

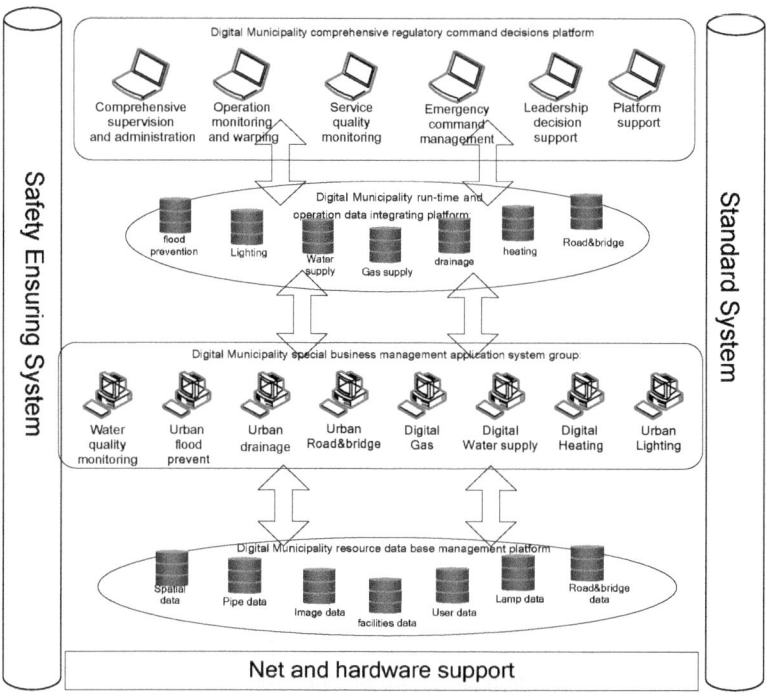

Fig. 1. The total frame of Jinan Digital Municipality,there is four layer in this hierarchy architecture

and in the standard platform, has the data, service, emergency, scheduling, decision-making[2], analysis, service and so on different stratification information integrated and shared, alternately with the integrated type management, by the smallest funding invested and the most highly effective information sharing mechanism, realizes united management, scheduling, and service of whole city various municipal industry with a safety precondition.

At the same time, based on the IOT technology, uses the scientific innovation method to carry on the real-time monitoring and the supervision to the run-time situation of various municipal facility, and integrate the data distributed in each special operational channel's independent operation and service data into the unified central data warehouse according to the unified data standard and operation flow rule, effectively manage and dynamically integrate city municipal run-time and operation data, provides the policy-making basis for the municipal entire business management and the plan.

3.3   Digital Municipality Resource Data Base Management Platform

The Digital Municipality resource data base management platform, is the entire digital municipality foundation. Using middleware, metadata, data warehouse, distributed computing technologies and so on, according to the unification information resource idea integrate the entire urban public foundation municipal domain's information, constructs the unified municipal resource base data management platform, forms digital municipal data information store management system [3], take the coordinated operational mechanism and the science management pattern as the foundation, take the complete technical standard and the standard system as the basis, take the effective system integration and the application support platform as the method, realizes the highly effective network data exchange and the sharing access mechanism, construct many kinds of special facilities system, can support flood preventing, lighting, drainage, bridge, gas, water supply, heat supply, and the application requirement, provide comprehensive, multi-dimensional, multi-scale, multi-resolution, much time Shared information service for global departments and the subordinate unit.

The Digital Municipality resource data base management platform is constructed in accordance with resources integration, resource management, resources sharing the overall thinking. The diagram below:

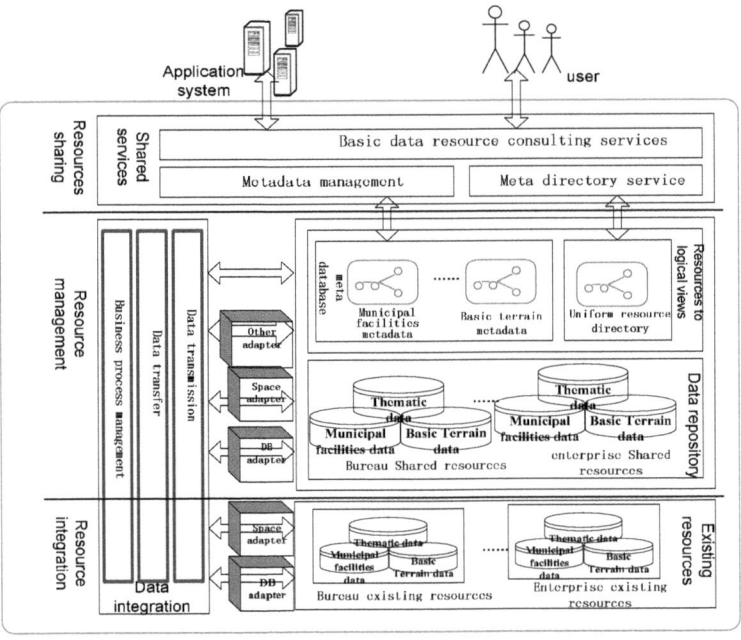

Fig. 2. The architecture of Digital Municipality resource data base management platform

The lowest level is the resource integration layer that is basis of construction data. according to specified data standards of Jinan Municipal Bureau, the bureau and subordinate unit carries on the extraction, transformation and conformity from the existing municipal administration resource base data, forms the shared resource storehouse; Uses the unified plan, the unification purchase, the unification construction regarding the unperfect municipal resource base data, and defers to the data standard which assigns through the resources conformity, merges to the municipal resource base data management platform. The resources integration realizes through the integration service and the different resources' adapter.

The intermediate level is the resource management level: The municipal resource base data is distributed in the Jinan Municipal Bureau and its subordinate unit, satisfies municipal resource base data management, routine maintenance and renewal through the using of synthesis pipe network as well as resource management function of the special data management system, realizes the municipal resource base data unity and uniform [4]. Through establishing unified municipal resource base data [5] directory storehouse, provides description to the entire municipal resource base data, provide the connection for the database retrieval and the inquiry.

The most top layer is the resource sharing level: Provides the foreign municipal data information sharing function through the resource sharing. In view of different object-oriented (the public, enterprise drawn game level government), open different level resource base data sharing and function sharing.

3.4   Digital Municipality Run-Time and Operation Data Integrating Platform

By establishing Digital Municipality run-time and operation data integrating platform, real-time or quantitatively extracts and compiles the operation data and run-time data of various special application system in subordinate unit which belongs to the public utilities bureau according to the unified data standard and the business process management rules, forms the municipal business data mass memory, provides the data source for supervision and management platform.

Digital Municipality run-time and operation data integrating platform uses the data extract integrated way while multi-storehouse heterogeneous data being integrated, does not affect each special application system's independent run-time. This platform construction both has guaranteed that we can ensure data accessing instantaneity demand of specialized application in the short-term application [6], and has guaranteed the various data edition immediate renewal and the unification. Moreover, guaranteed that the government is grasping throughout from the overall situation to the urban public foundation supply system's newest condition, may rationally makes the scientific judgment to the urban development and the adjustment.

Digital Municipality run-time and operation data integrating platform uses the data extract integrated way while multi-storehouse heterogeneous data being integrated, does not affect each special application system's independent run-time. This platform construction both has guaranteed that we can ensure data accessing instantaneity demand of specialized application in the short-term application, and has guaranteed the various data edition immediate renewal and the unification[7]. Moreover, guaranteed that the government is grasping throughout from the overall situation to the urban public foundation supply system's newest condition, may rationally makes the scientific judgment to the urban development and the adjustment.

Data sources: Is the data warehouse system's foundation, and overall system's data fountainhead. The run-time and operation platform mainly extracts business data from various platforms and various types data source such as the flood prevention, the illumination (lighting), drainage, heating, water supply, fuel gas, road & bridge and so on in each special business management application group, after cleaning and loading, data is loaded to run-time and operation data warehouse according to certain rules and standards.

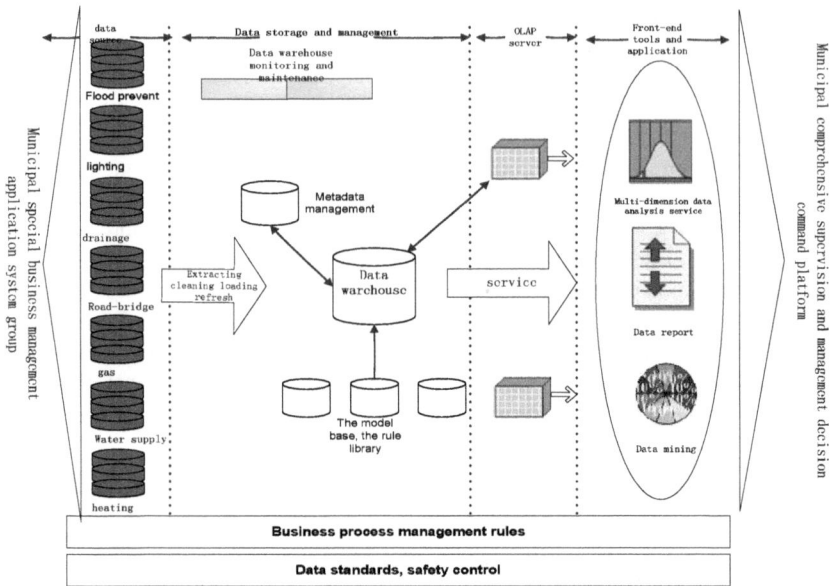

Fig. 3. The architecture of Digital Municipality run-time and operation data integrating platform

The data storage and management: is the core of the whole data warehouse system. on the basis of the existing each business system, carries on the extraction, the cleaning up to the data, and the effective integration, reorganizes these data according to the subject, finally define the physical memory structure of the specified data warehouse, simultaneously organizes metadata of storage data warehouse (specific include the data dictionary definition of data warehouse [8], recording system definition, data conversion rule, data load frequency as well as service rule). According to the data coverage area, the data warehouse storage may be divided into the enterprise data warehouse and the department level data warehouse (usually is called Data Mart). Data warehouse's management includes the data security, filing-up, backup, maintenance, recovery and so on work. These functions and current DBMS are basically consistent.

OLAP Server: reorganizes the data needed for analysis according to the multidimensional data model again, to support the user multi-angle, multi-level analysis, and finds data trends. Its realization can be divided into: MOLAP and HOLAP ROLAP, ROLAP basic data and polymerization data are stored in RDBMS; MOLAP basic data and polymerization data is stored in the multidimensional database; And the MOLAP is synthesized by ROLAP and HOLAP [9], the basic data is stored in the RDBMS, polymerization data is stored in multidimensional database.

Front-end tools and application: front-end tools include various data analysis tools, statements tools, inquires tools, data mining tools and various applications based on data warehouse or data mart development. The data analysis tools mainly aim at OLAP server, reporting tools, data mining tool aims at data warehouse and OLAP server. Through these tools data mining analysis results are provided for comprehensive regulatory command decisions platform.

3.5   Digital Municipality Comprehensive Regulatory Command Decisions Platform

Based on municipal resources data management platform and municipal run-time operation data integrating platform, Digital Municipality comprehensive regulatory command decisions platform analyses and mines global data, centrally shows global current run-time situation in unified operation surveillance platform, overall supervises the problems of system running situation, and realizes the overall management of basic data and the comprehensive analysis for pipeline data[10].

Based on the precise understanding to municipal industry, in order to improve management effectiveness, scientific decision, emergency command, service quality as the goal, through the circular developement and perfection, gradually forms the city decision-making management informationization integration including monitoring and management, emergency command, decision integration, above is the key of that digital municipality play a big role. Cooperate with remote control techniques for building up the city basic municipal areas of monitoring system, will gradually implement emergency command scheduling, energy saving optimization and simulation control, fault diagnosis warning etc. artificial intelligence auxiliary decision and management application.

3.6   Digital Municipality Special Business Management Application System Group

Based on Digital Municipality resource data base management platform, unified complete technical standard and standard hierarchy of municipal public utilities, establishes various special application system of enterprise respectively, satisfy both the cunrrent operation management requirements, and IT development trend, integrates department level management system step by step, ultimately all will be brought in the Digital Municipality comprehensive regulatory command decisions platform.

Municipal special business management application system include:

Urban flood prevention integrated management information system; Urban green illumination intelligence management system; urban drainage integrated management information system; Urban road & bridge facility management information system; Urban fuel gas facility management information system; Urban water supply facility management information system; Urban heating facility management information system.

4   Conclusion

After more than a year of construction began in March 2010, main frame of Jinan Digital Municipality has been build, main function system is also developed completely, it can roundly be online in August,2011. Centering on municipal utilities safety, management and service three aspects, Jinan Digital Municipality will implement construction object, to ensure the safe operation of Jinan city, and promote the healthy development and scientific development of the municipal public utilities, promote the urban management service level and constructing harmonious Jinan. This project has pass through the strict review of country Ministry of Construction (MOC) experts, will be expanded toward the national municipal services industry.

References

1. Jiang, H.-y.: Study on Strategic Development Plan of Water Supply in Jinan City, pp. 28–29. TianJin University (2005)
2. Zhang, D.-p.: Study on Evaluation System and Algorithm for China's Digital City Developing Status, pp. 70–71. Huazhong University of Science and Technology (2004)
3. Tao, W.-x.: Database Construction Practice and Spatial Analysis Application Research on Digital Municipality, pp. 65–67. China University of Petroleum (EastChina) (2009)
4. Yang, J.: The Research for Spatial ontoloies Database in CyberCity, pp. 101–103. Southwest Jiaotong University (2007)
5. Gao, S.: Study on Digital Municipal Management System Architecture with BPM and SOA Technology, pp. 24–25. ChongQing University (2009)
6. Yang, X.-w.: On Research of Urban Informatization In China' s Modernization Drive, pp. 56–56. Central China Normal University (2003)
7. Li, Z.: Research on the Construction and Application of Spatial Data Infrastructure of Digital City, pp. 43–45. Wuhan University (2005)
8. Du, H.-m.: Study on the Management of the Digital Urban Govermance, pp. 23–25. Chongqing University (2009)
9. Tang, L.: Analysis of the Development of Chinese IOT Industry and the status of the Industrial chain, pp. 66–67. Beijing University of Posts and Telecommunications (2010)
10. Gong, J.-r.: Research on Modern Digital City, pp. 46–47. Beijing University of Posts and Telecommunications (2009)

The Research of Weighted-Average Fusion Method in Inland Traffic Flow Detection

Zhong-zhen Yan[1,2], Xin-ping Yan[1,2], Lei Xie [1,2], and Zheyue Wang [1,2]

[1] Intelligent Transport System Research Center,
Wuhan University of Technology,Wuhan, 430063, P.R. China
[2] Engineering Research Center for Transportation Safety (Ministry of Education),
Wuhan University of Technology, Wuhan, 430063, P. R. China
fulianla@sohu.com, xpyan@whut.edu.cn,
xl_for_paper@163.com, 327431103@qq.com

Abstract. Inland waterway traffic flow statistical data is an important foundation for water transportation planning, construction, management, maintenance and safety monitoring. Proposing a multi-sensor data fusion algorithm based on the weighted average estimation method is used to deal with the vessel traffic flow data, and the optimal weight ratio is inducted. Data fusion method is on the basis of weighted average estimation theory, using distributing map of detection technology to test the consistency of data, checking the data to exclude abnormal ones and record missing data, fusing effective data to improve data accuracy. With MATLAB simulation, this example show that the weighed average estimate data fusion method is simple, with high reliability, can effectively improve the robustness of the system measurements, it can get accurate test results. For inland river ships traffic flow testing various sensing device for the collected data format is not the same, weighted average estimate data fusion method is suitable for the situation.

Keywords: traffic flow detection, multisensor, data fusion, weighted average.

1 Introduction

A rapid development of water transportation makes the "golden waterway" of coast and the Yangtze River in China. On the one hand, these important water transportation hubs have made great contributions to China's economic and social development. On the other hand, because of the increasement in vessel traffic flow, maritime accidents have occured thick and fast, brought huge economical losses. Therefore, the detection of ship traffic flow for the channel or navigable waterway is very essential for planning and navigation management. Ship traffic flow data acquisition and processing provide a strong basis for maritime decision makers [1]. It is an important foundation for inland marine monitoring platform. Traffic flow data processing is mainly applied for to gathering and analysing island river vessel traffic flow, piloting vessel, chosing the best channel for the real-time ship in flight, as well as providing strong confirmation marine decision-makers. Single data acquisition method can not meet the needs of data processing; multi-sensor information fusion technology is very effective to improve

B. Liu and C. Chai (Eds.): ICICA 2011, LNCS 7030, pp. 89–96, 2011.
© Springer-Verlag Berlin Heidelberg 2011

efficiency and accuracy of traffic flow data processing, then reaching a more accurate estimates conclusion. Conducting the consistency test of collected data, using the optimal ratio of weighted methods for weighted average estimation, applying MATLAB simulation methods of the algorithm ,can help verify the reliability of weighted average estimation of data fusion.

2   Island River Vessel Traffic Flow Acquisition

Inland river vessel traffic flow arguments are ship traffic flow, speed, vessel traffic density, etc. on this major research in this article is vessels traffic flow. Traffic flow is usually obedient to Poisson distribution or binomial distribution, because the weighted average data processing of data when the distribution is not strict requirements, so here ,vessel traffic flow is subject to any traffic distribution. Ship detection method is divided into two kinds of active detection and passive detection, the monitoring ship set point that some sensing device on the Yangtze River observing vessel underway as Figure 1 to Figure 4 shows [2].

Fig. 1. Radar monitoring

Fig. 2. RFID monitoring

Fig. 3. CCTV monitoring

Fig. 4. AIS berth equipment

3   Multi-sensor Data Fusion Based on Optimal Weight

Among many data fusion methods, the weighted average estimate of fusion algorithm is the eastest. It weights average redundant information provided by multiple sensors inorder to get fused data. It's able to handle the static and dynamic original data. Weight selection is critical, this paper introduces the concept of the optimal ratio of weight, conducts overall weighted multi-sensor fusion, optimized weighted average estimate fusion algorithm [3-5].

3.1   Target Identification Method of Consistency Test

When parameters on the traffic flow is measured, there must be negative influence such as the sudden strong interference on the scene , and   the failure of the measuring device itself,which will inevitably produce negligent errors, and take negative affect on the consistency of measurement data[6]. Therefore, prior to the measurement data fusing, negligent errors should be removed. Inland ship traffic flow is much smaller than road traffic and inland waterways is not as heavy as road traffic, inland waterway has its own inherent characteristics. Inland navigable ability is closely related to vessel traffic flow, the level of water and navigable   environment.

We use the distribution method to detect the consistency of data.The method of distribution map is not limited to the data distribution, capable of enhancing the adaptability of uncertainty in data processing.The method of distribution map does not require complex data structures, easy to program,with less computation[7].

Here is the distribution map algorithm:

An array of data is obtained by different sensors collecting vessel traffic flow information at the same time.We sort 10 acquisitions in ascending, and get the sequences:x1, x2,......x10 ,x1 is the Lower limit,x10 is the higher limit. Definite the mediam as:

$$x_m = \frac{x_5 + x_6}{2} \tag{1}$$

Upper quartile x9 is the mediam of interval[xm, x10],lower quartile x3 is the mediam of interval[x1, xm],quartile dispersion is:

$$dF = x_8 - x_3 \tag{2}$$

Where dF is the dispersion,it reflects the physical quantity of data dispersion. Small data dispersion is considered normal, and large data is considered abnormal. It provides a quantitative criteria for judging whether the data is normal or not.

The data whose distance from median is lager thanγdF is affirmed abnormal.That is to say judgment intervals of invalid data is[y1,y2]:

$$y_1 = x_3 + \frac{\gamma}{2} dF, \quad y_2 = x_8 + \frac{\gamma}{2} dF \tag{3}$$

In the arithmetic expression(3),γ is a const,Its size depends on the system testing accuracy,usually take 1,2 and so on. Data belong to the interval [y1,y2] is considered invalid data of consistency measurement. Taking advantage of this interval to select data, we can eliminate interference of abnormal value at least half of the whole.After eliminating,the remaining datas are considered valid,then we can fuse these consistency data.

3.2   Multi-sensor Weighted Fusion Algorithm

Weighted multi-sensor data fusion is that multiple sensors take test on the same parameters data in a particular environment, taking into account the local estimates of each sensor, according to the same principle to give the weight of each sensor, and finally get the best estimate of global by weighting all the local estimates.

We usually adopt a single sensor to collecte insland vessel traffic data.The detection parameters is relative slow compare to the system sampling frequency,it is subjected to normal distribution. At some time when variance is smallest, we can get the optimal weight ratio that is the state estimate value devides a number (the sum of detected and estimated values) [8-10]. We can utilize the weight to regulate the volatilization or estimates of large divergence, to be convergent. In multi-sensor data fusion, it can provide good and stable source. The purpose of multi-sensor data fusion is to achieve a higher accuracy of target estimation, because the sensor variance is fixed, we need to consider the sensor variance influence on weight of the Integration during the fusion.

Assuming in the information fusion system, among some periods of time the number of ships testedare x1,x2, x3, ...xn, the weights of each sensorare h1,h2,h3, ...hn , variances are σ_1, σ_2,...σ_n, we weighted average processing of measurement data coming from different sensors by multiplied the corresponding weighted in accordance with its precision weights, then obtained the synthetic data,mean output is $\hat{x}$, the model is shown as Figure.5.

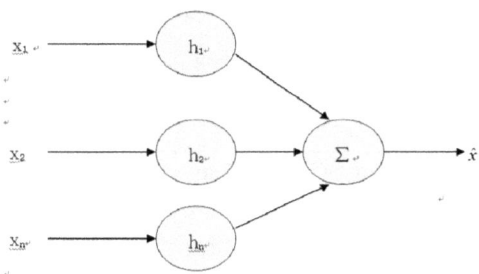

Fig. 5. Weighted average data fusion model

The purpose of multi-sensor data fusion is to enable target estimates more accurate. However, because the variance of the sensor is fixed,we only need to consider the sensor variance influence to fusion weight. According to the data fusion Synthetic model shown in Figure 5,if n is the number of sensors ,output of data and weight should satisfy such condition:

$$\hat{x}_k = \sum_{i=1}^{n} h_i x_k^i = hx^T \ (i = 1,2,3,...n \ ) \tag{4}$$

In the arithmetic expression(4),weight should satisfy such condition:

$$\sum_{i=1}^{n} h_i = 1 \tag{5}$$

Because x_k is subjected to standard normal distribution,based on the average weight of each sensor expectations,accuracy of variance is:

$$\sigma^2 = E[h_i^2 (x - x_i)^2] = \sum_{i=1}^{n} h_i^2 \sigma_i^2 \tag{6}$$

According to $\sum_{i=1}^{n} h_i = 1$, multi-sensor fusion weight can be concluded as follow:

$$h_i = \frac{1}{\sigma_i^2} \bigg/ \sum_{i=1}^{n} \frac{1}{\sigma_i^2} \tag{7}$$

Single sensor weight is:

$$\hat{x}_j = h_k \alpha_j \ (j=1,2,\ ...p) \tag{8}$$

Supposing $\alpha_j = \hat{x}_k + Z_k$, optimum weight ratiois:

$$h_k = \hat{x}_k \bigg/ \alpha_j \tag{9}$$

Multi-sensor fusion weight is presented by:

$$\beta_i \ (i=1,2,\ ...n) \tag{10}$$

In the arithmetic expression(7), α_j is the sum of detection and estimation valueof value at certain time. On the basis of(6)and(8),improved formula for multiple data fusion algorithm weight is:

$$\hat{x}_j = \sum_{i=1}^{n} \alpha_i h_k^i \beta_j^i \tag{11}$$

The antecedent condition to minimum the value of $\hat{x}_j$ is:

$$\sum_{i=1}^{n} \beta_i = 1 \tag{12}$$

4   MATLAB Simulation

This is the procedure of arithmetic operations:

(1) Capturing an 10-dimensional array of data whose mean output is 20,excluding abnormal and error ones,consistency data is obtained;

Table 1. The data collected at a certain time

1	2	3	4	5	6	7	8	9	10
5	26	6	5	9	20	9	31	13	13
13	5	16	18	18	31	17	14	23	46
16	19	25	28	5	17	14	39	32	24
12	26	29	38	35	13	42	10	21	21
28	6	10	6	20	39	27	37	5	5
24	33	36	9	16	24	18	7	40	7
30	22	5	5	5	16	32	22	10	18
18	32	26	46	41	28	8	6	24	42
32	25	23	18	20	5	27	12	8	10
22	6	24	27	31	7	6	22	24	14

Table 2. The consistency data after removing the errors

1	2	3	4	5	6	7	8	9	10
0	26	0	0	9	20	9	0	13	13
12	26	25	38	5	13	0	39	24	21
13	0	0	18	41	31	17	14	23	46
18	32	26	46	35	28	27	31	32	42
22	6	24	27	31	7	6	10	24	14
24	33	36	0	16	24	18	0	40	7
26	19	29	28	18	17	14	22	21	24
28	6	0	0	20	39	27	37	0	0
30	22	0	0	5	16	32	12	10	18
32	25	23	18	20	0	8	0	0	10

Table 1 shows the original data collected by sensor at a particular time,Table 2 shows the appliance of distribution method to analyse error and exclude abnormal data,consistency data is arose.

(2) When each sensor is in the case of minimum variance,calculate the optimal proportion weight;

(3) According to the fixed variance of each sensor,calculate the integration weight.

Figure.6 shows the comparison between weighted average algorithm and actual data values, Figure.7 shows the comparison of state filtering error curve between improved fusion algorithm and weighted average fusion algorithm.

It can be seen that target vessel tracking accuracy via the multi-sensor fusion is much higher than the single sensor. The data go through testing the consistency,it's fusion result is berrer both at estamate value and variance,it has higher precision and accuracy. The algorithm doesn't require knowing any priori knowledge of sensor data, just relying on the test data provided by sensors, could obtain fusion data of minimum MSE (mean square error).It is simple and convenient, and can effectively improve the speed of computing, effectiveness as well.

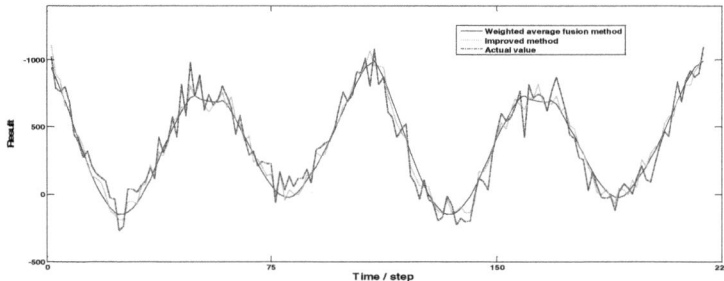

Fig. 6. The comparison between the two estimated fusion value and true value

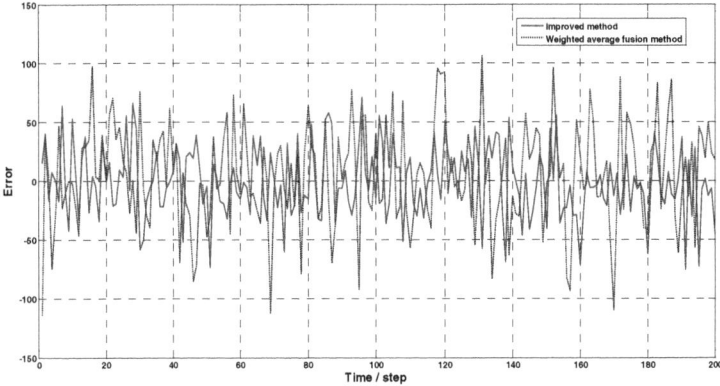

Fig. 7. The comparison of the two fusion state filtering error curve

5   Conclusion

Manipulation checked data and forecast real-time date though method of weighted mean are convenience and high-accuracy.It is particularly applicable to be used for handling vessel flux and the speed of ships in vessel traffic flow,this method can satisfy real-time detection,and have no require of data distribution.Both little vessel flux accord with Poisson distribution and large vessel flux accord with binomial distribution can use this method.Combine method of weighted with distribution map method can improve data accuracy and precisely control the flow of assignment furthermore,and is beneficial for planning in real time vessel traffic flow in the future.

Acknowledgments. This work was supported by Western Traffic Construction Technology Projects (2009328811064) and Independent innovation research fund projects of Wuhan University of Technology (2010-IV-063).The authors are indebted to the reviewers for helpful comments and suggestions.

References

1. Yan, Z., Yan, X., Ma, F., et al.: Green Yangtze River, Intelligent Shipping Information System and Its Key Technologies. Journal of Transport Information and Safety, 76–81 (2010)
2. Yan, X., Ma, F., Chu, X., et al.: Key Technology of Collecting Traffic Flow on the Yangtze River in Real-Time. Navigation of China, 40–45 (2010)
3. Bataillou, E., Thierry, E., Rix, H.: Weighted averaging with adaptive weight estimation. Computers in Cardiology, Proceedings, 37–40 (1991)
4. Fu, H., Du, X.: Multi-sensor Optimum Fusion Based on the Bayes Estimation. Techniques of Automation and Applications, 10–12 (2005)
5. Liu, H., Gong, F.: An Improved Weighted Fusion Algorithm of Multi-sensor. Electronic Engineering & Product World, 19–21 (2009)

6. Shozo, M., Barker, W.H., Chong, C.Y.: Track association track fusion with nondeterministic dynamic target. IEEE Trans. on Aerospace and Electronic System, 659–668 (2002)
7. Spalding, J., Shea, K., Lewandowisk, M.: Intelligent Waterway System and the Waterway Information Network. The Institute of Navigation National Technical Meeting, 484–487 (2002)
8. Pilcher, C., Khotanzad, A.: Nonlinear Classifier Combination for a Maritime Target Recognition Task. In: IEEE 2009 Radar Conference, pp. 873–877. IEEE Press, California (2009)
9. Zhang, M., Hu, J., Zhou, Y.-f.: Research of Improved Particle Filtering Algorithm. Ordnance Industry Automation, 61–63 (2008)
10. Zhang, J., Wang, W., Wu, Q.: Federated Kalman Filter Based on Optoelectronic Tracking System. Semiconductor Optoelectronics, 602–605 (2008)

Distance-Related Invariants of Fasciagraphs and Rotagraphs*

Fuqin Zhan[1], Youfu Qiao[1,**], and Huiying Zhang[2]

[1] Department of Mathematics, Hechi University,
Yizhou, Guangxi 546300, China
[2] Department of Chemistry and Life Science, Hechi University,
Yizhou, Guangxi 546300, China
{zhanfq06, qiaoyf78}@yahoo.com.cn, zhy5158@126.com
http://www.hcnu.edu.cn/

Abstract. The Szeged index, edge Szeged index and GA_2 index of graphs are new topological indices presented very recently. In this paper, a definition approach to the computation of distance-related invariants of fasciagraphs and rotagraphs is presented. Using those formulas, the Szeged index, edge Szeged index and GA_2 index of several graphs are computed.

Keywords: Graph invariant, Szeged index, Edge Szeged index, Geometric-arithmetic index.

1 Introduction

The notion of a polygraph was introduction in chemical graph theory as a generalization of the chemical notion of polymers[1]. Fasciagraphs and rotagraphs form an important class of polygraphs. They describe polymers with open ends and polymers that are closed upon themselves, respectively. It was shown in [2] how the structure of fasciagrapgs and rotagraphs can be used to obtain efficient algorithms for computing the Wiener index of such graphs.

Besides the Wiener index, we will consider several related indices. We First introduce some notation. Throughout this paper, graphs are finite, undirected, simple and connected, the vertex and edge-shapes of which are represented by $V(G)$ and $E(G)$, respectively.

Suppose that x and y are two vertices of G, by $d(x, y)$ we mean the number of edges of the shortest path connecting x and y. We call two vertices u and v to be neighbors if they are the end points of an edge e and we denote it by $e = uv$. The degree of vertex v is the number of its neighbor vertices. Suppose that $e = uv$ is an edge of G connecting the vertices u and v, the distance of e

---

* Supported by the Scientific Research Foundation of the Education Department of Guangxi Province of China (201010LX471;201010LX495;201106LX595; 201106LX608); the Natural Science Fund of Hechi University (2011YBZ-N003).
** Corresponding author.

B. Liu and C. Chai (Eds.): ICICA 2011, LNCS 7030, pp. 97–104, 2011.

to a vertex $w \in V(G)$ is the minimum of the distances of its ends to w, that means $d(e, w) := min\{d(w, u), d(w, v)\}$. Suppose that $W \subseteq V(G)$ by $d(e, w)$ we mean $min\{d(e, w) : w \in W\}$. The number of vertices of G whose distance to the vertex u is smaller than the distance to the vertex v is denoted by $n_u(e)$. We also denote the number of edges of G whose distance to the vertex u is smaller than the distance to the vertex v by $m_u(e)$. In the other words, $n_u(e) := |\{x \in V(G)|d(x, u) < d(x, v)\}|$ and $m_u(e) := |\{f \in E(G)|d(f, u) < d(f, v)\}|$. The vertices and the edges of G with the same distance to u and v are not counted. The Szeged index of the graph G is defined as

$$Sz(G) = \sum_{e=uv \in E(G)} n_u(e)n_v(e).$$

The edge Szeged index of the graph G is defined as

$$Sz_e(G) = \sum_{e=uv \in E(G)} m_u(e)m_v(e).$$

In a recent paper [8] the so-called the second geometric-arithmetic index was conceived, defined as

$$GA_2(G) = \sum_{e=uv \in E(G)} \frac{\sqrt{n_u(e)n_v(e)}}{\frac{1}{2}[n_u(e) + n_v(e)]}.$$

Let $G_1, G_2, ..., G_n$ denote the set of arbitrary, mutually disjoint graphs, and let $X_1, X_2, ..., X_n$ be a sequence of sets of edges such that an edge of X_i joins a vertex of $V(G_i)$ with a vertex of $V(G_{i+1})$. For convenience we also set $G_0 = G_n$, $G_{n+1} = G_1$ and $X_0 = X_n$. This in particular means that edges in X_n join vertices of G_n with vertices of G_1. A polygraph

$$\Omega_n = \Omega_n(G_1, G_2, ..., G_n; X_1, X_2, ..., X_n)$$

over monographs $G_1, G_2, ..., G_n$ is defined in the following way:

$$V(\Omega_n) = V(G_1) \bigcup V(G_2) \bigcup ... \bigcup V(G_n),$$

$$E(\Omega_n) = E(G_1) \bigcup X_1 \bigcup E(G_2) \bigcup X_2 \bigcup ... \bigcup E(G_n) \bigcup X_n.$$

Denote by Γ_i the subgraph of Ω_n with $V(\Gamma_i) = V(G_1) \bigcup V(G_2) \bigcup ... \bigcup V(G_i)$ and $E(\Gamma_i) = E(G_1) \bigcup X_1 \bigcup ... \bigcup E(G_i)(i \in \{1, 2, .., n\})$. If $G_1 = G_2 = ... = G_n = G$ and $X_1 = X_2 = ... = X_n = X$, we denote Ω_n by ω_n and Γ_i by ψ_n; ω_n and ψ_n are called a rotagraph and a fasciagraph , respectively. P_n and C_n will denote the path on n vertices and the cycle on vertices, respectively. Let F_{nm} or M_{nm} be a graph which is obtained from a path P_m or a cycle C_m by replacing each vertex by a star graph S_n and replacing each edge by a fixed set of edges joining the corresponding copies of S_n (see Figure 1).

In this paper we compute the Szeged index, edge Szeged index and GA_2 index of F_{nm} or M_{nm} and apply this result to determine the Szeged index and edge Szeged index of several graphs.

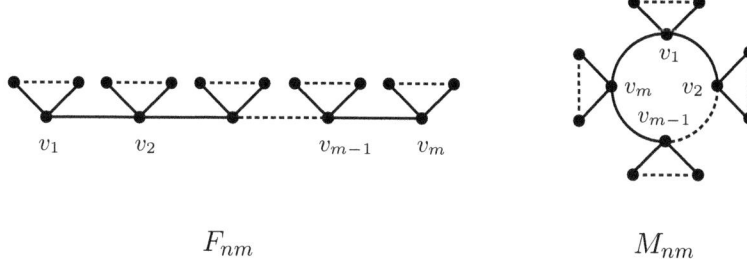

$$F_{nm} \qquad\qquad\qquad M_{nm}$$

Fig. 1. Graphs F_{nm} and M_{nm}

2 The Szeged Index of F_{nm} and M_{nm}

The Szeged index is closely related to the Wiener index and is a vertex-multiplicative type that takes into account how the vertices of a given molecular graph are distributed and coincides with the Wiener index on trees.

Theorem 1. The Szeged index of F_{nm} is given by

$$Sz(F_{nm}) = m\left[(n-1)(nm-1) + \frac{1}{6}n^2(m^2-1)\right].$$

Proof. Let $G = F_{nm}$. From the definitions we see that

$$Sz(G) = \sum_{e=uv\in E(G)} n_u(e)n_v(e)$$

$$= \sum_{i=1}^{m} \sum_{e=uv\in E(S_n)} n_u(e)n_v(e) + \sum_{i=1}^{m-1} n_{v_i}(v_iv_{i+1})n_{v_{i+1}}(v_iv_{i+1})$$

$$= \sum_{i=1}^{m}(n-1)(nm-1) + \sum_{i=1}^{m-1} in(nm-in)$$

$$= m(n-1)(nm-1) + \sum_{i=1}^{m-1}(n^2mi - n^2i^2)$$

$$= m(n-1)(nm-1) + \frac{m(m-1)}{2}n^2m - n^2\cdot\frac{1}{6}m(m-1)(2m-1)$$

$$= m(n-1)(nm-1) + \frac{1}{6}n^2m(m-1)(2m-1)$$

$$= m\left[(n-1)(nm-1) + \frac{1}{6}n^2(m^2-1)\right].$$

Chemically relevant fasciagraphs F_{nm} correspond to the cases $n = 2$ (monomer unit S_2) and $n = 3$ (monomer unit S_3) (see Figure 2). As a corollary of Theorem 1 we have the following result.

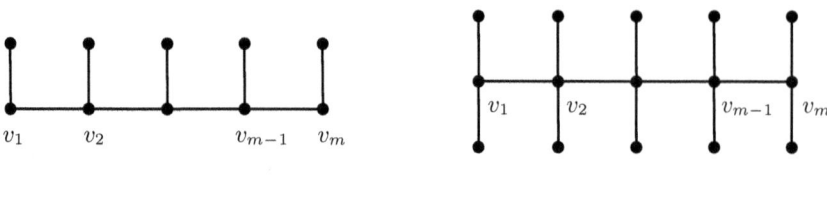

$$F_{2m}$$

$$F_{3m}$$

Fig. 2. Graphs F_{2m} and F_{3m}

Corollary 2. $(1) Sz(F_{2m}) = \frac{1}{3}m(2m^2 + 6m - 5)$.

$$(2) Sz(F_{3m}) = \frac{1}{2}m(3m^2 + 12m - 7).$$

Theorem 3. The Szeged index of rotagraphs M_{nm} is given by

$$
Sz(M_{nm}) =
\begin{cases}
m\left[(n-1)(nm-1) + \frac{1}{4}m^2n^2\right], & \text{for even } m. \\[2ex]
m\left[(n-1)(nm-1) + \frac{1}{4}(m-1)^2n^2\right], & \text{for odd } m.
\end{cases}
$$

Proof. Let $G = M_{nm}$. From the definitions we see that

$$Sz(G) = \sum_{e=uv \in E(G)} n_u(e)n_v(e)$$

$$= \sum_{i=1}^{m} \sum_{e=uv \in E(S_n)} n_u(e)n_v(e) + \sum_{i=1}^{m} n_{v_i}(v_i v_{i+1}) n_{v_{i+1}}(v_i v_{i+1})$$

$$= \sum_{i=1}^{m} (n-1)(nm-1) + \sum_{i=1}^{m} n_{v_i}(v_i v_{i+1}) n_{v_{i+1}}(v_i v_{i+1}).$$

If m is even, then

$$\sum_{i=1}^{m} n_{v_i}(v_i v_{i+1}) n_{v_{i+1}}(v_i v_{i+1}) = \sum_{i=1}^{m} \frac{1}{4}m^2n^2$$

$$= \frac{1}{4}m^3n^2.$$

If m is odd, then

$$\sum_{i=1}^{m} n_{v_i}(v_i v_{i+1}) n_{v_{i+1}}(v_i v_{i+1}) = \sum_{i=1}^{m} \frac{1}{4}(m-1)^2n^2$$

$$= \frac{1}{4}m(m-1)^2n^2.$$

Hence, the Szeged index of the graph G is given by

$$
Sz(G) = \begin{cases}
m\left[(n-1)(nm-1) + \frac{1}{4}m^2n^2\right], & \text{for even } m. \\
\\
m\left[(n-1)(nm-1) + \frac{1}{4}(m-1)^2n^2\right], & \text{for odd } m.
\end{cases}
$$

3   The Edge Szeged Index of Fasciagraph and Rotagraph

Since the Szeged index takes into account how the vertices are distributed, it is natural to introduce an index that takes into account the distribution of edges. The edge Szeged index are recently defined graph invariants.

Theorem 4. The edge Szeged index of fasciagraph F_{nm} is given by

$$
Sz_e(F_{nm}) = (m-1)\left[\frac{1}{6}n^2m(m+1) - (nm-1)\right].
$$

Proof. Let $G = F_{nm}$. From the definitions we see that

$$
Sz_e(G) = \sum_{e=uv \in E(G)} m_u(e)m_v(e)
$$

$$
= \sum_{i=1}^{m-1} m_{v_i}(v_iv_{i+1})m_{v_{i+1}}(v_iv_{i+1})
$$

$$
= \sum_{i=1}^{m-1} \left(ni-1\right)\left[nm - 2 - (ni-1)\right]
$$

$$
=(nm-2)\sum_{i=1}^{m-1}(ni-1) - \sum_{i=1}^{m-1}(ni-1)^2
$$

$$
=n^2m\sum_{i=1}^{m-1}i - (nm-1)(m-1) - n^2\sum_{i=1}^{m-1}i^2
$$

$$
=\frac{1}{2}n^2m^2(m-1) - (nm-1)(m-1) - \frac{1}{6}n^2m(m-1)(2m-1)
$$

$$
=m(m-1)(\frac{1}{6}n^2m + \frac{1}{6}n^2) - (nm-1)(m-1)
$$

$$
=(m-1)\left[\frac{1}{6}n^2m(m+1) - nm + 1\right].
$$

Corollary 5. $(1) Sz_e(F_{2m}) = \frac{1}{3}(m-1)(2m^2 - 4m + 3).$

$(2) Sz_e(F_{3m}) = \frac{3}{2}(m-1)(m^2 - m + \frac{2}{3}).$

Theorem6. The edge Szeged index of rotagraph M_{nm} is given by

$$Sz_e(M_{nm}) = \begin{cases} \frac{1}{4}mn^2(m-2)^2, \text{ for } n \text{ even }. \\ \frac{1}{4}mn^2(m-1)^2, \text{ for } n \text{ odd }. \end{cases}$$

Proof. Let $G = M_{nm}$. From the definitions we see that

$$Sz_e(G) = \sum_{e=uv\in E(G)} m_u(e)m_v(e)$$

$$= \sum_{i=1}^{m} m_{v_i}(v_iv_{i+1})m_{v_{i+1}}(v_iv_{i+1})$$

If m is even, then

$$\sum_{i=1}^{m} m_{v_i}(v_iv_{i+1})m_{v_{i+1}}(v_iv_{i+1}) = \sum_{i=1}^{m} \frac{1}{4}n^2(m-2)^2$$

$$= \frac{1}{4}mn^2(m-2)^2.$$

If m is odd, then

$$\sum_{i=1}^{m} m_{v_i}(v_iv_{i+1})m_{v_{i+1}}(v_iv_{i+1}) = \sum_{i=1}^{m} \frac{1}{4}n^2(m-1)^2$$

$$= \frac{1}{4}mn^2(m-1)^2.$$

Hence , the edge Szeged index of the graph G is given by

$$Sz_e(G) = \begin{cases} \frac{1}{4}mn^2(m-2)^2, \text{ for } n \text{ even }. \\ \frac{1}{4}mn^2(m-1)^2, \text{ for } n \text{ odd }. \end{cases}$$

4   The GA_2 Index of Fasciagraph and Rotagraph

The GA_2 index is named as geometrical-arithmetic index because, as it can be seen from the definition, it consists from geometrical mean of $\sqrt{n_u n_v}$ as numerator and arithmetic mean of $\frac{1}{2}(n_u + n_v)$ as denominator.

Theorom 7. The GA_2 index of F_{nm} is given by

$$GA_2(F_{nm}) = \frac{2(n-1)}{n}\sqrt{nm-1} + \frac{2}{nm}\sum_{i=1}^{m-1} \sqrt{(ni-1)(nm-ni+1)} \ .$$

Proof. Let $G = F_{mn}$. From the definitions we see that

$$GA_2(G) = \sum_{e=uv\in E(G)} \frac{\sqrt{n_u(e)n_v(e)}}{\frac{1}{2}[n_u(e) + n_v(e)]}$$

$$= \sum_{i=1}^{m} \sum_{e=uv\in E(S_n)} \frac{\sqrt{n_u(e)n_v(e)}}{\frac{1}{2}[n_u(e) + n_v(e)]}$$

$$+ \sum_{i=1}^{m-1} \frac{\sqrt{n_{v_i}(v_iv_{i+1})n_{v_{i+1}}(v_iv_{i+1})}}{\frac{1}{2}[n_{v_i}(v_iv_{i+1}) + n_{v_{i+1}}(v_iv_{i+1})]}$$

$$= \sum_{i=1}^{m} \frac{2(n-1)\sqrt{nm-1}}{nm} + \sum_{i=1}^{m-1} \frac{2\sqrt{(ni-1)(nm-ni+1)}}{nm}$$

$$= \frac{2(n-1)}{n}\sqrt{nm-1} + \frac{2}{nm} \sum_{i=1}^{m-1} \sqrt{(ni-1)(nm-ni+1)} \ .$$

As a corollary of Theorem 7 we have the following result.

Corollary 8. $(1)GA_2(F_{2m}) = \sqrt{2m-1} + \frac{1}{m} \sum_{i=1}^{m-1} \sqrt{(2i-1)(2m-2i+1)}$.

$(2)GA_2(F_{3m}) = \frac{4}{3}\sqrt{3m-1} + \frac{2}{3m} \sum_{i=1}^{m-1} \sqrt{(3i-1)(3m-3i+1)}$.

Theorom 9. The GA_2 index of M_{nm} is given by

$$GA_2(M_{nm}) = \frac{2(n-1)}{n}\sqrt{nm-1} + m.$$

Proof. Let $G = M_{nm}$. From the definitions we see that

$$GA_2(G) = \sum_{e=uv\in E(G)} \frac{\sqrt{n_u(e)n_v(e)}}{\frac{1}{2}[n_u(e) + n_v(e)]}$$

$$= \sum_{i=1}^{m} \sum_{e=uv\in E(S_n)} \frac{\sqrt{n_u(e)n_v(e)}}{\frac{1}{2}[n_u(e) + n_v(e)]}$$

$$+ \sum_{i=1}^{m} \frac{\sqrt{n_{v_i}(v_iv_{i+1})n_{v_{i+1}}(v_iv_{i+1})}}{\frac{1}{2}[n_{v_i}(v_iv_{i+1}) + n_{v_{i+1}}(v_iv_{i+1})]}$$

$$= \sum_{i=1}^{m} \frac{2(n-1)}{nm}\sqrt{nm-1} + \sum_{i=1}^{m} \frac{\sqrt{n_{v_i}(v_iv_{i+1})n_{v_{i+1}}(v_iv_{i+1})}}{\frac{1}{2}[n_{v_i}(v_iv_{i+1}) + n_{v_{i+1}}(v_iv_{i+1})]}$$

If m is even, then

$$\sum_{i=1}^{m} \frac{\sqrt{n_{v_i}(v_iv_{i+1})n_{v_{i+1}}(v_iv_{i+1})}}{\frac{1}{2}[n_{v_i}(v_iv_{i+1}) + n_{v_{i+1}}(v_iv_{i+1})]} = \sum_{i=1}^{m} \frac{\frac{mn}{2}}{\frac{mn}{2}} = m.$$

If m is odd, then

$$\sum_{i=1}^{m} \frac{\sqrt{n_{v_i}(v_iv_{i+1})n_{v_{i+1}}(v_iv_{i+1})}}{\frac{1}{2}\left[n_{v_i}(v_iv_{i+1}) + n_{v_{i+1}}(v_iv_{i+1})\right]} = \sum_{i=1}^{m} \frac{\frac{n(m-1)}{2}}{\frac{n(m-1)}{2}} = m.$$

Hence, the GA_2 index of the graph G is given by

$$GA_2(G) = \frac{2(n-1)}{n}\sqrt{nm-1} + m.$$

References

1. Babic, D., Graovac, A., Mohar, B., Pisanski, T.: The matching polynomial of a polygraph. Discrete Appl. Math. 15, 11–24 (1986)
2. Juvan, M., Mohar, B., Graovac, A., Klavzar, S., Zerovnik, J.: Fast computation of the Wiener index of fasciagraphs and rotagraphs. J. Chem. Inf. Comput. Sci. 35, 834–840 (1995)
3. Klavzar, S., Zerovnik, J.: Algebraic approach to fasciagraphs and rotagraphs. Discrete Appl. Math. 68, 93–100 (1996)
4. Mekenyan, O., Dimitrov, S., Bonchev, D.: Graph-theoretical approach to the calculation of physico-chemical properties of polymers. European Polym. J. 19, 1185–1193 (1983)
5. Chang, T.Y., Clark, W.E.: The domination numbers of the $5 \times n$ and $6 \times n$ grid graphs. J. Graph Theory 17, 81–107 (1993)
6. Dolati, A., Motevavian, I., Ehyaee, A.: Szeged index, edge Szeged index, and semi-star trees. Discrete Appl. Math. 158, 876–881 (2010)
7. Mansour, T., Schork, M.: The vertex PI index and Szeged index of bridge graphs. Discrete Appl. Math. 157, 1600–1606 (2009)
8. Fath-Tabar, G., Furtula, B., Gutman, I.: A new geometric-arithmetic index. J. Math. Chem. 47, 477–486 (2010)
9. Zhou, B., Gutman, I., Furtula, B., Du, Z.: On two types of geometric-arithmetic index. Chem. Phys. Lett. 482, 153–155 (2009)
10. Eliasi, M., Iranmanesh, A.: On ordinary generalized geometric-arithmetic index. Appl. Math. Lett. (in press)
11. Hao, J.: Some graphs with extremal PI index. MATCH Commun. Math. Comput. Chem. 63, 211–216 (2010)

Application of the Maximum Real Roots
of Matching Polynomial*

Youfu Qiao and Fuqin Zhan**

Department of Mathematics, Hechi University,
Yizhou, Guangxi 546300, China
{qiaoyf78,zhanfq06}@yahoo.com.cn
http://www.hcnu.edu.cn/

Abstract. To discuss the matching uniqueness of the simple undirected graph G. To find the necessary and sufficient conditions for the matching uniqueness of $T(a, b, c)$. Use the maximum real roots, and the properties of matching polynomials to compute. For $n \geq 5$, $T(1, 5, n)$ and its complement are matching uniqueness if and only if $n \neq 5, 8$ and 15.

Keywords: Matching polynomial, Matching equivalence, Matching uniqueness, The maximum real roots.

1 Introduction

All the graphs considered here are finite, undirected and simple. A matching of graph G is a spanning subgraph of G in which each component is isolated vertex or isolated edge. $r-$ matching is a matching with r edges. In Ref.[1], Farrell defines the matching polynomial of graph G with order n as follows:

$$M(G, W) = \sum_{t \geq 0} p(G, t) x^{n-2t} y^t, \qquad (\dagger)$$

where $p(G, t)$ denotes the number of $t-$ matching, and vector $W = (x, y)$. x, y, respectively, being the weights of vertex and edge. If $y = -1$, we get the matching polynomial of graph G defined in [2], i.e.,

$$\mu(G, x) = \sum_{t \geq 0} (-1)^t p(G, t)(G) x^{n-2t}. \qquad (\ddagger)$$

It is clear that $(\dagger)$ and $(\ddagger)$ are determined by each other. Hence, for (1) and (2), the following notions are identical. Two graphs G and H are said to be matching equivalent, symbolically $G \sim H$, if $\mu(G, x) = \mu(H, x)$. A graph G is matching unique if $\mu(G, x) = \mu(H, x)$ implies that $H \cong G$.

* Supported by the Scientific Research Foundation of the Education Department of Guangxi Province of China (201010LX471;201010LX495;201106LX595; 201106LX608); the Natural Science Fund of Hechi University (2011YBZ-N003).
** Corresponding author.

B. Liu and C. Chai (Eds.): ICICA 2011, LNCS 7030, pp. 105–112, 2011.

Since Farrell [2] introduced the notion of the matching polynomial of graphs, many researchers have been studying matching unique and matching equivalence of graphs, using comparison with the coefficients or maximum real roots of the matching polynomial of graphs [4-9], but this work is difficult, hence there are many unsolved problems. In this paper, the matching uniqueness of a class of the dense graphs $\overline{T(1,5,n)}$ is proved by the algebraic properties and maximum real roots of matching polynomials.

2   Preliminaries

For a graph G, Let $V(G)$, $E(G)$, $p(G)$, $q(G)$, $\overline{G}$, respectively, denote vertex set, edge set, order, size and the complement of G. $G \bigcup H$ denotes the union of two graphs G and H which have no common vertices. nG stand for the disjoint union of n copies of G. $M(G, x)$ denotes the maximum real roots of matching polynomials of graph G. For convenience, we simply denote $\mu(G, x)$ by $\mu(G)$, $M(G, x)$ by $M(G)$, $\mu(T(a, b, c), x)$ by $\mu(a, b, c)$, $\mu(T(a, b, n, c, d), x)$ by $\mu(a, b, n, c, d)$, respectively. F is said to be the special component of graph H, if F is a connected component of graph H and $M(F, x) = M(H, x)$. $\phi(G, x)$ denotes the characteristic polynomials of graph G. $\lambda_1(G)$ denotes the maximum eigenvalues of graph G. In this paper, the maximum roots of matching polynomials are calculated by Mathematica4.0.

We also define the following family of graphs, which will be used throughout the paper. Undefined notation and terminology can be found in [3]:

C_n(resp. P_n) denotes the cycle (resp. the path) of order n.

$T(a, b, c)(a \le b \le c)$ denotes a tree with a vertex v of degree 3 and three vertices of degree 1 such that the distances from v to the three vertices of degree 1 are a, b, c, respectively (see Fig. 1).

$D_{m,n}$ denotes the graph obtained from C_m and P_{n+1} by identifying a vertex of C_m with an end-vertex of P_{n+1} (see Fig. 1).

$T(a, b, n, c, d)$ denotes the tree with a path of length n, and from each end-vertex of this path we draw two pathes of length a, b and two pathes of length c, d. If $a = b = c = d = 1$, then it denote $U_n = T(1, 1, n, 1, 1)$ (see Fig. 1).

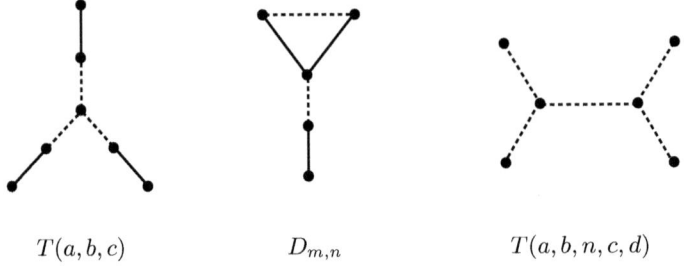

$T(a, b, c)$ $D_{m,n}$ $T(a, b, n, c, d)$

Fig. 1.   Graph $T(a, b, c)$, $D_{m,n}$ and $T(a, b, n, c, d)$

Lemma 1. [2] (1)Let G be any graph with k components $G_1, G_2, ..., G_k$, then

$$\mu(G, x) = \prod_{i=1}^{k} \mu(G_i, x).$$

(2) If $uv \in E(G)$, then

$$\mu(G, x) = \mu(G - uv, x) - \mu(G - \{u, v\}, x).$$

Lemma 2. [2] Let $u \in V(G)$, $e \in E(G)$ and G be a connected graph, then

(1) $M(G) > M(G - u)$.
(2) $M(G) > M(G - e)$.

Lemma 3. [3] Let G be a connected graph, $u \in V(G)$, then $M(G)$ is a single root of $\mu(G, x)$ and $M(G) > M(G - u)$.

Lemma 4. [3] Let G be a connected graph, then
(1) $M(G) < 2$ if and only if

$$G \in \Omega_1 = \{K_1, P_n, C_n, T(1, 1, n), T(1, 2, i)(2 \le i \le 4), D_{3,1}\}.$$

(2) $M(G) = 2$ if and only if

$$G \in \Omega_2 = \{K_{1,4}, T(2, 2, 2), T(1, 3, 3), T(1, 2, 5), U_n, D_{3,2}, D_{4,1}\}.$$

Lemma 5. [4] Let G be a connected graph. Then $M(G) \in \left(2, \sqrt{2 + \sqrt{5}}\right]$ if and only if G is one of the following graphs:
(1) $T(a, b, c)$, for $a = 1, b = 2, c > 5$, or $a = 1, b > 2, c > 3$, or $a = b = 2, c > 2$, or $a = 2, b = c = 3$;
(2) $T(1, a, b, c, 1)$, for $(a, b, c) \in \{(1, 1, 2), (2, 4, 2), (2, 5, 3), (3, 7, 3), (3, 8, 4)\}$, or $a \ge 1, b \ge b^*(a, c), c \ge 1$, where $(a, c) \ne (1, 1)$ and

$$b^*(a, c) = \begin{cases} a + c + 2 & \text{for } a \ge 3 \\ 3 + c & \text{for } a = 2 \\ c & \text{for } a = 1 \end{cases}$$

(3) $D_{3,n}(n \ge 3)$, or $D_{m,1}(m \ge 5)$, or $D_{4,2}$.

Lemma 6. [5] Let G be a graph with n vertices and $n - 1$ edges. Let degree sequences of G be $\pi(G) = (1, 1, 1, 3, 2, ..., 2)$, if $\mu(G, x) = \mu(H, x)$, then degree sequences of H is $\pi(H) = (1, 1, 1, 3, 2, ..., 2)$ or $\pi(H) = (0, 2, ..., 2)$.

Lemma 7. Let $m \ge 3$ be integers, then

$$M(D_{m,1}) = M(D_{3,m-2}).$$

Proof. By Lemma 1(2), we have

$$\mu(D_{m,1}) = \mu(P_{m+1}) - x\mu(P_{m-2}),$$

$$\mu(D_{3,m-2}) = \mu(P_{m+1}) - x\mu(P_{m-2}),$$

then

$$\mu(D_{m,1}) = \mu(D_{3,m-2}).$$

Therefor $M(D_{m,1}) = M(D_{3,m-2})$.

Lemma 8. [1] Let graph G be a forest, then

$$\mu(G, x) = \phi(G, x).$$

Lemma 9. Let T be a tree, then

$$M(G) = \lambda_1(G).$$

Proof. By Lemma 8, the Lemma 9 holds.

A internal $x_1 x_k - path$ of a graph G is a path $x_1 x_2 ... x_k$ (possibly $x_1 = x_k$) of G such that $d(x_1)$ and $d(x_k)$ are at least 3 and $d(x_2) = d(x_3) = ... = d(x_{k-1}) = 2$(unless $k = 2$).

Lemma 10. [7] Let $xy \in E(G)$, G_{xy} is the graph obtained from G by inserting a new vertex on the edge xy of G, then

(1) If xy is not an edge on an internal path of G and $G \neq C_n$, then

$$\lambda_1(G_{xy}) > \lambda_1(G).$$

(2) If xy is an edge on an internal path of G and $G \neq U_n$, for all $n \geq 6$, then

$$\lambda_1(G_{xy}) < \lambda_1(G).$$

Lemma 11. Let G be a tree, $uv \in E(G)$, G_{uv} be the graph obtained from G by inserting a new vertex on the edge uv of G, then

(1) If uv is not an edge on an internal path of G, then

$$M(G_{uv}) > M(G).$$

(2) If uv is an edge on an internal path of G and $G \neq U_n$, for all $n \geq 6$, then

$$M(G_{uv}) < M(G).$$

Proof. By Lemma 9 and Lemma 10, the Lemma 11 holds.

Lemma 12. [7] A graph G is matching unique if and only if $\overline{G}$ is matching unique.

3   Main Results

Lemma 13. Let $t \leq n_1$ be integers, if $max\{b, d\} \geq t$, then

$$M(1, t, n_1) < M(1, b, n_2, d, 1).$$

Proof. Without loss of generality, we can assume that $max\{b, d\} = b \geq t$. By Lemma 2 and Lemma 11, we have

$$
\begin{aligned}
M(1, t, n_1) &< M(1, b, n_1) \\
&< M(1, b, n_1 + n_2) \\
&< M(1, b, n_1 + n_2, d, 1) \\
&< M(1, b, n_2, d, 1).
\end{aligned}
$$

Theorem 1. Let $n \geq 5$ be integers, tree $T(1, 5, n)$ is matching unique if and only if $n \neq 5, 8$ and 15.

Proof. By

$$
\mu(1, 5, 5) = \mu(P_5)\mu(D_{3,4}) = \mu(P_5)\mu(D_{6,1}),
$$

$$
\mu(1, 5, 8) = \mu(P_7)\mu(D_{3,5}) = \mu(P_7)\mu(D_{7,1}),
$$

$$
\mu(1, 5, 15) = \mu(C_7)\mu(1, 6, 7),
$$

So, the necessity of the condition is obvious.

The following the proof of the sufficiency of the condition:

Let $H \sim T(1, 5, n)$. By Lemma 6, we have

$$
H \in \left\{ K_1 \bigcup \left(\bigcup_{i \in A} C_i \right), T(a, b, c) \bigcup \left(\bigcup_{i \in A} C_i \right), P_m \bigcup D_{s,t} \bigcup \left(\bigcup_{i \in A} C_i \right) \right\},
$$

where A is multiset with integer numbers of more than or equal to 3 as its elements.

Case 1. If $T(1, 5, n) \sim H = K_1 \bigcup \left(\bigcup_{i \in A} C_i \right)$, then

$$
M(1, 5, n) = M\left(K_1 \bigcup \left(\bigcup_{i \in A} C_i \right) \right).
$$

By Lemma 4 and Lemma 5, we have $M(1, 5, n) > 2$, $M\left(K_1 \bigcup \left(\bigcup_{i \in A} C_i \right) \right) < 2$, contradiction.

Case 2. If $T(1, 5, n) \sim H = T(a, b, c) \bigcup \left(\bigcup_{i \in A} C_i \right)$, then

$$
M(1, 5, n) = M\left(T(a, b, c) \bigcup \left(\bigcup_{i \in A} C_i \right) \right).
$$

By Lemma 4 and Lemma 5, we have

$$
M(1, 5, n) = M(a, b, c) \in \left(2, \sqrt{2 + \sqrt{5}} \right].
$$

Then

$$T(a, b, c) \in \left\{ T(1, 2, i)(i > 5), T(1, b, c)(2 < b \leq c, c \neq 3), T(2, 2, c)(c > 2), \right.$$
$$\left. T(2, 3, 3) \right\}.$$

Case 2.1. Let $T(1, 5, n) \sim H = T(1, b, c) \bigcup \left(\bigcup_{i \in A} C_i \right) (2 \leq b \leq 3, b \leq c)$.

Since $M(1, 5, 5) = M(1, 1, 2, 2, 1) = M(1, 1, 6, 3, 1)$, by Lemma 13 and Lemma 2 we have

$$M(1, 2, c) < M(1, 1, 2, 2, 1) = M(1, 5, 5) \leq M(1, 5, n),$$
$$M(1, 3, c) < M(1, 1, 6, 3, 1) = M(1, 5, 5) \leq M(1, 5, n).$$

Therefore $T(1, 2, c)$ and $T(1, 3, c)$ are not the special component of H.

Case 2.2. Let $T(1, 5, n) \sim H = T(1, 4, c) \bigcup \left(\bigcup_{i \in A} C_i \right)$.

By calculating, we obtain

$$2.04734 = M(1, 1, 9, 4, 1) < M(1, 5, 7) = 2.04781.$$

From Lemma 13, we have

$$M(1, 4, c) < M(1, 1, 9, 4, 1) < M(1, 5, 7) < M(1, 5, 7 + m).$$

Noticing $M(1, 4, 7) = M(1, 5, 5)$, $M(1, 4, 13) = M(1, 5, 6)$, but $q\big(T(1, 4, 7)\big) > q\big(T(1, 5, 5)\big)$, $q\big(T(1, 4, 13)\big) > q\big(T(1, 5, 6)\big)$.
   Therefore $T(1, 4, c)$ is not the special component of H.

Case 2.3. Let $T(1, 5, n) \sim H = T(1, 5, c) \bigcup \left(\bigcup_{i \in A} C_i \right)$.

   By Lemma 4 and Lemma 5, we have $c = n$. Then $A = \phi$.
   Therefore $H \cong T(1, 5, n)$.

Case 2.4. Let $T(1, 5, n) \sim H = T(1, 6, c) \bigcup \left(\bigcup_{i \in A} C_i \right)$.

By calculating, we obtain

$$2.05218 = M(1, 6, 8) > M(1, 1, 10, 5, 1) = 2.05185.$$

From Lemma 13, we have

$$M(1, 5, n) < M(1, 1, 10, 5, 1) < M(1, 6, 8) < M(1, 6, 8 + m).$$

Noticing $M(1, 6, 6) = M(1, 5, 8)$, $M(1, 6, 7) = M(1, 5, 15)$, but from the condition of theorem we have $n \neq 8$ and 15.
   Therefore $T(1, 6, c)$ is not the special component of H.

Case 2.5. Let $T(1, 5, n) \sim H = T(1, b, c) \bigcup \left(\bigcup_{i \in A} C_i \right) (7 \leq b \leq c)$.

Since $M(1,7,7) = M(1,6,9)$, by case 2.4 and Lemma 2 we have

$$M(1,5,n) < M(1,7,7) \leq M(1,b,c)(7 \leq b \leq c).$$

Therefore $T(1,b,c)(7 \leq b \leq c)$ is not the special component of H.

Case 2.6. Let $T(1,5,n) \sim H = T(2,2,c) \bigcup (\bigcup_{i \in A} C_i)$.

Since $M(2,2,3) = M(1,4,4)$, $M(2,2,6) = M(1,7,7)$, by case 2.2 and case 2.5 we have

$$M(2,2,3) = M(1,4,4) < M(1,4,7) = M(1,5,5),$$

$$M(1,5,n) < M(1,7,7) = M(2,2,6) < M(2,2,6+m).$$

Noticing $M(2,2,4) = M(1,5,5)$, $M(2,2,5) = M(1,5,8)$, but from the condition of theorem we have $n \neq 5$ and 8.

Therefore $T(2,2,c)$ is not the special component of H.

Case 2.7. Let $T(1,5,n) \sim H = T(2,3,3) \bigcup (\bigcup_{i \in A} C_i)$.

Since $M(2,3,3) = M(2,2,6)$, by case 2.6 we have $T(2,3,3)$ is not the special component of H.

Case 3. If $T(1,5,n) \sim H = P_m \bigcup D_{s,t} \bigcup (\bigcup_{i \in A} C_i)$, then

$$M(1,5,n) = M\left(P_m \bigcup D_{s,t} \bigcup (\bigcup_{i \in A} C_i) \right).$$

By Lemma 4 and Lemma 5, we have

$$M(1,5,n) = M(D_{s,t})$$

and

$$D_{s,t} \in \{ D_{3,t}(t \geq 3), D_{s,1}(s \geq 5), D_{4,2} \}.$$

Case 3.1. Let $T(1,5,n) \sim H = P_m \bigcup D_{3,t} \bigcup (\bigcup_{i \in A} C_i)$.

Since $M(D_{3,3}) = M(1,4,4)$, $M(D_{3,6}) = M(1,7,7)$, by case 2.2 and case 2.5 we have

$$M(D_{3,3}) = M(1,4,4) < M(1,4,7) = M(1,5,5),$$

$$M(1,5,n) < M(1,7,7) = M(D_{3,6}) < M(D_{3,6+m}).$$

Noticing $M(D_{3,4}) = M(1,5,5)$, $M(D_{3,5}) = M(1,5,8)$, but from the condition of theorem we have $n \neq 5$ and 8.

Therefore $D_{3,t}(t \geq 3)$ is not the special component of H.

Case 3.2. Let $T(1,5,n) \sim H = P_m \bigcup D_{s,1} \bigcup (\bigcup_{i \in A} C_i)$.

Since $M(D_{s,1}) = M(D_{3,s-2})(s \geq 5)$, by case 3.1 we have $D_{s,1}(s \geq 5)$ is not the special component of H.

Case 3.3. Let $T(1,5,n) \sim H = P_m \bigcup D_{4,2} \bigcup (\bigcup_{i \in A} C_i)$.

Since $M(D_{4,2}) = M(D_{3,6})$, by case 3.1 we have $D_{4,2}$ is not the special component of H.

By Lemma 5, case 1, case 2 and case 3, we have the special component of $T(1,5,n)$ is itself.

Therefore, $T(1,5,n)$ is matching unique.

Theorem 2. Let $n \geq 5$ be integers, graph $\overline{T(1,5,n)}$ is matching unique if and only if $n \neq 5, 8$ and 15.

Proof. By Lemma 12 and Theorem 1, it is easy to prove Theorem 2.

References

1. Godsil, C.D.: Algebraic Combinatorics. Chapman and Hall, New York (1993)
2. Bondy, J.A., Murty, U.S.R.: Graph Theory with Applications. North-Holland, Amsterdam (1976)
3. Ma, H.: Graphs Characterized by the Roots of Matching. Journal of Qufu Normal University (Natural Science) 27(1), 33–36 (2001)
4. Ma, H.: The Graphs G with $2 < M(G) \leq \sqrt{2 + \sqrt{5}}$. Acta Scientiarum Naturalium Universitatis Neimongol 36(5), 485–487 (2005)
5. Sheng, S.: The Matching Uniqueness of $T-$shape Trees. Journal of Mathematical Study 32(1), 86–91 (1999)
6. Cvetkovic, D., Rowlinson, P.: The largest eigenvalue of a graph: A survey. Linear and Multilinear Algebra 28(1,2), 3–33 (1990)
7. Cvetkovic, D.M., Doob, M., Gutman, I., Torgaser, A.: Recent result in the theory of graph spectra. Elsevier Science Publishers, New York (1988)
8. Ma, H., Zhao, H.: The Matching Unique Graphs with Large Degree or Small Degree. Journal of Mathematical Research and Exposition 24(2), 369–373 (2004)
9. Ma, H., Ren, H.: The new methods for constructing matching-equivalence Graphs. Discrete Mathematics 307, 125–131 (2007)
10. Qiao, Y., Zhan, F.: Matching Uniqueness of $T(1,4,n)$ and its Complement. Natural Science Journal of Hainan University 26(3), 220–224 (2008)
11. Zhan, F., Qiao, Y., Zhao, L.: The Matching Uniqueness of a Class of New Graphs. Journal of Southwest China Normal University (Natural Science Edition) 35(3), 7–11 (2010)
12. Shen, S.: On matching characterization of a Caterpillars. Pure and Applied Mathematics 26(4), 541–545 (2010)
13. Shen, S.: On Matching Characterization of $K_1 \bigcup T(1,3,n)$ and Its Complement. Journal of Southwest China Normal University (Natural Science Edition) 34(3), 5–9 (2009)
14. Zhang, G., Li, Y.: A Lower Bound for the Second Largest Eigenvalue of a Class of Tree Matching. Journal of North University of China (Natural Science Edition) 32(1), 1–3 (2011)

Applying Backtracking Heuristics for Constrained Two-Dimensional Guillotine Cutting Problems

Luiz Jonatã, Piresde Araújo, and Plácido Rogério Pinheiro

University of Fortaleza(UNIFOR)-Graduate Program in Applied Informatics,
Av.Washington Soares, 1321-J30,
60.811-905 Fortaleza-Brazil
ljonata@gmail.com, placido@unifor.br

Abstract. The Backtracking Heuristic (BH) methodology consists in to construct blocks of items by combination beetween heristics, that solve mathematical programming models, and backtrack search algorithm to figure out the best heuristics and their best ordering. BH has been re- cently introduced in the literature in order to solve three-dimensional Knapsack Loadin Problems, showing promising results. In the present Work we apply the same methodology to solve constrained two-dimensional Guillotine cutting problems. In order to assess the potentials of this novel ersion also for cutting problems, we conducted computational experiments on a set of difficult and well known benchmark instances.

Keywords: Cutting Problems, Backtracking, Integer Programming.

1 Introduction

The Backtracking Heuristic (BH) methodology has been recently introduced in literature in order to solve three-dimensional knapsack loading problems[1, 2]. It consists in to construct blocks of items by combination beetween heristics, that solve mathematical programming models, and backtrack search algorithm to figure out the best order of heuristics. The comparinson with other methodologies using well known benchmark instances showed promising results. In the present work we apply the same methodology to solve unconstrained and constrained two-dimensional cutting problems.

In two-dimensional cutting problems a rectangular plate sized (L,W), the plate's lenght and width, is to be ortogonally cut into determined and smaller rectangular pieces sized (l_i, w_i). To each piece is associated a utility value v_i. The cutting problem objectives an optimal cutting pattern that optimizes the sum of utility values. If $v_i = l_i w_i$, the problem is equivalent to minimize the waste (trim loss) or maximizing the used area of the [3]. This problem's definition fits on 2/B/O classification, according Dyckhoff's typology of cutting and packing problems [13]. An improved typology was proposed in [30].

B. Liu and C. Chai (Eds.): ICICA 2011, LNCS 7030, pp. 113–120, 2011.

According the cargo we make some considerations:

- When there's a limit b_i on the quantity of type of piece i the problem is called unconstrained. Otherwise, when there is not upper bound on the quantity of pieces, the problem is called constrained;
- When we deal only identical demanded pieces, we call the problem as pallet loadin problem. Some works adress specifically the pallet loadin problem: [7]

We still can classify the two-dimensional cutting problems according the type of cutting, guillotine or unguillotine cutting. The first one occurs when pieces (rectangles) are to be obtained by successive edge-to-edge cuts of the stock sheet. Figure 1-b shows a guillotine cutting and Figure 1-c illustrates the successive steps to produce each piece. Some works adress this problem as Morabito and Morabito [21, 5]. When there's a rectangle that can not be obtained by after a successive edge-to-edge cuts, we say it is a non-guillotine cutting. Figure 1-a illustrates an example of this type of cutting.

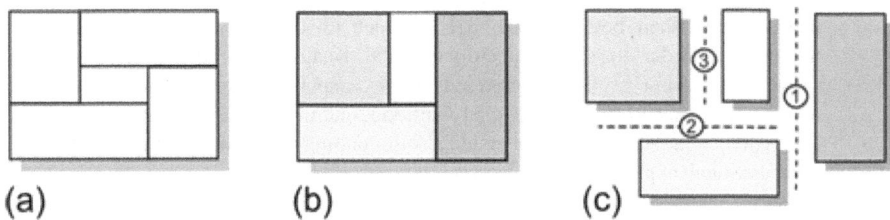

Fig. 1. (a) An instance of non-guillotined cutting problem. (b) A guillotined cut and (c) the successive edge-to-edge cuts to produce its pieces

Some adittional requirements in literature:

- $-90°$ rotation of a piece is allowed;
- Non-rectangular pieces have to be cut;
- Cutting of plates that present some defected areas. Carnieri et al. [10] approach a heuristic procedure that deal this requirement.

In this work we aplly the Backtracking Heuristic from [1, 2] to constrained two-dimensional guillotine cutting problems, that is a NP-hard problem [15, 12].

Due to this characteristic, few methodologies approach exact methods to solve it. Beasley approached an integer linear programming model to solve the general problem 2/B/O/M [4]. However, the model has a large number of variables and restrictions, mainly for big dimensions. It turns more difficult to be optimally solved. Gilmore-Gomory elaborated a technique for overcoming the difficulty in the linear programming formulation of the Cutting Stock Problem[16, 17].

Development of heuristic methods helped to figure out acceptable approximated solutions. Wang presented a combinatoric algorithm that finds guillotines cutting patterns by successively add rectangles to each other, forming horizontal and vertical builds [29]. An AND/OR graph problem representation to solve guillotine cutting problems were proposed by Morabito et al [20, 21]. Wall building heuristic for generating cutting patterns was proposed by Gilmore and Gomore[17].

Recently has approached methods that consist of to combine methods coming from metaheuristics, such as simulated annealing, tabu search and evolutionary algorithms, and more conventional exact methods into a unique optimization framework. This kind of methodology has been referred to by "hybrid meta- heuristics" [8, 9]. For example Mahfoud and Goldberg combined simulated an- nealing to improve the population obtained by a genetic algorithm [19]. Moscato presented various kinds of local search embedded in an evolutionary algorithm for locally improving candidate solutions obtained from variation operators [22]. Nepomuceno combined a metaheuristic engine (genetic algorithm) that works as a generator of reduced instances for the original optimization problem, which are formulated as mathematical programming models. These instances, in turn, are solved by an exact optimization technique (solver). It was applyed for cutting problems in [26, 24].

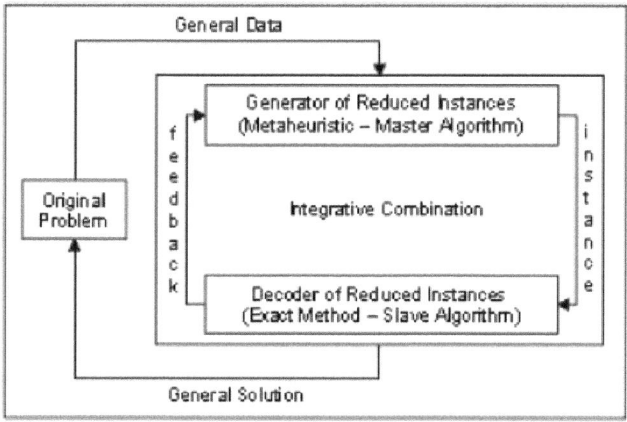

Fig. 2. The hybrid framework investigated in [23]

Our methodology fits on the last category of approaches. It combines a search algorithm, the backtracking, with heuristics which solve integer linear programming models. Next section we present the methodology. Then we use benchmark tests to compare our computational results with others methodologies. Finally, we draw conclusions regarding the quality of the solutions provided by our algorithm and make some considerations concerning future development.

2   Methodology

The present methodology was firstly proposed in [1, 2], which was based on a wall building heuristic [27] and a backtracking algorithm applying in order to choose the best order of heuristics. Each heuristic, when succesfull, pack a block of boxes. After this step, a Genetic Algorithm chooses the best rotations for the set of packed blocks. The figure 3-a illustrates the methodology. Figure 3-b the steps executed by each heuristic. Firstly, the algorithm firstly receives the input problem that includes the list of avaiable types of boxes and their quantities, and the avaiable avaiable space to be

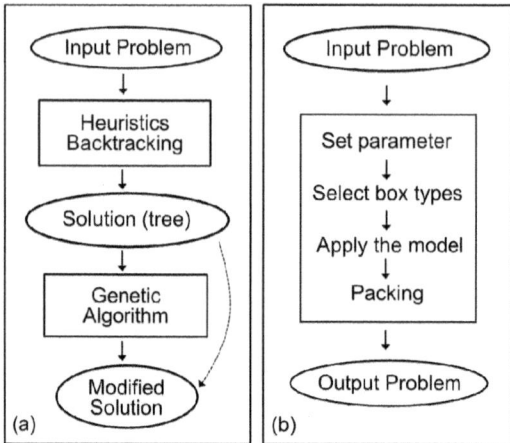

Fig. 3. (a) Methodology's overview. (b) Steps executed by each heuristic.

cut. So, the algorithm selects the types of boxes that attend the filter criterious for the heuristic, according the table 1. Then, using the selected types of boxes, the algorithm builds and solves the specific integer programming model for that heuristic. The model's solution defines the list of blocks to be packed. For example, for 'X Layer' heuristic in figure 4-a, the model's solution achieved the result $nx_\alpha = 3$ for the type α (the most near to origin) and $nx_\beta = 5$,with calculated constan $ny_\alpha = 2$ and $ny_\beta = 4$. The algorithm pack a block of $3x2$ boxes and updates the avaiable space to be cut and avaiable types of boxes, generating an output problem.For the constrained two-dimensional guillotine cutting problem we adaptedthe used heristics in [1, 2]. They are: X Mixed Layer (a), Y Mixed Layer (b),Partition on X (c) and Partition on Y (d). They are illustrated in figure 4.

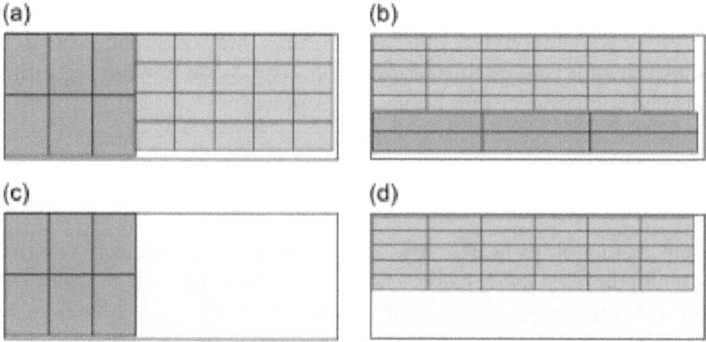

Fig. 4. The heuristics used in the algorithm

The table 1 shows the characteristics of the integer models solved by the heuristics.

Table 1. Comparison between used heurisitcs in the present methodology

	Filter	Decision Variable	Calculated Constant
X Mixed Layer	$(H\%h_s)/H \leq 0.02$	nx_s	$ny_s = H/h_s$
Y Mixed Layer	$(W\%w_s)/W \leq 0.02$	ny_s	$nx_s = H/w_s$
Partition on X	$(H\%h_s)/H \leq 0.02$	nx_s	$ny_s = H/h_s$
Partition on Y	$(W\%w_s)/W \leq 0.02$	ny_s	$nx_s = H/w_s$

Most heuristics aims to maximize the total packed relevance. To each box type we associate a constant, which we call relevance. The higher the volume, the greater the relevance. We adopted a particular non-linear function. The relevance coefficient value for for the ith greatest box type in a list of n box types measures $r(i,n) = 2^{n-i}$. The ratio beetween relevance coefficient and volume of a box is not linear. So, maximizing the relevance is not (necessarily) the same that maximizing the volume. This constant will allow us to prioritize the large packing boxes, leaving the smaller boxes (easier to be allocated) to the end of the process, when residual space is small.

We present the integer model to be solved by 'X Mixed Layer' heuristic, according its characteristics described in table 1.

The MIP that this strategy tries to solve.

$$\text{Maximize} \sum_{s \in S'} (nx_s ny_s r_s)$$

Subject to:

$$\left(\sum_{s \in T} (nx_s ny_s)\right) \leq qT, \forall T \qquad (C1) // \text{ avaiable quantity}$$

$$\sum_{s \in S} ny_s \geq 1, \forall T \qquad (C1) // \text{ avaiable quantity}$$

$$\sum_{s \in S} (ny_s h_s) \geq H(1-0.02), \forall T \qquad (C3.1) // \text{ few space waste in H}$$

$$\sum_{s \in S} (ny_s h_s) \leq H, \forall T \qquad (C3.1) // \text{ fits on plate}$$

Where:

$[W, H]$ Weight, height avaiable in container

S Set of all selected subtypes

T Set of original type of boxes from we obtain S

h_s Height of selected subtype s

r_s Relevance coefficient of selected subtype s

0.02   Wasting space margim

nx_s Number of strips on x-axis (constant)

ny_s Variable on the model. Number of strips on y-axis

Figure 4-a presents a X Mixed Layer compoused by two Partitions on X, blocks of boxes compoused by only one type of box that extends from edge-to-edge. The first partition (nearest to origin) we call α. The second one is β. So, $nx_\alpha = 3$ and $ny_\alpha = 2$. For β we have that $nx_\beta = 5$ and $ny_\beta = 4$.

The objective funciont aims to maximize the total packed relevance, thatis, trying firstly to pack bigger rectangles. (C1) constraint garantee a minimal quantity of boxes for create the partition. (C2) constraint garantees that at least one subtype of box have be used. A subtype is a type of box obtained after a 90o rotation. (C3.1) and (C3.2) constraints garantee the space waste in Y-axis at most equals to 0.02 and the partitions have to fit in container. If the model has solution, the algorithm process the packing. After packing, it updates the list of avaiable boxes and avaiable space. Then, the algorithm goes on.

3   Computational Results

In order to test the approached methodology we tested some benchmarks libraries and compare the found results with other ones found in literatue.The computational results were obtained using an Intel Core 2 Duo 2.1 GHz with 3 GB of RAM. The operating system is Windows Vista Home Edition. The development platform is Java 6.0 and Eclipse 3.1.1 tool, while the solver utilized was CPLEX 9.0.

In computational tests we were concerned with using space, avoiding the waste space. We present the results for various benchmark tests.

	W	CW1	CW2	CW3	OF1
Sol.	97.21%	100%	97.75%	97.21%	96.50%
Time	0.48	0.75	1	0.76	0.53
	OF2	CHW1	CHW2	TH1	TH2
Sol.	95.57%	97.75%	97.21%	98.45%	99.57%
Time	0.73	1.01	0.78	0.68	0.56

W: Example from [29]; CW1-3: from [11]; OF1-2: from [25]; CHW1-2: from[11]; TH1-2: from [28].

Another results from examples in [14].

	FS1	FS2	FS3	FS4	FS5
Sol.	99.60%	96.89%	96.85%	99.75%	100%
Time	1.43	3.84	7.92	28.87	0.48

We notice that HB is a good methodology to achieve a good space using and [28] little waste space for all benchmarks tests, in acceptable execution time.

Acknowledgements. This work has been financially supported by CNPq/Brazil via Grants #308249/2008-9, #473454/2008-4, and #312934/2009-2.The authors also acknowledge IBM for making the IBM ILOG CPLEX Optimization Studio available to the academic community.

References

1. Aráujo, L.J.P., Pinheiro, P.R.: Combining Heuristics Backtracking and Genetic Algorithm to Solve the Container Loading Problem withWeight Distribution. Advances in Intelligent and Soft Computing 73, 95–102 (2010)
2. Aráujo, L.J.P., Pinheiro, P.R.: Heuristics Backtracking and a Typical Genetic Algorithm for the Container Loading Problem withWeight Distribution. Communications in Computer and Information Science 16, 252–259 (2010)
3. Arenales, M.N., Morabito, R.N.: An And/Or-Graph Approach To The Solution Of Two-Dimensional non-Guillotine Cutting Problems. European Journal of Operational Research 84(1), 599–617 (1995)
4. Beasley, J.E.: An Exact Two-dimensional Non-guillotine Cutting Tree Search Procedure. Operations Research 33(1), 49–64 (1985)
5. Beasley, J.E.: Algorithms for Unconstrained Two-dimensional Guillotine Cutting. Journal of Operational Research Society 36(4), 297–306 (1985)
6. Beasley, J.E.: OR-Library: distributing test problems by electronic mail. Journal of the Operational Research Society 41(11), 1069–1072 (1990)
7. Bischoff, E., Dowsland, W.B.: An Application of the Micro to Product Design and Distribution. Journal of the Operational Research Society 33(3), 271–280 (1982)
8. Blesa, M.J., Blum, C., Roli, A., Sampels, M.: HM 2005. LNCS, vol. 3636, pp. VI–VII. Springer, Heidelberg (2005)
9. Raidl, G.R.: A Unified View on Hybrid Metaheuristics. In: Almeida, F., Blesa Aguilera, M.J., Blum, C., Moreno Vega, J.M., Pérez Pérez, M., Roli, A., Sampels, M. (eds.) HM 2006. LNCS, vol. 4030, pp. 1–12. Springer, Heidelberg (2006)
10. Carnieri, C., Mendoza, G.A., Lupold, W.G.: Optimal Cutting of Dimension Parts from Lumber with defect: a Heuristic Solution Procedure. Forest Products Journal, 66–72 (1993)
11. Christofides, N., Whitlock, C.: An algorithm for two-dimensional cutting problems. Operations Research 25(1), 30–44 (1977)
12. Dowsland, K.A., Dowsland, W.B.: Packing Problems. European Journal of Operational Research 56(1), 2–14 (1992)
13. Dyckhoff, H.: A typology of cutting and packing problems. European Journal of Operational Research 44, 145–159 (1990)
14. Fekete, S.P., Schepers, J.: A New Exact Algorithm for General Orthogonal D-dimensional Knapsack Problems. In: Burkard, R.E., Woeginger, G.J. (eds.) ESA 1997. LNCS, vol. 1284, pp. 144–156. Springer, Heidelberg (1997)
15. Garey, M.R., Johnson, D.S., Sethi, R.: Computers and intractability: a guide to the theory of NP-completeness. Freeman, New York (1979)
16. Gilmore, P., Gomory, R.: A Linear Programming Approach to the Cutting Stock Problem. Operations Research 9, 849–859 (1961)
17. Gilmore, P., Gomory, R.: A Linear Programming Approach to the Cutting Stock Problem - Part II. Operations Research 11, 863–888 (1963)
18. Gilmore, P., Gomory, R.: Multistage Cutting Stock Problems of Two and MoreDimensions. Operations Research 14, 94–120 (1965)

19. Mahfoud, S.W., Goldberg, D.E.: Parallel Recombinative Simulated Annealing: A Genetic Algorithm. Parallel Computing 21(1), 1–28 (1995)
20. Morabito, R.N., Arenales, M.N., Arcaro, V.F.: An And-Or Graph Approach For Two-Dimens ional Cutting Problems. European Journal of Operational Research 58(2), 263–271 (1992)
21. Morabito, R., Arenales, M.: An and/or-graph approach to the container loading problem. International Transactions in Operational Research 1, 59–73 (1994)
22. Moscato, P.: Memetic algorithms: A short introduction. In: New Ideas in Optimization, pp. 219–234. McGraw-Hill (1999)
23. Nepomuceno, N., Pinheiro, P.R., Coelho, A.L.V.: Tackling the Container Loading Problem: A Hybrid Approach Based on Integer Linear Programming and Genetic Algorithms. In: Cotta, C., van Hemert, J. (eds.) EvoCOP 2007. LNCS, vol. 4446, pp. 154–165. Springer, Heidelberg (2007)
24. Nepomuceno, N.V., Pinheiro, P.R., Coelho, A.L.V.: A Hybrid Optimization Framework for Cutting and Packing Problems: Case Study on Constrained 2D Nonguillotine Cutting. In: Cotta, C., van Hemert, J. (eds.) Recent Advances in Evolutionary Computation for Combinatorial Optimization. SCI, vol. 153, ch. 6, pp. 87–99. Springer, Heidelberg (2008) ISBN:978-3-540-70806-3
25. Oliveira, J.F., Ferreira, J.S.: An improved version of Wang's algorithm for two-dimensional cutting problems. European Journal of Operational Research 44, 256–266 (1990)
26. Pinheiro, P.R., Coelho, A.L.V., Aguiar, A.B., Bonates, T.O.: On the Concept of Density Control and its Application to a Hybrid Optimization Framework: Investigation into Cutting Problems. Computers & Industrial Engineering (to appear, 2011)
27. Pisinger, D.: Heuristc for the Conteiner Loading Problem. European Journal of Operational Research 141, 382–392 (2000)
28. Tschoke, S., Holthofer, N.: A new parallel approach to the constrained two-dimensional cutting stock problem. Preprint, University of Paderbon, D.C.S. 33095 Paderborn, Germany (1996)
29. Wang, P.Y.: Two Algorithms for Constrained Two-Dimensional Cutting Stock Problems. Operations Research 31, 573–586 (1983)
30. Wascher, G., Hausner, H., Schumann, H.: An improved typology of cutting and packing problems. European Journal of Operational Research 183(3), 1109–1130 (2007)

Improvement and Application of TF * IDF Algorithm

Ji-Rui Li[1], Yan-Fang Mao[2], and Kai Yang[3]

[1] Information Engineer Department,
Henan Vocational and Technical Institute,
Lecturer, Zhengzhou, China
[2] School of Computer and Communication Engineering,
Zhengzhou University of Light Industry,
Lecturer, Zhengzhou, China
[3] Computer Department,
Armed Police Command College in Zhengzhou,
Lecturer, Zhengzhou, China

Abstract. The traditional TF-IDF probability model is a relatively simple formula. For a few words which are commonly used and not stop words in a paper,it is lack of better differentiate and is not suitable for many specific cases, such as news advertising service module, about extraction of key words of the article, according to the deficiencies and the demand of news advertising service module, on the basis of the original algorithm, presents a new probability model------MTF-IDF, it greatly improves the accuracy of news information data retrieval.

Keywords: TF * IDF, MI, MTF-IDF.

1 TF*IDF

Vector space model arises because the original form of natural language is not suitable for using directly mathematical approach, and therefore difficult to achieve automatic processing of natural language. thought of Vector space model is to use the form of vectors to describe the document, the document that will be shaped such as (W1, W2, W3, ...) form, Wi is the weight of each word. Now commonly use the TF * IDF formula[1] to calculate the weight, the formula in various forms, most commonly used are as follows:

$$W(t,d) = \frac{tf(t,d) \times \log(N/n_t + 0.01)}{\sqrt{\sum_{t \in d}[tf(t,d) \times \log(N/n_t + 0.01)]^2}} \qquad (1)$$

Where, W (t, d) is the weight of word t in the text d, and tf (t, d) is the word frequency of the word t in the text d, N is number of the training text, n1 is the text number of appearing word t in the training of the text, the denominator is the normalization factor.

B. Liu and C. Chai (Eds.): ICICA 2011, LNCS 7030, pp. 121–127, 2011.

2   The Mutual Information(MI)

In information theory, mutual information formula[2] is:

$$MI(W,C_j) = \log \frac{P(W \cap C_j)}{P(W) \times P(C_j)} \qquad (2)$$

Which, $P(W \cap C_j)$ refers to conditional probability of the term W and Cj class occurring simultaneously,P (W) is occurrence probability of term W, P (Cj) is the probability of class Cj. In order to facilitate the calculation, other literature also shows the approximate formula for calculating the above formula[4]:

$$MI(W,C_j) = \log \frac{A \times N}{(A+C) \times (A+B)} \qquad (3)$$

Where, A is the number of occurrences of term W in the Cj class document, B is the number of term W which does not occur in Cj class document, C is the number of Cj class documentation and non-occurrence of words W, and N is the number of all documents of the set of documents.

3   Improved Feature Selection Algorithm for TF*IDF------ MTF-IDF

The definition of information in information theory can be expressed in the following formula[3]:

$$I(x) = \log \left[\frac{1}{p(x)} \right] \qquad (4)$$

Which, p (x) is the probability of event x occurs.it can easily be seen from the formula, in information theory ,The smaller the probability of event x occurs, when x event occurs it is to provide the greater amount of information.

Applied to text classification feature words selected, that is, means that lower frequency words will be given greater weight. We are a little deformed mutual information formula for this trend can be seen:

$$I(x, y) = \log \frac{p(xy)}{p(x)p(y)} = \log p(xy) - \log p(x) - \log p(y) \qquad (5)$$

p(y) is determined for the determination and document class y, p (xy), the joint probability of words x and y type document, it is little difference for many word, then - log p (x) more large, the word x, the greater the mutual information, ie, low frequency words x can get higher mutual information. the reality is that, although there are some really low-frequency words should have higher importance, in general, the frequency of the most important word in the Chinese articles is moderate level, the emphasis of which is not enough, this is the shortcomings of mutual information.

For the TF * IDF formula, the term frequency (TF, Term Frequency) tf(t,d) and inverse document frequency (IDF, Inverse Document Frequency) log (N/n1 +0.0.1.)is the main part, although the formula taking into account two factors: word frequency and document frequency,the formula itself is a bit too simple, the lack of a better discrimination, such as for some of the more commonly used but not stop words, they will appear in large Most of the articles, therefore, in the formula the value of N/n1 items is less the same for them,the frequency of words in Chinese article is generally not high, This makes weight difference obtained from the TF * IDF formula is very small for a lot of words. Therefore, this paper proposes a new feature extraction method - Word Frequency Differentia Based Feature Selection, WFDBFS, and on this basis, giving the improved TF * IDF formula (MTF-IDF).

The first step, and the mutual information similarity, we calculate the type weight W(WK, Cj)of word WK on Cj, is calculated as:

$$W(w_k, c_j) = \exp\left[\frac{1}{f_k^c \times (N-1)} \sum_{\substack{i,j=1 \\ i+j}}^{N} (f_k^{c_i} - f_k^{c_j})\right] \times \left[1 + \lambda \frac{\sqrt{D(c_i, f_k^{dj})}}{E(c_i, f_k^{dj})}\right] \qquad (6)$$

Which, WK is the first K words, Ci is the i-class article, $f_k^{c_i}$ is the frequency of word WK in class Ci, f_k^{dj} is the frequency of word WK in the document dj , $D(c_i, f_k^{dj})$, $\overline{f_k^c} = E(f_k^{c_i})$ are respectively the variance and the mean of f_k^{dj} in the Ci, N is the number of total categories, λ is the proportional coefficient. two parts of before and after the multiplication formula are called between-class weight and the weight class.

The second step, average weight for calculation of all types of categories:

$$\overline{W_k} = \sum_{i=1} W(w_k, c_i) \qquad (7)$$

The third step, giving the threshold W of $\overline{W_k}$, words which the average weight category more than a threshold be retained as feature words. Or giving M that is the number of feature words, word which value of $\overline{W}_k$ before the first M-bit as feature words retained. On this basis, giving the modified TF * IDF formula:

$$W(t,d) = \frac{tf(t,d) \times \log(N/n_t + 0.01) \times W(w_k, c_i)}{\sqrt{\sum_{t \in d} [tf(t,d) \times \log(N/n_t + 0.01) \times W(w_k, c_i)]^2}} \qquad (8)$$

That is based on the original formula, we multiply the weight categories of words as the original formula adjustments. The reason to do so having regard to the fact the same word in different types of articles should have different degrees of importance, with not the same as the amount of information, and the original TF * IDF formula does not reflect this.

4   News Advertising Modular Framework

Press advertising service module consists mainly of two small modules: news advertising module and data module. The main process: extract the network news, analysis content of the article, extract the expressing keywords of the context, classify press advertising, calculate news content and advertising content, matching the most relevant ads for each news; matching of ads is relations of the keyword match and correspondence relations of classification system, every 1 hour, ad data updates. The system framework shown in Fig. 1:

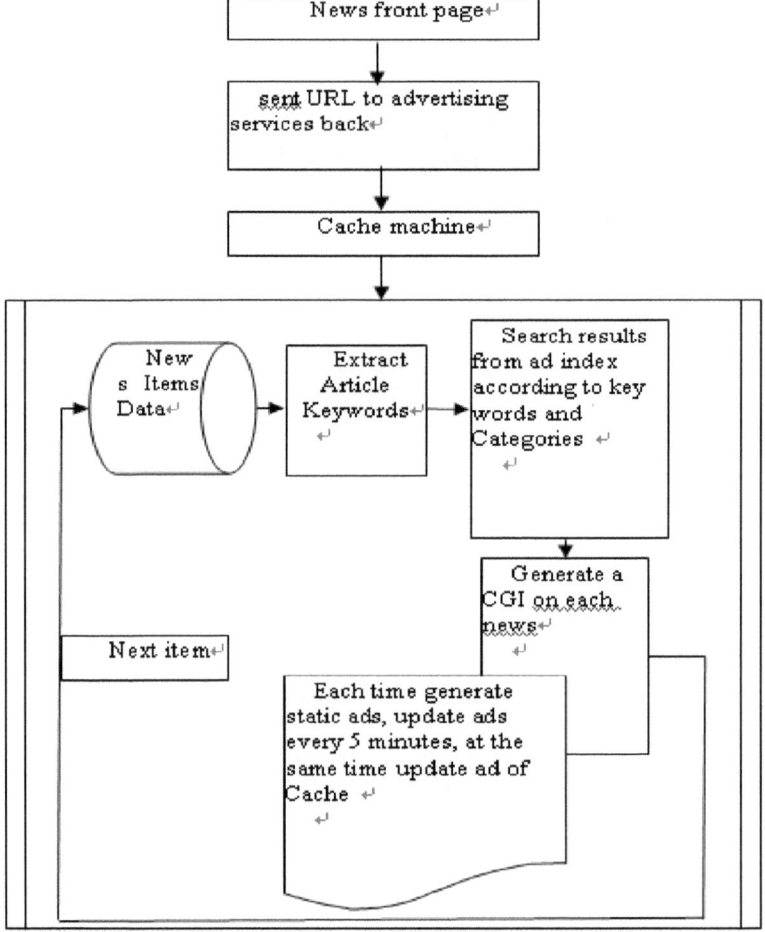

Fig. 1. System framework

4.1   The News Data Module

Follow is News data processing procedures.

(1) News data file exists, if it does not exist, then every 5 minutes duplication of 1.
(2) The entry reads the input data format is the document number, document title, document text, document URL, the document category.
(3) Keywords extracted document.
(4) And their weights according to key words, document type and weight to the advertising in the search indexing service.
(5) Extraction of the results of the former 20, in accordance with the document number modulo / 50 to create the directory, generates a dynamic CGI, CGI generated each time you run three ads data. Each 5-minute runs, generate different ads. At the same time update the cache.
(6) Repeat (1)-(5), to handle all the data items.
(7) Delete the data entry.

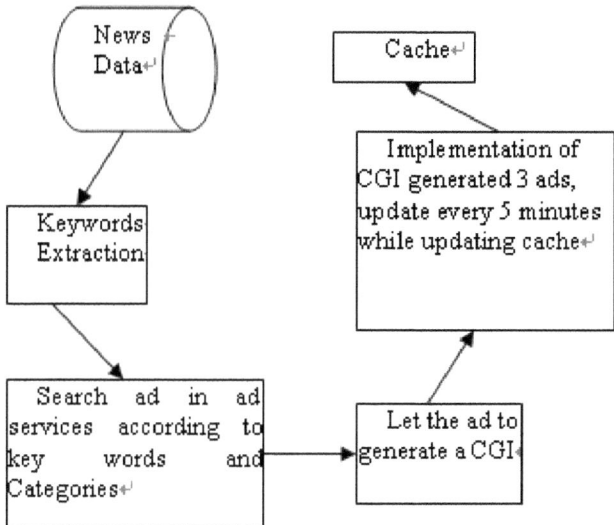

Fig. 2. News data processing procedures

4.2   Advertising Module

Adsense establish index for ad through the lucene, and provide search advertising for each news service, the specific process is:

(1) Read news data from the news file, extract the article keywords and categories
(2) Applying keywords to the ad system, search ads, generate static pages
(3) Delete the data
(4) Uploading the advertising data to polling pool.

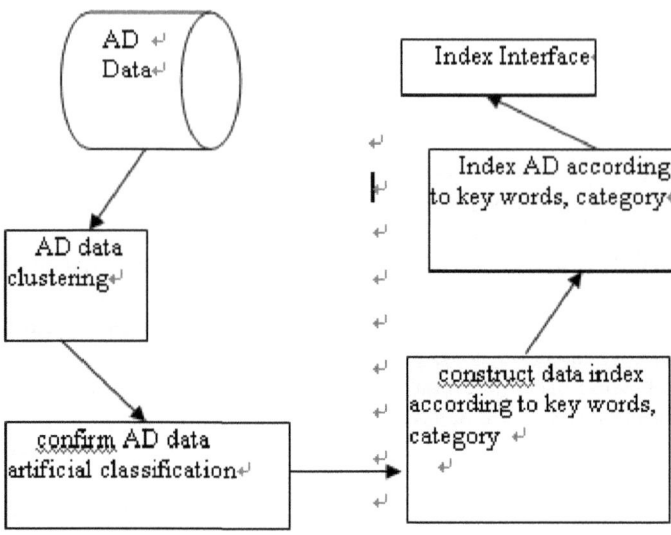

Fig. 3. Ad data processing

4.3   Keyword Extraction

Keyword extraction module is the title and content of a text is calculated to extract the key words used to summarize the text and given weight. Keyword extraction process are: segmentation, filtering, calculation of the value of segmentation, identifying specific reference to the word, identifying phrase recognition and effective, the results sorted. If the sentence is the sentence-level units will have to carry out the scoring calculation. Among them, the scoring of the word use basically improved the probability of TF * IDF scoring model. The basic score of each word is weight = idf * tf (where idf means the information content of the word, with a large corpus in the previous training. tf is the number of the word that appears in the article), based on the points in the participation rate of other strategies are as follows:

(1) On limiting the high value of IDF: IDF, probably because of the high value of the corpus of data sparseness caused so IDF value of more than 24 words should be reduced by 20%.
(2) TF limit: TF is likely to be more than a certain amount of noise than some of the text itself, so the maximum size of the TF is set to 10, and the use of TF as the final value of the square root of the square root of the role of TF.
(3) New words in the assignment using a high-value approach.
(4) On names, place names, organization names are weighted.
(5) Title weighted, in the words of the title are weighted.
(6) Appear in the title but the word does not appear in the body to reduce points, to highlight the strong correlation between full word.
(7) To consider the sentence score, making the text in scoring over the sentence, consider the sentence itself on word scores.

References

1. Shi, C.-y., Xu, C.-j., Yang, X.-j.: Documents cluster summary. Chinese Information Journal, 106–109 (2006)
2. Wei, J., Chang, C.-w.: Sub-dictionary of single array and full map. Computer Engineering and Applications, 184–186 (2007)
3. Zhai, W.-b., Zhou, Z.-l., Jiang, Z.-m., et al.: Design of Chinese Word Dictionary. Computer Engineering and Applications, 1–2 (2007)
4. Lin, Y.-m., Lu, Z.-y., Zhao, S., Zhu, W.-d.: Analysis and Improvement of Text Feature weighted method TFIDF. Computer Engineering and Design, 2923–2926 (2008)
5. Gao, X.-d., Wu, L.-y.: Chinese keywords extraction algorithm Based on high dimensional clustering technique. China Management Information, 9–12 (2011)
6. Li, P.: Text classification research Based on the improved the weights of the words. Northeast Normal University, 251–255 (2010)
7. Zhou, Y.-b., Chen, X.-s., Wang, W.-x.: The topic crawler research based on Bayes classifier. Computer Application Research, 33–35 (2009)
8. Zhang, X.-y., Wu, X.-q., Zhang, P.-y.: Study of garbage filter method in Agriculture website page. Network Security Technology and Application, 102–105 (2011)
9. Shi, C.-y., Xu, C.-j., Yang, X.-j.: TFIDF in algorithm. Computer Application, 321–324 (2009)
10. Xu, Z.-y.: Analysis and Improvement of Feature selection method in Text classification. Computer and Modernization, 28–30 (2010)
11. Sun, Q.-h.: Key extraction method based on spatial distribution and information entropy. Dalian University of Technology, 77–79 (2010)

Theory and Algorithm Based on the General Similar Relationship between the Approximate Reduction

Baoxiang Liu[*] and Hui Guo

College of Science, Hebei United University,
Tangshan Hebei 063009,China
{Liubx5888,guomaonao111}@126.com

Abstract. On fundamental aspect of variable precision rough approximate reduction is an important mechanism for knowledge discovery. This paper mainly deals with attribute reductions of an inconsistent decision information system based on a dependence space. Through the concept of inclusion degree, a generalized decision distribution function is first constructed. A decision distribution relation is then defined. On the basis of this decision distribution relation, a dependence analogy relation representation of VPRS data space is proposed, and an equivalence congruence based on the attribute sets is also obtained. Applying the congruence on a dependence space, new approaches to find a distribution consistent set are formulated. The theorems for judging distribution consistent sets are also established by using these congruences and the decision distribution relation.

Keywords: Variable precision rough set, Similarity relation, Approximate reduction.

1   Introduction

Rough set theory (RS) proposed by Pawlak (1982) is an extension of conventional set theory that supports approximations indecision making. It has also been conceived as a mathematical approach to analyze and conceptualize various types of data, especially to deal with vagueness or uncertainty. Rough set theory has been successfully used in diverse fields such as decision support systems, machine learning, and automated knowledge acquisition and so on.

RS was introduced more than twenty years ago [1-4] and had emerged as a powerful technique for automatic classification of objects [5]. It has provided a powerful tool for data analysis and for knowledge discovery from uncertain and vague data. RS has been successfully applied to machine learning, forecasting, knowledge acquisition, decision analysis, knowledge discovery from database, pattern recognition, and data mining.

The variable precision rough set model (VPRS) by Ziarko(1993) is an expansion of the basic rough set model. The choices of b values to be applied to find the

---

[*] Project supported by National Natural Science Foundation of China(No. 61170317) and Natural Science Foundation of Hebei Province of China(No.A2011209046).

B. Liu and C. Chai (Eds.): ICICA 2011, LNCS 7030, pp. 128–135, 2011.

β -reducts in VPRS for an system are somewhat arbitrary. The VPRS model considers some objects of the given data set as misclassified or uncertain. In a first step, the full attribute set is used to evaluate which objects are regarded as misclassified. The class information (decision) of these objects is changed resulting in a data set that can be handled by methods of the original RST. For the above problems, the initial definition of misclassified objects is preserved throughout the entire process. Although other approximation methods could be used in principle, we simply chose similar relationship.In summary,he proposed method bridges the approximation methodology and similar relationship of RS to solve the threshold value β determination problem for VPRS.

However, the success of RS techniques in correctly classifying a dataset relies upon all the collected data being correct and certain.In other words, performing a classification function with a controlled degree of uncertainty or misclassification error falls outside the realm of the RS approach (Ziarko, 1993). In an attempt to extend the applicability of the RS method, Ziarko developed the VPRS theory (Ziarko, 1993). In contrast to the original RS model, the aim of the VPRS approach is to analyze and identify data patterns which represent statistical trends rather than functional patterns (Ziarko, 2001) [6]. VPRS deals with uncertain classification problems through the use of a precision parameter β. Essentially, β represents a threshold value which determines the portion of the objects in a particular conditional class which are assigned to the same decision class. (Note that in conventional RS theory, β has a value of one.) Thus, determining an appropriate value of β is essential in deriving knowledge from partially-related data objects.

2   The Related Concept and Conclusion

Knowledge expression is the relationship between the form of data table, said the relation table elements corresponding to the object of study, and all the corresponding is the study of the attribute, object information is designated by the object each attribute to express.

Ziarko professor consider is classified the error rate β, The classification is allowed errors, thus $\beta \in [0,0.5)$. An.A make Understanding for classification accuracy make $\beta \in (0.5,1]$. VPRS model is the expansion of the rough set model. When β =0,VPRS degenerate into RS. Includes the introduction of degrees that decision about condition attribute set of the lower, relatively positive field, classification, the attribute importance and so on some concepts will be closely related with β value, rough set theory of many of the concepts and properties and further expansion of generalization. Variable precision rough set by two concepts: An information table is the following tuple:(1) where $U = \{x_1, x_2, \cdots x_n\}$ is a finite nonempty set of objects, $\Omega = \{A_1, A_2 \cdots A_m\}$ is a finite nonempty set of attributes, V=$\{V_1, V_2 \cdots V_m\}$is a nonempty set of values ,and $f : U \times \Omega \rightarrow V$, $f(x_i, A_j) \in V_j$ is an information function that maps an object in U to exactly one value in V. In classification problems, we consider an information table of the form $I =< U, \Omega, V, f >$ where $\Omega = \{A_1, A_2 \cdots A_m\}$ is a set of condition attributes describing the objects, and $\{A_1, A_2 \cdots A_m\}$ is a decision attribute that indicates the classes of objects. In general, we may have a set of decision attributes. A table with multiple decision attributes can be easily transformed into a table with a single decision attribute by considering the Cartesian product of the original decision attributes.

2.1   VPRS of Similarity Relationship

In this section, we will review several basic concepts in rough set theory. Throughout this paper, we suppose that the universe U is a finite nonempty set.

In VPRS, the β value represents a threshold value of the portion of objects in a particular conditional class being classified into the same decision class. In processing the information system K=(U,R) using a VPRS model with $0.5 < \beta \leq 1$, the data analysis procedure hinges on two basic concepts, namely the β-lower and β-upper approximations of a set. The β-lower approximation of sets $X \subseteq U$ and $R \subseteq C$ can be expressed as follows.Let U be a finite and nonempty set called the universe and R=U×U an equivalence relation on U .Then K=(U,R) is called an approximation space [7]. The equivalence relation R partitions the set U into disjoint subsets. This partition of the universe is called a quotient set induced by R, denoted by Y= U/R. It represents a very special type of similarity between elements of the universe. If two elements $X, Y \in U, X \neq Y$ belong to the same equivalence class, we say that X and Y are indistinguishable under the equivalence relation R, i.e., they are equal in R. We denote the equivalence class including X by Y=U/R. Each equivalence class Y=U/R may be viewed as an information granule consisting of indistinguishable elements [8]. The granulation structure induced by an equivalence relation is a partition of the universe. Given an approximation space K=(U,R) and an arbitrary subset X⊆U, one can construct a rough set of the set on the universe by elemental information granules in the following definition:

$$\underline{R}_\beta X = \cup\{E \in U/R \Big| X \overset{\beta}{\supseteq} E\}. \tag{1}$$

$$\overline{R}_\beta X = \cup\{U/R \Big| p(E, X) < 1 - \beta\}. \tag{2}$$

Where (2) and (3) are called β-lower approximation and β-upper approximation with respect to R, respectively. The order pair ($\underline{R}X, \overline{R}X$) is called a rough set of X with respect to the equivalence relation R. Equivalently, they also can be written a:

$$bnr_\beta(X) = \cup\{E \in U/R \Big| \beta < p(E, X) < 1 - \beta\}. \tag{3}$$

$$negr_\beta(X) = \cup\{E \in U/R \Big| p(E, X) \geq 1 - \beta\}. \tag{4}$$

Each of feature selection method preserves a particular property of a given information system, which is based on a certain predetermined heuristic function. In rough set theory, attribute reduction is about finding some attribute subsets that have the minimal attributes and retain some particular properties. For example, the dependency function keeps the approximation power of a set of condition attributes. To design a heuristic attribute reduction algorithm, three key problems should be considered, which are significance measures of attributes,search strategy and stopping (termination) criterion. As there are symbolic attributes and numerical attributes in real-world data, one needs to proceed with some preprocessing. Through attribute discretization,it is easy to induce an equivalence partition. However, the existing heuristic attribute reduction methods are computationally intensive which become infeasible in case of large-scale data. As already noted, we do not reconstruct significance measures of attributes and design new stopping criteria, but improve the search strategies of the existing algorithms by exploiting the proposed concept of positive approximation.

In the above expressions the unions are taken for U/R. Essentially, the value $\gamma^\beta(P,Q)$ indicates the proportion of objects in the universe U for which a classification based on decision attribute D is possible at the specified value of β. In other words,it involves combining all β-positive regions and summing up the number of objects involved in such a combination. The mea-surement (quality of classification) is used operationally to define and extract reducts, which is the kernel part of RS theory and VPRS in the application to data mining and rule construction.

2.2 β-Approximate Reduction

This section introduces some basic concepts of VPRS in order to present our notions conveniently. More detailed description of VPRSM can be found in[9] .

Formally, a decision information system S can be taken as a system S=(U,R), where U is the universe (a finite and nonempty set of objects)and R is the set of attributes(C is condition attributes set and D is decision attribute set).Each attribute $r \subseteq R$ defines an information function U/R , where $r \subseteq R$ is the value set of r, called the domain of attribute r. U/R is the union of attribute do mains.With equivalence relation C, referred to as an indiscernibility relation ,universe U can be partitioned into a collection of equivalence classes U/R Similarly, equivalence relation D partitions U into another collection of equivalence classes U/R. Each element of U/C is called condition class and each element of U/D is called decision class.

The VPRS approach to data analysis hinges on two important concepts, namely, the β-lower and β-upper approximation of a set. Given a decision in formation system S and inclusion degree $\beta \in (0.5,1]$,the β-lower and β-upper approximation of decision class U/D are presented as follows. Based on Ziarko's notions, the measure of classification quality in VPR is defined as:

$$\gamma^{\beta}(P,Q) = \frac{|POS(P,Q,\beta)|}{|U|}. \tag{5}$$

$$POS(P,Q,\beta) = \underset{r \in U/Q}{U} pos_p^{\beta}(Y). \tag{6}$$

Each region (positive region, negative region, boundary region) is composed of condition classes in classical rough set model as well as in VPRS. For consistent decision information system, the quality of classification always equals one, no matter what the β is. However, decision in formation system sometimes is in consistent in reality .The inconsistency of information leads to the change of classification quality. From the view of statistical probability β value determines the quality of classification. There are more than one β value in an inconsistent decision information system. The dynamic changes of β value should be considered in VPR. Each region is affected by β value directly, and the quality of classification will be in fluenced accordingly. Indeed, the quality of classification and the β value are related in versely.The lower the β value can be prepared to accept, the higher the quality of classification $\gamma^{\beta}(P,Q)$ will be attained. On one hand, the value of $\gamma^{\beta}(P,Q)$ holds the line for a specific β, on the other hand, the value of $\gamma^{\beta}(P,Q)$ will be decreased monotonically as β value is increased.

Reduct is one of the most important notions in rough set theory as well as in VPRS. Formally in VPRS from Ziarko,an approximate reduct can be defined as follows .

An approximate reduct of condition attribute set with respect to decision attribute set Q is a minimal subset $red(P,Q,\beta)$ of P which satisfies the following two criteria :

$$\gamma(P,Q,\beta) = \gamma(red(P,Q,\beta),Q,\beta). \tag{7}$$

no attributes can be eliminated from $red(P,Q,\beta)$ without affecting the requirement(7).

Ziarko's reduct definition depicts knowledge reduction for a specific β value, so β value must be given beforehand. Ziarko discussed how to choose optimal β value for approximate reduct [10]. However, it is very difficult to find optimal β value. It often depends on our subjectivity indeed or on some prior knowledge. For the same $\gamma^{\beta}(P,Q)$ value, the β value, which satisfies the criteria of approximate reduct, is often a specific value interval. Some reduction anomalies may occur if only two criteria are considered in the process of reduction.

3   Structure Similarity Relationship

Incomplete information is $S = (U, A = C \cup D, V, f)$, $U = \{X_1, X_2, \cdots X_{|n|}\}$, $C = \{C_1, C_2, \cdots C_{|n|}\}$, $D = \{d\}$, $x_i, x_j \in U$, about $C_k \in B \subseteq C$, similarity $R_{\{c_k\}}\{x_i, x_j\}$, $|V_{c_k}|$ Said domain of all objects in the attribute U take A different number of known value ,be defined as :

$$R_{\{c_k\}}\{x_i, x_j\} = \begin{cases} 1, i=j \\ 1, f(x_i, c_k) = f(x_j, c_k), i \neq j \\ 1/|V_{c_k}|, f(x_i, c_k) = *, i \neq j \\ 0, other \end{cases} \tag{8}$$

Incomplete information is $S = (U, A = C \cup D, V, f)$, $B \subseteq C$, $\alpha \in (0,1)$, and $\tau \in [0.5, 1]$, (α, τ), $\beta(0.5 < \beta \leq 1)$ be Similarity relation defined as:

$$R_\alpha^{\beta^\tau}(X) = \left| x_i \in U \left| \frac{\left| R_B^{\alpha^\tau}(x_i) \cap X \right|}{\left| R_B^{\alpha^\tau}(x_i) \right|} \geq \beta \right| \right. \tag{9}$$

$$R_\beta^{\alpha^\tau}(X) = \left| x_i \in U \left| \frac{\left| R_B^{\alpha^\tau}(x_i) \cap X \right|}{\left| R_B^{\alpha^\tau}(x_i) \right|} \geq 1 - \beta \right| \right. \tag{10}$$

Obviously, in complete information system, when $\alpha = 1, \tau = 1$ VPRS; it is called the In incomplete information system when $\tau = 1$, α full hour that is based on the a symmetric relationship for the RS. Therefore, this model is classical rough set model based on the asymmetric relationship in the incomplete information system of the development.Incomplete. Condition attribute is $c_k \in C (k = 1, 2, \cdots, |C|)$, For decision attribute the importance degree definition for D:

$$\sigma_D(c_k) = \frac{card(U) - card(POS^\beta_{\{C/c_k\}}(D))}{card(U)}. \tag{11}$$

Type of $card(\bullet)$ set of said potential, is the number of objects in the set:

$$POS^\beta_{\{C/c_k\}}(D) = \bigcup_{D_I \in U/R_D} \left| x \in U \left| \frac{\left\| [x]_{\{C/c_k\}} \cap D_i \right\|}{\left\| [x]_{\{C/c_k\}} \right\|} \geq \beta \right| = \bigcup_{D_I \in U/R_D} B^\beta_{\alpha \tau}(D_i). \tag{12}$$

Type of $POS^{\beta}_{\{C/c_k\}}(D)$ Said in (α, τ) Similar relationship between the D relative to the $\{C/c_k\}$ is domain, which is the U all objects based on the attribute set $\{C/c_k\}$ division, can use β correctly divide to D the equivalence relation of objects in the set.

4   Algorithm

Algorithm based on the similarity relation: approximate reduction algorithm is proposed.

Input A incomplete information system $S = (U, A = C \cup D, V, f)$, approximation β.
Output an $RED^{\beta}_{\alpha\tau}(C, D)$ approximate reduction set.

Step 1 to condition attribute C find out all objects in U similar class $RED^{\beta}_{\alpha\tau}(C, D)$;
Step 2 calculation approximation classified quality $\gamma^{\beta}_{\alpha\tau}(C, D)$;
Step 3 calculation all attributes in c_k condition attributes C set the important degree $\sigma_D(c_k)$, and the descending order

Make $B \neq \phi$;

Step 4 makes $B = B \cup \{c_K\}$, will be C/B the biggest in the attribute $\sigma_D(c_k)$ join B;

Step 5 calculation approximation classified quality is $\gamma^{\beta}_{\alpha\tau}(C, D)$, if $\gamma^{\beta}_{\alpha\tau}(B, D) = \gamma^{\beta}_{\alpha\tau}(C, D)$ retreat to cycle, or turn step 4;

Step 6 output.

5   Summary

In has been argued in this paper that approximate reduction is essential to the discovery of decision rules in information system .Based on a generalized decision distribution function, we have proposed a decision distribution relation to construct a congruence and then obtain a dependence space. By the equivalent congruence on the dependence space formed by the indiscernibility sets,we have constructed a new method to search for the distribution consistent sets.

In brief, this paper mainly studies similarity relation based on a dependence space in an inconsistent approximate reduction.These inconsistent approximate reduction are all complete. However, approximate reduction obtained from the real world might be incomplete, and attribute reductions are more complicated and difficult to make. Approaches to search for attribute reductions in incomplete information system under different requirements are thus necessary. The proposed approaches can be extended to solve attribute reduction problems in more general and complicated information system in further research.

References

1. Pawlak, Z.: Rough Sets. International Journal of Information and Computer Seienee 11, 13–34 (2008)
2. Kattan, M., Copper, R.: The Predietive accuracy of computer-based classification decision techniques. A Review and Research Directionso, MEGA 26(4), 452–467 (1998)
3. Zattan, W.: Variable preision rought set model. Journal of Computer and Systerm Science 46(1), 39–59 (1993)
4. Katzberg, J.D., Ziarko, W.: Vatriable Preeision rough sets with asymmetric bounds, in Banff, Alberta, Can (2001)
5. Tseng, T., Kwon, Y.J., Erterkin, Y.M.: Feature-baseed ruld inductin inmachining opration using rough set theory for quality assurance. Robotics and Computer-Intergrated Manufasctuing 21, 559–567 (2008)
6. Li, R.P., Wang, Z.O.: Mining classification rules using set and neural networks. European Journal of Operational Reaseach 157, 439–449 (2004)
7. Gong, Z.T., Sun, B.Z., Shao, Y.B., et al.: Variable Preeision rough set model based on general relation. In: Proeeedings of 2004 International Conference on Machine Learning and Cybemetics, Shanghai, China, pp. 2490–2494 (2004)
8. Tsumoto, S., Ziarko, W., Shan, N.: Kowledge diseovery in clinieal databases based on variable precision rough set model. In: Proc. Annu. Symp. Comput. APPI Med. Care, p. 270 (2007)
9. Mieszkowicz-Rolka, A., Rolka, L.: Variable Precision Fuzzy Rough Sets Model in the Analysis of Process Data. In: Ślęzak, D., Wang, G., Szczuka, M.S., Düntsch, I., Yao, Y. (eds.) RSFDGrC 2005. LNCS (LNAI), vol. 3641, pp. 354–363. Springer, Heidelberg (2005)
10. Nishino, T., Nagamaehi, M., Tanaka, H.: Variable Precision Bayesian rough set model and its applieation to human evaluation data. In: Proeeeding so FSPIE-The International Soeiety for Optical Engineering, San Jose, United States, pp. 294–303 (2010)

Effective Software Fault Localization
by Statistically Testing the Program Behavior Model

Azam Peyvandi-Pour and Saeed Parsa

az_peyvandypour@comp.iust.ac.ir, parsa@iust.ac.ir

Abstract. Existing Statistical fault localization approaches locate bugs by testing statistical behavior of each predicate and propose fault relevant predicates as nearest points to faults. In this paper, we present a novel statistical approach employing a weighted graph, elicited from run-time information of a program. The predicates are considered as nodes; an edge is denoting a run-time path between two predicates and its label is the number of simultaneous occurrence of connected predicates in the run. Firstly, a typical graph, representing failed run is contrasted with whole graphs of passed runs to find the two most similar graphs of the passed runs and failed runs and discriminative edges are chosen as suspicious edges. In next phase, we statistically test the distribution of the suspicious edges to find the most fault relevant edges; to this end, we apply a normality test on the suspicious edges and based on the test result, we use a parametric or non-parametric hypothesis testing to discover the most fault relevant edges. We conduct the experimental study based on Siemens test suite and the results show the proposing approach is remarkable.

Keywords: Statistical Fault Localization, Weighted Graph, Parametric and Non-parametric Hypothesis testing, Normality test.

1 Introduction

In spite of many efforts at software debugging, some faults have been revealing after the release which may cause disturbing crashes [4][15]. This motivates researchers to propose an automated fault localization technique to fix the problem [5][13][1][2].

Among fault localization techniques, predicated-based techniques [1][2][5] were proposed and reported to be promising. They statistically find the predicates which are very likely in close proximity to the faults. In these techniques some extra codes, namely predicates are injected at some program positions such as branch statements or function returns to encode run-time behavior of the program [4]. After collecting the evaluation of predicates via executing the program over the test cases, the statistical model of the predicate evaluations is built and analyzed to detect the fault relevant predicates.

Liblit et al. [5] consider the difference between the probability of a predicate to be evaluated to be true in all failed runs and this probability in all runs, as a measure of how much the predicate correlates with the fault and ranks the most fault relevant predicates based on this increase. As this approach has not distinguished when the

B. Liu and C. Chai (Eds.): ICICA 2011, LNCS 7030, pp. 136–144, 2011.
© Springer-Verlag Berlin Heidelberg 2011

predicate is evaluated to be true once from when it is evaluated to be true in more than once in each run; Liu et al. [2] proposed the concept of evaluation bias which indicates the probability that the predicate is evaluated to be true in each run. Let n_t be the number of times that predicate P is evaluated to be true, and n_f be the number of times it is evaluated to be false, π (P)$=\frac{n_t}{n_t+n_f}$ estimates the evaluation bias of predicate P. $\pi(P)$ changes from 0 to 1, hence is regarded as a random variable and the differences between the distribution of this random variable in the passed and failed runs is considered as an indicator of the fault relevance. According to the central limit theorem, a null hypothesis test based on means of the two distributions of the mentioned random variable is applied. Rejecting the null hypothesis means that the two distributions are different from each other and the predicate is fault relevant.

Although, the approach proposed by [2] resolved the mentioned problems proposed by Liblit et al. [5], but the result of the null hypothesis test is not reliable when the random variable X does not follow a normal distribution. So faced to this problem, one first has to figure out whether or not random variable X forms a normal distribution and how much the degree of normality is.

On the other hand, existing statistical approaches do not regard the association of the predicates and test the behavior of each predicate separately to detect the fault relevant predicate. In this paper, we study the correlation between the predicates over the run-time paths using failed and passed test cases. We encode control flow of the program representing the correlation of the predicates as a weighted graph which consists of a sub set of predicates observed in the run as the nodes of the graph; transitions from one predicate to another in each run are considered as edges and the number of the transitions among the predicates is considered as the label of each edge. According to the result of the runs the weighted graphs are divided into two sets, the passed and failed sets. Whereas, some slices of the failed graph make it discriminate from its similar passed graph; the aim is to find such suspicious slices (or edges) by contrasting the failed weighted graphs with the passed weighted ones. In fact, we want to find transitions which are more related to the faulty statements. We propose a difference function to fulfill this threat. Then, in next phase, we test the suspicious edges to find the most fault relevant edges and raise the accuracy. As the weight of each edge changes from one execution to another, the random values assigned to it constitute a statistical model. Suppose X is a random variable standing for the weight of the edge E, then the aim is to test how much two statistical models of the random variable X, related to the passed and failed graphs are different from each other.

In summary, we make the following contributions in this paper:

We propose a special kind of control flow graph as a weighted graph which encodes the association between predicates and exploit it to detect anomalous edges.

We offer a difference function to find suspicious edges which distinguish failed weighted graphs from the passed weighted ones.

The random values of each suspicious edge which constitute a statistical distribution are tested to determine normality nature of the distribution. And based on the test result a hypothesis testing is applied to measure the difference between two distributions of suspicious edges in the failed and passed graphs.

2   A Motivating Example

In this section, we present a code fragment from the program "Print_tokens" of Siemens test suite to show the problem in detail. In this code, the predicates are labeled as p_{17} to p_{44}. As shown in figure1, The statement "ch=get_char(tstream_ptr->ch_stream);" in "case16:" is commented out and two other faults are because of the wrong statements in the codes related to "case32:" and "case25".

The observed predicates during the run time from the starting point of the program to the end one constitute a chain of predicates namely a path. A sample run-time path of a typical test case (t_{271}) of the first version of the program "Print_tokens" is "p_2, p_5, p_9, p_{11}, p_{17}, p_{18}, p_{79}, p_{81}, p_{21}, ... , p_{84}, p_{109}". Using the different test cases leads to the different paths. The paths traversing the faulty statement/statements end in failure and the others end in success. The goal is to find predicates (e.g. p_{31}) which are close to the faulty statement by contrasting the failed paths with the passed ones.

```
P17:   while(!token_found)
P18:       {   if(token_ind < 80)
{token_str[token_ind++]=ch
next_st=next_state(cu_state,ch);
}
.
.
switch(next_st)
{
P26:            default : break;
P27:            case 6  :
P28:            case 9  :
P29:            case 11 :
P30:            case 13 :
P31:            case 16 :    /* ch=get_char(tstream_ptr->ch_stream);*/
P32:            case 32 : ch=get_char(tstream_ptr->ch_stream);    /*Bug*/
P33:                        if(check_delimiter(ch)==TRUE){
.
.
P44: case 25 : token_ptr->token_id=special(next_st);            /*Bug*/
        token_ptr->token_string[0]='\0';
        return(token_ptr);
```

Fig. 1. A fragment of code from the 1th version of print_tokens program

Let us take a short look at the two statistical approaches which had significant improvements in results. In first and powerful approach [2], the behavior of each predicate is inspected in the passed and failed runs. In other words, mean and variance values of the evaluation bias distribution of each predicate is tested based on a null hypothesis as "$\mu_p = \mu_f$ and $\sigma_p = \sigma_f$" and the central limit theorem. The predicate will be fault relevant if the null hypothesis is rejected. But the test result will be reliable if the evaluation bias for each predicate in all runs follows a normal distribution. These challenges are considered in the approach proposed by [1]. Such that, the distribution evaluation bias of each predicate is tested and according the test result, the normal or non-normal distribution, the parametric or non-parametric hypothesis testing is applied.

As to be discussed later, we look at the problem in different view. In fact, instead of analyzing the behavior of each predicate individually, we discover the anomalous relations between two predicates by contrasting the distribution of their mutual relations in the failed and passed runs of the program. In addition, we investigate the normality of the considering distribution and then apply the hypothesis testing as well.

3   Proposing Approach

3.1   Weighted Graph of Predicates

In this section, we introduce the first phase of our approach. Like previous statistical approaches, the subjected program is instrumented at first. To address this treat, the branches and return values are chosen to be instrumented [4]. Therefore, the sequence of observing predicates is profiled by running the program over test cases. Hence, a path consists of the appeared predicates is profiled, the transitions of two contiguous predicates and the frequency of their transitions are collected to be analyzed.

Let $P = \{p_1, p_2, \ldots, p_i, p_n\}$ be a set of predicates, each program run is a permutation of the predicates such as $\Gamma : p_1, p_4, p_{10}, p_3, p_1, p_4, \ldots, p_k$, while the predicates and their arrangements may be different in each run. The weighted graph G is built based on the predicates which are listed in Γ, such that the transitions from predicate p_i to p_j is modeled as an edge in graph and its weight is the occurrence frequency of $p_i\, p_j$ in Γ or the number of transitions from predicate p_i to predicate p_j. We represent a weighted graph G by the set $G = \{((p_i, p_j), w) \mid p_i, p_j \in \Gamma, w \in Z^+\}$, which (p_i, p_j) representing the edge connecting the predicate p_i and p_j and w is the weight of the edge(p_i, p_j) in the graph G.

Given a test suite $T = \{t_1, t_2, \ldots, t_n\}$ for the program P; Each test case t_i leads the program P to be failed if the observed output is not the expected output and it passes the test case t_i if and only if it gives the output which is identical to the expected output. Suppose R_p and R_f be the sets of the passed and failed runs by executing the program P over two sets of test cases, respectively. So, the weighted graphs related to the runs are split to two sets G_p and G_f (the passed and failed graph respectively). After the instrumentation of the faulty program and collecting the information of the program runs by running the instrumented program over the test cases, in the form of the passed and failed graphs, we apply proposed approach to discover the edges which are most probable to be anomalous. In next section we note the next phase of debugging approach.

3.2   Finding the Suspicious Edges

In this phase, we apply a difference function to find suspicious sub paths which are more approximately fault relevant. To address this concern, two sets of the passed and failed graphs G_p and G_f which manifest the execution paths, are used.

Suppose we have a failed run r_i, we want to find a passed run r_j that is very close to r_i. In other words, we need to find some slices of execution path of r_i that make it distinguish from the passed run r_j(the run which is most similar to r_i). Such sub paths are our starting point to approach exact defective positions of the faulty program. So, we exploit the weighted graph manifesting the run for the comparison process.

Let g_f be a failed graph of the set G_f, since we want to find the most similar passed graph g_p in the set G_p, we contrast the graph g_f with all passed graphs in the set G_p. A difference measure is determined based on the edges and their corresponding weights which make two graphs different. Let the set $G= \{((p_i ,p_j),w)| \ p_i, \ p_j \in \Gamma, \ w \in Z^+\}$ denoting the graph G. At first, we compute the set $g_{slice}=g_f-g_p$ for all the passed graph g_p in G_p and then select such g_p that causes the smallest set g_{slice}. In other words, we find the most similar passed graph g_p to the failed graph g_f regarding the difference between them.

If we have more than one passed graph g_p as the most similar passed graph to r_i, we consider all the result passed graphs g_p and regard the union of the sets g_{slice} as the suspicious edges which should be analyzed in last phase of the approach. So we preprocess the edges to relieve the investigating time of finding the most fault relevant edges.

3.3   Non-parametric/Parametric Hypothesis Testing

After finding the edges of the failed graphs that are more probable to be fault relevant, we try to detect most fault relevant edges by testing them statistically. By considering the fact that the program runs are independent of each others, given a random run $r_f \in R_f$ or $r_p \in R_p$ represented by a failed or passed weighted graph (g_f or g_p), we want to statistically test the behavior of the suspicious edges in all passed and failed graphs to detect the most fault relevant edges.

Let X be a random variable representing the weight of a suspicious edge E_s in the failed weighted graph g_f, we use a hypothesis testing procedure to test the distribution of the random variable X in two sets of the failed and passed runs. As we don't have any assumption about the normality or non-normality of the distribution of the random variable X, at first, we conduct Jarque-Bera test to determine whether the random variable X follows a normal distribution.

In fact, Jarque-Bera test is a goodness-of-fit measure of departure from the normality, based on the kurtosis and skewness of the sample. The p-value of Jarque-Bera test is used to determine how much the random variable X forms a normal distribution. For example p-value 0.01 means that the null hypothesis can be rejected at 0.01 significance level or in other words, the probability of getting an observation accepting the null hypothesis (being normal distribution) is less than 0.01. Generally, the smaller the p-value, the more certain the null hypothesis is rejected and the random variable X does not form a normal distribution.

In this study, we apply Jarque-Bera test on each random variable assigned to the resultant suspicious edges to find whether they form a normal distribution. The null hypothesis is rejected for 71% of the 1442 random variables (of 132 faulty versions of Siemens programs).

After applying the normality test on random variable X, we use a parametric or non-parametric hypothesis testing method according to the test result as follow:

Parametric test:

Consider the test result on the distribution of random variable X tells us, it conforms a normal distribution (based on the p-value). Then, we use statistical parameters mean and variance to estimate the similarity between the distribution of random variable X in two sets of the weighted graphs G_p and G_f (passed and failed graphs). Let μ_p and μ_f be the means, σ_p and σ_f be the variances of the random variable X in passed and failed graphs respectively, the null hypothesis is determined as "$\mu_p = \mu_f$ and $\sigma_p = \sigma_f$" which means μ_f should conform to $N(\mu_p, \frac{\sigma_p^2}{m})$, a normal distribution with mean μ_p and variance $\frac{\sigma_p^2}{m}$ and m is the number of failed graphs. Rejecting the null hypothesis means that the distribution of random variable X in the failed graphs is different from the passed weighted graphs. In other words, this statistical test shows that the behavior of the random variable X is different in two sets of passed and failed graphs and this anomalous behavior means that it is more probable to be closer to the defective location of faulty program and the examined edge is fault relevant. Let $X=(X_1, X_2, \ldots, X_m)$ be an independent and identically distributed random sample of the weight of the suspicious edge E_s in the set of failed graphs G_f, μ_p and σ_p are the mean and variance of the population of the weight of suspicious edges in the passed graphs set G_p. Under the null hypothesis, the following statistic:

$$Y = \frac{\sum_{i=1}^{m} X_i}{m}.$$

Conforms to $N(\mu_p, \frac{\sigma_p^2}{m})$, a normal distribution with the mean μ_p and the variance $\frac{\sigma_p^2}{m}$ (i.e. μ_p and σ_p are the mean and the variance of the population of the weights of the considering edge in population of the passed runs) and m is the number of failed runs. So, by applying the null hypothesis test and achieving the procedure on the suspicious edges resulting from step 3.2, the most fault relevant edges are detected.

Nonparametric test:

Consider the opposite condition, the random variable X does not conform a normal distribution. We do not use parametric method, because it depends on the assumption of the normality. To this end, we use Pearson correlation test, to assess the strength and direction of the linear relationship between two populations of random variable X (the weight of suspicious edge as a random variable X) in passed and failed weighted graphs G_f and G_p. The Pearson correlation is 1 in the case of perfect positive linear relationship, -1 in the case of a perfect decreasing linear relationship and some values between -1 and 1 in all other cases, indicating the degree of linear dependence between the populations. As it approaches zero there is less of the relationship. The closer the coefficient is to either -1 or 1, the stronger the correlation between the populations. If the populations are independent, Pearson's correlation coefficient is 0. Therefore, for each suspicious edge E_s resulting from the step 3.2, which does not resemble a normal distribution, we use Pearson correlation test to evaluate how much two populations of the random variable X are linearly correlated. In this study, based on the result of the test on the faulty versions of 132 program of Siemens test suite, we chose 0.7 as the threshold for rejecting the test.

4   Experimental Results

In this section, we evaluate the proposed approach by Siemens Test suite which contains 132 faulty versions of seven subjected programs and only one fault has been injected in each faulty version. There are some test cases for each faulty version, ending in failure or success [9].

As previous works [2][14], we use the T-score metric which estimates the percentage of the code should be investigated before finding the fault location. Certainly, presenting the statements where are closer to the faulty statement means proposing approach is strong in locating faults. Another criterion is the number of detected faults which means what percentages of the fault can be detected by the proposing approach.

In this paper, for the first time, we manifest the run-time behavior of the program by collecting the sequence of the predicates which has been visited in each run. So, an extended control flow graph, namely the weighted graph is constructed. Then, we detect anomalous transitions (edges) by applying the hypothesis testing on the weighted graphs of two sets of the failed runs and passed runs.

To achieve this, firstly we have instrumented the faulty programs of Siemens test suite manually. The control flow of the faulty versions of the programs as sequences of the appeared predicates has been collected by executing the programs over the passing and failing test cases. Then, the weighted graph has been constructed based on the collected data. After constructing the weighted graphs, we find the suspicious edges by applying a difference function and choose the edges which are more probable to constitute the faulty statement. Finally, we have applied our proposing statistical approach to find the most fault relevant predicates by considering the distribution of the suspicious edges. As shown later, the proposed approach excels in fault localization because of the appropriate method of modeling the run-time behavior of the program and especially the presented hypothesis testing approach for locating the faulty statements.

In figure 4, we have compared our proposed approach with some well-known existing approaches according to the average amount of code which should be manually inspected for locating the faults. As shown in figure 2 and 3, our approach considerably has achieved in fault localization.

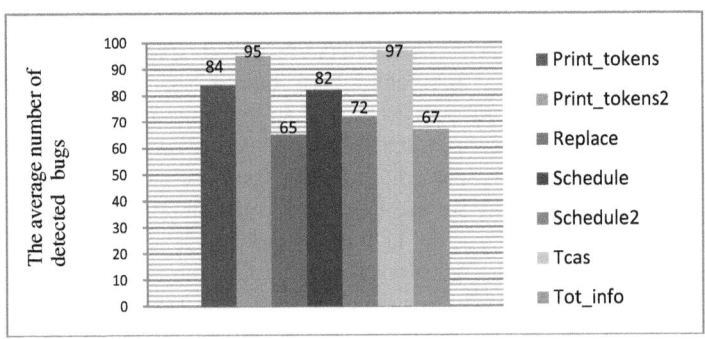

Fig. 2. The percentage of detected bugs in Siemens Test Suite

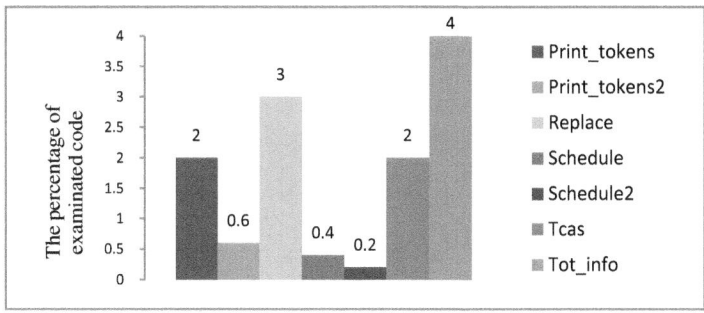

Fig. 3. The percentage of investigated code in Siemens Test Suite

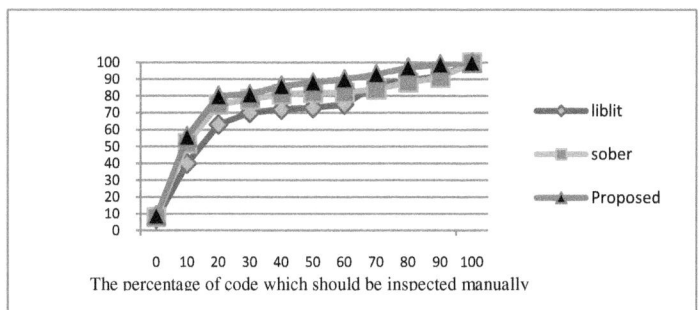

Fig. 4. The comparison of the proposed approach with two other approaches

5   Concluding Remarks

Recent fault localization approaches locate fault relevant predicates by contrasting the evaluation bias of the predicates between passing and failing runs. However, these techniques overlook examining the relationship of predicates with each other such that special kind of this relationship can lead the program to be successful or failed. Furthermore, they apply a parametric hypothesis testing without investigating the distribution of evaluation bias. In this paper, we propose the weighted graph of run-time behavior of the predicates which manifests the correlation of the predicates well. Then, we statistically test the behavior of the suspicious edges in two sets of the weighted graphs of failed and passed the program runs. As we don't have any assumption about the distribution of the weight of the edge, at first we use a normality test and based on the test result we apply a hypothesis testing to find the most fault relevant edges.

References

1. Zhenyu, Z., Chan, W.K., Tse, T.H., Hu, P., Wang, X.: Is non-parametric hypothesis testing model robust for statistical fault localization? Journal of Information and Software Technology 51(11), 1573–1585 (2009)

2. Liu, C., Yan, X., Fei, L., Han, J., Midkiff, S.P.: Sober: Statistical model-based bug localization. In: 10th European Software Eng. Conf./13th ACM SIGSOFT Int'l Symposium Foundations of Software Engineering, Lisbon, pp. 286–295 (2005)
3. Arumuga Nainar, P., Chen, T., Rosin, J., Liblit, B.: Statistical debugging using compound Boolean predicates. In: International Symposium on Software Testing and Analysis, pp. 5–15. ACM Press, London (2007)
4. Liblit, B.: Cooperative Bug Isolation. PhD thesis, University of California, Berkeley (2004)
5. Liblit, B., Naik, M., Zheng, A., Aiken, A., Jordan, M.: Scalable Statistical Bug Isolation. In: Int'l Conference Programming Language Design and Implementation, Chicago, pp. 15–26 (2005)
6. Fei, L., Lee, K., Li, F., Midkiff, S.P.: Argus: Online Statistical Bug Detection. In: Baresi, L., Heckel, R. (eds.) FASE 2006. LNCS, vol. 3922, pp. 308–323. Springer, Heidelberg (2006)
7. Zheng, A.X., Jordan, M.I., Liblit, B., Naik, M., Aiken, A.: Statistical debugging: simultaneous identification of multiple bugs. In: ICML 2006: Proceedings of the 23rd International Conference on Machine Learning, pp. 1105–1112. ACM Press, NY (2006)
8. Freund, J.E., Miller, I., Miller, M.: Mathematical statistics with applications, 7th edn. Prentice Hall (2004)
9. Software-artifact infrastructure repository, http://sir.unl.edu/portal
10. Liu, C., Lian, Z.: How Bayesian Debug. In: Proceedings of the 6th IEEE International Conference on Data Mining ICDM, Hong Kong, China (2006)
11. Chilimbi, T.H.: HOLMES: Effective Statistical Debugging via Efficient Path Profiling. In: Proceedings of the International Conference on Software Engineering ICSE, Canada (2009)
12. Casella, G., Berger, R.: Statistical Inference, 2nd edn., Duxbury (2001)
13. Jones, J.A., Harrold, M.J.: Empirical evaluation of the Tarantula automatic fault localization technique. In: Proceedings of the 20th IEEE/ACM International Conference on Automated Software Engineering (ASE 2005), pp. 273–282. ACM Press, New York (2005)
14. Renieris, M., Reiss, S.P.: Fault localization with nearest neighbor queries. In: Proceedings of the 18th IEEE International Conference on Automated Software Engineering (ASE 2003), pp. 30–39. IEEE Computer Society Press, Los Alamitos (2003)
15. Zeller, A.: Why Programs Fail: A Guide to Systematic Debugging. Morgan Kaufmann, San Francisco (2006)

Duplicate Form of Gould-Hsu Inversions and Binomial Identities

Chuanan Wei[1], Dianxuan Gong[2,*], and Jianbo Li[3]

[1] Department of Information Technology,
Hainan Medical College, Haikou 571101, China
[2] College of Sciences Hebei Polytechnic University,
Tangshan 063009, China
dxgong@heut.edu.cn
[3] Department of Statistics,
The Chinese University of Hong Kong,
Hong Kong, China

Abstract. It is well known that inversion techniques have an important role in the development of combinatorial identities. In 1973, Gould and Hsu [6] offered a pair of surprising inverse series relations. Then it was utilized by Chu [3,4] to study systematically hypergeometric series identities. By applying the duplicate form of Gould-Hsu inversions to a terminating $_4F_3-$series identity form Saalscütz's theorem, we shall establish a family of binomial identities implying numerous interesting hypergeometric series identities.

Keywords: Duplicate form of Gould-Hsu inversions, Binomial identity, Hypergeometric series identity.

1 Introduction

In 1973, Gould and Hsu [6] offered a pair of surprising inverse series relations. According to the parity of k, the duplicate form of it can be displayed as follows.

Lemma 1 *For two complex variables $\{y, z\}$ and four complex sequences $\{a_i, b_i, c_i, d_i\}_{i \geq 0}$, define two polynomial sequences by*

$$\varphi(y; 0) \equiv 1 \quad and \quad \varphi(y; n) = \prod_{i=0}^{n-1}(a_i + yb_i) \quad when \quad n \in \mathbb{N},$$

$$\psi(z; 0) \equiv 1 \quad and \quad \psi(z; n) = \prod_{i=0}^{n-1}(c_i + zd_i) \quad when \quad n \in \mathbb{N}.$$

Then the system of equations

$$\Omega_n = \sum_{k \geq 0} \binom{n}{2k} \frac{c_k + 2kd_k}{\varphi(n; k)\psi(n; k+1)} F(k)$$

---

* Corresponding author.

B. Liu and C. Chai (Eds.): ICICA 2011, LNCS 7030, pp. 145–152, 2011.
© Springer-Verlag Berlin Heidelberg 2011

$$-\sum_{k\geq 0}\binom{n}{1+2k}\frac{a_k+(1+2k)b_k}{\varphi(n;1+k)\psi(n;k+1)}G(k) \tag{1}$$

is equivalent to the system of equations

$$F(n)=\sum_{k=0}^{2n}(-1)^k\binom{2n}{k}\varphi(k;n)\psi(k;n)\,\Omega_k, \tag{2}$$

$$G(n)=\sum_{k=0}^{1+2n}(-1)^k\binom{1+2n}{k}\varphi(k;n)\psi(k;n+1)\,\Omega_k. \tag{3}$$

Combining the decomposition relation:

$$
{}_4F_3\left[\begin{matrix}-\frac{n}{2},\ \frac{1-n}{2},\ a-c,\ 1+x\\ 1/2-c-n-\lambda n,\ 2+a+\lambda n,\ x\end{matrix}\middle|\,1\right]
$$
$$
={}_3F_2\left[\begin{matrix}-\frac{n}{2},\ \frac{1-n}{2},\ a-c\\ 1/2-c-n-\lambda n,\ 1+a+\lambda n\end{matrix}\middle|\,1\right]\frac{(x+c-a)(1+a+\lambda n)}{x(1+c+\lambda n)}
$$
$$
+{}_3F_2\left[\begin{matrix}-\frac{n}{2},\ \frac{1-n}{2},\ 1+a-c\\ 1/2-c-n-\lambda n,\ 2+a+\lambda n\end{matrix}\middle|\,1\right]\frac{(1-x+a+\lambda n)(a-c)}{x(1+c+\lambda n)}
$$

with Saalschütz's theorem(cf. Bailey [2]p. 9):

$$
{}_3F_2\left[\begin{matrix}a,\ b,\ -n\\ c,\ 1+a+b-c-n\end{matrix}\middle|\,1\right]=\frac{(c-a)_n(c-b)_n}{(c)_n(c-a-b)_n},
$$

we derive the following hypergeometric series identity:

$$
{}_4F_3\left[\begin{matrix}-\frac{n}{2},\ \frac{1-n}{2},\ a-c,\ 1+x\\ 1/2-c-n-\lambda n,\ 2+a+\lambda n,\ x\end{matrix}\middle|\,1\right]
$$
$$
=\frac{(1/2+a+\lambda n)_n(1+2c+2\lambda n)_n}{(1/2+c+\lambda n)_n(1+2a+2\lambda n)_n}\frac{1+a+\lambda n}{1+c+\lambda n}U(x,n) \tag{4}
$$

where $U(x,n)=1+(c-a)\dfrac{2(1+2a+2n+2\lambda n)x+n^2-n}{x(1+2a+2\lambda n+n)(2+2a+2\lambda n+n)}$.

The purpose of this paper is to establish a family of binomial identities implying numerous interesting hypergeometric series identities by applying Lemma 1 to (4) and using the linear method.

2   Binomial Identities with Four Parameters

It is obvious that (4) reads as

$$
\sum_{k\geq 0}\binom{n}{2k}\frac{1+a+k+2\lambda k}{(1/2-c-n-\lambda n)_k(1+a+\lambda n)_{k+1}}\frac{(x+k)(1/2)_k(a-c)_k}{x(1+a+k+2\lambda k)}
$$
$$
=\frac{(1/2+a+\lambda n)_n(1+2c+2\lambda n)_n}{(1/2+c+\lambda n)_n(1+2a+2\lambda n)_n}\frac{U(x,n)}{1+c+\lambda n}
$$

which matches with (1) perfectly under the following specifications:

$$\varphi(y; n) = (1/2 - c - y - \lambda y)_n, \quad \psi(z; n) = (1 + a + \lambda z)_n,$$

$$F(n) = \frac{(x + n)(1/2)_n(a - c)_n}{x(1 + a + n + 2\lambda n)}, \quad G(n) = 0,$$

$$\Omega_n = \frac{(1/2 + a + \lambda n)_n(1 + 2c + 2\lambda n)_n}{(1/2 + c + \lambda n)_n(1 + 2a + 2\lambda n)_n} \frac{U(x, n)}{1 + c + \lambda n}.$$

Then (2) and (3) give the following pair of dual relations:

$$\sum_{k=0}^{2n} (-1)^k \binom{2n}{k} (1/2 - c - k - \lambda k)_n (1 + a + \lambda k)_n \Omega_k = F(n),$$

$$\sum_{k=0}^{1+2n} (-1)^k \binom{1 + 2n}{k} (1/2 - c - k - \lambda k)_n (1 + a + \lambda k)_{n+1} \Omega_k = G(n)$$

which are equivalent to the two summation formulas with four parameters.

Theorem 1 (Binomial identities)

$$\sum_{k=0}^{2n} \left(\frac{1}{4}\right)^k \binom{2n}{k} (1 + 2a + 2\lambda k + k)_k (1 + 2c + 2\lambda k)_k (1 + a + \lambda k + k)_{n-k}$$

$$\times (1/2 - c - \lambda k)_{n-k} \frac{U(x, k)}{1 + c + \lambda k} = \frac{(x + n)(1/2)_n(a - c)_n}{x(1 + a + 2\lambda n + n)},$$

$$\sum_{k=0}^{1+2n} \left(\frac{1}{4}\right)^k \binom{1+2n}{k} (1+2a+2\lambda k+k)_k (1+2c+2\lambda k)_k (1+a+\lambda k+k)_{1+n-k}$$

$$\times (1/2 - c - \lambda k)_{n-k} \frac{U(x, k)}{1 + c + \lambda k} = 0.$$

Letting $x \to a - c - 1$, $a \to a - 1$ for Theorem 1, we obtain the two equations.

Proposition 1 (Binomial identities)

$$\sum_{k=0}^{2n} \left(\frac{1}{4}\right)^k \binom{2n}{k} (1 + 2a + 2\lambda k + k)_{k-1} (1 + 2c + 2\lambda k)_k (a + \lambda k + k)_{n-k}$$

$$\times (1/2 - c - \lambda k)_{n-k} = \frac{(1/2)_n(a - c)_n}{2(a + 2\lambda n + n)},$$

$$\sum_{k=0}^{1+2n} \left(\frac{1}{4}\right)^k \binom{1+2n}{k} (1+2a+2\lambda k+k)_{k-1}(1+2c+2\lambda k)_k(a+\lambda k+k)_{1+n-k}$$

$$\times (1/2 - c - \lambda k)_{n-k} = 0.$$

Letting $\lambda \to -1/2$, $a \to a/2$, $c \to -a/2$ for Proposition 1, we recover the two known results.

Corollary 1 (Finite Kummer identities: Bailey [[2] p. 27])

$$
{}_2F_1\left[\begin{matrix} -2n,\ a \\ 1-a-2n \end{matrix}\ \middle|\ -1\right] = \frac{(1/2)_n(a)_n}{(a/2)_n(a/2+1/2)_n},
$$

$$
{}_2F_1\left[\begin{matrix} -1-2n,\ a \\ -a-2n \end{matrix}\ \middle|\ -1\right] = 0.
$$

Letting $x \to \frac{a-c-1}{(1+2\lambda)^2}$, $a \to a-1$ for Theorem 1, we get the two equations.

Proposition 2 (Binomial identities: $u = \frac{(a-1/2)(c+1)}{1+c+(3/2+a+c)\lambda+\lambda^2}$)

$$
\sum_{k=0}^{2n} \left(\frac{1}{4}\right)^k \binom{2n}{k}(1+2a+2\lambda k+k)_{k-2}(1+2c+2\lambda k)_k(a+\lambda k+k)_{n-k}
$$
$$
\times\ (1/2-c-\lambda k)_{n-k}\frac{u+k}{1+c+\lambda k} = \frac{\{a-c-1+n(1+2\lambda)^2\}(1/2)_n(a-c)_{n-1}}{4\{1+c+(3/2+a+c)\lambda+\lambda^2\}(a+2\lambda n+n)},
$$
$$
\sum_{k=0}^{1+2n} \left(\frac{1}{4}\right)^k \binom{1+2n}{k}(1+2a+2\lambda k+k)_{k-2}(1+2c+2\lambda k)_k(a+\lambda k+k)_{1+n-k}
$$
$$
\times\ (1/2-c-\lambda k)_{n-k}\frac{u+k}{1+c+\lambda k} = 0.
$$

Letting $\lambda \to -1/2$, $a \to 1+a/2$, $c \to -1-a/2$ for Proposition 2, we gain the two equations.

Corollary 2 (Kummer-type identities)

$$
{}_3F_2\left[\begin{matrix} -2n,\ 1+a/2,\ a \\ a/2,\ -1-a-2n \end{matrix}\ \middle|\ -1\right] = \frac{(1/2)_n(1+a)_n}{(1+a/2)_n(a/2+3/2)_n},
$$

$$
{}_3F_2\left[\begin{matrix} -1-2n,\ 1+a/2,\ a \\ a/2,\ -2-a-2n \end{matrix}\ \middle|\ -1\right] = 0.
$$

Letting $x \to \frac{1+c-a}{3+4c+2\lambda+4a\lambda+4c\lambda}$, $a \to a-1$ for Theorem 1, we achieve the two equations.

Proposition 3 (Binomial identities)

$$
\sum_{k=0}^{2n} \left(\frac{1}{4}\right)^k \binom{2n}{k}(1+2a+2\lambda k+k)_{k-2}(1+2c+2\lambda k)_k(a+\lambda k+k)_{n-k}
$$
$$
\times\ (1/2-c-\lambda k)_{n-k}\frac{u+k^2}{1+c+\lambda k} = \frac{\{a-c-1-n(3+4c+2\lambda+4a\lambda+4c\lambda)\}(1/2)_n(a-c)_{n-1}}{4\{1+c+(3/2+a+c)\lambda+\lambda^2\}(a+2\lambda n+n)},
$$
$$
\sum_{k=0}^{1+2n} \left(\frac{1}{4}\right)^k \binom{1+2n}{k}(1+2a+2\lambda k+k)_{k-2}(1+2c+2\lambda k)_k(a+\lambda k+k)_{1+n-k}
$$
$$
\times\ (1/2-c-\lambda k)_{n-k}\frac{u+k^2}{1+c+\lambda k} = 0.
$$

Letting $\lambda \to -1/2, a \to 1 - a, c \to a - 1$ for Proposition 3, we attain the two equations.

Corollary 3 (Kummer-type identities)

$$
{}_4F_3\left[\begin{array}{c} -2n,\, 1+\sqrt{a},\, 1-\sqrt{a},\, -2a \\ \sqrt{a},\, -\sqrt{a},\, 2a-1-2n \end{array}\Bigg| -1\right] = \frac{(3/2)_n(1-2a)_n}{(1-a)_n(3/2-a)_n},
$$

$$
{}_4F_3\left[\begin{array}{c} -1-2n,\, 1+\sqrt{a},\, 1-\sqrt{a},\, -2a \\ \sqrt{a},\, -\sqrt{a},\, 2a-2-2n \end{array}\Bigg| -1\right] = 0.
$$

3   Binomial Identities with Five Parameters

Splitting the factor

$$
1 + a + (1+2\lambda)k = \frac{d+2k}{d+n}(1+a+\lambda n+k) - \frac{\lambda d - a - 1 - k}{d+n}(n-2k),
$$

we can reformulate (4) as

$$
\sum_{k\geq 0}\binom{n}{2k}\frac{-1/2-c-k-2\lambda k}{(1+a+\lambda n)_k(-1/2-c-n-\lambda n)_{k+1}}\,\frac{(x+k)(d+2k)(1/2)_k(a-c)_k}{x(1+a+k+2\lambda k)(1/2+c+k+2\lambda k)}
$$

$$
-\sum_{k\geq 0}\binom{n}{1+2k}\frac{1+a+\lambda+k+2\lambda k}{(1+a+\lambda n)_{k+1}(-1/2-c-n-\lambda n)_{k+1}}\,\frac{(x+k)(1+a-d\lambda+k)(3/2)_k(a-c)_k}{x(1+a+k+2\lambda k)(1+a+\lambda+k+2\lambda k)}
$$

$$
= \frac{(1/2+a+\lambda n)_n(1+2c+2\lambda n)_n}{(1/2+c+\lambda n)_{n+1}(1+2a+2\lambda n)_n}\,\frac{d+n}{1+c+\lambda n}U(x,n)
$$

which fits into (1) ideally under the following specifications:

$$
\varphi(y;n) = (1+a+\lambda y)_n, \qquad\qquad \psi(z;n) = (-1/2-c-z-\lambda z)_n,
$$

$$
F(n) = \frac{(x+n)(d+2n)(1/2)_n(a-c)_n}{x(1+a+n+2\lambda n)(1/2+c+n+2\lambda n)}, \quad G(n) = \frac{(x+n)(1+a-d\lambda+n)(3/2)_n(a-c)_n}{x(1+a+n+2\lambda n)(1+a+\lambda+n+2\lambda n)},
$$

$$
\Omega_n = \frac{(1/2+a+\lambda n)_n(1+2c+2\lambda n)_n}{(1/2+c+\lambda n)_{n+1}(1+2a+2\lambda n)_n}\,\frac{d+n}{1+c+\lambda n}U(x,n).
$$

Then (2) and (3) create the following pair of dual relations:

$$
\sum_{k=0}^{2n}(-1)^k\binom{2n}{k}(1+a+\lambda k)_n(-1/2-c-k-\lambda k)_n\,\Omega_k = F(n),
$$

$$
\sum_{k=0}^{1+2n}(-1)^k\binom{1+2n}{k}(1+a+\lambda k)_n(-1/2-c-k-\lambda k)_{n+1}\,\Omega_k = G(n)
$$

which are equivalent to the two summation formulas with five parameters.

Theorem 2 (Binomial identities)

$$\sum_{k=0}^{2n} \left(\frac{1}{4}\right)^k \binom{2n}{k}(1 + 2a + 2\lambda k + k)_k(1 + 2c + 2\lambda k)_k(1 + a + \lambda k + k)_{n-k}$$

$$\times (1/2 - c - \lambda k)_{n-k-1}\frac{d+k}{1+c+\lambda k}U(x,k) = \frac{-(x+n)(d+2n)(1/2)_n(a-c)_n}{x(1+a+n+2\lambda n)(1/2+c+n+2\lambda n)},$$

$$\sum_{k=0}^{1+2n} \left(\frac{1}{4}\right)^k \binom{1+2n}{k}(1 + 2a + 2\lambda k + k)_k(1 + 2c + 2\lambda k)_k(1 + a + \lambda k + k)_{n-k}$$

$$\times (1/2 - c - \lambda k)_{n-k}\frac{d+k}{1+c+\lambda k}U(x,k) = \frac{(x+n)(d\lambda - a - 1 - n)(3/2)_n(a-c)_n}{x(1+a+n+2\lambda n)(1+a+\lambda+n+2\lambda n)}.$$

It should be pointed out that Theorem 1 is only a special case of Theorem 2. The first equation of Theorem 2 reduces to the first equation of Theorem 1 when $d = \frac{1/2+c-n}{1+\lambda}$ and the second equation of Theorem 2 reduces to the second equation of Theorem 1 when $d = \frac{1+a+n}{\lambda}$.

Letting $x \to a - c - 1$ and $a \to a - 1$ for Theorem 2, we obtain the extension of Proposition 1.

Proposition 4 (Binomial identities)

$$\sum_{k=0}^{2n} \left(\frac{1}{4}\right)^k \binom{2n}{k}(1 + 2a + 2\lambda k + k)_{k-1}(1 + 2c + 2\lambda k)_k(a + \lambda k + k)_{n-k}$$

$$\times (1/2 - c - \lambda k)_{n-k-1}(d+k) = \frac{-(d/2+n)(1/2)_n(a-c)_n}{(a+n+2\lambda n)(1/2+c+n+2\lambda n)},$$

$$\sum_{k=0}^{1+2n} \left(\frac{1}{4}\right)^k \binom{1+2n}{k}(1 + 2a + 2\lambda k + k)_{k-1}(1 + 2c + 2\lambda k)_k(a + \lambda k + k)_{n-k}$$

$$\times (1/2 - c - \lambda k)_{n-k}(d+k) = \frac{(d\lambda - a - n)(1/2)_{n+1}(a-c)_n}{(a+n+2\lambda n)(a+\lambda+n+2\lambda n)}.$$

Letting $\lambda \to -1/2, a \to a/2, c \to -a/2$ for Proposition 4, we derive the extension of Corollary 1.

Corollary 4 (Kummer-type identities)

$$_3F_2\left[\begin{matrix} -2n, a, 1+d \\ 2-a-2n, d \end{matrix}\middle| -1\right] = \frac{(d+2n)(1/2)_n(a)_n}{d(a/2)_n(a/2-1/2)_n},$$

$$_3F_2\left[\begin{matrix} -1-2n, a, 1+d \\ 1-a-2n, d \end{matrix}\middle| -1\right] = \frac{(d+a+2n)(1/2)_{n+1}(a)_n}{-d(a/2)_n(a/2-1/2)_{n+1}}.$$

Letting $x \to \frac{a-c-1}{(1+2\lambda)^2}, a \to a - 1$ for Theorem 2, we get the extension of Proposition 2.

Proposition 5 (Binomial identities: $\beta = \frac{1+c-a-n(1+2\lambda)^2}{1+c+(3/2+a+c)\lambda+\lambda^2}$)

$$\sum_{k=0}^{2n} \left(\frac{1}{4}\right)^k \binom{2n}{k}(1+2a+2\lambda k+k)_{k-2}(1+2c+2\lambda k)_k(a+\lambda k+k)_{n-k}$$

$$\times (1/2-c-\lambda k)_{n-k-1}\frac{(d+k)(u+k)}{1+c+\lambda k} = \frac{\beta(d+2n)(1/2)_n(a-c)_{n-1}}{4(a+n+2\lambda n)(1/2+c+n+2\lambda n)},$$

$$\sum_{k=0}^{1+2n} \left(\frac{1}{4}\right)^k \binom{1+2n}{k}(1+2a+2\lambda k+k)_{k-2}(1+2c+2\lambda k)_k(a+\lambda k+k)_{n-k}$$

$$\times (1/2-c-\lambda k)_{n-k}\frac{(d+k)(u+k)}{1+c+\lambda k} = \frac{\beta(a-d\lambda+n)(3/2)_n(a-c)_{n-1}}{4(a+n+2\lambda n)(a+\lambda+n+2\lambda n)}.$$

Letting $\lambda \to -1/2$, $a \to 1+a/2$, $c \to -1-a/2$ for Proposition 5, we gain extension of Corollary 2.

Corollary 5 (Kummer-type identities)

$${}_4F_3\left[\begin{array}{c} -2n,\, 1+a/2,\, a,\, 1+d \\ a/2,\, -a-2n,\, d \end{array}\middle| -1\right] = \frac{(d+2n)(1/2)_n(1+a)_n}{d(1/2+a/2)_n(1+a/2)_n},$$

$${}_4F_3\left[\begin{array}{c} -1-2n,\, 1+a/2,\, a,\, 1+d \\ a/2,\, -1-a-2n,\, d \end{array}\middle| -1\right] = \frac{(2+a+d+2n)(3/2)_n(2+a)_{n-1}}{-d(1+a/2)_n(3/2+a/2)_n}.$$

Letting $x \to \frac{1+c-a}{3+4c+2\lambda+4a\lambda+4c\lambda}$, $a \to a-1$ for Theorem 2, we achieve the extension of Proposition 3.

Proposition 6 (Binomial identities: $\gamma = \frac{1+c-a+(3+4c+2\lambda+4a\lambda+4c\lambda)n}{1+c+(3/2+a+c)\lambda+\lambda^2}$)

$$\sum_{k=0}^{2n} \left(\frac{1}{4}\right)^k \binom{2n}{k}(1+2a+2\lambda k+k)_{k-2}(1+2c+2\lambda k)_k(a+\lambda k+k)_{n-k}$$

$$\times (1/2-c-\lambda k)_{n-k-1}\frac{(d+k)(u+k^2)}{1+c+\lambda k} = \frac{\gamma(d+2n)(1/2)_n(a-c)_{n-1}}{4(a+n+2\lambda n)(1/2+c+n+2\lambda n)},$$

$$\sum_{k=0}^{1+2n} \left(\frac{1}{4}\right)^k \binom{1+2n}{k}(1+2a+2\lambda k+k)_{k-2}(1+2c+2\lambda k)_k(a+\lambda k+k)_{n-k}$$

$$\times (1/2-c-\lambda k)_{n-k}\frac{(d+k)(u+k^2)}{1+c+\lambda k} = \frac{\gamma(a-d\lambda+n)(3/2)_n(a-c)_{n-1}}{4(a+n+2\lambda n)(a+\lambda+n+2\lambda n)}.$$

Letting $\lambda \to -1/2$, $a \to 1-a$, $c \to a-1$ for Proposition 6, we attain the extension of Corollary 3.

Corollary 6 (Kummer-type identities)

$${}_5F_4\left[\begin{array}{c} -2n,\, 1+\sqrt{a},\, 1-\sqrt{a},\, -2a,\, 1+d \\ \sqrt{a},\, -\sqrt{a},\, 2a-2n,\, d \end{array}\middle| -1\right] = \frac{(d+2n)(3/2)_n(1-2a)_n}{d(1/2-a)_n(1-a)_n},$$

$${}_5F_4\left[\begin{array}{c} -1-2n,\, 1+\sqrt{a},\, 1-\sqrt{a},\, -2a,\, 1+d \\ \sqrt{a},\, -\sqrt{a},\, 2a-1-2n,\, d \end{array}\middle| -1\right] = \frac{(1+2n)(a-1-d/2-n)(3/2)_n(1-2a)_n}{d(1-a)_n(1/2-a)_{n+1}}.$$

Let $\mathcal{A}_n$ and $\mathcal{B}_n$ stand, respectively, for the first and second equations of Theorem 2. Then the linear combinations of $\mathcal{A}_n$ with $\mathcal{B}_{n-1}$ and $\mathcal{A}_n$ with $\mathcal{B}_n$ can produce binomial identities with six parameters. Due to the limit of space, the details will not be mentioned here.

Remark: The equivalent forms of Propositions 1 and 4 have been displayed in Wei [10] that is the Master theses of the first author.

References

1. Andrews, G.E., Askey, R., Roy, R.: Special Functions. Cambridge University Press, Cambridge (2000)
2. Bailey, W.N.: Generalized Hypergeometric Series. Cambridge University Press, Cambridge (1935)
3. Chu, W.: Inversion techniques and combinatorial identities. Boll. Un. Mat. Ital. B-7, 737–760 (1993)
4. Chu, W.: Inversion techniques and combinatorial identities: Strange evaluations of basic hypergeometric series. Compositio Math. 91, 121–144 (1994)
5. Chu, W., Wei, C.: Lengendre inversions and balanced hypergeometric series identities. Discrete Math. 308, 541–549 (2008)
6. Gould, H.W., Hsu, L.C.: Some new inverse series relations. Duke Math. J. 40, 885–891 (1973)
7. Ma, X.: An extension of Warnaar's matrix inversion. Proc. Amer. Math. Soc. 133, 3179–3189 (2005)
8. Riordan, J.: Combinatorial Identities. John Wiley & Sons, Inc., New York (1968)
9. Warnaar, S.O.: Summation and transformation formulas for elliptic hypergeometric series. Constr. Approx. 18, 479–502 (2002)
10. Wei, C.: Applications of Inversions Techniques in Combnatorial Identities. Dalian University of Technology, Dalian (2006) (in Chinese)

Schur-Convexity on Generalized Information Entropy and Its Applications

Bo-yan Xi, Shu-hong Wang, and Tian-yu Zhang

College of Mathematics, Inner Mongolia University for Nationalities,
Tongliao City, Inner Mongolia Autonomous Region, 028043, China
baoyintu68@sohu.com, baoyintu78@qq.com

Abstract. The information entropy has general applications in different subjects, such as information theory, linear algebra, signal processing, dynamical systems, ergodic theory, probability and statistical. Then the study of inequality on the information entropy has important signification in theory. Schur-convexity and Schur-geometric convexity and Schur-harmonic convexity entropy are studied for the generalized information based on the well-known Schur's condition. As applications, some inequalities of the entropy are established by use of majorization.

Keywords: Information entropy, Schur-convexity, inequality, application.

1 Introduction

Throughout this paper, let $\mathbb{R} = (-\infty, +\infty)$, and $\mathbb{R}_+ = (0, +\infty)$.

Let $p_i \geq 0 (i = 1, 2, \cdots, n)$ and $\sum_{i=1}^{n} p_i = 1$, the function(see [1]: p.101):

$$H(p_1, \cdots, p_n) = -\sum_{i=1}^{n} p_i \log p_i \qquad (1)$$

(Here $x \log x = 0$ for $x = 0$) is called the entropy of p, or the Shannon information entropy of p. Then $H(p)$ is a strict Schur-concave, consequently, $H(p) \geq H(q)$ whenever $p \prec q$, and in particular

$$H(1, 0, \cdots, 0) \leq H(p) \leq H(1/n, \cdots, 1/n). \qquad (2)$$

A more general entropy function, known as Kapur's entropy of order 1 and type t(see [2]) is defined as

$$H_t(p_1, \cdots, p_n) = -\frac{1}{\sum_{k=1}^{n} p_k^t} \sum_{i=1}^{n} p_i^t \log p_i^t. \qquad (3)$$

for $t > 0$.

B. Liu and C. Chai (Eds.): ICICA 2011, LNCS 7030, pp. 153–160, 2011.

When $t = 1$, this reduces to the usual entropy function. Consider the inequality $H_t(p) \leq \ln n$ for every probability vector p. This inequality holds for $t = 1$ as remarked in (2). It does not hold for every $t > 0$. Stolarsky(see [3]) shows that it holds only for $t \geq t_0(n)$, where $t_0(n)$ is constantly depending on n and $t_0(2) = \frac{1}{2}$. Subsequently, Clausing(see [4]) verified that if $n > 3$ and $t = t_0(n)$, then equality holds in (3) for a probability vector $p = (1/n, \cdots, 1/n)$. Thus for this value of t, $H_t(p)$ is not strictly Schur-convex.

The inequality on generalized entropy see [5]-[7].

Now, let $t \in \mathbb{R}$, and $t \neq 0$, the function is defined as follows:

$$H_t(p_1, p_2) = -\frac{p_1^t \log p_1^t + p_2^t \log p_2^t}{p_1^t + p_2^t}, \tag{4}$$

for $(p_1, p_2) \in \mathbb{R}_+^2$.

Particularly, if $p_i > 0 (i = 1, 2)$, and $p_1 + p_2 = 1$, then $H_t(p_1, p_2)$ is information entropy. The purpose of this paper is discussing Schur-convexity and Schur-geometric convexity and Schur-harmonic convexity of $H_t(p_1, p_2)$ on $\mathbb{R}_+^2$. As applications, some inequalities are obtained.

2   Definitions and Lemmas

The Schur-convex function was introduced by I. Schur in 1923(see [1]: p.12;p.80), and it has many important applications in analytic inequalities, linear regression, graphs and matrices, combinatorial optimization, information-theoretic topics, Gamma functions, stochastic orderings, reliability, and other related fields (see [1, 4, 7,15-18]).

In 2004, Zhang Xiao-ming first propose concepts of "Schur-geometric convex function" which is extension of "Schur-convex function" and establish corresponding decision theorem [8]. Since then, Schur-geometric convex has evoked the interest of many researchers and numerous applications and extensions have appeared in the literature, see [9] and [12]-[14].

In order to verify our Theorems, the following Definitions and Lemmas are necessary.

Definition 2.1([1]:p.12). Let $x = (x_1, x_2, \cdots, x_n), y = (y_1, y_2, \cdots, y_n) \in \mathbb{R}^n$. x is said to be majorized by y(in symbols $x \prec y$) if

$$\sum_{i=1}^{k} x_{[i]} \leq \sum_{i=1}^{k} y_{[i]}, k = 1, 2, \cdots, n - 1, \quad \sum_{i=1}^{n} x_{[i]} = \sum_{i=1}^{n} y_{[i]},$$

where $x_{[1]} \geq x_{[2]} \geq \cdots \geq x_{[n]}$ and $y_{[1]} \geq y_{[2]} \geq \cdots \geq y_{[n]}$ are rearrangements of x and y in a descending order.

Definition 2.2. (i) Let $\Omega \subset \mathbb{R}^n$ is symmetric set. Ω is called a convex set if

$$(\lambda x_1 + (1 - \lambda)y_1, \lambda x_2 + (1 - \lambda)y_2, \cdots, \lambda x_n + (1 - \lambda)y_n) \in \Omega$$

for every $x, y \in \Omega$, where $\lambda \in [0, 1]$.

(ii) Let $\Omega \subset \mathbb{R}_+^n$ is symmetric set. Ω is called a geometric convex set if

$$\left(x_1^\lambda y_1^{1-\lambda}, x_2^\lambda y_2^{1-\lambda}, \cdots, x_n^\lambda y_n^{1-\lambda}\right) \in \Omega$$

for every $x, y \in \Omega$, where $\lambda \in [0, 1]$.

(iii) Let $\Omega \subset \mathbb{R}_+^n$ is symmetric set. Ω is called a harmonic convex set if

$$\left(\left(\tfrac{\lambda}{x_1} + \tfrac{1-\lambda}{y_1}\right)^{-1}, \left(\tfrac{\lambda}{x_2} + \tfrac{1-\lambda}{y_2}\right)^{-1}, \cdots, \left(\tfrac{\lambda}{x_n} + \tfrac{1-\lambda}{y_n}\right)^{-1}\right) \in \Omega$$

for every $x, y \in \Omega$, where $\lambda \in [0, 1]$.

Definition 2.3 (i)([1:p.80]). Let $\Omega \subset \mathbb{R}^n$. The function $\varphi : \Omega \to R$ be said to be a Schur-convex function on Ω if $x \prec y$ on Ω implies $\varphi(x) \leq \varphi(y)$. $\varphi(x)$ is said to be a Schur-concave function on Ω if and only if $-\varphi(x)$ is Schur-convex.

(ii)([8]) Let $\Omega \subset \mathbb{R}_+^n$. The function $\varphi : \Omega \to \mathbb{R}_+$ be said to be a Schur-geometric convex function on Ω if $(\ln x_1, \cdots, \ln x_n) \prec (\ln y_1, \cdots, \ln y_n)$ on Ω implies $\varphi(x) \leq \varphi(y)$.

(iii)([10]) Let $\Omega \subset \mathbb{R}_+^n$. The function $\varphi : \Omega \to \mathbb{R}_+$ be said to be a Schur-harmonic convex function on Ω if $1/x := \left(\tfrac{1}{x_1}, \cdots, \tfrac{1}{x_n}\right) \prec 1/y := \left(\tfrac{1}{y_1}, \cdots, \tfrac{1}{y_1}\right)$ on Ω implies $\varphi(x) \leq \varphi(y)$.

Lemma 2.1(Schur-Ostrowski Theorem)([1:p.84]). Let $\Omega \subset \mathbb{R}^n$ be a symmetric convex set with nonempty interior, and $\varphi : \Omega \to R$ be a continuous symmetric function on Ω. If φ is differentiable in Ω°. Then φ is Schur-convex (Schur-concave) on Ω if and only if

$$(x_1 - x_2)\left(\frac{\partial \varphi(x)}{\partial x_1} - \frac{\partial \varphi(x)}{\partial x_2}\right) \geq 0(\leq 0)$$

for all $x \in \Omega^\circ$.

Lemma 2.2([8]). Let $\Omega \subset \mathbb{R}_+^n$ be symmetric with a nonempty interior geometric convex set, and $\varphi : \Omega \to \mathbb{R}_+$ be a continuous symmetric function on Ω. If φ is differentiable in Ω°. Then φ is Schur- geometric convex(Schur- geometric concave) on Ω if and only if

$$(\ln x_1 - \ln x_2)\left(x_1\frac{\partial \varphi(x)}{\partial x_1} - x_2\frac{\partial \varphi(x)}{\partial x_2}\right) \geq 0(\leq 0)$$

for all $x \in \Omega^\circ$.

Lemma 2.3([10]). Let $\Omega \subset \mathbb{R}_+^n$ be symmetric with a nonempty interior harmonic convex set, and $\varphi : \Omega \to \mathbb{R}_+$ be a continuous symmetric function on Ω. If φ is differentiable in Ω°. Then φ is Schur-harmonic convex(Schur-harmonic concave) on Ω if and only if

$$(x_1 - x_2)\left(x_1^2\frac{\partial \varphi(x)}{\partial x_1} - x_2^2\frac{\partial \varphi(x)}{\partial x_2}\right) \geq 0(\leq 0)$$

for all $x \in \Omega^\circ$.

The well-known Logarithmic means is defined by(see [1: p.141]):

$$L(a,b) = \begin{cases} \dfrac{a-b}{\log a - \log b}, & a \neq b, \\ \sqrt{ab}, & a = b. \end{cases} \quad (a,b) \in \mathbb{R}_+^2. \qquad (5)$$

Lemma 2.4([1: p.141]) Let $(a,b) \in \mathbb{R}_+^2$, then

$$L(a,b) \leq \frac{a+b}{2}.$$

Lemma 2.5([11]). Let $a \leq b, u(t) = ta + (1-t)b$ and $v(t) = (1-t)a + tb$, if $\frac{1}{2} \leq t_2 \leq t_1 \leq 1$, or $0 < t_1 \leq t_2 \leq \frac{1}{2}$, then

$$\left(\frac{a+b}{2}, \frac{a+b}{2}\right) \prec (u(t_2), v(t_2)) \prec (u(t_1), v(t_1)) \prec (a,b).$$

3   Main Results

In this section, we discuss the Schur-convexity and the Schur-geometric convexity and the Schur-harmonic convexity of $H_t(p_1, p_2)$, our main results are the following.

Theorem 3.1 Let $(p_1, p_2) \in \mathbb{R}_+^2$, and $t < 0$, or $t \geq \frac{1}{2}$. Then $H_t(p_1, p_2)$ is Schur-concave on $\mathbb{R}_+^2$.

   Proof. For

$$H_t(p_1, p_2) = -\frac{p_1^t \log p_1^t + p_2^t \log p_2^t}{p_1^t + p_2^t},$$

we have

$$\frac{\partial H_t}{\partial p_1} = \frac{-t}{(p_1^t + p_2^t)^2}\left[p_1^{2t-1} + p_1^{t-1}p_2^t \log p_1^t + p_1^{t-1}p_2^t - p_1^{t-1}p_2^t \log p_2^t\right], \qquad (6)$$

$$\frac{\partial H_t}{\partial p_2} = \frac{-t}{(p_1^t + p_2^t)^2}\left[p_2^{2t-1} + p_1^t p_2^{t-1} \log p_2^t + p_1^t p_2^{t-1} - p_1^t p_2^{t-1} \log p_1^t\right] \qquad (7)$$

and

$$(p_1 - p_2)\left(\frac{\partial H_t}{\partial p_1} - \frac{\partial H_t}{\partial p_2}\right) = \frac{p_1 - p_2}{(p_1^t + p_2^t)^2}[(p_1^{2t-1} - p_2^{2t-1}) + p_1^{t-1}p_2^{t-1}(p_2 - p_1)$$

$$+ p_1^{t-1}p_2^{t-1}(p_1 + p_2)(\log p_1^t - \log p_2^t). \qquad (8)$$

If $t < 0$, we have

$$(p_1 - p_2)\left(\frac{\partial H_t}{\partial p_1} - \frac{\partial H_t}{\partial p_2}\right) \leq 0.$$

If $t \geq \frac{1}{2}$, from Lemma2.4, we have

$$-\frac{(p_1 - p_2)}{(p_1^t + p_2^t)^2}(p_2 - p_1) = \frac{(p_1 - p_2)(\log p_1 - \log p_2)}{(p_1^t + p_2^t)^2} \cdot \frac{p_1 - p_2}{\log p_1 - \log p_2}$$

$$\leq \frac{(p_1 - p_2)(\log p_1 - \log p_2)}{(p_1^t + p_2^t)^2} \cdot \frac{p_1 + p_2}{2}. \tag{9}$$

Thereforeby (8) and (9), we obtain

$$(p_1 - p_2)\left(\frac{\partial H_t}{\partial p_1} - \frac{\partial H_t}{\partial p_2}\right)$$

$$= -\frac{t(p_1 - p_2)}{(p_1^t + p_2^t)^2}\left((p_1^{2t-1} - p_2^{2t-1}) + p_1^{t-1}p_2^{t-1}(p_2 - p_1)\right.$$

$$\left. + p_1^{t-1}p_2^{t-1}(p_1 + p_2)(\log p_1^t - \log p_2^t)\right)$$

$$\leq -\frac{t(p_1 - p_2)}{(p_1^t + p_2^t)^2}\left((p_1^{2t-1} - p_2^{2t-1}) - p_1^{t-1}p_2^{t-1}\frac{(p_1 + p_2)(\log p_1 - \log p_2)}{2}\right.$$

$$\left. + p_1^{t-1}p_2^{t-1}(p_1 + p_2)(\log p_1^t - \log p_2^t)\right)$$

$$= -\frac{t(p_1 - p_2)}{(p_1^t + p_2^t)^2}\left((p_1^{2t-1} - p_2^{2t-1})\right.$$

$$\left. + (t - 1/2)p_1^{t-1}p_2^{t-1}(p_1 + p_2)(\log p_1 - \log p_2)\right) \leq 0,$$

by the Lemma 2.1, it follows that $H_t(p_1, p_2)$ is Schur-concave on $\mathbb{R}_+^2$.
   Thus the proof of Theorem 3.1 is complete.

Remark. The condition $t \geq \frac{1}{2}$ in Theorem 3.1 and the condition $t_0(2) = \frac{1}{2}$ in [3] are consistent, here we give proof of majorization.

Theorem 3.2. Let $(p_1, p_2) \in \mathbb{R}_+^2$, and $t \in \mathbb{R}, t \neq 0$. Then $H_t(p_1, p_2)$ is Schur-geometric concave on $\mathbb{R}_+^2$.
   Proof. By (6) and (7), we have

$$(\log p_1 - \log p_2)\left(p_1\frac{\partial H_t}{\partial p_1} - p_2\frac{\partial H_t}{\partial p_2}\right)$$

$$= -\frac{t(\log p_1 - \log p_2)}{(p_1^t + p_2^t)^2}\left((p_1^t - p_2^t)(p_1^t + p_2^t) + 2p_1^t p_2^t(\log p_1^t - \log p_2^t)\right) \leq 0,$$

by the Lemma 2.2, it follows that $H_t(p_1, p_2)$ is Schur-geometric concave on $\mathbb{R}_+^2$.
   Thus the proof of Theorem 3.2 is complete.

Theorem 3.3. Let $(p_1, p_2) \in \mathbb{R}_+^2$, and $t > 0$, or $t < -\frac{1}{2}$. Then $H_t(p_1, p_2)$ is Schur-harmonic concave on $\mathbb{R}_+^2$.
   Proof. By (6) and (7), we have

$$(p_1 - p_2)\left(p_1^2\frac{\partial H_t}{\partial p_1} - p_2^2\frac{\partial H_t}{\partial p_2}\right) = -\frac{t(p_1 - p_2)}{(p_1^t + p_2^t)^2}\left((p_1^{2t+1} - p_2^{2t+1})\right.$$

$$\left. + p_1^t p_2^t(p_1 - p_2) + p_1^t p_2^t(p_1 + p_2)(\log p_1^t - \log p_2^t)\right). \tag{10}$$

If $t > 0$, we have

$$(p_1 - p_2)\left(p_1^2 \frac{\partial H_t}{\partial p_1} - p_2^2 \frac{\partial H_t}{\partial p_2}\right) \le 0.$$

If $t < -\frac{1}{2}$, from Lemma2.4, we have

$$(p_1 - p_2)(p_1^t + p_2^t)^2(p_1 - p_2) = \frac{(p_1 - p_2)(\log p_1 - \log p_2)}{(p_1^t + p_2^t)^2} \cdot \frac{p_1 - p_2}{\log p_1 - \log p_2}$$

$$\le \frac{(p_1 - p_2)(\log p_1 - \log p_2)}{(p_1^t + p_2^t)^2} \cdot \frac{p_1 + p_2}{2}. \qquad (11)$$

Therefore, by (10) and (11), we obtain

$$(p_1 - p_2)\left(p_1^2 \frac{\partial H_t}{\partial p_1} - p_2^2 \frac{\partial H_t}{\partial p_2}\right)$$
$$\le -\frac{t(p_1 - p_2)}{(p_1^t + p_2^t)^2}\left((p_1^{2t+1} - p_2^{2t+1}) + (t + \tfrac{1}{2})p_1^t p_2^t(p_1 + p_2)(\log p_1 - \log p_2)\right) \le 0,$$

by the Lemma 2.3, it follows that $H(p_1, p_2)$ is Schur-harmonic concave on $\mathbb{R}_+^2$. Thus the proof of Theorem 3.3 is complete.

4   Applications

Theorem 4.1. Let $(p_1, p_2) \in \mathbb{R}_+^2$, $t \in \mathbb{R}, t \ne 0$, and $\lambda \in [0, 1]$.

(i) If $t < 0$, or $t \ge \frac{1}{2}$, then

$$H_t(p_1, p_2) \le H_t(\lambda p_1 + (1 - \lambda)p_2, (1 - \lambda)p_1 + \lambda p_2) \le -t \log\left(\frac{p_1 + p_2}{2}\right). \quad (12)$$

(ii) If $t \in \mathbb{R}, t \ne 0$, then

$$H_t(p_1, p_2) \le H_t(p_1^\lambda p_2^{1-\lambda}, p_1^{1-\lambda} p_2^\lambda) \le -t\left(\frac{\log p_1 + \log p_2}{2}\right). \qquad (13)$$

(iii) If $t > 0$, or $t < -\frac{1}{2}$, then

$$H_t(p_1, p_2) \le H_t\left(\left(\frac{\lambda}{p_1} + \frac{1-\lambda}{p_2}\right)^{-1}, \left(\frac{1-\lambda}{p_1} + \frac{\lambda}{p_2}\right)^{-1}\right) \le t \log\left(\frac{1}{2}\left(\frac{1}{p_1} + \frac{1}{p_2}\right)\right).$$
$$(14)$$

Proof. From Lemma 2.5, we have

$$\left(\frac{p_1 + p_2}{2}, \frac{p_1 + p_2}{2}\right) \prec (\lambda p_1 + (1 - \lambda)p_2, (1 - \lambda)p_1 + \lambda p_2) \prec (p_1, p_2),$$

$$(\ln \sqrt{p_1 p_2}, \ln \sqrt{p_1 p_2}) \prec \left(\ln(p_1^\lambda p_2^{1-\lambda}), \ln(p_1^{1-\lambda} p_2^\lambda)\right) \prec (\ln p_1, \ln p_2),$$

$$\left(\frac{1}{2}\left(\frac{1}{p_1}+\frac{1}{p_2}\right),\frac{1}{2}\left(\frac{1}{p_1}+\frac{1}{p_2}\right)\right) \prec \left(\frac{\lambda}{p_1}+\frac{1-\lambda}{p_2},\frac{1-\lambda}{p_1}+\frac{\lambda}{p_2}\right) \prec \left(\frac{1}{p_1},\frac{1}{p_2}\right),$$

where $\lambda \in [0,1]$.

By Theorem 3.1-3.3, the function $H_t(p_1,p_2)$ is Schur-concave on $\mathbb{R}_+^2$ ($t < 0$, or $t \geq \frac{1}{2}$), Schur-geometric concave on R_+^2 ($t \in R, t \neq 0$) and Schur- harmonic concave on $\mathbb{R}_+^2$ ($t > 0$, or $t < -\frac{1}{2}$), so we have (12)-(14).

Thus the proof of Theorem 4.1 is complete.

Let X be a random variable, and its distribution are

$$P(X=0)=p_1, P(X=1)=p_2, p_i > 0 (i = 1,2),$$

then, the generalized entropy of X is

$$H_t(p_1,p_2) = -\frac{p_1^t \ln p_1^t + p_2^t \ln p_2^t}{p_1^t + p_2^t}, t > 0.$$

Corollary 4.1. Let $p_i > 0 (i = 1,2)$, $p_1 + p_2 = 1$, $t \in R, t \neq 0$, and $\lambda \in [0,1]$, then

(i) If $t \geq \frac{1}{2}$, then

$$0 \leq H_t(p_1,p_2) \leq H_t(\lambda p_1 + (1-\lambda)p_2, (1-\lambda)p_1 + \lambda p_2) \leq t \log 2.$$

(ii) If $t > 0$, then

$$0 \leq H_t(p_1,p_2) \leq H_t(p_1^\lambda p_2^{1-\lambda}, p_1^{1-\lambda} p_2^\lambda) \leq -t\left(\frac{\log p_1 + \log p_2}{2}\right).$$

(iii) If $t > 0$, then

$$0 \leq H_t(p_1,p_2) \leq H_t\left(\left(\frac{\lambda}{p_1}+\frac{1-\lambda}{p_2}\right)^{-1}, \left(\frac{1-\lambda}{p_1}+\frac{\lambda}{p_2}\right)^{-1}\right)$$

$$\leq t \log\left(\frac{1}{2}\left(\frac{1}{p_1}+\frac{1}{p_2}\right)\right).$$

Let $p_i > 0, \lambda_i > 0, i = 1, \cdots, n, n \geq 2)$, and $\sum_{i=1}^n p_i = 1, \sum_{i=1}^n \lambda_i = 1$. Denote

$$q_k = \sum_{i=1}^n \lambda_{k+i-1} p_i \ (k = 1, \cdots, n), \ \lambda_{k+n} = \lambda_k (k = 1, \cdots, n-1),$$

then

$$(1/n, 1/n, \cdots, 1/n) \prec (q_1, q_2, \cdots, q_n) \prec (p_1, p_2, \cdots, p_n) \prec (1,0,\cdots,0)$$

and by [1, p.101], the function $H(p_1, \cdots, p_n) = -\sum_{i=1}^n p_i \log p_i$ is Schur-concave on $\mathbb{R}_+^n$, we have

Corollary 4.2 Let $p_i > 0, \lambda_i > 0 (i = 1, \cdots, n, n \geq 2)$, $\sum_{i=1}^n p_i = 1$, $\sum_{i=1}^n \lambda_i = 1$, then

$$0 \leq H(p_1, p_2, \cdots, p_n) \leq H(q_1, q_2, \cdots, q_n) \leq \log n,$$

where $q_k = \sum\limits_{i=1}^{n} \lambda_{k+i-1} p_i \ (k = 1, \cdots, n)$.

Acknowledgements. The present investigation was supported, in part, by the *National Natural Science Foundation of the People's Republic of China* under Grant No. 10962004.

References

1. Marshall, A.M., Olkin, I., Arnold, B.C.: Inequalities: Theory of Majorization and its Application, 2nd edn. Springer, New York (2011)
2. Kapur, J.N.: On Some Properties of Generalised Entropies. Indian J. Math. 9, 427–442 (1967)
3. Stolarsky, K.B.A.: Stronger Logarithmic Inequality Suggested by the Entropy Inequality. SIAM J. Math. Anal. 11, 242–247 (1980)
4. Clausing, A.: Type T Entropy And Majorization. SIAM J. Math. Anal. 14, 203–208 (1983)
5. Nanda, A.K., Paul, P.: Some Results on Generalized Residual Entropy. Information Sciences 176, 27–47 (2006)
6. Bebiano, N., Lemos, R., da Providência, J.: Inequalities for Quantum Relative Entropy. Linear Algebra and its Appl. 401, 159–172 (2005)
7. Madiman, M., Barron, A.: Generalized Entropy Power Inequalities and Monotonicity Properties of Information. IEEE Transactions on Information Theory 53, 2317–2329 (2007)
8. Zhang, X.-M.: Geometrically Convex Functions, pp. 107–108. An'hui University Press, Hefei (2004) (in Chinese)
9. Chu, Y.-M., Zhang, X.-M.: The Schur Geometrical Convexity of the Extended Mean Values. J. Convex Analysis 15, 869–890 (2008)
10. Xia, W.-F., Chu, Y.-M.: Schur-Convexity for a Class of Symmetric Functions and its Applications. J. Inequal. Appl. Article ID 493759, 15 pages (2009)
11. Shi, H.-N., Jiang, Y.-M., Jiang, W.-D.: Schur-Convexity and Schur-Geometrically Concavity of Gini Mean. Comp. Math. Appl. 57, 266–274 (2009)
12. Guan, K.-Z.: A Class of Symmetric Functions for Multiplicatively Convex Function. Math. Inequal. Appl. 10, 745–753 (2007)
13. Niculescu, C.P.: Convexity According to the Geometric Mean. Math. Inequal. Appl. 2, 155–167 (2000)
14. Stepniak, C.: An Effective Characterization of Schur-Convex Functions with Applications. J. Convex Anal. 14, 103–108 (2007)
15. Elezovic, N., Pecaric, J.: Note on Schur-Convex Functions. Rocky Mountain J. Math. 29, 853–856 (1998)
16. Sándor, J.: The Schur-Convexity of Stolarsky and Gini Means. Banach J. Math. Anal. 1, 212–215 (2007)
17. Chu, Y.-M., Long, B.-Y.: Best Possible Inequalities between Generalized Logarithmic Mean and Classical Means. Abstract Applied Anal. Article ID 303286, 13 pages (2011)
18. Niezgoda, M.: Majorization and relative concavity. Linear Algebra Appl. 434, 1968–1980 (2011)

Reduced 4th-Order Eigenvalue Problem

Shu-hong Wang[1], Bao-cai Zhang[2], and Zhu-quan Gu[2]

[1] College of Mathematics, Inner Mongolia University for Nationalities,
Tongliao, 028043 Inner Mongolia, China
[2] Department of Mathematics and Physis,
Shijiazhuang Tiedao University, Shijiazhuang, 050043 Hebei, China
shuhong7682@163.com

Abstract. The technique of the so-called nonlinearization of Lax pairs has been developed and applied to various soliton hierarchies, and this method also was generalized to discuss the nonlinearization of Lax pairs and adjoint Lax pairs of soliton hierarchies. In this paper, by use of the nonlinearization method, the reduced 4th-order eigenvalue problem is discussed and a Lax representation was deduced for the system. By means of Euler-Lagrange equations and Legendre transformations, a reasonable Jacobi-Ostrogradsky coordinate system has been found, and the Bargmann system have been given. Then, the infinite-dimensional motion system described by Lagrange mechaics is changed into the Hamilton cannonical coordinate system.

Keywords: Eigenvalue problem, Reduced System, Jacobi-Ostrogradsky coordinate, Bargmann system.

1 Introduction

Since the first of the 1990s, the technique of the so-called nonlinearization of Lax pairs[1-2] has been developed and applied to various soliton hierarchies. Recently, this method was generalized to discuss the nonlinearization of Lax pairs and adjoint Lax pairs of soliton hierarchies [3-5, 11-13]. However, the 4th-order eigenvalue problem

$$L\varphi = (\partial^4 + q\partial^2 + \partial^2 q + p\partial + \partial p + r)\varphi = \lambda\varphi$$

has been discussed, except its reduced system

$$L\varphi = (\partial^4 + \partial u\partial + v)\varphi = \lambda\varphi .$$

In this paper, we consider the reduced 4th-order eigenvalue problem

$$L\varphi = (\partial^4 + \partial u\partial + v)\varphi = \lambda\varphi \tag{1.1}$$

here $\partial = \partial/\partial x$,eigenparameter $\lambda \in R$, and potential u, v is a actual function in (x,t) .

By means of Euler-Lagrange equations and Legendre transformations, a reasonable Jacobi-Ostrogradsky coordinate system has been found, and the Bargmann system have been given. Then, the infinite-dimensional motion system described by Lagrange mechaics is changed into the Hamilton cannonical coordinate system.

B. Liu and C. Chai (Eds.): ICICA 2011, LNCS 7030, pp. 161–168, 2011.

2   Conceptions

In order to verify our Theorems, the following definitions are necessary.

Definition 1[6][9]. Assume that our linear space is equipped with a L_2 scalar product

$$(f,g)_{L_2(\Omega)} = \int_\Omega fg^* dx < \infty$$

symbol * denoting the complex conjugate.

Definition 2 [6][9]. Operator $\overline{A}$ is called dual operator of A , if $(Af,g)_{L_2(\Omega)} = (f,\overline{A}g)_{L_2(\Omega)}$, and A is called a self-adjioint operator, if $\overline{A} = A$.

Definition 3 [7][8]. The commutator of operators W and L is defined as follows

$$[W,L] = WL - LW$$

3   Main Results

3.1   Evolution Equations and Lax Pairs

Now, suppose Ω is the basic interval of (1.1), if u,v,φ in the (1.1) and their derivates on x are all decay at infinity, then $\Omega = (-\infty, +\infty)$; if they are all periodic T functions, then $\Omega = [0, 2T]$.

We consider the following 4th-order operator L in the interval Ω :

$$L = \partial^4 + \partial u \partial + v ,$$

here u, v is potential function of the eigenvalue problem (1.1). λ is called eigenvalue of the eigenvalue problem (1.1), and φ is called eigenfunction to eigenvalue λ ,if $L\varphi = \lambda\varphi, \varphi \in L_2(\Omega)$ is non-trivial solution of the eigenvalue problem (1.1) [6][9].

Theorem 1. $L = \partial^4 + \partial u \partial + v$ is a self-adjioint operator on Ω .

Proof. By Definition 2, we easily derive Theorem1.

If $L' = \partial/\partial\varepsilon\big|_{\varepsilon=0} L(u + \varepsilon\delta u)$ is called the derived function of differential operator $L = L(u)$ [8][10], here u is potential function of L , then we have

Theorem 2. If φ is an eigenfunction corresponding to the eigenvalue λ of (1.1), then functional Gradient [7][8]

$$\nabla\lambda = \left(\delta\lambda/\delta u \quad \delta\lambda/\delta v\right)^T = \left(\int_\Omega \varphi^2 dx\right)^{-1} \left(-\varphi_x^2 \quad \varphi^2\right)^T \tag{3.1}$$

Set

$$\begin{cases} W_m = \sum_{j=0}^m [\frac{1}{4}b_j\partial^3 - \frac{1}{8}b_{jx}\partial^2 + \frac{1}{8}(2a_j + b_{jxx} + 2ub_j)\partial - \frac{1}{8}(3a_{jx} + b_{jxxx} + ub_{jx})]L^{m-j} \\ G_j = \left(a_j \quad b_j\right)^T \end{cases} \tag{3.2}$$

here $m = 0,1,2\cdots$, $j = 0,1,2,\cdots m$.

$$J = \begin{pmatrix} 0 & \partial \\ \partial & \frac{3}{4}\partial^3 + \frac{1}{2}u\partial + \frac{1}{4}u_x \end{pmatrix} \tag{3.3}$$

$$K = \begin{pmatrix} K_{11} & K_{12} \\ K_{21} & K_{22} \end{pmatrix} \tag{3.4}$$

here

$K_{11} = \frac{4}{5}\partial^3 + \frac{1}{2}u\partial + \frac{1}{4}u_x, \quad K_{12} = \frac{3}{8}\partial^5 + \frac{3}{8}u\partial^3 + \frac{3}{4}u_x\partial^2 + (\frac{3}{8}u_{xx} + v)\partial + \frac{3}{4}v_x,$

$K_{21} = \frac{3}{8}\partial^5 + \frac{3}{8}u\partial^3 + \frac{3}{8}u_x\partial^2 + v\partial + \frac{1}{4}v_x,$

$K_{22} = \frac{1}{8}\partial^7 + \frac{1}{4}u\partial^5 + \frac{5}{8}u_x\partial^4 + (\frac{3}{4}u_{xx} + \frac{1}{8}u^2 + \frac{3}{4}v)\partial^3 + (\frac{1}{2}u_{xxx} + \frac{3}{8}uu_x + \frac{9}{8}v_x)\partial^2$

$\quad + (\frac{7}{8}v_{xx} + \frac{1}{8}u_{xxxx} + \frac{1}{8}uu_{xx} + \frac{1}{2}uv + \frac{1}{8}u_x^2)\partial + (\frac{1}{4}uv_x + \frac{1}{4}vu_x + \frac{1}{4}v_{xxx}).$

Theorem 3. (1) operators J and K are the bi-Hamilton operators[6][9] ;
(2) if λ and φ are the eiegenvalue and eigenfunction of (1.1) , then

$$JG_{m+1} = KG_m \qquad m = 0,1,2\cdots \tag{3.5}$$

$$K\nabla\lambda = J\lambda\nabla\lambda \tag{3.6}$$

Set $JG_0 = 0$, then

$$G_0 = \begin{pmatrix} 4 & 0 \end{pmatrix}^T \tag{3.7}$$

or

$$G_0 = \begin{pmatrix} -u & 4 \end{pmatrix}^T \tag{3.8}$$

So, we have

$$L_{t_m} = [W_m, L] = \sum_{j=0}^{m} [L_*(KG_j) - L_*(JG_j)L]L^{m-j} = L_*(KG_m) = L_*(JG_{m+1})$$

here $L_*(\varsigma) = d/d\varepsilon_{\varepsilon=0}\left((q \quad p)^T + \varepsilon\varsigma\right): R^2 \to R$ is one to one.

Theorem 4. Set $JG_0 = 0$, then the m-th-order evolution equation of (1.1) is

$$q_{t_m} = X_m = JG_{m+1} = KG_m, q = (u,v)^T, m = 0,1,2\cdots \tag{3.9}$$

and (3.9) become the isospectral compatible condition of the following Lax pairs

$$L\varphi = \lambda\varphi, \quad \varphi_{t_m} = W_m\varphi, \quad m = 0,1,2\cdots \tag{3.10}$$

So, we have the first evolution equation

$$u_{t_0} = u_x, v_{t_0} = v_x \tag{3.11}$$

and its Lax pairs

$$L\varphi = \lambda\varphi, \quad \varphi_{t_0} = \varphi_x \tag{3.12}$$

or the first evolution equation

$$u_{t_0} = -\frac{5}{4}u_{xxx} - \frac{3}{4}uu_x + 3v_x, v_{t_0} = -\frac{3}{8}u_{xxxxx} - \frac{3}{8}uu_{xxx} - \frac{3}{8}u_xu_{xx} + v_{xxx} + \frac{3}{4}uv_x \tag{3.13}$$

and its Lax pairs

$$L\varphi = \lambda\varphi, \varphi_{t_0} = \frac{3}{8}u_x\varphi + \frac{3}{4}u\varphi_x \tag{3.14}$$

Remark 1. (1) Condition (3.7) generates the Bagmann system for the 4th-order eigenvalue problem (1.1);

(2) Condition (3.8) generates the C.neumann system for the 4th-order eigenvalue problem (1.1).

3.2 Jacobi-Ostrogradsky Coordinates and Hamilton Canonical Forms of the Bargmann System

Now, suppose $\{\lambda_j, \varphi_j\}(j = 1, 2, \cdots, N)$ are eigenvalues and eigenfunctions for the 4th-order eigenvalue problem (1.1), and $\lambda_1 < \lambda_2 < \cdots < \lambda_N$, then

$$L\Phi = (\partial^4 + \partial u \partial + v)\Phi = \Lambda\Phi \tag{3.15}$$

here $\Lambda = diag(\lambda_1, \lambda_2, \cdots, \lambda_N)$, $\Phi = diag(\varphi_1, \varphi_2, \cdots, \varphi_N)^T$.

From (3.6), we have

$$K\left(\sum_{j=1}^{N} \lambda_j^k \nabla \lambda_j\right) = J\left(\sum_{j=1}^{N} \lambda_j^{k+1} \nabla \lambda_j\right), \quad k = 0, 1, 2, \cdots$$

so

$$K\left(-\langle \Lambda^k \Phi_x, \Phi_x \rangle \quad \langle \Lambda^k \Phi, \Phi \rangle\right)^T = J\left(-\langle \Lambda^{k+1} \Phi_x, \Phi_x \rangle \quad \langle \Lambda^{k+1} \Phi, \Phi \rangle\right)^T, \quad k = 0, 1, 2, \cdots \tag{3.16}$$

Set $G_0 = (4, 0)^T$, using (3.5), we have

$$G_1 = \left(-\tfrac{3}{4} u_{xx} - \tfrac{3}{8} u^2 + v \quad u\right)^T$$

Now we define the constrained system to the Bargmann system

$$G_1 = \left(-\tfrac{3}{4} u_{xx} - \tfrac{3}{8} u^2 + v \quad u\right)^T = \left(-\langle \Phi_x, \Phi_x \rangle \quad \langle \Phi, \Phi \rangle\right)^T \tag{3.17}$$

then we have the relation between the potential (u, v) and the eigenvector Φ as follows

$$(u \quad v)^T = \left(\langle \Phi, \Phi \rangle \quad \tfrac{3}{2}\langle \Phi, \Phi_{xx} \rangle + \tfrac{1}{2}\langle \Phi_x, \Phi_x \rangle + \tfrac{3}{8}\langle \Phi, \Phi \rangle^2\right)^T \tag{3.18}$$

From (3.5), (3.16), (3.17) and (3.18), we have

$$G_k = \left(-\langle \Lambda^{k-1}\Phi_x, \Phi_x \rangle \quad \langle \Lambda^{k-1}\Phi, \Phi \rangle\right)^T, \quad k = 1, 2 \cdots$$

Based on the constrained system (3.18), the eigenvalue problem

$$L\Phi = \Lambda\Phi \tag{3.19}$$

is equivalent to the following system

$$\Phi_{xxxx} + \langle \Phi, \Phi \rangle \Phi_{xx} + 2\langle \Phi, \Phi_x \rangle \Phi_x + (\tfrac{3}{2}\langle \Phi, \Phi_{xx} \rangle + \tfrac{1}{2}\langle \Phi_x, \Phi_x \rangle + \tfrac{3}{8}\langle \Phi, \Phi \rangle^2)\Phi = \Lambda\Phi \tag{3.20}$$

Remark 2. we call the equation system (3.20) to be the Bargmann system for the 4th-order eigenvalue problem (1.1).

In order to obtain the Hamilton canonical forms that is equivalent to the Bargmann system (3.20), the Lagrange function $\hat{I}$ [6][9] is defined as follows :

$$\hat{I} = \int_{\Omega} I dx \tag{3.21}$$

here

$$I = \tfrac{1}{2}\langle \Phi_{xx},\Phi_{xx}\rangle + \tfrac{1}{2}\langle \Phi,\Phi\rangle\langle \Phi,\Phi_{xx}\rangle + \tfrac{1}{4}\langle \Phi,\Phi_x\rangle^2 + \tfrac{3}{48}\langle \Phi,\Phi\rangle^3 - \tfrac{1}{2}\langle \Lambda\Phi,\Phi\rangle \qquad (3.22)$$

Theorem 5. The Bargmann system (3.20) for the 4th-order eigenvalue problem (1.1) is equivalent to the Euler-Lagrange equations[1][4]

$$\delta \hat{I}\big/\delta\Phi = 0 \qquad (3.23)$$

Set $y_1 = -\Phi, y_2 = \Phi_x$, and $h = \sum_{j=1}^{2}\langle y_{jx},z_j\rangle - I$, our aims is to find the coordinates z_1,z_2 and the Hamilton function h , that satisfy the following Hamilton canonical equations

$$y_{jx} = \{y_j,h\} = \partial h/\partial z_j\,, \quad z_{jx} = \{z_j,h\} = -\partial h/\partial y_j\,, \quad j=1,2 \qquad (3.24)$$

here, the symbol $\{\bullet,\bullet\}$ is the Poisson bracket in the symplectic $\left(\omega = \sum_{j=1}^{2} dy_j \wedge dz_j, R^{4N}\right)$, and the Poisson bracket of Hamilton functions F,H in the symplectic is defined as follows

$$\{F,H\} = \sum_{j=1}^{2}\sum_{k=1}^{N}\left(\frac{\partial F}{\partial y_{jk}}\frac{\partial H}{\partial z_{jk}} - \frac{\partial H}{\partial y_{jk}}\frac{\partial F}{\partial z_{jk}}\right) = \sum_{j=1}^{2}\left(\langle F_{y_j},H_{z_j}\rangle - \langle F_{z_j},H_{y_j}\rangle\right)$$

By $h = \sum_{j=1}^{2}\langle y_{jx},z_j\rangle - I$,then $dh = \sum_{j=1}^{2}(\langle y_{jx},dz_j\rangle + \langle z_j,dy_{jx}\rangle) - dI$. On other hand, $h = h(y_j,z_j),(j=1,2)$, so

$$dh = \sum_{j=1}^{2}\left(\langle \partial h/\partial y_j,dy_j\rangle + \langle \partial h/\partial z_j,dz_j\rangle\right) = \sum_{j=1}^{2}\left(-\langle z_{jx},dy_j\rangle + \langle y_{jx},dz_j\rangle\right)$$

so that

$$dI = \langle z_1,dy_{1x}\rangle + \langle z_2,dy_{2x}\rangle + \langle z_{1x},dy_1\rangle + \langle z_{2x},dy_2\rangle = -\langle z_1,d\Phi_x\rangle + \langle z_2,d\Phi_{xx}\rangle - \langle z_{1x},d\Phi\rangle + \langle z_{2x},d\Phi_x\rangle$$
$$= \langle z_{1x},d\Phi\rangle + \langle z_{2x} - z_1,d\Phi_x\rangle + \langle z_2,d\Phi_{xx}\rangle$$

we have the relations

$$z_1 = \Phi_{xxx} + \tfrac{1}{2}\Phi_x\langle \Phi,\Phi\rangle + \tfrac{1}{2}\Phi\langle \Phi,\Phi_x\rangle = \Phi_{xxx} + 1/2\,\Phi_x u + 1/2\,\Phi u_x$$
$$z_2 = \Phi_{xx} + \tfrac{1}{2}\Phi\langle \Phi,\Phi\rangle = \Phi_{xx} + \tfrac{1}{2}\Phi u$$

Theorem 6. The Jacobi-Ostrogradsky coordinates are as follows

$$y_1 = -\Phi,\quad y_2 = \Phi,\quad z_1 = \Phi_{xxx} + \tfrac{1}{2}u\Phi_x + \tfrac{1}{4}u_x\Phi_x,\quad z_2 = \Phi_{xx} + \tfrac{1}{2}u\Phi \qquad (3.25)$$

and the Bargmann system (3.20) for the 4th-order eigenvalue problem (1.1) is equivalent to the Hamilton canonical system

$$y_{jx} = \{y_j,h\} = \partial h/\partial z_j\,, \quad z_{jx} = \{z_j,h\} = -\partial h/\partial y_j\,, \quad j=1,2 \qquad (3.26)$$

here, the Hamilton function h is

$$h = \tfrac{1}{2}\langle \Lambda y_1,y_1\rangle - \langle y_2,z_1\rangle + \tfrac{1}{2}\langle z_2,z_2\rangle - \tfrac{1}{4}\langle y_1,y_2\rangle^2 + \tfrac{1}{2}\langle y_1,y_1\rangle\langle y_1,y_2\rangle + \tfrac{1}{16}\langle y_1,y_1\rangle^3 \qquad (3.27)$$

Remark 3. Based on the Jacobi-Ostrogradsky coordinate system (3.25), the Bargmann constrained equations associated with the 4th-order eigenvalue problem (1.1) is

$$u = \langle y_1, y_1 \rangle, v = -\tfrac{3}{2}\langle y_1, z_2 \rangle + \tfrac{1}{2}\langle y_2, y_2 \rangle - \tfrac{3}{8}\langle y_1, y_1 \rangle^2 \tag{3.28}$$

3.3   Hamilton Equation of Bargmann System

Theorem 7. (1) The Bargmann system (3.20) for the 4th-order eigenvalue problem (1.1) is equivalent to

$$Y_x = MY \tag{3.29}$$

there

$$Y = (y_1, y_2, y_3, y_4)^T = (-\Phi, \Phi_x, \Phi_{xx} + \tfrac{1}{2}u\Phi, \Phi_{xxx} + \tfrac{1}{2}u\Phi_x + \tfrac{1}{4}u_x\Phi)^T \tag{3.30}$$

$$M = \begin{pmatrix} 0 & -E & 0 & 0 \\ \tfrac{1}{2}uE & 0 & E & 0 \\ -\tfrac{1}{4}u_xE & 0 & 0 & E \\ -\Lambda + (v - \tfrac{1}{4}u_{xx} - \tfrac{1}{4}u^2)E & -\tfrac{1}{4}u_xE & -\tfrac{1}{2}uE & 0 \end{pmatrix} \tag{3.31}$$

(2) The system of the Lax pairs for the evolution equation hierarchy (3.8) is equivalent to

$$Y_{t_m} = W_m Y \qquad m = 0,1,2,\cdots \tag{3.32}$$

there

$$W_m = (\omega_{ij}^m)_{4\times4} \qquad m = 0,1,2\cdots \tag{3.33}$$

$$\omega_{11}^m = \sum_{j=0}^{m}(\tfrac{1}{8}b_{j-1xxx} + \tfrac{1}{16}ub_{j-1x} + \tfrac{1}{16}u_xb_{j-1} + \tfrac{3}{8}a_{j-1x})\Lambda^{m-j},$$

$$\omega_{12}^m = \sum_{j=0}^{m}(\tfrac{1}{8}b_{j-1xx} + \tfrac{1}{8}ub_{j-1} + \tfrac{1}{4}a_{j-1})\Lambda^{m-j}, \quad \omega_{13}^m = \sum_{j=0}^{m}(-\tfrac{1}{8}b_{j-1x})\Lambda^{m-j}, \quad \omega_{14}^m = \sum_{j=0}^{m}(\tfrac{1}{4}b_{j-1})\Lambda^{m-j},$$

$$\omega_{21}^m = \sum_{j=0}^{m}(-\tfrac{1}{8}b_{j-1xxxx} - \tfrac{1}{8}ub_{j-1xx} - \tfrac{5}{32}u_xb_{j-1x} - \tfrac{1}{4}vb_{j-1} - \tfrac{3}{8}a_{j-1xx} - \tfrac{1}{8}ua_{j-1})\Lambda^{m-j},$$

$$\omega_{22}^m = \sum_{j=0}^{m}(-\tfrac{1}{16}ub_{j-1x} + \tfrac{1}{8}a_{j-1x})\Lambda^{m-j}, \quad \omega_{23}^m = \sum_{j=0}^{m}(-\tfrac{1}{4}a_{j-1})\Lambda^{m-j}, \quad \omega_{24}^m = \sum_{j=0}^{m}(-\tfrac{1}{8}b_{j-1x})\Lambda^{m-j},$$

$$\omega_{31}^m = \sum_{j=0}^{m}[\tfrac{1}{16}b_{j-1xxxx} + \tfrac{3}{32}u_xb_{j-1x} + (-\tfrac{1}{32}u^2 + \tfrac{1}{8}v + \tfrac{1}{16}u_{xx})b_{j-1x}$$

$$+ (\tfrac{1}{8}v_x - \tfrac{1}{32}uu_x)b_{j-1} + \tfrac{1}{4}a_{j-1xxx} + \tfrac{1}{16}u_xa_{j-1}]\Lambda^{m-j} - \sum_{j=0}^{m}(\tfrac{1}{8}b_{j-1x})\Lambda^{m-j+1},$$

$$\omega_{32}^m = \sum_{j=0}^{m}[\tfrac{1}{8}b_{j-1xxx} + \tfrac{1}{8}u_xb_{j-1x} + (\tfrac{1}{4}v - \tfrac{1}{16}u^2)b_{j-1} + \tfrac{1}{2}a_{j-1xx}]\Lambda^{m-j} - \sum_{j=0}^{m}(\tfrac{1}{4}b_{j-1})\Lambda^{m-j+1},$$

$$\omega_{33}^m = \sum_{j=0}^{m}(\tfrac{1}{16}ub_{j-1x} - \tfrac{1}{8}a_{j-1x})\Lambda^{m-j}, \quad \omega_{34}^m = (-\tfrac{1}{8}b_{j-1xx} - \tfrac{1}{8}ub_{j-1} - \tfrac{1}{4}a_{j-1})\Lambda^{m-j},$$

$$\omega_{41}{}^{m} = \sum_{j=0}^{m} [-\frac{1}{32}ub_{j-1xxxxx} - \frac{1}{32}ub_{j-1xxx} - \frac{5}{32}u_{x}b_{j-1xx} - (\frac{3}{32}u_{xx} + \frac{1}{4}v)b_{j-1xx}$$

$$-(\frac{1}{32}uu_{x} + \frac{3}{16}v_{x} + \frac{1}{32}u_{xxx})b_{j-1x} - (\frac{1}{16}v_{xx} + \frac{1}{64}u_{x}{}^{2})b_{j-1} - \frac{1}{16}a_{j-1xxx}$$

$$+\frac{1}{8}ua_{j-1xx} + (\frac{1}{16}u^{2} - \frac{1}{4}v + \frac{1}{16}u_{xx})a_{j-1}]\Lambda^{m-j} + \sum_{j=0}^{m}(\frac{1}{4}b_{j-1xx} + \frac{1}{8}a_{j-1})\Lambda^{m-j+1},$$

$$\omega_{42}{}^{m} = \sum_{j=0}^{m} [\frac{1}{16}b_{j-1xxxx} + \frac{3}{32}u_{x}b_{j-1xx} + (-\frac{1}{32}u^{2} + \frac{1}{8}v + \frac{1}{16}u_{xx})b_{j-1x}$$

$$+(\frac{1}{8}v_{x} - \frac{1}{32}uu_{x})b_{j-1} + \frac{1}{4}a_{j-1xxx} + \frac{1}{16}u_{x}a_{j-1}]\Lambda^{m-j} - \sum_{j=0}^{m}(\frac{1}{8}b_{j-1x})\Lambda^{m-j+1},$$

$$\omega_{43}{}^{m} = \sum_{j=0}^{m} [\frac{1}{8}b_{j-1xxxx} + \frac{1}{8}ub_{j-1xx} + \frac{5}{32}u_{x}b_{j-1x} + \frac{1}{4}vb_{j-1} + \frac{3}{8}a_{j-1xx} + \frac{1}{8}ua_{j-1}]\Lambda^{m-j} - \sum_{j=0}^{m}(\frac{1}{4}b_{j-1})\Lambda^{m-j+1},$$

$$\omega_{44}{}^{m} = \sum_{j=0}^{m} [-\frac{1}{8}b_{j-1xx} - \frac{1}{16}ub_{j-1x} - \frac{1}{16}u_{x}b_{j-1} - \frac{3}{8}a_{j-1x}]\Lambda^{m-j}.$$

Now, substituting (3.17)-(3.19) into (3.29) and (3.30), then

$$Y_{x} = MY \quad Y_{t_{m}} = W_{m}Y \qquad m = 0,1,2,\cdots \tag{3.34}$$

there

$$M = \begin{pmatrix} 0 & -E & 0 & 0 \\ \frac{1}{2}\langle y_{1}, y_{1} \rangle E & 0 & E & 0 \\ \frac{1}{2}\langle y_{1}, y_{2} \rangle E & 0 & 0 & E \\ -\Lambda - \langle y_{1}, z_{2} \rangle - \frac{3}{8}\langle y_{1}, y_{1} \rangle^{2} E & \frac{1}{2}\langle y_{1}, y_{2} \rangle E & -\frac{1}{2}\langle y_{1}, y_{1} \rangle E & 0 \end{pmatrix} \tag{3.35}$$

and

$$W_{m} = (\omega_{ij}{}^{m})_{4\times4} \qquad m = 0,1,2\cdots \tag{3.36}$$

$$\omega_{11}{}^{m} = \sum_{j=1}^{m}(-\frac{1}{4}\langle \Lambda^{j-1}y_{1}, z_{1} \rangle)\Lambda^{m-j}, \quad \omega_{12}{}^{m} = -\sum_{j=1}^{m}(\frac{1}{4}\langle \Lambda^{j-1}y_{1}, z_{2} \rangle)\Lambda^{m-j} + \Lambda^{m},$$

$$\omega_{13}{}^{m} = \sum_{j=1}^{m}(\frac{1}{4}\langle \Lambda^{j-1}y_{1}, y_{2} \rangle)\Lambda^{m-j}, \quad \omega_{14}{}^{m} = \sum_{j=1}^{m}(\frac{1}{4}\langle \Lambda^{j-1}y_{1}, y_{1} \rangle)\Lambda^{m-j},$$

$$\omega_{21}{}^{m} = -\sum_{j=1}^{m}(\frac{1}{4}\langle \Lambda^{j-1}y_{2}, z_{1} \rangle)\Lambda^{m-j} - \frac{1}{2}\langle y_{1}, y_{1} \rangle \Lambda^{m}, \quad \omega_{22}{}^{m} = -\sum_{j=1}^{m}(\frac{1}{4}\langle \Lambda^{j-1}y_{2}, z_{2} \rangle)\Lambda^{m-j},$$

$$\omega_{23}{}^{m} = \sum_{j=1}^{m}(\frac{1}{4}\langle \Lambda^{j-1}y_{2}, y_{2} \rangle)\Lambda^{m-j} - \Lambda^{m}, \quad \omega_{24}{}^{m} = \sum_{j=1}^{m}(\frac{1}{4}\langle \Lambda^{j-1}y_{2}, y_{1} \rangle)\Lambda^{m-j},$$

$$\omega_{31}{}^{m} = -\sum_{j=1}^{m}(\frac{1}{4}\langle \Lambda^{j-1}z_{2}, z_{1} \rangle)\Lambda^{m-j} - \frac{1}{2}\langle y_{1}, y_{2} \rangle \Lambda^{m}, \quad \omega_{32}{}^{m} = -\sum_{j=1}^{m}(\frac{1}{4}\langle \Lambda^{j-1}z_{2}, z_{2} \rangle)\Lambda^{m-j},$$

$$\omega_{33}{}^{m} = \sum_{j=1}^{m}(\frac{1}{4}\langle \Lambda^{j-1}y_{2}, z_{2} \rangle)\Lambda^{m-j}, \quad \omega_{34}{}^{m} = \sum_{j=1}^{m}(\frac{1}{4}\langle \Lambda^{j-1}y_{1}, z_{2} \rangle)\Lambda^{m-j} - \Lambda^{m},$$

$$\omega_{41}{}^{m} = -\sum_{j=1}^{m}(\frac{1}{4}\langle \Lambda^{j-1}z_{1}, z_{1} \rangle)\Lambda^{m-j} + \Lambda^{m+1} + \langle y_{1}, z_{2} \rangle \Lambda^{m} + \frac{3}{8}\langle y_{1}, y_{1} \rangle^{2} \Lambda^{m},$$

$$\omega_{42}{}^{m} = -\sum_{j=1}^{m}(\frac{1}{4}\langle \Lambda^{j-1}z_{2}, z_{1} \rangle)\Lambda^{m-j} - \frac{1}{2}\langle y_{1}, y_{2} \rangle \Lambda^{m},$$

$$\omega_{43}{}^m = \sum_{j=1}^{m} (\tfrac{1}{4}\langle \Lambda^{j-1}y_2, z_1 \rangle)\Lambda^{m-j} + \tfrac{1}{2}\langle y_1, y_1 \rangle \Lambda^m, \quad \omega_{44}{}^m = \sum_{j=1}^{m} (\tfrac{1}{4}\langle \Lambda^{j-1}y_1, z_1 \rangle)\Lambda^{m-j}.$$

So that, the following theorem holds.

Theorem 8. On the Bargmann constrained equation (3.28), the evolution equation hierarchy (3.09) of the 4th-order eigenvalue problem (1.1) are nonlinearized as following Hamilton canonical equation system

$$Y_x = \partial h/\partial Z, Y_{t_m} = \partial h_m/\partial Z, Z_x = \partial h/\partial Y, Z_{t_m} = -\partial h_m/\partial Y, \quad m = 0,1,2,\cdots$$

there

$$h_m = -\tfrac{1}{2}\langle \Lambda^{m+1}y_1, y_1 \rangle + \langle \Lambda^m y_2, z_1 \rangle + \tfrac{1}{4}\langle \Lambda^m y_1, y_2 \rangle^2 - \tfrac{1}{2}\langle y_1, y_1 \rangle \langle \Lambda^m y_1, z_2 \rangle$$

$$- \tfrac{1}{16}\langle \Lambda^m y_1, y_1 \rangle^3 - \tfrac{1}{2}\langle \Lambda^m z_2, z_2 \rangle + \tfrac{1}{8}\sum_{j=1}^{m} (\langle \Lambda^{m-j}z_1, z_1 \rangle \langle \Lambda^{j-1}y_1, y_1 \rangle$$

$$+ \langle \Lambda^{m-j}z_2, z_2 \rangle \langle \Lambda^{j-1}y_2, y_2 \rangle - \langle \Lambda^{m-j}y_1, z_1 \rangle \langle \Lambda^{j-1}y_1, z_1 \rangle - \langle \Lambda^{m-j}y_2, z_2 \rangle \langle \Lambda^{j-1}y_2, z_2 \rangle)$$

$$+ \tfrac{1}{4}\sum_{j=1}^{m} (\langle \Lambda^{m-j}z_2, z_1 \rangle \langle \Lambda^{j-1}y_1, y_2 \rangle - \langle \Lambda^{m-j}y_2, z_1 \rangle \langle \Lambda^{j-1}y_1, z_2 \rangle).$$

References

1. Cao, C.W., Geng, X.G.: Research Reports in Physics. In: Nonlinear Physics. Springer, Heidelberg (1990)
2. Cao, C.W.: A Classica Integrable Systems and the Involutive of Solutions of the KDV Equation. Acta Math. Sinica, New Series 7, 5–15 (1991)
3. Ma, W.X.: Binary Nonlinearization of Spectrial Problems of the Perturbation AKNS Systems. Chaos Solitions & Fractals B, 1451–1463 (2002)
4. Gu, Z.Q.: Two New Completely Integrable Systems Related to the KDV Equation Hierarchy. IL Nuove Cimento 123B(5), 605–622 (2008)
5. Gu, Z.Q.: A New Constrained Flow for Boussinesq-Burgers' Hierarchy. IL Nuove Cimento 122B(8), 871–884 (2007)
6. Huang, X., Bai, Z.J., Su, Y.F.: Nonlinear Rank-one Modification of the Symmetric Eigenvalue Problems. Journal of Computational Mathematics 28(2), 218–234 (2010)
7. Adler, M.: On a trace functional for formal pseudo-Differential operators and the symplectic structure of the KDV type equations. Inventions Mathematical, 19–248 (1979)
8. Wang, D.S.: Complete Integrability and the Miura Transformation of a Coupled KDV Equation. Appl. Math. Lett. 03, 665–669 (2010)
9. Gu, Z.Q.: The Neumann system for the 3rd-order eigenvalue problems related to the Boudssinesq equation. IL Nuovc Cimento 117B(6), 615–632 (2002)
10. Gu, Z.Q.: Complex confocal involution system associated with the solutions of the AKNS evolution equations. J. Math. Phys. 32(6), 1498–1504 (1991)
11. Wang, D.S., Zhang, D.J., Yang, J.K.: Integrable Properities of the General Completed Nonlinear Schodinger Equations. J. Math. Phys. 51, 023540 (2010)
12. Zeng, X., Wang, D.S.: A Generalized Extended Rational Expansion Method and its Application to (1+1)-dimensional Dispersive Long Wave Equation. Appl. Math. Comput. 212, 296–304 (2009)
13. Cao, Y.H., Wang, D.S.: Prolongation Structures of a Generelized Coupled Korteweg-de Vries Equation and Miura Transformation. Commun. in Nonl. Sci. and Num. Simul. 15, 2344–2349 (2010)

Numerical Analysis of Interface Crack Propagation in Reinforced Concrete Beams Strengthened with FRP by Extended Isoparametric Finite Element Method

Hong Wang[1,2], Hong Yuan[1], and Shuai Yang[2]

[1] Moe Key Lab of Disaster Forecast and Control in Engineering,
Institute of Applied Mechanics, Jinan University,
Guangzhou, China
[2] College of Civil and Transportation Engineering,
Guangdong University of Technology,
Guangzhou, China
{wanghong.cn,s_yang54}@163.com,
tyuanhong@jnu.edu.cn

Abstract. Extended isoparametric finite method (XIFEM) consists in enriching the basis of the classical finite element method and taking into account the discontinuity of the displacement field across the crack by a discontinuous function along the crack line. It simulates the discontinuous character resulted from discontinuity such as crack or joint and by some trigonometric basis functions around the crack tip to embody singularity at the end of discontinuity. With the improved XIFEM, the tracking of crack propagation in reinforced concrete beams strengthened with FRP is simulated and the failure model is analyzed. Compared with the traditional finite element method, the XIFEM allows crack surface to be in any position of finite element mesh without dense mesh near the crack tips and without re-meshing, therefore crack growth is traced and modeled effectively. The results show the effectiveness and superiority of the improved XIFEM.

Keywords: Extended isoparametric finite element, FRP, interface crack propagation, discontinuous function.

1 Introduction

Interface Crack Propagation in FRP-to-Concrete is difficult to simulate [1-2]. Based on the displacement model of traditional finite element, the enriched discontinuous interpolating function is introduced and the crack propagation can be modeled without re-meshing to embody the discontinuities of displacement at crack edge. At present, the enriched generalized Heaviside function has been adopted to model the displacement jump at the crack in XFEM [3-4]. But the jump function is not fine enough. In fact, the effect of crack on different point is also not same which is weak with the distance from the crack. So the exponent discontinuous function is adopted to embody the discontinuous character resulted from the discontinuities [5-8].

B. Liu and C. Chai (Eds.): ICICA 2011, LNCS 7030, pp. 169–176, 2011.

The principle of XFEM consists in enriching the basis of the classical finite element method and taking into account the discontinuity of the displacement field across the crack by a discontinuous function along the crack line to simulate the discontinuous character resulted from discontinuity such as crack or joint and by some trigonometric basis functions around the crack tip to embody singularity at the end of discontinuity [9-10]. In addition, the application of XFEM only involves the numerical simulation of single material contained in element. For the simulation of interface crack in which the extended element contains two different materials has involved at present. Finally, the improved XFEM will be adopted to model the development of crack in the FRP-to-concrete beam and the failure mode will be discovered subsequently [11].

2   Displacement Function

For plane problem, the extended finite element function is defined as follows[12]

$$
\left\{ \begin{matrix} u_x \\ u_y \end{matrix} \right\} = \sum_{i=1}^{4} N_i(x) \left(\left\{ \begin{matrix} u_x^i \\ u_y^i \end{matrix} \right\} + \varphi_j(x) \left\{ \begin{matrix} a_x^j \\ a_y^j \end{matrix} \right\} \right)
\tag{1}
$$

Where i is the node of element, j the enrichment degree of freedom of those node, and $\varphi(x)$ the enrichment displacement function, here the exponent discontinuous function is adopted as follows

$$
\phi_i(x) = h(d(x))
\tag{2}
$$

The introduction of the sign function in $d(x)$ is to distinguish those points with the same distance located at the two sides of crack. For those elements with interface, the exponent discontinuous function is modified as follows in terms of the rigidity ratio of the material distributing at the two sides of interface crack.

$$
\phi_i(x) = h(d(x)) = \begin{cases} \eta e^{-d(x)} & d(x) > 0 \\ -e^{d(x)} & d(x) < 0 \end{cases}
\tag{3a}
$$

Otherwise,

$$
\phi_i(x) = h(d(x)) = \begin{cases} e^{-d(x)} & d(x) > 0 \\ -\eta e^{d(x)} & d(x) < 0 \end{cases}
\tag{3b}
$$

Where η is the rigidity ratio defined as $\eta = \dfrac{\|E_1 - E_2\|}{\max(E_1, E_2)}$, and E_1, E_2 are corresponding respectively to the elastic module of the two materials. The modified exponent discontinuous function is presented as fig.1.

Where $d(x)$ is the minimum distance of the point x to the crack. If the crack curve equation is $f(x) = 0$, then the $d(x)$ can be defined as follows

$$d(x) = \min_{\overline{x} \in f(x)=0} \|x - \overline{x}\| sign(f(x) \cdot f(x_i))$$ (4)

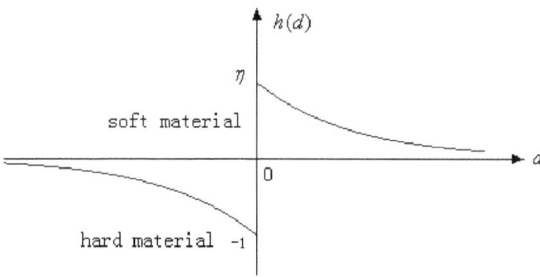

Fig. 1. The enrichment function of interface crack

For the element containing the crack tip, the displacement functions which are I type crack and II type crack problem are presented respectively can be chosen as follows in terms of the crack tip displacement field function.

$$u_x = \frac{K_I}{2\mu}\sqrt{\frac{r}{2\pi}}\cos\frac{\theta}{2}(k-1+2\sin^2\frac{\theta}{2}) + \frac{K_{II}}{2\mu}\sqrt{\frac{r}{2\pi}}\sin\frac{\theta}{2}(k+1+2\cos^2\frac{\theta}{2})$$

$$u_y = \frac{K_I}{2\mu}\sqrt{\frac{r}{2\pi}}\sin\frac{\theta}{2}(k+1-2\cos^2\frac{\theta}{2}) + \frac{K_{II}}{2\mu}\sqrt{\frac{r}{2\pi}}\cos\frac{\theta}{2}(k-1-2\sin^2\frac{\theta}{2})$$ (5)

Where, $k = \begin{cases} 3-4\mu & \text{plane strain} \\ \dfrac{3-4\mu}{1+\mu} & \text{plane stress} \end{cases}$

3   Extended Isoparametric Finite Element Method

The whole stiffness equation can be deduced as follows in term of the minimum potential energy principal [13].

$$\begin{pmatrix} K_{uu} & K_{ua} & K_{uI} \\ K_{au} & K_{aa} & 0 \\ K_{Iu} & 0 & K_{I\cdot I} \end{pmatrix} \begin{Bmatrix} u \\ a \\ K_{I\cdot II} \end{Bmatrix} = \begin{Bmatrix} f^{ext} \\ q^{ext} \\ fq^{ext} \end{Bmatrix}$$ (6)

Where, $K_{uu} = \sum_e g_u^T K_{uu}^e g_u$, $K_{ua} = \sum_e g_u^T K_{ua}^e g_a$,

$$K_{aa} = \sum_e g_a^T K_{aa}^e g_a \quad K_{ul} = \sum_e g_u^T K_{ul}^e g_l \quad ,$$

$$K_{l \cdot l} = \sum_e g_l^T K_{l \cdot l}^e g_l \quad ,$$

$$f^{ext} = \sum_e \left(\int_{\Omega_e} N^T b d\Omega + \int_{\Gamma_t^{te}} N^T \bar{t} d\Gamma \right)$$

$$q^{ext} = \sum_e \left(\int_{\Omega_e} (N\phi)^T b d\Omega + \int_{\Gamma_t^{te}} (N\phi)^T \bar{t} d\Gamma \right) ,$$

$$fq^{ext} = \sum_{e \in tip} \left(\int_{\Omega^e} \left(N \begin{bmatrix} f_{11} & f_{12} \\ f_{21} & f_{22} \end{bmatrix} \right)^T b d\Omega + \int_{\Gamma_t^e} \left(N \begin{bmatrix} f_{11} & f_{12} \\ f_{21} & f_{22} \end{bmatrix} \right)^T \bar{t} d\Gamma \right) .$$

The enrichment function of interface crack is simulated by extended isoparametric finite element. The displacement function for four nodes is

$$N_i = \frac{1}{4}(1 + \xi_0)(1 + \eta_0) \tag{7}$$

where, $\xi_0 = \xi_i \xi, \eta_0 = \eta_i \eta, \varsigma_0 = \varsigma_i \varsigma$ ($i = 1,2,3,4$).

If the crack curve equation is $f(x, y) = 0$, then the $d(x)$ can be defined as follows

$$d(x) = \sqrt{(x - \bar{x})^2 + (y - \bar{y})^2} sign(f(x, y) \cdot f(x_i, y_i)) \tag{8}$$

$$x = \sum_i N_i x_i , y = \sum_i N_i y_i \ (i = 1,2,3,4)$$

Then,

$$\begin{cases} f(\bar{x}, \bar{y}) = 0 \\ (y - \bar{y}) \dfrac{\partial f}{\partial x} \Big|_{(\bar{x}, \bar{y})} = (x - \bar{x}) \dfrac{\partial f}{\partial y} \Big|_{(\bar{x}, \bar{y})} \ (i = 1,2,3,4) \end{cases} \tag{9}$$

If $x(\xi, \eta)$ is on the crack curve, then $f(x, y) = 0$, and $d(x) = 0$.

Sitting mark partial derivative for the enrichment displacement function, it can be obtained

$$\begin{Bmatrix} \dfrac{\partial\phi}{\partial x} \\[2mm] \dfrac{\partial\phi}{\partial y} \end{Bmatrix} = \dfrac{1}{\det J} \begin{bmatrix} \displaystyle\sum_i \dfrac{\partial N_i}{\partial\eta} y_i & -\displaystyle\sum_i \dfrac{\partial N_i}{\partial\xi} y_i \\[4mm] -\displaystyle\sum_i \dfrac{\partial N_i}{\partial\eta} x_i & \displaystyle\sum_i \dfrac{\partial N_i}{\partial\xi} x_i \end{bmatrix} \begin{Bmatrix} \dfrac{\partial\phi}{\partial\xi} \\[2mm] \dfrac{\partial\phi}{\partial\eta} \end{Bmatrix} \quad (i=1,2,3,4) \qquad (10)$$

where, $J = \begin{bmatrix} \displaystyle\sum_i \dfrac{\partial N_i}{\partial\xi} x_i & \displaystyle\sum_i \dfrac{\partial N_i}{\partial\xi} y_i \\[4mm] \displaystyle\sum_i \dfrac{\partial N_i}{\partial\eta} x_i & \displaystyle\sum_i \dfrac{\partial N_i}{\partial\eta} y_i \end{bmatrix}$ ..

4   Example

The extended isoparametric finite element with the modified enriched exponent discontinuous function and the enriched triangle function modeling the crack tip are used to simulate the extension of the interface crack in concrete beam with FRP gusset thin plate. There is a crack with 44cm long at the bottom centre of the beam with 10m long, 60cm high and 30cm thick presented as Fig3. There are steels ($4\phi12$) lay on the bottom of the beam. The elastic modulus of concrete is 32.5GPa, Poisson's ratio 0.2, fracture toughness 2.395MPa $\sqrt{m}$, tension strength 2.4MPa and shear strength 3.5MPa. For steel, the elastic modulus is 210GPa, Poisson's ratio 0.3. For FRP material, the elastic modulus is 235GPa, Poisson's ratio 0.23, fracture toughness 30MPa $\sqrt{m}$ and tension strength 3.35GPa. The fracture toughness of the bond between FRP and concrete beam is 4MPa $\sqrt{m}$ and gusset strength 20MPa. The simulated results are presented as table1 and Fig.2-3.

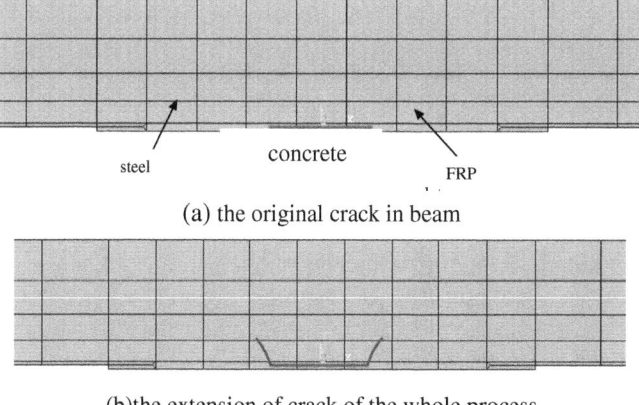

(a) the original crack in beam

(b)the extension of crack of the whole process

Fig. 2. The schematic diagram of crack propagation of beam

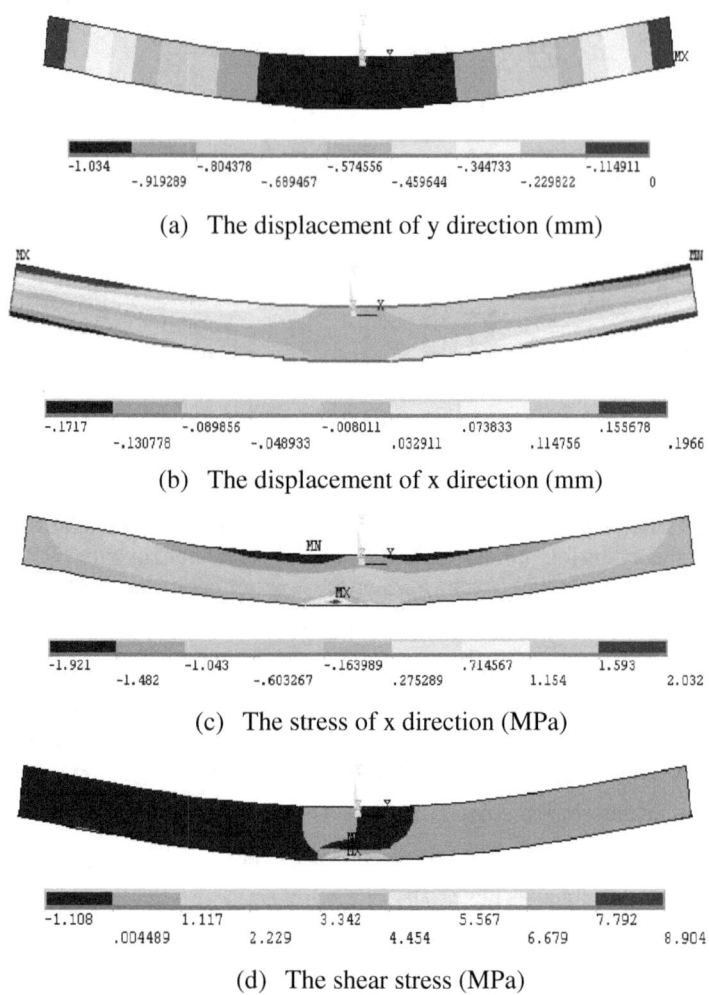

(a) The displacement of y direction (mm)

(b) The displacement of x direction (mm)

(c) The stress of x direction (MPa)

(d) The shear stress (MPa)

Fig. 3. The displacement and stress distribution for forth load step (suffer from uniform pressure with 0.01kN/cm at the top beam basing on the 3rd step)

The gusset strength of the bond between the FRP thin plate and concrete beam is bigger than the strength of concrete material, so, after reinforcing with FRP thin plate, the interface crack should be developed along the inner of concrete, and results in desquamation failure. From the development diagram of crack presented as Fig.3, the nearer crack stretch the end of beam, the more apparent what's the action of shear force, which agrees with the practice. In order to show the effectively of XIFEM, The displacement and stress distribution for first load step with FEM is done as Fig.4.

(a) The displacement of y direction (m)

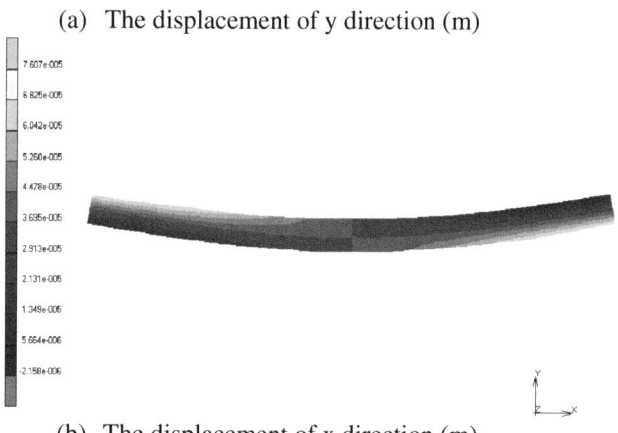

(b) The displacement of x direction (m)

Fig. 4. The displacement and stress distribution for first load step with FEM

Compared with the traditional finite element method, the XFEM allows crack surface to be in any position of finite element mesh without dense mesh near the crack tips and without re-meshing, therefore crack growth is traced and modeled effectively, which are shown in Table1.

Table 1. The results comparison of FEM and XFEM

		FEM	XFEM
y	Max	-0.377	-0.418
(mm)	Min	0.000	0.000
x	Max	0.76	0.079
(mm)	Min	-0.021	0.000
transverse	Max	1.602	0.609
stress (Mp)	Min	-0.383	-0.824

5   Conclusions

This paper analyzed crack propagation and failure mode in a FRP-to-concrete beam basing on the improved XFEM. Considering that hardness material in one side of interface had little influence, this article introduced a concept of the ratio of rigidity and modified the enrichment function properly to make it incorporate the different reactions of the materials of two sides in response to the interface cracks in the units. With the improved XFEM, the tracking of crack propagation in steel reinforced concrete beams is simulated and the failure model is analyzed. The results show that the XFEM can model crack propagation well.

References

1. Chen, J.F., Yuan, H., Teng, J.G.: Debonding failure along a softening FRP-to-concrete interface between two adjacent cracks in concrete members. Engineering Structures 29(2), 259–270 (2007)
2. Coronado, Lopez, C.A., Maria, M.: Numerical modeling of concrete-FRP debonding using a crack band approach. Journal of Composites for Construction 14(1), 11–21 (2010)
3. Iarve, E.V.: Mesh independent modeling of cracks by using higher order shape functions. Int. J. Numer. Meth. Eng. 56, 869–882 (2003)
4. Stazi, F.L., Budyn, E., Chessa, J., Belytschko, T.: An extended finite element method with higher-order elements for curved cracks. Comput. Mech. 31(38-48), 38 (2003)
5. Chessa, J., Wang, H., Belytschko, T.: On the construction of blending elements for local partition of unity enriched finite elements. International Journal for Numerical Method in Engineering 57(7), 1015–1038 (2003)
6. Hansbo, A., Hansbo, P.: A finite element method for the simulation of strong and weak discontinuities in solid mechanics. Comput. Meth. Appl. Mech. Eng. 193, 3523–3540 (2004)
7. Godat, A., Labossière, P., Neale, K.W.: Numerical modeling of shear crack angles in FRP shear-strengthened reinforced concrete beams. Australian Journal of Structural Engineering 11(2), 87–101 (2010)
8. Belytschko, T., Moes, N., Usui, S., Parimi, C.: Arbitrary discontinuities in finite elements. International Journal for Numerical Methods in Engineering 50, 993–1013 (2001)
9. Xia, X.-z., Zhang, Q.: Discontinuous finite element method for simulation of discontinuities. Journal of Hohai University 33(6), 682–687 (2006) (in Chinese)
10. Khoei, A.R., Biabanaki, S.O.R., Anahid, M.: Extended finite element method for three-dimensional large plasticity deformations on arbitrary interfaces. Computer Methods in Applied Mechanics and Engineering 197(9-12), 1100–1114 (2008)
11. Zhou, J.-M., Qi, L.-H.: Treatment of discontinuous interface in liquid-solid forming with extended finite element method. Transactions of Nonferrous Metals Society of China (English Edition) 20(suppl. 3), s911–s915 (2010)
12. Liu, Borja, F., Ronaldo, I.: Finite deformation formulation for embedded frictional crack with the extended finite element method. International Journal for Numerical Methods in Engineering 82(6), 773–804 (2010)
13. Yang, S., Wang, H., Xia, X.-z., Yuan, H.: The Numerical Simulation of Crack Propagation in a FRP-to-concrete Beam without Remeshing. In: ICIECS 2009, vol. (3), pp. 1810–1812 (2009)

An Algorithm of Improving the Consistence
of the Positive Reciprocal Matrix Based on Relative Error

Qiuhong Fan[1], Baoxiang Liu[1], Yuhuan Zhang[2], and Ruiying Zhou[2]

[1] College of Science, Hebei United University, 063009 Tangshan, Hebei, China
[2] College of Light Industry, Hebei United University, 063000 Tangshan, Hebei, China
qhf1026@163.com

Abstract. By analyzing the inconsistent relationship between the small triangular matrix and positive reciprocal matrix, a convenient correction method based on the relative error is brought up. This method can fully retain the effective information of original positive reciprocal matrix. It helps people to solve practical problems effectively and enriches the theories and methods of decision analysis.

Keywords: positive reciprocal matrix, consistency examination, ER correction method, small triangular matrix.

1 Introduction

In view of the integer characteristics of 1-9 scale system, that decimals or the fractional number case that numerators are not one when adjusting positive reciprocal matrix does not meet the psychological of judges obviously.[1,2,3] In this regard, we can get a calculation algorithm with a high degree of clarity which not only can achieve the purpose of adjustment but does not need complex calculation through careful comparison and repeated experiments.[4] We adjust the maximum relative error of corresponding elements between the elements in positive reciprocal matrix A and complete consistency matrix W to improve its consistency reasonably under the condition of retaining fully the effective information of original positive reciprocal matrix. The above-mentioned method is called ER (Error Reduction) correction method. Examples show that the ER correction method is the quick and easy improvement method for the positive reciprocal matrix consistence.

2 Definitions and Theorems

To facilitate our narrative we introduce the following notation $\Omega = \{1,2,\ldots,n\}$

Supposing $A = (a_{ij})_{n \times n}$ is a positive reciprocal matrix. If the elements of it satisfy

$a_{ij} > 0$, $a_{ji} = \dfrac{1}{a_{ij}}$, $a_{ii} = 1$, $i, j \in \Omega$, we consider that A has a satisfactory consistency

when $C.R = \dfrac{\lambda_{\max} - n}{(n-1)R.I} < 0.1$. [5] In general, the positive reciprocal matrix experts give

B. Liu and C. Chai (Eds.): ICICA 2011, LNCS 7030, pp. 177–183, 2011.

is difficult to meet the conditions of complete consistency.[6,7]Therefore, it is necessary to adjust the positive reciprocal matrix without satisfying consistency appropriately.

Definition 1: Supposing x^* is the approximation value of accuracy x, we call the ratio of absolute error and exact value as the relative error of approximation, e_r

$$e_r = \frac{e}{x} = \frac{x - x^*}{x}$$

Definition 2: We call the matrix $C = \begin{pmatrix} c_{11} & c_{12} & \cdots & c_{1n} \\ & c_{22} & \cdots & c_{2n} \\ & & \cdots & \cdots \\ & & & c_{nn} \end{pmatrix}$ as the small triangular

matrix of positive reciprocal matrix A.

Among them, $c_{ij} = \dfrac{w_{ij} - a_{ij}}{w_{ij}}$, $i, j \in \Omega$

Theorem: The necessary and sufficient condition that positive reciprocal matrix A is completely consistent matrix is all the elements in C are 0. That is

$$C = \begin{pmatrix} 0 & \cdots & \cdots & 0 \\ & 0 & \cdots & 0 \\ & & \cdots & \\ & & & 0 \end{pmatrix}$$

Proof

(1) Proof for necessity

If the positive reciprocal matrix A is a completely consistent matrix, then the normalization vector of each column vector of A is equal. So it is equal to the ordering vector obtained by sum product method, that is, $w_{ij} = a_{ij}$, $i, j \in \Omega$. So

$c_{ij} = 0$. That is, $C = \begin{pmatrix} 0 & \cdots & \cdots & 0 \\ & 0 & \cdots & 0 \\ & & \cdots & \\ & & & 0 \end{pmatrix}$.

(2) Proof for sufficiency

If $C = \begin{pmatrix} 0 & \cdots & \cdots & 0 \\ & 0 & \cdots & 0 \\ & & \cdots & \\ & & & 0 \end{pmatrix}$, that is $c_{ij} = 0$. Thus, $w_{ij} = a_{ij}$, $i, j \in \Omega$. That is, the

normalization vector of each column is equal. So A is a completely consistent matrix.[8,9,10]

From the Theorem we know that if there is certain element $c_{ij} \neq 0$ in a small triangular matrix C, then it shows the positive reciprocal matrix A is not a completely consistent matrix. What's more, for c_{ij} the larger the deviation from 0, it shows the greater the influence of a_{ij} to the inconsistency of A. When $c_{ij} < 0$, a_{ij} tends to larger which should be reduced appropriately. When $c_{ij} > 0$, a_{ij} tends to smaller which should be increased appropriately. [11,12] Taking into account the authority of expert and the integer features of 1-9 scale system, the specific adjustment method is as follows. First we find the position of the largest element deviated from 0 of C. Then we compare the corresponding position element of the original positive reciprocal matrix A and 1. If $c_{ij} < 0$ of the corresponding position, then it shows a_{ij} is rather larger than w_{ij} in completely consistent matrix W. If $a_{ij} > 1$, then we order $a_{ij}' = a_{ij} - 1$; otherwise we order $a_{ij}' = 1 \Big/ (\frac{1}{a_{ij}} + 1)$. If $c_{ij} > 0$, then it indicates a_{ij} is rather smaller than w_{ij} in completely consistent matrix W. If $a_{ij} > 1$, then we order $a_{ij}' = a_{ij} + 1$; otherwise we order $a_{ij}' = 1 \Big/ (\frac{1}{a_{ij}} - 1)$.

3   Improvement Method of Positive Reciprocal Matrix Consistency

The specific steps of ER correction method to positive reciprocal matrix consistency improvement are as follows:

Step1: Sum product method is used to evaluate $\lambda_{\max}(A)$ for positive reciprocal matrix A. The main feature vector is $v = (v_1, v_2 \ldots v_n)$ and the consistency index of A is $C.R$. If $C.R < 0.1$, it ends. Otherwise, turn to Step2.

Step2: By V structure $W_A(w_{ij})$, among them, $w_{ij} = \dfrac{v_i}{v_j}, i, j \in \Omega$, we calculate $c_{ij} = \dfrac{w_{ij} - a_{ij}}{w_{ij}} i, j \in \Omega$ and obtain C.

Step3: Order $c_{kl} = \max\limits_{1 \le i, j \le n} \{|c_{ij}|\}, k, l \in \Omega$

Step4: If $c_{kl} < 0$ and a_{kl} is an integer, then we order $a_{kl}' = a_{kl} - 1$. Or we order $a_{kl}' = 1 \Big/ (\frac{1}{a_{kl}} + 1)$. If $c_{kl} > 0$ and a_{kl} is an integer, then we order $a_{kl}' = a_{kl} + 1$.

Or we order $a_{kl}' = 1 \Big/ (\frac{1}{a_{kl}} - 1)$.

Step5: Order $a'_{kl} = \dfrac{1}{a'_{kl}}, a'_{ij} = a_{ij}, i, j \neq k, l$

Step6: If $A' = (a'_{ij})$ has a satisfactory agreement, then it stops and A' is the positive reciprocal matrix obtained with a satisfactory consistency. Otherwise, we order $c'_{fg} = \max\limits_{1 \leq i, j \leq n}\{|c_{ij}|\} f, g \in \Omega$ and $f, g \neq k, l$ and turn to Step4.

4   Example

$$\text{Case } A = \begin{pmatrix} 1 & \frac{1}{9} & 3 & \frac{1}{5} \\ 9 & 1 & 5 & 2 \\ \frac{1}{3} & \frac{1}{5} & 1 & \frac{1}{2} \\ 5 & \frac{1}{2} & 2 & 1 \end{pmatrix}$$

$C.R = 0.1874 > 0.1$, A does not possess a satisfactory consistency.
From ordering vector obtained by sum product method we can get:

$$W = \begin{pmatrix} 1 & 0.2126 & 1.2670 & 0.4302 \\ 4.7046 & 1 & 5.9609 & 2.0239 \\ 0.7892 & 0.1678 & 1 & 0.3395 \\ 2.3245 & 0.4941 & 2.9453 & 1 \end{pmatrix}$$

So the small triangular matrix of A

$$C = \begin{pmatrix} 0 & 0.4770 & -1.3680 & 0.5351 \\ & 0 & 0.1612 & 0.0118 \\ & & 0 & -0.4728 \\ & & & 0 \end{pmatrix}$$

The largest element deviated from 0 of C is $c_{13} = -1.3680$ and $a_{13} = 3$ is an integer. So a_{13} must be reduced by 1, that is $a'_{13} = 2, a'_{31} = \frac{1}{2}$. So

$$A' = \begin{pmatrix} 1 & \frac{1}{9} & 2 & \frac{1}{5} \\ 9 & 1 & 5 & 2 \\ \frac{1}{2} & \frac{1}{5} & 1 & \frac{1}{2} \\ 5 & \frac{1}{2} & 2 & 1 \end{pmatrix}.$$

Further examination: $C.R = 0.10944 > 0.1$ and A' does not still obtain a satisfactory consistency. Re-examining the small triangular matrix C we get $c_{14} = 0.5351$ and

$a_{14} = \frac{1}{5}$ is a fraction. So the denominator of a_{14} needs to be decreased by 1, that is $a'_{14} = \frac{1}{4}, a'_{41} = 4$. So

$$A'' = \begin{pmatrix} 1 & \frac{1}{9} & 2 & \frac{1}{4} \\ 9 & 1 & 5 & 2 \\ \frac{1}{2} & \frac{1}{5} & 1 & \frac{1}{2} \\ 4 & \frac{1}{2} & 2 & 1 \end{pmatrix}.$$

Further examination: $C.R = 0.08747 < 0.1$ is with the satisfaction of consistency. Then the correction is over.

The following narratives are results of other two correction methods of positive reciprocal matrix in examples.

(1) The correction method of the concept of introduction vector deviation. After one correction we get

$$A' = \begin{pmatrix} 1 & \frac{1}{9} & \frac{1}{2} & \frac{1}{5} \\ 9 & 1 & 5 & 2 \\ 2 & \frac{1}{5} & 1 & \frac{1}{2} \\ 5 & \frac{1}{2} & 2 & 1 \end{pmatrix}.$$

(2) The correction method of definition induced matrix. After two corrections we get

$$A'' = \begin{pmatrix} 1 & \frac{1}{9} & 1 & \frac{1}{5} \\ 9 & 1 & 5 & 2 \\ 1 & \frac{1}{5} & 1 & \frac{1}{2} \\ 5 & \frac{1}{2} & 2 & 1 \end{pmatrix}.$$

The changes to the original matrix in (1) are $a_{13} : 3 \to \frac{1}{2}; a_{31} : \frac{1}{3} \to 2$.

The changes to the original matrix in (2) are $a_{13} : 3 \to 1; a_{31} \frac{1}{3} \to 1$.

The changes to the ER correction method are

$$a_{13} : 3 \to 2; a_{31} : \frac{1}{3} \to \frac{1}{2}, a_{14} : \frac{1}{5} \to \frac{1}{4}; a_{41} : 5 \to 4$$

We can see that although (1) ends only through one calculation correction, its correction result is obviously not easy to be accepted by judges. At first the 1st element is more important than the 3rd one by the experts' judgment, but the 3rd element is more important than the 1st one after correction. That is a "message reversal" phenomenon, which is a reversal of experts' opinions. In the correction of (2), that the 1st element is relatively important than the 3rd one is corrected to that the 1st element is equally important with the 3rd one. Although there is no reversal phenomenon of experts' opinions, the actual meaning between slightly important and important is of some distance and changes the experts' opinions relatively largely. But ER modification method to the control of correction amplitude can avoid larger changes, and it will not reverse the information but carry on correction on the basis of retaining the experts' opinions. On the other hand, we can see by example comparison that the calculation quantity of ER correction is also significantly less than other methods. The advantages of ER correction can be summarized as the following:

(1) It is visual with less calculation quantity, that is, it's simple and quick.
(2) It retains fully the effective information of original positive reciprocal matrix ensuring the correction quality.

5   Conclusion

By the above discussion, we give an intuitive consistency modification method with less calculation quantity.   And on the basis of accurate understanding to relative error and in-depth understanding to positive reciprocal matrices, we fully consider the mental capacity of decision makers and try best to reach the purpose of correcting the matrix which cannot satisfy the consistency on the basis of keeping the original positive reciprocal matrix information. ER correction method brings up the practical limit that each couple of weights can be corrected once at most. We fully respect for the views of experts and can get satisfying results ensuring that the correction results will not appear decimal and avoids the skipping correction. Therefore, the algorithm is a distinctive method of improvement for 1-9 scale system. It helps people to solve practical problems effectively, enriches the theories and methods of decision analysis and reduces blindness greatly in the decision analysis process, and it is easy to implement on a computer and has a quick speed.

References

1. Huang, D.-c., Xu, L.: Proportion Criteria and Method for Building Comparison Matrices in the Analytic Hierarchy Process. Control and Decision 04, 484–486 (2002)
2. Chen, J.-x., Luo, W.-q., Pang, S.-l.: A New Standard of Consistent for Comparison Matrix in the AHP. Mathematics in Practice and Theory 20, 213–216 (2010)
3. Wu, Z.-k., Qu, S.-j.: A Valid Method for Adjusting Inconsistency Judgment Matrix in AHP. Journal of Qingdao Agricultural University (Natural Science) 02, 160–162 (2008)
4. Guo, Z.-m.: A New Method for Improving the Consistency of the Judgement Matrix in AHP. Journal of Qiqihar University (Natural Science Edition) 06, 84–86 (2010)
5. Han, L.-l., Li, J.-q.: The New Method of Comparison Matrix Consistency Checking. Journal of Qufu Normal University (Natural Science) 03, 44–46 (2002)
6. Wu, S.-h., Guo, N.-l.: A Method for Improving the Consistency Check in AHP. Journal of Projectiles, Rockets, Missiles and Guidance S9, 1059–1060 (2006)
7. Guo, P., Zheng, W.-w.: Certain Improvements in Application of AHP. Systems Engineering 01, 28–31 (1995)
8. Yue, L.-z.: Another Method of Testing Judgement Matrix Uniformity. Journal of Shenyang Institute of Chemical Technology 02, 139–144 (1991)

9. Zhou, X.-h., Zhang, J.-j.: Generalized Consistency Transformation of Judgement Matrix and a Algorithm of Ranking. Journal of Zhejiang University (Science Edition) 02, 157–162 (2011)
10. Zhang, G.-q., Chen, Y.-h.: A New Method for Improving the Consistency of the Comparison Matrix in AHP. Mathematics in Practice and Theory 23, 140–146 (2009)
11. Cai, M.-y., Gong, S.-w., Li, X.-b.: Technique of Human Error Failure Analysis Based on Analytic Hierarchy Process. Journal of Safety Science and Technology 02, 74–77 (2008)
12. Zhang, X., Li, G., Xiong, W.-q.: Particle Swarm Optimization for Correcting Judgment Matrix Consistency in Analytic Hierarchy Process. Computer Engineering and Applications 36, 43–47 (2010)

The Generalized Mann Iterative Process with Errors for Strongly Pseudocontractive Mappings in Arbitrary Banach Spaces

Cheng Wang, Hai-E Zhang, and Zhi-Ming Wang

Department of Basic Science, Tangshan College, Tangshan, 063000 P.R. China
149960953@qq.com, haiezhang@126.com,
wangzhiming.wzm@163.com

Abstract. Let E be a real Banach space and D be a nonempty closed convex subset of E. Suppose that $T : D \to D$ is a uniformly continuous and strongly pseudocontractive mapping with bounded range. It is proved that the generalized Mann iterative process with errors converges strongly to the unique fixed point of T. It is also to establish the convergence theorems of the new iterative methods for strongly pseudocontractive and strongly accretive operators in Banach spaces. The related results deal with the approximation of the solutions of nonlinear equation for strongly accretive operators.

Keywords: Generalized Mann iterative process with errors, Strongly pseudocontractive mapping, Strongly accretive operator, Banach space.

1 Introduction

Let E be a real Banach space and E^* be its dual space. The normalized duality mapping $J : E \to 2^{E^*}$ is defined by

$$Jx = \left\{ f \in E^* : < x, f > = \|x\| \cdot \|f\| = \|f\|^2 \right\}, x \in E,$$

where $< \cdot, \cdot >$ denotes the generalized duality pairing.

Definition 1.1. A mapping T with domain $D(T)$ and range $R(T)$ in E is said to be strongly pseudocontractive if for all $x, y \in D(T)$, there exist $j(x - y) \in J(x - y)$ and a constant $k \in (0,1)$ such that

$$< Tx - Ty, j(x - y) > \le k \|x - y\|^2. \tag{1.1}$$

Closely related to the class of strongly pseudocontractive mappings are those of strongly accretive types. It is well known that T is strongly pseudocontractive if and only if $I - T$ is strongly accretive, where I is the identity mapping.

Definition 1.2. The mapping $T : E \to E$ is called strongly accretive if for each $x, y \in E$, there exist $j(x - y) \in J(x - y)$ and a constant $k \in (0,1)$ such that

$$< Tx - Ty, j(x - y) > \ge k \|x - y\|^2. \tag{1.2}$$

B. Liu and C. Chai (Eds.): ICICA 2011, LNCS 7030, pp. 184–191, 2011.

The accretive operators were introduced independently by Browder [1] and Kato [2] in 1967. An early fundamental result in the theory of accretive mapping, due to Browder, states that the initial value problem

$$du(t)/dt + Tu(t) = 0, u(0) = u_0$$

is solvable if T is locally Lipschitzian and accretive on E. In recent years, much attention has been given to solving nonlinear operator equations in Banach spaces by using Mann and Ishikawa iterative processes.

Definition 1.3. Let D be a nonempty closed convex subset of Banach space E and $T: D \to D$ be a map. The Mann iterative process [3] is defined by the sequence $\{x_n\}_{n=0}^{\infty} \subset D$

$$\begin{cases} x_0 \in D, \\ x_{n+1} = (1 - a_n)x_n + a_n T x_n, n \geq 0, \end{cases} \quad (1.3)$$

where $\{a_n\}_{n=0}^{\infty}$ is a real sequence in $[0,1]$. In 1998, Xu [4] introduced the iterative scheme with errors defined by

$$\begin{cases} x_0 \in D, \\ x_{n+1} = a_n x_n + b_n T x_n + c_n u_n, n \geq 0, \end{cases} \quad (1.4)$$

where $\{a_n\}_{n=0}^{\infty}, \{b_n\}_{n=0}^{\infty}, \{c_n\}_{n=0}^{\infty}$ are sequences in $[0,1]$ such that $a_n + b_n + c_n = 1$ and $\{u_n\}_{n=0}^{\infty}$ is a bounded sequence in D. Clearly, this iterative scheme contains (1.3) as its special case. Furthermore, we define the generalized Mann iteration process and the generalized Mann iteration process with errors, respectively

$$\begin{cases} x_0 \in D, \\ x_{n+1} = a_n' x_n + b_n' T x_n + c_n' T^2 x_n, n \geq 0, \end{cases} \quad (1.5)$$

$$\begin{cases} x_0 \in D, \\ x_{n+1} = a_n x_n + b_n T x_n + c_{\partial n} T^2 x_n + d_n u_n, n \geq 0, \end{cases} \quad (1.6)$$

where $\{a_n'\}_{n=0}^{\infty}, \{b_n'\}_{n=0}^{\infty}, \{c_n'\}_{n=0}^{\infty}$ and $\{a_n\}_{n=0}^{\infty}, \{b_n\}_{n=0}^{\infty}, \{c_n\}_{n=0}^{\infty}, \{d_n\}_{n=0}^{\infty}$ are sequences in $[0,1]$ such that $a_n' + b_n' + c_n' = 1$, $a_n + b_n + c_n + d_n = 1$ and $\{u_n\}_{n=0}^{\infty}$ is a bounded sequence in D. If $c_n' = 0$, then (1.5) becomes (1.3). If $c_n = 0$, then (1.6) reduces to (1.4).

It is our purpose in this paper to establish the convergence theorems of the new iterative methods for strongly pseudocontractive and strongly accretive operators in Banach spaces. For this, we need to introduce the following Lemmas.

Lemma 1.4 [10]. Let E be a real Banach space. Then for all $x, y \in E$ and

$$j(x + y) \in J(x + y), \|x + y\|^2 \leq \|x\|^2 + 2 < y, j(x + y) >.$$

Lemma 1.5 [6]. Suppose that $\{\rho_n\}_{n=0}^{\infty}$, $\{\sigma_n\}_{n=0}^{\infty}$, and $\{\lambda_n\}_{n=0}^{\infty}$ are nonnegative sequences suh that

$$\rho_{n+1} \leq (1 - \lambda_n)\rho_n + \sigma_n, \forall n \geq n_0,$$

where n_0 is a positive integer and $\{\lambda_n\}_{n=0}^{\infty} \subset (0,1)$, $\sum_{n=1}^{\infty} \lambda_n = \infty$ and $\sigma_n = o(\lambda_n)$.

Then $\rho_n \to 0$ as $n \to \infty$.

Lemma 1.6 [7]. Let $\{\rho_n\}_{n=0}^{\infty}$, $\{\sigma_n\}_{n=0}^{\infty}$, and $\{\delta_n\}_{n=0}^{\infty}$ be nonnegative real sequences satisfying the condition:

$$\rho_{n+1} \leq (1 - t_n)\rho_n + \sigma_n + \delta_n, \forall n \geq n_0,$$

where n_0 is a positive integer and $\{t_n\}_{n=0}^{\infty} \subset (0,1)$, $\sum_{n=1}^{\infty} t_n = \infty$, $\sigma_n = o(t_n)$ and

$\sum_{n=1}^{\infty} \delta_n < \infty$. Then $\rho_n \to 0$ as $n \to \infty$.

2   Main Results

In the sequel, we always assume that E is a real Banach space, D is a nonempty closed convex subset of E.

Theorem 2.1. Let $T : D \to D$ be a uniformly continuous and strongly pseudo-contractive mapping with a bounded range $R(T)$. Let q be a fixed point of T. Assume that $\{u_n\}_{n=0}^{\infty}$ is an arbitrary bounded sequence in D and $\{a_n\}_{n=0}^{\infty}$, $\{b_n\}_{n=0}^{\infty}$, $\{c_n\}_{n=0}^{\infty}$, $\{d_n\}_{n=0}^{\infty}$, are any sequences in $[0,1]$ such that $a_n + b_n + c_n + d_n = 1$ satisfying the conditions: (i) $b_n \to 0$ as $n \to \infty$; (ii) $\sum_{n=0}^{\infty} b_n = \infty$; (iii) $c_n = o(b_n)$; (iv) $d_n = o(b_n)$. Suppose that $\{x_n\}_{n=0}^{\infty}$ is the sequence generated from arbitrary $x_0 \in D$ by (1.6). Then the sequence $\{x_n\}_{n=0}^{\infty}$ converges strongly to the unique fixed point q of T.

Proof. Since T is strongly pseudocontractive, it is evident to verify that q is the unique fixed point of T.
    Put

$$M = \|x_0 - q\| + \sup_{x \in D} \|Tx - q\| + \sup_{n \geq 0} \|u_n - q\|. \tag{2.1}$$

It follows from (1.6) and (2.1) that

$$\|x_n - q\| \le M, \forall n \ge 0. \tag{2.2}$$

By using induction. Indeed, it is clear that $\|x_0 - q\| \le M$. Let $\|x_n - q\| \le M$ hold. Next we will show that $\|x_{n+1} - q\| \le M$. From (1.9), it follows that

$$
\begin{aligned}
\|x_{n+1} - q\| &= \|a_n(x_n - q) + b_n(Tx_n - q) + c_n(T^2 x_n - q) + d_n(u_n - q)\| \\
&\le a_n\|x_n - q\| + b_n\|Tx_n - q\| + c_n\|T^2 x_n - q\| + d_n\|u_n - q\| \\
&\le a_n M + b_n M + c_n M + d_n M = M.
\end{aligned}
\tag{2.3}
$$

So the sequence $\{x_n\}$ is bounded.

Define a real convex function $f : [0, \infty) \to [0, \infty)$ with $f(t) = t^2$. Then, for all $\lambda_1, \lambda_2, \lambda_3, \lambda_4 \in [0,1]$ with $\lambda_1 + \lambda_2 + \lambda_3 + \lambda_4 = 1$ and $t_1, t_2, t_3, t_4 \ge 0$, we have

$$f(\lambda_1 t_1 + \lambda_2 t_2 + \lambda_3 t_3 + \lambda_4 t_4) \le \lambda_1 f(t_1) + \lambda_2 f(t_2) + \lambda_3 f(t_3) + \lambda_4 f(t_4). \tag{2.4}$$

From (1.6), (2.2) and (2.4), we obtain that

$$
\begin{aligned}
\|x_{n+1} - q\|^2 &\le (a_n\|x_n - q\| + b_n\|Tx_n - q\| + c_n\|T^2 x_n - q\| + d_n\|u_n - q\|)^2 \\
&\le a_n\|x_n - q\|^2 + b_n\|Tx_n - q\|^2 + c_n\|T^2 x_n - q\|^2 + d_n\|u_n - q\|^2 \\
&\le (1 - b_n - c_n - d_n)\|x_n - q\|^2 + b_n M^2 + c_n M^2 + d_n M^2 \\
&\le (1 - b_n)\|x_n - q\|^2 + b_n M^2 + c_n M^2 + d_n M^2.
\end{aligned}
\tag{2.5}
$$

Using Lemma 1.4 and (1.6), we have

$$
\begin{aligned}
\|x_{n+1} - q\|^2 &\le (1 - b_n - c_n - d_n)^2\|x_n - q\|^2 + 2b_n <Tx_n - Tq, j(x_{n+1} - q)> \\
&\quad + 2c_n <T^2 x_n - T^2 q, j(x_{n+1} - q)> + 2d_n <u_n - q, j(x_{n+1} - q)> \\
&\le (1 - b_n - c_n - d_n)^2\|x_n - q\|^2 + 2b_n <Tx_{n+1} - Tq, j(x_{n+1} - q)> \\
&\quad + 2b_n <Tx_n - Tx_{n+1}, j(x_{n+1} - q)> + 2c_n M^2 + 2d_n M^2 \\
&\le (1 - b_n - c_n - d_n)^2\|x_n - q\|^2 + 2b_n(1 - k)\|x_{n+1} - q\|^2 \\
&\quad + 2b_n\|Tx_n - Tx_{n+1}\| \cdot \|x_{n+1} - q\| + 2c_n M^2 + 2d_n M^2 \\
&\le (1 - b_n)^2\|x_n - q\|^2 + 2b_n(1 - k)\|x_{n+1} - q\|^2 \\
&\quad + 2b_n A_n M + 2c_n M^2 + 2d_n M^2
\end{aligned}
\tag{2.6}
$$

with $A_n = \|Tx_n - Tx_{n+1}\| \to 0$ as $n \to \infty$. Indeed, we easily conclude from (1.6) and (2.1) that

$$\|x_n - x_{n+1}\| = \|b_n(x_n - Tx_n) + c_n(x_n - T^2x_n) + d_n(x_n - u_n)\|$$
$$\leq b_n(\|x_n - q\| + \|Tx_n - q\|) + c_n(\|x_n - q\| + \|T^2x_n - q\|) \tag{2.7}$$
$$+ d_n(\|x_n - q\| + \|u_n - q\|)$$
$$\leq 2(b_n + c_n + d_n)M \to 0$$

as $n \to \infty$. Since T is uniformly continuous on D,

$$A_n = \|Tx_n - Tx_{n+1}\| \to 0$$

as $n \to \infty$. Substituting (2.5) in (2.6), we get

$$\|x_{n+1} - q\|^2 \leq ((1 - b_n)^2 + 2b_n(1 - k))\|x_n - q\|^2$$
$$+ 2(b_n + c_n + d_n)M^2 + 2b_n A_n M + 2(c_n + d_n)M^2$$
$$\leq (1 - b_n(k - b_n))\|x_n - q\|^2 \tag{2.8}$$
$$+ 2b_n(b_n + c_n + d_n)M^2 + 2b_n A_n M + 2(c_n + d_n)M^2.$$

Note that $\lim_{n \to \infty} b_n = 0$, which implies that there exists a natural number N such that, for all $n \geq N$, $b_n \leq \dfrac{k}{2}$. Therefore, (2.8) can be written as

$$\|x_{n+1} - q\|^2 \leq (1 - \frac{kb_n}{2})\|x_n - q\|^2 + 2b_n(b_n + c_n + d_n)M^2 + 2b_n A_n M + 2(c_n + d_n)M^2 \tag{2.9}$$

$$\rho_n := \|x_n - q\|^2, \lambda_n := \frac{kb_n}{2}, \sigma_n := 2b_n(b_n + c_n + d_n)M^2 + 2b_n A_n M + 2(c_n + d_n)M^2 = o(\lambda_n)$$

and using Lemma 1.5, we obtain that $\lim_{n \to \infty}\|x_n - q\| = 0$. This completes the proof.

Theorem 2.2. Let T, D, q, $\{u_n\}$ be as in Theorem 2.1. Suppose that $\{a_n\}_{n=0}^{\infty}, \{b_n\}_{n=0}^{\infty}, \{c_n\}_{n=0}^{\infty}, \{d_n\}_{n=0}^{\infty}$ are sequences in [0, 1] such that $a_n + b_n + c_n + d_n = 1$ satisfying (i), (ii) and (iii) in Theorem 2.1 and the condition $\sum_{n=1}^{\infty} d_n < \infty$. Then the conclusion of Theorem 2.1 holds.

Proof. From the reasoning course of Theorem 2.1, we obtain

$$\|x_{n+1} - q\|^2 \leq \left(1 - \frac{kb_n}{2}\right)\|x_n - q\|^2 + 2b_n(b_n + c_n + d_n)M^2 + 2b_n A_n M + 2(c_n + d_n)M^2 \tag{2.10}$$

where A_n, M are as in Theorem 2.1.

Let

$$\rho_n := \|x_n - q\|^2, \lambda_n := \frac{kb_n}{2}, \sigma_n := 2b_n(b_n + c_n + d_n)M^2 + 2b_n A_n M + 2c_n M^2 = o(t_n)$$

and $\delta_n = 2d_n M$ with $\sum_{n=1}^{\infty} \delta_n < \infty$. By Lemma 1.6, we see that $\lim_{n\to\infty} \|x_n - q\| = 0$. This completes the proof.

Theorem 2.3. Let $D, q, \{u_n\}$ and the parameters $\{a_n\}_{n=0}^{\infty}, \{b_n\}_{n=0}^{\infty}, \{c_n\}_{n=0}^{\infty}$, $\{d_n\}_{n=0}^{\infty}$ be as in Theorem 2.1. Assume that $T : D \to D$ is a Lipschitz and strongly pseudocontractive mapping. Then the conclusion of Theorem 2.1 holds.

Proof. From (1.6) and Lipschitz condition, we get

$$\|x_{n+1} - q\| \le a_n \|x_n - q\| + b_n \|Tx_n - q\| + c_n \|T^2 x_n - q\| + d_n \|u_n - q\|$$

$$\le a_n \|x_n - q\| + b_n \|Tx_n - Tq\| + c_n \|T^2 x_n - T^2 q\| + d_n \|u_n - q\|$$

$$\le a_n \|x_n - q\| + b_n L \|x_n - q\| + c_n L^2 \|x_n - q\| + d_n \|u_n - q\|$$

$$\le (a_n + b_n L + c_n L^2) \|x_n - q\| + d_n \|u_n - q\| \tag{2.11}$$

and

$$\|Tx_n - Tx_{n+1}\| \le L\|x_n - x_{n+1}\| \le L\|b_n(x_n - Tx_n) + c_n(x_n - T^2 x_n) + d_n(x_n - u_n)\|$$

$$\le Lb_n(\|x_n - q\| + \|Tx_n - q\|) + Lc_n(\|x_n - q\| + \|T^2 x_n - q\|) + Ld_n(\|x_n - q\| + \|u_n - q\|) \tag{2.12}$$

$$\le L\left[((1+L)b_n + (1+L^2)c_n + d_n)\|x_n - q\| + d_n\|u_n - q\|\right].$$

By virtue of (2.11) and (2.12), we conclude that,

$$2b_n \|Tx_n - Tx_{n+1}\| \cdot \|x_{n+1} - q\|$$

$$\le 2b_n L(((1+L)b_n + (1+L^2)c_n + d_n)(a_n + b_n L + c_n L^2)\|x_n - q\|^2$$

$$+ d_n \|u_n - q\|(((1+L)b_n + (1+L^2)c_n + d_n)$$

$$+ (a_n + b_n L + c_n L^2)) + d_n \|u_n - q\|(((1+L)b_n + (1+L^2)c_n + d_n) \tag{2.13}$$

$$+ (a_n + b_n L + c_n L^2))\|x_n - q\|^2 + d_n^2 \|u_n - q\|^2)$$

$$= 2b_n L B_n \|x_n - q\|^2 + 2b_n L D_n,$$

where

$$B_n = \left((1+L)b_n + (1+L^2)c_n + d_n\right)\left(a_n + b_n L + c_n L^2\right)$$
$$+ d_n \|u_n - q\|\left((1+L)b_n + (1+L^2)c_n + d_n + (a_n + b_n L + c_n L^2)\right),$$
$$D_n = d_n \|u_n - q\|\left((1+L)b_n + (1+L^2)c_n + d_n + (a_n + b_n L + c_n L^2)\right) + d_n^2 \|u_n - q\|^2.$$

In view of (2.4) and (1.6), we have

$$\|x_{n+1} - q\|^2 \le \left(a_n \|x_n - q\| + b_n \|Tx_n - q\| + c_n \|T^2 x_n - q\| + d_n \|u_n - q\|\right)^2$$
$$\le a_n \|x_n - q\|^2 + b_n \|Tx_n - Tq\|^2 + c_n \|T^2 x_n - T^2 q\|^2 + d_n \|u_n - q\|^2$$
$$\le a_n \|x_n - q\|^2 + b_n L^2 \|x_n - q\|^2 + c_n L^4 \|x_n - q\|^2 + d_n \|u_n - q\|^2$$
$$\le (a_n + b_n L^2 + c_n L^4)\|x_n - q\|^2 + d_n \|u_n - q\|^2$$

(2.14)

By Lemma 1.4 and (2.6), (2.13), (2.14), we obtain

$$\|x_{n+1} - q\|^2 \le (1 - b_n - c_n - d_n)^2 \|x_n - q\|^2 + 2b_n(1-k)\|x_{n+1} - q\|^2$$
$$+ 2b_n \|Tx_n - Tx_{n+1}\| \cdot \|x_{n+1} - q\| + 2c_n L^2 \|x_n - q\| \cdot \|x_{n+1} - q\| + 2d_n \|u_n - q\| \cdot \|x_{n+1} - q\|$$
$$\le (1 - b_n)^2 \|x_n - q\|^2 + 2b_n(1-k)((a_n + b_n L^2 + c_n L^4)\|x_n - q\|^2 + d_n \|u_n - q\|^2)$$
$$+ 2b_n LB_n \|x_n - q\|^2 + 2b_n LD_n + 2c_n L^2 \|x_n - q\|((a_n + b_n L + c_n L^2)\|x_n - q\| + d_n \|u_n - q\|)$$
$$+ 2d_n \|u_n - q\|((a_n + b_n L + c_n L^2)\|x_n - q\| + d_n \|u_n - q\|)$$

(2.15)

$$\le ((1 - 2b_n + b_n^2 + 2b_n(1-k) + 2b_n(1-k)(b_n L^2 + c_n L^4) + 2b_n LB_n$$
$$+ 2c_n L^2(a_n + b_n L + c_n L^2) + c_n L^2 d_n \|u_n - q\| + d_n \|u_n - q\|(a_n + b_n L + c_n L^2))\|x_n - q\|^2$$
$$+ 2b_n(1-k)d_n \|u_n - q\|^2 + 2b_n LD_n + c_n L^2 d_n \|u_n - q\| + d_n \|u_n - q\|(a_n + b_n L + c_n L^2) + 2d_n \|u_n - q\|^2$$
$$\le (1 - 2b_n k + E_n)\|x_n - q\|^2 + F_n$$

where

$$E_n = b_n^2 + 2b_n(1-k)(b_n L^2 + c_n L^4) + 2b_n LB_n + 2c_n L^2(a_n + b_n L + c_n L^2)$$
$$+ c_n L^2 d_n \|u_n - q\| + d_n \|u_n - q\|(a_n + b_n L + c_n L^2)$$
$$F_n = 2b_n(1-k)d_n \|u_n - q\|^2 + 2b_n LD_n + c_n L^2 d_n \|u_n - q\|$$
$$+ d_n \|u_n - q\|(a_n + b_n L + c_n L^2) + 2d_n \|u_n - q\|^2.$$

From the known conditions, we obtain that $E_n = o(b_n)$ and $F_n = o(b_n)$, thus there exists a natural number N such that for all $n \ge N$, we have $E_n \le kb_n$. From (2.15), we get

$$\left\| x_{n+1} - q \right\|^2 \le \left(1 - kb_n \right) \left\| x_n - q \right\|^2 + 2F_n \tag{2.16}$$

It follows from Lemma 1.4 and (2.16) that $\lim_{n \to \infty} \left\| x_n - q \right\| = 0$. This completes the proof.

References

1. Browder, F.E.: Nonlinear mappings of nonexpansive and accretive type in Banach spaces. Bull. Amer. Math. Soc. 73, 875–882 (1967)
2. Kato, T.: Nonlinear semigroup and evolution equations. J. Math. Soc. Japan 19, 508–520 (1967)
3. Mann, W.R.: Mean value methods in iteration. Proc. Amer. Math. Soc. 4, 506–510 (1953)
4. Xu, Y.G.: Ishikawa and Mann iteration process with errors for nonlinear strongly accretive operator equations. J. Math. Anal. Appl. 224, 91–101 (1998)
5. Takahashi, W.: Nonlinear Functional Analysis, Kindaikagakusha, Tokyo (1998)
6. Weng, X.: Fixed point iteration for local strictly pseudocontractive mapping. Proc. Amer. Math. Soc. 113, 727–731 (1991)
7. Liu, L.S.: Ishikawa and Mann iterative process with errors for nonlinear strongly accretive mappings in Banach spaces. J. Math. Anal. Appl. 194, 114–125 (1995)
8. Ishikawa, S.: Fixed points by a new iteration method. Proc. Amer. Math. Soc. 44, 147–150 (1974)
9. Berinde, V.: On the convergence of Ishikawa iteration in the class of quasicontractive operators. Acta. Math. Univ. Comenianae LXXIII, 119–126 (2004)
10. Xue, Z.Q., Zhou, H.Y.: Iterative approximation with errors of fixed point for a class of nonlinear operator with a bounded range. Applied Mathematics and Mechanics 20(1), 99–1041 (1999)

Parallel Fourth-Order Runge-Kutta Method to Solve Differential Equations

Chunfeng Liu, Haiming Wu, Li Feng, and Aimin Yang

College of Science, Hebei United University, Tangshan Hebei 063009, China
{liucf403,wuhaiming-08}@163.com, fengli3@126.com

Abstract. Through research for the method of serial classic fourth-order Runge-Kutta and based on the method, we construct Parallel fourth-order Runge-Kutta method in this paper, and used in the calculation of differential equation, then under the dual-core parallel, research the Parallel computing speedup and so on. By compared the results of traditional numerical algorithms and parallel numerical algorithms, the results show parallel numerical algorithms have high accuracy and computational efficiency in the dual-core environment.

Keywords: Differential equations, Runge-Kutta method, Parallel computing.

1   Introduction

With the gradual deepening of Parallel studies,domestic and foreign scholars have achieved some results,and parallel algorithms has become an important subject[1][2], differential equations numerical parallel algorithms have a wide range of research value and application prospects[3]. All fields of engineering and science in many of the problems can be described by differential equations. With the development of various fields, such as energy, meteorology, military, medicine, artificial intelligence and some basic research, the need for more and more complex system for digital simulation, this is a very time-consuming calculations, if the use of fast and efficient parallel computing to achieve, will save some time. For parallel computing, need to construct parallel algorithms and suitable for running on a computer system[4]. In the field of parallel computing, parallel algorithm is the core technology and bottlenecks. This article is based on the serial classic fourth-order Runge-Kutta method, and constructed parallel fourth-order Runge-Kutta method, thus speeding up of the calculation is more efficient and save computing time, this method is verified in the dual-core environment.

2   Classical Fourth-Order Runge-Kutta Method

Runge-Kutta algorithm are generated in the solution of differential equations, it has more than 200 years, it is a mature way in computational mathematics. Compared with other methods of solving differential equations (for example, the differential method, finite element method, finite volume method), the method has the high precision and flexibility, especially the classic Runge-Kutta method [5]-[7].

B. Liu and C. Chai (Eds.): ICICA 2011, LNCS 7030, pp. 192–199, 2011.

Fourth-order Runge-Kutta method is structured by the Taylor expansion for the function, and this method has good stability and convergence. The following formula:

$$y_{n+1} = y_n + \frac{h}{6}(k_1 + 2k_2 + 2k_3 + k_4)$$

where $k_1 = f(x_n, y_n)$

$$k_2 = f\left(x_n + \frac{h}{2}, y_n + \frac{h}{2}k_1\right)$$

$$k_3 = f\left(x_n + \frac{h}{2}, y_n + \frac{h}{2}k_2\right)$$

$$k_4 = f(x_n + h, y_n + hk_3)$$

This method is often used, but it has a large amount of computation, the loss of time will be more.

3   Parallel Fourth-Order Runge-Kutta Method

3.1   Construction of Parallel Method

Consider the differential equation

$$y' = f(x, y), \quad y(x_0) = y_0$$

For the numerical solution of this problem, based on the classic fourth-order Runge-Kutta method, construct a parallel four-order Runge-Kutta method, as follows:

$$\begin{cases} y_{n+1} = y_n + \frac{1}{6}(K_{1,n} + 2K_{2,n} + 2K_{3,n} + K_{4,n}) \\ \\ K_{1,n} = hf(x_n, y_n) \\ K_{2,n} = hf\left(x_n + \frac{h}{2}, y_n + \frac{1}{2}\sum_{j=1}^{4} b_{2j} K_{jn-1}\right) \\ K_{3,n} = hf\left(x_n + \frac{h}{2}, y_n + \frac{1}{2}\sum_{j=1}^{4} b_{3j} K_{jn-1}\right) \\ K_{4,n} = hf\left(x_n + h, y_n + \sum_{j=1}^{4} b_{4j} K_{jn-1}\right) \end{cases}$$

3.2   Convergence of the Method

We define $\Delta K_{i,n} = K_i(x_n, h) - K_{i,n}$, $\Delta y_{n+1} = y(x_{n+1}) - y_{n+1}$ [8], where f satisfy the Lipschitz condition on y, and the coefficient is L,

We get

$$\left|\Delta K_{1,n}\right| = \left|K_1(x_n, h) - K_{1,n}\right| = \left|hf(x_n, y(x_n)) - hf(x_n, y_n)\right|$$
$$\leq hL\left|y(x_n) - y_n\right| = hL\left|\Delta y_n\right|$$

$$\left|\Delta K_{2,n}\right| = \left|hf(x_n + \frac{h}{2}, y(x_n) + \frac{1}{2}\sum_{j=1}^{4}b_{2j}K_{jn-1} - hf(x_n + \frac{h}{2}, y_n + \frac{1}{2}\sum_{j=1}^{4}b_2 K_{jn-1})\right|$$
$$\leq hL\left|y(x_n) - y_n\right| + hL * \frac{1}{2}\sum_{j=1}^{4}b_{2j}\left|\Delta K_{jn-1}\right|$$
$$\leq hL\left|\Delta y_n\right| + \frac{hL}{2}\sum_{j=1}^{4}b_{2j}\left|\Delta K_{jn-1}\right|$$

$$\left|\Delta K_{3,n}\right| = \left|hf(x_n + \frac{h}{2}, y(x_n) + \frac{1}{2}\sum_{j=1}^{4}b_{3j}K_{jn-1}) - hf(x_n + \frac{h}{2}, y_n + \frac{1}{2}\sum_{j=1}^{4}b_{3j}K_{jn-1})\right|$$
$$\leq hL\left|y(x_n) - y_n\right| + hL * \frac{1}{2}\sum_{j=1}^{4}b_{3j}\left|\Delta K_{jn-1}\right|$$
$$\leq hL\left|\Delta y_n\right| + \frac{hL}{2}\sum_{j=1}^{4}b_{3j}\left|\Delta K_{jn-1}\right|$$

$$\left|\Delta K_{4,n}\right| = \left|hf(x_n + h, y(x_n) + \sum_{j=1}^{4}b_{4j}K_{jn-1}) - hf(x_n + h, y_n + \sum_{j=1}^{4}b_{4j}K_{jn-1})\right|$$
$$\leq hL\left|y(x_n) - y_n\right| + hL\sum_{j=1}^{4}b_{4j}\left|\Delta K_{jn-1}\right|$$
$$\leq hL\left|\Delta y_n\right| + hL\sum_{j=1}^{4}b_{4j}\left|\Delta K_{jn-1}\right|$$

$$\left|\Delta y_{n+1}\right| = \left|y(x_{n+1}) - y_{n+1}\right|$$
$$= \left|y(x_n) + \frac{1}{6}(K_1(x_n, h) + 2K_2(x_n, h) + 2K_3(x_n, h) + K_4(x_n, h))\right|$$
$$- \left|y_n + \frac{1}{6}(K_{1,n} + 2K_{2,n} + 2K_{3,n} + K_{4,n})\right|$$
$$\leq \left|y(x_n) - y_n\right| + \frac{1}{6}(\left|\Delta K_{1,n}\right| + 2\left|\Delta K_{2,n}\right| + 2\left|\Delta K_{3,n}\right| + \left|\Delta K_{4,n}\right|)$$

Let $R_n = \max\{|\Delta y_n|, |\Delta K_{jn-1}|, j = 2,3,4\}$

Then $|\Delta y_{n+1}| \le L(1+h)R_n + \dfrac{hL}{6}\sum_{i=2}^{4}\sum_{j=1}^{4}b_{ij}R_n$

So, $K_{1,n}, K_{2,n}, K_{3,n}, K_{4,n}$ satisfy the Lipschitz condition on y, according to theorem can know the method is convergent .

3.3   Coefficient of Solving

Let

$$K_i(x,h) = hf(y(x) + \sum_{j=1}^{4}b_{ij}K_j(x-h,h)), \qquad i = 1,2,3,4$$

For $K_{i,n}$, we conduct the Taylor expansion, then

$$K_i(x,h) = hy^{(1)} + h^2 p_i y^{(2)} + h^3[q_i f_y y^{(2)} + \frac{1}{2}p_i^2 f_{yy}(y^{(1)})^2] + h^4[\frac{1}{6}p_i^3 f_{yyy}(y^{(1)})^3 +$$

$$p_i q_i ff_y y^{(1)} + (\frac{1}{2}\sum_{j=1}^{4}b_{ij}p_j^2 - q_i - \frac{1}{2}p_i)f_y f_{yy}(y^{(1)})^2 + (\sum_{j=1}^{4}b_{ij}q_j - q_i - \frac{1}{2}p_i)$$

$$f_y f_y f_y y^{(1)}] + o(h^5) \qquad i = 1,2,3,4$$

$$\text{where, } p_i = \begin{cases} \dfrac{1}{2}\sum_{j=1}^{4}b_{ij}, & i = 2,3 \\[2mm] \sum_{j=1}^{4}b_{ij}, & i = 4 \end{cases}, q_i = \begin{cases} \dfrac{1}{2}\sum_{j=1}^{4}b_{ij}(p_j - 1), & i = 2,3 \\[2mm] \sum_{j=1}^{4}b_{ij}(p_j - 1), & i = 4 \end{cases},$$

and $p_1 = q_1 = 0, \quad b_{11} = b_{12} = b_{13} = b_{14} = 0$.

We may replace y_{n+1} and y_n by $y(x_{n+1})$ and $y(x_n)$ in the formula, and conduct the Taylor expansion, we obtained :

$$y(x_{n+1}) = y(x_n) + hy^{(1)} + \frac{h^2}{2}y^{(2)} + \frac{h^3}{6}y^{(3)} + \frac{h^4}{24}y^{(4)} + o(h^5) \qquad (5)$$

And

$$y(x_{n+1}) = y(x_n) + \frac{1}{6}[K_1(x,h) + 2K_2(x,h) + 2K_3(x,h) + K_4(x,h)] \qquad (6)$$

$K_i(x,h)$ substitute into (6), comparing the coefficients with a exponent of h on sides of the equation, there was established the following formula:

$$\frac{1}{3}p_2 + \frac{1}{3}p_3 + \frac{1}{6}p_4 = \frac{1}{2} \qquad (7)$$

$$\frac{1}{3}q_2 + \frac{1}{3}q_3 + \frac{1}{6}q_4 = \frac{1}{6} \tag{8}$$

$$\frac{1}{6}p_2^2 + \frac{1}{6}p_3^2 + \frac{1}{12}p_4^2 = \frac{1}{6} \tag{9}$$

$$\frac{1}{18}p_2^3 + \frac{1}{18}p_3^3 + \frac{1}{36}p_4^3 = \frac{1}{24} \tag{10}$$

$$\frac{1}{3}p_2 q_2 + \frac{1}{3}p_3 q_3 + \frac{1}{6}p_4 q_4 = \frac{1}{24} \tag{11}$$

$$\frac{1}{3}\left(\frac{1}{2}\sum_{j=1}^{4} b_{2j} p_j^2 - q_2 - \frac{1}{2}p_2\right) + \frac{1}{3}\left(\frac{1}{2}\sum_{j=1}^{4} b_{3j} p_j^2 - q_3 - \frac{1}{2}p_3\right) +$$
$$\frac{1}{6}\left(\frac{1}{2}\sum_{j=1}^{4} b_{4j} p_j^2 - q_4 - \frac{1}{2}p_4\right) = \frac{1}{24} \tag{12}$$

$$\frac{1}{3}\left(\sum_{j=1}^{4} b_{2j} q_j - q_2 - \frac{1}{2}p_2\right) + \frac{1}{3}\left(\sum_{j=1}^{4} b_{3j} q_j - q_3 - \frac{1}{2}p_3\right) +$$
$$\frac{1}{6}\left(\sum_{j=1}^{4} b_{4j} q_j - q_4 - \frac{1}{2}p_4\right) = \frac{1}{24} \tag{13}$$

By the formula (7), (9), (10) can be obtained,

$$p_2 = \frac{1}{2}, \ p_3 = \frac{1}{2}, \ p_4 = 1$$

By the formula (8) and (11) can be obtained:

$$q_4 = -\frac{1}{2}$$

Let $q_2 = \frac{1}{2}$, then $q_3 = \frac{1}{4}$,

Based on the above formula, we can get some relations:

$$b_{21} + b_{22} + b_{23} + b_{24} = 1$$
$$b_{31} + b_{32} + b_{33} + b_{34} = 1$$
$$b_{41} + b_{42} + b_{43} + b_{44} = 1$$
$$b_{21} + \frac{1}{2}b_{22} + \frac{1}{2}b_{23} = -1$$

$$b_{31} + \frac{1}{2}b_{32} + \frac{1}{2}b_{33} = -\frac{1}{2}$$

$$b_{41} + \frac{1}{2}b_{42} + \frac{1}{2}b_{43} = \frac{1}{2}$$

$$\frac{1}{3}(\frac{1}{4}b_{22} + \frac{1}{4}b_{23} + b_{24}) + \frac{1}{3}(\frac{1}{4}b_{32} + \frac{1}{4}b_{33} + b_{34}) +$$

$$\frac{1}{6}(\frac{1}{4}b_{42} + \frac{1}{4}b_{43} + b_{44}) = \frac{11}{12}$$

$$\frac{1}{3}(\frac{1}{2}b_{22} + \frac{1}{4}b_{23} - \frac{1}{2}b_{24}) + \frac{1}{3}(\frac{1}{2}b_{32} + \frac{1}{4}b_{33} - \frac{1}{2}b_{34}) +$$

$$\frac{1}{6}(\frac{1}{2}b_{42} + \frac{1}{4}b_{43} - \frac{1}{2}b_{44}) = \frac{11}{24}$$

Let $b_{22} = 1, b_{23} = 0, b_{33} = -1, b_{34} = 0$, then we can obtain:

$b_{21} = -3/2, b_{24} = 3/2, b_{31} = -2, b_{32} = 4, b_{41} = 1/2, b_{42} = 0, b_{43} = 0, b_{44} = 1/2$

Then we can get the parallel fourth-order Runge-Kutta formula,as follows:

$$\begin{cases} y_{n+1} = y_n + \frac{1}{6}(K_{1,n} + 2K_{2,n} + 2K_{3,n} + K_{4,n}) \\[2mm] K_{1,n} = hf(x_n, y_n) \\[2mm] K_{2,n} = hf(x_n + \frac{h}{2}, y_n - \frac{3}{4}K_{1n-1} + \frac{1}{2}K_{2n-1} + \frac{3}{4}K_{4n-1}) \\[2mm] K_{3,n} = hf(x_n + \frac{h}{2}, y_n - K_{1n-1} + 2K_{2n-1} - \frac{1}{2}K_{3n-1}) \\[2mm] K_{4,n} = hf(x_n + h, y_n + \frac{1}{2}K_{1n-1} + \frac{1}{2}K_{4n-1}) \end{cases}$$

4   Calculation Example

Consider the differential equation

$$y' = -20 \, y \, , y_0 = 1, h = 0.1$$

We apply the classic fourth-order Runge-Kutta method and parallel fourth-order Runge-Kutta method to solve the initial value problem differential equation, the calculation as following table1:

Table 1. We obtained the results when $h = 0.1$

Node coordinates	0.1	0.2	0.3	0.4	0.5
Classical fourth-order R-K formula	0.3333	0.1111	0.0370	0.0123	0.0041
Parallel fourth-order R-K formula	0.4000	0.1600	0.0640	0.0256	0.0102
Node coordinates	0.6	0.7	0.8	0.9	1.0
Classical fourth-order R-K formula	0.0014	4.5725e-004	1.5242e-004	5.0805e-005	1.6935e-005
Parallel fourth-order R-K formula	0.0041	0.0016	6.5536e-004	2.6214e-004	1.0486e-004

It used the time 2.114545 seconds by the classical fourth-order Runge-Kutta formula, it used the time 0.033066 seconds seconds by parallel fourth-order Runge-Kutta formula and on the dual-core computing environments, time is obviously reduced.

5   Conclusion and Outlook

Parallel Runge-Kutta formula can be effectively implemented on a multi-processor system, this article is achieved in the dual-core computer environment. Based on the classic fourth-order Runge-Kutta formula, construct a parallel show fourth-order Runge-Kutta formula, and gives the calculation method of the parametersthe for formula, compare the serial and parallel methods, parallel algorithm saves computation time and improve the computational efficiency. Based on this paper, the need to further study the optimization method of the parameters, enable the results of the calculation to higher speed ,the computational efficiency, accuracy and stability.

Acknowledgment. This work is supported by the National Nature Science Foundation of China (No.61170317) and the Nature Science Foundation of Hebei Province of China (No.A2009000735). The authors are grateful for the anonymous reviewers who made constructive comments.

References

1. Ding, X.-h., Geng, D.-h.: The convergence theorem of parallel Runge-Kutta methods for delay differential equation. Journal of Natural Scinece of Heilongjiang University 21(1) (March 2004)

2. Fei, J.: A Class of parallel Runge-Kutta Methods for differential-algebraic sytems of index 2. Systems Engineering and Electronics 4 (2000)
3. Zou, J.-y., Ding, X.-h., Liu, M.-z.: Two-stage and three-stage continuous Runge-Kutta-Nystrom method of orders two and three. Journal of Natural Science of Heilongjiang University 20(2) (June 2003)
4. Gay, D., Galenson, J., Naik, M., Yelick, K.: Straightforward parallel progrmmng. Parallel Computing (2011)
5. Ninomiya, M., Ninomiya, S.: A new higher-order weak approximation scheme for stochastic differential equationsand the Runge–Kutta method. Finance Stoch. 13, 415–443 (2009)
6. Abbasbandy, S., Allahviranloo, T., Darabi, P.: Numerical solution of N-order fuzzy differential equations by Runge-Kutta method. Mathematical and Computational Applications 16(4), 935–946 (2011)
7. Gan, S.Q., Zhang, W.M.: Stability of Multistep Runge-Kutta Methods for Systems of Functional-Differential and Functional Equations. Applied Mathematics Letters 17, 585–590 (2004)
8. Fei, J.: A class of parallel Runge-Kutta formulas. Computer Engineering and Design 3 (1991)

The Two-Phases-Service M/M/1/N Queuing System with the Server Breakdown and Multiple Vacations

Zeng Hui and Guan Wei

School of Sciences, Yanshan University,
Qinhuangdao, China
zenghui@ysu.edu.cn

Abstract. An two-phases-service M/M/1/N queuing system with the server breakdown and multiple vacations was considered. Firstly, equations of steady-state probability were derived by applying the Markov process theory. Then, we obtained matrix form solution of steady-state probability by using blocked matrix method. Finally, some performance measures of the system such as the expected number of customers in the system and the expected number of customers in the queue were also presented.

Keywords: two phases of service, the server breakdown, multiple vacations, steady-state probability, blocked matrix method.

1 Introduction

Repairable queuing is the expansion of the classical queuing theory, which service stations may be breakdown and can be repaired during the service period. Kell Avi-Itzhak and Thiuvengadam respectively investigate an M/M/1/N queuing system and an M/M/1/N queuing system with the server breakdown. Gray[1] studied an queueing model in which the server takes a vacatioan when the system becomes idle and obtain the queue length distribution and the mean queue length of the model.

The models above study the server can only provide once service for each customer. In fact, in our daily life, we often encouter a server offer different services for the same customer. In such queuing models, all the customers should accepted the fisrt phase service and only of them may be ask the server provide a second phase service, which is the two-phases-service queuing system. Recently, there have been serveral conributions considering queueing system in which the server may provide a second phase service. Madan[2] studied an M/G/1 queue with the second optional service in which first essential service time follows a general disribution but second optional service is assumed to be exponetialy disributed. Medhi[3] generalized the model by considering that the second optional is also governed by a general disribution. Yue[4-10] studied an M/M/1/N queue with the multiple vacations, they obtained the matrix form solution of steady-state probability.

In this paper, we consider a model in which the server can provide two phases service, and the server takes a vacatioan when the system become idel. Once service begins, the service mechanism is subject to breakdowns which occur at a constant.

B. Liu and C. Chai (Eds.): ICICA 2011, LNCS 7030, pp. 200–207, 2011.

2   System Model

We consider the M/M/1/N queueing system with the following assumptions.

(1) Customers arrive according to Poisson process with different rates. Arrival rate during vatiaon, active service, breakdown are λ_0, λ_1 and λ_2. The first service rate is μ_1, and the second service rate is μ_2, and have exponential distribution. As soon as the first service of a customer is completed, he may opt for the second service with probability $\theta(0 < \theta < 1)$,or leave with probability $1-\theta$.

(2) The server goes on vacation at the instant when the queue becomes empty, and continues to take vacations of exponential length until, at the end of a vacation, cusomers are found in the queue. The vacatioan rate is v , vacatioan time is exponential.

(3) The service mechanism breakdowns occur only during the first active service, the breakdown rate is $b, (0 < b < 1)$.

(4) The service mechanism goes through a repair process of random duration, and then repair is completed, the server returns to the customer whose service was interrupted,the repair rate is r .

(5) Various stochastic processes involved in the system are assumed independent of each other.

3   Steady-State Probability Equations

Let $X(t)$ be the number of customers in the system at time t . Define $C(t)$ as the state of the server at the time t . And define the state as follows:

$$C(t) = \begin{cases} 0 , & (\text{ The server is on vacation at the time } t \ ) \\ 1 , & (\text{ The server is on the first service at the time } t \ ) \\ 2 , & (\text{ The server is on the second service at the time } t) \\ 3 , & (\text{ The server is on breakdown process at the time } t) \end{cases}$$

Then, $\{X(t), C(t), t \geq 0\}$ is a Makov process with state space as follows:

$$\Omega = \{(n,0) : 0 \leq n \leq N\} \bigcup \{(n, j) : 1 \leq n \leq N, j = 1,2,3\}$$

The steady-state probability of the system are defined as follows:

$$p_0(n) = \lim_{t \to \infty} p(X(t) = n, C(t) = 0), (0 \leq n \leq N)$$

$$p_j(n) = \lim_{t \to \infty} p(X(t) = n, C(t) = j), (1 \leq n \leq N)$$

By applying the Makov process theory, we can obtain the following set of steady-state probability equaions:

$$\lambda_0 p_0(0) = \mu_1(1-\theta)p_1(1) + \mu_2 p_2(1), \tag{1}$$

$$(v + \lambda_0) p_0(n) = \lambda_0 p_0(n-1), \ (1 \le n \le N-1), \tag{2}$$

$$vp_0(N) = \lambda_0 p_0(N-1), \tag{3}$$

$$(\mu_1 + \lambda_1 + b) p_1(1) = vp_0(1) + \mu_1(1-\theta) p_1(2) + \mu_2 p_2(2) + rp_3(1), \tag{4}$$

$$(\mu_1 + \lambda_1 + b) p_1(n) = vp_0(n) + \lambda_1 p_1(n-1) + \mu_1(1-\theta) p_1(n+1)$$
$$+ \mu_2 p_2(n+1) + rp_3(n), (2 \le n \le N) \tag{5}$$

$$(\mu_1 + b) p_1(N) = vp_0(N) + \lambda_1 p_1(N-1) + rp_3(N), \tag{6}$$

$$(\mu_2 + \lambda_1) p_2(1) = \mu_1 \theta p_1(1), \tag{7}$$

$$(\mu_2 + \lambda_1) p_2(n) = \mu_1 \theta p_1(n) + \lambda_1 p_2(n-1), (2 \le n \le N-1), \tag{8}$$

$$\mu_2 p_2(N) = \mu_1 \theta p_1(N) + \lambda_1 p_2(N-1), \tag{9}$$

$$(r + \lambda_2) p_3(1) = bp_1(1), \tag{10}$$

$$(r + \lambda_2) p_3(n) = bp_1(n) + \lambda_2 p_3(n-1), (2 \le n \le N-1), \tag{11}$$

$$rp_3(N) = bp_1(N) + \lambda_2 p_3(N-1), \tag{12}$$

$$\sum_{n=0}^{N} p_0(n) + \sum_{n=1}^{N} p_1(n) + \sum_{n=1}^{N} p_2(n) + \sum_{n=1}^{N} p_3(n) = 1. \tag{13}$$

4   Matrix Form Solution

In the following, we derive the steady-state probability by using the partitioned block marix method.

Let $P = (p_0(0), P_0, P_1, P_2, P_3)$ be the steady-state probability vector of the transition rate matrix Q, where $P_0 = (p_0(1), p_0(2), \cdots, p_0(N))$, $P_i = (p_i(1), p_i(2), \cdots, p_i(N)), (1 \le i \le 3)$. Then, the steady-state probability equations above can be rewritten in the matrix form as

$$\begin{cases} PQ = 0 \\ Pe = 1 \end{cases} \tag{14}$$

Where e is a column vector with $4N+1$ components, and each component of e equal to one, and the transition rate matrix Q of the Markov process has the following blocked matrix structure:

$$Q = \begin{pmatrix} \lambda_0 & \eta & 0 & 0 & 0 \\ 0 & A_0 & B_0 & 0 & 0 \\ \alpha & 0 & B_1 & C_1 & D_1 \\ \beta & 0 & B_2 & C_2 & 0 \\ 0 & 0 & B_3 & 0 & D_3 \end{pmatrix}$$

where λ_0 is a constant, $\eta = (-\lambda_0, 0, \cdots, 0)$ is a $1 \times N$ row vector, $\alpha = \left[-\mu_1(1-\theta), 0, \cdots, 0\right]^T$, $\beta = (-\mu_2, 0, \cdots, 0)^T$ are $1 \times N$ column vectors. $A_0, B_i \ (0 \le i \le 2), C_i \ (1 \le i \le 2), D_i \ (1 \le i \le 3)$ are square matrices.

Eq. (14) is rewritten as follows:

$$\lambda_0 p_0(0) + P_1 \alpha + P_2 \beta = 0, \tag{15}$$

$$p_0(0)\eta + P_0 A_0 = 0, \tag{16}$$

$$P_0 B_0 + P_1 B_1 + P_2 B_2 + P_3 B_3 = 0, \tag{17}$$

$$P_1 C_1 + P_2 C_2 = 0, \tag{18}$$

$$P_1 D_1 + P_3 D_3 = 0, \tag{19}$$

$$p_0(0) + P_0 e_N + P_1 e_N + P_2 e_N + P_3 e_N = 1, \tag{20}$$

Where e_N is column vector with N components, and component of e_N to one.

Form Eq. (2) we get

$$p_0(1) = \frac{\lambda_0}{v + \lambda_0} p_0(0) \tag{21}$$

$$p_0(k) = \left(\frac{\lambda_0}{v + \lambda_0}\right)^k p_0(0), \ (1 \le k \le N-1) \tag{22}$$

$$p_0(N) = \frac{\lambda_0}{v} \left(\frac{\lambda_0}{v + \lambda_0}\right)^{N-1} p_0(0) \tag{23}$$

Form Eq. (18) we get

$$P_2 = -P_1 C_1 C_2^{-1}$$

(24)

Form Eq. (19) we get

$$P_3 = -P_1 D_1 D_3^{-1}$$

(25)

Substituting Eq. (24) and (25) into (17), we get

$$P_1 \left[B_1 - C_1 C_2^{-1} B_2 - D_1 D_3^{-1} B_3 \right] = v P_0$$

(26)

Let $A = B_1 - C_1 C_2^{-1} B_2 - D_1 D_3^{-1} B_3$, after some algebraic mvnipulation we find the component of the A as follows:

$$
a_{ij} = \begin{cases}
\mu_1 + \lambda_1 + b - \mu_1 \mu_2 \theta \dfrac{\lambda_1}{(\mu_2 + \lambda_1)} + \dfrac{br}{r + \lambda_2} , i = j \neq N \\[2mm]
\mu_1 , i = j = N \\[2mm]
-\mu_1 (1 - \theta) - \dfrac{\mu_2}{\mu_2 + \lambda_1} , j = i + 1, j \neq N \\[2mm]
-\mu_1 (1 - \theta) , i = N, j = N - 1 \\[2mm]
-\lambda_1 - \dfrac{\mu_2 \lambda_1^{j-i+1}}{(\mu_2 + \lambda_1)^{j-i+2}} + \dfrac{br \lambda_2}{(r + \lambda_2)^2} , i < j \leq N - 1 \\[2mm]
\dfrac{br \lambda_2}{(r + \lambda_2)^2} , i = 1, j = N \\[2mm]
-\lambda_1 + \dfrac{br \lambda_2}{(r + \lambda_2)^2} , i = N - 1, j = N \\[2mm]
0, others
\end{cases}
$$

Let $A = \begin{pmatrix} r_1 & 0 \\ \tilde{A} & r_2^T \end{pmatrix}$, where $\tilde{A}$ is a upper triangular matrix $r_1 = (a_{11}, a_{12}, a_{13}, 0, \ldots, 0)$ is a

$1 \times (N - 1)$ row vector, $r_2^T = \left(0, \ldots, 0, a_{(N-1)N}, a_{NN} \right)^T$ is a $1 \times (N - 1)$ column vector.

Let $P_1 = \left(p_1(1), \tilde{P}_1 \right)$, Eq. (14) is rewritten as follows:

$$p_1(1) r_1 + \tilde{P}_1 \tilde{A} = v \left(p_0(1), p_0(2), \ldots, p_0(N-1) \right) = v \sigma p_0(0)$$

(27)

$$\tilde{P}_1 r_2^T = v p_0(N) = \lambda_0 \left(\frac{\lambda_0}{v + \lambda_0} \right)^{N-1} p_0(0)$$

(28)

where $\sigma = \left(\dfrac{\lambda_0}{v + \lambda_0}, \left(\dfrac{\lambda_0}{v + \lambda_0} \right)^2, \ldots, \left(\dfrac{\lambda_0}{v + \lambda_0} \right)^{N-1} \right).$

Theorem 4.1. $\tilde{A}$ is an invertible matrix, the determinant is

$$\left|\tilde{A}\right| = \prod_{i=2}^{N} a_{i(i-1)} \neq 0$$

Proof. $\tilde{A}$ is a upper triangular matrix, the determinant is equal to the product of diagonal elements, that is $\left|\tilde{A}\right| = \prod_{i=2}^{N} a_{i(i-1)}$.

For $\lambda_0, \lambda_1, \lambda_2, \mu_1, \mu_2, v, b, r > 0$, $0 < b < 1, \lambda_1, \lambda_2 \geq 0$, According to the above expression of the a_{ij} , that $a_{i(i-1)} < 0, i = 2,3,\ldots,N$, so $\left|\tilde{A}\right| \neq 0$.

Form theorem 4.1 and Eq. (26) we get

$$\tilde{P}_1 = v\sigma p_0 (0)\tilde{A}^{-1} - p_1(1)r_1\tilde{A}^{-1} \tag{29}$$

Substituting Eq. (29) into (28), we get

$$p_1(1) = cp_0(0) \tag{30}$$

where $c = \dfrac{1}{r_1\tilde{A}^{-1}r_2^T}\left[v\sigma\tilde{A}^{-1}r_2^T - \lambda_0\left(\dfrac{\lambda_0}{v+\lambda_0}\right)^{N-1}\right]$ is a constant.

Substituting Eq. (30) into (29), we get

$$\tilde{P}_1 = p_0(0)\left(v\sigma\tilde{A}^{-1} - cr_1\tilde{A}^{-1}\right) \tag{31}$$

Let $q = v\sigma\tilde{A}^{-1} - cr_1\tilde{A}^{-1}$ is a $1\times(N-1)$ row vector, that $\tilde{P}_1 = p_0(0)q$. So

$$P_1 = \left(p_1(1), \tilde{P}_1\right) = p_0(0)(c,q) \tag{32}$$

Substituting Eq. (32) into (24), we get

$$P_2 = \mu_1\theta p_0(0)(c,q)C_2^{-1} \tag{33}$$

Substituting Eq. (32) into (25), we get

$$P_3 = bp_0(0)(c,q)D_3^{-1} \tag{34}$$

Substituting Eq. (16) ,(33)and(34) into (20), we get

$$p_0(0) = \left[1+\eta A_0^{-1}e_N +(c,q)e_N +\mu_1\theta(c,q)C_2^{-1}e_N +b(c,q)D_3^{-1}e_N\right]^{-1} \tag{35}$$

Substituting Eq.(35) into (16),(33)and(34), we get the matrix solution of P_0, P_2, P_3 . In summary, probability matrix of the steady state solution is:

$$p_0(0)=\delta \quad P_0 = -\delta\eta A_0^{-1} \quad P_1 = \delta(c,q) \quad P_2 = \mu_1\theta\delta(c,q)C_2^{-1} \quad P_3 = b\delta(c,q)D_3^{-1}$$

Where $\delta = m^{-1}, m = 1+\eta A_0^{-1}e_N +(c,q)e_N +\mu_1\theta(c,q)C_2^{-1}e_N +b(c,q)D_3^{-1}e_N$

5   Performance Measures of System

(1) The probability that the system service station during busy period

$$P_B = \delta \sum_{n=1}^{N} \left[(c,q) + \mu_1 \theta(c,q) C_2^{-1} \right]$$

(2) The probability that the system service station during vacation period

$$P_V = \delta \left(1 - \sum_{n=0}^{N-1} \eta A_0^{-1} \varepsilon_{n+1} \right)$$

(3) The average waiting queue length of the system

$$E(L_q) = \delta \sum_{n=1}^{N-1} n \left[(c,q) \varepsilon_{n+1} + \mu_1 \theta(c,q) C_2^{-1} \varepsilon_{n+1} + b(c,q) D_3^{-1} \varepsilon_{n+1} - \eta A_0^{-1} \varepsilon_{n+1} \right] + N\delta$$

(4) The average queue length of the system

$$E(L) = -\delta \sum_{n=1}^{N-1} n\eta A_0^{-1} \varepsilon_{n+1} + N\delta + \delta \sum_{n=1}^{N} n \left[(c,q) \varepsilon_n + \mu_1 \theta(c,q) C_2^{-1} \varepsilon_n + b(c,q) D_3^{-1} \varepsilon_n \right]$$

6   Numerical Examples

Obtained by the above analysis, we obtain the average waiting queue length and the the average queue length of the system, and some other state indicators. But as a management decision makers should understand some of the parameters on the impact of these state indicators of the system, so that the queuing system as optimal. We take $N = 5$ for example.

In Fig.1, we fix $\lambda_0 = \lambda_1 = \lambda_2 = 1$, $\mu_2 = 1, \theta = 0.5$, $v = 1$, $r = 1$, and μ_1 from 0.5 to 2.5, b from 0 to 1. Looking at Fig.1, with the μ_1 increase will find that attendant faster and faster, steady state system in reducing the number of customers. When μ_1 is fixed, with the b increases, the captain gradually increasing.

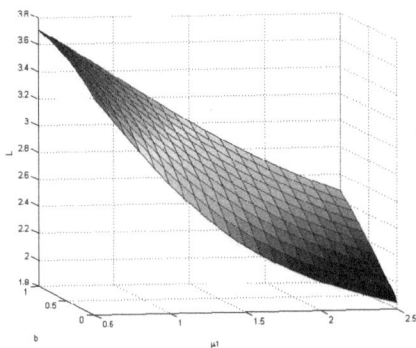

Fig. 1. The expected queue length $E(L)$ vs.the service rate μ_1 and b

Through the above analysis, a clearer understanding of the system, some of the parameters on the performance of queuing systems. Using this result, service providers can design a reasonable rate and holiday vacation service rate, so that the queuing system as optimal.

References

1. Gray, W.J., Wang, P.P., Scot, M.: A Vacation Queueing Model with Service Breakdowns. Applied Mathematics Modelling 24, 391–400 (2000)
2. Madan, K.C.: An M/G/1 Queue with Second Optional Service. Queue. Syst. (34), 37–46 (2000)
3. Medhi, J.: A Single Server Poisson input Queue with a Second Optional Channel. Queue. Syst. (42), 239–242 (2002)
4. Yue, D., Zhang, Y., Yue, W.: Optimal Performance Analysis of an M/M/1/N Queue System with Balking, Reneging and Server Vacation. International Journal of Pure and Applied Mathematics 28, 101–115 (2006)
5. Tian, R., Yue, D., Hu, L.: M/ H2 / 1 Queuing System with Balking, N-Policy and Multiple Vacations. Operation Research and Management Science 4, 56–60 (2007)
6. Yue, D., Sun, Y.: The Waiting Time of the M/M/1/N Queuing System with Balking Reneging and Multiple Vacations. Chinese Journal of Engineering Mathematics 5, 943–946 (2008)
7. Yue, D., Sun, Y.: The Waiting Time of MPMPCPN Queuing System with Balking, Reneging and Multiple Synchronous Vacations of Partial Servers. Systems Engineering Theory & Practice 2, 89–97 (2008)
8. Wu, J., Yin, X., Liu, Z.: The M/G/1 retrial G-queues with N-policy, Feedback, Preemptive Resume and Unreliable Server. ACTA Mathematicae Applicatae Sinica 32, 323–335 (2009)
9. Tang, Y.: Some Seliability Indices in Siscrete Geomx/G/1 Repairable Queueing System with Delayed Multiple Vacations. System Engineering—Theory & Practice 29, 135–143 (2009)
10. Feng, Y.: A Repairable M/G/1 Queue System with Negative Arrivals Under Multiple Vacation Policy. College Mathema TICS 25, 86–90 (2009)

Improving Data Availability in Chord p2p System[*]

Rafał Kapelko

Institute of Mathematics and Computer Science,
Wrocław University of Technology,
Poland
`rafal.kapelko@pwr.wroc.pl`

Abstract. In this paper we investigate the information availability in the direct union of 2 copies of chord which is the improved version of popular P2P Chord system. We present precise asymptotic formula describing the resistance to the loss of documents stored in them.

After unexpected departures of $k = o(\sqrt{n})$ nodes from the system no information disappears with high probability and $\sqrt{n} = o(k)$ unexpected departures lead to loss some data with high probability.

Our modification of Chord is very soft and requires only a small interference in the programming code of the original Chord protocol.

Keywords: peer-to-peer network, Chord, information, threshold, order statistics, incomplete Beta function.

1 Introduction

In the last couple of years, P2P systems have been popular and powerful network for successful sharing of certain resources. Despite of the advantages of P2P systems there is problem with unexpected departure of nodes from the system (see [1]). In the Chord P2P system each such as departure of node effects in losing data stored in the node.

To increase data availability we present the direct union of 2 copies of chord which is a very soft modification of the original Chord protocol. The first preliminary result for information availability in the direct union of 2 copies of Chord was obtained in [2]. It states that after unexpected departures of $k \leq \sqrt{\frac{n}{\log n}}$ nodes the system is safe with high probability and for $k \geq \sqrt{n \log n}$ it is unsafe with high probability.

In this paper we generalize the result from [2] and show that $k = \sqrt{n}$ is a threshold for the information availability (see Remark 1). Moreover, we give a very precise asymptotic formula (see Theroem 1). In the proof we use order statistics, asymptotic techniques and present the new inequalities for incomplete Beta function.

There are several previous works related to the subject [3], [4]. In [3] authors modify original Chord to Chordet - the replication scheme. The efficiency is confirmed by the series of numerical experiments. In [4] authors present theoretical model of nodes with long session time.

---

[*] Supported by grant nr 2011/342643 of the Institute of Mathematics and Computers Science of the Wrocław University of Technology.

B. Liu and C. Chai (Eds.): ICICA 2011, LNCS 7030, pp. 208–215, 2011.

The paper is organised as follows. In Section 2 we discuss the basic facts and notation. In Section 3 we describe the direct union of 2 copies of Chord and estimate threshold for information availability. In Section 4 we present several inequalities for incomplete beta function. Finally, Section 5 summaries and concludes our work.

2   Basic Facts and Notation

The classical Chord protocol defined in [5] and developed in [6], [2] and many other papers from the formal point of view may be described as a structure

$$\text{Chord} = (\{0,1\}^{160}, H, H_1),$$

where H is a hash function assigning position to each node and H_1 is a hash function assigning position of descriptors of documents. The space $\{0,1\}^{160}$ is identified with the set $\{0, 1, \ldots, 2^{160} - 1\}$ considered as the circular space with the ordering $0 < 1 < \ldots < 2^{160} - 1 < 0 < \ldots$. Each new node X obtains a position $H(Id)$ (where Id is an identifier of the node) in the space $\{0,1\}^{160}$ and is responsible for the interval starting at point $H(Id)$ and ending at the next point from the set $\{H(Id') : Id' \neq Id\}$. This node is called the successor of the node X. Each document with a descriptor doc is placed at point $H_1(doc)$ in the space $\{0,1\}^{160}$ and the information about this document is stored by the node which is responsible for the interval into which $H_1(doc)$ falls.

We shall identify the space $\{0,1\}^{160}$, after a proper scaling, with the unit interval $[0,1)$ and we shall interpret positions of nodes of a Chord as elements of $[0,1)$. Moreover, we may assume that one node is at point 0. The random sets corresponding to nodes of Chord are generated in the following way: we generate independently n random points $X_1, \ldots, X_n$ from $[0,1)$ using the uniform distribution on $[0,1)$ and then we sort them in increasing order and obtain a sequence $X_{1:n} \leq \ldots \leq X_{n:n}$ This construction will be used as a formal model of the Chord protocol with $n+1$ nodes. We call the segment $[X_{i:n}, X_{i+1:n})$ the interval controlled by the node i.

By $\Gamma(z)$ we denote the standard generalization of the factorial function. The following identities hold: $n! = \Gamma(n+1), \Gamma(z+1) = z\Gamma(z)$. We will also use several times the Eulerian function

$$\text{B}(a,b) = \int_0^1 x^{a-1}(1-x)^{b-1}dx$$

and the incomplete beta function

$$\text{B}(t; a, b) = \frac{1}{\text{B}(a,b)} \int_0^t x^{a-1}(1-x)^{b-1}dx \tag{1}$$

which are defined for all complex numbers a, b such as $\Re(a) > 0$ and $\Re(b) > 0$.

We will use the following basic identity $\text{B}(a,b) = \Gamma(a)\Gamma(b)/\Gamma(a+b)$.

Let $X_1, \ldots, X_n$ be independent random variables with the uniform density on $[0,1)$. The order statistics $X_{1:n}, \ldots X_{n:n}$ are the random variables obtained from $X_1, \ldots, X_n$ by sorting each of their realizations in the increasing order. The probabilistic density $f_{k:n}(x)$ of the variable $X_{k:n}$ equals

$$f_{k:n}(x) = \frac{1}{\mathrm{B}(k, n - k + 1)} x^{k-1}(1 - x)^{n-k} \quad . \tag{2}$$

(see e.g. [7]). Let us recall that these kinds of probabilistic distributions are called Beta distributions.

Let X be a random variable. We denote its expected value by $\mathbf{E}\,[X]$.

3   Direct Union of Chord

In this section we analyse the maximal number of nodes which can be removed to system stay in safe configuration and minimal number of nodes which can be removed to system stay in unsafe configuration. The modyfication called direct union of Chord and the concept of safety (see Definiton 1) were introduced in [2].

The *direct union of 2 copies of Chord* is the structure

$$\mathrm{Chord} \oplus \mathrm{Chord} = (\Omega \times \{1, 2\}, \{H_1, H_2\}, H) \ ,$$

where H_1, H_2, H are independent hash functions. Each new node with an identifier Id uses the value $H_i(Id)$ to calculate its position in the i-th copy $C_i = \Omega \times \{i\}$ of the Chord and each document doc is placed at point $H(doc)$ in each copy C_i of the Chord. Notice that we use the same function H to place new documents in distinct copies of the Chord.

Definition 1. *Let $A = \{n_1, \ldots, n_k\} \subseteq \{1, \ldots, n + 1\}$ be a random subset of nodes from the structure* $\mathrm{Chord} \oplus \mathrm{Chord}$ *with $n + 1$ nodes. We denote by $K_{A,1}$ the unions of intervals controlled by nodes from A in the first circle and we denote by $K_{A,2}$ the unions of intervals controlled by nodes from A in the second circle. We say that the set A is **safe** if $K_{A,1} \cap K_{A,2} = \emptyset$. We say that the set A is **unsafe** if A is not safe.*

Notice that if the set A is safe then no information disappears from the system after unexpected departure of all nodes from A.

Theorem 1. *Let $n + 1 \geq 257$ be the number of nodes in the structure* $\mathrm{Chord} \oplus \mathrm{Chord}$ *and let A denote a set of nodes of cardinality k.*

1. If $n^{\frac{3}{8}} < k < \sqrt{n}$ then $\Pr[A$ is safe$] > 1 - \left(\frac{2k^2}{n} + \frac{k^2}{n-k} + \frac{3e}{2\sqrt{2\pi k}} \left(\frac{2}{e^{\frac{7}{8}}} \right)^k \right)$.

2. If $\sqrt{n} < k < n^{\frac{5}{8}}$ then $\Pr[A$ is unsafe$] > \left(1 - \frac{1}{e^{\frac{k^2}{2n}}} \right) \left(1 - \frac{3e}{\sqrt{2\pi k}} \left(\frac{e^{\frac{17}{32}}}{2} \right)^k \right)$.

Notice that the bounds in Theorem 1 do not depend on the number of documents put into the system and the functions: $\left(\frac{2}{e^{\frac{7}{8}}} \right)^k$, $\left(\frac{e^{\frac{17}{32}}}{2} \right)^k$ are exponentially small.

Remark 1. *Theorem 1 shows that $k = \sqrt{n}$ is a threshold for the information availability in the* $\mathrm{Chord} \oplus \mathrm{Chord}$. *This means that after unexpected departure of $k = o(\sqrt{n})$ nodes from the system no information disappears with high probability and for $\sqrt{n} = o(k)$ the system losses some data with high probability.*

Proof of Theorem 1. Suppose that $n^{\frac{3}{8}} < k < \sqrt{n}$. Let us recall that $K_{A,1}$ is the union of intervals controlled by nodes from A in the first circle and $K_{A,2}$ is the union of intervals controlled by nodes from A in the second circle. Let L_A be the area of the set $K_{A,1}$, where $|A| = k$. Notice L_A is a random variable with the same density as the density of the k-th order statistics, i.e. equals $f_{k:n}$ (see formula (2)) therefore $\Pr[L_k < t] = \int_0^t f_{k:n}(x)dx$. From Theorem 2 we deduce that if A is a random set of nodes of cardinality k then

$$\Pr\left[L_A < \frac{2k}{n}\right] > 1 - \frac{3e}{2\sqrt{2\pi k}}\left(\frac{2}{e}e^{\frac{2(k-1)}{n}}\right)^k > 1 - \frac{3e}{2\sqrt{2\pi k}}\left(\frac{2}{e^{\frac{7}{8}}}\right)^k.$$

Fix a random set $A = \{n_1, \ldots, n_k\}$ of cardinality k such that $L_A < \frac{2k}{n}$. Notice that the positions of nodes in the second copy of Chord are chosen independently from the position of nodes from the first copy. Let $Y_1, \ldots, Y_n$ denotes the positions of nodes in the second copy of Chord $\oplus$ Chord.

There are two reasons why an interval controlled by a variable Y_i may have a nonempty intersection with the set $K_{A,1}$. The first one is caused by the fact $Y_i \in K_{A,1}$. The second reason is that $Y_i \notin S_1$ but $[Y_i, Y_i^*) \cap K_{A,1} \neq \emptyset$, where Y_i^* is the end of the interval controlled by the i-th node. Therefore

$$\Pr[K_{A,1} \cap K_{A,2} \neq \emptyset] \leq$$
$$\Pr[(\exists i)(Y_{n_i} \in K_{A,1})] + \Pr[(\exists i)(Y_{n_i} \notin K_{A,1} \wedge [Y_{n_i}, Y_{n_i}^*) \cap K_{A,1} \neq \emptyset)].$$

Notice that

$$\Pr[(\exists i)(Y_{n_i} \in K_{A,1})] \leq kL_A < \frac{2k^2}{n}.$$

Notice also that the set $K_{A,1}$ is a union of k intervals, so

$$|\{i : Y_i \notin K_{A,1} \wedge [Y_i, Y_i^*) \cap K_{A,1} \neq \emptyset\}| \leq k .$$

Therefore

$$\Pr[(\exists i)(Y_{n_i} \notin K_{A,1} \wedge [Y_{n_i}, Y_{n_i}^*) \cap K_{A,1} \neq \emptyset)] \leq 1 - \frac{\binom{n-k}{k}}{\binom{n}{k}}.$$

Since $n^{\frac{3}{8}} < k < \sqrt{n}$, so we have

$$\frac{\binom{n-k}{k}}{\binom{n}{k}} = \prod_{i=1}^{k}\left(1 - \frac{k}{n-k+i}\right) \geq \left(1 - \frac{k}{n-k}\right)^k \geq 1 - \frac{k^2}{n-k} .$$

Hence

$$\Pr\left[K_{A,1} \cap K_{A,2} \neq \emptyset | L_A < \frac{2k}{n}\right] \leq \frac{2k^2}{n} + \left(1 - \frac{\binom{n-k}{k}}{\binom{n}{k}}\right) \leq \frac{2k^2}{n} + \frac{k^2}{n-k}.$$

Therefore the inequality $\Pr[A] \leq \Pr[A|B] + \Pr[B^c]$ gives us

$$\Pr[K_{A,1} \cap K_{A,2} \neq \emptyset] \leq \Pr\left[K_{A,1} \cap K_{A,2} \neq \emptyset | L_A < \frac{2k}{n}\right] + \Pr\left[L_A \geq \frac{2k}{n}\right] <$$

$$\frac{2k^2}{n} + \frac{k^2}{n-k} + \frac{3e}{2\sqrt{2\pi k}} \left(\frac{2}{e^{\frac{7}{8}}}\right)^k ,$$

so the first part of the Theorem is proved.

Suppose now that $\sqrt{n} < k < n^{\frac{5}{8}}$ and that $A = \{n_1, \ldots, n_k\}$ is a random subset of $\{1, \ldots, n\}$ of cardinality k. From Theorem 2 we deduce that $\Pr[L_A < \frac{k}{2n}] \leq \frac{3e}{\sqrt{2\pi k}} \left(\frac{\sqrt{e}}{2} e^{\frac{k}{2n}}\right)^k$, so

$$\Pr[L_A \geq \frac{k}{2n}] > 1 - \frac{3e}{\sqrt{2\pi k}} \left(\frac{\sqrt{e}}{2} e^{\frac{k}{2n}}\right)^k > 1 - \frac{3e}{\sqrt{2\pi k}} \left(\frac{e^{\frac{17}{32}}}{2}\right)^k .$$

Suppose hence that $L_A \geq \frac{k}{2n}$. Let, as in the previous part of the proof, $Y_1, \ldots, Y_n$ denote the positions of nodes in the second copy. Then $\Pr[Y_i \in K_{A,1}] = L_A$ for each i and the random variables $Y_1, \ldots, Y_n$ are independent and uniformly distributed. Therefore

$$\Pr[\{Y_{n_1}, \ldots, Y_{n_k}\} \cap K_{A,1} \neq \emptyset] = 1 - (1 - L_A)^k ,$$

so we see that

$$\Pr[\{Y_{n_1}, \ldots, Y_{n_k}\} \cap K_{A,1} \neq \emptyset | L_A \geq \frac{k}{2n}] \geq 1 - (1 - \frac{k}{2n})^k .$$

From the inequality $\left(\left(1 - \frac{1}{\frac{2n}{k}}\right)^{\frac{2n}{k}} < \frac{1}{e} \text{ for } \frac{2n}{k} \geq 2\right)$ we get

$$\Pr[\{Y_{n_1}, \ldots, Y_{n_k}\} \cap K_{A,1} \neq \emptyset | L_A \geq \frac{k}{2n}] \geq 1 - \frac{1}{e^{\frac{k^2}{2n}}} .$$

Hence

$$\Pr[\{Y_1, \ldots, Y_{n_k}\} \cap K_{A,1} \neq \emptyset] \geq$$

$$\Pr[\{Y_{n_1}, \ldots, Y_{n_k}\} \cap K_{A,1} \neq \emptyset | L_A \geq \frac{k}{2n}] \Pr[L_A \geq \frac{k}{2n}] >$$

$$\left(1 - \frac{1}{e^{\frac{k^2}{2n}}}\right) \left(1 - \frac{3e}{\sqrt{2\pi k}} \left(\frac{e^{\frac{17}{32}}}{2}\right)^k\right) ,$$

so the second part of the Theorem is proved. $\square$

4   Inequalities for Incomplete Beta Function

In this section we present the inequalities for incomplete Beta function (see formula (1)) which are new to the best author knowledge. Some related works have been reported in the literature [8], [9].

Theorem 2. *Let* $n^{\frac{3}{8}} < k < n^{\frac{5}{8}}$ *and* $n \geq 256$. *Then*

1. $B(\frac{k}{2n}; k, n - k + 1) < \frac{3e}{\sqrt{2\pi k}} \left(\frac{\sqrt{e}}{2} e^{\frac{k}{2n}} \right)^k,$

2. $B(\frac{2k}{n}; k, n - k + 1) > \left(1 - \frac{3e}{2\sqrt{2\pi k}} \left(\frac{2}{e} e^{\frac{2(k-1)}{n}} \right)^k \right).$

Notice that for $n^{\frac{3}{8}} < k < n^{\frac{5}{8}}$ and $n \geq 256$ we have $\frac{2}{e} e^{\frac{k}{2n}} < 1$ and $\frac{\sqrt{2}}{e} e^{\frac{2(k-1)}{n}} < 1$, so the functions $\left(\frac{\sqrt{e}}{2} e^{\frac{k}{2n}} \right)^k$ and $\left(\frac{2}{e} e^{\frac{2(k-1)}{n}} \right)^k$ are exponentially small.

Proof. The first part of the Theorem follows directly from Lemma 2 and Lemma 3.
The identity $B(\frac{2k}{n}; k, n-k+1) = B(k, n-k+1) - \int_{\frac{2k}{n}}^{1} x^{k-1}(1-x)^{n-k} dx$, Lemma 2 and Lemma 4 give the second part of the Theorem.   □

Lemma 2. *Let* $n^{\frac{3}{8}} < k < n^{\frac{5}{8}}$ *and* $n \geq 256$. *Then*

$$B(k, n - k + 1) > \frac{\sqrt{2\pi}}{e} \left(\frac{1}{e} \right)^k \left(\frac{k}{n} \right)^k \sqrt{\frac{1}{k}}.$$

Proof. The Stirling's approximation inequality

$$x^{x+\frac{1}{2}} e^{-x} \sqrt{2\pi} e^{\frac{1}{12x+1}} \leq \Gamma(x+1) \leq x^{x+\frac{1}{2}} e^{-x} \sqrt{2\pi} e^{\frac{1}{12x}}$$

for $x \geq 2$ (see e.g. [10]) gives us

$$B(k, n - k + 1) = \frac{\Gamma(k)\Gamma(n - k + 1)}{\Gamma(n + 1)} \geq$$

$$\sqrt{2\pi} \frac{e^{1 + \frac{1}{12(n-k)+1}}}{e^{\frac{1}{12n} - \frac{1}{12(k-1)+1}}} \left(\left(1 - \frac{1}{k} \right)^{k-1} \right)^{\frac{k - \frac{1}{2}}{k-1}} \left(\left(1 - \frac{1}{n} \right)^{\frac{n}{k}-1} \right)^{\frac{n-k+\frac{1}{2}}{\frac{n}{k}-1}} \left(\frac{k}{n} \right)^k \sqrt{\frac{1}{k}}.$$

From the inequality $\left(\left(1 - \frac{1}{x} \right)^{x-1} > \frac{1}{e} \text{ for } x \geq 2 \right)$ we get

$$B(k, n - k + 1) > \sqrt{2\pi} \frac{e^{\frac{1}{12(n-k)+1} + \frac{1}{12(k-1)+1}}}{e^{\frac{1}{12n} + \frac{1}{2(k-1)} + \frac{k}{2(n-k)}}} \left(\frac{1}{e} \right)^k \left(\frac{k}{n} \right)^k \sqrt{\frac{1}{k}}.$$

Notice that

$$\frac{e^{\frac{1}{12(n-k)+1} + \frac{1}{12(k-1)+1}}}{e^{\frac{1}{12n} + \frac{1}{2(k-1)} + \frac{k}{2(n-k)}}} \geq \frac{e^0}{e^1}.$$

Therefore the Lemma is proved.   □

Lemma 3. *Let $n^{\frac{3}{8}} < k < n^{\frac{5}{8}}$ and $n \geq 256$. Then*

$$\int_0^{\frac{k}{2n}} x^{k-1}(1-x)^{n-k}dx < \frac{3}{k}\left(\frac{k}{2n}\right)^k \left(\frac{1}{e}\right)^{\frac{k}{2}} e^{\frac{k^2}{2n}}.$$

Proof. Notice that the function $f(x) = x^{\frac{2}{3}k}(1-x)^{n-k}$ is increasing on the interval $\left[0, \frac{2k}{3n-k}\right]$, decreasing on the interval $\left[\frac{2k}{3n-k}, 1\right]$ and $\frac{k}{2n} < \frac{2k}{3n-k}$. Hence

$$\int_0^{\frac{k}{2n}} x^{k-1}(1-x)^{n-k}dx \leq \max_{x\in[0,\frac{k}{2n}]} \left(x^{\frac{2}{3}k}(1-x)^{n-k}\right) \int_0^{\frac{k}{2n}} x^{\frac{1}{3}k-1}dx =$$

$$\frac{3}{k}\left(\frac{k}{2n}\right)^k \left(\left(1 - \frac{1}{\frac{2n}{k}}\right)^{\frac{2n}{k}}\right)^{\frac{k}{2n}(n-k)}.$$

From the inequality $\left(\left(1 - \frac{1}{x}\right)^x < \frac{1}{e} \text{ for } x \geq 2\right)$ we get

$$\int_0^{\frac{k}{2n}} x^{k-1}(1-x)^{n-k}dx < \frac{3}{k}\left(\frac{k}{2n}\right)^k \left(\frac{1}{e}\right)^{\frac{k}{2}} e^{\frac{k^2}{2n}}.$$

$\square$

Lemma 4. *Let $n^{\frac{3}{8}} \leq k \leq n^{\frac{5}{8}}$ and $n \geq 256$. Then*

$$\int_{\frac{2k}{n}}^1 x^{k-1}(1-x)^{n-k}dx < \frac{3}{2k}\left(\frac{2k}{n}\right)^k \left(\frac{1}{e}\right)^{2k} e^{\frac{2k(k-1)}{n}}.$$

Proof. Notice that the function $f(x) = x^{k-1}(1-x)^{\frac{2}{3}n-k+1}$ is increasing on the interval $\left[0, \frac{3k-3}{2n}\right]$, deacreasing on the interval $\left[\frac{3k-3}{2n}, 1\right]$ and $\frac{3k-3}{2n} < \frac{2k}{n}$. Hence

$$\int_{\frac{2k}{n}}^1 x^{k-1}(1-x)^{n-k}dx \leq \max_{x\in[\frac{2k}{n},1]} \left(x^{k-1}(1-x)^{\frac{2}{3}n-k+1}\right) \int_{\frac{2k}{n}}^1 (1-x)^{\frac{1}{3}n-1}dx =$$

$$\frac{3}{2k}\left(\frac{2k}{n}\right)^k \left(\left(1 - \frac{1}{\frac{n}{2k}}\right)^{\frac{n}{2k}}\right)^{\frac{2k}{n}(n-k+1)}.$$

From the inequality $\left(\left(1 - \frac{1}{x}\right)^x < \frac{1}{e} \text{ for } x \geq 2\right)$ we get

$$\int_{\frac{2k}{n}}^1 x^{k-1}(1-x)^{n-k}dx < \frac{3}{2k}\left(\frac{2k}{n}\right)^k \left(\frac{1}{e}\right)^{2k} e^{\frac{2k(k-1)}{n}}.$$

$\square$

5   Conclusions and Further Works

In the paper the author proved that $\sqrt{n}$ is a threshold for the information availability in the Chord $\oplus$ Chord. Theorem 1 gives a very precise influence of parameter k on keeping stored data in the system.

In the paper the author focuses on the precise theoretical description of data availability in the Chord $\oplus$ Chord. The technical aspects of the shown results along with the simulations we shall present in the next paper.

Furthermore, we plan to generalize our results to the direct union of l chords ($l \geq 2$) and extend our study to other P2P systems such us CAN (see [11]), KADEMLIA (see [12]), PASTRY (see [13]) and others.

References

1. Derek, L., Zhong, Y., Vivek, R., Loguinov, D.: On Lifetime-Based Node Failure and Stochastic Resilience of Decentralized Peer-to-Peer Networks. IEEE/ACM Transactions on Networking 15, 644–656 (2007)
2. Cichoń, J., Jasiński, A., Kapelko, R., Zawada, M.: How to Improve the Reliability of Chord? In: Meersman, R., Herrero, P. (eds.) OTM-WS 2008. LNCS, vol. 5333, pp. 904–913. Springer, Heidelberg (2008)
3. Park, G., Kim, S., Cho, Y., Kook, J., Hong, J.: Chordet: An Efficient and Transparent Replication for Improving Availability of Peer-to-Peer Networked Systems. In: 2010 ACM Symposium on Applied Computing (SAC), Sierre, Switzerland, pp. 221–225 (2010)
4. Hong, F., Li, M., Yu, J.: SChord: Handling Churn in Chord by Exploiting Node Session Time. In: Zhuge, H., Fox, G.C. (eds.) GCC 2005. LNCS, vol. 3795, pp. 919–929. Springer, Heidelberg (2005)
5. Stoica, I., Morris, R., Karger, D., Kaashoek, M.F., Balakrishnan, H.: Chord: A Scalable Peer-to-Peer Lookup Service for Internet Applications. In: SIGCOMM 2001, San Diego, California, USA, pp. 149–160 (2001)
6. Liben-Nowell, D., Balakrishnan, H., Karger, D.: Analysis of the Evolution of Peer-to-Peer Systems. In: ACM Conference on Principles of Distributed Computing, Monterey, California, USA, pp. 233–242 (2002)
7. Arnold, B., Balakrishnan, N., Nagaraja, H.: A First Course in Order Statistics. John Wiley & Sons, New York (1992)
8. Kechriniotis, A.I., Theodorou, Y.A.: New Integral Inequalities for n-Time Differentiable Functions with Applications for pdfs. Applied Mathematical Sciences 2, 353–362 (2008)
9. Grinshpan, A.Z.: Weighted Integral and Integro-differential Inequalities. Advances in Applied Mathematics 41, 227–246 (2008)
10. Flajolet, F., Sedgewick, R.: Analytic Combinatorics. Cambridge University Press, Cambridge (2009)
11. Ratnasamy, S., Francis, P., Handley, M., Karp, R., Shenker, S.: A Scalable Content-addressable Network. In: SIGCOMM 2001, San Diego, California, USA, pp. 161–172 (2001)
12. Maymounkov, P., Mazières, D.: Kademlia: A Peer-to-Peer Information System Based on the XOR Metric. In: Druschel, P., Kaashoek, F., Rowstron, A. (eds.) IPTPS 2002. LNCS, vol. 2429, pp. 53–65. Springer, Heidelberg (2002)
13. Rowstron, A., Druschel, P.: Pastry: Scalable, Decentralized Object Location, and Routing for Large-Scale Peer-to-Peer Systems. In: Guerraoui, R. (ed.) Middleware 2001. LNCS, vol. 2218, pp. 329–350. Springer, Heidelberg (2001)

No Regret Learning for Sensor Relocation in Mobile Sensor Networks

Jin Li[1,2,3], Chi Zhang[2], Wei Yi Liu[1], and Kun Yue[1]

[1] School of Information Science, Yunnan University,
Kunming, China
[2] School of Software, Yunnan University,
Kunming, China
[3] Key laboratory in Software Engineering of Yunnan Province,
Kunming, China
ljatynu@gmail.com

Abstract. Sensor relocation is a critical issue because it affects the coverage quality and capability of a mobile sensor network. In this paper, the problem of sensor relocation is formulated as a repeated multi-player game. At each step of repeated interactions, each node uses a distributed no-regret algorithm to optimize its own coverage while minimizing the locomotion energy consumption. We prove that if all nodes adhere to this algorithm to play the game, collective behavior converges to a pure Nash equilibrium. Simulation results show that a good coverage performance can be obtained when a pure Nash equilibrium is achieved.

Keywords: mobile sensor networks; sensor relocation; repeated game; pure Nash equilibria; no-regret learning.

1 Introduction

With the improvements in mobile network techniques and reduced expense of individual sensing devices, there has been a large amount of research and practical applications [1-5] in mobile sensor networks (MSN). In many application scenarios, such as target tracking in some hostile environments, it is difficult to achieve an exact sensor placement through manual means. Instead, sensors may be self-deployed somewhat randomly from a initial position, and then reposition themselves to provide the required sensing coverage. Furthermore, as the network condition or application requirements change, sensor nodes may need to be redeployed to recover failures or respond to occurring events. Therefore, it is a challenging and crucial tasks for a MSN that designing a sensor relocation algorithm to achieve the required coverage level or response to new events in a dynamical way.

Towards the problem of optimal sensor relocation, there has been a few research works. For example, Meguerdichian and Koushanfar [6] present a formal definition of the worst and best coverage, then proposed polynomial-time algorithms for solving each case. Bai and Yun [8] presented results on optimal

B. Liu and C. Chai (Eds.): ICICA 2011, LNCS 7030, pp. 216–223, 2011.
© Springer-Verlag Berlin Heidelberg 2011

two-coverage deployment patterns in sensor networks. Maredn [7] proposed a general framework for the noncooperative resource allocation problems. However, all the results above do not consider the problem of sensor's movement, which is realistic constrained in mobile sensor networks. Wang and Cao [9] proposed a framework of relocating mobile sensors in a centralized manner. They defined a grid head, which is responsible for information collecting for its members, and for deter-mining sensors redundancy. However, this requires a highly computing resource for the grid head sensor to control the whole network. Moreover, there also exists the risks of head node failure, which may lead to the broken down of the whole sensor network.

In this paper, different from the exist approaches, we address the problem of sensor relocation using a game theoretic approach [14], in which the sensor nodes in MSNs are self-deployed. Our goal is to maximize the total coverage value by relocating sensor nodes from an initial configuration to optimal positions in a distributed way. While the energy constraint and balanced energy expense are also considered as well. Specifically, we model the problem of sensor relocation as a repeated game, namely sensor relocation game (SRG). In our game model, a payoff function is carefully set up. The payoff of each position for each player is a function of the number of players covering it. This in turn depends on the number of players which cover the same position. By constructing a potential function, SRG is proved as a potential game [10] . Thus, a desirable property possessed by SRG: the existence of a pure Nash equilibrium [13], which ensures that a distributed algorithm based on our repeated game approach can be designed for sensor relocating in MSNs.

In a SRG, starting with a randomly deployed positions, sensors learn to play a game through repeated interactions. More precisely, at each time step, based on the decision histories over finite previous stage, a sensor learns the regret of each relocation strategy and make a decision, either taking the same strategy as in last step with a probability β or selecting, with a a probability $1 - \beta$, the strategy which corresponds to a high aggregate regrets. We prove that if each sensor adheres to this distributed learning algorithm to play a SRG, the group behavior then converges to a pure Nash equilibrium.

As for the coverage performance, our objective is to achieves an optimal sensors relocation configuration which maximize the total coverage value under the constrains of energy consumption. This goal is refer to as a social optimum. In our game, since the social function is approximated by the potential function, the no regret dynamics converges within a neighborhood of social optimal point, thereby inducing near-optimal performance in a dynamical sense. Additionally, we demonstrate through simulations that a good coverage performance can be obtained using our proposed distributed relocation algorithm.

The organization of this paper is as follows. In Section 2, we introduce the game model and the design of payoff function. We also prove that SRG is a potential game by constructing a potential function. In Section 3, a distributed no-regret relocation algorithm is proposed with limited demand in communication and computing capabilities to reach a pure equilibrium. Simulation tests are performed in Section 4, and we conclude the paper in Section 5.

2   Problem Formulations

We divide the mission area into grids, each of which has a detecting value. The higher detecting value the grid has, the more valuable it is. Each grid is indexed by a tuple $\langle m, n \rangle$, where m, n represent the row and column respectively. Assuming that once the mission starts, every node will be in an exact grid $g(m, n)$. For simplicity, say $g_k = g(m, n)$. The grid index of sensor s_i can be represented as p_i. r_i represents the detecting radius of s_i. The coverage area is presented as follows:$c_i = \{g_k | d(p_i, g_k) \leq r_i\}$. In addition, energy consumption is considered in sensor movements. We define the cost of locomotion from grid $g_1 = g(m_1, m_1)$ to $g_2 = g(m_2, m_2)$ as:$m(g_1, g_2) = \mu \cdot (|m_1 - m_2| + |n_1 - n_2|)^{\eta}$. where $\mu > 0$ and $\eta \geq 1$. Now our relocation problem can be defined as follows: A mobile sensor network $S = \{s_i | 1 \leq s_i \leq n\}$, containing n mobile sensors deployed in a mission space, which is divided by a grid $G = \{g_k | g_k = g(i, j), 0 \leq i \leq M, 0 \leq j \leq N\}$. $V = \{v_i | 0 \leq i \leq M, 0 \leq j \leq N\}$ is the evaluated detecting value in each grid, which indicates the importance of monitoring in this section. Our goal is to design a distributed protocol to control sensors' locomotion for covering G, reaching a maximum global profit and balancing the cost of locomotion at the same time.

Sensor nodes may fail or be blocked due to lack of power, which will cause physical damage or environmental interference. The reliability of a sensor is represented as $\gamma(0 < \gamma < 1)$, then the probability that a sensor s_i correctly detect a section g_k is: $\alpha_{i,k} = \begin{cases} \gamma, g_k \in c_i \\ 0, g_k \notin c_i \end{cases}$. When there are n_k sensors covering a grid g_k, we are able to calculate the probability of correctly detecting the section: $\beta_k = 1 - (1 - \gamma)^{n_k}$. After all the sensor nodes have chosen a position to relocate, the global welfare W can be defined as follows, which we want to optimize in this mission: $W = \sum_{g_k \in G} v_k \cdot \beta_k - \sum_{s_i \in S} m_i$. m_i represents the locomotion expense of s_i from its initial section to the target area in the game.

Now, we formulate the sensor relocation problem as a so called *sensor relocation game* (SRG). A SRG is a triple $SRG = \langle N, (A_i)_{i \in N}, (u_i)_{i \in N} \rangle$, where N is a set of mobile sensors. For every sensor i, its action a_i is to move from its initial position p_i to any grid in mission space, say p_i'. In particular, the payoff function $u_i(\cdot)$ should at least satisfy the following properties in our sensor relocation game:(1) For the same distance, a higher valued section is considered to gain more; (2) For the same distance and evaluated section value, an increased degree of coverage results in a lower profit; (3) At least one pure Nash equilibrium should be guaranteed in the game; (4) Result in an optimum relocation strategy for the global welfare.

Naturally, sharing the welfare in average to all sensor nodes that covering a section is the simplest approach to achieve the requirements. However, a better payoff function with some other good properties, which are specified in the following part, can be designed for players. For each section g_k in the coverage c_i of sensor i, we define the detecting profit:

$$u_i(\boldsymbol{a}, \lambda) = \left[\sum_{g_k \in c_i} v_k \cdot \left(\frac{n_k}{n_k + \lambda} - \frac{n_k - 1}{(n_k - 1) + \lambda} \right) \right] - m_i \qquad (1)$$

where $n_k(n_k \geq 1)$ is the coverage degree of section g_k, λ is a parameter, which can be controlled by the system designer. This payoff function satisfies all first four properties above. More importantly, we can construct a potential function which guarantees that at least one pure Nash equilibrium exists in this game, for the sharing rule.

Theorem 1. *SRG where the payoff function defined by equation (1) is an exact potential game with the potential function* $\Phi(\boldsymbol{a}, \lambda) = \sum_{g_k \in G} v_k \cdot \frac{n_k}{n_k + \lambda} - \sum_{s_i \in S} m_i$.

3   A Distributed No-regret Relocation Algorithm

3.1   The Description of the Distributed Relocation Algorithm

No regret learning is a class of learning algorithm which has been widely used in distributed control problem of multi-agent systems [11-12]. In the more general regret based algorithms, each player makes a decision using only information regarding the regret for each of his possible actions. If an algorithm guarantees that a players maximum regret asymptotically approaches zero then the algorithm is referred to as a no-regret algorithm. In any game, if all players adhere to a no-regret learning algorithm, then the group behavior will converge asymptotically to a set of points of no-regret. Moreover, a Nash equilibrium is a special case of no regret points, which motivates us to develop a distributed algorithm based on no regret learning to deal with the sensors relocation problem.

Particularly, in a *SRG*, at time t, suppose the length of profile histories is l and denoted the factor of decay by $\delta \in [0, 1]$. Let $g(l) = 1/(1 + \delta + \dots + \delta^{l-1})$, the average regret of a strategy $a_i(t)$ for s_i is defined as: $R_i^{a_i}(t; l) = g(l) \sum_{k=1}^{l} \delta^{l-k} \Delta_{t-l-1+k}^{a_i}$, where $\Delta_{t-l-1+k}^{a_i} = u_i(a_i, \boldsymbol{a}_{-i}(t - l - 1 + k)) - u_i(\boldsymbol{a}(t - l - 1 + k))$. A sensor s_i has an average regret vector $[R_i^{a_i}(t)]_{a_i \in A_i}$, each element of which corresponds to the average regret of every strategy. At each step, s_i updates regrets using the following recurrence formula: $R_i^{a_i}(t+1) = \frac{t-1}{t} R_i^{a_i}(t) + \frac{1}{t}[u_i(a_i, \boldsymbol{a}_{-i}(t)) - u_i(\boldsymbol{a}(t))]$. Generally speaking, s_i selects a particular strategy at each step with probability proportional to the average regret for not choosing that particular strategy in the past steps. The main steps of the distributed relocation algorithm based on no regret learning is described as follows.

Algorithm 1. A distributed no-regret relocation algorithm

1 At step $t \leq l$ a sensor s_i randomly selects a strategy from A_i, thus, we get a l
  length profile histories: $\boldsymbol{a}(1), \boldsymbol{a}(2), ..., \boldsymbol{a}(l)$;
2 Let $t_0 = l + 1$, s_i computes regrets according to the following way:
$$R_i^{a_i}(t; l) = g(l) \sum_{k=1}^{l} \delta^{l-k} \Delta(t - l - 1 + k), \ R_i(t) = [R_i^{a_i}(t; l)]_{a_i \in A_i};$$
3 **while** $t \geq t_0$, s_i and $\exists i \in N$, $[R_i(t)]^+ \neq 0$ **do**
4 $\quad$ s_i selects a strategy according to the following rules:
5 $\quad$ With β probability, $a_i(t) = a_i(t-1)$. With $1 - \beta$ probability, s_i selects a
  $\quad$ strategy based on the following probability distribution:
  $\quad$ $$B_i^j(R_i(t)) = \frac{[R_i^j(t)]^+}{\sum\limits_{k=1}^{|A_i|}[R_i^k(t)]^+};$$
6 $\quad$ if $[R_i(t)]^+ \neq 0$ then choose a $a_i(t) \in A_i$ according to distribution $B_i^j(R_i(t))$;
7 $\quad$ otherwise $a_i(t) = a_i(t-1)$;
8 $\quad$ Updates the regrets using the following formula :
  $\quad$ $R_i^{a_i}(t; l) = \delta R_i^{a_i}(t - 1; l) + g(l)[\Delta(t-1) - \delta^l \Delta(t - l - 1)];$
9 $\quad$ $t = t + 1$;
10 **end**

3.2 Convergence Result

Generally speaking, a no regret learning algorithm converges to a no regret point
does not implies that it also converges to a pure Nash equilibrium. Now, we
prove that in a SRG, if all sensors adhere to the algorithm 1, collective behavior
converges to a pure Nash equilibrium.

Theorem 2. *The algorithm 1 converges to a pure Nash equilibrium.*

Proof. **Firstly**, we prove that at time t_0, once no regret dynamics moves to a
strict pure Nash equilibrium $\boldsymbol{a}(t_0)$, $\forall t > t_0$, $\boldsymbol{a}(t) = \boldsymbol{a}(t_0)$ then occurs with a positive probability $\beta^{n(l-1)}$. $\forall t_0 > l$, suppose the profile $\boldsymbol{a}(t_0) = (a_1(t_0), ..., a_n(t_0))$
is a strict Nash equilibrium. $\boldsymbol{a}(t_0)$ is repeated l consecutive steps, i.e. $\boldsymbol{a}(t_0) =$
$\boldsymbol{a}(t_0 + 1) = ... = \boldsymbol{a}(t_0 + l - 1)$, which occurs with at least probability $\beta^{n(l-1)}$. This
conclusion holds in both following cases: (1). Based on the concept of pure Nash
equilibria, for any strategy $a_i(t_0)$ of equilibrium profile $\boldsymbol{a}(t_0)$, we have: $\Delta_{t_0}^{a_i(t_0)} =$
$u_i(a_i(t_0), \boldsymbol{a}_{-i}(t_0)) - u_i(\boldsymbol{a}(t_0)) = \Delta_{t_0+1}^{a_i(t_0)} = ... = \Delta_{t_0+l-1}^{a_i(t_0)} = 0$. (2). Since, $\boldsymbol{a}(t_0)$ is
a strict Nash equilibrium, for any sensor $s_i \in S$ and $\forall a_i \in A_i$, and $a_i \neq a_i(t_0)$, we
have $u_i(a_i, \boldsymbol{a}_{-i}(t_0)) < u_i(\boldsymbol{a}(t_0))$. Therefore, $\Delta_{t_0}^{a_i} = u_i(a_i, \boldsymbol{a}_{-i}(t_0)) - u_i(\boldsymbol{a}(t_0)) < 0$,
and then $\Delta_{t_0+1}^{a_i} < 0, ..., \Delta_{t_0+l-1}^{a_i} < 0$. So, at the time $t_0 + l$, for a sensor s_i and
$\forall a_i \in A_i$, we have $R_i^{a_i}(t_0 + l; l) \leq 0$. Hence, based on the distributed relocation
algorithm, if $\boldsymbol{a}(t_0)$ is not a strict Nash equilibrium, $\forall t > t_0$, $\boldsymbol{a}(t) = \boldsymbol{a}(t_0)$ condi-
tioned on $\boldsymbol{a}(t_0) = \boldsymbol{a}(t_0 + 1) = ... = \boldsymbol{a}(t_0 + l - 1)$ which occurs with a positive
probability $\beta^{n(l-1)}$.

Secondly, we prove that at time t_0, if $\boldsymbol{a}(t_0)$ is not a strict Nash equilibrium,
$\exists i \in S$ and $a_i' \in A_i$, the dynamics moves from $\boldsymbol{a}(t_0)$ to $\boldsymbol{a}'$ with a positive

probability, where $a' = (a'_i, a_{-i}(t_0))$ and $u_i(a) < u_i(a')$. $\forall t_0 > l$, suppose $a(t_0) - (a_1(t_0), ..., a_n(t_0))$ is **not** a strict pure Nash equilibrium, that implies that $\exists s_i \in S$,$a'_i \in A_i$, $u_i(a(t_0)) < u_i(a'_i, a_{-i}(t_0))$. Let a' be $(a'_i, a_{-i}(t_0))$. Suppose that is $a(t_0)$ repeated l consecutive steps, i.e. $a(t_0) = a(t_0 + 1) = ... = a(t_0 + l - 1)$, which occurs with at least probability $\beta^{n(l-1)}$. Since $u_i(a(t_0)) < u_i(a'_i, a_{-i}(t_0))$, we have $\Delta^{a'_i}_{t_0} = u_i(a'_i, a_{-i}(t_0)) - u_i(a(t_0)) > 0$. Also, we have $\Delta^{a'_i}_{t_0} = ... = \Delta^{a'_i}_{t_0+l-1} > 0$ due to $a(t_0) = ... = a(t_0 + l - 1)$. Thus, we have $R^{a'_i}_i(t_0 + l; l) = \frac{1}{1+...+\delta^{l-1}}\{\delta^{l-1}\Delta^{a'_i}_{t_0} + ... + \delta^0 \Delta^{a'_i}_{t_0+l-1}\} > 0$.

Based on the above two claims, we can conclude that starting from any vertex $a(l+1)$, if $a(l+1)$ is a strict Nash equilibrium, then the dynamics will stay on $a(l+1)$ with a positive probability. if $a(l+1)$ is not a strict Nash equilibrium, the dynamics then moves from $a(l+1)$ to a Nash equilibrium with a positive probability along a directed path in profile state graph.

4   Simulation Experiments

Consider a SRG in which the sensors start from compact initial configuration. The inceptive evaluated grids' values are followed Zipf distribution, namely, $v_i = 1/i^{\theta}$ for a parameter $0 < \theta \leq 1$.

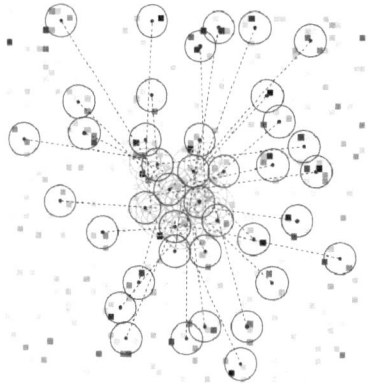

Fig. 1. Coverage performance of the distributed relocation algorithm

In Figure 1, 50 sensors are randomly deployed in a central region of 60*60 grid space with the parameters: detect radius of sensors $r = 2.8$, $\lambda = 0.2$, $\mu = 1.0$, $\eta = 1.5$, $\theta = 1.2$, the length of decision histories $l = 2$ and $\beta = 0.7$. The gray circles represent the initial positions of sensors, and the blue circles are final coverage positions when a pure NE is achieved in this game. Each dash line connects between the initial and final positions of a sensor node. We can see that grids with high values (dark squares) are covered by more sensors than those with relative low values.

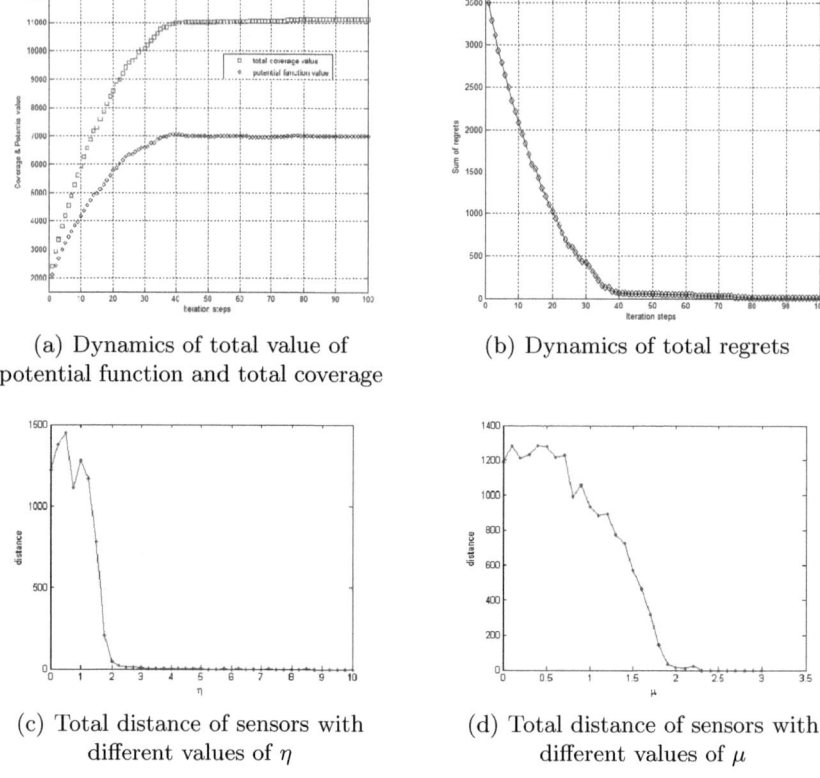

(a) Dynamics of total value of
potential function and total coverage

(b) Dynamics of total regrets

(c) Total distance of sensors with
different values of η

(d) Total distance of sensors with
different values of μ

Fig. 2. The performances of the distributed no-regret relocation algorithm

Dynamics of the total coverage value and total potential function value are shown in Figure 2(a). The potential function value increases in every iteration until it reaches the Nash equilibrium point, while the total coverage value is significantly improved. From Figure 2(b), we observe that the total regret converges monotonically close to 0, which means a Nash equilibrium, after around 80 iterations.

In our game, the cost of locomotion greatly affects the actions of sensor players. Figure 2(c) and Figure 2(d) indicates the relationship of total distance and two parameters μ and η which are associated to the cost of locomotion. When the moving cost increases, the total distance will decrease, as every sensor prefer to a nearer section with higher profit.

5   Conclusion

In this paper, the problem of sensors optimal relocation is formulated as a repeated game and a distributed no regret learning algorithm is also proposed for sensors to play the game. In the repeated game dynamics, the potential function value increases, and collective behavior will finally converge to a an acceptable equilibrium point where total coverage value are optimized. Simulation results have demonstrated the convergence and coverage performance of the proposed algorithms.

Acknowledgments. This work was supported by the Research Foundation of the Educational Department of Yunnan Province (2010Y251), the Foundation of National Innovation Experiment Program for University Students (101067301), the Key Discipline Foundation of Software School of Yunnan University (2010KS01).

References

1. Younisa, M., Akkaya, K.: Strategies and techniques for node placement in wireless sensor networks: A survey. Journal Ad Hoc Networks Archive 6(4) (June 2008)
2. Bartolini, N., Calamoneri, T., La Porta, T., Massini, A., Silvestri, S.: Autonomous deployment of heterogeneous mobile sensors. In: ICNP 2009, pp. 42–51 (2009)
3. Ghosh, A., Das, S.K.: Coverage and connectivity issues in wireless sensor networks: A survey. Pervasive and Mobile Computing 4(3), 303–334 (2008)
4. Bartolini, N., Calamoneri, T., La Porta, T.F., Silvestri, S.: Mobile Sensor Deployment in Unknown Fields. In: INFOCOM 2010, pp. 471–475 (2010)
5. Wang, G., Irwin, M.J., Fu, H., Berman, P., Zhang, W., La Porta, T.: Optimizing sensor movement planning for energy efficiency. TOSN 7(4), 33 (2011)
6. Meguerdichian, S., Koushanfar, F., Potkonjak, M., Srivastava, M.B.: Coverage Problems in Wireless Ad-hoc Sensor Networks. IEEE INFOCOM 3, 1380–1387 (2001)
7. Marden, J.R., Wierman, A.: Distributed welfare games with applications to sensor coverage. In: Proc. CDC, pp. 1708–1713 (2008)
8. Bai, X., Yun, Z., Xuan, D., Chen, B., Zhao, W.: Optimal Multiple-Coverage of Sensor Networks. In: Proc. of IEEE International Conference on Computer Communications (INFOCOM) (to appear, April 2011)
9. Wang, G., Cao, G., La Porta, T.F., Zhang, W.: Sensor relocation in mobile sensor networks. In: INFOCOM 2005, pp. 2302–2312 (2005)
10. Monderer, D., Shapley, L.S.: Potential games. Games and Economic Behavior 14, 124–143 (1996)
11. Banerjee, B., Peng, J.: Efficient No-Regret Multiagent Learning. In: AAAI 2005, pp. 41–46 (2005)
12. Greenwald, A., Jafari, A.: A General Class of No-Regret Algorithms and Game-Theoretic Equilibria. In: Proceedings of the 2003 Computational Learning Theory Conference, pp. 1–11 (August 2003)
13. Nash, J.F.: Equilibrium points in n-person games. Proc. of National Academy of Sciences 36, 48–49 (1950)
14. Fudenberg, D., Tirole, J.: Game Theory. MIT Press (1991)

Research on Decentralized Message Broadcasting for Multi-emergent Events

Haoming Guo, Feng Liang, and Yunzhen Liu

Beihang University Computer School,
New Main Building G1134,Xueyuan street 37#,
Haidian district, Beijing, China
guohm@nlsde.buaa.edu.cn

Abstract. Many emergency information systems are built on decentralized architectures for time efficiency. Due to lack of centralized information analyzing, message loop and identity confusion affect their performance. An approach called Decentralized Message Broadcasting Process is introduced to address the issue. In the approach, messages are specified to carry information about source event and node path it has been passed through. Distributed nodes are involved initiatively to organize message exchange. In the decentralized message broadcasting, time efficiency and message filtering is achieved. Upon the approach, a platform is built for CEA' SPON to support decentralized emergency information processing applications.

Keywords: Decentralized System, Message Exchange, Sensor, IOT.

1   Introduction

One primary objective of IOT web is to watch and report exceptional event's development in real world[1][2]. For this purpose, large number of sensor resource is deployed. Information collected from the sensor resources are processed to analyze events[3][4]. Time efficiency is core of the systems. As result, many systems are developed on decentralized architecture. In many cases, while one event takes place, sensors around may catch the information early or late depending on their distance to its location or something else. One event's information will be exchanged by sensors once they detect it for alert or further proceedings. If system processes the information without distinguishing, the information could be viewed as motivated by multi individual events so that chaos and confusion will be produced. This problem could be addressed by centralized information process system easily. In the centralized systems, sensors are separated in work. Information about event won't be exchange between sensors but aggregated in system center where they could be filtered by specific computation[4][5]. However, in many distributed emergency information control system cases, there is no time for unnecessary message dispatching and centralized processing. Sensor resources should be initiatively coordinated together to spread message as fast as possible[6]. Under such circumstances, it will be hard for system to organize right message exchange between sensor resources.[7]

B. Liu and C. Chai (Eds.): ICICA 2011, LNCS 7030, pp. 224–232, 2011.
© Springer-Verlag Berlin Heidelberg 2011

Earthquake-Early-Warning-Application(EEWA) is developed for in China Earth-quake Administration(CEA). EEWA is to send out alert message for areas where earthquake takes place. For this purpose, CEA deploys a large number of sensors all over the country to watch exceptional geological fault vibration. Once earthquake takes place, the sensor that is located nearest may catch and produce the event's information and broadcast it first. Because electric signal runs faster than earthquake wave's spread speed, the neighboring sensors receive the message and broadcast it to others before they really catch the earthquake wave. While these sensors are driven by the messages to repeat broadcasting, the earthquake wave may reach them. At that moment, they work as the initial one to produce earthquake information and broadcasting initiatively.

EEWA is a typical distributed emergency information control system. In DEICS, time efficiency is essential. To spread information as quick as possible, there is no time to aggregate message for centralized information processing and dispatching because the unnecessary data transferring between resource and centralized service may affect system's performance. Message should be forwarded between resources directly. However, this approach leads to problems:

Messages flooding and looping the system: to spread message as quick as possible. Sensors broadcast information as they received from others. This repeated message broadcasting may lead to message loop. As result, while all sensors begin to receive and broadcast message about the event, the system will be flooded by repeated messages.

Events' identity confusion: as shown in EEWA, sensors produce information about event once they caught it. In DEICS, there is no centralized organization to assert identity of events by the information. As result, it's hard for system to distinguish event while multiply information are reported.

The message exchange mechanism of DEICS is built upon decentralized approaches for requirement over time efficiency. To spread message of emergency as quick as possible, resources not only work as information producer of events but also as message relay station. In this process, a message exchange group will be formed in dynamic for a caught event. In this dynamic message exchange group, one sensor works as information producer and others as message relay station that broadcasts message as it receive without knowing who will receive. The involvement of sensors is determined by whether it received message from others. For uncertainty of events' occurrences, it's impossible to predict membership of one group and message exchange order. There is no uniform approach for the group to filter repeatedly broadcasted messages to reduce unnecessary flooding. So the reliability of DEICS may be affected.

Meanwhile, in the group, following the beginning sensor, the adjacent sensor may catch the event when event is developing. Because there is no centralized component in the system to assert identity of events, the system could not distinguish whether the newly caught event is a new one or not. If the very sensor produces new information about the event and broadcast, the system may be confused as many events by multiply information produced for one same event. Senor may ban out the information production simply if it has been involved in one message exchange group during the period. However, in many case, a number of events may take place in same space and period. This approach results in lost of information about multi-happening events that really take place in same time. The accuracy of DEICS may be affected.

As introduced above, DEICS needs a decentralized message broadcasting mechanism. In this approach, there is no centralized message exchange organizer. Sensors receive and broadcast message as equal peer. Message is the only reference for sensors to organize self-direct message exchange action and filter unnecessary looping and flooding. Meanwhile, with this reference, identity of events is to be distinguished by sensor locally so that event's information could be produced and broadcasted rightly. The mechanism is called Decentralized Message Broadcasting Process (DMBP). In the message exchange process, sensor receives and broadcasts message. The message will be cached locally for lifetime for filtering message and event distinguishing. It is composed of two parts: event information core and message path. Event information core is created by initial information producer and utilized as reference by others who receive the message. While the sensor catches an event, it compares the information with local cached messages' event information core. Through specific processing, identity of event will be clarified. If it's newly found event, the sensor produce information and broadcast accordingly or ban out the information if it's not. While sensor receives a message, it check its own local path information in message path and broadcast message before add local path information in its message path if no path information is found in message path or ban out the message if it's found. Through the approach, DEICS is about to address issues of identity assertion and unnecessary message flooding for decentralized message exchange circumstance.

DMBP is to enhance reliable message exchange in DEICS. Through DMBP, sensors could exchange message initiatively without centralized information processing. As event takes place, the initial sensor that catches it produces information and broadcast. The message will be exchanged among sensors who receive it. While in message exchanging, sensors use the message as reference to filter sent messages and distinguish identity of events. Through this approach, message loop and confusion of multi-events' identity could be addressed for decentralized message exchange environment.

2   Related Works

Message broadcasting is widely applied in peer-to-peer applications, such as Gnutella. Gnutella is a fully decentralized and unstructured peer-to-peer network. It uses the distributed message broadcasting mechanism. By Combining standard protocol mechanisms of time-to-live (TTL) with unique message identification (UID), paper[8] introduced an approach to govern flooding message transmissions that potentially have a devastating effect on the reachability of message broadcast for Gnutella. By analysis Gnutella's broadcasting mechanism, paper[9]introduced   an approach for connection management. In the approach, an discarding connection management algorithm is provided, which discards the redundant connection. The researches introduced above focused on how to manage connectivity and spread message as efficiency as possible. The messages were not utilized as reference to influence broadcasting activities. As result, repeated message can't be filtered.

In paper [10],An approach is introduced to filter out unnecessary messages to avoid impact on bandwidth of network. In the approach, each node maintains a table of parameters for associated terminal in the network. The parameters in that table are compared to the parameters in message frames received by the node and only those messages having parameters found in the table are put on the network. The parameters in the table are controllable to adjust level of filtering. The approach is designed to filter valued message to be transmitted. Message loop can't be addressed. However, the mechanism of maintaining parameter table locally can be utilized to realize goal of filtering repeated messages.

3   Definition of DMBP Message

As introduced above, DMBP is message for decentralized exchange. It is created by the node that catch event first and exchanged by others. As result, DMBP consists of two parts: information core and message path. It's used as reference to avoid message loop and distinguish of events. The definition of DMBP message is shown as below:

DMBPMsg = (GUID,   infoCore, msgPath)

GUID is identity of the message. In DEICS, GUID is used for system to identify event information for further analysis and operation. "infoCore" is created by the initial node who catch event first. In "infoCore", the sensor production detailed information about the event. It's definition is shown as below:

infoCore = (producerID, eventDsp, timeClock, lifetime}

In infoCore's definition, "producerID" defines the identity of the node who catch the event first and production information. "eventDsp" is created by the producer to describe detailed information of the event. "timeClock" is the right time of the event being caught. "lifetime" is created to define life of the message. While the event is caught by initial node, the node creates message and writes information about the event in it. After that, the node broadcast the message. Others who receives the message checks "msgPath" of the message and determined whether it should be continued to broadcast or ban. "msgPath" is used to keep path information of the nodes who has broadcasted the hosted message. It's definition is shown as below:

msgPath = {nodePathi | i=1,2,…….n };

nodePathi is the information about node who has broadcasted the message. It consists of three parts:

nodePath = (nodeID, inTime, outTime);

"nodeID" is identity of the node that has broadcasted the message. "inTime" is created to define when message is received by the node. "outTime" is created to define when the message is broadcasted by the node. In message exchange, the msgPath is used to specify nodes through whom the message has been exchanged. By the specification, a node may decide whether the message to be broadcasted continuingly or not to kill loop in decentralized message exchange.

3.1   Definition of DMBP Node

DMBP node is core of whole system. DMBP node functions as sensor wrapper, information producer/publisher, message relay station and other roles. Its definition is shown as below:

Node = (ID, MsgCache ,NodeOp);
NodeOp = MsgReceive⊕ MessageFilter ⊕ MsgBroadcast ⊕ Publish ⊕   EvtCatch

In node's definition, ID is to specify identity of the node. "msgcache" is the node's local message cache which is created to cache received message. It consists of an array of DMBPMsg and its definition is shown as below:

MsgCache   = (msgArr , cacheOp) ;
msgArr = { DMBPMsg i | i=1,2,....n}
cacheOp = findMsg ⊕ addMsg ⊕ killMsg⊕ MsgLife

MsgCache's primary operation includes getMsg, addMsg and killMsg. "findMsg" is created to enable check message's existence in local message cache. It's definition is :

findMsg : infoCore@ DMBPMsg × DMBPMsg * → DMBPMsg

"addMsg" is create to add message into local message cache. It's definition is:

addMsg: DMBPMsg× DMBPMsg * → DMBPMsg *

"killMsg" is created to enable remove message from local message cache. It's definition is:

killMsg: DMBPMsg× DMBPMsg * → DMBPMsg *

one node's primary operation includes MsgReceive, MsgProc, MsgBroadcast, InfoPublish, MsgLife and EvtCatch

MsgReceive is created to enable node to listen and receive message from others.

3.2   Message Process for Node

In DMBP, Nodes are deployed to watch events and listen messages from others. When one node receive message, it needs to check whether the message has been exchanged by current node. In message's "msgPath", the information about nodes that the message has been passed through is specified. By the information, the node could filter the message or continue to broadcast it. Before the message is about to be broadcasted, the node append its path information in message's "msgPath" and the message is cached in local message cache. The whole process is shown as below:

MessageProcess : Node → MsgCache
MessageProcess:(node)   =   let msg = MsgReceive()   in ;
                       let cache = node(-,msgcache) ;
                       if FilterMessage(msg , cache) = false
                       then   AppendPath(node, msg) ;
                           MsgBroadcast(msg) ;
                           addMsg (msg , cache) ;

AppendPath: Node × DMBPMsg   → DMBPMsg
AppendPath:(node , msg) = let nid = node(id,-)   in ;
                  let p = path(nid, - , -) in;
                  let parr = msg(-,-, pa) in ;
                  p → parr ;

3.3   Filter Message Loop

In the decentralized system, node broadcast messages while they receive initiatively and automatically. Because there is no centralized unit in the system to organize message exchange order, the message will be mutually broadcasted and received between two nodes. As result, message loop could be motivated. To enhance the DEICS, one primary objective of DMBP is filtering unnecessary mutually motivated message loop for one event.

In node, all received and broadcasted message are cached in local message cache-"MsgCache". Once the node receives a new message, it load existed message from local message cache with reference of the newly received message's information core. If there is no such message found in local message cache, the newly received message is viewed as valid message and continues to be broadcasted. If there is a related message found in local message cache, the operation check the node's path information in the newly received message's msgPath. If in the message's msgPath there is the node's path information, the message is viewed as invalid message and filtered. If no such information is found in message's msgPath, the message is view as valid message and continued to be broadcasted. The whole process is defined as below:

FilterMessage : DMBPMsg× MsgCache $\rightarrow$ Boolean
FilterMessage: ((inMsg), (cache)) = let   info = inMsg(-,infocore,,-) in;
$\qquad$ let   cached = cache. findMsg (info , cache.
msgArray) in;

$\qquad$ if(cached = null) $\rightarrow$ false,
$\qquad$ {   let pathArr = cached(-, - , path*) ; in
$\qquad$ let result = isNodeInPath(ID, pathArr) $\rightarrow$ true,
false ;
$\qquad$ }

isNodeInPath:ID × nodePath *$\rightarrow$ Boolean
isNodeInPath:(ID , path) = let nID = path(nodeId,-,-) in ;
$\qquad$ if(ID = nID) $\rightarrow$ true, false ;
isNodeInPath:(ID , path:paths) = isNodeInPath:(ID , path) $\rightarrow$ true, isNodeInPath:(ID , paths) ;

3.4   Distinguish Identity of Events

As introduced, nodes may catch one event's information at different time. On the other hand, events may take place simultaneously. In DMBP, there is no centralized organization to distinguish and assert event's identity. The Message is only reference to realize that purpose.

In DMBP, infoCore is created and appended in message by node who catch an event initially. In inforCore, the event's key information is produced and may utilized by other nodes as reference to distinguish. In message broadcasting, the information is cached locally by nodes. while one node catches a event, it search the local message cache by certain rules with reference of newly caught event's information. If there is relevant message found in the cache, the event is viewed as already caught and the message won't be broadcasted. If there is no relevant message found in the cache, the event is viewed as newly caught and the message will be broadcasted. The whole process is defined as below:

EvtCatch: Node × DMBPMsg   → DMBPMsg
EvtCatch: (node , msg): let cache = node(-,msgcache) in ;
                     if ChkInfo(cache, msg) = false
              then AppendPath(node, msg) ;
                   MsgBroadcast(msg) ;
                   addMsg (msg , cache) ;
                   Publish(msg) ;
ChkInfo: DMBPMsg*× DMBPMsg → Boolean
ChkInfo:(m   , msg) =   let newDsp = m (-,infocore(-,dsp,-,-), -) in ;
                  let oldDsp = msg (-,infocore(-,dsp,-,-), -) in ;
                  if newDsp = oldDsp → true, false;

ChkInfo:(m:ms   , msg) = ChkInfo(m,msg) → true, ChkInfo(ms,msg);

4   Application and Test

DMBP is developed to address issues of message loop and identity confusion for decentralized message exchange applications. Through message specification and broadcasting mechanism, the purpose is realized. With DMBP, a platform, called Emergency Message Publish System(EMPS), is created for CAE's SPON. In test, a centralized message system is built for comparison. In the system, once a sensor catches sign of event, it reports it to system center. The system center will send out the message to all nodes. In the process, every 50 nodes will be grouped and transmitted parally. In comparison, DMBP system spread the message by decentralized broadcasting approach. The graphic below shows comparison of total time for all nodes to receive the message.

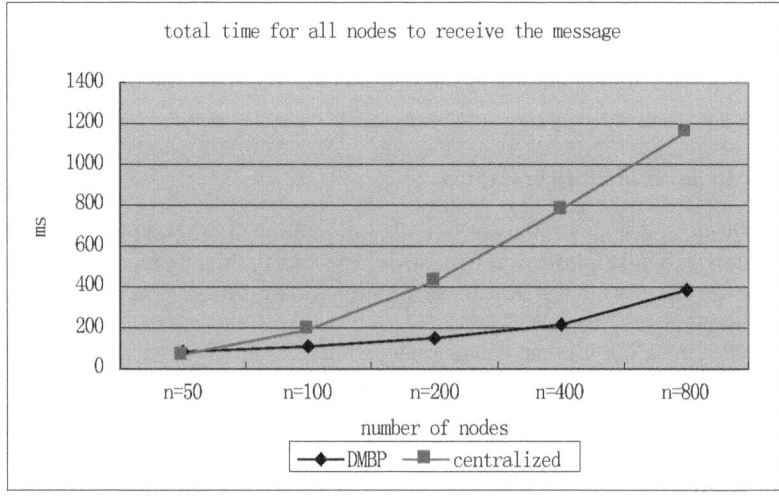

Fig. 1. Time for all nodes to receive the message

As shown by the graphic, the DMBP approach takes less time than the centralized approach. The graphic below shows result of test that simulates 5 events taking place simultaneously

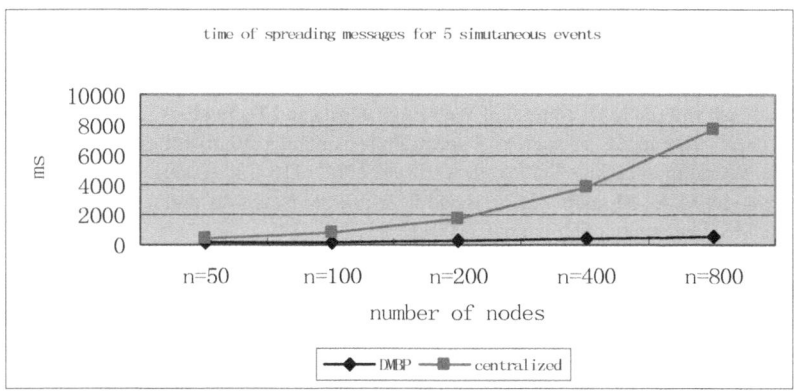

Fig. 2. Spreading messages for 5 simultaneous events

In the test, more events takes place, the centralized system need to dispatch more messages. In contrary, DMBP takes much less time than the centralized approach. The tests shows effectiveness of DMBP

5   Conclusion

In distributed emergency information control system, the message exchange mechanism is different from centralized counterparts. Message is only exchange among nodes for time efficiency. There is no centralized organization in the system to organize order of message exchange and distinguish event's identity. As result, message loop and confusion will affect distributed emergency information control system's performance. For the problem, DMBP is developed to address the issue. By specific message protocol, nodes filter message loop and distinguish events' identity.

Acknowledgement. This research work was supported by China Earthquake Administration's program for Seism-Scientific Research "Research in Online Processing Technologies for Seismological Precursory Network Dynamic Monitoring and Products" (NO. 201008002).

References

1. Chaouchi, H.: The Internet of Things: Connecting Objects. In: Wiley-ISTE (2010)
2. Yan, L., Zhang, Y., Yang, L.T., Ning, H.: The Internet of Things: From RFID to the Next-Generation Pervasive Networked Systems (Wireless Networks and Mobile Communications) Auerbach Publications (2008)

3. Hu, W., Bulusu, N., Chou, C.T., Jha, S., Taylor, A., Tran, V.N.: Design and evaluation of a hybrid sensor network for cane toad monitoring. ACM Trans. Sen. Netw. 5, 1–28 (2009)
4. Schelp, J., Winter, R.: Business application design and enterprise service design: a comparison. International Journal of Services Sciences 1, 206–224 (2008)
5. Demirkan, H., Kauffman, R.J., Vayghan, J.A., Fill, H.-G., Karagiannis, D., Maglio, P.P.: Service-oriented technology and management: Perspectives on research and practice for the coming decade. Electronic Commerce Research and Applications 7(4), 356–376 (2008)
6. Papazoglou, M.P., Van Den Heuvel, W.-J.: Service-oriented design and development methodology. International Journal of Web Engineering and Technology 2(4), 412–442 (2006)
7. Yucek, T., Arslan, H.: A survey of spectrum sensing algorithms for cognitive radio applications. Communications Surveys & Tutorials 11(1), 116–130 (2009)
8. Jovanovic, M.A.: Modeling Large-scale Peer-to-Peer Networks and a Case Study of Gnutella. MS, University of Cincinnati (2001)
9. Zhuang, L., Pan, C.-J., Guo, Y.-Q., Wang, C.-Y.: Connection Management Based on Gnutella Network. Ruan Jian Xue Bao 16(1), 158–164 (2005)
10. Baker, M.C., Cheung, R.Y.M., Bhattacharya, P.P., Kobo, R.M., Kolbe, E.M., Naghshineh, M.: Broadcast/multicast filtering by the bridge-based access point. Application Number:08/443793 (1996)

Research in Intrusion Detection System
Based on Mobile Agent

Zhisong Hou, Zhou Yu, Wei Zheng, and Xiangang Zuo

School of Information Engineering,
Henan Institute of Science and Technology,
Xinxiang, China, 453003
forhouor@gmail.com,yyuzhou@tom.com,
{karl777,zuoxg2002}@163.com

Abstract. According to the problems of traditional intrusion detection system, there was a design of intrusion detection system based on mobile agent. The internal function of mobile agent was divided. Moreover, the communication manner and interactive process were applied in the paper. It is can be seen from the result that the system could reduce the network load and shorten the waiting time of network. Then, the dynamic adaption of the system could be implemented while false alarm rate and false negative rate would be reduced.

Keywords: Mobile Agent, Intrusion Detection, Intrusion Detection System, Architecture, Information Exchange.

1   Introduction

As the Internet user increases rapidly, intrusion activities on computer system became popular. Intrusion detection system, an effective method for defending intrusion, developed rapidly in recent years.

According to the increase of network size and complexity, in huge heterogeneous network, traditional intrusion detection system could not complete the work for multiple information processing, and the safety of the whole system could not be achieved as the condition that intrusion detection system was partly failure [1]. At the same time, intrusion activity could be found before losses because current intrusion detection system is lack of early warning mechanism. When network is busy, the characteristic data of intrusion activities would be loss as the network traffic cannot be processed by the system, and there would be security vulnerabilities. Therefore, current intrusion detection system has loss the function for defending the intrusion activities.

For solving the problems of traditional intrusion detection system, the design is to apply mobile agent for intrusion detection system. Then, a new architecture of intrusion detection system would be produced which is based on the mobile agent.

2   Research of Related Technologies

Mobile Agent is the intelligent Agent with mobility. It could independently move between each node in heterogeneous network. Mobile agent is a software entity which

B. Liu and C. Chai (Eds.): ICICA 2011, LNCS 7030, pp. 233–240, 2011.
© Springer-Verlag Berlin Heidelberg 2011

could represent other entities (including person or other agents). It could independently choose the option place and time to interrupt the current execution according to different circumstance, then, it could move to another equipment and return the related result. The purposes of the movements are to make the execution closer for data source, reduce network overhead, save bandwidth, balance load, accelerate the process. All these could be contribute to enhancing the efficiency of distributed system.

The main advantages of mobile agent include several points: network bandwidth saved and network delay reduced as the data processed in local node; interoperability of heterogeneous networks achievement based on the decrease of components coupling as bottom network protocol encapsulated; dynamic adaption implemented as dynamic adjustment of mobile agent depending on the current network configuration, topology architecture and flow [2].

Intrusion detection system evaluates the user or system behaviors in suspicious level, and identifies the legitimacy of the behavior based on the evaluation. Then system administrator could manage system safely and cope system attacking. Intrusion detection system includes data acquisition, data analysis and response module. Specifically, data acquisition module collects the status and behaviors of system, network, file data and users. The collected information would be applied for the analysis if there is intrusion. Module of data analysis assembles the collected information, and applied pattern matching, statistical analysis or integrity analysis for detection. When intrusion was detected, corresponding response would be applied for terminating intrusion or attacking, and the system would try to recover influenced service and loss data. The architecture of the system is shown in figure 1.

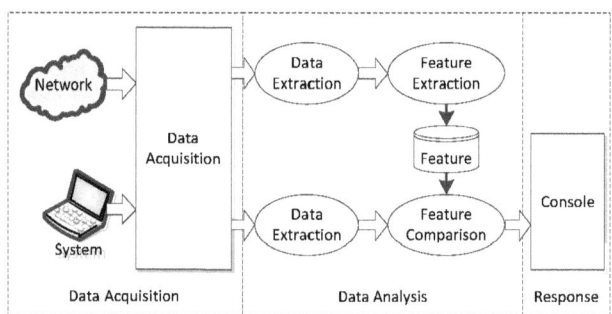

Fig. 1. The structure of traditional intrusion detection system

There are three architectures of traditional intrusion detection system: integral structure, hierarchal structure and distributed structure. Integral structure integrated all functions, but it is weak for system prevention; hierarchal structure is a tree structure which is composed with detector and controller, and it shares information of all subsystems for detection the intrusion. The disadvantage of the structure is single-point failure in the structure.

In distributed structure, intrusion detection system is divided into several modules. The modules distribute in the heterogeneous network environment, and each module receives different input information, finishes different missions. Then, they report the result to the high function units until integrated console. The distribution of the modules is matched with the characters of mobile agent. Therefore, intrusion detection system applies technology of mobile agent. As a result, mixed structure could be implemented based on the technical advantages of mobile agent [3, 4]. This could compensate the defects of traditional intrusion detection system.

3   Intrusion Detection System Based on Mobile Agent

3.1   The Architecture of the System

Intrusion detection system based on mobile agent is developed on mobile agent platform. The architecture is shown in figure 2.

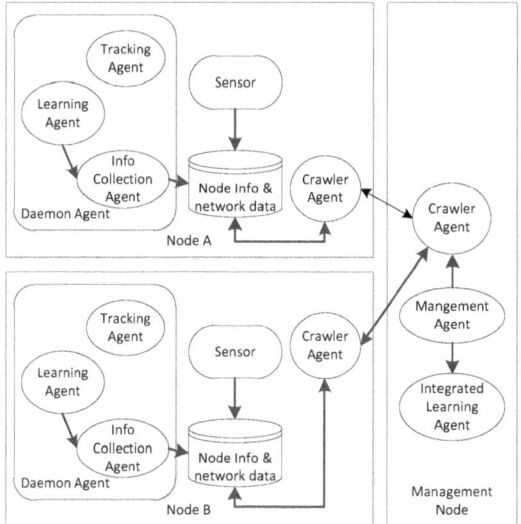

Fig. 2. The Architecture of intrusion detection system

As can be seen from figure 2 that: the intrusion detection system based on mobile agent composes Management Agent, Integrated Learning Agent, Crawler Agent, Sensor, Daemon Agent, Tracking Agent, Learning Agent and Info Collection Agent. The functions of them are shown below:

(1) Management Agent: Management Agent is the agent which is in the highest level. There is one management agent at least in a system. In complex distributed network, several management agents are applied to be network management mode of hierarchal structure. This could avoid single-point failure.

Integrated learning agent of Management Agent analyses the learning result from learning agent and responses; at the same time, management agent provides the human machine interface between administrator and system.

Management agent records the status of all agents in network, and completes the detection and management of the agents regularly. Management agent has control over all agents in system. Management agent could send related agents to a target system according to network status. Management agent notices daemon agent suspend, resume or transmit agents when agent is running. At last, management agent is responsible for agent recycling as mission complemented.

For hidden distributed attacks, sensor of single node could not detect intrusion, and this could lead the intrusion detection failure. Therefore, management agent has to send tracking agent to all network nodes in timing, and collect the information to analysis or judge whether there is intrusion.

(2) Crawler agent: crawler agent is sent to whole network by management agent in timing which is for information collection. Crawler agent is responsible for traversing the entire network. It is necessary as that premeditated network intrusion with planning and technical preparation is span a large space and time. As a result, reducing the rate of intrusion depends on traversal whole network regularly.

(3) Daemon agent: Daemon agent is the basis of system. Each daemon agent is processing on each node. It provides applicable environment of operation and transmission between internal agents.

Daemon agent is controlled by management agent, and it could manage the agents on node. Daemon agent can start, suspend or terminate agent, and transmit agents in target networks.

(4) Sensor: Each node installs sensor. Sensor monitors network information and system audit log, and then it formats the monitored data and saves them into intrusion detection database.

In a fixed period, the saved information in database is analyzed in a certain rule. If distributed attack was found, there are two ways to deal with: firstly, sending info collection agent and learning agent to other nodes to collect and learn for judging that if distributed attack occurred; secondly, noticing management agent, and management agent sends info collection agent and learning agent. At last, collect detection information to management agent for analysis.

(5) Info collection: Info collection agent can move in network, and collects the information related to intrusion from database of target node.

(6) Learning agent: Learning agent can move in network. It can apply data mining algorithm to extract characters of user behavior and related pattern rules.

(7) Integrated learning agent: Integrated learning agent is on management node. It is responsible for analysis and synthesis of information from collection agent and result from learning agent. Then, comprehensive safe mode and character of user could be achieved by that. According to the analysis, management agent could judge whether there is intrusion and decide to send warning or not.

(8) Tracking agent: Tracking agent is to track the path of intrusion, judge the source node of intrusion and the node position where invaders log on the network. At the same time, tracking agent tracks the nodes which invaders reached to.

In common, a tracking agent could not finish the task of tracking intrusion path. Then, several tracking agents work together to track the intrusion path. Therefore, when several intrusion behaviors are found on a target node, daemon agent on the node sends several tracking agents which are corresponded to the intrusion behavior. The 8 functional agents work corporately and complete intrusion detection for network system. The system is distributed on all network nodes. Each node has tracking agent, sensor, info collection agent and learning agent. Tracking agent tracks intrusion path and judge the intrusion node; info collection agent collects information of target node, sends the information to learning agent for analysis, extract the pattern mode related safety and character of user behavior. Crawler agent collects information of all nodes, and sends the information to the integrated learning agent which stays on the management node. At last, management agent detects intrusion according to the analysis from integrated learning agent.

3.2   Communication between Agents

There are many agents in the system. Each agent takes specific functions. Several agents need to work corporately and exchange information for intrusion detection.

There would be conflicts in agents as the autonomy of mobile agent[5]. The communication between agents is very important as that if there are several tracking agents, info collecting agent and learning agent for intrusion. Moreover, the communication of integrated learning agent, info collection agent and learning agent is necessary as that management agent could receive the operational status of system. The system sets specific public information exchange area on each node to exchange information for communication between agents. The area is bulletin board which is used by learning agent; it is message board used by tracking agent.

All agents can visit the public information exchange area. Learning agents exchange rule information via bulletin board. Tracking agent finds the information that if a path was tracked by other tracking agent via message board for avoiding conflict of path tracking.

There are bulletin boards and message boards installed on all management nodes, which is used for collecting messages from nodes and the tracked paths.

In the system, the path of information exchange between agents is shown if figure 3.

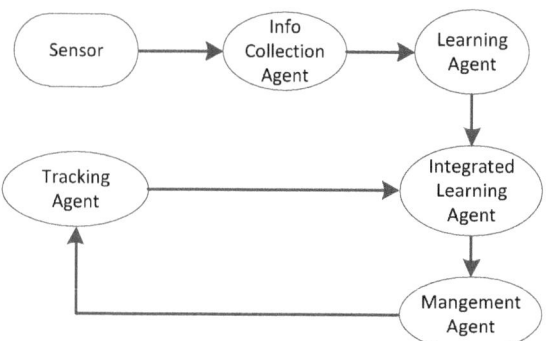

Fig. 3. The Path of Information Exchange between Agents

3.3   Process of Intrusion Detection

Intrusion detection is completed by both network node and management node. The agent of network node collects information, extracts safe pattern rules and character of user behavior, and sends the information to management node; the agent of management node travels information of all network nodes, receives information from network node, judges and responds as analysis is done.

According to the information collected ways of network agents, process of intrusion detection is composed of information collection sub-process and intrusion detection and response of management node sub-process.

The sub-process of information collection is illustrated in figure 4.

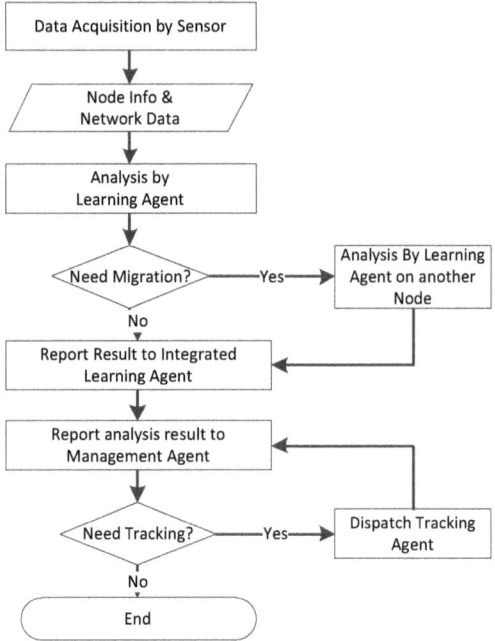

Fig. 4. The Process of Node Info Collection

As seen from the figure, it is necessary to collect suspected acts of intrusion for intrusion detection. The detailed steps are shown below:

(1) Sensor on network node monitors network data and node data, collects information, then sends the information to info collection agent.

(2) The formatted information is sent to learning agent for analysis, and then judged that if the process has to be transmitted to other node for collection information and learning. Otherwise, the process turns to step (4).

(3) If the transmission is necessary, select a certain node and transmit the agent to this node. The process of collecting and analyzing are finished by the selected node as well.

(4) The result of learning would be reported to integrated learning agent which is on the management agent as learning agent completed mission.
(5) Management agent decides whether tracking agent is sent according to the information from integrated agent. If intrusion behavior is confirmed, the collection of information completed. Otherwise, tracking agent is sent.
(6) Tracking agent tracks the suspected intrusion behavior until the source node, and then sends the tracking information to management agent.

The information collection sub-process only reports the information of network node. To achieve the whole network defense, it is necessary to collect the information of the whole network nodes. The process is achieved by intrusion detection and response on management node sub-process. The process of that is shown in figure 5.

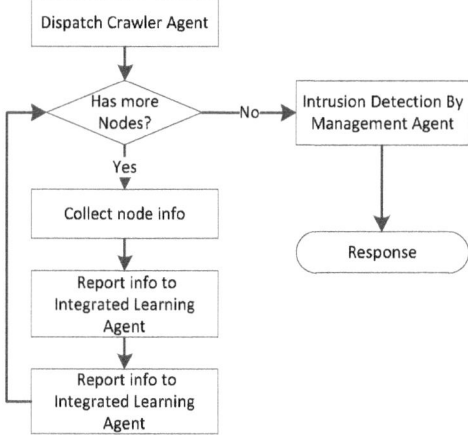

Fig. 5. The Process of Node Traversal and Intrusion Detection

In figure 5, intrusion detection and response process sends crawler agent in timing. The crawler agent travels the whole network nodes, collects information of target node and returns with the information to management agent. At last, the information is transmitted to integrated learning agent for analysis.

Management agent detects and judges intrusion according to the analysis of integrated learning agent. If intrusion is confirmed, the warning would be alarmed and certain protection measures would be taken to deal with the intrusion consequence, then the system turns to response process.

4   Conclusion

This design applies mobile agent in intrusion detection system. The architecture of the intrusion detection system based on mobile agent was implemented in the paper. Moreover, internal agents function was distributed. The information exchange was given and the process of intrusion detection was discussed. The system enhanced the ability of defending attack and intrusion detection as the single-point failure and false alarm of tradition intrusion detection system were eliminated; adapting dynamically and executing synchronously and autonomously are achieved by mobile agent. This contributes on robustness and fault tolerance of system from system architecture.

References

1. Zakim, Sobh, T.S.: A cooperative agent based model for active security systems. Journal of Network and Computer Application 27(4), 201–220 (2004)
2. Lange, D., Oshima, M.: Seven Good Reasons for Mobile Agents. Communications of the ACM 42(3), 88–89 (1999)
3. Jansen, W.A.: Counterm easures for mobile agent security. Computer Communications 23, 1667–1676 (2000)
4. Gowadia, V., Farkas, C., Valtorta, M.: PAID.: a probabilistic agent-based intrusion detection system. Computers & Security 24(7), 529–545 (2005)
5. Balasubramaniyan, J.S.: An Architecture for Intrusion Detection using Autonomous Agent. Coast TR 98-05, Department of Computer Sciences, Purdue University (1998)
6. Aksyonov, K.A., Smoliy, E.F., Popov, M.V., Dorosinskiy, L.G.: Development of decision support system BPsim3: Multi-service telecommunication networks design and modeling application. In: Proceedings of 10th International PhD Workshop on Systems and Control, Hluboka nad Vltavou, Czech Republic, pp. 112–117 (2009)
7. Fok, C., Roman, G.: Mobile Agent Middleware for Sensor Networks: An Application Case Study. In: Proc. Fourth International Symp. Information Processing in Sensor Networks, Los Angeles, California, USA, pp. 382–287 (2005)
8. Aksyonov, K.A., Bykov, E.A., Smoliy, E.F., Khrenov, A.A.: Industrial enterprises business processes simulateon with BPsim.MAS. In: Proceedings of the 2008 Winter Simulation Conference, Miami, FL, United States of America, pp. 1669–1677 (2008)
9. Aksyonov, K.A., Sholina, I.I., Sufrygina, M.: Multi-agent resource conversion process object-oriented simulation and decision support system development and application. In: Scientific and Technical Bulletin. Informatics. Telecommunication. Control. St.Petersburg, Russia, pp. 87–96 (2009)
10. Helmer, G.: Lightweight agents for intrusion detection. The Journal of Systems and Software. 67, 109–122 (2003)

Research on Network Information Search System Based on Mobile Agent

Zhisong Hou, Zhou Yu, Wei Zheng, and Xiangang Zuo

School of Information Engineering, Henan Institute of Science and Technology
Xinxiang, China, 453003
forhouor@gmail.com, yyuzhou@tom.com,
karl777@163.com, zuoxg2002@163.com

Abstract. According to the disadvantages of traditional network information search system, there is a new network information search system designed which is based on mobile agent. The architecture of the system is provided; the function of mobile agent and the search process of search agent are discussed in this paper. It is can be seen from the result that the system could reduce the network load and the dynamic adaption of route planning could be implemented. As a result, search efficiency and reliability of search executing in unstable network are improved.

Keywords: Mobile Agent, Network, Information Search, Search Engine.

1 Introduction

Accompany with the increase of network information on Internet, management of network resources became complicated. The main problem of people using Internet is how to find the information accurately in the mass of resources.

Search engine is the effective tool for network information search which provides convenience for users. However, structure of traditional search engine is a kind of client/server structure. In the structure, client receives the pages from searching, then creates local index database. The database provides the search service which solves the problem of information resource location. The efficiency of the structure is hardly to improve for the internal defects. These defects include: in searching process, pages of remote site has to be downloaded to local site, and then index process could be operated. The network communication resource wastes and the index speed decreased as there are useless and temporary information in downloaded pages. At the same time, the information is poor in real time, because the search engine could be hardly to refresh with the document update in website. That means the traditional search engine could not provide the latest information. Moreover, the structure of traditional search engine is based on the message passing and remote procedure call. The search process is heavily depended on the function of network. If the network communication is unreliable in search, the efficiency and effectiveness of search could not be achieved.

For solving the problems of traditional network information search system, the design is to apply mobile agent in the system. According to distributed computing and artificial intelligence of mobile agent, a network information search system based on mobile agent is designed.

B. Liu and C. Chai (Eds.): ICICA 2011, LNCS 7030, pp. 241–248, 2011.

2   Research in Related Technologies

Mobile Agent is the intelligent Agent with mobility. It could independently move between each node in heterogeneous network. Mobile agent is a software entity which could represent other entities (including person or other agents) [1]. It could independently choose the option place and time to interrupt the current execution according to different circumstance, then, it could move to another equipment and return the related result. The purposes of the movements are to make the execution closer for data source, reduce network overhead, save bandwidth, balance load, accelerate the process. All these could be contribute to enhancing the efficiency [2]. Usually, mobile agent communicates and woks with other agents coordinately to complete mission.

The main advantages of mobile agent include several points:

(1) Mobile agent could choose moving routine independently according to its mission and current network status. The moving routine of agent changes with network status. Therefore, mobile agent has the strain ability to unexpected network status or events which is fit for the construction of complex distributed information system. This could enhance the robustness and fault tolerance of system.

(2) Network bandwidth is saved and network delay is reduced as the decrease of remote call. Mobile agent packs the problem solving demands, and sends them to target host. Then, the result would be computed in host and returned. It avoids the data transmitting process of computing on network which decreases the demand of network bandwidth. At the same time, the process of computing in host is non-connecting which contributes the decrease of network dependence, and increase of reliability in unstable network.

(3) Interoperability of heterogeneous networks and seamless connection between distributed information system are achieved based on the decrease of components coupling as bottom network association encapsulated.

(4) Mobile agent divides the complex mission into several smaller, easy to solve subtask, and sends them into network nodes to computing which enhances parallel computing ability of system.

The structure of traditional information search system is shown in figure 1.

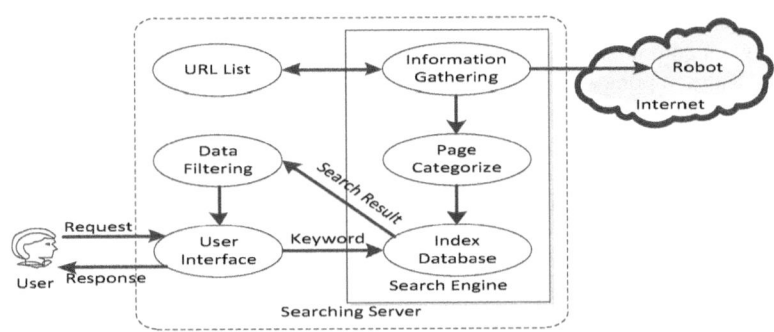

Fig. 1. The Structure of Traditional Information Search System

As can be seen from figure 1, the core of the network search system is the network search engine. The structure is two-tier client/server architecture. Firstly, a user which works as client accesses search engine and submits the search request. Then, search engine is the server. It searches its index database and returns the result to client according to a certain format; secondly, search engine sends search robot. The robot collects data on Internet. Then, search engine is the client while the information site is the server. The server returns the information that is searching for to search engine.

Search engine includes three modules: Information gathering module, page categorize module and Index database module. Specifically, network information collection, key words extraction on web site, page classification and construction of index database are implemented respectively. These functions are the key to search engine and determine the reliability and efficiency of search engine [3]. Structurally, the functions are relatively independent, high cohesion and belong to distributed computing. These are matched with independence, distribution of mobile agent.

As a result, mobile agent technology applied in network search system could overcome the defects of traditional network information search system.

3   Network Information Search System Based on Mobile Agent

3.1   Architecture

The network information search system based on mobile agent is constructed on mobile agent platform. Its architecture is shown in figure 2.

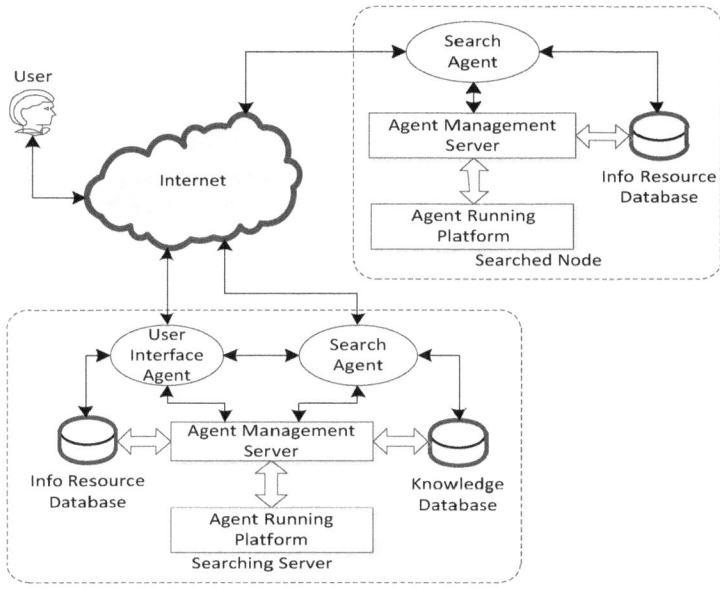

Fig. 2. The Architecture of Information Search System

As shown in figure 2, the system includes User Interface Agent, Search Agent, Agent Management Server, Info Resource Database, Knowledge Database and Agent Running Platform. Specifically, agent running platform is the basement of the system.

(1) User interface agent: user interface agent is the interaction interface between system and user which is used for receiving search demand and display the search result.

The core of User interface agent is analysis and inference machine which includes algorithm and rule for analysis and inference. Analysis and inference machine collects the users' search interest and habits, combines with database to be the domain knowledge of search.

(2) Search agent: search agent is the key of the system, which is responsible for information search service.

Search agent receives the search request from user interface agent. Then, combined with the domain knowledge, search agent generates several sub-searches by partitioning the search request. At last, information search is achieved by combination of local and remote search.

The next host should be determined when search agent finished search in a host, because search agent has to move between hosts. The search routine is determines by the internal routing list.

For the implementation of search independent, there is a monitor object for every search agent. The moving status of search agent is monitored, and the next behavior of search agent is issued according to the instruction of the status. Then, search flow is controlled.

(3) Agent management server: agent management server is the agent which is in the highest level. There is one agent management server at least in a system. Several agent management servers are applied to be network management mode of hierarchal structure. This could avoid single-point failure. Agent management server records the statuses of all agents, detects and manages the agents in timing. Agent management server controls all agents and sends search agent according to the search request and network status.

Agent management server determines search mode according to the information and domain knowledge of info resource database and knowledge database when agent management server received search request from user interface agent. Different search modes set different agent parameters and search tasks. When search results are returned from search agent, interest models and personalized data are compared with search results. If the similarity is greater than the reserved threshold value, the results are the information which is user searching for. Then, the information is displayed by user interface agent. In the system, push technology is used for sending search result to user by user interface agent. As a result, the real time is improved.

The agent management server is divided into service spaces and anonymous spaces. Specifically, service space is composed of search agent and user interface agent; anonymous space provides an independent running environment for the agent which finishes the search task. Service space and anonymous space are running on search server and searched node specifically.

(4) Info resource database: info resource database saves the information which user is interested in. In the system, there are three kinds of info resource database:

Local info resource database is composed of index database which is responsible for recording information resource from sites and classifying the information.

Node info resource database includes the information resource index of current site.

Internal info resource database of agent saves the search results from search agent.

(5) Knowledge database: knowledge database saves the domain knowledge related with user. In searching process, domain knowledge is used for search task generating, information analysis and information response.

(6) Agent running platform: agent running platform is the basement of system operating. It provides a running platform for highly distribution heterogeneous network environment. In addition, agent running platform provides dynamic communication mechanism and reliable security mechanism for the agents in the system.

Above 6 components work coordinately to accomplish the search task on Internet.

Process of information search. In the system, network information search service is provided by the cooperation of all agents. In the searching process, information resource databases of network internal nodes are traveled, and then the resource which meets the search demand is provided. The process is illustrated in figure 3.

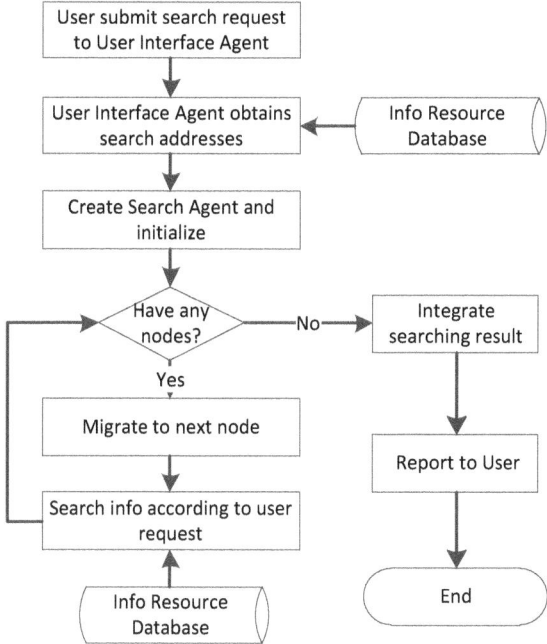

Fig. 3. The process of information searching

Network information search process is the process of collecting information resource for user demand. The detailed steps are as follow:

(1) User submits search request to user interface agent via browser.

(2) When user interface agent received user request, it analyses the request and obtains domain knowledge which is related with user from knowledge database. Then, the agent searches local info resource database and acquires search address list.

(3) Agent management server creates search agent, and transmits the request and address list to the search agent. The search task would be operated y search agent.

(4) Search agent is initialized according to the parameters. Then it moves independent in the range of search address list and searches information which is related to the user request.

(5) Search agent chooses next searched host as determined rule and moves to the host.

(6) When search agent reaches the host, the node information database is searched and the demanded search results are saved in internal information resource database.

(7) Search agent examines address list for finding the hosts without search. If there were, step (5) and (6) would be repeated. Otherwise, step (8) would be executed.

(8) Search agent returned to search server and submits its internal info resource database.

(9) Search server processes search results, and display them via user interface agent.

3.2   Routine Planning of Search Agent

The core of the system is the process that the search agent searches the host according to search address list. In the process, system determines a best route based on the search address list and current network status. This could finish the search task with lowest consuming [4].

In the system, network monitoring module and route planning modules are applied. They monitors network in real-time and plans moving routine of search agent. When search agent is initialized, the search address list is submitted to route planning module; the hosts addresses in search list are sent to network monitoring module; network monitoring module monitors the network status, sends the information which includes host running status, network delay and bandwidth to route planning module. At last, network planning module computes a best route according to the information.

The information interaction of search agent, network monitoring module and route planning module is illustrated in figure 4.

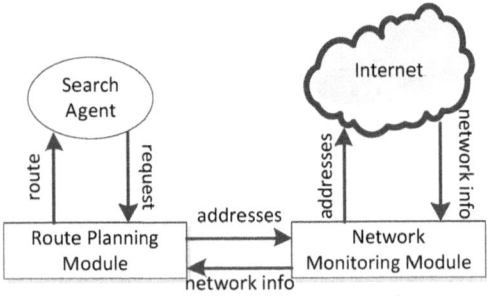

Fig. 4. Route Planning of Search Agent

3.3 Implementation and Testing of the System

The system is implemented with JADE (Java Agent Development Framework). JADE is a software framework fully implemented in Java language and supplies a stable, reliable platform [5]. MySQL database is applied for info resource database and knowledge database as the system characters that are large data, high performance and multi-thread programming. JADE supplies the running platform and manages internal mobile agent. Firstly, when system is initialized, user interface is created. It receives user request, searches local info resource database and obtains the address list which is saved in MySQL database. Secondly, search agent is created and initialized. At last but not least, search agent is started and sent to the searched hosts. In searching process, JADE could be suspended or stopped current agent. JADE destroys current search agent when search is completed.

For the compare with traditional information search system, an information search system is implemented which is based on client/server. At the same time, traditional system and system based on mobile agent are run on LAN for compare.

The testing environment is composed of 10 PC servers which are running FreeBSD. Servers are connected with that Fast LAN. The finish time of them are almost same, 5.7s and 4.6s, with same search request when search document is less than 1M; the difference became large with the document increases. When the document was 10MB, the search time is 243.6s and 121.5s respectively.

In traditional system, searched node sends the searched document to server, and the transmitting needs longer time while the document is larger; the system based on mobile agent search in local area which saves the time for transmitting. That means the efficiency of search is enhanced. Moreover, the search computing is allocated on the search node which decreases the computing load of search server.

4 Conclusion

Compared to traditional network search system, the system has the capability of asynchronous computing. A lot of data transmission is avoided and the dependence on network bandwidth is reduced as the search process is dynamically executed on searched node. Meanwhile, load balance among network nodes is achieved for the mobile agent can choose the best route dynamically according to network status. In addition, parallel execution of the search is implemented by creating more search agents for more searched node according to search task.

References

1. Lange, D.B.: Introduction to special issue on mobile agents. Autonomous Agents and Multi-Agent Systems 5(1), 5–6 (2002)
2. Lange, D., Oshima, M.: Seven Good Reasons for Mobile Agents. Communications of the ACM 42(3), 88–89 (1999)
3. Chen, J., Liu, L., Yu, X.: An intelligent information retrieval system model. Intelligent Control and Automation 34(4), 450–455 (2006)

4. Cabri, G., Leonardi, L., Zambonelli, F.: Mobile-agent coordination models for Internet applications. IEEE Computer 33(2), 82–89 (2000)
5. JADE-A White Paper,
   http://jade.tilab.com/papers/2003/WhitePaperJADEEXP.pdf
6. Gurel, U., Parlaktuna, O.: Agent based route planning for a mobile robot. In: 5th International Advanced Technologies Symposium, Karabuk, Turkey (2009)
7. Djekoune, A.O., Achour, K.: A Sensor Based Navigation Algorithm for a Mobile Robot using the DVFF Approach. International Journal of Advanced Robotic Systems 6, 97–108 (2009)
8. Fok, C., Roman, G.: Mobile Agent Middleware for Sensor Networks: An Application Case Study. In: Proc. Fourth International Symp. Information Processing in Sensor Networks, Los Angeles, California, USA, pp. 382–287 (2005)
9. Webots User Guide, http://www.cyberbotics.com
10. Ma, Y., Ding, R.: Mobile Agent technology and Its Application in Distributed Data Mining. In: 2009 First International Workshop on Database Technology and Applications, pp. 151–155. IEEE Computer Society, Los Alamitos (2009)

Research on Portable Frequency Hopping Wireless Data Transmission System

Ai-ju Chen[1] and Yuan-juan Huang[2]

[1] School of Electronic and Electrical Engineering, Wuhan Textile University, Wuhan, China 430073
[2] Department of Gynecology and Obstetrics, Qianjiang Municipal Hospital, Qianjiang City, Hubei Province, China 433100
aijuchen@126.com

Abstract. To remedy the complexity and sophistication of the Bluetooth equipments, a low-cost, high reliability, low power consumption and user friendly point to point wireless frequency hopping communication system was established. The system applies wireless communications chips nRF24L01 chips, DS/FH-SS techniques and adaptive frequency hopping technology, and enjoys the merits of excellent anti-jamming performance, good confidentiality, strong anti-interference, good mobility, and high accessibility. In addition, the system realized a low power consumption of around 70mW, which promises a good application to modern handheld wireless devices.

Keywords: Wireless Communication Systems, DS/FH-SS technique, Adaptive Frequency Hopping.

1 Introduction

With the recent development of various industries, people are increasingly demanding data transmission, enterprises in different regions departments hope to achieve a lower cost but high-speed networking, network operators and police hope to achieve in some places Reliable multi-point wireless transmission, or regional coverage. These reflect that people want to have such a wireless communication system can be very convenient, safe and economical to achieve these functions.

Bluetooth is a open global standard for wireless data and voice communications, establishing a special connection to fixed and mobile devices communications environment, The program written in a 9 x 9 mm microchip, the Bluetooth work in the global generic 2.4GHz ISM (ie industrial, scientific, medical) band.Bluetooth's data rate is 1Mb / s. Time division duplex transmission scheme is used to achieve full-duplex transmission, using the IEEE802.15 protocol. Bluetooth specially designed to quickly identify and frequency hopping scheme to ensure link stability, making it more stable than other systems. Use of FEC (Forward Error Correction, FEC) inhibited long-distance link random noise. Frequency-hopping transceiver (application binary frequency modulation (FM) technology) is used to suppress interference and prevent fading.Bluetooth can support an asynchronous data channel, three simultaneous

B. Liu and C. Chai (Eds.): ICICA 2011, LNCS 7030, pp. 249–257, 2011.

synchronous voice channels. It can also use a simultaneous transmission of asynchronous data channel and synchronous voice. Each voice channel supports 64kb/s synchronous voice links. Asynchronous channel can support one end of the maximum rate of 721kb / s while the other end of the rate of 57.6kb / s asymmetric connection, its also support the 43.2kb / s symmetric connection.

These characteristics make extensive use of Bluetooth in the mouse, keyboard, printer, laptop computer, headphones and speakers and other occasions. Not only increase the beauty of the office area, but also for interior decoration provides more creativity and freedom. Bluetooth, whether in the office, at the dinner table, or on the way, has more than a simple technology, but a concept. It can reduce the word-processing, shortening the waiting time, and seamless transaction for customers; But its complex configuration process, high cost and large power equipment, these problems exist.

To meet the industrial sector's request of wireless data transmission with high reliability, low cost, long working hours, This topic has developed a basic data transmission only, but high reliability, low power consumption, low cost, easy to use portable wireless data transmission system.

2   System Design

2.1   Hardware Design

The Wireless Communication System is mainly based on the chip nRF24L01. Modular design, with great flexibility and scalability, as needed, add other related data acquisition modules, System CPU uses ST company's STM32F103, which is 32bit, and 72MHz frequency, can efficiently process the data and nRF24L01's hopping control. The data can be collected and real-time packet transmission in time to achieve fast data transmission requirements. Figure 1 shows the specific block diagram of the hardware.

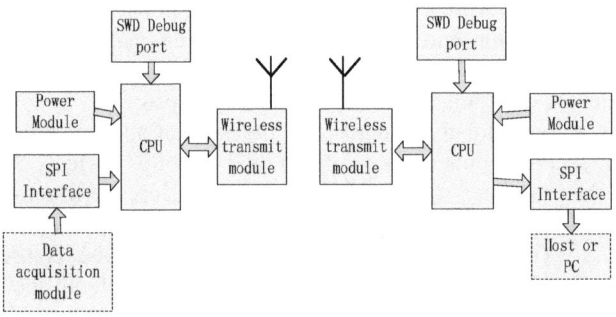

Fig. 1. System block diagram of the hardwar

RF part uses Nordic company's nRF24L01 chip, with features such as fast transmission speed, low power consumption, short state switching time, moderate distance, suitable for short-range frequency hopping communications. Meanwhile, the system is designed to a separate send module and receiving module, with good independence and portability, only need to transfer the data to module if necessary.

2.2   nRF24L01 Circuit Structure

Except NORDIC company's RF chip nRF24L01's power pins, The remaining pins are all can connect with MCU's generate I/O port,nRF24L01 itself has a SPI Interface with maximum speed 10MHz, STM32F103 is through SPI3 and nRF24L01 transmit data to each other. Figure 2 shows the connection circuit nRF24L01 module.

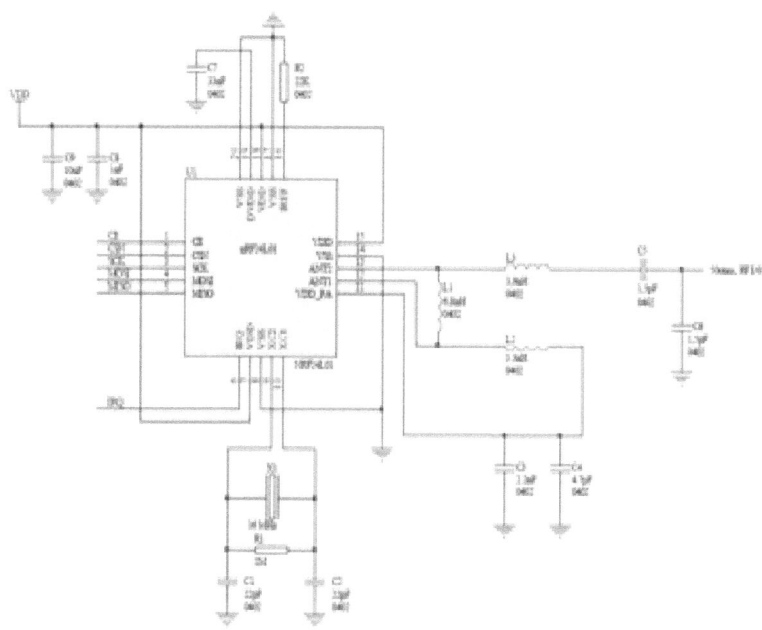

Fig. 2. nRF24L01 50 ohm single-ended RF output circuit

2.3   Software Design

This design is to develop an adaptive frequency hopping communication protocol which based on the nRF24L01 chip. Including: ARM's firmware development, nRF24L01 driver development, wireless communication protocol development, particularly in the hopping part and the adaptive part, which is the core idea of this design. Fig.3. is the overall structure.

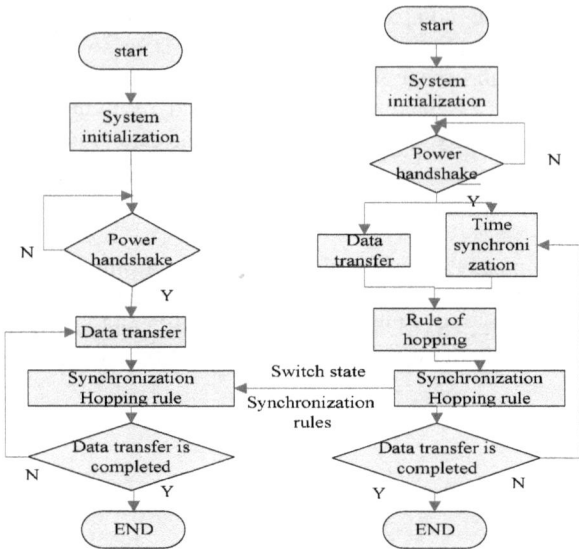

Fig. 3. The overall structure

2.3.1   Power Handshake

Shaking hands four times: First, the before two shaking hands is to establish an initial path, the equivalent of Pathfinder; Second, the after two shaking hands is to transport the synchronization information, the two sides start the timer synchronization. The two shaking hands sides connected on the frequency hopping 16 channels, sender 16 times per frequency-hopping, the receiver had just jumped again. Prevent the single frequency interference.

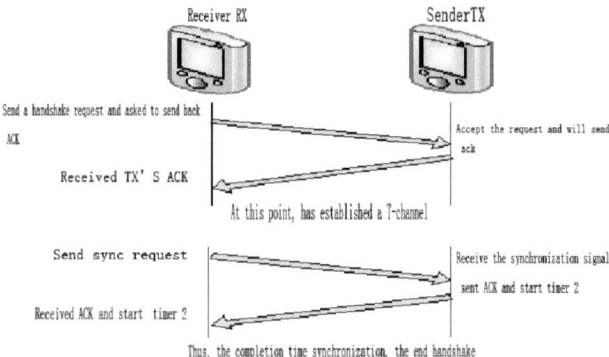

Fig. 4. Schematic diagram of power-handshake

2.3.2   Time Synchronization

Traditional methods: independent pathway, header method and self-synchronization method. The system uses data synchronization method. Its advantages are: Data need not to sync mark; before read Rx data can be in processing synchronization function.

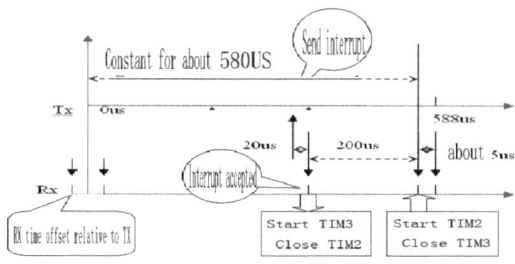

Fig. 5. Schematic Diagram of Time ActiveSync

The key point of Synchronous: the time difference the sender starts sends data to the receiver read is always constant.

2.3.3 Frequency Hopping Spread Spectrum and Direct Sequence Spread Spectrum (FH / DSSS)

Frequency hopping is to divide band into a number of hopping channels (hop channel), in a connection, the radio transceiver according to a certain code sequence (is a certain law, being called a "pseudo random code" technically, is a "false" random code), continuously "jump" from one channel to another channel. Only the two transceiver sides communicate according to the law, while others can not interfere with the interference by the same law, so the anti-interference ability is strong.

Frequency Hopping Spread Spectrum: Anti-interference ability, Low bit error rate

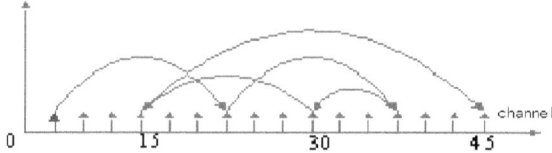

Fig. 6. Schematic diagram of frequency hopping spread spectrum

Direct Sequence Spread Spectrum: Excellent anti-jamming performance, Good Confidentiality.

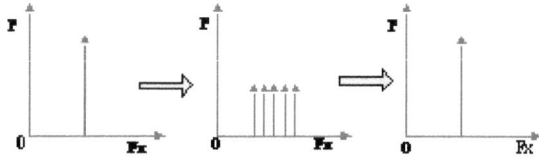

Fig. 7. Schematic diagram of direct sequence spread spectrum

2.3.4   Adaptive Frequency Hopping

Adaptive Frequency Hopping (AFH) belongs to avoid the interference sources actively, on the basis of a fixed frequency hopping law, it increases the frequency of adaptive control, that is, the law of the signal carrier frequency is variable. Firstly determine the channel quality, then formulate frequency hopping synchronization rule according to channel quality.

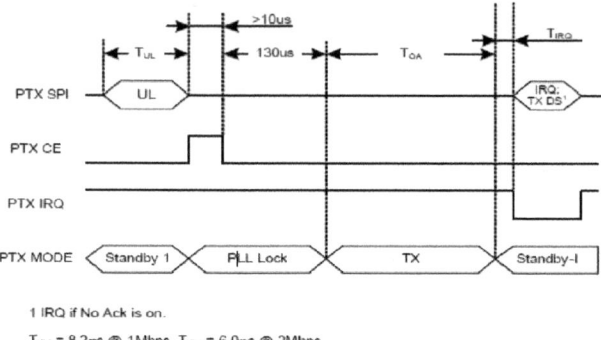

Fig. 8. Sending timing with no automatic response

Advantages are to automatically avoid frequency interference, strong anti-interference and very low bit error rate; fit for the complex environment of the crowded electromagnetic environment.

3   Test

3.1   The State Transition

Since it is adaptive frequency hopping, RX and TX need to interact with communications, this step is critical, because only the interactive communication succeeds, RX is possible to send the new law enacted FH to TX, in the interactive communication process:

Receiver: Rx->> Tx->> Rx
  Sender: Tx->> Rx->> Tx

Interactive communication will involve 4 times of Tx and Rx state conversion, in order not to affect the upload of data in real time, the whole process of interactive communication transmission frequency hopping law is not occupation of too much time, we must ensure that after the change of state can do may soon enter the normal communication state. This requires to ensure a good time synchronization when sending to the receiver and receiving to the sending converses.

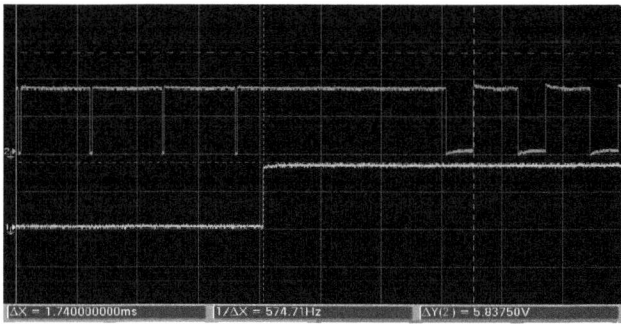

Fig. 9. State switch timing measured diagram

Figure 9 is the waveform measured by the Agilent oscilloscope. It can be observed from the chart, the send state is the first half part of the green line, the latter part of the green wave signifies the state to receive, that is the state transition occurred, and can be observed from the state transition moment to the communication system back to normal circumstances, only for three periods of time. 1740/588 = 2.959 = 3S, fully consistent with the oretical projections.

3.2   Frequency Hopping Test

Test Environment: 12 * 12 laboratory, normal phase (is EMI serious / more serious cases).Bluetooth interference sources: Bluetooth transmitter and receiver open at the same time. Transport normal data. Bluetooth power 0dB. Test plate: launching board STM32 + NRF24L01, Receiver board STME32 + NRF24L01, nRF power 0dB. The program of using: adaptive frequency hopping. The distance between transmit and receive Plate: 6 meters. The Antenna: transmitter: PCB antenna; the receiving end: cylindrical antenna. Obstacles: None

Table 1. Hopping Communications Test

Communication Rate Frames / sec	total number (Frame)	Data sent	Dropped frames	Bluetooth interference dropped frames XX cm distance from the XX					
				R 10	R 50	Middle	T 10	T 50	Middle
340	2*10e5	(1)	0	0	0	0	0	0	0
		(1)	0	0	0	0	0	0	0
		(2)	0	0	0	0	0	0	0
		(2)	0	0	0	0	0	0	0

Note: During testing, there are pillars in the interval between the transmitter and receiver because of the selection of some point, and some points are or even 10 meters. Test Description: A sideway The NRF launching is basically in the same position with receiving R 10: refers to the interference source 10cm away from the NRF areas receiving end. T 10: refers to the interference source 10cm away from the place where the NRF sender. Transmission time: a 200,000 transfer time approximately 9.8 minutes. Data sent:

(1) 0x68, 0x55, 0xAA, 0x55, 0xAA, 0x55, 0xAA, 0x55, 0xAA 0x55, 0xAA
(2) 0x00, 0x11, 0x22, 0x33, 0x44, 0x55 ... ... 0xEE, 0xFF 0xFF, 0XFF

Transport 200,000, a transfer takes approximately 9.8 minutes, cumulative about 274.4 minutes. In the process of the testing does not appear packet loss rate, the reliability of frequency hopping communications is relatively high. Table 1 also shows that frequency hopping communication is good for some occasional interference; In addition, the transmitter uses the PCB antenna, so comparing the frequency of other communications such as fixed communications, advantage is obvious.

4   Conclusion

The system uses the NORDIC company's wireless communications chips nRF24L01, with easy installation and maintenance, network structure flexibility, a wide coverage range and so on. Couple with the DS / FH-SS (combination of direct sequence spread spectrum and frequency hopping spread spectrum) technology and adaptive frequency hopping technology to control the error rate more than in the 10E-5, making the system be highly resistant to electromagnetic interference, be fit for crowded electromagnetic environment and a place wireless data transmission has the demanding requirements; The system can be carried, has good mobility, and transport data at any time. In addition, the system power consumption is very low, about 70mW, can use battery for power up to the time, can meet the modern handheld wireless devices for low power requirements well. The portable wireless data transmission system, as a rapid and efficient mobile information transmission system, will have very broad application prospects in people's daily life and work.

References

1. Chen, L.-f., Han, Y.-j., Xu, K.-h.: Multi-points dynamic temperature measurement and control system based on wireless communication technology. Transducer and Microsystem Technologies (2011-02)
2. Zhao, J., Wang, B.-w., Tang, C., Yang, T.-l.: Collision Avoidance System of Railway Maintenance Machineries Based on GPS and Wireless Communication Technologies. Microcomputer Information (2011-03)
3. Gong, H.: Wireless Centralized Meter Reading System Based on Short Distance Wireless Communication and 3G. Computer Engineering (2011-02)
4. Li, Y.-j.: Simulation Design of Frequency-hopping Communication System Suitable for Teaching. Communications Technology (2011-04)
5. Li, S.: Design of Wireless Data Transmission System Based on nRF24L01. Journal of Hubei University of Education (2011-02)

6. Qiao, T.-x.: Design of DS/FH signal transmission system based on FPGA. Electronic Design Engineering (2011-02)
7. Hao, W.: Wireless Environment Monitoring System Based on DSP and nRF24L01. Microcontrollers & Embedded Systems (2011-03)
8. Pan, Y., Guan, X., Zhao, R.: Design of smart wireless temperature measurement system based on NRF24L01. Electronic Measurement Technology (2010-02)
9. Tian, Z.-w.: Design of data communication system in wireless sensor networks based on Bluetooth. Transducer and Microsystem Technologies (2010-09)
10. Zhou, Y.: Implementation of RFID Wireless Communication System Based on FPGA. Modern Electronics Technique (2010-17)

Research on Improving Throughput of Wireless MIMO Network

Yongli An[1,2], Yang Xiao[2], Pei Wu[1], and Xuefei Gao[1]

[1] College of Information Engineering, Hebei United University,
Tangshan Hebei 063004, China
[2] Institute of Information Science, Beijing Jiaotong University, Beijing 100044, China
any1@heut.edu.cn

Abstract. To improve the throughput of wireless MIMO network, this paper proposed two new methods. One is called interference aligment, the other is called interference separation based on transmission time delay. These two methods based on transmiter and receiver precoding design are all of linear complexity. By adoptting these proposed methods, the wireless MIMO network can transmitt more packets or support more users simultaneously. Simulation result show that these proposed methods are not affected by channel matrix correlation.

Keywords: WLAN, MIMO, Virtual MIMO, Throughput.

1 Introduction

The Multiple-inputs Multiple-output technology uses multiple transmission antennas and multiple receiving antennas to resist the influence of wireless fading channel. Under the superiority of MIMO technology, WLAN may be used for transmitting multimedia applications which do not allow for time delay and need massive bandwidth, for example, the HDTV wireless real-time dissemination. It also provids for the enterprise or the family more reliable and higher speed transmission in bigger coverage area. Moreover , WLAN enables the unceasingly promoted network connection speed to display its advantage fully. The MIMO technology may improve the covering scope and the transmission speed of WLAN largely[1-3]. 802.11 LAN that using MIMO technology, it's client side and access point all use 2 antennas. This causes the system to be possible to transmit two data packets simultaneously. In theoretical, the throughput of the system   is 2 times of the existing wireless local area network without using MIMO technology.

In this work we provide two new interference Cancellation schemes for the MIMO WLAN through joint transmiter and receiver design[4,5]. First, we will discuss the transmitting signal space precoding technique. It is used to align the multi-user interferences, so the throughput of 802.11 LAN may further enhance by using this algorithm. Secondly, we will propose a scheme based on transmission time delay. It separates access point's receiving signal by odd time slot and even time slot, thus expected signal and unexpected signal are separated for clients. So this scheme can support more clients to transmit data packts to Aps.

B. Liu and C. Chai (Eds.): ICICA 2011, LNCS 7030, pp. 258–265, 2011.

2   Improve Throughput of MIMO WLAN

Currently, we suppose a local area network has two APs and two users, as shown in Figure 1. If we do not use the special algorithm or the anti-interference mechanism, these two users cannot transmit data packets simultaneously[6-8]. As Figure 1, if this time user 1 transmits two data packets, and user 2 also transmits a data packet, then the AP 1 will receive these three data packets simultaneously. Similarly, the AP 2 will also receive these three data packets simultaneously. But these two APs have only two antennas separately. It is impossible for them to decode three data packets p_1, p_2 and p_3 all together. Namely, two equations include three unknowns can not be decoded.

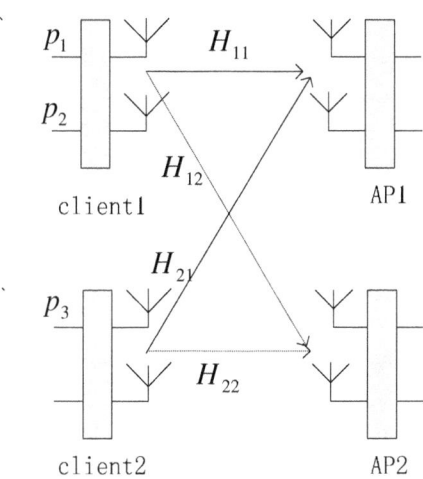

Fig. 1. Traditional multiuser MIMO LAN

Then we may consider one kind of new algorithm which is called interference alignment[9,10]. We reconsider the MIMO system, when the transmitting side transmits the data p_1 using the first antenna, it is equal to multiply the unit vector $[1,0]^T$ with the data packet sampling. After transmission, receiving side's receiving vector is $H[1,0]^T p_1$. If the transmitting side is multiplied by another vector for example $\bar{v}$, then AP receives $H\bar{v}p_1$. Therefore, we let p_i multiply $\bar{v}_i$ to replace transmitting a data on each antenna independently. We called $\bar{v}_i$ is the decoding vector of p_i. As shown in Figure 2, we still consider the example above, that has two APs and two users currently. First, select two unequal vectors $\vec{v}_1$ and $\vec{v}_2$ randomly, as Figure 2, we must make p_3 and p_2 combine at AP 1, namely request $H_{11}\vec{v}_2 = H_{21}\vec{v}_3$. From this we may extract $\vec{v}_3$, namely $\vec{v}_3 = H_{21}^{-1}H_{11}\vec{v}_2$.

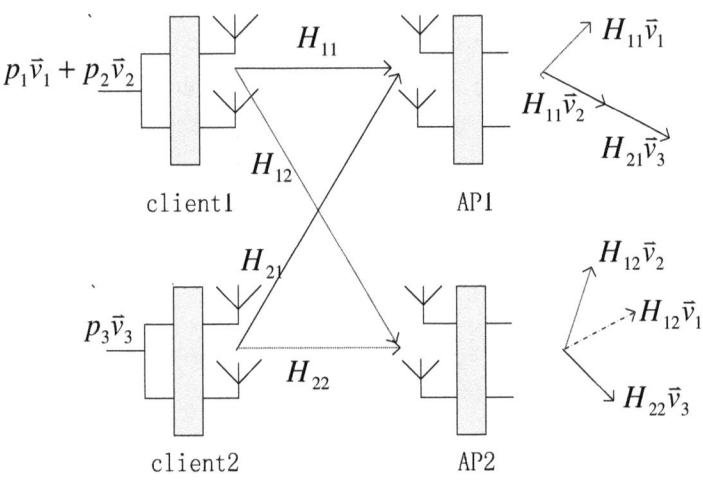

Fig. 2. MIMO LAN using new algorithm

Therefore, at AP 1, we may decode the data p_1 as long as founding a vector which is orthogonal with the vector $H_{11}\vec{v}_2$ or $H_{21}\vec{v}_3$. When decoding of p_1 is completed, transmit p_1 from AP 1 to AP 2 through ethernet, then the AP 2 will use interference cancellation to subtract the packet p_1. So two antennas' receiving equations with two unknown data packet p_2 and p_3 might be completly decoded.

3   Improving Client Quantity of MIMO WLAN

Consider the 3×2 uplink MIMO WLAN system shown in Fig 4, that is 3 users with single antenna and one base station with two antennas. In this wireless network there is a propagation delay associated with each channel. In particular, let us assume that the propagation delay is equal to one symbol duration for all desired signal paths and two symbol durations for all paths that carry interference signals[6-10]. The channel output at AP's receiving antenna are defined separately as (1), (2).

$$y_1(n) = h_{11}p_1(n-1) + h_{21}p_2(n-2) + h_{31}p_3(n-2) + z_1(n) \ . \tag{1}$$

$$y_1(n) = h_{12}p_1(n-2) + h_{22}p_2(n-2) + h_{32}p_3(n-1) + z_3(n) \ . \tag{2}$$

Where during the nth time slot (symbol duration) transmitter $j \in \{1,2,3\}$ sends symbol $p_j(n)$ and $z_j(n)$ is i.i.d. zero mean unit variance Gaussian noise (AWGN). All inputs and outputs are complex. Now, with all the interferers present, suppose each client side transmits only during odd time slots and is silent during the even time slots. Let us consider what happens at antenna 1of the AP. The symbols sent from its desired transmitter (transmitter 1) are received free from interference during the even time slots and all the undesired (interference) transmissions are received simultaneously during the odd time slots. Thus, we separate the receiving single to even time slots receiving single and odd time slots receiving single. For antenna 1 of the AP, we can define them as (3),(4)

$$y_1(n_e) = h_{11}p_1(n-1) \ . \tag{3}$$

$$y_1(n_o) = h_{21}p_2(n-2) + h_{31}p_3(n-2) \ . \tag{4}$$

As the same, even time slots and odd time slots receiving signle at antenna 2 of the AP can also be defined as (5), (6)

$$y_2(n_e) = h_{32}p_3(n-1) \ . \tag{5}$$

$$y_2(n_o) = h_{12}p_1(n-2) + h_{22}p_2(n-2) \ . \tag{6}$$

We may see from the formula above that it is easy to decode the receiving signal when we separate the receiving signal time slot into odd time slot and even time slot.

From the formula (3) and (5), we may decode the sending date packts P_1 and P_3 directly, certainly, this can be worked out only in the situation that the channel parameters h_{11} and h_{32} are already estimated. Then, we substitute P_3 into formula (4) or substitute P_1 into formula (6). Choosing one formulas from (4) and (6) can decode the date packt P_2. Therefore, decoding three client's data packets simultaneously also only need to estimate 4 channel parameters then, namely, $h_{11}, h_{32}, h_{21}, h_{31}$ or $h_{11}, h_{32}, h_{12}, h_{22}$.

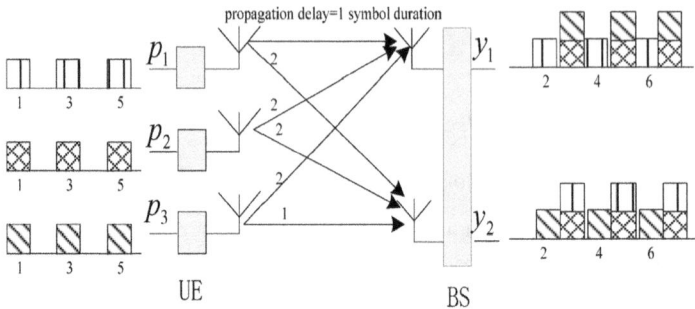

Fig. 3. 3×2 MIMO using propagation delay

Compared with the traditional 2×2 MIMO, this kind of method which separates the time slot into odd time slot and even time slot by using time delay will be wonderful to separate the wanted signal and the interference signal. This method has these following merits. First, it can support more users to transmit data packets simultaneously, and also achieve 3×2 virtual MIMO(V-MIMO). Secondly, each client has only one antenna. Thus, it reduces the complexity of client side's equipment.Thirdly, supporting more users at the same time has not increased the quantity of channel parameter estimation. 3 users still only needed to estimate 4 channel parameters. It is possible to see form the analysis above, in fact 3 users have only used the odd time slot to transmit data packets, but the even time slot was idle, namely, average every two time slots altogether transmits 3 data packets. But the traditional 2×2 MIMO is every two time slots altogether transmits 4 data packets. Although this method has merits as stated above, it causes the throughput of the system in fact dropped. In order to solve this problem, we make a improvement to the algorithm as follows.

4   Improved Algorithm

Let the number of uplink clients increase to 4, namely, compose 4×2 MIMO system.As the front analysis, we express the receiving signal at AP's two antennas separately as(7),(8)

$$
\begin{aligned}
y_1(n) = h_{11}p_1(n-1) + h_{21}p_2(n-2) + h_{31}p_3(n-2) + \\
h_{41}p_4(n-2) + z_1(n)
\end{aligned}
\tag{7}
$$

$$
\begin{aligned}
y_2(n) = h_{12}p_1(n-2) + h_{22}p_2(n-2) + h_{32}p_3(n-2) + \\
h_{42}p_4(n-1) + z_4(n)
\end{aligned}
\tag{8}
$$

Then we separate the receiving signal by odd time slot and even time slot. We write:

$$y_1(n_e) = h_{11} p_1(n-1) \quad . \tag{9}$$

$$y_1(n_o) = h_{21} p_2(n-2) + h_{31} p_3(n-2) + h_{41} p_4(n-2) \quad . \tag{10}$$

$$y_2(n_e) = h_{42} p_4(n-1) \quad . \tag{11}$$

$$y_2(n_o) = h_{12} p_1(n-2) + h_{22} p_2(n-2) + h_{32} p_3(n-2) \quad . \tag{12}$$

Obviously, we can decode p_1 and p_4 from formula (9) and (11) directly. Substitute the decoded data packets p_1 and p_4 into formula (10) and (12) to Cancel directly, then we get:

$$y_{1'}(n_o) = h_{21} p_2(n-2) + h_{31} p_3(n-2) \quad . \tag{13}$$

$$y_{2'}(n_o) = h_{22} p_2(n-2) + h_{32} p_3(n-2) \quad . \tag{14}$$

The above equation is similar with traditional 2×2 MIMO, so long as rank of channel matrix $H \begin{bmatrix} h_{21} & h_{31} \\ h_{22} & h_{32} \end{bmatrix}$ is 2, we might decodes p_2 and p_3 simultaneously. These four single antenna client side and the AP with two antennas will compose a MIMO system which can decode 4 data packets simultaneously. Compared with the traditional 2×2 MIMO, this algorithm has supported more users and has not reduced system's average throughput.

5   Simulation and Conclusion

Because this proposed algorithm does not have the request for pairing user's channel matrix correlation. So we simulate the difference between the traditional V-MIMO and the proposed V-MIMO algorithm, when the pairing users are selected randomly, namely the channel matrix is random. In Figure 5, we simulate the real transmitting signal value; restorative transmitting signal value when selecting the pairing users randomly, namely the channel matrix is random; and the restorative transmitting signal value when the channel matrix is high correlated, namely orthogonal factor is small. Obviously channel matrix's correlation has affected the signal restoration. When channel matrix has high correlation, the difference between real transmitting signal value and restorative transmitting signal value is also big.

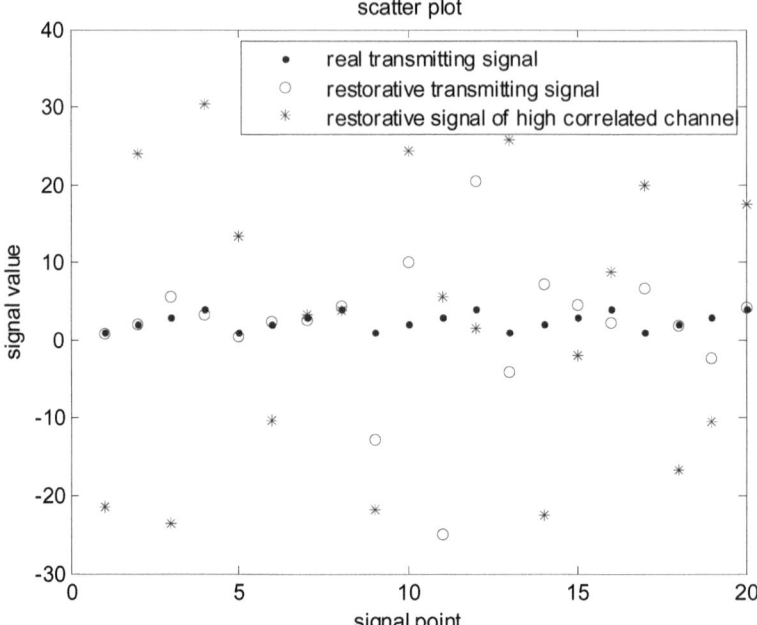

Fig. 5. Influence of channel matrix correlation

Thus we may know, our proposed algorithm is more optional to user's selection and the signal restoration quality does not rely on pairing users' channel matrix correlation, namely it does not need pairing strategy or only some partial users need to apply pairing strategy.

In this paper, we have investigated how to enhance the performance of WLAN using MIMO technology. We have proposed two interference cancellation approaches from different angles. One algorithm is based on transmitting signal space precoding. This algorithm has increased the average throughput of the MIMO WLAN system. And the other algorithm is based on transmission time delay. It can support more single antenna clients to reduce the complexity of the client side's equipment. All these two algorithm have shown better performance in WLAN uplink system using MIMO technology.

References

1. Xiao, Y., Zhao, Y., Lee, M.H.: MIMO Precoding of CDMA Systems. In: Proc. 8th International Conference on Signal Processing, Beijing, vol. 1, pp. 397–401 (2006)
2. Xiao, Y., Lee, M.H.: MIMO Multiuser Detection for CDMA Systems. In: ICSP 2006 Proceedings (2006)
3. Li, L., Jing, Y., Jafarkhani, H.: Multi-user transmissions for relay networks with linear complexity (July 2010), http://arxiv.org/abs/1007.3315
4. Jing, Y., Jafarkhani, H.: Interference cancellation in distributed spacetime coded wireless relay networks. In: Proc. IEEE ICC (2009)

5. Chen, W.J., Xiao, Y., Zhao, Y.: The Algorithm Implement of WCDMA Channel Estimation. In: Proc. 7th International Conference on Signal Processing, vol. 3, pp. 1894–(1897)
6. Cadambe, V.R., Jafar, S.A.: Interference alignment and the degrees of freedom for the K user interference channel. IEEE Trans. Inf. Theory IT-54(8), 3425–3441 (2008)
7. Cadambe, V.R., Jafar, S.A., Shamai, S.: Interference alignment on the deterministic channel and application to fully connected AWGN interference networks. IEEE Trans. Inf. Theory IT-55(1), 269–274 (2009)
8. Peters, S.W., Heath Jr., R.W.: Interference Alignment Via Alternating Minimization. In: Proc. of ICASSP 2009 (2009)
9. Tresch, R., Guillaud, M.: Cellular Interference Alignment with Imperfect Channel Knowledge. In: Proc. of ICC 2009 (2009)
10. Thukral, J., Bölcskei, H.: Interference Alignment with Limited Feedback. In: Proc. of ISIT (2009)

An Efficient and Lightweight User Authentication Scheme for Wireless Sensor Networks

Xiancun Zhou[1,2] and Yan Xiong[1]

[1] School of Computer Science, University of Science and Technology of China, Hefei
230026, China
[2] Department of Information and Engineering, West Anhui University, Lu'an 237012, China
{zhouxcun,yxiong}@mail.ustc.edu.cn

Abstract. As wireless sensor networks(WSNs) are susceptible to attacks and sensor nodes have limited resources, it is necessary to design a secure and lightweight user authentication protocol for WSNs. Over the past few years, a few user authentication schemes for WSNs were proposed, in which the concept of timestamps were employed without exception. Analysis shows that it brings new security flaws. An efficient and lightweight user authentication scheme for WSNs is proposed in this paper. In proposed scheme, we employ one-way hash function with a secret key. With the help of smart card, mutual authentication is performed using a challenge-response handshake between user and gateway node in both directions, and it avoids the problem of time synchronization between smart card and the gateway node. An analysis of the scheme is presented and it shows its resilience against classical types of attacks, and it adds less computational overhead than other schemes.

Keywords: Wireless sensor network, user authentication, smart card.

1   Introduction

Recent advances in electronics and wireless communication technologies have enabled the development of large-scale wireless sensor networks (WSNs), which consist of many low-power, low-cost, and small-size sensor nodes. WSNs hold the promise of facilitating large-scale and real-time data processing in complicated environments. The applications of WSNs include military sensing and tracking, environmental monitoring, health care, intelligent household control, wildlife monitoring and so on[1]. It is a challenging task to secure WSNs as it presents a uncontrollable environment with constrained resources[2]. Each sensor node carries a limited, generally irreplaceable energy source. Therefore, energy conservation is the most important performance consideration for extending network lifetime. Security mechanisms designed for WSNs must be lightweight and efficient[3]. User authentication is necessary and important if users want to access the real-time data through WSNs. For example, the user might use his mobile devices to query the WSN about the average humidity or temperature, or soldiers might use a small computer to query the WSN about images of the surrounding area in battlefield. It's well known that the user authentication is developed in traditional network. But it is so different

B. Liu and C. Chai (Eds.): ICICA 2011, LNCS 7030, pp. 266–273, 2011.
© Springer-Verlag Berlin Heidelberg 2011

from the traditional network because computation, storage, and battery power of WSNs are quite limited. Given the resource-constrained, it is difficult to apply traditional user authentication solutions in WSNs. For the last decade, a plenty of literatures and related schemes have been proposed on user authentication. However, only few researches focus on user authentication for WSNs. The problem of user authentication in WSNs was firstly identified only in 2004 by Benenson et al.[4]. In order to resistance to replay attack, most schemes based on timestamps have been put forward[7]-[11]. However, these schemes all have security flaws. In this paper, we design a secure and lightweight user authentication scheme for WSNs using smart card, it is based on one-time random numbers(nonces) instead of timestamps, mutual authentication is performed using a challenge-response handshake between user and gateway node in both directions, and it avoids the problem of synchronism between smart card and the gateway node.

The rest of the paper is organized as follows. In Section 2, we review some related works in this area; In Section 3, we define the model of sensor network. We describe the design of the proposed user authentication scheme in Section 4. And then, we discuss the security and computation costs of the proposed schemes in Section 5. Finally, the conclusions of this paper is shown in Section 6.

2 Related Works

In 2006, Wong et al. [7] proposed a password-based dynamic user authentication scheme for WSNs. The scheme is a password-based dynamic user authentication scheme with lightweight computational load, it requires only hash function and exclusive-OR operations. But, there are some security flaws in the scheme [8][9]. For example, it cannot protect against the replay and forgey attacks; passwords can be revealed by any of the sensor nodes; it is vulnerable to many logged-in users with the same login-id threat; the protocol also suffers from stolen-verifier attack, because both the gateway node and login-node maintain the lookup table of registered users' credentials[9]. In 2007,Tseng et al. [8] proposed an improved user authentication scheme that is modification of Wong et al.'s scheme. It claimed that it not only fixes the weaknesses but also enhances the security of Wong et al.'s scheme. However, Tseng et al. 's scheme brings other problems[9]. In 2009, Das[9] proposed a two-factor user authentication scheme based on smart card and password. He claimed that the approach can defend the replay, impersonation, stolen-verifier and guessing attacks. However, in 2010, Khan and Algahathbar [10] pointed out that Das's scheme is vulnerable to gateway node bypassing attack and privileged-insider attack. Recently, B Vaidya et al.[11] pointed out that several security pitfalls remain in both Das's and Khan-Algahathbar's schemes, and proposed an improved scheme. It was claimed that the scheme can protect WSNs from stolen smart card attacks and other common types of attacks. But the overall computational overhead of the scheme is higher comparing with those of Das's scheme and Khan-Algahathbar's scheme.

As a matter of fact, a potential Denial of Service (Dos) attack exists in all schemes mentioned in the above because of employing the concept of timestamps. Since these schemes ensure validity of authentication request by timestamps, the adversary can intercept login request, delay for a period of time, then replay it; or the adversary can modify the timestamp of login request and replay it. So, the login requests can't pass the verification of timestamp, users requests will be denied. In addition, when network congestion happening, Dos attack occurs probably. In order to avoid the disadvantages above, we propose a new scheme.

3   Network Model

We consider a WSN organized in a star topology, and is managed by a special sensor node called gateway node(GWN). Sensor nodes can use, for example, the ZigBee/ 802.15.4 standard as a communication protocol. GWN is a sensor node that will play the role of relay between the user and the rest of the WSN: users   receive data through GWN. Thus, GWN represents the point of access to the WSN. Therefore, the access control process is implemented at GWN and also the authentication of users is made by GWN, which is responsible for maintaining the network security, managing the network, responding to user queries and sending commands to the other sensor nodes.

4   The Proposed Authentication Scheme

In this section, we propose an efficient and lightweight user authentication scheme for WSNs. Before issuing any queries to access data from sensor network, the user has to register with the gateway node. Upon successful registration, a user will receive a personalized smart card, and then, with the help of user's password and the smart card, the user can login to GWN and access data from the network. The scheme is divided into four phases: registration phase, login phase and mutual authentication phase and password change phase.

The notations used throughout this paper is shown in Table 1.

Table 1. The Symbols and Notations

symbols and notations	Description
U	user
IDi	identity of user
pwi	password chosen by user
GWN	gateway node.
y	secret key of gateway node
h( )	a secure one-way hash function
hK( )	hash function with secret key K
$\oplus$	exclusive-OR operation
=?	verification operation
‖	bit-wise concatenation operator
$\rightarrow$	a common channel

4.1   Registration Phase

When a user wants to access the wireless sensor network, he has to register with GWN of the network firstly.

GWN selects a secret parameter y and it is securely stored. IDi is Ui's identity, which is unique. The following steps are performed to complete the registration phase:

1) Ui chooses his password pwi.
2) Ui submits his IDi and pwi to GWN via a secure channel.
3) GWN checks the registration information of Ui, confirms the identity of Ui.
4) GWN computes $Ri=h(y\|IDi)\oplus pwi$.
5) GWN write $h(\bullet)$, IDi, Ri and h(pwi) into a smart card. Then,GWN removes pwi, Ri from his memory.
6) GWN issues smart card to Ui over a secure channel.

4.2   Login Phase

When the user Ui wants to acquire the sensed data, he needs to insert his smart card into a terminal, and input IDi and pwi. The smart card computes the value of h(pwi), and compare the result with the stored one in it. The steps to be performed in this phase are detailed as follows:

1) Ui inserts his smart card into his mobile devices and enters his identity IDi and password pwi.
2) The smart card computes $p^*= h(pwi)$, and check whether p^* is equal the value of h(pwi) stored in the card or not. If yes, the legitimacy of the user can be assured and proceed   to the next step. Otherwise, end this phase.
3) Smart card generates a one-time random number(called nonce) Nu in order to challenge GWN to prove its identity.
4) Ui sends the login message { IDi, Nu} to GWN for the authentication process.

4.3   Mutual Authentication Phase

when receives login message { IDi, Nu}, GWN will perform as follows:

1) GWN verifies ID. validates if IDi is valid. GWN has a ID table which includes user's ID with counter, Initial value of which is zero. Upon receiving the login request message { IDi, L1}, GWN checks the table , validates if IDi is a valid user's ID then performs further operations, otherwise terminates the operation and informs Ui about it . Meanwhile, GWN checks the value of the counter ,if the value is bigger than a threshold value, it can be judged that an illegal login or   dos attack ,the login request of Ui will be refused; otherwise, GWN   accepts the login request of Ui, and the value of the counter will be increased by one .

2) GWN Computes $S= h(y\|IDi)$.
3) GWN generates a one-time random number(nonce) Ns.
4) GWN Computes $Vs= h_S(Nu, Ns)$.
5) $GWN \rightarrow Ui:Ns, Vs$.

Upon receiving the mutual authentication message, Ui Computes S'= Ri⊕pwi and Vu=hₛ·(Nu, Ns). verifies Vu=?Vs, if not, Ui rejects this message and terminates the operation; otherwise, Ui authenticates GWN and following steps are performed by Ui :

    6) Ui Computes Vu= hₛ·(Ns).
    7) Ui→GWN:Vu.

When GWN receives{ Ns, Vu}, computes and verifies Vu=? hₛ(Ns). If it is true, Ui has been authenticated by GWN; if not, Ui will be rejected, GWN send Reject message to the user Ui.

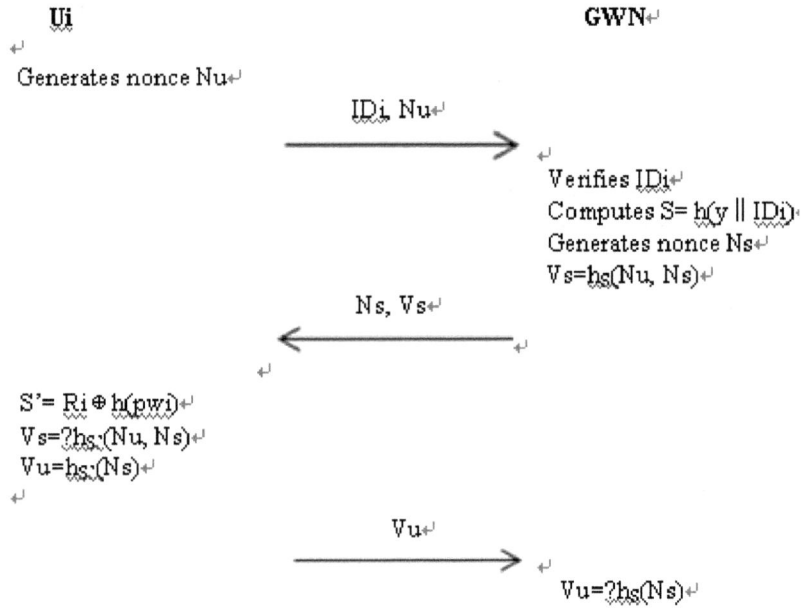

Fig. 1. Login/Mutual Authentication Phase

4.4   Password Change Phase

When the authenticated user Ui wants to update his password, he inserts his smart card into the terminal and keys his ID and old password pwi. The smart card computes h(pwi), and check whether the value is equal h(pwi) stored in it or not. If not, end the password change phase; otherwise, the legitimacy of the user can be assured and proceed to the next. Ui selects new password pwi*. The smart card performs the following operations without interacting with GWN:

    1) Computes Ri*=Ri⊕h(pwi)⊕h(pwi*), where pwi* is the new password.
    2) Replaces the old value of Ri with the new value Ri*.

5   Analysis

In this section, we present the analysis of security features of the proposed protocol and the comparison of the cost overhead.

5.1   Security Analysis

(1) Resist Impersonation Attack

In our scheme, if the intruder wants to impersonate the user to pass the verification of GWN, he must launch the attack by choosing a random number Nu' and sending {IDi , Nu' } as a login request message to GWN. Note that from GWN's point of view, Nu'is perfectly indistinguishable from Nu of an honest execution since both are simply random numbers. However, the intruder cannot compute the valid Vu in the step 6 of mutual Authentication phase due to not knowing S'= Ri$\oplus$pwi. Without not knowing pwi, the intruder can't compute S'. What's more, the intruder can't impersonation of GWN, because he can't compute Vs without knowing the secret key y.

(2) Prevent from Stolen Smart Card Attack

Although it is generally assumed that the smart card is safe and cannot be stolen or be lost, there is a risk of smart card being stolen. Here it is demonstrated that our proposed scheme can prevent from stolen smart card attack. The details are as follows: Firstly, if the intruder steal a smart card and cracks it, he can extract h($\bullet$), IDi, Ri and h(pwi) stored in the user's smart card. However, the intruder cannot obtain pwi from h(pwi), the security is based on one-way property of the hash function. He cannot use h($\bullet$), IDi, Ri and h(pwi) that he has obtained.Without knowing the pwi, it is impossible to compute the correct value of Vu. Therefore, the intruder cannot masquerade as the legal user to to pass GWN's authentication.

(3) Resist Replay Attack

A replay attack (replaying an intercepted message) can't work in our scheme because mutual authentication is performed using a challenge-response handshake in both directions. The scheme employs a nonce as the challenge to ensure that every challenge-response sequence is unique. This protects against a replay attack.

(4) Resist Dos Attack

The whole authentication process requires only need small amounts of computation, so it doesn't occupy too much energy, memory and computational resources on the gateway node. What's more, in our scheme, the attack is considered and has been taken measures to protect from it when the intruder initiates Dos attack in the phase of login. In our scheme, Upon receiving the login request message, GWN checks the ID table. It checks the value of the counter in the table, which is used to record the login times of the same ID. If the value is bigger than a threshold, Dos attack can be judged, the login request will be refused.

5.2   Performance Analysis

In this section, we give the performance comparisons between other schemes and our proposed scheme. Some notations are defined as follows:

th: the time for performing a one-way hash function;
tXOR :the time for performing a XOR operation;
tRAN: the time for performing a randomized algorithm.

Table 2. Comparison with Other Related Schemes

Scheme	Computation Cost		Mutual Authent-ication	Synchro-nization Indepe-ndence
	Registration phase	Login/authentication phase		
Our scheme	$1t_h+1t_{XOR}$	$6t_h+1t_{XOR}+2t_{RAN}$	Yes	Yes
Vaidya et al's scheme	$5t_h+3t_{XOR}$	$13t_h+6t_{XOR}$	No	No
Das's scheme	$3t_h+1t_{XOR}$	$5t_h+3t_{XOR}$	No	No

Table 2 shows the overall performance comparison between other schemes and our scheme. Apparently, our proposed scheme adds less computational overhead than Das's scheme and Vaidya et al's scheme. Moreover, our scheme provids mutual authentication between user and gateway node in WSNs and it need not synchronized clocks.

6   Conclusion

In this paper, we have proposed an efficient user authentication scheme based on smart card and a challenge-response mechanism. The proposed scheme not only need not synchronized clocks, but also is efficient in terms of message exchanges and computation burdens, so it can be easily implemented in the resource-constrained wireless sensor networks. As a part of our future work, we will go on a detailed simulation and implementation for our proposed scheme.

Acknowledgments. The work presented in this paper is supported by the Nature Science Foundation of Anhui Education Department under Grant No. KJ2011B204 and directional entrusting projects of Lu'an under Grant No. 2008LWA002.

References

1. Xiangqian, C., Makki, K., Kang, Y., Pissinou, N.: Sensor network security: a survey. IEEE Communications Surveys & Tutorials 11(2), 52–73 (2009)
2. Li, N., Zhang, N., Das, S.K., Thuraisingham, B.: Privacy preservation in wireless sensor networks: A state-of-the-art survey. Ad Hoc Networks 7(8), 1501–1514 (2009)
3. Sharifi, M., Kashi, S.S., Ardakani, S.P.: Lap: A lightweight authentication protocol for smart dust wireless sensor networks. In: International Symposium on Collaborative Technologies and Systems, pp. 258–265 (2009)

4. Benenson, Z., Gartner, F., Kesdogan, D.: User Authentication in Sensor Networks (Extended Abstract). In: Lecture Notes in Informatics Proceedings of Informatik 2004, Workshop on Sensor Networks, Ulm, Germany (2004)
5. Benenson, Z., Gedicke, N., Raivio, O.: Realizing Robust User Authentication in Sensor Networks. In: Workshop on Real-World Wireless Sensor Networks, Sweden (2005)
6. Jiang, C.M., Li, B., Xu, H.: An Efficient Scheme for User Authentication in Wireless Sensor Networks. In: Proc. of 21st International Conference on Advanced Information Networking and Applications Workshops, AINAW 2007 (2007)
7. Wong, K.H.M., Zheng, Y., Cao, J., Wang, S.: A dynamic user authentication scheme for wireless sensor networks. In: Proc. of the IEEE International Conference on Sensor Networks, Ubiquitous, and Trustworthy Computing (SUTC 2006), pp. 244–251 (2006)
8. Tseng, H.R., Jan, R.H., Yang, W.: An Improved Dynamic User Authentication Scheme for Wireless Sensor Networks. In: IEEE Global Telecommunications Conference (GLOBECOM 2007), pp. 986–990 (2007)
9. Das, M.L.: Two-Factor User Authentication in Wireless Sensor Networks. IEEE Trans. Wireless Comm., 1086–1090 (2009)
10. Khan, M.K., Alghathbar, K.: Cryptanalysis and Security Improvements of Two-Factor User Authentication in Wireless Sensor Networks. Sensors 10(3), 2450–2459 (2010)
11. Vaidya, B., Markrakis, D., Mouftah, H.T.: Improved Two-factor user authentication in wireless sensor networks. In: 2nd International Workshop on Network Assurance and Security Services in Ubiquitous Environments, pp. 600–605 (2010)

Performance Analysis of Location Based Service Using TPEG&MBMS in 3G Mobile Networks

Lu Lou[1], Xin Xu[2], Zhili Chen[3], and Juan Cao[1]

[1] College of Information Science and Engineering, Chongqing Jiaotong University,
Chongqing, China, 400074
cloudlou@163.com
[2] Library of Chongqing Jiaotong University, Chongqing, China, 400074
xx1771@163.com
[3] Faculty of Information and Control Engineering, Shenyang Jianzhu University,
Shenyang, China, 110168
chenzhili@sjzu.edu.cn

Abstract. Location-based services (LBS) provide content that is dynamically customized according to the user's location. These services are commonly delivered to mobile devices. This paper proposed a novel LBS application based on the TPEG over Multimedia Broadcast and Multicast Service and explained the implement of that using the stream delivery method and download delivery method. For download service, Results of performance tests for FLUTE Protocol (File Delivery under unidirectional Transport) and FEC (Forward error correction) are also presented.

Keywords: Performance Analysis, Location based Service, TPEG, MBMS, FLUTE, FEC.

1 Introduction

A Location based Service (LBS) is an information and entertainment service, accessible with mobile devices through the mobile network and utilizing the ability to make use of the geographical position of the mobile device. On the driving or walking around, consumers want to find an optimal path to destination or information of POI (point of interest) such as park, restaurant, hotel, cinema, gas station, traffic jam , and so forth. Location based services usually also rely on real-time traffic information that be often encoded in TPEG protocol. TPEG (Transport Protocol Expert Group) standard designed by EBU (European Broadcasting Union) is a new traffic information transfer protocol which has three major characteristics, language independent, bearer independent, and multi-modal application [1][2][3]. Broadcasting based transmission technology is one of main methods to provide dynamic traffic information or public emergency service in recent years, which is used by countries all over the world.

The Third Generation Partnership Project (3GPP) suggests an enhancement of current cellular networks to support Multimedia Broadcast and Multicast

B. Liu and C. Chai (Eds.): ICICA 2011, LNCS 7030, pp. 274–281, 2011.

Services (MBMS). MBMS is a broadcasting service that can be offered via cellular networks using point-to-multipoint links instead of the usual point-to-point links. Because of its operation in broadcast or multicast mode, MBMS can be used for efficient streaming or file delivery to mobile phones [4][5].

In this paper, we present a novel POI application specification based on TPEG over MBMS, which is satisfied to LBS applications in mobile devices, and explain the practicability and feasibility of that.

2   Preparation of Information for MBMS

MBMS is an IP datacast type of service that can be offered via existing GSM and UMTS cellular networks, which has been standardized in various of 3GPP, and the first phase standards are to be finalized for UMTS release6 [5].

The MBMS is a unidirectional point-to-multipoint bearer service in 3GPP cellular network in which data are transmitted from a single source entity to multiple mobiles. Various MBMS user services can be made up of these MBMS bearer services. To support the MBMS, Broadcast and Multicast-Service Center (BM-SC) is newly added to the network, and MBMS controlling functions are added to the existing network entities such as UMTS Terrestrial Radio Access Network (UTRAN), Serving GPRS Support Node (SGSN), and Gateway GPRS Support Node (GGSN). For user equipments (UEs) (or mobiles) to support the MBMS, much additional functionality are also required to be added. Some of them, for example, the interaction among protocol entities which is beyond of the standardization should be defined in detail. New protocols such as File Delivery over Unidirectional Transport (FLUTE) and media codecs are also needed to be implemented [5][6][7]. The content is distributed via unicast (point-to-point) connections, forcing the network to process multiple requests for the same content sequentially and therefore wasting resources in the radio access and core networks. With the expected increase of high bandwidth applications, especially with a large number of UEs(User Equipments) receiving the same high data rate services, efficient information distribution becomes essential. Broadcast and multicast are methods for transmitting datagrams from a single source to multiple destinations. Thus, these techniques decrease the amount of transmitted data within the network. This results in a dramatic cost reduction.

MBMS provides two modes of services —broadcast and multicast. The broadcast mode enables multimedia data (e.g. text, audio, picture, video) to be transmitted to all users within a specific service area. The broadcast mode needs neither service subscription nor charging. The multicast mode enables multimedia data to be transmitted from one source entity to a specific multicast group. For a user to receive a multicast mode service, it should be subscribed to the multicast group in advance. Multicast mode services are charged, and therefore should be protected from illegal users [8].

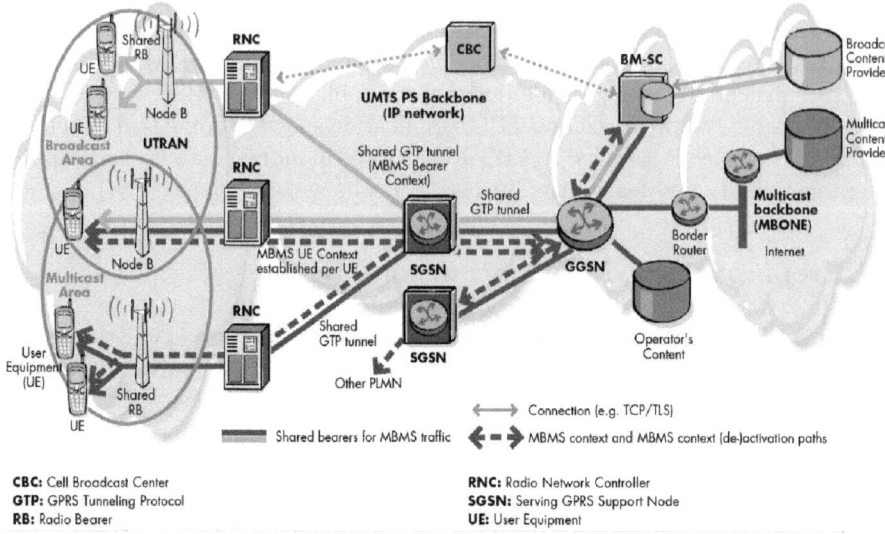

Fig. 1. The MBMS architecture designed in 3GPP

3   The Implemention of TPEG over MBMS

3.1   The Design of POI Message Using TPEG

In order to provide the greatest flexibility, the TPEG system was designed to allow a number of different service providers to deliver a number of different types of information for ITS, without needing to rely on any facilities provided by the bearer system. This was done to void compromising the bearer independence of the protocol. For this reason, the first level of TPEG framing – *the 'Transport Frame'* – carriers a multiplex of TPEG *'Application Frames'* carrying potentially different types of information, and a TPEG stream is constructed from a sequence of transport frames from potentially different service providers [2][3].

We design a new POI application specification, which is satisfied to be interoperable with TPEG protocol. The hierarchical transport frame structure including the POI message made up three data fields is embedded into the existing TPEG applications. Each data field is called container. The first one of the message management container is used to manage the POI information in the receiving side. The second one, POI event container, consists of the four items such as classification, description, reservation, time information. The POI information is divided into more than ten categories and each category is also classified into several sub-categories. For example, restaurant category is consist of Chinese, western and fast-food restaurant, and so forth. The last one, TPEG-location container, represents the exact position of POI by using the WGS84 co-ordinate or descriptor.

We use standard XML tools to generate tpegML that in the final phase of stream generation will be encoded for the stream. An example message: *A hotel can accommodate up to fifty peoples.it's located on A10 and apart from only 5 kilometers of Chongqing international airport. Rate for standard room is 180 yuan per day. discounting only on 8:30AM -18:00PM.* Expressed in tpegML, this message looks like:

```
<?xml version="1.0" encoding="ISO-8859-1"?>
<!DOCTYPE tpeg_document PUBLIC "ITS/tpegML/EN" "tpegML_05.dtd">
<tpeg_document>
<POI message_id="123"
version_number="1"
message_generation_time="2009-02-03T13:03:00Z"
severity_factor="&rtm31_4;">\\
<!-- Location is on A10, Chongqing international airport.-->
<tpeg_loc_container language="&loc41_30;">
<location_coordinates location_type="&loc1_5;">
<WGS84 longitude="105.6212" latitude="29.3864"/>
<descriptor descriptor_type="&loc3_7;"
descriptor="A10"/>
<descriptor descriptor_type="&loc3_8;"
descriptor=" Chongqing "/>
<direction_type direction_type="&loc2_2;"/>
</location_coordinates>
</tpeg_loc_container>
<!-- A hotel can accommodate up to fifty peoples. -->
<hotel number_of="1">
<position position="&rtm10_37;"/>
<people number_of="50">
</hotels>

......

</POI_message>
</tpeg_document>
```

The values of the tables are naturally transformed into XML entities, also the document type definition (DTD) of tpegML is divided so that all of these entities are defined in separate files.

3.2   The Accessing of MBMS Services

Broadcast mode enables the unidirectional point-to-multipoint transmission of multimedia data and enriched multimedia services. Broadcast mode necessitates neither subscription nor joining by the user. All users located in the broadcast service area as defined by the mobile network operator can receive the data. However, since not all users may wish to receive those messages, the user has the possibility to disable their reception by configuring the UE. Unlike the broadcast mode, the multicast mode requires the user to subscribe for the general reception of MBMS services. End users monitor service announcements and can decide to join one or more available services. Charging is possible either on subscription or on purchasing keys enabling access to the transmitted data. Fig.2 shows the typical flow of MBMS sessions [9].

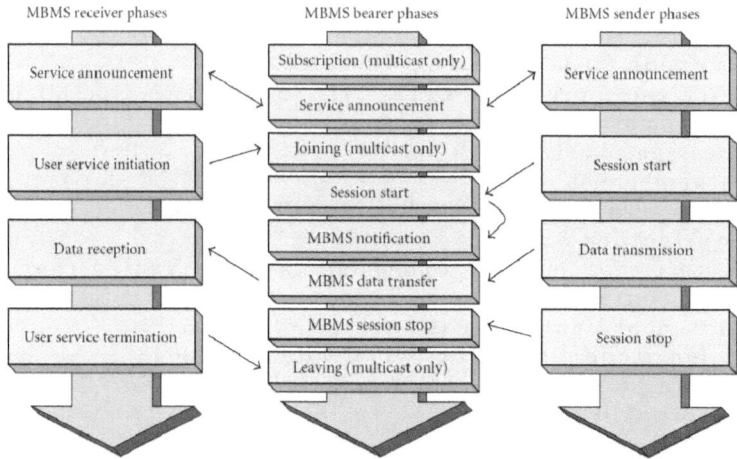

Fig. 2. MBMS Session Flow

When announcing a particular service, the BM-SC informs the devices about the nature of the service and about the parameters required for the activation of the service (e.g. IP multicast address).

Subscription (Multicast Mode Only). Service subscription establishes the relationship between the user and the service provider, which allows the user to receive the related MBMS multicast service. Service subscription is the request of a user to receive the service(s) offered by the operator and that he agrees to a potential payment associated with the service [5][6].

Service Announcement. In this phase the terminal receives all information needed for the reception of a specific service, for example IP multicast addresses and associated port numbers, used codecs and codec configurations.

User Service Initiation. In this phase the receiver initiates the reception of a certain MBMS service. The MBMS service may use one or several MBMS Bearer Services. In case of an MBMS Multicast bearer, the initiation on the receiver triggers the "Joining" phase in the network. During this "Joining" phase the receiver becomes a member of themulticast group.

3.3   Performance Analysis of MBMS Protocols and Codecs

MBMS defines two methods for service announcements: pull (the initiative comes from the receiving UE) and push (the initiative arises from the service itself). In the case of a pull method, the devices fetch the announcements (HTTP or WAP) from a web server. As push method, SMS (Short Message Service) cell broadcast,

SMS-PP (point-to-point), WAP-PUSH, MBMS broadcast, MBMS multicast and MMS (Multimedia Message Service) are used [5][6].

Essentially, MBMS offers a scalable mechanism for delivering multimedia content over 3G networks. MBMS User services use the protocol stack shown in Fig.3. A general distinction between two delivery methods exists: the download delivery method and the streaming delivery method. Streaming uses RTP (Realtime Transport Protocol) is used, which in turn uses UDP (User Datagram Protocol). For downloading, the FLUTE (File Delivery over Unidirectional Transport) protocol applies.

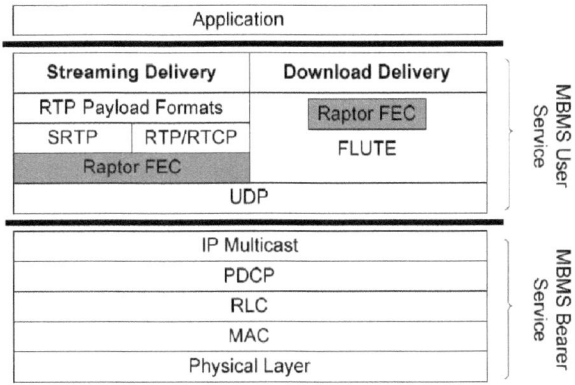

Fig. 3. MBMS Protocol Stack

The FLUTE packets are encapsulated in UDP/IP packets and are forwarded to the Packet Data Convergence Protocol (PDCP) layer. The packets are then sent to the Radio Link Control (RLC) layer. The RLC layer functions in unacknowledged mode. The RLC layer is responsible for mapping IP packets to RLC SDUs. The Media Access Control (MAC) Layer adds a 16 bit header to form a PDU, which is then sent in a transport block on the physical layer[10]. The protocol stack introduces the following headers, reducing the Maximum Transmission Unit (MTU) payload from 1500 bytes for an Ethernet maximum frame size to a payload of 1444 bytes:

- FLUTE header: 16 bytes
- UDP/IP header: 8+20 bytes
- GTP-User traffic header: 12 bytes

For FLUTE, the file is partitioned in one or several source blocks, as shown in Fig.4. Each source block is split into source symbols of a fixed size. Each encoding symbol is assigned a unique encoded symbol ID (ESI). If the ESI is smaller than the number of source symbols, then it is a source symbol; otherwise it is a repair symbol. Symbols are either transmitted individually or concatenated and mapped to a FLUTE packet payload.

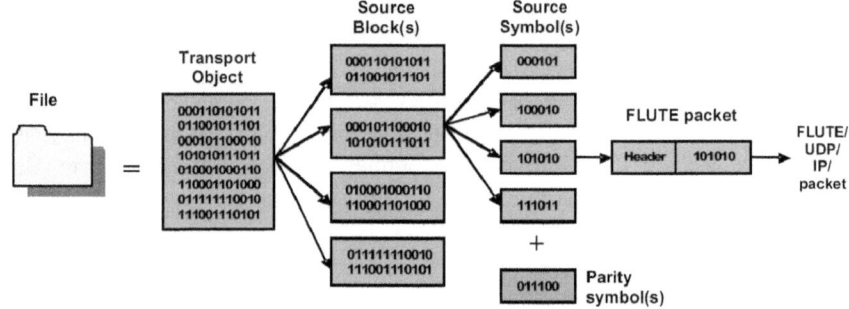

Fig. 4. Construction of FLUTE packets

Raptor Codes are in use for FEC. The aim of Raptor codes is to recover the lost symbols using encoding symbols. Raptor codes achieve very close to ideal performance for a wide range of parameters and Raptor codes are effectively recover the source block with a probability even if the number of lost symbols is close to the number of repair symbols. The small inefficiency of the Raptor code can be modeled as [11]:

$$p_f(m, k) = \begin{cases} 1 & if \ m < k, \\ 0.85 \times 0.567^{m-k} & if \ m \geq k. \end{cases}$$

For LBS applications, the file download Delivery is very important way to Multicast POI message. Because FLUTE uses unreliable transport protocol, packet losses must be handled at higher layers. Data carousel is one option to recover from packet losses so that missing packets are tried to be caught in the next loop(s). Two types of carousels were used, data carousel and FEC data carousel. Better results are received by using FEC data carousel, which includes parity data to recover from packet losses. We studied how many loops were needed to receive the whole file with different amount of Reed-Solomon FEC data. The results of the data carousel simulation experiments are shown as Table 1. We can notice that performance gets quite well with low average packet loss.

4   Conclusions

Open standards will be a necessary catalyst for LBS growth. We propose a novel POI application based on the TPEG over Multimedia Broadcast and Multicast, and explain the implement of that using the stream delivery method and download delivery method. The results of performance analysis show that FLUTE and FEC can work well to recover from packet losses using data carousel for download services.

Table 1. FEC DATA CAROUSEL WITH UNIFORMLY DISTRIBUTED ERRORS

Loss[%]	FEC RS[%]	Avg loops	Min loops	Max loops	Num of Experiment
1	5	1.1	1	2	2000
5	5	2.2	2	3	2000
10	5	2.4	2	3	2000
10	10	2.0	2	2	2000
25	5	3.4	3	4	2000
25	10	3.0	2	4	2000
50	10	5.0	4	6	2000
50	25	3.5	3	4	2000

Acknowledgment. This work was supported in part by Natural Science Foundation Project of CQ CSTC (No. cstc2011jjA40030).

References

1. Transport Protocol Experts Group (TPEG) Specifications. Part 4: Road (2006)
2. Transport Protocol Experts Group (TPEG) Specifications. Part 5: Public (2006)
3. Transport Protocol Experts Group (TPEG) Specifications. Part 6: Location referencing for applications (2006)
4. Digital Audio Broadcasting. Data roadcasting transparent data channel. ETSI TS 101 759 v1.2.1 (2005)
5. 3GPP TS 23.246. Multimedia broadcast/multicast service (mbms):architecture and functional description (2008)
6. 3GPP TS 26.346. Multimedia broadcast/multicast service (mbms): Protocols and codecs (2008)
7. de Vriendt, J., Gmez Vinagre, I., Van Ewijk, A.: Multimedia broadcast and multicast services in 3g mobile networks. Alcatel Telecommunications Review 1 (2004)
8. Shin, J., Park, A.: Design of mbms client functions in the mobile. Proceedings of World Academy of Science, Engineering and Technology 18 (2006)
9. 3GPP TR 26.946. Multimedia broadcast/multicast service (mbms)-user service guidelines (2008)
10. Thorsten Lohmar, U.: Scalable push file delivery with mbms. Ericsson Review 1, 12–16 (2009)
11. Luby, Watson, M., Gasiba, M., Stockhammer, T., Wen Xu, T.: Raptor codes for reliable download deliveryin wireless broadcast systems. In: 3rd IEEE Consumer Communications and Networking Conference, vol. 1, pp. 192–197 (2006)

On the Dividend Problem in a Risk Model with Delayed Claims

Wei Zou* and Jie-Hua Xie

Department of Science, NanChang Institute of Technology,
330099 NanChang, P.R.China
{fion_zou,jiehuaxie}@163.com

Abstract. In this paper, a discrete time risk model with dividend payments is considered, in which the occurrences of the claims may be delayed. A system of differential equations with certain boundary conditions for the expected present value of dividend payments prior to ruin is derived and solved. Moreover, the closed form expressions are given when the claim size distributions belong to the rational family. The qualitative properties of ruin probabilities for this risk model are also obtained. Numerical results are provided to illustrate the applicability of the main result.

Keywords: Discrete time risk model, Delayed claim, Dividend.

1   Introduction

In recent years, risk models with dividend payments have been one of the major interests in the risk theory literature. The barrier strategy was initially proposed by De Finetti [1] for a binomial risk model. More general barrier strategies for a risk process have been studied in a number of papers and books. Related works can be found in Gerber, Shiu and Yang [2], Frosting [3], Liu, Li and Ameer [4], Dassios and Wu [5], Yin, Song and Yang [6] and references therein. The main focus of these papers is on optimal dividend policies and the time of ruin under various barrier strategies and other economic conditions.

In reality, insurance claims may be delayed due to various reasons. Since the work by Waters and Paratriandafylou [7], risk models with this special feature have been discussed by many authors in the literature. For example, Yuen and Guo [8] studied a compound binomial model with delayed claims and obtained recursive formulas for the finite time ruin probabilities. Xiao and Guo [9] also studied this risk model, they derived an upper bound for the ruin probability. Xie and Zou [10,11] studied the ultimate ruin probability and expected discounted function for a risk model with delayed claims. Exact analytical expressions for the ruin functions were obtained. Yuen, Guo and Kai [12] studied a risk model with delayed claims, in which the time of delay for the occurrence of a by-claim is assumed to be exponentially distributed. A framework of delayed claims is

---

* Corresponding author.

B. Liu and C. Chai (Eds.): ICICA 2011, LNCS 7030, pp. 282–289, 2011.

built by introducing two kinds of individual claims, namely main claims and by-claims, and allowing possible delays of the occurrences of by-claims.

Motivated by these idea, in this paper, we consider a compound binomial model with delayed claims and dividend payments that are ruled by a constant barrier. It is obvious that the incorporation of the delayed claim and dividend payments makes the problem more interesting. It also complicates the analysis of the risk model. This paper is to devoted to calculating the expected present value of dividends for this risk model. In Section 2, we defines the model of interest, describes various payments, including the premiums, claims and dividends, and lists the notation. In Section 3, difference equations are developed for the expected present value of dividend payments. Then an explicit expression is derived, using the technique of generating functions. Moreover, closed-form solutions for the expected present value of dividends are obtained for K_n claim size distributions in Section 4. Numerical examples are also provided to illustrate the impact of the delay of claims on the expected present value of dividends.

2   The Model

Here, we consider a discrete-time model which involves two types of insurance claims; namely the main claims and the by-claims. Denote the discrete time units by $k = 0, 1, 2, \cdots$. It is assumed that each main claim induces a by-claim. In any time period, the probability of having a main claim is p, $0 < p < 1$, and thus the probability of no main claim is $q = 1 - p$. The by-claim and its associated main claim may occur simultaneously with probability θ, or the occurrence of the by-claim may be delayed to the next time period with probability $1 - \theta$. The main claim amounts $X_1, X_2, \cdots$, are independent and identically distributed. Put $X = X_1$. Then the common probability function of the X_i is given by $f_m = \Pr(X = m)$ for $m = 1, 2, \cdots$. The corresponding probability generating function and mean are $\tilde{f}(z) = \sum_{m=1}^{\infty} f_m z^m$ and $\mu_X = \sum_{n=1}^{\infty} n f_n$, respectively. Let $Y_1, Y_2, \cdots$ be the independent and identically distributed by-claim amounts and put $Y = Y_1$. Denote their common probability function by $g_n = \Pr(Y = n)$ for $n = 1, 2, \cdots$, and write the probability generating function and mean as $\tilde{g}(z) = \sum_{n=1}^{\infty} g_n z^n$ and μ_X, respectively.

Let U_k be the total amount of claims up to the end of the kth time period, $k \in \mathbb{N}^+$ and $U_0 = 0$. We define $U_k = U_k^X + U_k^Y$, where U_k^X and U_k^Y are the total main claims and by-claims, respectively, in the first k time periods.

Assume that premiums are received at the beginning of each time period with a constant premium rate of 1 per period, and all claim payments are made only at the end of each time period. We introduce a dividend policy to the company that certain amount of dividends will be paid to the policyholder instantly, as long as the surplus of the company at time k is higher than a constant dividend barrier $b(b > 0)$. It implies that the dividend payments will only possibly occur at the beginning of each period, right after receiving the premium payment. The surplus at the end of the kth time period, S_k, is then defined to be

$$S_k = u + k - U_k - UD_k. \tag{2.1}$$

Here the initial surplus is u, $u= 1, 2, \cdots, b$, $k = 1, 2, \cdots$. The positive safety loading condition holds if $p(\mu_X + \mu_Y) < 1$. We define UD_k as the sum of dividend payments in the first k periods, for $k = 1, 2, \cdots$, $UD_k = D_1 + D_2 + \cdots + D_k$, $(UD_0 = 0)$. Denote by D_n the amount of dividend paid out in period n, for $n= 1, 2, \cdots$, with the definition $D_n = \max\{S_{n-1} + 1 - b, 0\}$.

Define $T = \min\{k | S_k \leq 0\}$ to be the time of ruin, $\Psi(u; b) = \Pr(T < \infty | S_0 = u)$ to be the ruin probability, and $\Phi(u; b) = 1 - \Psi(u; b)$, to be the survival probability. Let v be a constant annual discount rate for each period. Then the expected present value of the dividend payments due until ruin is

$$W(u; b) = E\left[\sum_{k=1}^{T} D_k v^{k-1} | S_0 = u\right].$$

3   The Expected Present Value of Dividends

To study the expected present value of the dividend payments, $W(u; b)$, we need to study the claim occurrences in two scenarios. The first is that if a main claim occurs in a certain period, its associated by-claim also occurs in the same period. Hence there will be no by-claim in the next time period and the surplus process really gets renewed. The second is simply the complement of the first scenario. In other words, if there exists a main claim, its associated by-claim will occur one period later. Conditional on the second scenario, that is, the main claim occurred in the previous period and its associated by-claim will occur at the end of the current period, we define the corresponding process as

$$S_{1k} = u + k - U_k - UD_{1k} - Y, \quad k = 1, 2, \cdots, \tag{3.1}$$

with $S_{10} = u$, where UD_{1k} is the sum of dividend payments in the first k time periods, and Y is a random variable following the probability function g_n; $n= 1, 2, \cdots$, and is independent of all other claim amounts random variables X_i and Y_j for all i and j. The corresponding ruin probability is denoted by $\Psi_1(u; b)$ with $\Psi_1(0; b) = \Psi_1(1; b) = 1$, the survival probability is denoted by $\Phi_1(u; b) = 1 - \Psi_1(u; b)$, and the expected present value of the dividend payments is denoted by $W_1(u; b)$. Then conditioning on the occurrences of claims at the end of the first time period, we can set up the following differential equations for $W(u; b)$ and $W_1(u; b)$:

$$W(u; b) = v \left\{ qW(u+1; b) + p\theta \sum_{m+n \leq u+1} W(u+1-m-n; b) f_m g_n \right.$$

$$\left. +p(1-\theta) \sum_{m=1}^{u+1} W_1(u+1-m; b) f_m \right\}, \quad u = 1, 2, \cdots, b-1, \tag{3.2}$$

$$W_1(1; b) = vq \sum_{n=1}^{2} W(2-n; b) g_n, \tag{3.3}$$

$$W_1(u;b) = v\left\{q\sum_{n=1}^{u+1} W(u+1-n;b)g_n + p\theta \sum_{m+n+l\leq u+1} W(u+1-m-n-l;b)\right.$$

$$\left. \times f_m g_n g_l + p(1-\theta)\sum_{m+l\leq u+1} W_1(u+1-m-l;b)f_m g_l\right\}, \quad u = 2,3,\cdots,b-1,$$

$$(3.4)$$

with boundary conditions:

$$W(0;b) = 0; \quad W(b;b) = 1+W(b-1;b); \quad W_1(0;b) = 0; \quad W_1(b;b) = 1+W_1(b-1;b).$$

The second boundary condition holds because when the initial surplus is b, the premium received at the beginning of the first period will be paid out as a dividend immediately. Except the first dividend payment, the rest of the model is the same as that starting from an initial surplus $b-1$. The last condition can be explained similarly. From (3.2) and (3.4) one can rewrite $W_1(u;b)$ as

$$W_1(u;b) = \sum_{n=1}^{u} W(u-n;b)g_n, \quad u = 2,3,\cdots,b-1. \qquad (3.5)$$

Substituting (3.5) into (3.2) gives

$$W(1;b) = vqW(2;b) + v^2pq(1-\theta)W(1;b)f_1g_1, \qquad (3.6)$$

and for $u = 2,3,\cdots,b-1$,

$$W(u;b) = v\left\{qW(u+1;b) + p\sum_{m+n\leq u+1} W(u+1-m-n;b)f_m g_n\right\}, \qquad (3.7)$$

with a new boundary condition:

$$W(b;b) = 1 + W(b-1;b). \qquad (3.8)$$

To obtain an explicit expression for $W(u;b)$ from (3.6) and (3.7), we define a new function $A(u)$ that satisfies the following differential equation,

$$A(1) = vqA(2) + v^2pq(1-\theta)A(1)f_1g_1, \qquad (3.9)$$

$$A(u) = v\left\{qA(u+1) + p\sum_{m+n\leq u+1} A(u+1-m-n)f_m g_n\right\}, \quad u = 2,3,\cdots.$$

$$(3.10)$$

Apart from a multiplicative constant, the solution of (3.9) and (3.10) is unique. Therefore, we can set $A(1) = 1$. It follows from the theory of differential equations that the solution to (3.6) and (3.7) with boundary condition (3.8) is of the form

$$W(u;b) = A(u)B(b), \qquad (3.11)$$

where $B(b) = 1/[A(b) - A(b-1)]$.

Let the generating function of $A(u)$ be $\tilde{A}(u) = \sum_{u=1}^{\infty} A(u)z^u, -1 < \mathbf{R}(z) < 1$. Similarly, $\tilde{f}(z) = \sum_{m=1}^{\infty} f_m z^m$ and $\tilde{g}(z) = \sum_{n=1}^{\infty} g_n z^n$ are probability generating functions of $\{f_m\}_{m=1}^{\infty}$ and $\{g_n\}_{n=1}^{\infty}$, respectively. Furthermore, we construct two new generating functions $\tilde{h}(z,1) = q + p\tilde{f}(z)\tilde{g}(z)$ and $\tilde{h}(z,k) = [\tilde{h}(z,1)]^k$. We denote the probability function of $\tilde{h}(z,k)$ by $h(i,k)$. Yuen and Guo [8] have commented that $h(i,k)$ is the probability function of the total claims in the first k time periods in the compound binomial model with individual claim amount $X_1 + Y_1$. In the following theorem, we show that $W(u;b)$ can be expressed explicitly in terms of $h(i,k)$.

Theorem 1. For the expected present value of dividend payments, $W(u,b)$, of model (2.1), we have, for $u = 2, 3, \cdots, b$,

$$W(u;b) = B(b) \sum_{i=1}^{\infty} v^{i+1} q \left(vp(1-\theta)f_1 g_1 h(i+u-1,i) - h(i+u,i)\right), \quad (3.12)$$

where $B(b) = (\sum_{i=1}^{\infty} v^{i+1} q(vp(1-\theta)f_1 g_1[h(i+b-1,i) - h(i+b-2,i)] - [h(i+b,i) - h(i+b-1,i)]))^{-1}$ and $W(1,b) = B(b)$.

Proof. Multiplying both sides of (3.10) by z^u and summing over u from 2 to ∞, we get

$$\tilde{A}(z) - A(1)z = vqz^{-1}\left(\tilde{A}(z) - A(1)z - A(2)z^2\right) + vqz^{-1}\tilde{A}(z)\tilde{f}(z)\tilde{g}(z). \quad (3.13)$$

From (3.9) and the fact that $A(1) = 1$, the above equation simplifies to

$$\tilde{A}(z) = \frac{vq[vp(1-\theta)f_1 g_1 z - 1]}{1 - v\tilde{h}(z,1)z^{-1}}. \quad (3.14)$$

Rewriting $[1 - v\tilde{h}(z,1)z^{-1}]^{-1}$ in terms of a power series in z, we have

$$A(z) = \sum_{i=1}^{\infty} v^{i+1} q\left[vp(1-\theta)f_1 g_1 h(i+u-1,i) - h(i+u,i)\right].$$

The above result together with (3.11) gives us the explicit expression for $W(u;b)$ as in (3.12). This completes the proof.

To end this section, we show that the ruin is certain in the risk model described in (2.2). For $b = 1$; since $\Phi(1;1) = q\Phi(1;1)$ and $0 < q < 1$; then $\Phi(1;1) = 0$. For $b = 2$, $0 \le \Phi(1;2) \le \Phi(2;2) = q\Phi(2;2) + p(1-\theta)\Phi_1(1;2)f_1 = q\Phi(2;2)$, then $\Phi(1;2) = \Phi(2;2) = 0$. The following theorem shows that ruin is certain for $b \ge 3$ under certain conditions.

Theorem 2. The ruin probability in the risk model (2.1) is one, i.e., $\Psi(u;b) = 1$, for $u = 1, 2, \cdots, b$, provided that $\sum_{m+n \le b-1} f_m g_n < 1$, for $b \ge 3$.

Proof. Conditioning on the occurrences of claims at the end of the first time period gives

$$\Phi(b;b) = q\Phi(b;b) + p\theta \sum_{m+n\leq b-1} \Phi(b-m-n;b)f_m g_n + p(1-\theta)\sum_{m=1}^{b-1}\Phi_1(b-m;b)f_m, \tag{3.15}$$

$$\Phi_1(u;b) = \sum_{n=1}^{u-1}\Phi(u-n;b)g_n, \quad u = 2,3,\cdots,b. \tag{3.16}$$

Substituting (3.16) into (3.15) yields

$$\Phi(b;b) = q\Phi(b;b) + p \sum_{m+n\leq b-1}\Phi(b-m-n;b)f_m g_n.$$

It follows from the inequality $\Phi(b-m-n;b) \leq \Phi(b;b)$ that

$$\Phi(b;b) \leq q\Phi(b;b) + p\Phi(b;b) \sum_{m+n\leq b-1} f_m g_n. \tag{3.17}$$

Since $\sum_{m+n\leq b-1} f_m g_n < 1$ and $0 \leq \Phi(b;b) \leq 1$, then inequality (3.17) gives $\Phi(b;b) = 0$, this implies that $\Psi(b;b) = 1$.

4   Explicit Results for K_n Claim Amount Distributions

K_n distributions includes geometric, negative binomial, discrete phase-type, as well as linear combinations (including mixtures) of these.

For the two independent claim amount random variables X_1 and Y_1, if they have K_n distributions, so does their sum. Therefore, in this Section, we assume that $(f * g)_x = \Pr(X_1 + X_1 = x)$ is K_n distributed for $x = 2,3,\cdots$, and $n = 1,2,\cdots$, i.e., the probability generating function of $(f * g)$ is given by $\tilde{f}(z)\tilde{g}(z) = z^2 E_{n-1}(z)/\prod_{i=1}^{n}(1 - zq_i)$, $\mathbf{R}(z) < \min(1/q_i : 1 \leq i \leq n)$, where $0 < q_i < 1$; for $i = 1,2,\cdots,n$ and $E_{n-1}(z) = \sum_{k=0}^{n-1} z^k e_k$ is a polynomial of degree $n-1$ or less with $E_{n-1}(1) = \prod_{i=1}^{n}(1-q_i)$. Then $\tilde{A}(z)$ can be transformed to the following rational function

$$\tilde{A}(z) = \frac{z[v^2 pq(1-\theta)f_1 g_1 z - vq]\prod_{i=1}^{n}(1 - zq_i)}{z\prod_{i=1}^{n}(1 - zq_i) - vq\prod_{i=1}^{n}(1 - zq_i) - vpz^2 E_{n-1}(z)}.$$

Since the denominator of the above equation is a polynomial of degree $n+1$, it can be factored as $[(-1)^n \prod_{i=1}^{n+1} q_i - vp e_{n-1}]\prod_{i=1}^{n+1}(z - R_i)$, where $R_1, R_2, \cdots, R_{n+1}$ are the $n+1$ zeros of the denominator. We remark that $(-1)^n \prod_{i=1}^{n} q_i - vp e_{n-1} = (-1)^n vq/\prod_{i=1}^{n} R_i$. Inverting $\tilde{A}(z)$ gives $A(1) = 1$, and

$$A(u) = \left(\prod_{i=1}^{n+1} R_i\right)\sum_{i=1}^{n+1} r_i\left[1 - vp(1-\theta)f_1 g_1 R_i\right]R_i^{-u}, \quad u = 2,3,\cdots$$

where $r_i = \prod_{j=1}^{n}(R_i q_j - 1)/\prod_{j=1, j\neq i}^{n+1}(R_i - R_j)$, $i = 1, 2, \cdots, n+1$. Now that $B(b) = 1/[A(b) - A(b-1)]$, then finally we have

$$W(1; b) = \cfrac{1}{\left(\prod_{i=1}^{n+1} R_i\right) \sum_{i=1}^{n+1} r_i \left[1 - vp(1 - \theta)f_1 g_1 R_i\right](1 - R_i)R_i^{-b}}, \qquad (4.1)$$

and

$$W(u; b) = \cfrac{\sum_{i=1}^{n+1} r_i \left[1 - vp(1 - \theta)f_1 g_1 R_i\right] R_i^{-u}}{\sum_{i=1}^{n+1} r_i \left[1 - vp(1 - \theta)f_1 g_1 R_i\right](1 - R_i)R_i^{-b}}, \qquad u = 2, \cdots, b. \qquad (4.2)$$

Example 1. In this example, we assume that the main claim X_1 follows a geometric distribution with $f_x = \alpha(1 - \alpha)^{x-1}, 0 < \alpha < 1, x = 1, 2, \cdots$, and the by-claim Y_1 follows a geometric distribution with $g_x = \beta(1 - \beta)^{x-1}, 0 < \beta < 1, x = 1, 2, \cdots$, so that $n = 2, q_1 = \alpha, q_2 = \beta$ and $E_{n-1}(z) = (1 - \alpha)(1 - \beta)$. Then (4.1) and (4.2) simplify to

$$W(1; b) = \cfrac{1}{(R_1 R_2 R_3) \sum_{i=1}^{3} r_i \left(1 - vp(1 - \theta)\alpha\beta R_i\right)(1 - R_i)R_i^{-b}},$$

and

$$W(u; b) = \cfrac{\sum_{i=1}^{3} r_i \left(1 - vp(1 - \theta)\alpha\beta R_i\right) uR_i^{-u}}{\sum_{i=1}^{3} r_i \left(1 - vp(1 - \theta)\alpha\beta R_i\right)(1 - R_i)R_i^{-b}}, \qquad u = 2, \cdots, b,$$

Table 1. Values of $W(u; 10)$ for geometric distributed claims

$W(u; 10)$	$\theta = 0$	$\theta = 0.25$	$\theta = 0.5$	$\theta = 0.75$	$\theta = 1$
$u = 1$	0.05164	0.04968	0.04786	0.04617	0.04460
$u = 2$	0.07264	0.07252	0.07242	0.07232	0.07223
$u = 3$	0.11652	0.11637	0.11624	0.11612	0.11601
$u = 4$	0.18535	0.18519	0.18504	0.18490	0.18477
$u = 5$	0.29301	0.29284	0.29268	0.29253	0.29239
$u = 6$	0.46121	0.46104	0.46087	0.46072	0.46058
$u = 7$	0.72391	0.72373	0.72356	0.72340	0.72326
$u = 8$	1.13409	1.13391	1.13374	1.13358	1.13344
$u = 9$	1.77455	1.77437	1.77420	1.77404	1.77390
$u = 10$	2.77455	2.77437	2.77420	2.77404	2.77390

As an example, let $p = 0.35; v = 0.95; b = 10; \alpha = \beta = 0.8$. From the above results we get $R_1 = 0.64044, R_2 = 1.01812, R_3 = 1.47972, r_1 = 0.75021, r_2 = -0.19739$, and $r_3 = 0.08718$. The values of $W(u; 10)$ for $\theta = 0, 0.25, 0.5, 0.75, 1$, and $u = 1, \cdots, 10$ are listed in Table 1. We observe the same features as in Example, that $W(u; b)$ is an increasing function with respect to u, and a decreasing function over θ. Also, the impact of the delay of by-claims on the expected present value of dividends is reduced for a higher initial surplus of the company.

5   Concluding Remarks

In this paper, we study a discrete risk model involving dividend payments. In this risk model, the occurrences of the claims may be delayed. We study how to compute the expected discounted dividend payments prior to ruin in this risk model. The results illustrate the impact of the delay of claims on the expected present value of dividends. The formulas are readily programmable in practice.

References

1. De Finetti, B.: Su un′ Impostazione Alternativa Dell Teoria Collettiva Del Rischio. Transactions of the XVth International Congress of Actuaries (1957)
2. Gerber, H.U., Shiu, E.S.W., Yang, H.L.: An Elementary Approach to Discrete Models of Dividend Strategies. Insurance: Mathematics and Economics 46, 109–116 (2010)
3. Frostig, E.: Asymptotic Analysis of a Risk Process with High Dividend Barrier. Insurance: Mathematics and Economics 47, 21–26 (2010)
4. Liu, Z.M., Li, M.M., Ameer, S.: Methods for Estimating Optimal Dickson and Waters Modification Dividend Barrier. Economic Modelling 26, 886–892 (2009)
5. Dassios, A., Wu, S.: On Barrier Strategy Dividends with Parisian Implementation Delay for Classical Surplus Processes. Insurance: Mathematics and Economics 45, 195–202 (2009)
6. Yin, G., Song, Q.S., Yang, H.: Stochastic Optimization Algorithms for Barrier Dividend Strategies. Journal of Computational and Applied Mathematics 223, 240–262 (2009)
7. Waters, H.R., Papatriandafylou, A.: Ruin Probabilities Allowing for Delay in Claims Settlement. Insurance: Mathematics and Economics 4, 113–122 (1985)
8. Yuen, K.C., Guo, J.Y.: Ruin Probabilities for Time-Correlated Claims in the Compound Binomial Model. Insurance: Mathematics and Economics 29, 47–57 (2001)
9. Xiao, Y.T., Guo, J.Y.: The Compound Binomial Risk Model with Time-Correlated Claims. Insurance: Mathematics and Economics 41, 124–133 (2007)
10. Xie, J.H., Zou, W.: Ruin Probabilities of a Risk Model with Time-Correlated Claims. Journal of the Graduate School of the Chinese Academy of Sciences 25, 319–326 (2008)
11. Zou, W., Xie, J.H.: On the Ruin Problem in an Erlang(2) Risk Model with Delayed Claims. In: Zhu, R., Zhang, Y., Liu, B.X., Liu, C.F. (eds.) ICICA 2010, Part II. CCIS, vol. 105, pp. 54–61. Springer, Heidelberg (2010)
12. Yuen, K.C., Guo, J.Y., Kai, W.N.: On Ultimate Ruin in a Delayed-Claims Risk Model. Journal of Applied Probability 42, 163–174 (2005)

Discrete Construction
of Power Network Voronoi Diagram

Yili Tan[1], Ye Zhao[2], and Yourong Wang[3]

[1] College of Science, Hebei United University, Hebei Tangshan 063009, China
[2] Department of Mathematics and Physics, Shijiazhuang Tiedao University, Hebei Shijiazhuang 050043, China
[3] Department of Basic, Tangshan College, Hebei Tangshan 063000, China
{tanyili_2002,ye_box,yourong1214}@163.com

Abstract. Power Network Voronoi diagrams are difficult to construct when the position relation of road segments are complicated. In traditional algorithm, The distance between objects must be calculated by selecting the minimum distance to their shared borders and doubling this value. When road segments cross or coincide with each other, production process will be extremely complex because we have to consider separately these parts. In this paper, we use discrete construction of network Voronoi diagrams. The algorithm can get over all kinds of shortcomings that we have just mentioned. So it is more useful and effective than the traditional ones. We also construct model according the algorithm. And the application example shows that the algorithm is both simple and useful, and it is of high potential value in practice.

Keywords: Network voronoi diagram, discrete, Power network Voronoi diagrams.

1 Introduction

As a branch of Computational Geometry, Voronoi diagram has been quickly developed on account of the development of theory and the need of application. We have already noted that the concept of the Voronoi diagram is used extensively in a variety of disciplines and has independent roots in many of them. Voronoi diagram was appeared in meteorology, biology discipline and so on. [1-3] Network Voronoi diagram is an important concept of it. And it can be used for city designing . The multiplicatively power network Voronoi diagram has been used for construction of highway and railway. More and more people pay attention to the algorithm that can construct power network Voronoi diagram fast and effectively. [4-6] We investigate a discrete method for constructing, and it proved to be satisfactory by experiment.

2 Definitions

2.1 Definition 1 (A Planar Ordinary Voronoi Diagram) [7]

Given a finite number of distinct points in the Euclidean plane
$P = \{p_1, p_2, \cdots, p_n\} \subset R^2$, where $2 < n < +\infty$, $x_i \neq x_j$, for $i \neq j$, $i, j \in I_n$.

B. Liu and C. Chai (Eds.): ICICA 2011, LNCS 7030, pp. 290–297, 2011.
© Springer-Verlag Berlin Heidelberg 2011

We call the region given by

$$V(p_i) = \left\{ x \middle\| \|x - x_i\| \leq \|x - x_j\| \; for \; j \neq i, j \in I_n \right\} \tag{1}$$

the planar ordinary Voronoi polygon associated with p_i, and the set given by

$$\mathcal{V} = \{V(p_1), V(p_2), \cdots, V(p_n)\} \tag{2}$$

the planar ordinary generated by P (or the Voronoi diagram of P). [8] We call p_i of $V(p_i)$ the generator point or generator of the ith Voronoi, and the set $P = \{p_1, p_2, \cdots, p_n\}$ the generator set of the Voronoi diagram (in the literature, a generator point is sometimes referred to as a site), as shown in Figure 1.

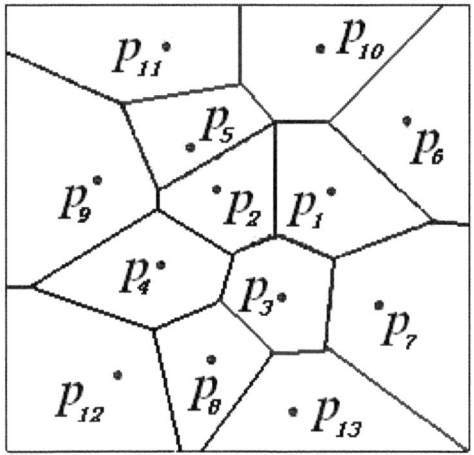

Fig. 1. A planar ordinary Voronoi diagram, there are 13 points in the plane, which are generator

2.2 Definition 2 (Network Voronoi Diagram) [9]

A Network Voronoi Diagram is a special kind of Voronoi diagram constructed on spatial networks [10]. The decomposition is based on the connection of the discrete objects rather than Euclidean distance. In the Network Voronoi Diagram, the Voronoi polygon changes to a set of road segments termed Network Voronoi polygon, and the edges of the polygons also shrink to some midpoints, termed border points, of the road network connection between two objects of interest Fig. 2. An order-2 Voronoi diagram in which the region with the symbol $\{i, j\}$ indicates the order-2 Voronoi polygon of $\{p_i, p_j\}$.

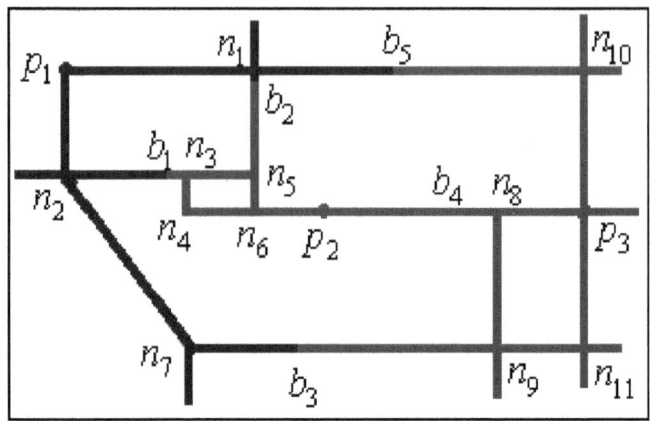

Fig. 2. Network Voronoi Diagram

Fig. 2 shows an example of Network Voronoi Diagram. Besides the objects of interest (p), a Network Voronoi Diagram also includes some road network intersections (n) and border points (b). According to the properties of a Voronoi diagram, from border points to a pair of adjacent objects is equidistant (e.g. dis (b7 , p1) = dis (b7 , p3)).

2.3   Definition 3 (Power Voronoi Diagram) [10]

Power Voronoi diagram is characterized by the weighted distance

$$d_{pw}\left(p,p_i,w_i\right)=\left\{\left\|x-x_i\right\|^2-w_i\right\} \tag{3}$$

Which we call the power distance. We call the region given by

$$V\left(p_i\right)=\left\{d_{pw}\left(x,x_i,w_i\right)\le d_{pw}\left(x,x_j,w_j\right) \ for \ j\ne i, j\in I_n\right\} \tag{4}$$

the power Voronoi polygon associated with p_i, and the set given by

$$\mathcal{V}=\left\{V(p_1),V(p_2),\cdots,V(p_n)\right\} \tag{5}$$

the power Voronoi diagram of P.

Fig. 3 shows an example of power Voronoi Diagram. There are 100 generators that were automatically generated by the computer. Every generator has power value respectively.

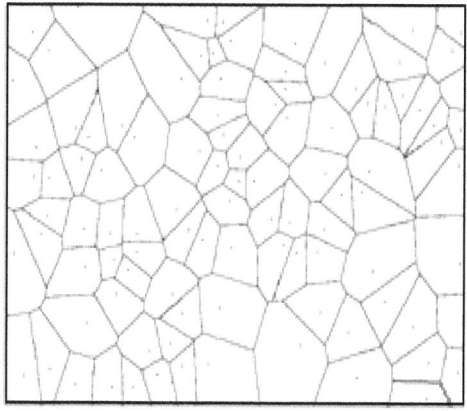

Fig. 3. Power Voronoi Diagram with 100 generators

2.4   Definition 4 (Power Network Voronoi Diagram)

A power network Voronoi Diagram is also a special kind of Voronoi diagram constructed on spatial networks. In the power network Voronoi Diagram, we use power distance in place of Euclidean distance. And every generator has their own power value. Fig. 4 shows an example of power Voronoi Diagram. Contrasting it with classical network Voronoi Diagram, we can observe that there is some difference between them. In fact, the difference was caused by the power values of generators.

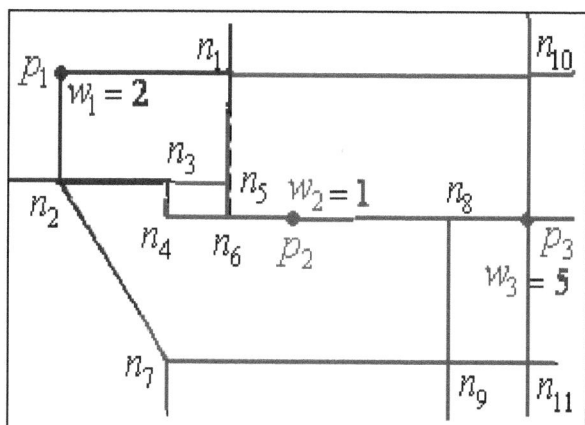

Fig. 4. Power network Voronoi Diagram

3   Discrete Construction of Power Network Voronoi Diagram

3.1   Outline of Discrete Algorithm

Suppose that there are n generator points in the Euclidean plane, and we will construct power network voronoi diagram. First, we assign different colors for different generator points, and black color for network. Then take generator points as the centre and draw circles. Spreading out from the center is of with the same power distance at a time. Unlike the discrete algorithm of planar ordinary Voronoi diagram, we only need to consider four points around generator point: the upper one, the lower one, the left one and the right one, and only need to consider those points on road segments. That is to say, we only assign color for a point with black color. This will greatly improve the efficiency of construction. The procedure end when all points with black color on screen are marked color. This time, we get the power network Voronoi diagram..

3.2   Algorithm

Input: $p_1, p_2, \cdots, p_n, w_1, w_2, \cdots, w_n$. Here, p_i is generator, w_i is its power.
Output: power network Voronoi diagrams generated by those generators.
Step 1: Built linked lists L that holds generators' data including: abscissa "x", ordinate "y", power "w_i", color
Step 2: for generator, square of minor diameter "d2", square of outside diameter "e2";
Step 3: Initialize screen as white color;
Step 4: Generate data sheet of Δx, Δy, and rw;
Step 5: k=1;
Step 6: define a pointer to L;
Step 7: When is not empty, do loop:
  {(1)Read data of row k: Δx, Δy, and rw;
  (2)read p->x, p->y, p->color;
  {SetPixel (p->x+Δx, p->y, p->color);
  SetPixel (p->x-Δx, p->y, p->color);
  SetPixel (p->x , p->y+Δx, p->color);
  SetPixel (p->x , p->y-Δx, p->color);}
   if one pixel above is black;
   hen p->d2=rw;
          else p->e2=rw;
   if p->e2 p->d2+2+1
    then delete the node which "p" pointed to from L;
   if "p" point to the end of L
   then k++, let "p" point to the first node of L;
   else p++;}
   End.

4   Practical Application

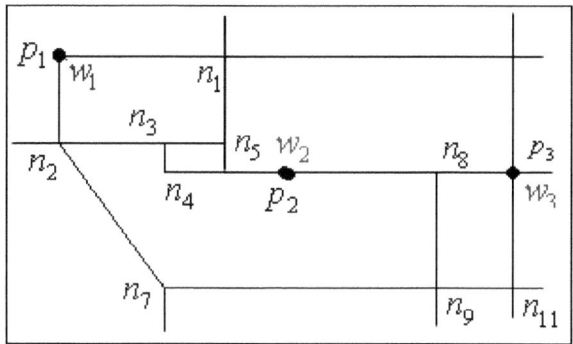

Fig. 5-1. p_1, p_2, p_3 are three generator points with power w_1, w_2, w_3 separately

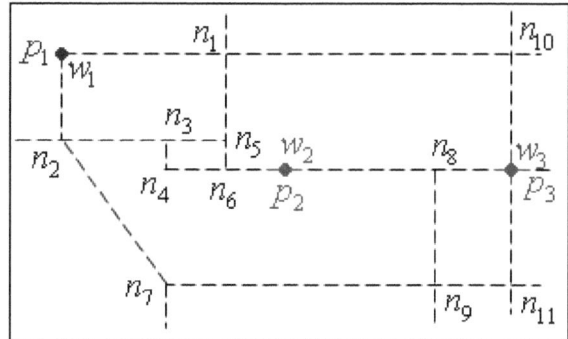

Fig. 5-2. Assign different colors for different generator point

Now we take 3 generator points as the example, and construct power network Voronoi diagram using discrete algorithm. p_1, p_2, p_3 are three generator points with power $w_1 = 2$, $w_2 = 1$, $w_3 = 5$ separately.   Figure 4 show us the generation of process.

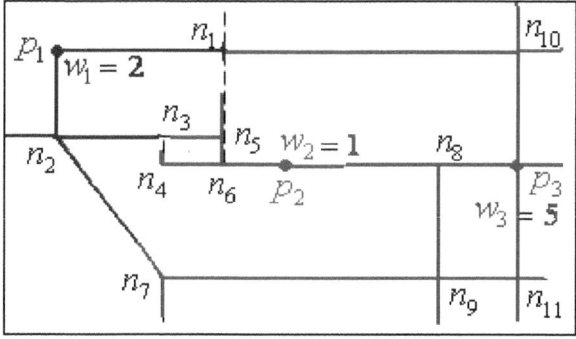

Fig. 5-3. take generator points as the centre and draw circles

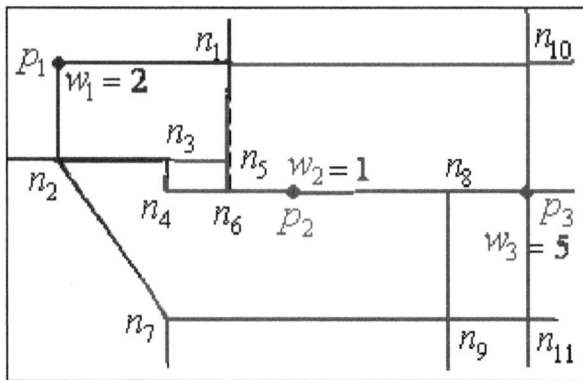

Fig. 5-4. Power network Voronoi diagram

Fig. 5. Discrete production process of an power network diagram with 3 generator points

Network distribution is shown in Fig. 5-1. First, we assign different colors for different generator points, and black color for network (in Fig. 5-2, network is represented by the dotted line). Then draw circles with taking generator points as the centre and power distance as radius (Fig. 5-3). At last, we get the network power Network Voronoi diagram (Fig. 5-4).

5   Conclusions

The Discrete construction of power network Voronoi diagrams can get over many kinds of shortcomings of traditional ones, because we need not to consider the situation of network distribution. So it has unique advantage in the construction of power network Voronoi diagrams, and it is of high potential value in practice.

References

1. Voronoi, G.: Nouvelles: applications des paramèters continus à la théorie des formes quadratiques. Premier Mémoire: Sur quelques Proprieteés des formes quadratiques positives parfaits. Reine Angew, Math. 133, 97–178 (1907)
2. Clarkson, K.L.: New applications of random sampling in computational geometry. Discrete and Computational Geometry 2, 195–222 (1987)
3. Sud, A., Govindaraju, N., Gayle, R., Dinesh Manocha, Z.: Interactive 3D distance field computation using linear factorization. In: Proceedings of the 2006 Symposium on Interactive 3D Graphics and Games, Redwood City, California, pp. 14–17 (2006)
4. Qian, B., Zhang, L., Shi, Y., Liu, B.: New Voronoi Diagram Algorithm of Multiply-Connected Planar Areas in the Selective Laser Melting. Tsinghua Science & Technology 14, 137–143 (2009)
5. Aurenhammer, F., Drysdale, R.L.S., Krasser, H.: Farthest line segment Voronoi diagrams. Information Processing Letters 100, 220–225 (2006)

6. Chen, J., Zhao, R., Li, Z.: Voronoi-based k-order neighbour relations for spatial analysis. ISPRS Journal of Photogrammetry and Remote Sensing 59, 60–72 (2004)
7. Lee, I., Lee, K.: A generic triangle-based data structure of the complete set of higher order Voronoi diagrams for emergency management. Computers, Environment and Urban Systems 33, 90–99 (2009)
8. Cabello, S., Fort, M., Sellarès, J.A.: Higher-order Voronoi diagrams on triangulated surfaces. Information Processing Letters 109, 440–445 (2009)
9. Chen, J., Zhao, R., Li, Z.: Voronoi-based k-order neighbour relations for spatial analysis. ISPRS Journal of Photogrammetry and Remote Sensing 59, 60–72 (2004)
10. Wu, Y., Zhou, W., Wang, B., Yang, F.: Modeling and characterization of two-phase composites by Voronoi diagram in the Laguerre geometry based on random close packing of spheres. Computational Materials Science 47, 951–996 (2010)

Research on Mathematical Model of Equipment Capability Planning Based on Resource Constraint

Xin Wang and Chongchong Ji

College of Mechanical Engineering, Hebei United University, 46 West Xinhua Road,
Tangshan 063009, Hebei, China
wx_wyzh@sina.com, cc520317@163.com

Abstract. In order to increase the utilization of equipments of the enterprise, taking flexible production planning for Mass Customization as the research background, the improved method of capacity planning is proposed and a mathematical model of equipment capability planning based on resource constraint is established, which aims at maximized flexibility to make the production cycle required in completing the tasks as little as possible. At the same time, the two methods of how to use the equipment idle time fully in the operation are also studied. And the methods which contain based on delivery and Sort-Insert for scheduling idle time are proposed.

Keywords: Mass customization, Equipment programming, Equipment idle time, Equipment scheduling.

1 Preface

As competition increasing in the modern markets, mass customization has become a new mode of production in the 21st century, whose purpose is providing personalized and customized products and services for customers by approaching the large scale production cost and speed [1]. The new manufacturing mode demands that enterprises have a fast and flexible manufacturing system to respond to unpredictable markets. Therefore, some demand factors about manufacturing system control are considered, which include the rapid increase in product variety, the increasing randomness of the order arrival, the equipment capacity of the enterprises itself and etc.

To meet customer's demand, enterprises must fully improve their management system and equipment layout as an important field of the whole enterprise management is the important component. It not only directly affects the current production and operation of enterprises, but also relates to long-term development and success or failure of enterprises. The task of equipment planning is to realize the goal of production and operations with good efficiency and well equipped investment efficiency, as a result, enterprise can obtain the best economic and social benefits [2].

Therefore, with the uncertainty and diversity of the customer demand, and market competition becomes increasingly fierce, equipment capability planning has become a key factor in an invincible position in the competition.

B. Liu and C. Chai (Eds.): ICICA 2011, LNCS 7030, pp. 298–304, 2011.

2   Equipment Planning Mathematical Model

2.1   Problem Description

M (the amount of equipments) equipments are supposed in the processing system. Processing products can be divided into different parts and components, which are supposed as independent workpieces. Parts are classified as purchased parts, outsourcing parts and homemade parts, among which there are N homemade parts. Each workpiece contains more than one procedure, which can be processed on different equipments. And with the different ability of different equipments, the processing time is various. Choosing the best processing equipment in the equipment capability constraint conditions is considered as the research goal. At the same time, the best processing sequence of the workpiece in each device is determined, in order to make the workspan between different workpieces shortest and make the punishment of the early / delayed lowest.

2.2   Mathematical Model

Number of workpieces: N; Amount of equipments: M; Part ID: i; Process ID: j; Equipment ID: m; The available ability of m among t time section : $C_m(t)$; The beginning processing time of the workpiece i in the m: S_{im}; The process period of i in the m: P_{im}; The process period of i in the m among t time section: $P_{im}(t)$; The completion time of the 'j' working procedure of i: E^{ij}; The process period of the 'j' working procedure of i: P^{ij}; The production period of i: t_i; The completion time of i: e_i; The advanced time of i: A_i; The delay time of i: D_i; The penalty coefficient of i in advance: α_i; The penalty coefficient of i in delay: β_i; Unit of all time variables is hour. The completion time of the 'j' working procedure of i in the m: E_{ijm}; The completion time of all parts in the m: E_{N_m}; The final completion time of all parts: E_N; When the 'j' working procedure of i and the 'l' working procedure of k are processed in the same equipment, in addition, l is ahead of j, $R_{klij} = 1$; Otherwise, $R_{klij} = 0$. When the 'j' working procedure of i is processed in the m, $X_{ijm} = 1$; Otherwise, $X_{ijm} = 0$.

The objective function and the constraints are as follows:

$$\begin{cases} f = \min \sum_{i=1}^{N} \left[\alpha_i \left(\max \left(0, t_i - e_i \right) \right) + \beta_i \left(\max(0, e_i - t_i) \right) \right] \\ = \min \sum_{i=1}^{N} \left(\alpha_i A_i + \beta_i D_i \right) \\ \min(E_N) = \max_{m=1,\dots,M} \left(E_{N_m} \right) \end{cases} \qquad (1)$$

S.t.:

$$E_{N_m} = \max(E_{ijm}) \qquad (2)$$

$$S_{im} - S_{(i+1)m} \le P_{im} \quad X_{klm} = X_{ijm} = 1 \quad R_{klij} = 1 \qquad (3)$$

$$E^{ij} - E^{i(j+1)} \le P^{ij} \quad X_{ijm} = X_{i(j+1)m} = 1 \qquad (4)$$

$$\sum_{i=1,2,\dots,N} P_{im}(t) \le C_m(t) \qquad ,m=1,\dots,M . \qquad (5)$$

The first formula (1) is objective function, namely each workpiece's punishment of advanced/delay is lowest and the completion time is shortest. The second formula (2) shows that the completion time of *m* depends on the longest completion time of all workpieces, which are processed in the *m*. The third one shows the constraints about resources. Whenever, the identical equipment can't be used for processing any two different workpieces and any two different working procedures simultaneously. The forth formula (4) shows the conditions of technological constraints, which determined the processing order of each workpiece's technique. For example, the '*j*+1' working procedure begins to be processed, after the '*j*' one is finished. The last formula expresses that the load of equipment can't exceed the inherent capability in a same period of time.

2.3   Model Analysis

Research on most flexible production scheduling problem in the shop, Companies often overlook planning for equipment, Simply, equipment capacity will be assumed to be constant, the study of equipment capacity is commenced in the paper, and a mathematical model is established. Analysis from the problem model above, the objective function is to make penalties in advance/extension penalties Minimum and the total task duration shortest. Among them, the duration defined here cannot simply sum the duration of all tasks, after all tasks in the implementation process may occur in parallel, and if sum the entire duration of the decomposition task and estimated duration will be too high. Therefore, based on the topology of the task decomposition, serial and parallel coexistence, using all available resources for production, and ultimately make advance/deferred punitive minimum and total task duration shortest.

The sum of advance/extension penalties are selected as the objective function which is based on the delivery scheduling indicators. Stressed the completion of advance/delay delivery will be punished, it is better to just complete in delivery or as close as possible in the delivery. Therefore, the mathematical model is solved by adopting a thought based on the reverse delivery. Firstly, reverse sort according to the delivery deadline, with each workpiece delivery for sorting starting point, arrange processing equipment for the workpiece in the last procedure. and its completion time is just equal to delivery, And then arrange the before procedure in sequence, until the completion of the first work arrangements, if the start time of the first procedure is less than equal to the minimum start time, then sorted in reverse order successfully, there is no advance/deferred penalty. Otherwise, positive ordering is sorted basing on the early completion.

Clearly, reverse ordering ideas to cope with the order change and other emergencies is more flexible than positive thoughts considering the completed order as early as possible.

3   Equipment Idle Time Scheduling

Due to the processing time of various finished production is different, working process will inevitably have idle time situation about equipment and workpieces are mutual waiting. How to reduce equipment free time is the effective means to improve the equipment utilization. Therefore, that how to improve the equipment utilization efficiency, and reduce equipment idle time in the existing production constraints has important theoretical significance and practical value to increase enterprise production efficiency.

3.1   The Method Based Delivery for Scheduling Idle Time

A supposed order is composed of N workpieces, which contains Q working procedures. The processing hour of a working procedure is t_j ; the idle time of equipment is t_m ; the total maintenance time of equipments is t_h ; the total service time of equipments is t_x ; the delivery cycle is D. Consequently, there is mathematical expression on the premise of not delay: $\sum_{j=1}^{Q} t_j + \sum_{m=1}^{M} t_m + t_h + t_x \leq D$ (D is standard delivery cycle). Form the above order, we know that the total idle time of equipments is $\sum_{m=1}^{M} t_m$ and the maintenance time is t_h . There is another order, of which equipment maintenance time is t_{h_2} ; a supposed task is scheduled according to a standard delivery time. Now each working hour of another order is t_{j_2} ; we use idle time of the first order to calculate $\sum_{m=1}^{M} t_m / t_{j_2}$ on the premise of using the idle time fully, when the equipment maintenance time of another order is invariant. By calculating mathematical expression, we can obtain the processing number of the second order. Finally, we get the shortest processing time of this order.

During the enterprises' process of production, the set-up time is essential. That how to use the idle time fully except for the set-up time of the same resource is an effective measure to increased efficiency of using resources. Using the example of mould changeover time, we can make a person monitor several equipments. So that, the time of people and equipments is utilized effectively. However, equipment utilization should be the largest at the planning level. That is to say, maximum equipment existing capacity should be used. When more than one order are arranged, technics arrangement of an order cause idle time which can be used to arrange the other orders before there are sufficient resources whether or not.

Accordingly, the concept of resource pooling is raised, which collects all available resources of enterprise to realize unified distribution. For example of above idle time scheduling, resource data during the idle time of equipment is reverted to resource pooling to wait calling for next task. After application of resource pooling, some problems can be considered generally at the system level and the resource utilization and the idle time obtain comprehensive measure in the current state of enterprise. Thus added orders should be treated correctly, but facing the order's temporary cancellation, enterprise makes a proper treatment by analyzing the finished parts and processing components. It not only improves the equipment utilization, but also increases the ability of answering order's changes.

3.2   *S-I* Method for Scheduling Idle Time

The explanation of *S-I* is sort-insert. Through the decomposition of the ordered product structure tree, assembly produced processes can be classified as relevant process and independent process, how to dispatch relevant process and independent process reasonably, is the key to shorten equipment idle time, fully improve equipment utilization rate. The scheduling problem mainly solved the product is in the process of production but new products in. This time, it just puts surplus process of these products and new product processing procedures into one hypothesized whole tree style flow diagram, and then put processes classification. As independent process has characters of parallel and short processing time and no process started processing time constraints, therefore we compare of man-hour time directly. If there is great free time more than independent working procedure time, it will put independent process into the closest length in leisure time, or put the maximum free time subsection necessarily.

Specific scheduling content:

With the above problems, one order: N workpieces, the number of total working procedures is Q, which are processed through M different equipments, in the different equipments between each process there are a certain sequencing constraint relation, including basic constraint requirements:

(1) A device only assemble a procedure at one time, a procedure can only be assembled in one device at one time.

(2) Once a device is assembling a procedure, it can assemble other processes after the completion of this equipment.

(3) After the completion of former procedure, it could be built next working procedure; Parallel two procedures are completed, it could be built next working procedure.

According to the principle of the shortest processing time priority, and process assembly constraints between processing, and considering the flexibility of the production planning, and according to production plan and delivery cycle's reverse recursive thought, it mainly aims at scheduling independence procedures in spare time, so as to solve the problem of using the leisure time.

(1) Put the independent working procedure in different equipment in ordering according to production hours.

(2) Put the devices' idle time in ordering according to the time.

(3) Compare device idle time with independent processing production time, eventually put the independent working procedure insert into the suited free time period; it will shorten the equipment free time and improve the utilization rate of equipment.

4   Conclusion

Based on the scheduling problem about the constraints of equipment resource capacities for mass customization, it put advanced or delayed punishment problem and the shortest production period as the target function, a mathematical model is established for equipment capacities planning. Put this model applied to practical investigation manufacturing enterprises, and planned to use linear programming knowledge for solving problem, it realizes the reasonable layout of machinery processing equipment, and improves the operation rate of equipment, increases the production efficiency, thus verifies the effectiveness of the model.

Through the research on equipment idle time, a rational utilization method of using idle time is proposed, which provides the basis for processing the temporary situation about increasing the order, cancelling the order and etc. increases the equipment utilization, improve the feasibility and flexibility of the main production planning, thereby enhancing the ability of enterprises responding to changes in market demand.

References

1. Haijia, L.: Research on production planning for mass customization in process industry. Shanghai Jiao Tong University, ShangHai (2006)
2. Yu, J.: Equipment management. Mechanical Industry Publishing (2001)

3. Zhao, X.: Research on the flexibility of manufacturing systems for the modern enterprise. Hebei University of Technology, TianJin (2009)
4. Xu, H.: Research and application on the flexible production planning for order. Graduate School of Zhejiang University, HangZhou (2004)
5. Xie, Z., Cong, J.: An algorithm of shortening idle time for a kind of special assembly problem. In: HangZhou: The International Technology and Innovation Conference 2006 (2006)
6. Peng, X., Jiang, Y.: Planning systems based on the combination of resource restrict and local search. Control Engineering of China 13(2), 185–189 (2006)
7. Zheng, C.L., Guan, X.G.: The calculation of equipment free time and its interval part waiting Time and its interval for given operating sequence. In: Proceedings of the First China-Japan International Symposium on Industrial Management (1991)
8. Bansal, N., Mahdian, M., Sviridenko, M.: Minimizing makespan in no-wait job shops. Mathematics of Operations Research 30, 817–831 (2005)
9. Qi, G., Gu, X., Li, R.: Research on mass customization and the model. CIMS 6(2), 41–44 (2000)
10. Janiak, A.: Single machine scheduling problem with a common deadline and resource dependent release dates. European Journal of Operational Research 53(3), 317–325 (1991)

Generalized Lax-Friedrichs Schemes for Linear Advection Equation with Damping

Yun Wu, Hai-xin Jiang*, and Wei Tong

College of Science, Jiujiang University
Jiujiang, China, 332005
jianghaixin@163.com

Abstract. To analyze local oscillations existing in the generalized Lax-Friedrichs(LxF) schemes for computing of the linear advection equation with damping, we observed local oscillations in numerical solutions for the discretization of some special initial data under stable conditions. Then we raised three propositions about how to control those oscillations via some numerical examples. In order to further explain this, we first investigated the discretization of initial data that trigger the chequerboard mode, the highest frequency mode. Then we proceeded to use the discrete Fourier analysis and the modified equation analysis to distinguish the dissipative and dispersive effects of numerical schemes for low frequency and high frequency modes, respectively. We find that the relative phase errors are at least offset by the numerical dissipation of the same order, otherwise the oscillation could be caused. The LxF scheme is conditionally stable and once adding the damping into linear advection equation, the damping has resulted in a slight reduction of the modes' height; We also can find even large damping, the oscillation becomes weaker as time goes by, that is to say the chequerboard mode decay.

Keywords: Finite difference schemes, low and high frequency modes, oscillations, chequerboard modes.

1 Introduction

Oscillation is an unsurprising phenomenon in numerical computations, but it is a very surprising phenomenon for computing of hyperbolic conservation laws using the Lax-Friedrichs(LxF) scheme. In this thesis, we consider the linear advection equation with damping.

$$u_t + au_x = -\alpha u, \quad x \in R, \alpha > 0, \tag{1}$$

In ([1],[11]), to compute the numerical solution of the hyperbolic conservation laws

$$u_t + f(u)_x = 0, \quad x \in R, t > 0, \tag{2}$$

---

* Corresponding author.

B. Liu and C. Chai (Eds.): ICICA 2011, LNCS 7030, pp. 305–312, 2011.
© Springer-Verlag Berlin Heidelberg 2011

where $u = (u_1, \cdots, u_m)^T$, and $f(u) = (f_1, \cdots, f_m)^T$, we consider the generalized LxF scheme of the viscosity form

$$u_j^{n+1} = u_j^n - \frac{\nu}{2}[f(u_{j+1}^n) - f(u_{j-1}^n)] + \frac{q}{2}(u_{j+1}^n - 2u_j^n + u_{j-1}^n), \tag{3}$$

where the mesh ratio $\nu = \tau/h$ is assumed to be a constant, τ and h are step sizes in time and space, respectively, u_j^n denotes an approximation of $u(jh, n\tau)$, the term $q \in (0, 1]$ is the coefficient of numerical viscosity. When $q = 1$, it is the classical LxF scheme.

With the flux function $f = au$, (1) is the linear advection equation accordingly

$$u_t + au_x = 0, \quad x \in R, t > 0, \tag{4}$$

and the scheme (3) turns into the generalized Lax-Friedrichs scheme of (4)

$$u_j^{n+1} = u_j^n - \frac{a\nu}{2}(u_{j+1}^n - u_{j-1}^n) + \frac{q}{2}(u_{j+1}^n - 2u_j^n + u_{j-1}^n). \tag{5}$$

By adding a damping term $-\alpha u$ (α is a positive constant) to the right of (4), we obtain the linear advection equation with damping(1). We can take account of the discretization of the damping term in three forms

$$u_j^{n+1} = u_j^n - \frac{a\nu}{2}(u_{j+1}^n - u_{j-1}^n) + \frac{q}{2}(u_{j+1}^n - 2u_j^n + u_{j-1}^n) - \alpha\tau u_j^n \tag{6}$$

$$u_j^{n+1} = u_j^n - \frac{a\nu}{2}(u_{j+1}^n - u_{j-1}^n) + \frac{q}{2}(u_{j+1}^n - 2u_j^n + u_{j-1}^n) - \alpha\tau \frac{u_{j+1}^n + u_{j-1}^n}{2} \tag{7}$$

$$u_j^{n+1} = u_j^n - \frac{a\nu}{2}(u_{j+1}^n - u_{j-1}^n) + \frac{q}{2}(u_{j+1}^n - 2u_j^n + u_{j-1}^n) - \alpha\tau u_j^{n+1} \tag{8}$$

For convenience, we take scheme (6) as an example in the first place, after that we investigate the rest.

2   Local Oscillations in the Solutions by the Generalized Lax-Friedrichs Scheme

We solve the linear advection equation with damping $u_t + au_x = -\alpha u$ by using (6) with the initial data:

$$u_j^0 = \begin{cases} 1, & j = M/2, \\ 0, & otherwise \end{cases}$$

where M is the number of the grid points.

As a matter of convenience, we will use a uniform grid, with grid spacing $\tau = 1/M$ and consider $[0, 1]$ as the space interval. We take $M = 50$, $a = 0.5$, $v = \alpha$, $n = 20$, the mesh ratio $\nu = \tau/h$ is assumed to be a constant c, and the boundary condition is set to be periodic just for simplicity.From these examples, we summarize as follows.

(1) The presence of oscillations is related to the numerical viscosity coefficient q. If q is large enough, the oscillations become weaker as q decrease, and finally the oscillations vanish. If q is small, the oscillations become weaker as q increase, and finally the oscillations vanish.

(2) The presence of oscillations is related to the damping coefficient α. The oscillations become weaker as α decrease, finally the oscillations vanish. And we also can find that if the damping coefficient is consistent, the oscillations become weaker as time steps n increase.

(3) The presence of oscillations is related to the width of mesh h. The oscillations become weaker as h decrease, and finally the oscillations vanish.

2.1 Linear Advection Equation with Damping with a Single Square Pulse Initial Data

We still use (6) to compute the linear advection equation with damping $u_t + au_x = -\alpha u$. The initial data now is a single square pulse:

$$u_j^0 = \begin{cases} 1, & j = M/2, M/2 + 1, \\ 0, & otherwise. \end{cases}$$

We can know that if we express the middle initial state with two points, i.e. an even number, the numerical solution is different from the case with one point. No oscillation is present.

3 Chequerboard Modes in the Initial Discretization

As observed in the numerical examples and also in [1,3], the numerical solutions display very distinct behaviors if the initial data are discretized in different ways. It turns out that chequerboard modes are present and affect the solutions if the initial data contains a square signal and are discretized with an odd number of grid points. So, we discuss the discretization of initial data

$$u(x, 0) = u_0(x), \qquad x \in [0, 1], \tag{9}$$

with M grid points and $h = 1/M$. For simplicity, we assume that M is even, and $u_0(0) = u_0(1)$. The numerical solution value at the grid point x_j is denoted by u_j^0. We express this grid point value u_j^0 by using the usual discrete Fourier sums, as in [1,5]. We use the scaled wave number $\xi = 2\pi k h$ and obtain

$$u_j^0 = \sum_{k=-M/2+1}^{M/2} c_k^0 e^{i\xi j}, \quad i^2 = -1, \quad j = 0, 1, \cdots, M-1, \tag{10}$$

where the coefficients c_k^0 are , in turn, expressed as

$$c_k^0 = \frac{1}{M} \sum_{j=0}^{M-1} u_j^0 e^{-i\xi j}, \quad k = -M/2+1, \cdots, M/2, \tag{11}$$

For the particular case that

$$c_k^0 = \begin{cases} 1, & k = M/2, \\ 0, & otherwise. \end{cases}$$

i.e. the initial data are taken to be just the highest Fourier mode which is a single chequerboard mode oscillation

$$u_j^0 = e^{i2\pi \frac{M}{2} jh} = e^{i\pi j} = (-1)^j$$

In accordance with the numerical examples, we start with the simple initial data of a square signal, i.e. an initial function of the following type

$$u(x,0) = \begin{cases} 1, & 0 < x^{(1)} < x < x^{(2)} < 1, \\ 0, & otherwise. \end{cases} \tag{12}$$

We take the following two ways to approximate the step function (12) as a grid function: One uses an odd number of grid points to take the value one of the square signal and the other uses the next smaller even number of grid points. The numerical results display oscillations for the former but not for the latter. We use the discrete Fourier sum to clarify the difference.

3.1   Discretization with an Odd Number of Grid Points

Take $j_1, j_2 \in N$ such that $j_1 + j_2$ is an even number. We set $x^{(1)} = (\frac{M}{2} - j_1)h$ and $x^{(2)} = (\frac{M}{2} + j_2)h$. We first discretize the square signal(12) with $p := j_1 + j_2 + 1$ nodes, i.e. an odd number of grid points, such that

$$u_j^0 = \begin{cases} 1, & if\ j = M/2 - j_1, \cdots, M/2 + j_2, \\ 0, & otherwise. \end{cases} \tag{13}$$

Substituting them into (11), we obtain by simple calculation,

$$c_k^0 = h \sum_{j=0}^{M-1} u_j^0 e^{-i\xi j} = \begin{cases} \frac{(-1)^k e^{i\xi j_1}(1-e^{-i\xi p})}{M(1-e^{-i\xi})}, & for\ k \neq 0, \\ ph, & for\ k = 0. \end{cases} \tag{14}$$

We give special attention to the term

$$c_{M/2}^0 = (-1)^{j_1 + M/2} h,$$

since M is even and p is odd. Hence the initial data (13) can be expressed in the form

$$u_j^0 = (-1)^{j+j_1+M/2} h + \sum_{k=-M/2+1}^{M/2-1} c_k^0 e^{i\xi j}$$

$$= (-1)^{j+j_1+M/2} h + ph + \sum_{k \neq 0, M/2} \frac{(-1)^k e^{i\xi(j+j_1)}(1 - e^{-i\xi p})}{M(1 - e^{-i\xi})} \tag{15}$$

3.2   Discretization with an Even Number of Grid Points

Rather than 3.1 above, we use $j_1 + j_2$(i.e. $p-1$) even number of grid points to express the square signal in (12) as follows

$$u_j^0 = \begin{cases} 1, & if \ j = M/2 - j_1 + 1, \cdots, M/2 + j_2, \\ 0, & otherwise. \end{cases}$$

Then we substitute these initial data into (13) to obtain

$$c_k^0 = \begin{cases} \frac{(-1)^k e^{i\xi(j_1-1)}[1-e^{-i\xi(p-1)}]}{M(1-e^{-i\xi})}, & for \ k \neq 0, \\ (p-1)h, & for \ k \ = \ 0. \end{cases}$$

When $k = M/2$, then $c_{M/2}^0 = 0$. And the initial data can be written by using the discrete Fourier sums as

$$u_j^0 = 0 \times (-1)^j + (p-1)h + \sum_{k \neq 0, M/2} \frac{(-1)^k e^{i\xi(j+j_1-1)}[1-e^{-i\xi(p-1)}]}{M(1-e^{-i\xi})} \tag{16}$$

Comparing (15) with (16), we observe an essential difference lies that a chequerboard mode $(-1)^{j+j_1+M/2}$ is present in (15), but it is filtered out in (16).

We summarize all of the above analysis in the following proposition.

Proposition 3.1. Suppose that the initial data (3.1) are given, we have two different types of discretization. If they are discretized with an odd number of grid points, the chequerboard, i.e. highest frequency, mode is present. In contrast, if they are discretized with an even number of grid points, this mode is suppressed.

Consider the generalized Lax-Friedrichs scheme (6) of the viscosity form for the linear advection equation with damping

$$u_j^{n+1} = u_j^n - \frac{a\nu}{2}(u_{j+1}^n - u_{j-1}^n) + \frac{q}{2}(u_{j+1}^n - 2u_j^n + u_{j-1}^n) - \alpha\tau u_j^n$$

where $\nu = \tau/h$. We assume that the solution to (6) has the standard form of discrete Fourier series,

$$u_j^n = \sum_{k=-M/2+1}^{M/2} c_k^n e^{i\xi j}, \tag{17}$$

The coefficients c_k^n are obtained successively and expressed as,

$$c_k^n = [1 - \alpha\tau - q(1-\cos\xi) - ia\nu\sin\xi]c_k^{n-1}$$
$$= [1 - \alpha\tau - q(1-\cos\xi) - ia\nu\sin\xi]^n c_k^0 \tag{18}$$

If we take

$$c_k^0 = \begin{cases} 1, & k = M/2, \\ 0, & otherwise. \end{cases}$$

Then we obtain

$$u_j^n = (1 - \alpha\tau - 2q)^n(-1)^j$$

This is a typical chequerboard mode solution.

In correspondence with the two kinds of discretization of a single square signal, the Fourier coefficients c_k^0 have different expressions, and the solutions are expressed, respectively, as follows

(1) Odd discretization case. We use (14),(17),(18), the solution of (6) is

$$
\begin{aligned}
u_j^n &= \frac{1}{M}(1 - \alpha\tau - 2q)^n(-1)^{j+j_1+M/2} + \sum_{k=-M/2+1}^{M/2-1} c_k^n e^{i\xi j} \\
&= \frac{1}{M}(1 - \alpha\tau - 2q)^n(-1)^{j+j_1+M/2} \\
&\quad + \sum_{k=-M/2+1}^{M/2-1} c_k^0 [1 - \alpha\tau - q(1 - \cos\xi) - ia\nu\sin\xi]^n e^{i\xi j},
\end{aligned} \tag{19}
$$

(2) Even discretization case. With the initial data (16), we have

$$
\begin{aligned}
u_j^n &= 0 \times (1 - \alpha\tau - 2q)^n(-1)^j + \sum_{k=-M/2+1}^{M/2-1} c_k^n e^{i\xi j} \\
&= 0 \times (1 - \alpha\tau - 2q)^n(-1)^j \\
&\quad + \sum_{k=-M/2+1}^{M/2-1} c_k^0 [1 - \alpha\tau - q(1 - \cos\xi) - ia\nu\sin\xi]^n e^{i\xi j},
\end{aligned} \tag{20}
$$

We emphasize that the Fourier coefficients c_k^n have different expressions in correspondence with the odd and even number of nodes taken for the discretization of the signal in the initial data. By comparing (19) and (20), we see that in the odd case the chequerboard mode does not vanish if it exists initially, unless $q = \frac{1-\alpha\tau}{2}$. However, the chequerboard mode will decay as time goes on, that is, the oscillations become weaker. We can see this phenomena . Therefore, a proper discretization of initial data (9) would be important to suppress these oscillations in the numerical solution of (1). This we can see in [1,3].

Remark: (1)If $q = 1 - \frac{\alpha\tau}{2}$, the solution oscillates between 1 and -1 alternately. The large numerical dissipation does not take effect on this chequerboard mode. (2)When $q \neq 1 - \frac{\alpha\tau}{2}$, we have $|1 - \alpha\tau - 2q| < 1$. Then the chequerboard mode is damped out. In particular, if $q = \frac{1-\alpha\tau}{2}$, the chequerboard mode is eliminated and has no influence on the solution at all. We will analyze this in Section 4.

4   Discussion and Conclusion

In the present paper, we are discussing the local oscillations in the particular generalized LxF scheme. We have individually analyzed the resolution of the low and high frequency modes $u_j^n = \lambda_k^n e^{ij\xi}$, $\xi = 2\pi kh$ in numerical solutions. Our approach is the discrete Fourier analysis and the modified equation analysis, which are applied to investigating the numerical dissipative and dispersive mechanisms as well as relative phase errors.

1. Relative phase error. For the low frequency modes, the error is of order $O(1)$, while for high frequency modes the error is of order $O(1)$ after each time step, which is generally independent of the parameter q.

2. Numerical dissipation. For the low frequency modes, the dissipation is usually of order $O(1)$ for the scheme (1.6), which closely depends on the parameter q. For high frequency modes, the scheme usually has the numerical damping of order $O(1)$ that becomes stronger as q is closer to $\frac{1-\alpha\tau}{2}$, unless it vanishes for the limit case $(q = 1 - \frac{\alpha\tau}{2})$.

Thus we obtain that the relative phase errors should be at least offset by the numerical dissipation of the same order. Otherwise the oscillation could be caused. We also obtain the following conclusions:

1. The GLxF scheme (1.6) is conditionally stable and $0 < \frac{a^2\nu^2}{1-\alpha\tau} \leq q \leq 1 - \alpha\tau$ is necessary and sufficient for stability.

2. Under the stable condition, the oscillation is connected with these factors:
(1) If $q > \frac{1-\alpha\tau}{2}$, the oscillation becomes weaker as q decrease. If $q > \frac{1-\alpha\tau}{2}$, the oscillation becomes weaker as q increase.
(2) The oscillation becomes weaker as α decrease.
(3) The oscillation becomes weaker as h decrease.
(4) If the initial data can be discretized as square signal, which be discretized with an odd number of grid points, the chequerboard mode (i.e. the oscillation) is present. In contrast, the chequerboard mode (i.e. the oscillation) is suppressed.

3. Once adding the damping into linear advection equation, it is clear that the damping has resulted in a slight reduction of the modes' height; We also can find even large damping, the oscillation becomes weaker as time goes by, that is to say the chequerboard mode decay.

4. When $q = \frac{1-\alpha\tau}{2}$, the oscillation vanish. When $q = 1 - \frac{\alpha\tau}{2}$, the oscillation is the strongest.

References

1. Li, J.-Q., Tang, H.-Z., Warnecke, G., Zhang, L.-M.: Local Oscillations in Finite Difference Solutions of Hyperbolic Conservation Laws. Mathematics of Computation, S0025-5718(09)02219-4 (2009)
2. Morton, K.W., Mayers, D.F.: Numerical Solution of Partial Differential Equations, 2nd edn. Cambridge University Press (2005)
3. Zhu, P., Zhou, S.: Relaxation Lax-Friedrichs sweeping scheme for static Hamilton-Jacobi equations. Numer. Algor. 54, 325–342 (2010)
4. Gomez, H., Colominas, I., Navarrina, F., Paris, J., Casteleiro, M.: A Hyperbolic Theory for Advection-Diffusion Problems. Mathematical Foundations and Numerical Modeling. Arch. Comput. Methods Eng. 17, 191–211 (2010)
5. Tang, H.-Z., Xu, K.: Positivity-Preserving Analysis of Explicit and Implicit Lax-CFriedrichs Schemes for Compressible Euler Equations. J. Sci Comput. 15, 19–28 (2000)
6. Breuss, M.: The correct use of the Lax-Friedrichs method. M2AN Math. Model. Numer. Anal. 38, 519–540 (2004)

7. Dou, L., Dou, J.: The Grid: Time-domain analysis of lossy multiconductor transmission lines based on the LaxCWendroff technique. Analog Integrated Circuits and Signal Processing 68, 85–92 (2011)
8. Breuss, M.: An analysis of the influence of data extrema on some first and second order central approximations of hyperbolic conservation laws. M2AN Math. Model. Numer. Anal. 39, 965–993 (2005)
9. Thomas, J.W.: Numerical Partial Differential Equations: Finite Difference Methods. Springer, Heidelberg (1995)
10. Warming, R.F., Hyett, B.J.: The modified equation approach to the stability and accuracy of finite difference methods. J. Comput. Phys. 14, 159–179 (1974)
11. Tadmor, E.: Numerical viscosity and the entropy condition for conservative difference schemes. Math. Comp. 43, 369–381 (1984)
12. Fernandez-Nieto, E.D., Castro Diaz, M.J., Pares, C.: On an Intermediate Field Capturing Riemann Solver Based on a Parabolic Viscosity Matrix for the Two-Layer Shallow Water System. J. Sci Comput. 48, 117–140 (2011)

Extension Limb Action Recognition
Based on Acceleration Median

Yonghua Wang[1], Lei Qi[2], Fei Yuan[3,*], and Jian Yang[1]

[1] Faculty of Automation, Guangdong University of Technology, Guangzhou, China
[2] Center of Campus Network&Modem Educational Technology, Guangdong University of Technology, Guangzhou, China
[3] School of Information Science and Technology, Sun Yat-sen University, Guangzhou, China
wangyonghua@gdut.edu.cn, lqi@gdut.edu.cn,
eric_f_y@foxmail.com, jmmyyoung@126.com

Abstract. An extension limb action recognition method based on acceleration median (EULAR-AM) is proposed in this paper. Stretch arm has the feature that the arm's acceleration increases firstly, and then decreases. So the EULAR-AM chooses the acceleration median of the arm outstretching process and the direction of acceleration at the initial moment of arm outstretching as its recognition characteristic values. It can reduce the affection of outstretching speed to the characteristic value of limb action, and can achieve the goal that the different outstretching speed actions having same direction could be described by the same characters. Then combining the extension recognition method, the EULAR-AM recognized the limb action. The experiment results show that the recognition accuracy rate of the EULAR-AM is 93.2 %.

Keywords: Upper limb action recognition, extension recognition, acceleration median.

1 Introduction

Human action recognition technology is an important research direction in Human-computer interaction[1]. Computer vision[2-3] and motion measurements[4-5] are two primary recognition methods. Acceleration sensor-based human action recognition technology belongs to the latter. Acceleration sensors are widely used in the interactive games. However, most interactive games do not use the acceleration sensor data to identify the action. So there is no real sense of interaction, and severely reducing the user's fun [6].

This paper proposed an extension upper limb action recognition method based on the acceleration median (EULAR-AM). Since the stretch arm has the feature that the arm's acceleration increases in the first, and then decreases. Therefore, the EULAR-AM selects the acceleration median of the arm outstretching process and the direction of acceleration at the initial moment of arm outstretching as its recognition characteristic values. Then it combines the extension recognition method[7-9] to

---

* Corresponding author.

B. Liu and C. Chai (Eds.): ICICA 2011, LNCS 7030, pp. 313–321, 2011.
© Springer-Verlag Berlin Heidelberg 2011

recognize the upper limb action. Experiments show that EULAR-AM has the high recognition rate to the upper limb movement recognition ,which means it could solve the interactive game action recognition question effectively.

2 Analysis on the Acceleration Characteristic of Stretch Arm

In this paper, the three-axis acceleration chip ADXL330 is used to measure the acceleration of outrigger action. When measuring a stretched forward arm, X-axis acceleration of the wrist position versus time waveform is shown in Figure 1.

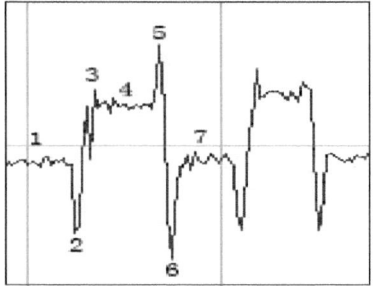

Fig. 1. Waveforms of x-axis acceleration when the arm stretched forward

The position 1 represents the static acceleration while the arm has not stretched out. The position 2 indicates that due to the sudden movement at the action start, the initial acceleration along the direction of movement is very large.When the arm continues the forward movement, the acceleration reduces slowly, while the speed increases continuously. At the last part of the forward arm movement, acceleration starts to increase reversely and the speed decreases. When arm reaching the farthest position, the outrigger action stops suddenly. At this time acceleration is large, the peak acceleration waveform (position 3) formed. While the arm in the furthest position, the acceleration is static(position 4), and only has a relationship with the accelerometer gesture.When the arm back, the process is the inverse of the above. At the beginning and ending of back arm, the waveform of acceleration will produces the same peak (position 5)and trough(position 6.) respectively. Reaching the end of a stretched arm cycle(position 7), the acceleration is the static acceleration value of back arm. It is similar to position 1.When the arm move to the opposite direction, the acceleration waveforms in position 2 will be upward peaks. It can be used as a feature to identify actions. However, this characteristic only can differentiate two kinds of movements in the opposite direction.

Figure 1 shows that: The acceleration median (the mean of maximum and minimum acceleration) of the arm outstretch is approximately equal to the acceleration median of the process of arm take back. Choosing the acceleration median to recognize the arm outstretch can avoid the case that same arm outstretch movements can't be distinguished because have different speeds.

For N kinds of stretch arm in different directions, the x-axis, y-axis, z-axis acceleration values measured by the accelerometer are different from each other and with certain of differentiability.Therefore, the acceleration median of arm outstretching can be used as recognition characteristic. Meanwhile, combined with the initial acceleration direction, it can very well distinguish two in opposite direction arm outstretch movements.

3   Analysis on Acceleration Medians of Stretch Arm

We install three three-axis accelerometers in different parts of the arm,so 9 acceleration median values and their direction characteristics can be measured. Through experiments, we obtained the acceleration waveforms of 4 kinds of stretch arm movements, as shown in Figure 2. The stretch arm movements are slow movement processes and when reached the farthest point the arm pause a moment.

Use "+ "and"- "indicate the direction of the initial moment of outrigger action."+"indicates the acceleration is less than the static acceleration and "-" is reverse. In order to the better description the outrigger process, we also measured the acceleration of the following outrigger movements. They are: Slow Continuous Outrigger (The outrigger speed is slow and don't stop at the farthest point), Fast Continuous outrigger (The outrigger speed is fast and don't stop at the farthest point), Fast Pause outrigger (The outrigger speed is fast and then pauses in a moment at the farthest point.).From Figure 2, the acceleration medians of above outrigger movements are got, as showed in Table 1,2.

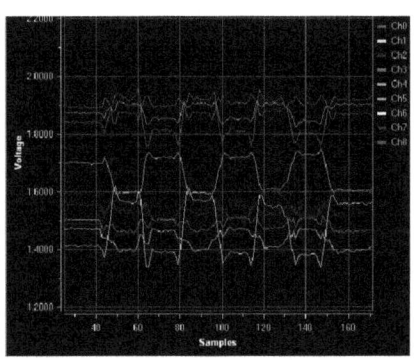

(a) Acceleration waveform of forward half outrigger

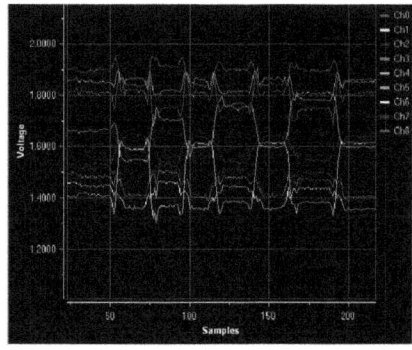

(b) Acceleration waveform of forward outrigger

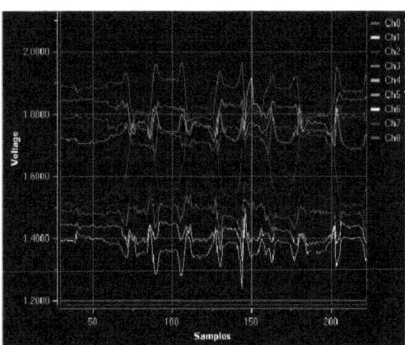

(c) Acceleration waveform of upward half outrigger

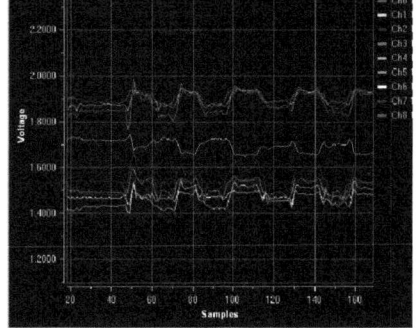

(d) Acceleration waveform of left half outrigger

Fig. 2. Acceleration waveforms of 4 kind of outrigger action

Table 1. The acceleration medians of Slow Pause, Slow Continuous Outrigger outrigger action

	Slow Pause Outrigger				Slow Continuous Outrigger			
	FHO	FO	UHO	LHO	FHO	FO	UHO	LHO
c_1	1.86	1.795	1.75	1.87	1.875	1.82	1.735	1.895
c_2	1.48	1.47	1.37	1.47	1.485	1.5	1.37	1.475
c_3	1.54	1.475	1.515	1.565	1.565	1.47	1.49	1.605
c_4	1.49	1.45	1.47	1.52	1.505	1.43	1.465	1.5
c_5	-1.67	-1.68	-1.81	-1.7	-1.725	-1.715	-1.81	-1.72
c_6	1.865	1.85	-1.79	1.895	1.87	1.84	-1.76	1.865
c_7	1.435	1.405	1.395	-1.47	1.45	1.405	1.415	-1.47
c_8	-1.55	-1.59	-1.67	-1.5	-1.5	-1.595	-1.675	-1.535
c_9	-1.89	-1.86	-1.87	-1.895	-1.915	-1.825	-1.84	-1.92

Where C_1, C_2, C_3 is the x-axis, y-axis, z-axis acceleration median of the accelerometer 1 respectively. C_4, C_5, C_6 is the x-axis,y-axis, z-axis acceleration median of the accelerometer 2 respectively. C_7, C_8, C_9 is the x-axis,y-axis,z-axis acceleration median of the accelerometer 3 respectively.FHO is the forward half outrigger.FO is the forward outrigger. UHO is the upward half outrigger. LHO is the left half outrigger.

Table 2. The acceleration medians of Fast Continuous Outrigger and Fast Pause Outrigger

	Fast Continuous Outrigger				Fast Pause Outrigger			
	FHO	FO	UHO	LHO	FHO	FO	UHO	LHO
c_1	1.815	1.65	1.71	1.88	1.74	1.71	1.69	1.865
c_2	1.565	1.535	1.33	1.495	1.505	1.565	1.47	1.5
c_3	1.47	1.435	1.485	1.525	1.505	1.5	1.51	1.46
c_4	1.475	1.405	1.485	1.465	1.47	1.515	1.385	1.495
c_5	-1.69	-1.73	-1.86	-1.73	-1.775	-1.72	-1.875	-1.67
c_6	1.805	1.73	-1.78	1.885	1.845	1.85	-1.83	1.875
c_7	1.45	1.465	1.38	-1.455	1.4	1.415	1.4	-1.46
c_8	-1.555	-1.515	-1.635	-1.535	-1.565	-1.585	-1.655	-1.505
c_9	-1.815	-1.865	-1.885	-1.895	-1.905	-1.94	-1.935	-1.9

Compared values in Table 1, 2, we can see that: for the same kind outrigger action, even if the movement form has the difference,the acceleration medians are basically consistent. For different Outrigger actions, acceleration median values have a certain differentiability. Combining the extension identification methods[7-9] for action recognition, and according to the size 0.1 extension rule, expand the values in the left part of table 3, the range of values used in the extension recognition are obtain ,as showed in wright part of Table 3.

Table 3. The acceleration medians of four outrigger actions

	Average value				The range of acceleration medians			
	FHO	FO	UHO	LHO	FHO	FO	UHO	LHO
c_1	1.823	1.744	1.721	1.878	<1.723, 1.923>	<1.644, 1.844>	<1.621, 1.821>	<1.778, 1.978>
c_2	1.509	1.518	1.385	1.485	<1.409, 1.609>	<1.418, 1.618>	<1.285, 1.485>	<1.385, 1.585>
c_3	1.52	1.47	1.5	1.539	<1.42, 1.62>	<1.37, 1.57>	<1.4, 1.6>	<1.439, 1.639>
c_4	1.485	1.45	1.451	1.495	<1.385, 1.585>	<1.35, 1.55>	<1.351, 1.551>	<1.395, 1.595>
c_5	-1.715	-1.711	-1.839	-1.705	<-1.815, -1.615>	<-1.811, -1.611>	<-1.939, -1.739>	<-1.805, -1.605>
c_6	1.846	1.818	-1.79	1.88	<1.746, 1.946>	<1.718, 1.918>	<-1.89, -1.69>	<1.78, 1.98>
c_7	1.434	1.423	1.398	-1.464	<1.334, 1.534>	<1.323, 1.523>	<1.298, 1.498>	<-1.564, -1.364>
c_8	-1.542	-1.571	-1.659	-1.519	<-1.642, -1.442>	<-.671, -1.471>	<-1.759, -1.559>	<-1.619, -1.419>
c_9	-1.881	-1.871	-1.881	-1.903	<-1.981, -1.781>	<-1.971, -1.771>	<-1.981, -1.781>	<-2.003, -1.803>

4   The Extension Recognition of Outrigger Action

4.1   Establish Classical Field Matter Elements and Segment Field Matter Elements

The classical field matter elements of outrigger action are the matter element model of all kinds of outrigger actions. Its basic form is:

$$R = (N,C,V) = \begin{pmatrix} N, & c_1, & v_1 \\ & c_2, & v_2 \\ & \vdots & \vdots \\ & c_n, & v_n \end{pmatrix} = \begin{pmatrix} N, & c_1, & <a_1,b_1> \\ & c_2, & <a_2,b_2> \\ & \vdots & \vdots \\ & c_n, & <a_n,b_n> \end{pmatrix} \qquad (1)$$

Where R -the outrigger action. N - the name. C - the feature names vector. V - the haracteristic value range vector. (a_i,b_i) - the characteristic value range. According Table 3, the classical field matter elements of the forward half outrigger actions is:

$$R_1 = \begin{bmatrix} \text{The forward} & & \\ \text{half outrigger,} & c_1, & v_{11} \\ & c_2, & v_{12} \\ & c_3, & v_{13} \\ & c_4, & v_{14} \\ & c_5, & v_{15} \\ & c_6, & v_{16} \\ & c_7, & v_{17} \\ & c_8, & v_{18} \\ & c_9, & v_{19} \end{bmatrix} = \begin{bmatrix} \text{The forward} & & \\ \text{half outrigger,} & c_1, & <1.723,1.923> \\ & c_2, & <1.409,1.609> \\ & c_3, & <1.420,1.620> \\ & c_4, & <1.385,1.585> \\ & c_5, & <1.615,1.815> \\ & c_6, & <1.746,1.946> \\ & c_7, & <1.343,1.543> \\ & c_8, & <1.443,1.643> \\ & c_9, & <1.781,1.981> \end{bmatrix} \tag{2}$$

And classical field matter elementsof the forward outrigger,the upward half outrigger, the left half outrigger are similarly the R_1 .Segment field matter elements are the characteristic value range of the Outrigger actions.

$$R_p = \begin{bmatrix} \text{Outrigger,} & c_1, & v_{P1} \\ & c_2, & v_{P2} \\ & c_3, & v_{P3} \\ & c_4, & v_{P4} \\ & c_5, & v_{P5} \\ & c_6, & v_{P6} \\ & c_7, & v_{P7} \\ & c_8, & v_{P8} \\ & c_9, & v_{P9} \end{bmatrix} = \begin{bmatrix} \text{Outrigger,} & c_1, & <1.621,1.978> \\ & c_2, & <1.285,1.625> \\ & c_3, & <1.379,1.639> \\ & c_4, & <1.351,1.595> \\ & c_5, & <-1.939,-1.601> \\ & c_6, & <-1.89,1.98> \\ & c_7, & <-1.664,1.543> \\ & c_8, & <-1.759,-1.443> \\ & c_9, & <-2.003,-1.759> \end{bmatrix} , \tag{3}$$

where $v_{pi} = v_{1i} \bigcup v_{2i} \bigcup v_{3i} \bigcup v_{4i}, (i = 1,2,\cdots,9)$

4.2   Determine the Correlation Function

$$\rho(x, X_0) = \left| x - \frac{1}{2}(a+b) \right| - \frac{1}{2}(b-a) \tag{4}$$

is the distance of point $x_0 \in (-\infty, +\infty)$ to interval $X_0 = \langle a, b \rangle$ .The correlation function is:

$$K_{ij}(v_i, v_{ji}) = \frac{\rho(v_i, v_{ji})}{\rho(v_i, v_{pi}) - \rho(v_i, v_{ji})} \tag{5}$$

Therein, v_i is the value of i characteristic of the object to be recognition $(i = 1,2,3\ldots9; j = 1,2,3,4)$ .When the denominator is zero,then

$$K_{ij}(v_i, v_{ji}) = -\rho(v_i, v_{ji}) - 1 \tag{6}$$

4.3   Determine the Weight Coefficient

According to the weight calculation method [10], and combined with the data in table 3, the weight coefficient vector can be calculated.

$$w = [1.1203, 0.1057, 0.0825, 0.0764, 0.1034, 0.1629, 0.1552, 0.0973, 0.0963] \quad (7)$$

4.4   The Extension Recognition of Action to be Identified

We measured the FHO, FO, UHO, LHO action 10 times and get 10 sets of acceleration information respetctively. Each set of information includes 9 acceleration medians and acceleration direction values in the initial movement time. Randomly selected one group data from the 10 groups of each type of action as the objects to be recognized, and carry on the recognition by the extension recognition methods. The random selected information of the actions to be recognized as shown in table 4.

Table 4. Characteristic values of the actions to be recognized

	FHO	FO	UHO	LHO
c_1	1.9	1.78	1.875	1.725
c_2	1.535	1.47	1.47	1.3675
c_3	1.515	1.4725	1.5725	1.515
c_4	1.5	1.4575	1.525	1.485
c_5	-1.7	-1.648	-1.7	-1.78
c_6	1.8725	1.85	1.9	-1.783
c_7	1.4425	1.4075	-1.48	1.41
c_8	-1.565	-1.588	-1.51	-1.675
c_9	-1.9	-1.855	-1.905	-1.87

Constructed the element model of the objects to be recognized based on its acceleration information firstly. It is:

$$R_o = \begin{bmatrix} \text{The object to} & & \\ \text{be recognized} & c_1, & v_1 \\ & c_2, & v_2 \\ & c_3, & v_3 \\ & c_4, & v_4 \\ & c_5, & v_5 \\ & c_6, & v_6 \\ & c_7, & v_7 \\ & c_8, & v_8 \\ & c_9, & v_{19} \end{bmatrix} = \begin{bmatrix} \text{The object to} & & \\ \text{be recognized} & c_1, & 1.9 \\ & c_2, & 1.535 \\ & c_3, & 1.515 \\ & c_4, & 1.5 \\ & c_5, & -1.7 \\ & c_6, & 1.8725 \\ & c_7, & 1.4425 \\ & c_8, & -1.565 \\ & c_9, & -1.9 \end{bmatrix} \quad (8)$$

Then Combined with the formula 4~6, calculated the interval v_{ji}, v_{pi} correlation degree $K_{ji} = v_i, v_{ji}$ of the characteristic value v_i of R_0. And constituted the association degree matrix K.

$$K = \begin{bmatrix} 0.0214 & -0.1613 & 0.0763 & -0.5032 \\ 0.0721 & 0.0891 & 0.0476 & -0.3571 \\ 0.0945 & 0.0618 & 0.0234 & 0.0837 \\ 0.0837 & 0.0659 & 0.0945 & 0.0486 \\ 0.0837 & 0.0989 & 0.0945 & -0.2826 \\ 0.0716 & 0.0613 & 0.0918 & -0.9707 \\ 0.0995 & 0.0747 & -0.9666 & 0.0531 \\ 0.0763 & 0.0827 & 0.0436 & 0.0055 \\ 0.0795 & 0.0567 & -0.0283 & 0.0795 \end{bmatrix} \tag{9}$$

Used the association degree matrix K and Weight Vector w, we can calculate the association degree vector K' of R_0 in relation to each outrigger type.

$$K' = w \times K = [0.0752, 0.0455, -0.1001, -0.2586] \tag{10}$$

According to Biggest Association Degree Criterion, we can judge that the R_o belongs to the left half outrigger action from Vector K'. This was consistent with the fact.Using the same method to recognize the other three sets of data in Table 4. The recognition results are all correct. Many experimental results show that the average recognition rate of this method is 93.2%. It is quite high.

5   Conclusion

This paper proposed an upper limb action recognition method based on acceleration median (EULAR-AM). EULAR-AM choosing the acceleration median of the arm outstretching process and the direction of acceleration at the initial moment of arm outstretching as the recognition characteristic values. It can reduce the affection of outstretching speed to the characteristic value of upper limb action. According to the request of extension recognition method, EULAR-AM expands the selected characteristics to interval form. The experimental results show that the EULAR-AM having the very good recognition rate.

References

1. Mitra, S., Acharya, T.: Gesture Recognition: A Survey. IEEE Transactions on Systems, Man, and Cybernetics-Part C: Applications and Reviews 37(3), 311–324 (2007)
2. Rashid, O., Al-Hamadi, A., Michaelis, B.: A framework for the integration of gesture and posture recognition using HMM and SVM. In: IEEE International Conference on Intelligent Computing and Intelligent Systems, ICIS 2009, pp. 572–577 (2009)
3. Duda, R.O., Hart, P.E.: Pattern Classification and Scene Analysis. Wiley, New York (1973)
4. Akl, A., Valaee, S.: Accelerometer-based gesture recognition via dynamic-time warping, affinity propagation & compressive sensing. In: 2010 IEEE International Conference on Acoustics Speech and Signal Processing (ICASSP), pp. 2270–2273 (2010)

5. Junqi, K., Hui, W., Guangquan, Z.: Gesture recognition model based on 3D accelerations. In: 4th International Conference on Computer Science & Education, ICCSE 2009, pp. 66–70 (2009)
6. Changxi, W., Xianjun, Y., Qiang, X.: Motion Recognition system for Upper Limbs Based on 3D Acceleration Sensors. Chinese Journal of Sensors and Actuators 23(6), 816–819 (2010)
7. Yang, C., Cai, W.: Extension Engineering. Science Press, Beijing (2007)
8. Yang, C., Cai, W.: Extension Engineering Methods. Science Press, Beijing (2003)
9. Cai, W., Yang, C., He, b.: Preliminary Extension Logic. Science Press, Beijing (2003)
10. Yuan, F., Cheng, T., Zhou, S.: Extension pattern recognition method based on interval overlapping degree. Moden Manufacturing Engineering (9), 139–142 (2010)

Research on Secure Multi-party Computational Geometry

Tao Geng[1,2,3], Shoushan Luo[1,2,3], Yang Xin[1,2,3], Xiaofeng Du[1,2,3], and Yixian Yang[1,2,3]

[1] Information Security Center, Beijing University of Posts and Telecommunications, Beijing
[2] National Engineering Laboratory for Disaster Backup and Recovery, Beijing
[3] Beijing Safe-Code Technology Co. Ltd., Beijing
taogeng@bupt.edu.cn

Abstract. With rapid growth of the internet, a plenty of collaboration opportunities exist when organizations and individuals share information or cooperate to compute in a distributed system. The cooperation may occur between mutually untrusted parties or competitors. This problem is referred to Secure Multi-party Computation (SMC) problem. Secure multi-party computational geometry (SMCG) is a new field of SMC. Due to the widely potential applications, the research of SMCG becomes more and more attractive. In this paper, research frameworks of SMCG are demonstrated. From an applied viewpoint, some basic and key techniques for building secure protocols to solve SMCG problems are described. At the same time, future research orientations of the SMCG are proposed in the end.

Keywords: Cryptograph, secure multi-party computation, computational geometry.

1 Introduction

With rapid growth of the internet, cooperative computation becomes more and more popular. The data used in computation are usually confidential or private. These computations may occur between trusted partners. But in the situation that the participants do not completely trust each other or even competitors, for privacy consideration, the participants just want to get the final result of the cooperative computation without revealing any private input. In fact, the situation in which the participants cooperate to compute but do not trust each other is very common. The secure multi-party computation (SMC for short) was introduced by Yao [1]. Goldwasser [2] once predicted that "the field of multi-party computations is today where public-key cryptography was ten years ago, namely an extremely powerful tool and rich theory whose real-life usage is at this time only beginning." Secure multi-party computational geometry (SMCG for short) is a new field of SMC. The main research of SMCG is how users take their private geometry information as input to gain the results through cooperative computation in distributed systems without leakage of their own input information. Problems about secure multi-party computational geometry are intersection problem, point-inclusion problem, range searching problem, closest pair problem and convex hulls problem, and so on [3].

B. Liu and C. Chai (Eds.): ICICA 2011, LNCS 7030, pp. 322–329, 2011.

The paper is organized in the following way: Section 2 demonstrate the research frameworks of SMCG. Section 3 discusses the work about SMCG. Section 4 shows some key building blocks or basic protocols for solving SMCG problems. Finally Section 5 conclusions the research aspects of SMCG and give out the future research orientations of SMCG.

2   Research Frameworks of SMCG

At present,the frameworks of the research of SMCG can be specified the following three types.

If trusted third party (TTP for short) exists, problems resolved by secure multi-party computation protocols can be easily solved: each party just gives his own private data to TTP, and then TTP computes the result and gives the result back to each party. As seen in Figure 1, there are n parties, $P_1, P_2, ..., P_n$, each one owning private input x_i which will be used in computation $f(x_1, x_2, ..., x_v)$. Using the anonymizer A_i to hide the identity and input of the party. TTP computes and announce the result of the computation. Each party can communicate with a trusted anonymizer which is a system that acts as an intermediately between the party and the TTP. But it is very hard to find such a TTP in reality.

In some multi-party computation methods, an untrusted third party (UTP for short) (seen in figure 1) is proposed for efficiency. An UTP is semi-honest, which means that one who follows the protocol properly with the exception that it keeps a record of all intermediate computations in an attempt to later derive additional information. With the aid of UTP, multi-party computation problems can be solved securely.

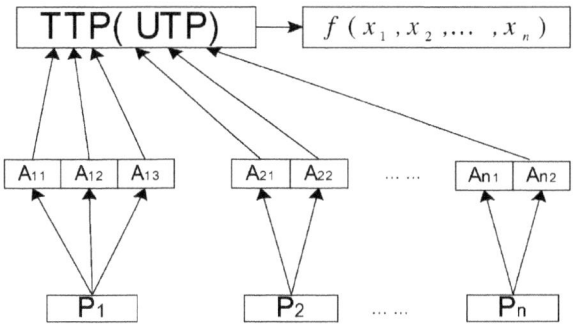

Fig. 1. TTP (UTP) architecture

Another discussed framework is transformed from normal multi-party computations, seen in Figure 2. In this framework, the input from each party is considered as private, and nobody wants to disclose its own input to other parties. Each party can use but not obtain other parties' inputs to compute the result. Under this framework, many solutions to SMCG problems have been developed.

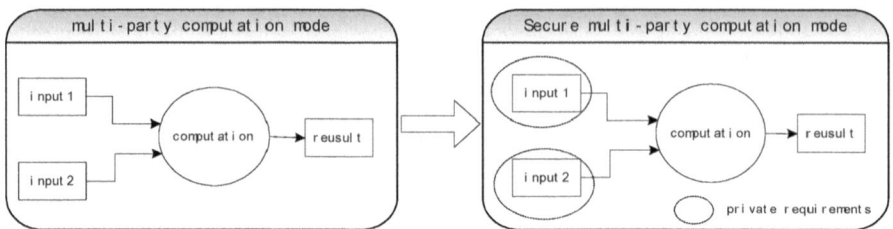

Fig. 2. Transformed framework

3   Related Work

SMCG is a special kind of SMC problem. It mainly investigates information security in computational geometry, especially preserving privacy of data between cooperative parties in distributed systems. Specifically, the aim of the research is to construct corresponding protocols to satisfy each party can use but not obtain the values of other parties such as point, line, segment and polygon and so on, during the computation.

Atallah *et al.* first presented the SMCG problem and developed a secure protocol for determining whether two polygons were intersectant [4]. At the beginning, they proposed two secure scalar product protocols by the oblivious transfer (OT for short) protocol and the homomorphic encryption scheme separately. But the communication cost is very high. As pointed by them, their solution could be further improved.

In [5], Luo *et al.* designed a protocol for determining whether two set of data were proportional correspondingly with the aid of secure scalar product protocol and secure equality-testing protocol, and solved the relative position determination problems for two spatial geometric objects such as point, line and plane and so on. In [6], they constructed a protocol for privacy-preserving intersect-determination of two line segments and an efficient probability algorithm for privacy-preserving intersect-determination of two polygons. A more secure and efficient protocol for the point-circle inclusion problem based on the homomorphic encryption scheme and private comparison protocol was proposed in [7]. By utilizing secure scalar product protocol and secure comparison protocol, a privacy-preserving cross product protocol was presented in [8], which is used to solve segment-intersection problem in computational geometry. Li *et al.* solved two private points distance, relations of point-circle and point-ellipse based on the Yao's millionaire problem and 1-out-of-n oblivious transfer protocol in [9]. Another solution to the problem of polygon-intersection was presented in [10].Based on secure comparison protocol, Monte Carlo approach and Cantor encoding, an approximate secure multi-party computation solution to graph-inclusion problem was solved with low communication complexity [11]. Applying Cantor encoding and the commutative encryption scheme, Li *et al.* proposed solutions to point-inclusion problem and intersection problem of rectangles [12]. With the aid of Homomorphic encryption scheme and secure scalar product protocol, a secure square of two real numbers sum protocol, a secure two real numbers relation determination protocol and a secure point's distance computation protocol were proposed to solve line-circle relation determination problem and circle-circle relation determination problem [13]. In [14], a privacy-preserving point-line relation determination protocol based on additive homomorphism encryption was developed. This protocol can be used to solve segments intersection problem and point inclusion problem. By using secure scalar product protocol, a privacy-preserving triangle inequality determination protocol was proposed in [15] to solve secure triangle shape determination problem and secure vector comparison problems. In [16], based on secure scalar product protocol and millionaire's protocol, a segment-ellipse intersect-determination protocol was presented to solve the determination problem of segment-ellipse intersection.

Table 1. Protocols and building blocks for specific SMCG problems

Problems	Protocols	Building blocks
Relative-position determination	Corresponding proportion protocol	Secure scalar product protocol; Secure equality-testing protocol
Polygon intersection	Intersect-determination protocol for segments and polygons	Secure scalar product protocol; Vector dominance protocol
	Intersection protocol for polygons	OT protocol
point-circle inclusion	Point-circle inclusion protocol	Homomorphic encryption scheme; private comparison protocol
segment-intersection	Privacy-preserving cross product protocol	Secure scalar product protocol; Secure comparison protocol
points distance	Distance protocol between two private points	Yao's millionaire problem; 1-out-of-n OT protocol
graph-inclusion	An approximate secure multi-party computation solution to graph-inclusion	Secure comparison protocol; Monte Carlo approach; Cantor encoding
Rectangle-intersection	Rectangle-intersection protocol	Cantor encoding; Commutative encryption scheme
Line- and circle-circle relation	Square of two real numbers sum protocol; Two real numbers relation protocol; Secure points distance computation protocol	Homomorphic encryption scheme; Secure scalar product protocol
Point-line relation determination	Privacy-preserving point-line relation determination protocol	Additive homomorphism encryption
triangle shape determination	Privacy-preserving triangle inequality determination protocol	Secure scalar product protocol
Segment-ellipse Intersection	Segment-ellipse intersect-determination protocol	Secure comparison protocol; Millionaire's protocol

4   Key Building Blocks

For a specific SMCG problem, the solution is to divide the problem into several steps, each of which can be considered as a sub-problem, and one may put some building blocks together to form a protocol to solve the sub-problems. In this section, we give out some key basic protocols used by solving SMCG problems.

4.1   The Circuit Evaluation Protocol

In a circuit evaluation protocol [17], each function is represented by a Boolean circuit and the construction takes the Boolean circuit and produces a protocol for evaluating it. The protocol scans the circuit from the input wires to the output wires, processing a single gate in each basic step. When entering each basic step, the parties hold shares of the values of the input wires and when the step is completed they hold shares of the output wire. Theoretically speaking, the general SMC problem can be solved by circuit evaluation protocol. But considering the communication complexity of the protocol, using the circuit evaluation protocol can be impractical.

4.2   Oblivious Transfer

The oblivious transfer (OT) is a fundamental primitive in secure protocols design, especially in designing SMCG protocols. The OT is a protocol typically involving two players, the sender and the receiver, and some parameters. The 1-out-of-2 OT proposed in [18] is a protocol where one party, the sender has input composed of two strings (M_0, M_1), and the input of another party, the receiver, which is a bit σ. The receiver should learn M_σ and nothing about $M_{1-\sigma}$ while the sender should gain no information about σ. In the most used form, the 1-out-of-N OT [19] which is also a straightforward generalization of 1-out-of-2 OT refers to a protocol where at the beginning of the protocol on party Alice has N inputs $x_1, x_2, \ldots x_n$, and at the end of the protocol the other party Bob learns one of the inputs x_i for some $1 \le i \le n$ for his choice, without learning any other inputs and allowing Alice to learn about i.

4.3   Homomorphic Encryption Schemes

A homomorphic encryption scheme (HES for short) means that someone can compute the encrypted data without knowing the value of the data. The formal description of HES is as follows:

Public key. The HES is public-key encryption scheme in that anyone with the public-parameter of the scheme can encrypt, but only one with the private-parameter can decrypt.

Semantically secure. The HES is semantically secure defined in [20]. That is, given the following game between a probabilistic polynomial time adversary A and a challenger C:

a. C creates a pubic-private key pair (E,D) for the system and sends to A.

b. A generates two messages M_0 and M_1, and sends back to C.

c. C picks a random bit $b \in \{0,1\}$, and sends $E(M_b)$ to A.

d. A outputs a guess $b' \in \{0,1\}$.

If $b' = b$, A wins the game. We define the advantage of A for a security parameter k to be $adv_A(k) = Pr[b = b'] - (1/2)$. If $adv_A(k)$ is negligible in k, we say the scheme is semantically secure.

Additive. Given the $E(x)$, $E(y)$ and the public-key of the additive HES, one can compute $E(x+y)=E(x)\otimes E(y)$, especially obtain $E(cx)=E(x)^c$.

Multiplicative. Given the $E(x)$, $E(y)$ and the public-key of the multiplicative HES, one can compute $E(xy)=E(x)\otimes E(y)$.

Re-encryption. When given an encryption $E(x)$, one can compute another encryption of x, by multiplying the original encryption by $E(1)$ in multiplicative HES or $E(0)$ in additive HES.

4.4   Scalar Product Protocol

Secure scalar product problem can be described as follows: Alice has a vector $X=(x_1,...,x_n)$ and Bob has a vector $Y=(y_1,...y_n)$. After secure computation, Alice obtains

$$w = XY + r = \sum_{i=1}^{n} x_i y_i + r ,$$

where r is a random number that only Bob knows. The secure scalar product protocol satisfies: Alice can not be able to derive the result of XY or any y_i from w, and the execution of the protocol; Bob can not be able to get the result of w (or XY) or any x_i and the execution of the protocol. There have been several protocols to solve this scalar product problem. One type of the solution is to use a UTP with the assumption that the UTP should not collude with either Alice or Bob. Another type of the solution is not to use a third party.

4.5   Comparison Problem

Generally speaking, comparison problem means Alice owns a private input x and Bob has a privacy input y. They want to compare which is larger or whether they are equal, without disclosing any information of their secret inputs. This problem is first proposed as the millionaires' problem [1], where the solution is exponential in the number of bits of the number involved. Thereafter, with a third party, a constant-round protocol based on the ϕ-hiding assumption is presented in [21]. Recently, many efficient comparison protocols such as [22] and [23] have been developed without third party, partly due to the useful applications as building blocks or sub-protocols.

5   Conclusions

For the widely potential application in various fields, the research of SMCG becomes more and more popular. Not limited to SMCG problems, it is impossible to give out a generally solution to solve all problems with considering the efficiency of the computation. So, for solving different problems, one must design different protocols.

Generally speaking, one can divide a whole SMCG problem into several sub-problems, each of which can be solved by putting some related building blocks together step by step. Among current research, many protocols overfull depend on the existed building blocks of SMC protocols, such as secure scalar product protocol, secure comparison protocol and so on. The computation and communication complexity of these protocols depend on the complexity of the existed building blocks. At the same time, the current research of SMCG are limited to designing protocols of privacy-preserving geometric elements relation determination, and not referred to complex geometric computation.

Not only for SMCG,but also for other SMC,the future research mainly focus on the following aspects: 1) Design an efficient and secure SMCG protocol which can compute arbitrary functions so that SMCG problems as many as possible can be solved by this protocol. 2) Modify the general SMCG protocols and get rid of the part which is meaningless to a specific application. The main research is to solve a specific problem by using SMCG theory and technology. 3) Construct some efficient and secure building blocks or sub- protocols as basic elements to solve SMCG problems. 4) The current research of SMCG is under the semi-honest situation, the future research may consider the SMCG under the malicious situation where any participator may not follow the procedure of the protocol, and may terminate the protocol randomly or modify any intermediate results.

Acknowledgments. This paper is supported by National Basic Research Program of China (973 Program) (2007CB311203), National Natural Science Foundation of China (No.61003285), National S&T Major project (2009ZX03004-003-03), and Ministry of S&T Major Project (2009BAH39B02).

References

1. Yao, A.-C.: Protocols for Secure Computation. In: 23rd IEEE Symposium on Foundations of Computer Science, pp. 160–164 (1982)
2. Goldwasser, S.: Multi-party Computations: Past and Present. In: 16th Annual ACM Symposium on Principles of Distributed Computing, pp. 1–6 (1997)
3. Du, W.-L., Atallah, J.M.: Secure Multi-party Computation Problems and Their Applications: A Review and Open Problems. In: New Security Paradigms Workshop, pp. 11–20 (2001)
4. Atallah, M.J., Du, W.: Secure Multi-party Computational Geometry. In: Dehne, F., Sack, J.-R., Tamassia, R. (eds.) WADS 2001. LNCS, vol. 2125, pp. 165–179. Springer, Heidelberg (2001)
5. Luo, Y.-L., Huang, L.-S., et al.: Privacy Protection in the Relative Position Determination for two Spatial Geometric Objects. Journal of Computer Research and Development 43, 410–416 (2006)
6. Luo, Y.-L., Huang, L.-S., et al.: A Protocol for Privacy-Perserving Intersect-Deterniation of Two Polygons. ACTA Electronica Sinica 35, 685–691 (2007)
7. Luo, Y.-L., Huang, L.-S., Zhong, H.: Secure Two-Party Point-Circle Inclusion Problem. Journal of Computer Science and Technology 22, 88–91 (2007)
8. Luo, Y.-L., Huang, L.-S., et al.: Privace-Preserving Cross Product Protocol and Its Applications. Chinese Journal of Computers 30, 248–254 (2007)

9. Li, S.-D., Dai, Y.-Q.: Secure Two-Party Computational Geometry. Journal of Computer Science and Technology 20, 258–263 (2005)
10. Li, S.-D., Dai, Y.-Q., et al.: Secure mutli-paty computations of geometric intersections. Journal of Tsinghua University 47, 1692–1695 (2007)
11. Li, S.-D., Si, T.-G., Dai, Y.-Q.: Secure Multi-party Computation of Set-Inclusion and Graph-Inclusion. Journal of Computer Research and Development 42, 1647–1653 (2005)
12. Li, S.-D., Dai, Y.-Q., et al.: A Secure Multi-party Computation Solution to Intersection Problems of Sets and Rectangles. Progress in Natural Science 16, 538–545 (2006)
13. Liu, W., Luo, S.-S., et al.: A Study of Secure Two-Party Circle Computation Problem. Journal of Beijing University of Posts and Telecommunications 32, 32–35 (2009)
14. Liu, W., Luo, S.-S., Chen, P.: Privacy-Preserving Point-Line Relation Determination Protocol and Its Applications. Journal of Beijing University of Posts and Telecommunications 32, 72–75 (2008)
15. Luo, S.-S., Liao, G.-C., Liu, W.: A Privacy-Preserving Triangle Inequality Determination Protocol and its Applications. Journal of Beijing University of Posts and Telecommunications 32, 47–51 (2009)
16. Tong, L., Luo, W.-J., et al.: Privacy-Preserving Segment-Ellipse Intersect-Determination Protocol. In: 2nd International Conference on E-Business and Infromation System Security, pp. 1–4 (2010)
17. Goldreich, O.: Secure Multi-party Computation (draft), http://www.wisdom.weizmann.ac.il/~oded/pp.html
18. Even, S., Goldreich, O., Lempel, A.: A Randomized Protocol for Signing Contracts. Communications of the ACM 28, 637–647 (1985)
19. Naor, M., Pinka, B.: Oblivious Transfer and Polynomial Evaluation. In: 31st ACM Symposium on Theory of Computing, USA, pp. 245–254 (1999)
20. Goldwasser, S., Micali, S.: Probabilistic encryption. Journal of Computer and System Sciences 28, 270–299 (1984)
21. Cachin, C.: Efficient private bidding and auctions with an oblivious third party. In: 6th ACM Conference on Computer and Communications Security, pp. 120–127 (1999)
22. Luo, Y.-L., Lei, S., Zhang, C.-Y.: Privacy-Preserving Protocols for String Matching. In: 4th International Conference on Network and System Security, pp. 481–485 (2010)
23. Liu, W., Luo, S.-S., Wang, Y.-B.: Secure Multi-Party Comparing Protocol Based on Muti-Threshold Secret Sharing Scheme. In: 6th International Conference on Wireless Communications Networking and Mobile Computing, pp. 1–4 (2010)

Grid Transaction-Level Concurrency Control Protocol

Jun Chen, Yan-Pei Liu, and Hong-Yu Feng

College of Information Engineering , Henan Institute of Science and Technology,
Xinxiang 453003, China
{47594409,32933415,332934328}@qq.com

Abstract. In this paper, the grid transaction-level (GTL) concurrency control protocol is proposed. According to the operating characteristics of the grid services, the grid transaction-level concurrency control protocol re-defines four types of transaction operations: read operation, write operation ,append operation, look operation. When the speculative shadows of transactions are created , the more redundancy resources in the grid, the higher the priority of transaction, the shorter the deadline of transaction, the more the number of speculative shadows that the transaction can establish; less and vice versa. If there is a new read - write conflict between the two services and a speculation shadow has to meet the needs of serialization order, it don't need create a new speculative shadow. Finally, the theory proof shows that the grid transaction-level concurrency control protocol ensures the serialization of transactions.

Keywords: Grid transaction management GTL concurrency control protocol.

1   Introduction

Grid with parallel, distributed, heterogeneous, scalable and dynamic self-adaptive characteristics such as the environment it needs in a dynamic combination of dynamic flexibility of resources and services[1]. The traditional distributed processing model and Web service transaction processing model to atomic and serializability as the basic features, in order to wait and give up as a fundamental means to no longer apply the database system in grid environment. So it is necessary to transform conventional transaction management mechanisms and expansion, so that it can meet the grid environment.

From the research status, the following problems exist in the grid real-time database[2,3]:

(1) Very few specialized studies to concurrency control and transaction processing models for real-time database in grid environment, to be further developed;

(2) Transaction processing model that isn't pertinent is relevant to the dynamic features in the grid unique environment;

(3) To existing concurrency control in improving success rates, there is still room for improvement.

B. Liu and C. Chai (Eds.): ICICA 2011, LNCS 7030, pp. 330–336, 2011.
© Springer-Verlag Berlin Heidelberg 2011

According to the dynamics in the grid, the grid transaction-level (GTL) concurrency control protocol is proposed. It aims to ensure real-time database transactions in the grid environment to achieve efficient and correct concurrent execution, and as far as possible reduce the number of transaction restart.

2   Grid Transaction-Level (GTL) Concurrency Control Protocol

2.1   Improved Thinking

According to the characteristics of real transaction in grid, this paper focuses on studying grid real-time transaction model and methods of concurrency control, and the GTL protocol that adapts the real-time transaction in the grid is proposed by improving speculative concurrency control protocol (SCC) [4].

Improvement ideas are as follows: before the projection, the transaction's type is refined to further expansion. In the implementation process, on the one hand according to resources' status in grid and the time information of real-time transaction, the protocol dynamiclly calculates of the maximum number of transaction's shadow; the other hand, it as concise as possible to set speculative shadow. In the submission, because of the introduction of a quasi-commit phase, the protocol as much as possible to reduce transaction restart without prejudice to submitting transaction.

The following describes the improvement of the SCC and the content of the protocol.

2.2   The Division of Transaction Operation

According to the traditional concurrency control protocol, the operations of transaction are divided into two categories:read operation and write operation.It's so likely to result in some unnecessary restarts of transaction. In order to improve the concurrency rate, the operations of transaction will be divided into four categories in this article[5]:

(1)Read operation (Read): data from the database to read out and applied in their own operation as an important parameter, so it needs to read the correct data;

(2) Write operation (Write): Some original data on the database is to be modified;

(3) Append operation (Append): It's no effect on original data in the database, and only to add new data;

(4) Look operation (Look): data from the database only need a general understanding, without affecting the correctness of operation.

Look operation can read the dirty data which is not submitted, so it does not cause any type of conflict. It can be directly implemented. When an append operation on the database, it means adding a new data into its database. No conflict with any type of transaction operations, it can be directly implemented.

2.3   Calculating the Number of Shadow

Sharing real-time database grid provides a large number of redundant resources, the agreement provides for the realization of the premise of ensuring SCC. The dynamics of the grid nodes can be allowed to join freely or withdraw from, and it may change the number of available resources. Therefore, the grid environment, the system can use redundant resources is also changing. In the real-time database, the system can use redundant resources to change constantly. At the same time, the various panels have different time limits and different importance. Therefore, when N is set to conduct a dynamic calculation, to make the N value with the transaction information and the system state varies [6]. Calculate the value of transaction T, a function of N is as follows:

$$N(T) = f(P(T), D(T), R(T)) = \left\lfloor \frac{P(T) + e^{\frac{R(T)}{R(T)_0}}}{K \times D(T)} \right\rfloor \quad (1)$$

Where, k is a constant coefficient, e is a constant, R0 is the average amount of redundant resources; P (T) on behalf of priority of transaction T, D (T) on behalf of deadline information of transaction T, R (T) on behalf of transaction T run-time can make use of redundant resource information.

As the GTL algorithm makes use of the system, redundant resources to build replacement schedule, so services can be established in the shadow of the number of redundancy and system resources, and the number has a direct relationship. The more redundant resources the system, the higher the priority; the shorter the deadline, the greater the N value, that is, the transaction can be established, the more the number of shadow; vice versa.

2.4   The Production Rule of Shadow

In this paper,a shadow can be in one of two modes: optimistic or speculative[7]. The establishment of shadow is actually to meet the different serialization order[4]. There are only two serialization order between Ti and Tj: Ti before Tj, or Tj before Ti. The optimistic shadow Tio of Ti corresponds to the first implementation of serialization order.For each conflict Services Tj, it only needs a speculative shadow of Ti which corresponds to the second serialized order. On the premise of ensuring transaction scheduling, it is as far as possible to reduce the number of shadow to enhance the system performance. Dealing with different situations, we have the production rule of shadow as follows.

Assume that a new read - write conflict is detected between transaction Ti and Tj, the first we should dynamically calculate the value of N (Ti). If N (Ti) is equal to 1, that is to say, the transaction is not allowed to set up the speculative shadow, the system 'll ignore the conflict to continue. Otherwise, according to the following steps[8,9]:

(1) If there is no the speculative shadow of Ti that corresponds to Tj, then:

① If there is a shadow Tik of Ti that does not execute the read operation of the conflict, the conflict is ignored to continue,as show in the figure 1;

② There is no shadow Tik of Ti that does not execute the read operation of the conflict. When the number of speculative shadow is equal to N (Ti), the conflict is ignored to continue; when the number of speculative shadow is not equal to N (Ti), a new speculative shadow of conflict is to be created,as show in the figure 2.

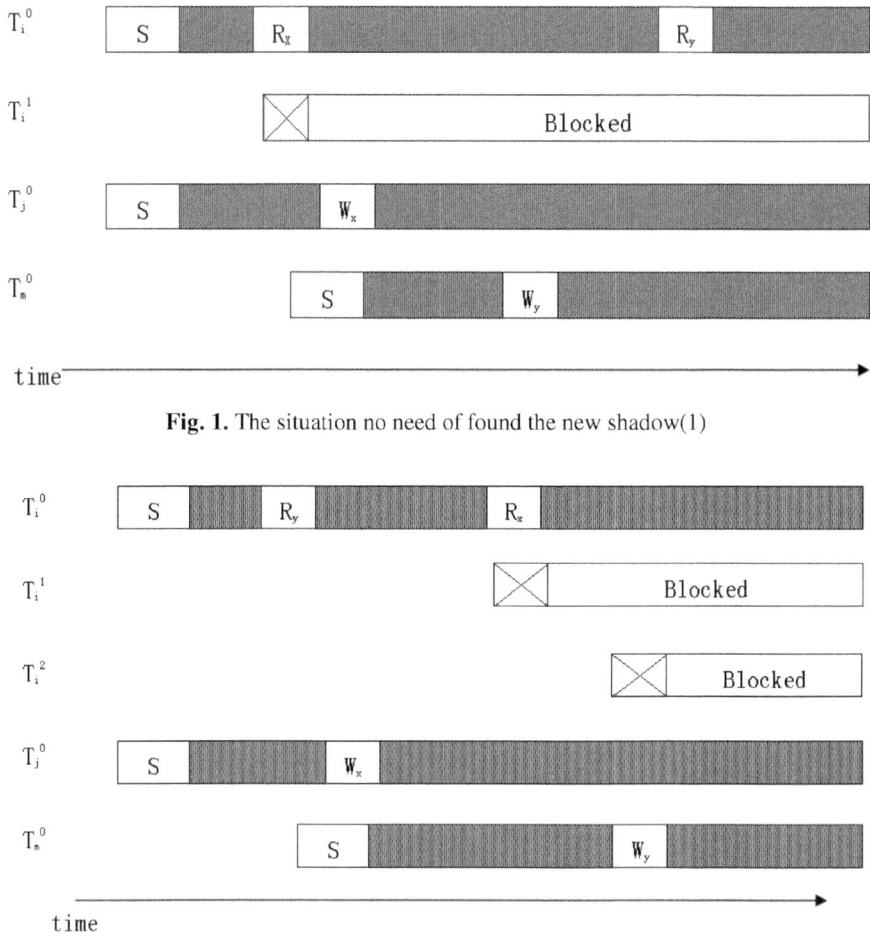

Fig. 1. The situation no need of found the new shadow(1)

Fig. 2. The situation need of found the new shadow(1)

(2) If there is a speculative shadow of Ti that corresponds to Tj, then:

① Before the first read - write conflict is detected and Ti does not execute the read operation of other conflicts, the new conflict is ignored to continue,as show in the figure3;

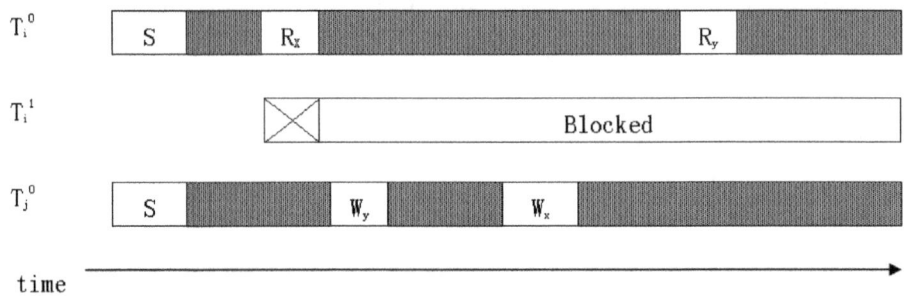

Fig. 3. The situation no need of found the new shadow(2)

② Before the first read - write conflict is detected and Ti has executed the read operation of other conflicts, the shadow performing other conflicts is interrupted ,and a new speculative shadow is to be created, ,as show in the figure 4.

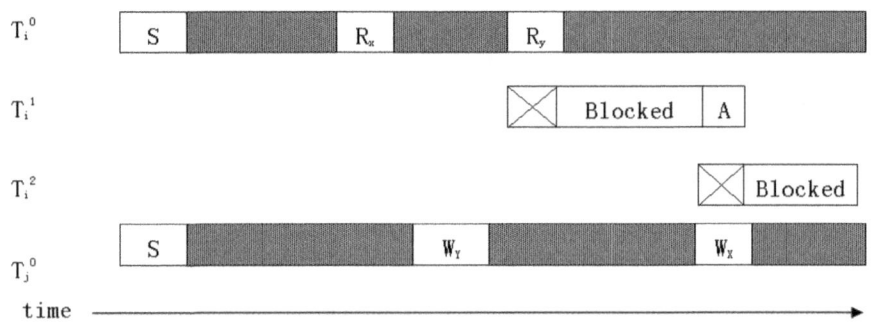

Fig. 4. The situation need of found the new shadow(2)

3   The Serializability Proof of GTL Algorithm

In GTL Algorithm ,the establishment of shadow is actually to meet the different serialization order. There are only two serialization order between Ti and Tj: Ti before Tj, or Tj before Ti. The optimistic shadow Tio of Ti corresponds to the first implementation of serialization order[10].For each conflict service Tj, it only needs a speculative shadow of Ti which corresponds to the second serialized order.

Theorem: The GTL algorithm ensures the serialization of transactions.

Proof: The method of induction and hypothesis.

Suppose a transaction group T={T1...TN} follows the GTL algorithm to perform[11]:

(1) When N = 1, the group which is only a single transaction T1 is serializable certainly.

(2) Suppose N = n, n transactions (T1...Tn) that follow the GTL algorithm to perform can produce a serializable sequence.Then, when N = n +1, there are two situations as following.

①If transaction Tn+1 doesn't conflict with other transactions (T1...Tn), they can produce a serializable sequence whenever Tn+1 commits.

②If transaction Tn+1 conflicts with other transactions (T1...Tn), there are two situations as following:

a) Optimistic shadow T0n+1 has committed. According to the implementation rule of optimistic shadow in GTL algorithm, it indicates that transaction Tn+1 has completed before other panels: T1 ... Tn. This can produce a serializable sequence.

b) Speculative shadow Tkn+1 has committed. According to the production rule of speculative shadow in GTL algorithm, every speculative shadow corresponds to a possible serializable sequence.The committing of Tkn+1 indicates that the serializable sequence comes true.

In a word, transactions that follow the GTL algorithm to perform can produce a serializable sequence, that is to say, the GTL algorithm ensures the serialization of transaction.

4   Conclusion

In this paper, the grid transaction-level concurrency control protocol is proposed. According to the operating characteristics of the grid services, the GTL re-defines four types of transaction operations: read operation, write operation ,append operation, look operation. When the speculative shadows of transactions are created , the more redundancy resources in the grid, the higher the priority of transaction, the shorter the deadline of transaction, the more the number of speculative shadows that the transaction can establish; less and vice versa. If there is a new read - write conflict between the two services and a speculation shadow has to meet the needs of serialization order, it don't need create a new speculative shadow. Finally, the theory proof shows that The GTL algorithm ensures the serialization of transactions.

References

1. Tang, F.-L., Li, M.-L., Huang, Z.-X., Wang, C.-L.: A Transaction Service for Service Grid and Its Correctness Analysis Based on Petri Net. Chinese Journal of Computers 28(04), 667–676 (2005)
2. Taniara, D., Goel, S.: Concurrency control issues in Grid databases. Future Generation Computer Systems 23, 154–162 (2007)
3. Goel, S., Sharda, H., Taniar, D.: Atomic Commitment in Grid Database Systems. In: Jin, H., Gao, G.R., Xu, Z., Chen, H. (eds.) NPC 2004. LNCS, vol. 3222, pp. 22–29. Springer, Heidelberg (2004)
4. Bestavros, A., Braoudakis, S.: Value-cognizant speculative concurrency control. In: Proceedings of the 21th International Conference on Very Large Data Bases, Zurich, Switzerland, pp. 122–133 (September 1995)

5. Ma, X., Li, T., Li, W.: Concurrency Control Method for Grid Transaction Based on Timestamp. Computer Engineering 33(20), 86–88 (2007)
6. Gray, J., Reuter, A.: Transaction Processing Concepts and Techniques, 1st edn. Morgan Kaufmann, San Francisco (1993)
7. Bestavros, A., Braoudakis, S.: A Family of Speculative Concurrency Control Algorithms. Technical Report TR-92-017, Computer Science Department, Boston University, Boston, pp. 1–21 (July 1992)
8. Chen, J., Gu, Y.-s.: A Grid Real-time Transaction Concurrency Control Protocol and The Analysis Based on Petri Net. Journal of Henan Normal University (Natural Science Edition) 39(01), 49–52 (2011)
9. Wang, Y., Wang, Q., Wang, H., Dai, G.: Dynamic Adjustment of Execution Order in Real-Time Databases. In: Proceedings of the 18th International Parallel and Distributed Processing Symposium, April 26-30, vol. (1), p. 87. IEEE Computer Society, Santa Fe (2004)
10. Xia, J.-l., Han, Z.-b., Chen, H.: Conflict-free Concurrency Control Protocol Based on Function Alternative Model. Computer Engineering 36(15), 57–59 (2010)
11. Chen, J., Mu, X.-x., Li, L.: The Transaction Model for Real-Time Database in Grid Environment. Journal of Henan Normal University 37(5), 29–32 (2009)

Wavelet Theory and Application Summarizing

Hong-Yu Feng, Jian-ping Wang, Yan-cui Li, and Jun Chen

Infomation Engineering College, Henan Institute of Science and Technology,
Xinxiang City, Henan Province, China
Feng_hongyu@126.com

Abstract. The wavelet analysis is the development based on Fourier transform. And it is a breakthrough of the Fourier analysis. The paper reviews the development history of wavelet analysis. Then it discusses the similarities and differences of Fourier transform and wavelet theory. As a new transform domain signal processing methods, wavelet transform is particularly good at dealing with non-stationary signal analysis. As wavelets are a mathematical tool, they can be used to extract information from many different kinds of data, including - but certainly not limited to - audio signals and images. In the end the paper summarizes and discusses applications of wavelet in various domains in detail.

Keywords: Wavelet, Analysis, Fourier, Time Domain, Frequency Domain.

1 Introduction

Wavelet analysis is a new subject rapidly developed from the 1980's of last century, and it is a breakthrough of the Fourier analysis. It also has double meaning of the theory deep and a wide range of application. In the application of wavelet, although not lack of good example, but widely application and the depth solve to the problem is developed along with the further development of wavelet analysis theory and algorithm. The solution of many practical problems promotes the development of the theory of wavelet. When we analyze our signal in time for its frequency content, unlike Fourier analysis, in which we analyze signals using sins and cosines, now we use wavelet functions.

As wavelets are a mathematical tool they can be used to extract information from many different kinds of data, including - but certainly not limited to - audio signals and images. Sets of wavelets are generally needed to analyze data fully. A set of "complementary" wavelets will deconstruct data without gaps or overlap so that the deconstruction process is mathematically reversible. Thus, sets of complementary wavelets are useful in wavelet based compression/decompression algorithms where it is desirable to recover the original information with minimal loss.

2 The Development History of Wavelet Analysis

The first known connection to modern wavelets dates back to a man named Jean Baptiste Joseph Fourier. In 1807, Fourier's efforts with frequency analysis lead to

B. Liu and C. Chai (Eds.): ICICA 2011, LNCS 7030, pp. 337–343, 2011.

what we now know as Fourier analysis. His work is based on the fact that functions can be represented as the sum of sins and cosines. Another contribution of Joseph Fourier's was the Fourier Transform. It transforms a function f that depends on time into a new function 'f hat,' which depends on frequency.

The next known link to wavelets came from a man named Alfred Haar in the year 1909. Haar's contribution to wavelets is very evident. The Haar wavelets are the simplest of the wavelet families. The concept of a wavelet family is easy to understand. The father wavelet is the starting point. By scaling and translating the father wavelet, we obtain the mother, daughters, sons, granddaughters etc.

After Haar's contribution to wavelets there was once again a gap of time in research about the functions until a man named Paul Levy. Levy's efforts in the field of wavelets dealt with his research with Brownian motion. He discovered that the scale-varying basis function – created by Haar (i.e. Haar wavelets) were a better basis than the Fourier basis functions. Unlike the Haar basis function, which can be chopped up into different intervals – such as the interval from 0 to 1 or the interval from 0 to ½ and ½ to 1, the Fourier basis functions have only one interval. Therefore, the Haar wavelets can be much more precise in modeling a function. If we were to project a function onto V3 it would be constant on eighths. On the other hand, we could project the same function onto V4, which is constant on sixteenths, leading to an even closer projection of the original function. Thus, the Haar basis functions seemed to be a better tool for Levy while dealing with the small details in Brownian motion.

Even though some individuals made slight advances in the field of wavelets from the 1930's to the 1970's, the next major advancements came from Jean Morlet around the year 1975. In fact, Morlet was the first researcher to use the term "wavelet" to describe his functions. More specifically, they were called "Wavelets of Constant Slope." Before 1975, many researchers had pondered over the idea of Windowed Fourier Analysis (mainly a man named Dennis Gabor). This idea allowed us to finally consider things in terms of both time and frequency. Windowed Fourier analysis dealt with studying the frequencies of a signal piece by piece (or window by window). These windows helped to make the time variable discrete or fixed. Then different oscillating functions of varying frequencies could be looked at in these windows.

The next two important contributors to the field of wavelets were Yves Meyer and Stephane Mallat. The two first met in the United States in 1986. They spent three days researching work being done on wavelets in many different applied fields. At the end of their research, multiresolution analysis for wavelets was born. This idea of multiresolution analysis was a big step in the research of wavelets. It was where the scaling function of wavelets was first mentioned, and it allowed researchers and mathematicians to construct their own family of wavelets using its criteria.

3   From the Fourier Transform to the Wavelet Analysis

3.1   Fourier Transform

In signal analysis, the basic characterization of the signal often takes two basic forms: time domain and frequency domain. The time or spatial location as independent variables, and the signal characteristics of a numerical signal as the dependent variable is commonly used to describe the way, so the range of variables is called time-domain. On the other hand, we often require the transform of frequency domain, which is Fourier transform.

$$f(\omega) = \frac{1}{\sqrt{2\pi}} \int_{-\infty}^{\infty} f(t) e^{-j\omega t} dt.$$ (1)

It identifies the spectrum characteristics of f(t) in the time domain. Fourier transform links the time domain features and frequency characteristics of signal together. In this way, signals can be observed from the time domain and frequency domain, but can't combine the two. Fourier transform is the time domain integration and the identified frequency at what time generated can't be known. So it is a global transformation, doesn't reflect the signal spectrum characteristics in a local time. So it can't reflect any resolution in the time domain. The analysis of the signal is facing a construction: the localized conflicts between time domain and frequency domain.

Fourier transform bring inconvenience and difficulties to analysis mutation signals, such as seismic waves, storms, floods and so on. This has prompted the search for a new time-frequency signal analysis method.

3.2   Short-time Fourier Transform

Short-time Fourier Transform, also known as the windowed Fourier transform, was proposed by the Gabor in 1946. The basic idea is: the signal is divided into many small time intervals, Fourier transform is used to analysis each time interval to determine the existing frequency, in order to achieve the purpose of time-frequency localization. Windowed Fourier transform can achieve a certain degree of time-frequency localization which is appropriate for deterministic smooth signal. Once the window function is determined, the window size and shape are fixed too, so its time and frequency resolution is single. As the frequency is inversely proportional with the period, it requires a higher time resolution (narrow time window) to reflect the high frequency signal components, a lower time resolution (wide time window) to reflect the low frequency signal components. Therefore, the windowed Fourier transform is not valid for the study of high-frequency signals and low frequency signals.

3.3   Wavelet Analysis

Wavelet analysis is a time-frequency localization signal analysis with fixed window size and changeable shape. It means that in the low frequency, it shows higher frequency resolution and lower time resolution. In the high frequency part, it shows higher time resolution and lower frequency resolution.

All the wavelet can be obtained by scale expansion and displacement of basic wavelet. Basic wavelet is a real-valued function with special nature; it is shock attenuation and often decays quickly. The condition of integral to zero is

$$\int_{-\infty}^{\infty} \psi(t) dt = 0.$$ (2)

and its spectrum is in meet of conditions as below.

$$c_\psi = \int_{-\infty}^{\infty} \frac{|\psi(s)|^2}{s} ds < \infty.$$ 　　　　　　(3)

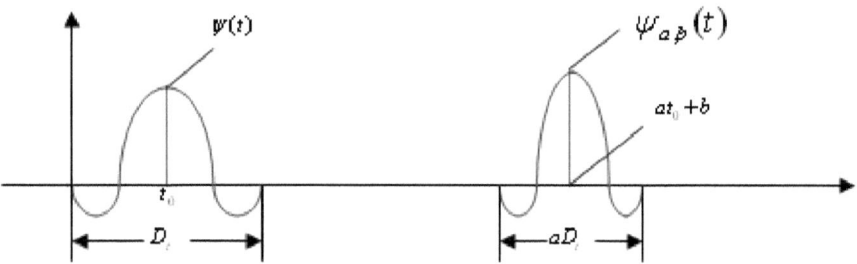

Fig. 1. If base wavelet $\psi(t)$ is function whose center is t0, valid width is Dt, then wavelet's center is at0+b and width is aDt.

The basic wavelet in the frequency domain also has good attenuation properties. In fact some basic wavelet is zero when it is outside a certain range, so this is a kind of wavelet which is decay fastest. A group of wavelet function is generated by scale factor and displacement factor from the basic wavelet.

$$\psi_{a,b}(X) = \frac{1}{\sqrt{a}} \psi\left(\frac{x-b}{a}\right).$$ 　　　　　　(4)

In this function, a is time axis scale expansion coefficient, b is time shift parameter. If the wavelet is considered as a time window with width changing with a and position changing with b, then the continuous wavelet transform can be seen as the collection of a set of short-term consecutive changing Fourier transform. These short-term Fourier transforms use different window function of different width for different signal frequencies. Specifically, high-frequency needs a narrow time domain window and low-frequency needs a wide time domain window. Wavelet transform with this valuable property is called "zoom" nature, "zoom" nature. Wavelet transform can change from wide (low frequency) and narrow (high frequency). That is to say wavelet transform can achieve time-frequency localization with fixed window size and variable shape, in this sense, wavelet transform is known as the mathematical "microscope".

Continuous wavelet has much redundancy, which permits sampling of scale and time shifting parameters. Then wavelet transformed value selected in scattered dot of time-scale plane describe signal, and they regain signal.

Wavelet Transform which is computed after discrete scale and time shifting parameters is called wavelet series coefficient.

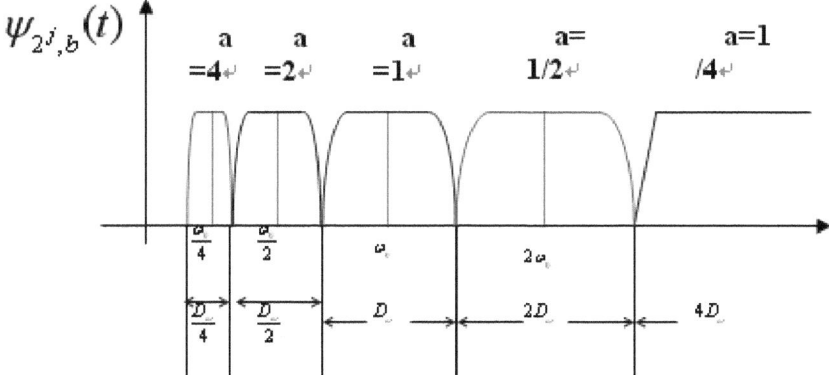

Fig. 2. when time shifting parameter b is sure, whether j is any Interger, wavelet transform keep all frequency spectrum that signal is in a tiny neighbourhood of b time.

3.4   Comparisons with Fourier Transform

The wavelet transform is often compared with the Fourier transform, in which signals are represented as a sum of sinusoids. The main difference is that wavelets are localized in both time and frequency whereas the standard Fourier transform is only localized in frequency. The Short-time Fourier transform (STFT) is more similar to the wavelet transform, in that it is also time and frequency localized, but there are issues with the frequency/time resolution trade-off. Wavelets often give a better signal representation using Multiresolution analysis, with balanced resolution at any time and frequency.

The discrete wavelet transform is also less computationally complex, taking O(N) time as compared to O(N log N) for the fast Fourier transform. This computational advantage is not inherent to the transform, but reflects the choice of a logarithmic division of frequency, in contrast to the equally spaced frequency divisions of the FFT(Fast Fourier Transform). It is also important to note that this complexity only applies when the filter size has no relation to the signal size. A wavelet without compact support such as the Shannon wavelet would require O(N^2). (For instance, a logarithmic Fourier Transform also exists with O(N) complexity, but the original signal must be sampled logarithmically in time, which is only useful for certain types of signals.)

4   The Applications of Wavelet

Fourier analysis, using the Fourier transform, is a powerful tool for analyzing the components of a stationary signal (a stationary signal is a signal that repeats). For example, the Fourier transform is a powerful tool for processing signals that are composed of some combination of sine and cosine signals.

The Fourier transform is less useful in analyzing non-stationary data, where there is no repetition within the region sampled. Wavelet transforms (of which there are, at least formally, an infinite number) allow the components of a non-stationary signal to be analyzed. Wavelets also allow filters to be constructed for stationary and non-stationary signals.

Although Haar wavelets date back to the beginning of the twentieth century, wavelets as they are thought of today are new. Wavelet mathematics is less than a quarter of a century old. Some techniques, like the wavelet packet transform are barely ten years old. This makes wavelet mathematics a new tool which is slowly moving from the realm of mathematics into engineering. For example, the JPEG 2000 standard is based on the wavelet lifting scheme.

The Fourier transform shows up in a remarkable number of areas outside of classic signal processing. Even taking this into account, I think that it is safe to say that the mathematics of wavelets is much larger than that of the Fourier transform. In fact, the mathematics of wavelets encompasses the Fourier transform. The size of wavelet theory is matched by the size of the application area. Initial wavelet applications involved signal processing and filtering. However, wavelets have been applied in many other areas including non-linear regression and compression. An offshoot of wavelet compression allows the amount of determinism in a time series to be estimated.

Generally, an approximation to DWT is used for data compression if signal is already sampled, and the CWT for signal analysis. Thus, DWT approximation is commonly used in engineering and computer science, and the CWT in scientific research.

Wavelet transforms are now being adopted for a vast number of applications, often replacing the conventional Fourier Transform. Many areas of physics have seen this paradigm shift, including molecular dynamics, ab initio calculations, astrophysics, density-matrix localization, seismology, optics, turbulence and quantum mechanics. This change has also occurred in image processing, blood-pressure, heart-rate and ECG analyses, DNA analysis, protein analysis, climatology, general signal processing, speech recognition, computer graphics and multifractal analysis. In computer vision and image processing, the notion of scale-space representation and Gaussian derivative operators is regarded as a canonical multi-scale representation.

One use of wavelet approximation is in data compression. Like some other transforms, wavelet transforms can be used to transform data, then encode the transformed data, resulting in effective compression. For example, JPEG 2000 is an image compression standard that uses biorthogonal wavelets. This means that although the frame is overcomplete, it is a tight frame (see types of Frame of a vector space), and the same frame functions (except for conjugation in the case of complex wavelets) are used for both analysis and synthesis, i.e., in both the forward and inverse transform. For details see wavelet compression.

A related use is that of smoothing/denoising data based on wavelet coefficient thresholding, also called wavelet shrinkage. By adaptively thresholding the wavelet coefficients that correspond to undesired frequency components smoothing and/or denoising operations can be performed.

Wavelet transforms are also starting to be used for communication applications. Wavelet OFDM is the basic modulation scheme used in HD-PLC (a powerline communications technology developed by Panasonic), and in one of the optional modes included in the IEEE P1901 draft standard. The advantage of Wavelet OFDM over traditional FFT OFDM systems is that Wavelet can achieve deeper notches and that it does not require a Guard Interval (which usually represents significant overhead in FFT OFDM systems).

5   Conclusion

Wavelet theory is a major breakthrough in the traditional Fourier transform, which is especially for a short time mutations and non-stationary signal and image processing, and now it has become a powerful tool for cutting-edge applied science. At present, the wavelet analysis theory is developing toward fast calculation in large-scale parallel scientific computing and real-time processing and its application in the production and life are increasingly in-depth and extensive.

References

1. Zhang, J., Zhang, Q.: the Application of Wavelet Analysis in Intrusion Detect System. Computer Safety, 3 (2009)
2. Zheng, W., Cui, Y.-l., Wang, F.: Study Summarizing on Image Compression Coding Based the Wavelet Transform. Communication Technology, 83–96 (2008)
3. Niu, Y.-r.: The Wavelet Analysis and Application Summarizing. Science and Technology Information, 3 (2009)
4. Wang, H.-s., Zhou, J.-h., Wu, W.-b.: The Application of the Wavelet Analysis in Remote Sensing Images Processing. Remote Information, 1 (2009)
5. Li, J.-h., Yan, t.-m., Ma, z.: Face Image Processing Based Wavelet Analysis. Metering and Testing Technology, 41 (2009)
6. Dong, W.-h.: The Realization of Wavelet Analysis in Imaging Compressing Based Matlab. Information Technology, 4 (2009)
7. Heng, T.: Study on the Wavelet Analysis and Application. Sichuan University Doctor Degree Paper (2003)
8. Chen, R.-p.: The Wavelet Theory and Application. Hunan Normal University Master Degree Paper (2006)
9. Yao, T.-r., Sun, H.: Modern Digital Signal Processing. HuaZhong University of Science and Technology Press (1999)
10. Li, X.-h., Zhang, G.-c., Wang, J.-t.: Watermarking Classified DWT Coefficients Algorithm Based on Blending. Computer Engineering and Applications 46, 139–141 (2010)

Research and Application on the Regular Hexagon Covering Honeycomb Model

Aimin Yang[1,2], Jianping Zheng[1], and Guanchen Zhou[1]

[1] Qinggong College, Hebei United University, Tangshan Hebei 063009, China
[2] College of Science, Hebei United University, Tangshan Hebei 063009, China
aimin_heut@163.com

Abstract. To ensure all the users can communicate at the same time, to solve the repeater coordination and the number of repeaters when the region, users and number of PLs are fixed, and to consider the two situations of users evenly distributed and random distributed, we build the regular hexagon honeycomb model based on the evenly distribution and the genetic algorithm model based on the random distribution in this paper. And then we separately get the repeater coordination and provide the improved Honeycomb model suitable for the signal transmission in the mountainous area; solve the problem of the number of the repeaters in the fixed area where a special mountain locates.

Keywords: Honeycomb Model, Repeaters, enetic algorithm model, Communication Transmission.

1 Introduction

As the over-exploitation of natural resources and the damage to the environment intensifies, it is recognized that resources are limited, environmental capacity is limited, and human activity and urban development will be subjected to the constraints of the resources and the environment. The concept of bearing capacity, which is originally a mechanical concept, has been gradually introduced into the study of urban development [1]. Bearing capacity can be better quantified. Therefore, it's been widely studied and used. Resource bearing capacity, environmental bearing capacity, population bearing capacity and other ecological bearing capacities become research hotspots. Tangshan, as a resource-based city, has been developing rapidly in recent years. However, the heavy-industry-based industrial structure exerts enormous pressure on the sustainable development of Tangshan's economy. A scientific evaluation of its ecological bearing capacity is of great practical significance for Tangshan, and also has important reference value for other similar cities in China. In this paper, bearing exponents, which are easily comparable, are used to study the basic condition of Tangshan's ecological bearing capacity. Hopefully, it can provide a scientific reference for the urban development of Tangshan.

Repeaters are the amplifier devices of the same frequency. It is a radio transmitting transshipment equipment, which can strengthen the signals during the wireless transmission. It can pick up weak signals, amplify them and retransmit them on a different frequency, which is shown in Fig. 1. In other words, the basic function of

B. Liu and C. Chai (Eds.): ICICA 2011, LNCS 7030, pp. 344–350, 2011.
© Springer-Verlag Berlin Heidelberg 2011

repeaters is to enhance the RF signal power. In the downlink relay, the repeaters pick up the signal from the donor antenna available in coverage area[4]. Through the isolation of the signals out of the band pass by the bandpass filter, amplify the filtering signal through a power amplifier and then re-launch to the required covering area. In the uplink path, signal of mobile stations in the coverage area is handled by uplink link by using the same way and then launch to the corresponding repeater, which achieves signal transmission from the repeater to the mobile station.

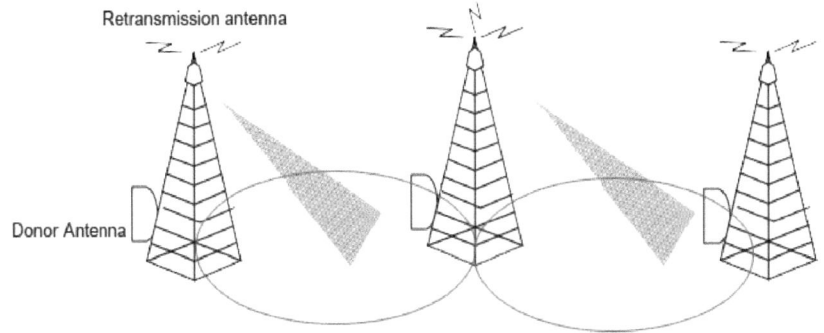

Fig. 1. Communication Transmission between the Repeaters

However, if the distance or frequency between repeaters is nearly, they will interference each other. In order to maintain the spacing between repeaters, we can use PL technology to reduce interference, and many repeaters can be installed to satisfy more users' communications needs[5]. So a new optimization scheme should be constructed. This scheme can satisfy the users' needs and optimize the utilization of resources.

Because of the different structures and functions, there are a lot of types repeaters. We generally divided them according to the direction of the signals sending and receiving, which can be divided into directional and non-stereo specific repeaters. The retransmission antenna of the directional repeater can only use directional radiation and deviate the direction of benefactor antenna. The site is required to be located in the edge of the covered area. The non-stereo specific repeaters regard itself as the circle center; the retransmission antenna radiate to arbitrary direction. If the gains from process are more than the loss of the links, the signal will by amplified continuously and finally produce the self-stimulation. So the isolation should necessarily considered by the repeaters[6]. In this paper, we choose the directional repeater, which shown in Fig.2. The benefactor antenna receives the rf signals sent from the base station and through the sending-receiving diplexer, amplify the low noise firstly, and then transform it to its original rf band to get the needed band width rf signals; finally amplify the power and then the retransmission antenna will send it to mobile station. The retransmission antenna receive the rf signals from mobile station and make the signals go through the sending-receiving diplexer, first amplify the low noise; then transform it to its original rf band to get the needed band width rf signals; finally amplify the power and then the benefactor antenna will send it to mobile station.

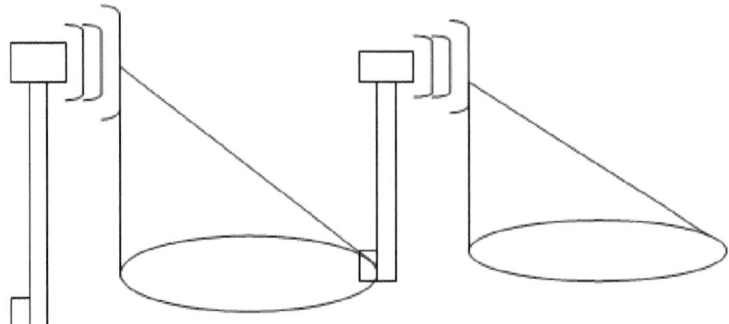

Fig. 2. The Schematic Diagram of the Directional Repeaters

One PL will be occupied when the repeater spread a signal, so some more factors should be considered. The midway repeaters will be occupied if two users far apart contact. The repeater in the middle position will be occupied with the maximum probability. When the users are fewer, we consider the repeaters combination with one frequency pair[7]. If there are more users, we should consider the repeaters combination with many frequency pairs.

With the progress of human science and technology, the network and communication technologies have been rapidly developed. At present, the users can develop the communication with each other from face to face to sending signals to the repeater, which can solve the problem of the users are far away. Of course, the usage of indispensable repeaters has become one of important topics in the present information age[1-3]. And one of the important unsolved mathematical problems is: when the geographic range and number of users are known, find the minimum number of required repeaters.

2   The Design of Hexagonal Honeycomb Model under the Even Distribution

First, based on the multiple repeaters covering the region of the users, ensure all users can communicate with each other, and then the situations of the square and hexagonal coverage were studied[8]. The arrangement is shown in Fig. 3:

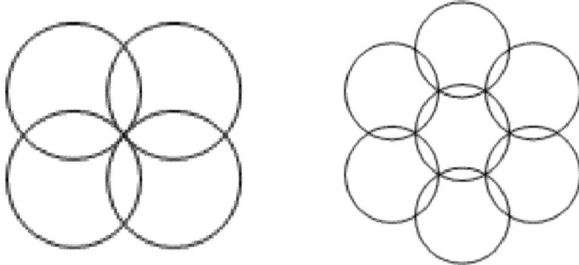

Fig. 3. The Arrangements of the Square and Regular Hexagon

The main features of the coverage are as follows:

(1) The interval of repeaters: set the coverage area's radius is r, and then we can get the center radiuses of the square and regular hexagon are $\sqrt{2}r$ and $\sqrt{3}r$ respectively. It can be seen that the repeaters of regular hexagon has the largest interval.

(2) The overlap area between repeaters: The unit coverage areas of repeaters and overlap areas vary with the regular polygon. Table 1 shows the area results of square and regular hexagon.

Table 1. The Analysis of Coverage of Square and Regular Hexagon

The Shape of Repeaters	Square	Regular Hexagon
The Interval of Repeaters	$\sqrt{2}r$	$\sqrt{3}r$
The Effective Area of Repeaters	$2r^2$	$\dfrac{3\sqrt{3}}{2}r^2 \approx 2.6r^2$
The Overlap Area of Repeaters	$(2\pi - 4)r^2 \approx 0.73\pi r^2$	$(2\pi - 3\sqrt{3})r^2 \approx 0.35\pi r^2$

As it can be seen from Table 2, the number of repeaters in the shape of regular hexagon is less than square when the coverage areas are the same. Therefore, we can build the repeater with the alveolate models which have the smaller overlap area, which is shown in Fig. 4.

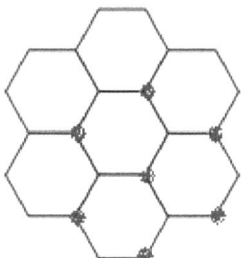

Fig. 4. The Repeater Coordination Picture

In the Fig.3, the black points in the lower right of each circle (regular hexagon) are expressed as the location of repeaters. In the alveolate coverage situation of regular hexagonal, there is a small area which has no signal, which can be ignored. In reality, there are few people in this area, so we also can ignore it. In the whole, the ignored area is very small.

For a repeater with the coverage radius r , the effective area is $\dfrac{3\sqrt{3}r^2}{2}$ in the coverage area. When the total area of n regular hexagons approximately equal to the given area, we can consider that n repeaters are needed. And n is the smallest number of the repeaters we need. The n repeaters are called the basic repeaters with the role of ensuring that the signal can fully cover the given area.

3   The Research of Repeaters Coordination and an Example in the Mountainous Environment

Considering the transmission of the repeaters signals in the environment of mountainous places, because the mountain of protuberant will increase the area, the number of repeaters will be increased if there are mountainous places located in certain places. Now think about increasing the repeaters in one mountainous area[9]. Suppose in a circle with the radius of 4 miles, there is a mountain with the height of 8 miles; for the convenience of research, we will regard it as a cone-shaped mountainous area, which is shown in Fig. 7. Now we calculate the surface area of the mountainous area and get $S = 71.55$ square miles, but the coverage area of one repeater is 41.57 square miles. So considering the special complicated environment, we need at least 3 repeaters to ensure that the signals can be found everywhere in this mountainous area.

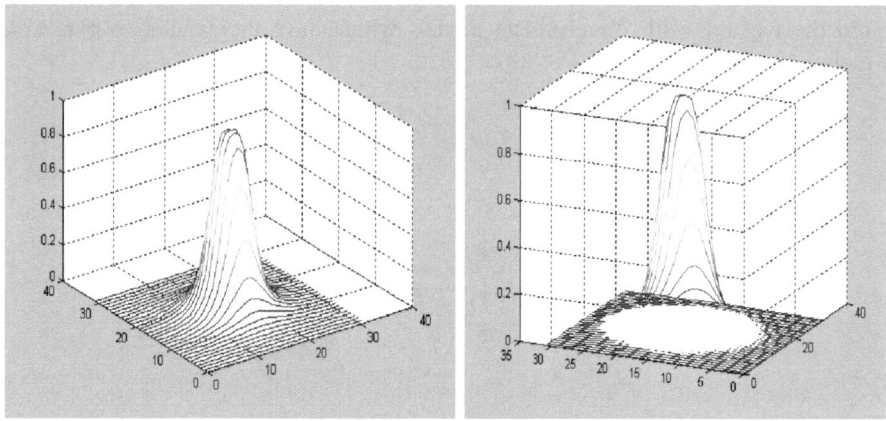

Fig. 5. Cone-shaped Mountainous Area

The result of the 3 repeaters shows that the number of the repeaters will be increased twice in the mountainous area. Now we will consider the location of the mountainous area and the number of the repeaters. When 1000 users make calls at the same time, because of fewer users, the basic repeaters haven't appeared saturation; the increasing 2 repeaters can satisfy the requirement. That is to say, if there is only one mountain in this area, we need build 129 repeaters to satisfy 1000 users making calls at the same time.

When 10000 users are making calls at the same time, which is far beyond the restriction of the PLs of the basic repeaters, we consider the relation between the mountainous area location and the number of the repeaters under the circumstance of the users approximately satisfying the normal distribution.

Table 2. The Relation Between the Mountainous Area Location and the Number of Repeaters

The Layer Where the Mountainous Area Locates	1	2	3	4	5	6	7
The Number of Repeaters	656	654	653	650	648	645	644

4   Assessment of Results and Conclusion

Under the circumstance of considering or not considering occupying the PLs when the repeaters communicate with each other, based on the users evenly distributed, the result from the regular hexagon honeycomb model is better than the result from the square one. And the detailed comparison is shown in table 3.

Table 3. The Comparison Between the Regular Hexagon and Square Repeaters Coordination Result

Number of Users	Occupying the PLs when Repeaters Communicating with Each Other			Not Occupying the PLs when Repeaters Communicating with Each Other		
	Regular Hexagon	Square	Difference	Regular Hexagon	Square	Difference
1000	127	164	37	17	24	7
10000	642	840	198	72	84	12

We use the compensation method to transform the raised using area to the above plane problem and also provide a common method. And then study the region with the special cone-shaped mountain. When the users are evenly distributed, we use the regular hexagon model to solve the problem; when considering occupying PLs, we get that we need at least 129 repeaters to satisfy 1000 users. When the users are random distributed, we use the genetic method model based on the random probability distribution to solve the problem; we get that we need at least 720 repeaters to satisfy 10000 users.

Compared with the results of the triangle honeycomb model, the square honeycomb model, and the traditional genetic algorithm model, we get that the number of the repeaters is smaller by using the above mentioned models, that is, we can use the resource more effectively. Besides the flat problems, we also successfully solve the problems of the areas with a mountain, even with more mountains.

Meanwhile, we the problems happening in the condition of occupying or not occupying PLs when the repeaters communicate with each other are also successfully solved. So our models and algorithm are successful in solving the problem of the repeater coordination and using number of the repeaters from the evidence shown above.

Acknowledgment. This work is supported by the National Nature Science Foundation of China (No.61170317) and the Nature Science Foundation of Hebei Province of China (No.A2009000735, A2010000908). The authors are grateful for the anonymous reviewers who made constructive comments.

References

1. Wang, B.: Honeycomb mobile telecommunication rf engineering. Posts & Telecom Press, Beijing (2005)
2. Luo, L.: Optimization and Practice of the Blindand Hot Spots Which Are Covered by Wireless Net. Fudan University (2006)
3. http://www.mc21st.com/techfield/systech/repeater/r03.htm#r30
4. Wan, X.: Probability Theory and Mathematical Statistics. Science Press, Beijing (2007)
5. Carey, D.: Small WiEx cellular phone repeater solves large problem. Electronic Engineering Times 10(19) (2009)
6. Oil Review Middle East Group. Radio repeater bends signal. Oil Review Middle East 12(2) (2009)
7. Choi, J., Lee, Y.: Design of an Indoor Repeater Antenna with High Isolation for WCDMA Systems. Antennas and Propagation 109(117) (2009)
8. Thanh Hiep, P., Kohno, R.: Examining exclusive transmits time for high propagation characteristic of Distributed MIMO Repeater System with amplify-and-forward. Radio Communication Systems 110(19) (2010)
9. Bakketeig, L.S.: Repeater's studies – Development of a new research field. Norsk Epidemiologic 17(2) (2009)
10. Tsai, L.-S., Shiu, D.-s.: Capacity scaling and coverage for repeater-aided MIMO systems in line-of-sight environments. IEEE Transactions on Wireless Communications 9(5) (2010)

Prediction for Water Surface Evaporation Based on PCA and RBF Neural Network

Wei Cao, Sheng-jiang Zhang, Zhen-lin Lu, and Zi-la Jia

Program Executive Office, Xinjiang Research Institute of Water Resources and Hydropower, Urumqi City, Xinjiang Province, China
caov100@163.com

Abstract. In order to build prediction model of the water surface evaporation so as to easily plan and manage water resources, authors presented a method with the principal component analysis(PCA) and radial basis function(RBF) neural network model for predicting the water surface evaporation. Firstly, the PCA was used to eliminate the correlation of the initial input layer data so that the problem of efficiency caused by too many input parameters and by too large network scale in neural network modeling could be solved. And then, the prediction model of water surface evaporation was built through taking the results of PCA as inputs of the RBF neural network. The research result showed that the model proposed had a better prediction accuracy that the average prediction accuracy reached 95.3%, and enhanced 5.5% and 5.0% compared with the conventional BP network and RBF network respectively, which met the requirements of actual water resources planning and provided a theoretical reference for other region of water surface evaporation forecasting.

Keywords: water surface evaporation, principal component analysis, radial basis function networks, neural networks.

1 Introduction

Evaporation is an important part of the the water cycle process, the rivers, lakes, reservoirs, ponds and other water bodies is an important part of the water loss.Water surface evaporation data are widely used in agricultural irrigation, water resources assessment, drought analysis, hydrological model identification, planning and design and management of water conservancy projects, four water(atmospheric water, surface water, soil, groundwater) into basic research.Arid regions far from the sea, little precipitation, high evaporation, water evaporation of surface water such as rivers, lakes, reservoirs and other water loss a major component, so the water evaporation in the arid area of surface water bodies and water resources research occupies a significant position ,also the basic of soil evaporation and ground water[1].Evaporation calculation for the prediction and rational use of limited water resources in arid areas to improve water use social, economic and ecological benefits, and promote the sustainable development of oasis agriculture is important.

In recent years, in order to establish a solid theoretical basis, the form of relatively simple parameters easily accessible and can meet the actual needs of the water surface

B. Liu and C. Chai (Eds.): ICICA 2011, LNCS 7030, pp. 351–358, 2011.

evaporation prediction models, many researchers use neural network prediction of evaporation have been studied, and achieved some results[2-5].Because neural networks do not need to establish an accurate mathematical model, we can achieve from input to output nonlinear mapping, so the evaporation has been widely applied.At present, the characteristics of these studies and the problems are: (1) Most are based on back propagation(BP) neural network predictive modeling, and BP neural network learning algorithm with slow convergence, the model length of the training machine, the total error in the training process such as lack of easy to fall into local minimum; (2) the factors that affect water evaporation are numerous and complex, traditional neural network prediction method is to simply put all the factors or man-made some of the major factors as the neural network input.However, due to the strong correlation between factors, and if people simply will simplify all factors or in combination, can cause a lot of useful information   missing or overlapping, thus affecting the prediction accuracy.

   In order to address the above situation, this paper sets up a model of water surface evaporation based on principal component analysis and RBF(Radial Basis Function) neural network, the first factors affecting the evaporation analysis to determine the impact parameter; then use principal component analysis eliminating the original input layer data correlation, not related to each other by a set of new input variables; Finally, the reconstruction of the training sample space as the RBF neural network input.

2   Main Influential Factors on Evaporation

2.1   Temperature

In the process of water surface evaporation, solar radiation is the primary energy, and through air and water temperatures to influence the increase of surface evaporation.Temperature determines the speed of diffusion of water vapor and how much capacity to accept water vapor, atmospheric temperature and humidity directly affect the stratification gradient, thereby affecting the water surface evaporation.High temperature, the water vapor pressure on the large difference of saturation .Surface temperature of the water vapor pressure determines the size and level of activity of water molecules, the water temperature is high, the water molecular motion energy, water evaporation rate is large.

2.2   Saturated Vapor Pressure

The water saturation vapor pressure is a function of surface temperature, air vapor pressure is a function of temperature and relative humidity.Therefore, the water temperature and air temperature on the rate of evaporation, from the saturation vapor pressure difference to be reflected.Vapor pressure deficit reflects the evaporation of moisture and surface evaporation of surface moisture gradient within a certain height, that affect the water evaporation rate of the main factors.According to diffusion theory, the evaporation rate and vapor pressure proportional to the change, the greater the vapor pressure, evaporation stronger, whereas smaller.

2.3   Air Saturation Deficit

Air saturation deficit reflects the evaporation of moisture in the air above the water situation, but does not reflect the evaporation of moisture on the surface size.Evaporation rate proportional to the difference between air saturation and air saturation deficit larger, stronger evaporation, whereas smaller.

2.4   Relative Humidity

Relative humidity is the actual vapor pressure in the air under the current temperature the ratio of the saturated vapor pressure, which reflects the extent of the air from far and near saturation.In general, the surface humidity (near saturation), the space above and external humidity is small, there is humidity difference.Therefore, the size of the relative humidity can be reflected on the water diffusion and exchange of water vapor out the speed.When the relative humidity, water vapor diffusion and exchange of more slowly outward, the water evaporation rate is small; hours when the relative humidity, water vapor diffusion and exchange of quick out, the water evaporation rate is large.

2. 5   Wind Speed

The size of wind speed, the performance of the role of turbulent diffusion in its strength and speed of wet and dry air exchange, the impact of the water evaporation rate is a major factor.No wind, the water depends mainly on molecular diffusion of water vapor, water vapor pressure decreased slowly, the saturation difference is small, so the water evaporates slowly.When the wind, turbulence enhanced the water vapor on the wind and turbulence quickly spread to the majority of the space, soon on the water vapor pressure decreases, saturation deficit bigger, the water evaporates faster.The greater the wind speed, the more intense turbulence, wet and dry faster air exchange, so the greater the evaporation.

3   PCA-RBF Neural Network Theory

PCA-radial basis function(PCA-RBF) neural network model is the principal component analysis and a combination of RBF neural network[6].First, using principal component analysis on the original set of variables for the main element modeling, and its role is to eliminate the relevance of the original variables, the cumulative contribution of more than 85% of the new set of variables as network inputs; Then, using the RBF neural networksimulation samples.

3.1   Principal Component Analysis

Principal component analysis is an effective method that the number of relevant variables into a few independent variables[7,8].Passes through a main element, you can keep the information on the basis of the main neural network reduces the input dimensions, reducing the size of neural networks, and related neural network input elements can enhance the elimination of the generalization performance of the network[6].

Suppose X is collected from a sample of $n \times m$ data matrix, where each column corresponds to a variable, each row corresponds to a sample, $X \in R^{n \times m}$, first standardized X as follows:

$$X_s = [X - (1,1,\cdots,1)^T M] diag \left| \frac{1}{s_1}, \frac{1}{s_2}, \cdots, \frac{1}{s_m} \right| \tag{1}$$

Where: M=[m$_1$, m$_2$, ..., m$_m$] is the mean of variable X, s=[s$_1$, s$_2$, ..., s$_m$] is the standard deviation of the variable X.

The principal component analysis of X can be:

$$X' = t_1 p_1^T + t_2 p_2^T + + t_m p_m^T \tag{2}$$

Where: $t_1, t_2, ..., t_m \in R_n$, the score vector; $p_1, p_2, ..., p_m \in R_m$, the load vector, X 'is also called the score vector X' of the principal component. Then we can get the main element model:

$$X' = t_1 p_1^T + t_2 p_2^T + ... + t_k p_k^T + E = X'_p + E \tag{3}$$

Where: E—error matrix.And X ' can be approximately expressed as

$$X'_p = t_1 p_1^T + t_2 p_2^T + ... + t_k p_k^T \tag{4}$$

3.2   RBF Neural Network

RBF neural network is a single hidden layer with 3-layer feed-forward network, input layer, hidden layer and output layer, can approximate any continuous function with arbitrary precision, the algorithm principlea certain extent, overcome the BP network learning speed and many parameters to adjust the disadvantages[9,10], is an ideal tool for the nonlinear calculation.

RBF neural network-based element of the input k-dimensional vector obtained $X'_p = \{X'_{p1}, X'_{p2}, ..., X'_{pk}\}$, that affect the surface evaporation k principal components expressed by X'_{pk}. Hidden layer for the l-dimensional vector $R = \{R_1, R_2, ..., R_m\}$, determine the number of hidden nodes has not theoretically perfect formula, the empirical formula of trial and error basis, until the error is satisfied.Output of the network is 1-dimensional vector f(X'), the output is:

$$f(X') = \sum_{i=1}^{l} W_{is} R_i(X') \tag{5}$$

Where: $R_i(X')$—hidden layer radial basis function; i—number of nodes in hidden layer, i=1,2, ..., l; s—the number of output neurons, s=1,2, ..., h; W_{is}—the first i-hidden units to output unit weights.

The role of the function of hidden layer radial basis function, we use the Gaussian kernel(Gaussian kernel function) to achieve information on the nonlinear transformation of input layer, the formula is as follows:

$$R_i(X') = \exp\{-\frac{1}{2}\left[\frac{\|X'-c_i\|}{\sigma_i}\right]^2\} \tag{6}$$

Where: c_i-i hidden units corresponding to the center of radial basis function ($c_i \in R^m$, i=1,2,..., l); $\acute{o}_i$-i units corresponding to the hidden layers are sensor center width, is used to adjust the sensitivity of the network;$\|X'-c_i\|$-the Euclidean distance of X 'and c_i.

4   Model Application

In this paper, PCA-RBF neural network, only need to analyze the impact of surface evaporation of the main meteorological factors, and mainly based on quantitative factors or values that group, select a broadly representative range of samples by principal component analysis, neural network model of Trained neural network model derived parameters of surface evaporation can be carried out.

4.1   Data Collection

This weather station at Yuli County,Xinjiang Province.Water surface evaporation data and meteorological factors related observations are collected. This choice of variables as follows: mid-averaged temperature(T), wind speed(V), vapor pressure(P), the relative humidity(H), sunshine duration(t), a total of five variables. Accordingly,a total of 21 sets of sample data obtained in this paper in mid-monthly water surface evaporation data as PCA-RBF neural network model prediction samples, the remaining 14 sets of data as training samples, the use of computer simulation data.

4.2   Principal Component Analysis and Modeling

First, 14 groups of training samples collected data principal component analysis. Table 1 shows the original sample data obtained through principal component analysis principal component coefficients, Table 2 shows the contribution of principal components, the cumulative contribution rate.

Table 1. Coefficent of PC Score

PC	Variable				
	T	V	P	H	t
PC1	0.529	0.392	0.442	-0.413	0.447
PC2	0.108	-0.557	0.488	0.543	0.380
PC3	0.006	0.561	0.451	0.504	-0.478
PC4	-0.388	0.469	-0.281	0.354	0.652
PC5	0.747	0.042	-0.533	0.395	-0.029

Table 2. Contribution Rate of PC

PC	Eigenvalue	Contribution rate/%	cumulative contribution rate/%
PC1	3.3847	67.7	67.7
PC2	0.9766	19.5	87.2
PC3	0.4206	8.4	95.6
PC4	0.1995	4.0	99.6
PC5	0.0185	0.4	100

Contribution rate means that the principal component principal component reflects the amount of information the original index, the cumulative contribution rate of the corresponding number of principal components that reflect the original index variables cumulative amount of information.Table 2 shows, the first a principal component can be explained by the original five indicator variables for 67.7% of the amount of information, the first two principal components can be explained by the original five indicator variables for 19.5% of the amount of information, the first three principal components can explainthe original five indicators 8.4% of the amount of information variables, we can see the first 3 principal components can be explained by the original five variables 95.6% of the amount of information, the basic indicators to retain the original five variables reflect the information.Therefore, we choose the first 3 principal components (ie, three composite indicators) on the surface of the water evaporation factors were analyzed, can be simplified and more objective indicators of the data structure.

Obtained by principal component analysis no correlation between the three comprehensive index model:

$$X'_{p1}=0.529x_1+0.392x_2+0.442x_30.413x_4+0.447x_5 \tag{7}$$

$$X'_{p2}=0.108x_10.577x_2+0.488x_3+0.543x_4+0.38x_5 \tag{8}$$

$$X'_{p3}=0.006x_1+0.561x_2+0.451x_3+0.504x_40.478x_5 \tag{9}$$

4.3   Simulation of RBF Neural Network Prediction Model

1)   RB Neural Network Design

The main element analysis of the three indicators as input samples, with the vector X'_p said, $X'p = \{X'_{p1}, X'_{p2}, X'_{p3}\}$, the surface evaporation Q(for normalization) as network output value, you can create a 3 input layer neurons, output layer neurons is a RBF neural network.

2)   Network Training

Taking 14 groups samples common factors as training samples, the neural network toolbox in Matlab function using newrb design the radial basis function network for function approximation with its, you can automatically   increase the radial basishidden layer neurons, until it reaches the mean square error so far, the format is:

net=newrb(P, T, GOAL, SPREAD, MN, DF)

Where: net-radial basis function network objects; newrb-radial basis function; P-network input sample vector; T-target vector; GOAL-mean square error; SPREAD-density of radial basis function; MN-the maximum number of neurons;DF-display frequency of the training   process.Network training, set GOAL 0.001; after testing, the time when the SPREAD of 0.3, RBF network error, and approaching the best.

Table 3. Results of Predction Model

Sample	Pritical value/ mm	PCA-RBF		BP		RBF	
		Prediction value/mm	Accurac y/%	Prediction value/mm	Accurac y/%	Prediction value/mm	Accura -cy/%
1	8.9	8.7	97.8	8.2	92.1	8.3	93.3
2	11.4	11.8	96.5	10.6	93.0	10.8	94.7
3	13.8	13.2	95.7	11.7	84.8	12.6	91.3
4	10.8	11.5	93.5	9.9	91.7	9.4	87.0
5	10.1	10.7	94.1	8.8	87.1	9.2	91.1
6	8	7.7	96.3	8.9	88.8	8.8	90.0
7	4.5	4.2	93.3	4.1	91.1	5.2	84.4

3)   Evaporation Prediction Models and Comparative Analysis

Taking other 7 groups samples as a prediction sample , with training after the completion of PCA-RBF neural network model for testing.Impact parameters of the test samples after PCA data input to the model to predict the network output and anti-standardized predicted by surface evaporation.Then, the PCA-RBF neural network model prediction results with all elements of the traditional BP neural network and RBF neural network prediction results were compared (Table 3).

Table 3 shows that the proposed PCA-RBF evaporation neural network prediction model, sample input for the prediction has a good effect, the average prediction accuracy is significantly higher than the element of the traditional BP neural network and RBF neural network prediction, the average predictionaccuracy of 95.3%, than the traditional BP neural network prediction accuracy increased by 5.5%, compared with RBF neural network prediction accuracy increased by 5%, fully in line with the actual demand for water resources planning, the model can accurately reflect changes in surface evaporation.

5   Conclusion

Evaporation is a complex nonlinear coupling system, it is difficult to predict the standard model.Surface evaporation is established a simplified prediction model modeling idea, proposed by using principal component analysis and neural network prediction model evaporation method.First, using principal component analysis to eliminate the original input layer data correlation, neural network modeling to solve too many input variables, network size is too large to decrease the efficiency of the problem; and then to enter the main element model results for the RBF to establish water surface evaporationneural network model.The method avoids the duplication of information, reducing the RBF neural network input dimensions, reducing the size of RBF neural network, RBF neural network while the correlation between input elements to eliminate the network's generalization performance can be enhanced. Taking Yuli County from April 2008 to October 2008 weather data as an example, application of PCA-RBF neural network prediction model on surface evaporation, the average prediction accuracy of 95.3%, fully in line with the actual needs of water resources planning, and contrast with total factor BP neuralnetwork and RBF neural network, prediction accuracy increased by 5.5% and 5.0%, indicating that the model has better prediction results, you can predict other areas of water surface evaporation of reference.

Acknowledgement. This research work was supported by public service   project for ministry of water resources,PRC(No.201101049);Research on Key Technologies of Agricultural Anti-drought in Xinjiang Province(No.200931105). We give the most heartfelt thanks to the reviewers for their great contributions on the early manuscript.

References

1. Zhou, J.-l., Yao, F.: Tianshan Mountains Plains Experimental Study of evaporation. Arid Zone Research 16(1), 41–43 (2010)
2. Wei, G.-h., Ma, L.: Prediction of water surface evaporation based on Gray Correlation Analysis and RBF Neural Network. Arid Meteorology 27(1), 73–77 (2009)
3. Min, Q.: Calculation model for water surface evaporation. Water Purification Technology 23(1), 41–44 (2009)
4. Wang, Y.-y.: Surface evaporation method of testing. Groundwater 28(2), 15–17 (2006)
5. Zhou, J.-l., Dong, X.-g., Chen, W.-j., et al.: Application of Penman-Monteith formula northern slope of Tianshan plain evaporation. Journal of Xinjiang Agricultural University 25(4), 35–38 (2002)
6. Guan, Z.-m., Chang, W.-b.: Prediction of aviation spare parts based on PCA-RBF neural network model. Journal of Beijing Technology and Business University: Natural Science 27(3), 60–64 (2009)
7. Ma, H.: Principal component analysis in the comprehensive evaluation of water quality. Journal of Nanchang Institute of Technology 25(1), 65–67 (2010)
8. Guo, T.-y., Li, H.-l.: Principal component analysis in the comprehensive evaluation of eutrophication of lakes in the application. Shaanxi Institute of Technology 18(3), 1–4 (2002)
9. Hagan, M.T., Demuth, H.B., Beale, M.H.: Neural Network Design. PWS Publishing, Boston (1996)
10. Qu, R., Liang, Y.: lycopene content in tomato fruit and Agronomic Traits and Path Analysis. Journal of Northwest Agricultural University 18(2), 233–236 (2009)

Decimal-Integer-Coded Genetic Algorithm for Trimmed Estimator of the Multiple Linear Errors in Variables Model

Fuchang Wang[1], Huirong Cao[2], and Xiaoshi Qian[1]

[1] Department of Basic Courses, Institute of Disaster Prevention of China Earthquake Administration, 065201 Sanhe, China
[2] College of Mathematics and Information Science, Langfang Teachers College, 065000 Langfang, China
{fzmath,huirongcao}@126.com, 8955563@qq.com

Abstract. The multiple linear errors-in-variables model is frequently used in science and engineering for model fitting tasks. When sample data is contaminated by outliers, the orthogonal least squares estimator isn't robust. To obtain robust estimators, orthogonal least trimmed absolute deviation (OLTAD) estimator based on the subset of h cases(out of n) is proposed. However, the OLTAD estimator is NP-hard to compute. So, an new decimal-decimal-integer-coded genetic algorithm(DICGA) for OLTAD estimator is presented. We show that the OLTAD estimator has the high breakdown point and appropriate properties. Computational experiments of the OLTAD estimator of multiple linear EIV model on synthetic data is provided and the results indicate that the DICGA performs well in identifying groups of high leverage outliers in reasonable computational time and can obtain smaller objective function fitness.

Keywords: decimal-integer-coded genetic algorithm, multiple linear error-in-variable model, orthogonal least trimmed absolute deviation estimator, robustness.

1 Introduction

It is well known that classical multiple linear regression is extremely sensitive to outliers in the data. This problem also holds in the case of multiple linear errors-in-variables(EIV) model and aroused many investigators' interest in recent decades [1]. Therefore, robust methods that can detect and resist outliers are needed so that reliable results can be obtained also in the presence of outliers. Substantial work has been done to develop methods. Zamar R.H proposed robust methods based on S-estimator [2] and M-estimator [3], Croux C. and Haesbroeck G proposed robust PCA method [4], Fekri M. and Ruiz-Gazen A. proposed weighted orthogonal regression method [5], Jung K.M. proposed least trimmed squares estimator [6], Hu T and Cui H.J. proposed t-type estimator [7]. All the estimators have good statistical properties, but they need greater computational effort. To obtain faster compute speed, evolutionary algorithm[8,9] and fast method [10] are proposed. In this paper, we proposed a new robust estimator and two compute methods for multiple EIV model.

B. Liu and C. Chai (Eds.): ICICA 2011, LNCS 7030, pp. 359–366, 2011.

The paper is organized as follows. In section 2, the definition, basic properties and computing complexity analysis of OLTAD estimator are presented. In section 3, we briefly discuss the genetic algorithm and develop a DICGA for OLTAD estimator. The objective function of the OLTAD estimator is continuous, non-differentiable and non-convex, showing multiple local minima. So the computation of the OLTAD estimator is laborious. We also proposed Fast-OLTAD for OLTAD estimator according to literature[6]. In section 4, simulations and a numerical example are given to illustrate the effectiveness of OLTAD estimator. Finally, in section 5, the conclusions are given.

2   The OLTAD Estimator for Multiple Linear EIV Model

2.1   The Definition of OLTAD Estimator

Errors-in-variables model arises from the study of regression models. We consider the multiple linear EIV model

$$\begin{cases} y_i = [1 \quad x_i^T]\boldsymbol{\beta} + \varepsilon_i \\ X_j = x_i + u_i \end{cases} \quad i = 1,2,\cdots,n. \tag{1}$$

where $\boldsymbol{\beta} = [\beta_0, \beta_1, \cdots, \beta_p]^T \in R^{p+1}$, β_0 is an intercept term, $(X_i^T, y_i)^T$ $(i = 1,2,\cdots,n.)$ are observations, The p dimensional vectors $x_i = (x_{i1}, x_{i2}, \cdots, x_{ip})^T$ are unknown design points and $(\varepsilon_i, u_i^T)^T$ $(i = 1,2,\cdots,n.)$ are independently distributed model statistical errors.

For estimated regression coefficients $\hat{\beta}_1, \hat{\beta}_2, \cdots, \hat{\beta}_p$ and an estimated intercept $\hat{\beta}_0$, we denote the orthogonal residuals

$$\hat{r}_i(\hat{\beta}_0, \hat{\beta}_1, \cdots, \hat{\beta}_p) = \frac{y_i - (\hat{\beta}_0 + \hat{\beta}_1 x_{i1} + \cdots + \hat{\beta}_p x_{ip})}{\sqrt{1 + \hat{\beta}_1^2 + \cdots, \hat{\beta}_p^2}} . \tag{2}$$

Further, let $u_{i:n}$ denote the i-th order statistic of n numbers $u_1, u_2, \cdots, u_n$. The high-breakdown estimator which we consider is defined as follows:

Definition 1. Let r_i be the i-th orthogonal residual determined by a Multiple linear EIV model with parameters $\hat{\beta}_0, \hat{\beta}_1, \cdots, \hat{\beta}_p$ and a given data set **Z**. The orthogonal least trimmed absolute deviation (OLTAD) estimator is given by

$$OLTAD(\mathbf{Z}) = \operatorname*{arg\,min}_{\hat{\beta}_0, \hat{\beta}_1, \cdots \hat{\beta}_p} \sum_{i=1}^{h} \{|r_1|, \cdots, |r_n|\}_{i:n} \tag{3}$$

where h with $[n/2] \le h \le n$ is a parameter influencing the estimation.

Let us consider the global robustness of the OLTAD estimator. This is the finite-sample version of breakdown point, introduced by Donoho and Huber (1983). The breakdown point of an estimator $T(W)$ at a sample W is defined as

$$\varepsilon^*(T) = \min\left\{\frac{m}{n}\left|\sup_{\tilde{w}}\left\|T(W) - T(\tilde{W})\right\|\right\}\right\} = \infty$$

Where $\tilde{W}$ is obtained by replacing m observations by arbitrary points. Roughly speaking, the breakdown point is the smallest fraction of the contaminated data to make the estimator useless. When the sample size is n, the breakdown point of the least trimmed square (LTS) estimator is $1/n$. So the asymptotic breakdown point of the LTS estimator is 0.

The OLTAD estimator is an h-sample estimator, so the following proposition is evident. Taking $h = [n/2]$ the OLTAD estimator has a 50% breakdown points, it means that the OLTAD estimator is a high breakdown estimator[11].

2.2   The Definition of OLTAD Estimator

Proposition 1. Assume also that any $p \times p$ submatrix of X is non-singular; in this case we say that observations are in general position. Then the finite sample breakdown point of the OLTAD estimator in the regression model (2) such that $\varepsilon^*(\hat{\beta}_{OLTAD}) = ([(n-p)/2]+1)/n$. Consequently it has a 50% asymptotic breakdown point.

Obtaining asymptotic theory for the least trimmed absolute deviation(OLTAD) estimator in linear EIV regression model is a very challenging problem, Cui(1997) gives some results of least absolute deviation(OLAD) estimator[12]. Thus, we use a Jackknife estimator for the asymptotic covariance matrix of the slope vector $\hat{\beta}_{OLTAD}$. The jackknife estimator can be written by

$$\frac{1}{n-1}\sum_{i=1}^{n}\left(\hat{\beta}_{OLTAD,(i)} - \hat{\beta}_{OLTAD,(\bullet)}\right)\left(\hat{\beta}_{OLTAD,(i)} - \hat{\beta}_{OLTAD,(\bullet)}\right)^T$$

where $\hat{\beta}_{OLTAD,(i)}$ is the OLTAD estimator with the i-th observation omitted and $\hat{\beta}_{OLTAD,(\bullet)} = \frac{1}{n}\sum_{i=1}^{n}\hat{\beta}_{OLTAD,(i)}$.

2.3   Computing Complexity Analysis of OLTAD Estimator

From the definition of OLTAD estimator, we can find that if we compute all the least absolute deviation estimators based upon all possible $\binom{n}{h}$ subset from data set Z, compare all the $\arg_{\hat{\beta}_0,\hat{\beta}_1,\cdots,\hat{\beta}_p}\sum_{i=1}^{h}\{|r_1|,\cdots,|r_n|\}_{i:n}$, then the best fitting estimator corresponding minimum of all $\arg_{\hat{\beta}_0,\hat{\beta}_1,\cdots,\hat{\beta}_p}\sum_{i=1}^{h}\{|r_1|,\cdots,|r_n|\}_{i:n}$ always exists. Obviously, the computation of the OLTAD is laborious by this method.

One of the practical attractions of our OLTAD estimator is the relative ease (compared with the OLTS estimator[6]) with which it can be computed. Because S.P. Feng [13] proves that an important property, just like least absolute deviation(LAD) estimator of multiple linear model, the OLTAD estimator corresponds to an exact fit to some subset of size $p+1$. So the OLTAD estimator is similarly characterized as a two-part problem - identifying the correct subset of size h to cover with the OLTAD fit, and determining the subset of size $p+1$ that minimizes the sum of orthogonal absolute deviations to these h cases. Denote the number of subsets of size h from a sample of size n. There are $\binom{n}{p}$ "elemental" subsets (subsets of size $p+1$) — a much smaller number than $\binom{n}{h}$ in typical applications — and one of these must provide an OLTAD solution for the full data set. By reversing the order of the two-part search therefore, we can dramatically reduce its computational complexity.

We propose the Fast-OLTAD algorithm by described as follows:

Step 1. Generate an initial set $H_1 \subset \{1,2,\cdots,n\}$ using the method described as follow: Generate a random $(p+1)$-subset J and compute the exact coefficient $\boldsymbol{\beta}_0$ from $y_i = [1 \quad x_i^T]\boldsymbol{\beta}$, $i \in J$ .If the data matrix X is not full rank, then add a random observation into J until it is. Then compute the residuals $r_{0i} = (y_i - [1 \quad x_i^T]\boldsymbol{\beta}_0)/\sqrt{1+\|\boldsymbol{\beta}'_0\|^2}$, $i = 1,2,\cdots,n$. wherein $\boldsymbol{\beta}'_0$ is a vector which $\boldsymbol{\beta}_0$ is removed the first element. Set H_1 such that $\{i \in H_1 \big| |r_{0(1:n)}| \le |r_{0(2:n)}| \le \cdots \le |r_{0(h:n)}|\}$.

Step 2. Compute the objective function, and repeat Step 1 5000 times.
Step 3. Set $\boldsymbol{\beta}_0$ corresponding to the minimum objective function as the OLTAD estimator of multiple EIV model.

Another important typical approach to tackle the problems is to use heuristics algorithm. Genetic Algorithm (GA)[14]is a well established search heuristic in computer science and steadily gaining importance in computational statistics[15]. In this paper, we concentrate on OLTAD estimator's computing and extend genetic algorithm to a framework that is applicable to it.

3   DICGA for OLTAD Estimator

As a first step, we have to appoint the code of the candidate solutions. In order to have a limited number of candidate solutions, we restrict ourselves to candidate solutions uniquely determined by a data subsample of fixed size. In the most common algorithms Fast-LTS[16] the subsamples are for sound reasons of size p. These reasons include the fact, that p linear independent points uniquely define a hyperplane. Additionally, smaller subsamples decrease the possibility of having outliers in the subsample. We will adopt this in letting for explanatory data $Z_e = \{(X_1, y_1),(X_2, y_2),\cdots,(X_n, y_n)\} \subset R^{p+1}$.

Let $S = \{s | s = (i_1, i_2, \cdots, i_{p+1}), i_k \in \{1, 2, \cdots, n\}, i_q = i_r, q \neq r, q, r, k = 1, 2, \cdots, p+1.\}$ be the code of our individuals that is mapped to its phenotype by the function

$$m : S \to R^{(p+1) \times (p+1)}$$

$$m(s) = m((i_1, i_2, \cdots, i_{p+1})) = \begin{bmatrix} X_{i_1} & y_{i_1} \\ X_{i_2} & y_{i_2} \\ \cdots & \cdots \\ X_{i_{p+1}} & y_{i_{p+1}} \end{bmatrix} \qquad (4)$$

Thus, we can obtain $\begin{pmatrix} n \\ p+1 \end{pmatrix}$ different possible individuals. The determination of the value of objective function of these individuals comprises two steps:

1. Compute a unique candidate solution hyperplane Hj from the given individual $s_j \in S$, where $j = 1, 2, \cdots, popsize, popsize \in N^+$.

2. Compute one of the objective function $f(s_j) = \sum_{i=1}^{h} \{|r_1|, \cdots, |r_n|\}_{i:n}$ (the residuals ri are determined by Hj and the given data) depending on the estimator chosen, h with $1 \leq h \leq n$ is a parameter influencing the estimation.

The algorithm that we propose is the following:

Algorithm (Integer-Coded Genetic Algorithm (DICGA) for Trimmed Estimators)

Step 0. (Initialization) Give the size of population popsize and randomly generate an initial population of chromosomes $s_j \in S, j = 1, 2, \cdots, popsize.$, compute a unique hyperplane Hj from sj and evaluate the fitness function for each of the chromosomes, let $fitness(s_j) = \dfrac{1}{1 + f(s_j)}, j = 1, 2, \cdots, popsize.$ for the OLTAD estimator. In general way, the population size popsize often take from 50 to 1000 depending on your demands and you computer hardware.

Step 1. (Parent Selection) Set if elitism strategy is not used; otherwise. Select with replacement parents from the full population (including the elitist elements). The parents are selected according to their fitness, with those chromosomes having a higher fitness value being selected more often. There are several schemes for the selection process: roulette wheel selection and its extensions: scaling techniques, tournament, elitist models, and ranking methods. A common selection approach assigns a probability of selection Pj to each individual, j based on its fitness value.

$$P_j(\text{individual } j \text{ is chosen}) = \frac{fitness(s_j)}{\sum_{j=1}^{popsize} fitness(s_j)}, j = 1, 2, \cdots, popsize.$$

Here, we adopt roulette wheel selection.

Step 2. (Crossover) For each pair of parents identified in Step 1, perform crossover on the parents at a randomly .If no crossover takes place (crossover probability pc(0.5 < pc <0.95), then form two offspring that are exact copies (clones) of the two parents. The crossover operator in details is as follows: Select chromosome(i1, i2,..., ip + 1) and (j1, j2, ..., jp + 1), union them and delete redundant same numbers. As a result, we acquire a long chromosome (k1, k2,..., km)quad (p < m ≤ 2p + 2). Subsequently, we attain two new chromosomes taking (i'1, i'2, ..., i'p + 1) and (j'1, j'2, ..., j'p + 1) from (k1, k2,..., km) randomly as offsprings.

Step 3. (Mutation) While retaining the best chromosomes from the previous generation, replace the remaining chromosomes with the current population of offspring from Step 2. For the integer-based implementations, mutate the individual integers with probability pm (0.01 < pm < 0.20). If chromosome (i1, i2,..., ip + 1) mutate,   randomly   select   ik   from   it   and   replace   ik   by   i'k, $i'_k \in \{1,2,\cdots,n\} \setminus \{i_1, i_2, \cdots, i_{p+1}\}$.

Step 4. (Fitness and termination criterion) Compute the fitness values for the new population of chromosomes. Terminate the algorithm if the stopping criterion is met or if the budget of fitness function evaluations is exhausted; else return to Step1. In general, generation number, running time and no improvement for some iterative times are often as stopping criterion.

The question how to compute a unique hyperplane from a subset of size p remains. As a first step, we compute the hyperplane Hj through the subset of data points. If it does not define a unique hyperplane, we try to add observations in fixed order (e.g. starting with (X_1, y_1) until it does.

4   Numerical Simulation

To get an idea of the performance of the overall algorithm, we start by applying the Fast-OLTAD and DICGA to synthetic data.

We simulate data with $n=500+20 \times p$ data points for $p =1,3,...,30$ from linear multiple regression model

$$ y_i = \beta_0 + \beta_1 x_{i1} + \cdots + \beta_p x_{ip} + \varepsilon_i $$

where β_0 is an intercept term and $\varepsilon_i \sim N(0,1)$ are statistical errors. The parameters $\beta_0, \beta_1, \cdots, \beta_p$ is set to 1, $x_{ij} \sim N(0,1)$, $j = 1,2,\cdots, p.$ are the nontrivial explanatory variables. We have introduced outliers in the x-direction and y-direction by replacing a 25 percent of the xij, (j=1,...,p) by that are normally distributed with mean 10 and variance 100.

The parameters of DICGA algorithm are set as follows: population size popsize = 50+30×p, crossover rate pc = 0.8, mutation rate pm = 0.2, generation number iters = 100+10×p. We also change parameters to other values, and obtain similar results and find the above parameters values can balance the algorithm's accuracy and efficiency in our test situation.

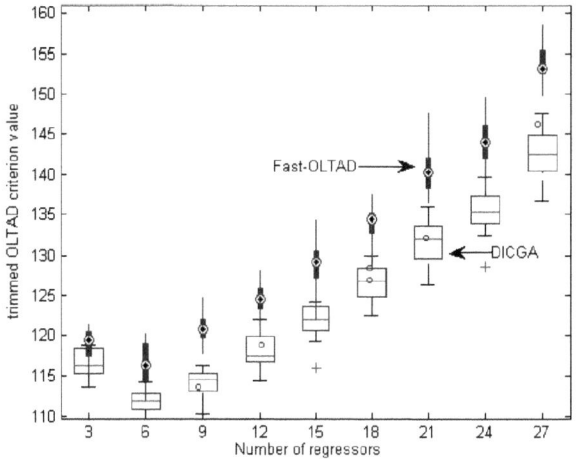

Fig. 1. The boxplots of Fast-OLTAD and DICGA algorithm

The results display in Fig. 1. In Fig. 1, the filled markers represent the results by Fast-OLTAD algorithm and null-box markers represent the results by DICGA. In nearly all conducted runs, our algorithm achieves better values. Of course, the DICGA requires more computation time. The result compute time is similar. When the p increase, the compute time mean of DICGA linearly increase from 1.3 second to 8.2 second, meanwhile, the compute time mean of DICGA linearly increase from 0.05 second to 0.08 second.

The simulation experiments were performed on Intel Core Duo 2.4GHz computer with 2GB of memory running MATLAB 2009a and Windows XP Sp2.

5   Conclusion

High breakdown estimator in large data sets or high dimensional regressor spaces is a challenging problem. Many high breakdown estimators are in all likelihood not computable exactly. The common heuristics to compute solutions in these cases is work with subsample versions of the estimators. By comparison with 0-1 integer-coded genetic algorithm and other evolution algorithms, the individuals of decimal-integer-coded genetic algorithm is shorter, especially when sample data number n is big, so decimal-integer-coded genetic algorithm occupies less computer's memory, the computing speed is faster.

Acknowledgments. The work was supported by the Hebei Provincial Natural Science Foundation of China(No.A2011408006),National Natural Science Foundation of China (No.40874022).

References

1. Ruiz-Gazenb, A., Fekria, M.: Robust estimation in the simple errors-in-variables model. Statistics & Probability Letters 76(16), 1741–1747 (2006)
2. Zamar, R.H.: Robust estimation in the errors-in-variables model. Biometrika 76(1), 149–160 (1989)
3. Zamar, R.H.: Bias robust estimation in the orthogonal regression. The Annals of Statistics 20(4), 1875–1888 (1992)
4. Croux, C., Haesbroeck, G.: Principal component analysis based on robust estimators of the covariance or correlation matrix: influence functions and efficiencies. Biometrika 87(3), 603–618 (2000)
5. Fekri, M., Ruiz-Gazen, A.: Robust weighted orthogonal regression in the errors-in-variables model. Journal of Multiple Analysis 88(1), 89–108 (2004)
6. Jung, K.M.: Least trimmed squares estimator in the errors-in-variables model. Applied Statistics 34(3), 331–338 (2007)
7. Hu, T., Cui, H.J.: T-type estimators for a class of linear errors-in-variables models. Statistica Sinica 19, 1013–1036 (2009)
8. Nunkessera, R., Morell, O.: An evolutionary algorithm for robust regression. Computational Statistics & Data Analysis 54(12), 3242–3248 (2010)
9. Akter, S., Khan, M.H.A.: Multiple-case outlier detection in multiple linear regression model using quantum-inspired evolutionary algorithm. Journal of Computers 5(12), 1779–1788 (2010)
10. Croux, Fekri, M., Ruiz-Gazen, A.: Fast and robust estimation of the multivariate errors in variables model. Test 19(2), 286–303 (2010)
11. Rousseeuw, P.J., Leroy, A.M.: Robust Regression and Outliers Detection. John Wiley and Son (1987)
12. Cui, H.J.: Asymptotic properties of Generalized LAD estimators in EV model (in Chinese). Science in China (Series A) 27(2), 119–131 (1997)
13. Feng, S.P.: Optimal solution on weighted total least absolute deviation (in Chinese). Journal of University of Science and Technology of China 39(12), 1260–1264 (2009)
14. Forrest, S.: Genetic algorithms: Principles of natural selection applied to computation. Science 261(5123), 872–878 (1993)
15. Chatterjeea, S., Laudatoa, M., Lynchb, L.A.: Genetic algorithms and their statistical applications: an introduction. Computational Statistics and Data Analysis 22(6), 633–651 (1996)
16. Rousseeuw, P.J., Van Driessen, K.: Computing lts regression for large data sets. Data Mining and Knowledge Discovery 12(1), 29–45 (2006)

Prediction of Wastewater Treatment Plants Performance Based on NW Multilayer Feedforward Small-World Artificial Neural Networks

Ruicheng Zhang and Xulei Hu

College of Electrical Engineering, Hebei United University,
Hebei Tangshan 063009, China
rchzhang@yahoo.com.cn

Abstract. In order to provide a tool for predicting wastewater treatment performance and form a basis for controlling the operation of the process, a reliable model is essential for any wastewater treatment plant. This would minimize the operation costs and assess the stability of environmental balance. For the multi-variable, uncertainty, non-linear characteristics of the wastewater treatment system, a NW multilayer feedforward small-world artificial neural network prediction model is established standing on the actual operation data in the wastewater treatment system. The model overcomes several disadvantages of the conventional BP neural network. Namely: slow convergence, low accuracy and difficulty in finding the global optimum. The results of model calculation show that the predicted value can better match measured value, played an effect of simulating and predicting and be able to optimize the operation status. The establishment of the predicting model provides a simple and practical way for the operation and management in wastewater treatment plant, and has good research and engineering practical value.

Keywords: NW small-world networks, Multi-layer forward neural networks, Wastewater plant, Modeling, Wastewater treatment.

1   Introduction

The increased concern about environmental issues has encouraged specialists to focus their attention on the proper operation and control of wastewater treatment plants (WWTPs). The characteristics of influent to the WWTPs are varied from one plant to another depending on the type of community lifestyle. Therefore, the performance of any WWTP depends mainly on local experience of a process engineer who identifies certain states of the plant[1]. The type of influent for any plant is also time-dependent and it is difficult to have a homogeneous influent to a WWTP[2]. This may result in an operational risk impact on the plant. Serious environmental and public health problems may result from improper operation of a WWTP, as discharging contaminated effluent to a receiving water body can cause or spread various diseases to human beings. Accordingly, environmental regulations set restrictions on the quality of effluent that must be met by any WWTP.

B. Liu and C. Chai (Eds.): ICICA 2011, LNCS 7030, pp. 367–374, 2011.
© Springer-Verlag Berlin Heidelberg 2011

A better control of a WWTP can be achieved by developing a robust mathematical tool for predicting the plant performance based on past observations of certain key parameters. However, modeling a WWTP is a difficult task due to the complexity of the treatment processes. The complex physical, biological and chemical processes involved in wastewater treatment process exhibit non-linear behaviors which are difficult to describe by linear mathematical models.

Owing to their high accuracy, adequacy and quite promising applications in engineering, artificial neural networks (ANNs) can be used for modeling such WWTP processes. The ANN can be used for better prediction of the process performance. It normally relies on representative historical data of the process. In a wastewater treatment plant, there are certain key parameters which can be used to assess the plant performance. These parameters could include biological oxygen demand (BOD), suspended solid (SS) and chemical oxygen demand (COD). Most of the available literature on the application of ANNs for modeling WWTPs utilized these parameters. For example, Oliveira-Esquerre et al. [3] obtained satisfactory predictions of the BOD in the output stream of a local biological wastewater treatment plant for the pulp and paper industry in Brazil. The principle component analysis was used to preprocess the data in the back propagated neural network. The Kohonen Self-Organizing Feature Maps (KSOFM) neural network was applied by Hong et al. to analyze the multidimensional process data and to diagnose the inter-relationship of the process variables in a real activated sludge process. The authors concluded that the KSOFM technique provides an effective analysis and diagnostic tool to understand the system behavior and to extract knowledge from multidimensional data of a large-scale WWTP. Hamed et al. [2,4]developed ANN models to predict the performance of a WWTP based on past information. The authors found that the ANN-based models provide an efficient and robust tool in predicting WWTP performance. But the large learning assignment, slow convergence, and local minimum in the neural network are observed.

For the problems in those wastewater treatment models mentioned above, in this study, a new wastewater treatment model, NW multilayer feedforward small-world artificial neural network model, is proposed, which integrates NW small-world networks and multi-layer forward neural networks, learns from others' strong points to offset one's weakness, and considerably improves precision, velocity, and anti-interference ability.

2   System Flow Description

A schematic diagram of the plant is shown in Fig. 1. The crude sewage (CS) from different pumping stations is collected and screened for floating debris and removal of grit is carried out by the grit collector and grit elevators. Primary settlement tanks (PST) are utilized to settle 65–75% of the solids. Settled solids are scrapped down in the hoppers of the PST with the help of mechanical drive scrappers. These settled solids are removed by the Hydro Valves which open in the Consolidation Sludge

Tank. Aerobic bacteria are activated by aeration and mixing with activated sludge to reduce the volume of mixed liquor. Primary treated effluent is mixed with the returned activated sludge from the secondary settlement tank and uniformly distributed in channels for aeration with the help of mechanically driven aerators. Mixed liquor out of the aeration tank is made to settle in the secondary settlement tanks. In the post-treatment, the secondary treated effluent is pre-chlorinated and lifted by screw pumps for uniform distribution to sand filters. The resulting stream, designated as final effluent (FE), flows down into the wet well.

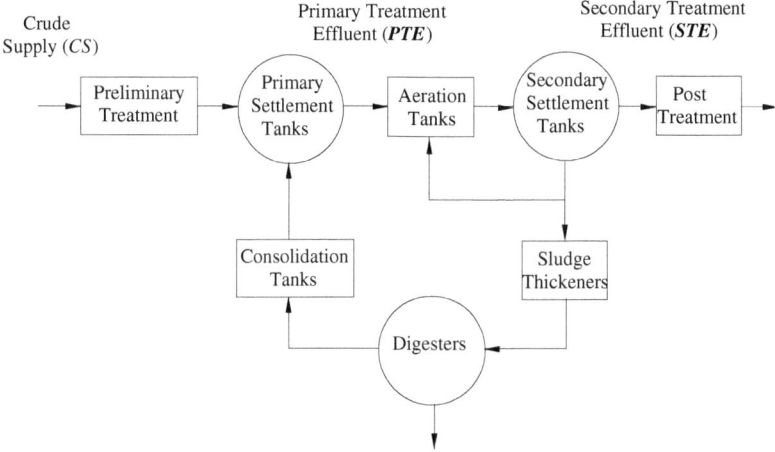

Fig. 1. Schematic diagram of the wastewater treatment system

3   NW Multilayer Feedforward Small-World Artificial Neural Networks for Wastewater Treatment

3.1   Model Generation Process

Based on the topology of the NW small-world networks, the model of NW multilayer feedforward small-world artificial neural networks is proposed. The model generation process is as follows.

(1) Initially, neurons are connected feedforward, i.e. each neuron of a given layer connects to all neurons of the subsequent layer.

(2) We make a random draw of two nodes which are connected to each other. We don't cut that "old" link. In order to create a "new" link, we make another random draw of two nodes. If those nodes are already connected to each other, we make further draws until we find two unconnected nodes. Then we create a "new" link between those nodes.

(3) In this way we create some connections between nodes in distant layers, i.e. short-cuts and the topology changes gradually (see Fig.[2]).

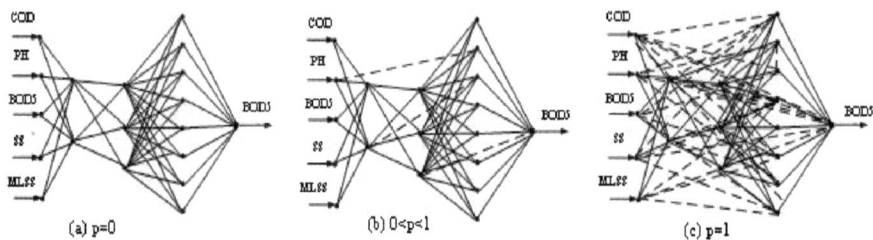

Fig. 2. The model of small world artificial neural network prediction of wastewater quality

3.2   The Flow of the Algorithm

The NW small-world artificial neural networks is a multilayered, feed forward neural network and is by far the most extensively used. Back Propagation works by approximating the non-linear relationship between the input and the output by adjusting the weight values internally. A supervised learning algorithm of back propagation is utilized to establish the neural network modeling. A normal NW small-world artificial neural networks model consists of an input layer, one or more hidden layers, and output layer. The input samples are $X = [X_1, X_2, \cdots, X_k, X_N]$, the any input sample is $X_k = [x_{k1}, x_{k2}, x_{kl}]$, the actual output of the network is $Y_k = [y_{k1}, y_{k2}, \cdots y_{ko}]$, Expected output is $d_k = [d_{k1}, d_{k2}, \cdots d_{ko}]$. So the NW small-world artificial neural networks program-training process as follows.

Step 1: Design the structure of neural network and input parameters of the network.
Step2: Get initial weights W and initial threshold values from randomizing.
Step 3: Input training data matrix X and output matrix d_k.
Step 4: Compute the output vector of each neural units.

(a) Compute the input and output vector of the No.1 hidden layer

$$u_{m_1}^{M_1} = \sum_{i=1}^{I} w_{im_1} x_{ki} \tag{1}$$

$$v_{m_1}^{M_1} = f\left(\sum_{i=1}^{I} w_{im_1} x_{ki}\right); m_1 = 1, 2, \cdots, M \tag{2}$$

Compute the input and output vector of the No.s hidden layer

$$u_{m_s}^{Ms} = \sum_{i=1}^{I} w_{im_s} x_{ki} + \sum_{t=1}^{s-1} \sum_{m_t=1}^{M_t} w_{m_t m_s} u_{m_t}^{M_t} \tag{3}$$

$$v_{m_s}^{Ms} = f(u_{m_s}^{Ms}); M_s = 1, 2, \cdots, M \tag{4}$$

(b) Compute the output vector of the output layer

$$u_o^{M_o} = \sum_{i=1}^{I} w_{io} x_{ki} + \sum_{t=1}^{s} \sum_{m_t=1}^{M_t} w_{m_t o} u_{m_t}^{M_t} \tag{5}$$

$$y_{ko} = f(u_o^{M_o}) \; ; o = 1,2,\cdots O \qquad (6)$$

(c) Compute the error signal of the Output layer neurons

$$E = \frac{1}{2}\sum_{o=1}^{O}[d_{ko}(n) - y_{ko}(n)]^2 \qquad (7)$$

Step 5: Compute the local gradient

(a) Compute the local gradient of the output layer M_o

$$\delta_o^{M_o}(n) = [(1 - y_{ko}(n))]^2 (d_{ko}(n) - y_{ko}(n)) \quad o = 1,2,\cdots,O \qquad (8)$$

(b) Compute the local gradient of the last one hidden layer

$$\delta_{m_s}^{M_s}(n) = v_{m_s}^{M_s}(n)(1 - v_{m_s}^{M_s}(n))\sum_{o=1}^{O}\delta_o^{M_o}(n)w_{m_s o}(n) \; m_s = 1,2,\cdots,M_S \qquad (9)$$

(c) Compute the local gradient of the hidden layer m_s

$$\delta_{m_s}^{M_s}(n) = v_{m_s}^{M_s}(n)(1 - v_{m_s}^{M_s}(n))[\sum_{t=s+1}^{S}\sum_{m_t=1}^{M_t}\delta_{m_t}^{M_t}(n)w_{m_s m_t}(n) \qquad (10)$$

$$\delta_{m_s}^{M_s}(n) = v_{m_s}^{M_s}(n)(1 - v_{m_s}^{M_s}(n))[\sum_{t=s+1}^{S}\sum_{m_t=1}^{M_t}\delta_{m_t}^{M_t}(n)w_{m_s m_t}(n)$$

$$+\sum_{o=1}^{O}\delta_o^{M_o}(n)w_{m_s o}(n)] \qquad (11)$$

$$m_s = 1,2,\cdots,M_s$$

Step 6: Renew W

$$\Delta w_{ji}^{S} = w_{ji}^{S}(n-1) + \eta\delta_j^{S} y_i^{S-1}(n-1) \qquad (12)$$

$$\Delta b_{ji}^{S} = \alpha b_{ji}^{S}(n-1) + \eta\delta_j^{S} y_i^{S-1}(n-1) \qquad (13)$$

Step 7: Repeat step 3 to step 6 until converge.

4   Simulations

Actual values of BOD in the training and testing data are compared to predicted values by the BP neural network models and NW multilayer feedforward small-world artificial neural network, to evaluate the models performance. Fig. 3 to 6 shows these comparisons for BOD outputs. Visual inspection indicates that the NW multilayer feedforward small-world artificial neural network models resulted in a good fit for the measured BOD data.

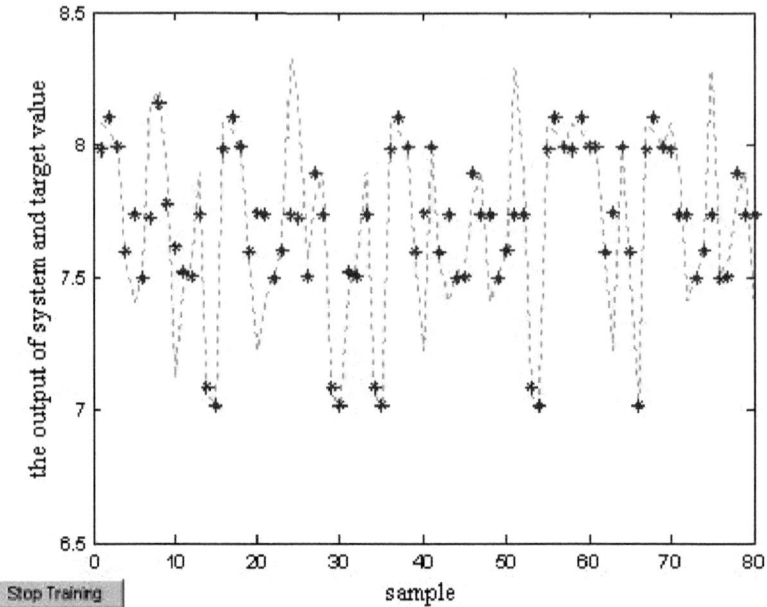

Fig. 3. The training result of effluent quality BOD5 of regular multi-layer forward neural networks

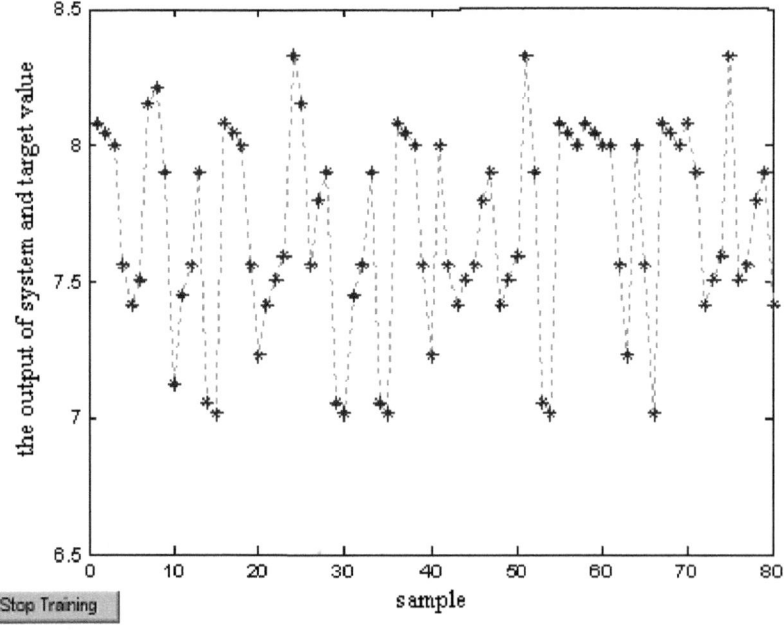

Fig. 4. The training result of effluent quality BOD5 of multi-layer forward small world l neural networks

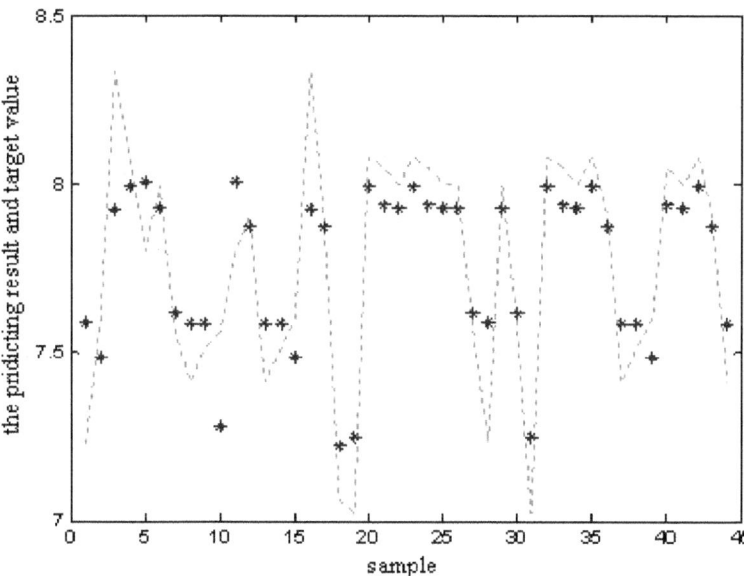

Fig. 5. Predicting results of regular multi-layer forward neural networks

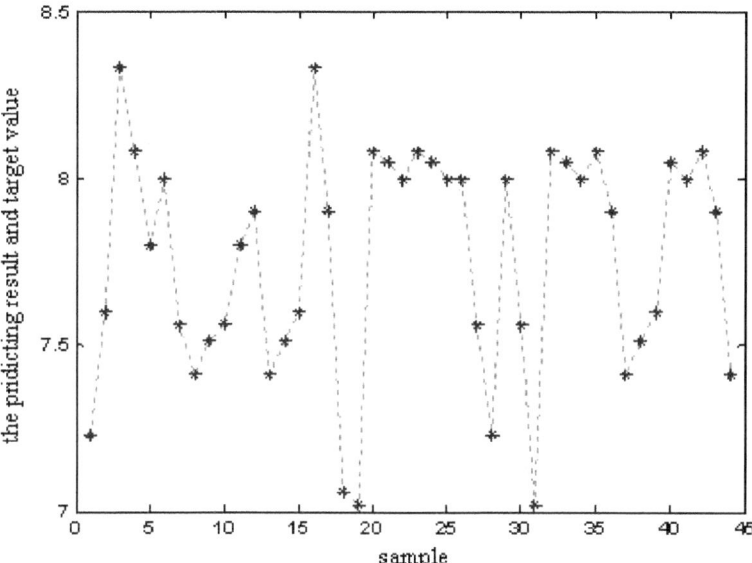

Fig. 6. Predicting results of multi-layer forward small-world artificial neural networks

5   Conclusions

In this paper, A models based on NW multilayer feedforward small-world artificial neural network were developed to predict the effluent concentrations of BOD for a WWTPs. The model is shown to fit the data precisely and to overcome several disadvantages of the conventional BP neural networks, namely: slow convergence, low accuracy and difficulty in finding the global optimum. A series of tests have been conducted based on the samples. It has been shown that the NW multilayer feedforward small-world artificial neural network model provided good estimates for the BOD data sets. After the network is trained, it becomes simple, fast, and precise, with strong self-adaptability and anti-interference ability to dispose data while predicting the wastewater treatment plant performance. The limitation in data, however, should be highlighted. If more data were collected, if the data were less noisy, this would have resulted in an improved predictive capability of the network.

Acknowledgments. This work was supported by the National Natural Science Foundation of China (grant number: 61040012).

References

1. Hong, Y.-S.T., Rosen, M.R., Bhamidimarri, R.: Analysis of a municipal wastewater treatment plant using a neural network-based pattern analysis. Water Research 37, 1608–1618 (2003)
2. Hamed, M., Khalafallah, M.G., Hassanein, E.A.: Prediction of wastewater treatment plant performance using artificial neural network. Environmental Modeling and Software 19, 919–928 (2004)
3. Oliveira-Esquerre, K.P., Mori, M., Bruns, R.E.: Simulation of an industrial wastewater treatment plant using artificial neural networks and principal components analysis. Brazilian Journal of Chemical Engineering 19, 365–370 (2002)
4. Mjalli, F.S., Al-Asheh, S., Alfadala, H.E.: Use of artificial neural network black-box modeling for the prediction of wastewater treatment plants performance. Journal of Environmental Management 83, 329–338 (2007)
5. Xia, Y.S., Wang, J.: A general methodology for designing globally convergent optimization neural networks. IEEE Transactions on Neural Networks 9(6), 1331–1343 (1998)
6. Watts, D.J., Strogatz, S.H.: Collective dynamics of small world networks. Nature 393, 440–442 (1998)
7. Newman, M.E.J., Watts, D.J.: Renormalization Group Analysis of the Small-world Network Model. Physics Letters A 263(4), 341–346 (1999)
8. Simard, D., Nadeau, L., Kroger, H.: Faster learning in small-world neural networks. Physics Letters A 336(1), 8–15 (2005)
9. Li, X., Du, H., Zhang, J.: Multilayer feedforward small-world neural networks and its function approximation. Control Theory & Applications 27(7), 836–842 (2010)
10. Wang, X., Li, X., Chen, G.: Complex network theory and its applications. Tsinghua University Press, Beijing (2006)
11. Yang, M., Xue, H.: Analysis of G Knowledge Communication Network Based on Complex Network. Computer Simulation 26(11), 122–123 (2009)

Hybrid Swarm-Based Optimization Algorithm of GA&VNS for Nurse Scheduling Problem

Zebin Zhang[1], Zhifeng Hao[1], and Han Huang[2]

[1] School of Computer, Guangdong University of Technology, China
[2] College of Business, City University of Hong Kong, China
bension8708@163.com, {mazfhao,hhan}@suct.edu.cn

Abstract. This paper presents a hybrid swarm-based optimization algorithm that combines genetic algorithm (GA) and variable neighborhood search (VNS) to deal with highly-constrained nurse scheduling problems in modern hospital environments. The problem is first divided into several sub-problems by the average principle. Then a genetic algorithm is used to solve the sub-problems including a subset of constraints. Better feasible solutions are built by a hybrid genetic algorithm; they are taken as the initial solution of the variable neighborhood search procedure. The proposed algorithm can be applied to other resource allocation problems with a large number of constraints. The experiment results show that our method can produce feasible solutions.

Keywords: Genetic algorithm, variable neighborhood search, Nurse Scheduling Problem.

1   Introduction

The nurse scheduling problem is essentially a multiple constrained resource scheduling problem. It requires improving the target in one hand and satisfying certain constraints on the other hand, which often causes conflict. The constraints mentioned here can be divided into hard constraints and soft constraints. Hard constraints are those that must be satisfied and not violated in order to have a feasible schedule. They are often generated by physical and requirements and legislation. Soft constraints are those that must be satisfied as more as possible but not all. They are often used to evaluate the quality of feasible schedules and generally decide by personal preference and fairness [5].

Most nurse scheduling problems in the real world are NP-hard [8] and were regarded as being more complex than the travelling salesman problem by [9]. Over the years, a wide variable of methodologics and models have been developed to deal with them. The available techniques can roughly be classified into two main categories: exact algorithm and (meta) heuristic. Mathematical programming is the traditional exact method [1], which guarantees to find an optimal solution. However, computational difficulties exist with this approach due to the enormous size of the search spaces that are generated. This is often ineffective in practice. Therefore, the method for solving such problem mainly focused in heuristic and meta-heuristic [2,6-9] which

B. Liu and C. Chai (Eds.): ICICA 2011, LNCS 7030, pp. 375–382, 2011.

can fine the better solution in a relatively short time. And it has achieved some results. Such as Burke et al. [1] proposed a hybrid multi-objective model that combines integer programming (IP) and variable neighborhood search (VNS) in modern hospital environments. Their method separated the hard constraints and the soft constraints into two phases of the problem. An IP is first used to solve the sub-problem which includes the full set of hard constraints and a subset of soft constraints. Then a basic VNS follows as a post-processing procedure to further improve the IP's resulting solutions. However, this method's solution is greatly depended on the first step (IP)'s solution. Kawaraka et al. [3] proposed an improving genetic algorithm with exchange strategy. Their algorithm used the exchange operation to meet the constraints both before and after genetic operators. But it is hard to extend this exchange method and easily lead to deadlock. Makoto et al. [4] proposed a genetic algorithm with simulative climbing operations. Their method first execute genetic algorithm in test data and the simulative climbing operations will be started when the algorithm fall into the local optimal. But its performance is poor in practical application.

In this paper, we present a new decomposition technique by combining genetic algorithm (GA) and variable neighborhood search (VNS) to deal with constraints and requirements. It can be divided into three phases. First phase, we apply throughout a separation strategy (average principle) splitting the problem into several sub-problems. This can greatly reduce the search space and speeding up the algorithm execution. Second phase, a genetic algorithm is used to solve the sub-problems which include a subset of constraints and better feasible solutions are built by a hybrid genetic algorithm with variable neighborhood search. This phase can obtain better feasible solution. Last phase, we proposed variable neighborhood search (VNS) to improve the better feasible solution.

2   Problem Description

The nurse scheduling problem we are tacking is derived from the literature [1]. It is provided by ORTEC, an international consultancy company specializing in planning, optimization and decision support solution. It involves assigning four types of shifts (i.e. shifts of early, day, late and light) within a scheduling period of 5 weeks to 16 nurses of different working contracts in a ward. Specific problem described as following: (HC is the hard constraints and SC is soft constraints)

The problem has 8 hard consrtaints: HC1-8, and 11 soft constraints: SC1-11.(see [1] for detail) If we consider the soft constraints as targets (f (xi)) and consider the hard constraints as constraints of object function (g (xi)). We find that the problem is a multiple constraints with multiple objective optimization problems. It is the essence of the nurse scheduling problem and we get a simple mathematical model as following (see [1] for details):

$$\text{Max } F(x) = [f1(x), f2(x), f3(x)\dots f8(x)] = w1*f1(x) + w2*f2(x) + \dots + w8*f8(x)$$
$$\text{s.t.   } gi(x) > 0; i = 1, 2, \dots, 11.$$

3   Hybrid Heuristic Algorithm

Our method has three phases which were metioned in the introduction above. And the following will introduce them in detail.

3.1   Splitting the Sub-problem

The practical nurse scheduling is a large scale problem in general. No matter what methods are used, its search space is very impressive and even lead to not obtain the feasible solution. To solve this problem, we propose a method that splitting the problem into several sub-problems. This will greatly reduce the search space, and reduce the time of execution.

There are many methods can split the problem and selecting which method is decided by the actual situation. As the scheduling tasks of the nurse scheduling problem we are talking is relatively homogeneous, we use an average principle to divided the problem. According to this principle, we can split the original problem into ten sub-problems with six nurses' scheduling task in one week and five sub-problems with four nurses' scheduling task in one week. The divided result is illustrated in table 1. As result of this, constraint for each sub-problem will change, because there are some constraints will play a role only in the overall scheduling process, such as HC4, SC6, and so on. These constraints will be satisfied in the step of the genetic algorithm with variable neighborhood search.

Table 1. The result of splitting

6 nurse's scheduling task in one week Constraint: HC1,2,6,7,8,910,11; SC1,2,4,5;	6 nurse's scheduling task in one week Constraint: HC1,2,6,7,8,910,11; SC1,2,4,5;	4 nurse's scheduling task in one week (Excluding Saturdays, Sundays) Constraint: HC1,2,6,7,8,910,11; SC1,2,4,5;
The same as up	The same as up	The same as up
The same as up	The same as up	The same as up
The same as up	The same as up	The same as up
The same as up	The same as up	The same as up

3.2   Hybrid Genetic Algorithm

This phase, a genetic algorithm is used to solve the sub-problems which include a subset of constraints and better feasible solutions are built by a hybrid genetic algorithm with variable neighborhood search, which is taken as the initial solution of the variable neighborhood search procedure.

Genetic algorithm for sub-problem. The numbers accorded to lemmas, propositions, and theorems etc. should appear in consecutive order, starting with the number 1, and not, for example, with the number 11.

Encoding. In this paper, we proposed a coding method which a sub-schedule problem corresponds to a chromosome. And the encoding rules are illustrated in Fig. 1. The benefit is that the constraints of scheduling tasks every can be satisfied in the coding stage, which can greatly enhance the efficiency of algorithm. To achieve this objective, we must do some special treatment when initializing the table. We first randomly assigned scheduling tasks to nurses until the scheduling tasks finished. Then the remaining nurses are random insert into the table where is blank. The following figure is one initial result which satisfied this requirement.

	Mon	Tue	Wed	Thu	Fri	Sat	Sun
Nurse1	day	night	late	null	early	late	day
Nurse2	early	null	night	day	late	night	night
Nurse3	null	early	early	night	day	day	null
Nurse4	night	day	day	late	null	day	late
Nurse5	null	late	early	early	night	null	early
Nurse6	late	early	null	early	null	early	null

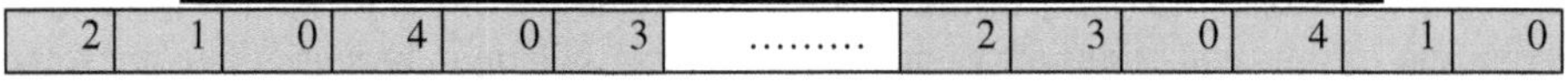

2	1	0	4	0	3	………	2	3	0	4	1	0

Fig. 1. Encoding of the chromosome for sub-problem

Selection. In this paper, we use the roulette wheel with elitist strategy as selecting operation. It is often required elitist strategy to ensure the convergence of the algorithm in multi-objective and multi-constrained optimization problem.

Crossover. In this paper, we use the crossover operator based on one day's scheduling tasks. The process is exchange one day's scheduling tasks from one chromosome to another chromosome, which is illustrated in Fig 2. This ensures that the vertical constraints are not destroyed in this process.

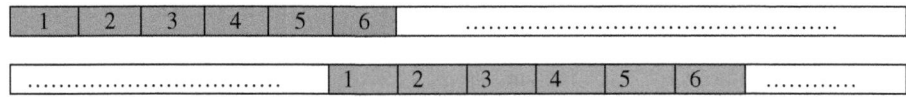

Fig. 2. The process of crossover

Mutation. In this paper, we use the mutation strategy with selection operator. The process is that first select one day's scheduling tasks from the chromosome, and then selects one gene from the scheduling tasks which dose on affect the vertical constraints for mutation. It is illustrated in Fig 3 and it also can ensure that the vertical constraints are not destroyed in this process.

……………	0	2	4	3	1	1	……………………

…………	0	2	4	3	0	1	……………………

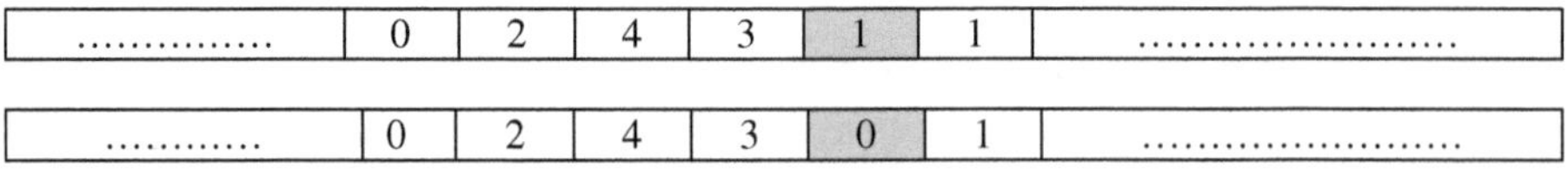

Fig. 3. The process of mutation

Evaluation. Max F(x) = [f1(x), f2(x), f3(x), f4(x), f5(x)] =w1*f1(x) + w2*f2(x) + w3*f3(x) + w4*f4(x) + w5*f5(x).

End conditions. When finding a feasible solution or the end of the loop condition.

Genetic algorithm with lcoal search. The numbers accorded to lemmas, propositions, and theorems etc. should appear in consecutive order, starting with the number 1, and not, for example, with the number 11.

Encoding. According to the result of splitting, we use one chromosome with the 15 genes to signify the nurse scheduling problem. Every gene indicates one sub-problem, and the sub-problems are obtained in above method (every sub-problem has 100 solutions).

Genetic operation. According Here, we also use the roulette wheel with elitist strategy as selecting operation. And the single point crossover and mutation is used as crossover operator and mutation operator.

Evaluation function. Max F(x) = [f1(x), f2(x), f3(x), f4(x), f5(x), f6(x), f7(x), f8(x)] = w1*f1(x) + w2*f2(x) + w3*f3(x) + w4*f4(x) + w5*f5(x) + w6*f6(x) + w7*f7(x) + w8*f8(x).

Neighborhood search. In order to improve the performance of the genetic algorithm, we proposed a variable neighborhood search based on 7 days. It is very likely the VNS which will be talking in following. The different is that the VNS based on 7 days consider 7 days (one week) as an element, and it can ensures that the vertical constraints are not destroyed in this process.

End conditions. When finding a feasible solution or the end of the loop condition.

3.3   Variable Neighborhood Search

Variable neighborhood search is a relatively recent meta-heuristic based on the simple idea of changing neighborhood within a local search to identify better local optima [10]. It has been applied to a wide variety in NP-hard problems such as the travelling salesman problem, the resource-allocation problem, the clustering problem, the linear ordering problem [12], vehicle routing, nurse scheduling and university course timetabling. An introduction can be seen in [11].

VNS is able to drive the search towards certain desirable objectives by defining the appropriate neighborhood structures associated with these objectives, although each result solution still needs to be evaluated by all the objectives in target function. As the VNS here is mainly used to make refinement on the GA's solution which has taken most constraints into account, the VNS's neighborhood structures should not be too complicated. In this paper, we apply the neighborhoods of swapping groups of consecutive shifts which are inspired by human scheduling processes of re-allocating sections of schedules. All possible swaps have been considered in these neighborhoods. That is shifts in a timespan from one day to the whole scheduling period can be switched between any set of two nurses in the schedule. To guarantee the satisfaction of the hard constraint HC1 of daily shift demands, swaps will only be allowed vertically. If any of the swaps results in an infeasible solution, this swap will be deemed to be invalid.

Table 2. One of possible moves in neighborhood between nurse1 and nurse3

	Mon	Tue	Wed	Thu
Nurse1	D	L	L	D
Nurse2	E	E	L	E
Nurse3	L		N	N

4   Experiment and Analysis

In this Section, we present the experimental results of the proposed algorithm and analyses its performance. All experiments were performed on a Windows XP machine with one 1.96 GHz Core2 processor and 2GB memory. And we use Java as programming languages to finish the algorithm.

To evaluable the performance of our algorithm, we used our method to solve the above nurse scheduling problem, and the example was executed 10 times. As show in table 1 and table 2, we find that our method can obtain the feasible solution and all the hard constraints are satisfied. As show in Fig. 4, it is clear that the variable neighbourhood search can effectively increase the performance. It plays a role in all the examples. One average, the VNS can improve the solution which is obtained by hybrid genetic algorithm 2.8%. Moreover, our method is stable. Fig. 4 illustrate that the standard deviation is 105.3 before VNS and 85.1 after VNS. In summary, our algorithm can obtain feasible solution in practice.

Table 3. 6 nurse's scheduling task in one week

	Mon	Tue	Wed	Thu	Fri	Sat	Sun
Nurse1	day	night	late	null	early	late	day
Nurse2	early	null	night	day	late	night	night
Nurse3	null	early	early	night	day	day	null
Nurse4	night	day	day	late	null	day	late
Nurse5	null	late	early	early	night	null	early
Nurse6	late	early	null	early	null	early	null

Table 4. 4 nurse's scheduling task in one week

	Mon	Tue	Wed	Thu	Fri
Nurse1	late	late	early	early	early
Nurse2	null	day	day	day	day
Nurse3	early	null	late	late	null
Nurse4	day	early	early	late	late

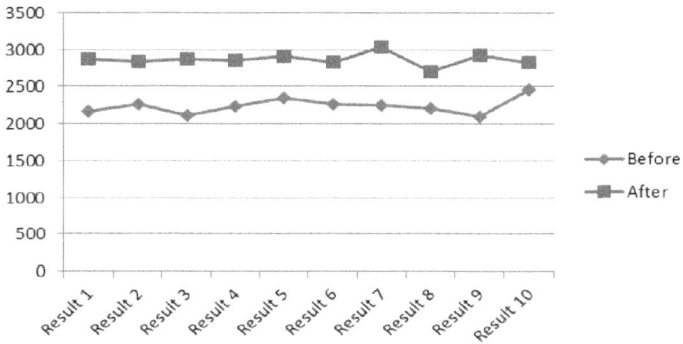

Fig. 4. The results before and after the VNS

5   Summary

It is a hot research fields that combination of the advantages of different heuristic and meta-heuristic to design more effective hybrid heuristic algorithm for combinatorial optimization problems. Combination of genetic algorithm and variable neighborhood search, this paper proposed a hybrid genetic algorithm to solving the Nurse scheduling problem. Make use of the rapid of the traditional heuristic, the search diversity of genetic algorithm and the strong local search ability of neighborhood search to improve the quality of the solution and speed up the convergence of the algorithm. In the experiment, we run our method in a practical nurse scheduling problem. The result illustrates that our algorithm can obtain better feasible solution.

Acknowledgement. This work is supported by Natural Science Foundation of China (61003066, 61070033), Doctoral Program of the Ministry of Education (20090172120035), Natural Science Foundation of Guangdong province (9251009001000005, 9151008901000165, 10151601501000015), Science and Technology Planning Project of Guangdong Province (2010B050400011, 2010B080701070, 2008B080701005, 2009B010800026), "Eleventh Five-year Plan" Program of the Philosophy and Social Science of Guangdong Province (08O-01), Opening Project of the State Key Laboratory of Information Security (04-01) and Fundamental Research Funds for the Central Universities, SCUT(2009ZM0052).

References

1. Burke, E.K., Li, J., Qu, R.: A hybrid model of integer programming and variable neighborhood search. European Journal of Operational Research 203, 439–484 (2010)
2. Aickelin, U., Dowsland, K.A.: Exploiting problem structure in a genetic algorithm approach to a nurse rostering problem. Journal of Scheduling 3(3), 139–153 (2000)
3. Kawanaka, H., Yamamoto, K., Yoshikawa, T., Shinogi, T., Tsuruoka, S.: Genetic Algorithm with the Constraints for Nurse Scheduling Problem. Evolutionary Computation 2, 1123–1130 (2001)

4. Ohki, M., Morimoto, A., Miyake, K.: Nurse Scheduling by Using Cooperative GA with Efficient Mutation and Mountain-Climbing Operators. Intelligent Systems, 164–169 (2006)
5. Burke, E.: Variable neighborhood search for nurse rostering problems. Metaheuristics: Computer Decision-Making, 153–172 (2003)
6. Goodman, M.D., Dowsland, K.A., Thompson, J.M.: A grasp-knapsack hybrid for a nurse-scheduling problem. Journal of Heuristics 15, 351–379 (2009)
7. Gutjahr, W.J., Rauner, M.S.: An ACO algorithm for a dynamic regional nurse-scheduling problem in Austia. Computers & Operations Research 34, 642–666 (2007)
8. Maenhout, B., Vanhoucke, M.: An electromagnetic meta-heuristic for the nurse scheduling problem. Journal of Heuristics 13, 359–385 (2007)
9. Maenhout, B., Vanhoucke, M.: Comparison and hybridization of crossover operators for the nurse scheduling problem. Annals of Operations Research 159, 333–353 (2008)
10. Karp, R.M.: Reducibility among combinatorial problems. Complexity of Computer Computations, 85–103 (1972)
11. Tien, J.M., Kamiyama, A.: On manpower scheduling algorithms. Society for Industrial and Applied Mathematics 24, 275–287 (1982)
12. Mladenovic, N., Hansen, P.: Variable neighbourhood search. Computers and Operations Research 24, 1097–1100 (1997)
13. Hansen, P., Mladenovic, N.: Variable neighbourhood search: Principles and applications. European Journal of Operational Research 130, 449–467 (1999)
14. Gonzáles, C.G., Pérez-Brito, D.: A variable neighbourhood search for solving the linear ordering problem. In: Proceedings of the Fourth Metaheuristics International Conference (MIC), pp. 181–185 (2001)

Evolutionary Computation of Multi-Band Antenna Using Multi-Objective Evolutionary Algorithm Based on Decomposition

Dawei Ding[1], Hongjin Wang[1], and Gang Wang[2]

[1] Department of Telecommunication Engineering, UJS, Zhenjiang, China
[2] Department of Electronic Engineering and Information Science, USTC, Hefei, China
gwang01@ustc.edu.cn

Abstract. Design of multi-band antenna involves multiple characteristics such as return loss in multiple operation bands. To apply MOEA/D (multi-objective evolutionary algorithm based on decomposition) to antenna structure optimization, it was introduced into this field for the first time. MOEA/D framework was demonstrated at first. Then it was in conjunction with electromagnetic solver, HFSS (high frequency structure simulator) to optimize and design tri-band bow-tie antenna to serve as an example. The evolutionary results showed that MOEA/D worked efficiently and generated multiple candidate structures at one single iteration, and that MOEA/D had lower computational overhead than NSGA-II (non-dominated sorting genetic algorithm II) for this problem. Therefore, MOEA/D shows great potential for antenna structure optimization and design.

Keywords: Bow-tie antenna, multi-objective evolutionary algorithm, multi-objective optimization.

1 Introduction

There have been many applications of optimization algorithms over the last several decades in the area of antenna engineering, which show great promise in handling the increasing demands on antenna design in terms of challenging performance, size, shape, and overall cost.

Recent years have witnessed that more and more new algorithms have emerged, such as Genetic Algorithm (GA), Bacterial Foraging Algorithm (BFA), Particle Swarm Optimization Algorithm (PSO), and Non-dominated Sorting Genetic Algorithm (NSGA) [1-4] in order to solve a multi-objective optimization problem (MOP). Moreover, these algorithms have been used successfully in synthesis and optimization problems. In [1], GA is employed to generate arbitrarily shaped planar monopole designs, which exhibits improved broadband performance and reduced size compared to reverse bow-tie. In [2], BFA is utilized to optimize the included angle of symmetrical V-dipole for higher directivity. The design of an antipodal Vivaldi antenna for ultra wideband (UWB) is implemented using PSO in [3]. And in [4], NSGA is used to design automobile conformal antenna considering the voltage standing wave ratio (VSWR) value at 100MHz and the average gain of the azimuthal gain pattern.

B. Liu and C. Chai (Eds.): ICICA 2011, LNCS 7030, pp. 383–390, 2011.

These algorithms attract much attention because of their diversity and flexibility, but each has its own advantage and disadvantage. For example, with the micro-particle swarm optimizer, a kind of modified version of PSO, improved optimization performances can be obtained especially for solving many structure parameters problems of antenna. It's advisable to consider GA firstly when the antenna shape in question refers to multi-dimension binary encoding. On the other hand, there are too many parameters and it's very difficult to select suitable operators in GA. PSO might prematurely converge to a local minimum. In addition, the majority of exiting multi-objective evolutionary algorithms (MOEAs), including NSGA, are based on Pareto dominance. This kind of Pareto dominance based algorithms are time-consuming to some extent for some antenna optimization problems. In view of the above, more effective optimization algorithms are appreciated by antenna design engineers.

A new kind of MOEA framework, referred to as multi-objective evolutionary algorithm based on decomposition (MOEA/D), was proposed by Zhang et al. in 2007. Unlike other optimization algorithms, especially Pareto dominance based evolutionary algorithm, MOEA/D has the following features (partly can be seen in [5-6]):

MOEA/D does not treat MOP as a whole. It converts a MOP into several scalar sub-problems by means of decomposition.

Many scalar optimization mechanisms, such as fitness assignment and diversity maintenance, can be utilized to solve MOP. And many optimization techniques, such as single objective local search, can be used in procedure.

Many sub-problems are simultaneously optimized using the information from its neighboring sub-problems, which has lower computational complexity.

Its search ability is not degraded using large population when the number of the objective function is more than three.

Objective normalization techniques can be incorporated into MOEA/D for solving disparately scaled objectives.

More recently, one version of MOEA/D, MOEA/D-DE (MOEA/D with differential evolution), has been proposed and utilized to solve some test instances with complicated Pareto set (PS) shapes. The experimental results have shown that it performs well [6]. Moreover, in the unconstrained MOEA competition (CEC) 2009, MOEA/D ranks first among 13 state-of-the-art MOEAs. In view of these, MOEA/D is promising in the area of antenna optimization design.

In this paper, we use MOEA/D-DE to optimize and design planar antenna for multi-band operation. For demonstration, we aim at the design of multi-band double-sided printed bow-tie due to its compact size, easy fabrication and simple structure. Double-sided printed dual-band and tri-band bow-tie antennas have been proposed by overlapping its two arms [7-8]. In our MOEA/D-based design, we follow the basic structure proposed in [8]. It is interesting to show what we will get if MOEA/D-based design is implemented. We expect more alternative structures for tri-band bow-tie antenna can be presented so that we may propose structures with high tolerance to the fabrication errors. To the best of our knowledge, it is the first paper that applies MOEA/D to this field.

The paper is organized as follows. In section 2, framework of MOEA/D-DE and basic steps for antenna structure optimization are introduced. In section 3, antenna optimization problems and MOEA/D-DE parameter setting are presented. Section 4 gives the evolutionary process and design results comparing with NSGA-II-DE.

2   MOEA/D-DE Framework

In MOEA/D implementation, a MOP is firstly decomposed into a number of single objective optimization sub-problems. Then each sub-problem is optimized using the information from its neighboring sub-problems. A sub-problem is essentially a weighted aggregation of all the individual objectives in the MOP. Neighborhood relations among these sub-problems are defined based on the Euclidean distance between their weight vectors [5]. MOEA/D-DE employs differential evolution (DE), one of the most powerful real parameter optimizers, as the search method [9]. DE operator and a polynomial mutation operator are utilized to generate a new solution from the selected solutions.

It's assumed that the optimization is for minimization, and the objective function of the i-th sub-problem is $g(x/\lambda^i)$. MOEA/D-DE works as follows [5-6]:

2.1   Algorithm Parameters

N: The number of the weight vectors and sub-problems;

$\lambda^1, \lambda^2, \cdots, \lambda^N$: A set of N weight vectors;

Gen: The number of generation;

Realb: The probability of selecting mating parents from neighborhood;

L: The maximal number of the updated solutions;

Niche: The number of the weight vectors in the neighborhood of each weight vector, and;

Rate: The probability of real-polynomial crossover and mutation.

It's noteworthy that Realb and L are set to maintain the diversity of population and in principle, Gen couldn't be determined beforehand for most real world optimization problems. N weight vectors represent different search direction.

2.2   Algorithm Steps

Step 1: Initialization

Step 1.1 For each i ($i = 1, 2, 3, \cdots, N$), compute the Niche closest weight vectors to λ^i in terms of Euclidean distance which are regarded as a set $T(i)$. Then store the information in the i-th individual.

Step 1.2 Initialize population $\{x^1, x^2, \cdots, x^N\}$ randomly, and ideal point.

Step 2: Update

For each $j = 1, 2, 3, \cdots, Gen$, we do the following work:

For each $i = 1, 2, 3, \cdots, N$, do:

Step 2.1 Selection of Mating/Update Range: Generate a random number rand, then set

$$G = \begin{cases} T(i) & \text{if (rand} < Realb) \\ \{1, 2, \cdots, N\} & \text{otherwise} \end{cases} \qquad (1)$$

Step 2.2 Reproduction: Select two different number m and n from G, produce a solution $\overline{y}$ by a DE operator, and then apply a real polynomial mutation operator on $\overline{y}$ to generate a new solution y. If y is out of the given decision space, then reset a value inside randomly.

Step 2.3 Update of population and ideal point: For each index $k \in T(i)$, if $g(y/\lambda^k) \leq g(x^k/\lambda^k)$, then set $x^k = y$. Then update the ideal point.

Step 3: Check

If the results meet performance requirements, stop and output the last population. Otherwise reset Gen to evaluate again.

3   Tri-band Antenna Design with MOEA/D-DE

3.1   Antenna Design Problem

To demonstrate the operation of MOEA/D in multi-band antenna design, we consider the optimization design of a double-sided printed bow-tie antenna shown in Fig. 1. The bow-tie antenna with overlapped arms shown in Fig. 1 has been proven to operate in the GPS band ranging from 1.57042GHz to 1.58042GHz and the two WLAN bands ranging from 2.4GHz to 2.5GHz and from 5.15GHz to 5.83GHz [8]. In [8], conventional parameter optimization method is employed, which was very time-consuming. We apply MOEA/D-DE to this optimization problem in this article.

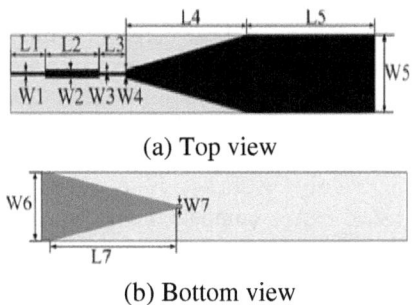

(a) Top view

(b) Bottom view

Fig. 1. Geometry of a multiband bow-tie antenna with overlapped arms

Table 1. Range of the design parameters for the tri-band bow-tie antenna

Parameter	L3	W3	L2	W2	W1
Range (mm)	0.1~10.91	0.1~11.2	0.1~10.91	0.1~11.2	0.1~11.2

Based on our previous analysis in [8], we may just take the length and width of the three stage impedance transformer, (L3,W3,L2,W2,L1,W1), into account. Due to the limitation on the length and width of exiting structure, L1 could be ignored and the parameters change in some decision space. Therefore, we consider five decision parameters, (L3,W3,L2,W2,W1) of which the decision space is shown in Table 1. In this paper, the objective is to assure the return loss of being less than -10dB over the GPS and two WLAN frequency bands. Therefore, we define

$$\text{objective function} = \max\left(10 - \min_{f\in[f_1,f_2]}|S_{11}|, 0\right). \tag{2}$$

All of the units are dB. $[f_1, f_2]$ in (2) may define the three bands respectively. The objective will reach 0dB when all the value of return loss over the given frequency range are more than 10dB, which is what we want through the optimization.

In addition, the function evaluation is realized by means of HFSS simulation. And a kind of technology of automatic modeling and simulation is applied to the function evaluation in order to improve the efficiency of multi-objective optimization. The interface of Visual C++ 6.0 for HFSS version 10.0 is developed for this aim.

3.2   Parameter Setting

Tchebycheff approach [5] is employed. We use one of the versions of DE, rand/1/bin [9], where crossover rate and scaling factor are set to be 0.1 and 0.5 respectively. N weight vectors correspond to different search direction. Any method which could spread all of the space is appreciated. Here, we employ weight vectors like those in [5]. N is 28. Realb is set to be 0.9. Niche is considered as 12. And Gen is set to be 18. It should be remarked that these parameters are set empirically, and the setting can affect the algorithm performance to some extent.

4   Evolutionary Process and Design Results

4.1   Evolutionary Process

As shown in Fig. 2, the evolutionary process can be displayed visually in terms of the population. Fig. 2 shows the 3D graph of the evolutionary process of the initial, 3rd, 6th, 9th, and 18th population where the plus, circle, star pentagon, rectangle, and asterisk signs mean corresponding population respectively.

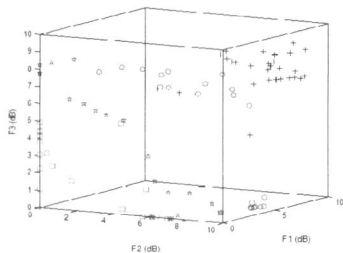

Fig. 2. 3D graph of the fitness of the initial, 3rd, 6th, 9th, 18th generation

4.2   Evolutionary Results

It took about 42.5 hours in a PC with Intel Core 2 at 2.99GHz with 2.0GB RAM running Windows XP to optimize the above tri-band bow-tie antenna. The optimization of five parameters and three objectives is completed after 18 generations. Finally, twelve solutions are available of which the objective vectors are (0dB,0dB,0dB). It's noteworthy that those twelve solutions are the expected points which satisfy with the performance requirements and they consist of one part of the Pareto set where its Pareto front is a point. The solutions are listed in Table 2. We note that the structure in [8] is out of the solutions we get after one run.

Table 2. Twelve candidate designs

Design	L3 (mm)	W3 (mm)	L2 (mm)	W2 (mm)	W1 (mm)
1	4.71398	0.230299	7.0236	1.11451	1.26001
2	3.00249	0.142959	7.35499	0.913356	1.19463
3	3.69229	0.130218	7.69602	1.06132	1.24756
4	5.2968	0.249436	7.27409	1.29293	1.29252
5	5.58836	0.274178	7.20046	1.32507	1.33221
6	3.05411	0.115892	6.70507	0.934141	1.18012
7	4.37677	0.192999	7.01878	1.08127	1.23812
8	4.10264	0.184065	6.90855	1.07697	1.23812
9	3.72011	0.100193	7.94942	0.89861	1.04554
10	3.4913	0.137832	6.7935	1.03942	1.21622
11	6.09905	0.309045	7.06313	1.40873	1.315
12	4.05703	0.194703	5.75494	1.14838	0.946857

4.3   Result Analysis

Twelve solutions are presented above. However, it's necessary to study the available physical size in the practical engineering application because the dimension can't be realized, such as 4.71398mm. To this aim, we analyzed the sensitivity of the five parameters and truncated the size parameters. Here, we take the 6th candidate for demonstration. Among the parameters, L3, L2, and W3 affect the bandwidth and its center frequency of the third operation band to some extend, W1 only tunes the center frequency of the third band, and W2 affects the bandwidth and the center frequency of the third band markedly. By increasing W2, the center frequency shifts to low frequency end and the bandwidth get wider. So we should pay much attention to W2 in the design and fabrication. For the precision of 0.1mm which can be met by commercial printed circuit board (PCB) engraving machines, we suggest a truncated solution as L3=3.1mm, W3=0.1mm, L2=6.7mm, W2=0.9mm, and W1=1.2 mm.

Fig. 3(a) shows the return loss and Fig. 3(b) shows the radiation pattern of the truncated one. From Fig. 3, we have the observation that the return loss over the frequency range of 1.5742GHz-1.5842GHz, 2.4GHz-2.5GHz, and 5.15GHz-5.83GHz are less than -10 dB. The radiation patterns are typical pattern of dipole-like antenna. It should be pointed out that the pattern at 5.49GHz shows down-tilted beam due to asymmetry of the two arms as defined in [8].

Therefore, MOEA/D-DE works well for our multi-band antenna design.

Comparison with NSGA-II

For comparison with NSGA-II-DE [6][10], NSGA-II in conjunction with DE, we also set the parameters the same as those in MOEA/D-DE. NSGA-II-DE does also work. It generated seven efficient designs after 30 generations. It's clearly seen that NSGA-II-DE took more time and generates less designs than MOEA/D-DE.

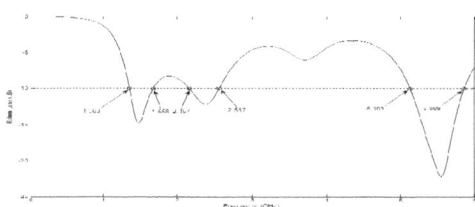

Fig. 3. (a) The simulated return loss of the 6th antenna (truncated)

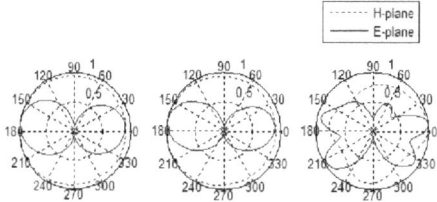

Fig. 3. (b) The simulated radiation pattern of the 6th antenna at 1.575GHz, 2.45GHz and 5.49 GHz (truncated)

5   Conclusion

An up-to-date multi-objective optimization evolutionary algorithm, called MOEA/D-DE, is employed to solve the optimization design of tri-band bow-tie antenna. Several alternative solutions are presented with lower computational overhead than NAGA-II. Therefore, MOEA/D is a promising optimization technology for some real world multi-objective antenna structure design.

Acknowledgements. The authors are grateful to Prof. Qingfu Zhang and Dr. Aimin Zhou of Essex University, UK, for their helpful suggestions. This work was supported in part by the Ministry of Science and Technology of China under Grant 2009GJC10042.

References

1. Kerkhoff, A.J., Rogers, R.L., Ling, H.: Design and Analysis of Planar Monopole Antennas Using a Genetic Algorithm Approach. IEEE Trans. Antennas Propag. 52, 2709–2718 (2004)

2. Mangaraj, B.B., Misra, I.S., Barisal, A.K.: Optimizing Included Angle of Symmetrical V-dipoles for Higher Directivity Using Bacteria Foraging Optimization Algorithm. Progr. Electromagn. Res. 3, 295–314 (2008)
3. Chamaani, S., Mirtaheri, S.A., Abrishamian, M.S.: Improvement of Time and Frequency Domain Performance of Antipodal Vivaldi Antenna Using Multi-objective Particle Swarm Optimization. IEEE Trans. Antennas Propag. 59, 1738–1742 (2011)
4. Kim, Y., Walton, E.K.: Automobile Conformal Antenna Design Using Non-dominated Sorting Genetic Algorithm (NSGA). IEE Proceedings: Microwaves, Antennas and Propagation, 579–582 (2006)
5. Zhang, Q., Li, H.: MOEA/D: A Multiobjective Evolutionary Algorithm Based on Decomposition. IEEE Trans. Evol. Comput. 11, 712–731 (2007)
6. Li, H., Zhang, Q.: Multiobjective Optimization Problems with Complicated Pareto Sets, MOEA/D and NSGA-II. IEEE Trans. Evol. Comput. 13, 284–302 (2009)
7. Wang, G., Liu, J., Xia, J., Yang, L.: Coaxial-Fed Double-Sided Bow-Tie Antenna for GSM/CDMA and 3G/WLAN Communications. IEEE Trans. Antennas Propag. 56, 2739–2742 (2008)
8. Zhang, Y., Wang, G.: Tri-band Bow-Tie Antenna with Overlapped Arms. Microw. Opt. Tech. Lett. 52, 1539–1542 (2010)
9. Price, K.V., Storn, R.M., Lampinen, J.A.: Differential Evolution: A Practical Approach to Global Optimization, New York (2005)
10. Deb, K., Pratap, A., Agarwal, S., Meyarivan, T.: A Fast and Elitist Multiobjective Genetic Algorithm: NSGA-II. IEEE Trans. Evol. Comput. 6, 182–197 (2002)

The Study of Improved Grid Resource Scheduling Algorithm

Qingshui Li[1], Yuling Zhai[2], Shanshan Han[1], and Binbin Mo[1]

[1] Computer Science and Technology College, Zhejiang
University of Technology Hangzhou, China
[2] Hangzhou Yuanjiang Technology Ltd. Hangzhou, China
mdusa@zjut.edu.cn, 303510628@qq.com

Abstract. With the computer technology and network technology development, great meet people's work and life needs, but a single computer can not meet the needs of computing or storage, grid resource scheduling strategy can achieve resource sharing, This paper introduces artificial school of fish algorithm to grid resource scheduling in order to further use the element of heuristic optimization method to find a more suitable high-performance grid computing environment, resource scheduling strategy. Through uses AFSA algorithm solving this kind of scheduling of resources question, seeks the new key to the situation for the scheduling of resources question, by enhances the scheduling of resources effectively the efficiency. And carried on the simulation experiment after the improvement algorithm in the Gridsim grid simulation software, and has carried on the contrast with other algorithms, finally indicated this article proposed the algorithm has the better search ability and the convergence rate.

Keywords: grid resources, Artificial School of Fish, DCC strategy, simulations, analysis.

1 Introduction

Grid is use high-speed Internet or private network to the earth widespread computing resources, storage resources, communication resources, network resources, software resources, data resources, information resources, knowledge and resources together into a logical whole, Elimination of the computer network composed of resource islands of information and resources, and ultimately provide users with a virtual and efficient information processing and resource sharing environment, and to the greatest extent to meet the various users of computing and resource sharing needs.

Grid architecture division of the main functions of the basic components of the system, specify the purpose and function of components, describe the interaction between the components, and integrate the various components is to help people be more accurate Understanding the mechanism of the grid to run and play grid advantage is the core part of the grid is the soul of the grid. Mature theory of the two current grid architecture is a 5-layer architecture of the hourglass, the first is out of Ian Foster Boshi Ti. The second is the Open Grid Services Architecture (referred to as OGSA) is Dr. Ian Foster and IBM, proposed by the Global Grid Forum 4 is an

B. Liu and C. Chai (Eds.): ICICA 2011, LNCS 7030, pp. 391–398, 2011.

important criterion for recommendations, while also following the five most important after the hourglass structure is one of the latest Kinds of grid architecture [1]. Here are the two architectures are introduced. Five hourglass architecture [2].

The architecture is a kind of agreement as the center of the structure, emphasizing service and the importance of API and SDK. Architecture is divided into: structural layer, link layer, resource layer, the convergence layer and application layer, as follows Fig. 1.

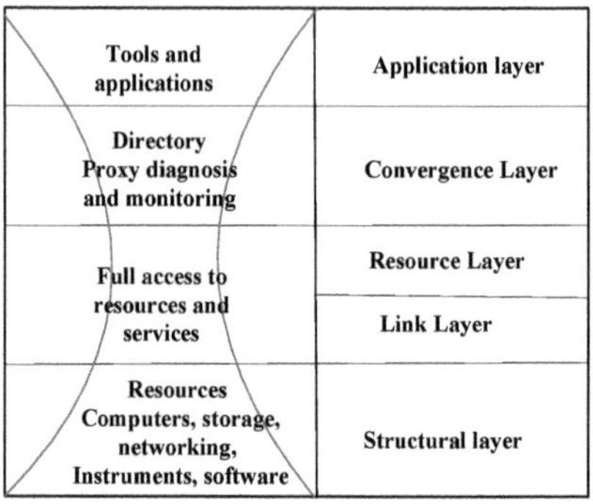

Tools and applications	Application layer
Directory Proxy diagnosis and monitoring	Convergence Layer
Full access to resources and services	Resource Layer
	Link Layer
Resources Computers, storage, networking, Instruments, software	Structural layer

Fig. 1. Five layer grid protocols

2   Grid Resources Scheduling

Grid resources scheduling simple said to a group of users submit task through certain strategy submitted to a group of resources up execution, Regarding the user, can in the price which may withstand obtain a more satisfactory service, looking from the dispatch way and the process, may divide into the dynamic scheduling and the static scheduling. Regarding the static scheduling, compared with typical has the Min-min algorithm, the Max-min algorithm, this kind of algorithm characteristic is the need beforehand forecast each duty running time, before the work execution on each task assignment correspondence's processor, forecast from this in advance the task execution order, the execution beginning time, the computing time as well as complete information and so on time, Under grid system environment, on the one hand the duty has the complexity, the duty form not solely is a Yuan duty, in many situations is compound the duty, also namely a duty the Yuan duty constitution which strict relates successively by many; On the other hand, the resources dynamic, is very unstable in the performance, when uses under the market economy model in the scheduling of resources, between the resources also has the competition relations, increased the scheduling of resources strategy complexity,   Therefore in this kind of situation proposed that can satisfy the strategy which this kind highly complex and the dynamic duty falls not to be very easy, a good scheduling strategy, enormous will reduce the task execution the price, enhances the grid system's volume of goods handled, thus raises the grid system's efficiency.

3   Improve of Artificial School of Fish Algorithm

3.1   DCC Strategy Introduces and the Algorithm Realizes

DCC strategy: Explains this strategy through the following Fig. 2:

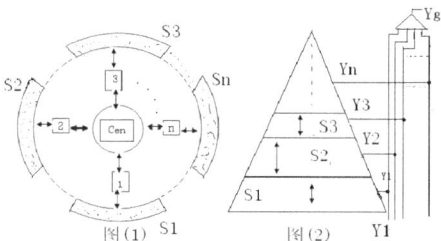

Fig. 2. DCC strategy

The DCC strategy profits from the multi-school of fish competition thought that in chart (1), $S_1,S_2,S_3....S_n(n=1,2,3...)$has the different search strategy artificial school of fish algorithm, digital 1, 2, 3...n respectively is various schools of fish correspondence call-board. Central call-board C_{en} when the search process is various schools of fish communication platform, the search termination may output the optimum value; it also played time the reduction communication data conflict role. Figure (2) is to Figure (1) longitudinal explanation, from the chart saw that in this system's each (here figure of school of fish) maintains the relative independence in the search process and in the result output, between each realizes the communication or the data updating through C_{en}.

$S_1,S_2,S_3....S_n(n=1,2,3...)$is the tendency hypothesis which increases gradually according to the algorithm order of complexity. The S1 algorithm order of complexity is low, single carries out the speed to be relatively quick; S2 is higher than the S1 order of complexity, the single execution has the high search efficiency algorithm $(S_1,S_2,S_3....S_n(n=1,2,3...)$ likewise), Meanwhile to avoid the data conflict which multiple-valued reads, each group reads in first own call-board value call-board a buffer, and disposes a condition flag bit for it, after various schools of fish each time iterate, besides renews this group in the call-board the value, must carry on one time with the C_{en} call-board in value quite to operate,   , if is more superior than the C_{en} in value, this crowd of call-board value to the specific buffer in the write data, and starts C_{en} to carry on the traversal, renews own value.

3.2   DCC Strategy Structural Design

In this article tests the system to suppose is three, uses three different algorithm strategies. The S1 level uses AFSA; In the S2 level joins the Gauss variation operator, but under the DCC strategy, is the most superior artificial fish's condition which uses in the C_{en} call-board guides; In the S3 level's algorithm is integrates the simulation annealing operator in the S2 foundation.

3.3   Experiment Simulations and Performance Analysis

In order to examine the algorithm the performance, selects following three to have the model multi-peak function (solution maximum value) to carry on the simulation to test [3].

$$f_1(x_1, x_2) = 0.5 - \frac{\sin^2 \sqrt{x_1^2 + x_2^2} - 0.5}{1.0 + 0.001*(x_2^2 + x_1^2)}, -100 \le x_1, x_2 \le 100 \qquad (1)$$

The SchafferF6 function has the intense shake multi-peak function, has the maximum value 1 in (0,0).

$$\max f(x, y) = 1 + x \times \sin(4\pi x) - y \times \sin(4\pi y + \pi)$$

$$+ \frac{\sin(6\sqrt{x^2 + y^2})}{6\sqrt{x^2 + y^2 + 10^{-15}}}, \quad x, y \in [-1,1] \qquad (2)$$

The function is many peak functions, there are four overall situation maxima 2.118, symmetrical railing distribution Rayleigh in (+0.64, +0.64), (-0.64, -0.64), (+0.64, -0.64), (-0.64, +0.64), exist a great deal of local to biggest be worth, betwixt the region has a to take value and overall situation's maximum the approximate local be biggest worth (2.077) convex stand.

$$f_3(x,y) = \left(\frac{3}{0.05 + (x^2 + y^2)}\right)^2 + (x^2 + y^2)^2, \quad x, y \in [-5.12, 5.12] \qquad (3)$$

This function is needle-in-a-haystack, optimal solution for 3600, approximate distribution in (0, 0); Four local extreme value point (+ 5.12, + 5.12), (5.12, - 5.12), (+ 5.12, - 5.12), (5.12, 5.12) for the function value in 2748.78.

For the full comparison and the examination this article algorithm strategy in seeks for the overall situation maximum value ability, separately with AFSA, in literature [4] the Gauss variation algorithm, the literature [5] variation operator and the simulation annealing algorithm carries on the simulation result contrast to the above four trial function, compares the peak value which and the value stable condition together they search. The algorithm USES Java programming realization, to eliminate random disturbance, the algorithm of this paper respectively, AFSA, literature [4] and [5] algorithm operated independently 20 times. The parameter establishment is: Each artificial school of fish scale is 50, each artificial fish attempt greatest number of times $Try\_number = 20$, crowded factor $\delta = 0.618$, field of vision Visual=4, the error range $\varepsilon = 10^{-4}$, length of stride Set p=0.50, iterates most greatly 100 times, annealing operation cooling coefficient C=0.90, initial temperature $T_0 = 100$, annealing operation biggest iteration number of times $K_{max} = 15$.

Below is uses three trial functions to carry on the test separately to four algorithms the result: (to express convenient, in next table based on variation operator and simulation annealing artificial fish algorithm simple form is SAGMAFSA).

Table 1. Four algorithms comparison on f_1 by 20 times

Algorithm	Maximum	Mean value	Optimum deviation value
AFSA	0.989794798110644	0.98337164644356256	1.6312e-005
AGMAFSA	0.9902070769633385	0.98391260539051147	1.2468e-004
SAGMAFSA	0.9901889328960866	0.98092636476070706	8.4470e-005
This algorithm	0.9998245930193874	0.989476153372777095	9.4777e-006

Table 2. Four algoriths comparison on f_2 by 20 times

Algorithm	Maximum	Mean value	Optimum deviation value
AFSA	2.118472927099663	2.11560309223124501	3.1899e-006
AGMAFSA	2.1182940024024353	2.11640564565168435	4.2385e-006
SAGMAFSA	2.1183667464864664	2.11545538598508767	1.2740e-005
This algorithm	2.118595965897675	2.11793559795461482	2.7084e-007

Table 3. Four algorithms comparison on f_3 by 20 times

Algorithm	Maximum	Mean value	Optimum deviation value
AFSA	3576.2712765630818	3148.567316010188495	1.1055e+005
AGMAFSA	3563.7778185298234	3167.278840555648061	1.0025e+005
SAGMAFSA	3561.818716254671	3156.34944626340971	9.3052e+004
This algorithm	3598.4147014766136	3275.265709959620475	1.2770e+005

From table 1, table 2 and table 3 can be seen, to 4 algorithm respectively using three function of test obtain the optimal value of the maximum and the average of the optimal value, this paper algorithm is better to other three algorithms; Especially the use f_1, f_2 carries on tests, the optimum value standard deviation obvious is smaller than other three algorithms, this explained that this article algorithm asks in the solution performance the stability to be quite good; In uses the function to test in separately the respective independence 40 tests to four algorithms, falls into the partial optimum value the number of times: AFSA falls into 15 times, the AGMAFSA algorithm 8 times, the literature [6-8] algorithm falls into 5 times, but this article algorithm falls into 2 times, this explained that this article algorithm has well compared to other three algorithms avoids falling into the partial optimal solution ability.

4 Grids Scheduling of Resources Strategy Based on AFSA

This article proposed the artificial school of fish algorithm's improvement strategy, applies after the improvement algorithm in the grid scheduling of resources, the grid scheduling of resources is a NP difficult problem, what current community intelligence algorithm solution NP kind of question mostly was uses in solving TSP, the combination optimization question, the experimental result has made a more remarkable progress.

4.1   The Basic Grids Scheduling of Resources Strategy Based on AFSA

The grid scheduling of resources' basic philosophy may use the following fig.3 indicate:

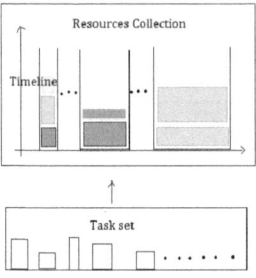

Fig. 3. Resource Matching

Task scheduling goal is assigns duty set inside all duties to the corresponding resources set, the final goal is causes between each resource the load to be basically balanced.

4.2   Basic Mentality of Grid Scheduling of Resources Based on AFSA

Applies AFSA in the grid scheduling of resources most important task is considered how unifies this algorithm and the grid scheduling of resources, chooses the appropriate code and the decoding rule[9]. The chromosome code uses the indirect code the way, namely the resources - duty's indirect encoding method, the chromosome indicated with the unvaried integer, its length is equal to the complete child duty the integer[10], like Fig. 4:

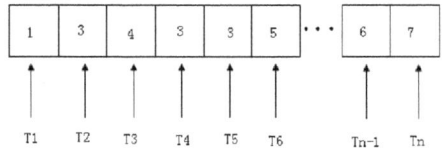

Fig. 4. Coding strategy

Thought analysis of grid scheduling of resources algorithm Based on AFSA

Initialization: Supposes the resources total is N, namely AR={R1,R2,R3,...,RN}, supposes the resources the performance factor respectively is r1,r2,r3,...,rr; The school of fish integer is M (duty integer), by AJ={J1,J2,J3...,Jm}, the fish individual's vision is away from Visual, the biggest permission attempt number of times is ntry-num, the length of stride is Step; The density marginal value is σ [11].

The duty just started, in the supposition each resources' duty number was 0 (relatively present need dispatch duty, then which one was did not have to enter resources), and the hypothesis this time each resources load was equal, then regarding some fish individual, enters the resources the way is stochastic; Like Fig. 5:

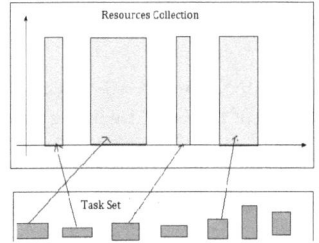
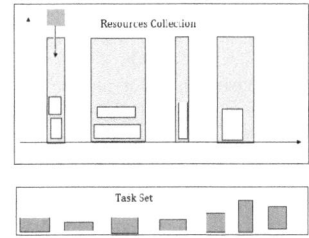

Fig. 5. Initial Matching of the Gridlets   **Fig. 6.** FIFO matching

Because the fish individual's number is more than far conforms to the condition resources integer, after therefore period of time, did not have the load for the spatial resources, then the fish individual was carrying on food (resources) chose, needs take the density (load) as is standard. Like Fig 6.

Gathers the group behavior: If we above enter from the crown stack's way regard are the AFSA algorithm follow behavior, then the resources exchange was belongs based on food density gathers the group behavior, In order to enable each fish to eat equally full, may let in the different subgroup in the fish other subgroup's individual exchange, after causing the exchange, these two fish both sides can obtain the warm and sufficient condition, namely realizes the affluent society, the common enrichment, avoids the obvious gap between rich and poor, like Fig. 7.

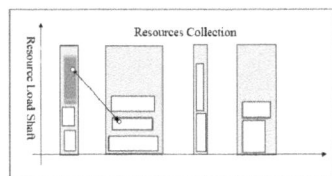

Fig. 7. Exchange of Gridlet

5   Summary

This article applies AFSA in the solution grid scheduling of resources question is a positive attempt, is applies to the artificial school of fish algorithm in the solution grid scheduling of resources question is a positive attempt, Is to artificial school of fish algorithm theory more thorough research and the application, has further developed the artificial school of fish algorithm application domain and the fundamental research scope, has the positive reference significance regarding the current grid computation's research and the application. Has carried on the experimental verification based on the AFSA grid scheduling of resources strategy which studies to this article, and has carried on the contrast with the classics algorithm Min-min, Max-min algorithm, has confirmed the feasibility which and the superiority this article calculated.

References

1. Xu, H.: Grid based on independent task scheduling algorithm. 5, 10–11 (2008)
2. Foster, I., Kesselmna, C., Tuecke, S.: The Anatomy of the Grid: Enabling Scalable Virtual Orgnaiztions. International J. Supercomputer Applications 15(3) (2001); Proceedings of the First IEEE/ACM Intenrational Symposium on Cluster Computing and the Grid (2001)
3. Du, H., Jiao, L., et al.: Immune optimization calculation, learning and recognition. Science and Technology Press 7, 401–402 (2007)
4. Lu, X.Y., Cai, F.: The improvement of artificial fish algorithm Based on competition. Journal of Wuzhou College 18(3), 66–72 (2008)
5. Du, H., Jiao, L., et al.: Immune optimization calculation, learning and recognition. Science and Technology Press 7, 401–402 (2007)
6. Vincenzo, D.M., Mililotti, M.: Sub-optimal scheduling in a grid usinggenetic algorithm. Parallel Computing (2004)
7. Stutzle, T., Dorigo, M.: A short convergence proof for a class of antcolony optimization algorithms. IEEE Transactions on Evolutionary Computation (2005)
8. Granvill, L.Z., Da rose, D.M., Panisson, A., et al.: Managing com2puter networks using peer2to2peer technologies. IEEE Communications Magazine 43 (2005)
9. Kun, W.X., Po, L.H.: Operate to computing a mesh operate to adjust one degree algorithm based onmisty shot excellent. Computer Science (2007)
10. The studying and imitate of the mesh task based on heredity algorithm. Master's thesis
11. Abraham, A., Buyya, R., Nath, B.: Natur heuristics forscheduling jobs on computational Grids. In: Proc. of the 8th IEEE International Conference on Advanced Computingand Communications, pp. 45–52. IEEE, Cochin (2000)

An Improved Genetic Algorithm
for Network Nodes Clustering*

Yong Li[1,2] and Zhenwei Yu[1]

[1] College of Mechanical and Electronic Engineering, China University of Mining and
Technology (Beijing), Beijing 100083, China
[2] School of Information Science and Technology, Yanchang Teachers College, Yancheng
224002, Jiangsu Province, China
yctcyy@163.com

Abstract. Nodes clustering is a useful way to construct an effective network
infrastructure for large-scale distributed network applications. In this paper,
network nodes are clustered by the K-medoids clustering algorithm according
to their coordinates. The coordinates of network nodes are gained by Vivaldi
which is a simple and lightweight network coordinates system. But K-medoids
algorithm is sensitive to the initial cluster centers and easy to get stuck at the
local optimal solutions. In order to improve the performance of K-medoids
algorithm, KCIGA(K-medoids clustering based on improved genetic algorithm)
is presented in this paper. The improved genetic algorithm that uses self-
adaptive genetic operator, dynamically adjusting the crossover rate and
mutation rate, can avoid premature and slow convergence phenomenon in
SGA(standard genetic algorithm). Experimental results show KCIGA has good
reliability and expansibility, and it is effective for clustering network nodes.

Keywords: Network nodes clustering, improved genetic algorithm, K-medoids,
Vivaldi.

1 Introduction

As the internet infrastructure becomes more pervasive and connects increasingly
powerful commodity resources, many diverse and complex applications attempt to
harvest the potential of that infrastructure. For many of these applications, being able
to identify nodes that are relatively close to each other in the internet, i.e., identify
node clusters, can provide ways to improve both performance and scalability [1]. For
example, peer-to-peer routing protocols need to maintain sets of nearby peers in order
to route queries more efficiently [2,3,4]. Other application domain for which
clustering is becoming an important issue is that of distributed computing. In order to
assign the geographic adjacent nodes to the same cluster, the topology information of
network nodes must be learned. The work in [5] assumes that each autonomous
system has a SNMP. The static network topology can be obtained through SNMP.

---

* The research is supported by "the Fundamental Research Funds for the Central Universities"
(grant No.:2011YJ).

B. Liu and C. Chai (Eds.): ICICA 2011, LNCS 7030, pp. 399–406, 2011.

But this method requires all routers, switches and end systems in the autonomous system to support SNMP protocol, so it is not practical. The study in [6] obtains topology information directly from the BGP router, nodes grouped into clusters based on their IP addresses. However, due to the end nodes not having the right to access BGP routing table, this method is not appropriate. Zhang et al. [7] proposes the method of nodes clustering based on static landmarks, but this method must designate some nodes as static landmarks. When the number of end nodes is large, landmark nodes will become the bottleneck [4]. Jiang et al. [8] calculates the network distances among the nodes according to their network coordinates, and uses the K-means algorithm to cluster nodes. Owning to randomly select the number of clusters and initial cluster centers in K-means algorithm, the clustering result is greatly impacted.

At present, there are many clustering methods. Among the clustering algorithms based on partition, K-means and K-medoids are two main algorithms [9,10]. These two algorithms are very similar. They use a distance measure for evaluating the similarity of object, then determine an objective function to evaluate the quality of clustering results. Giving the initial cluster centers, the best clustering result can be obtained when the objective function takes the extreme value through iterating. The disadvantage of these two algorithms is sensitive to the initial cluster centers and easy to get stuck at the local optimal solutions. The difference of these two algorithms is in that K-means algorithm uses the average value of objects as cluster centers which are abstract points, not the concrete objects. In addition, the K-means algorithm is sensitive to the isolated point. Comparatively, K-medoids algorithm uses the actual objects as cluster centers, then may solute the isolated point question of K-means algorithm and enormously enhance the precision of clustering algorithm [10].

In this paper, network nodes are mapped into the coordinate space by Vivaldi [11] which is a simple and lightweight network coordinates system. Because the network nodes are separate spots in coordinate space, K-medoids algorithm which has good robustness on the isolated point problem is used for network nodes clustering. But K-medoids algorithm is sensitive to the initial cluster centers and easy to get stuck at the local optimal solutions. To address this problem, KCIGA is presented in this paper. The rest of this paper is organized as follows. In Section 2, we present network nodes clustering based on K-medoids. KCIGA is presented in Section 3. Experiments and simulation results are presented in Section 4. Finally, we conclude this paper in Section 5.

2   K-medoids Clustering for Network Nodes

Network nodes are clustered based on K-medoids according to their network coordinates which are learned by the network coordinates system Vivaldi. Given a network having n nodes, each node i has an m-dimension real value vector i.e., $v_i = \{v_{i1}, v_{i2}, \cdots, v_{im}\}, i = \{1, 2, \cdots n\}$. Network nodes are mapped into the points of metric space R^m , therefore, $V = \{v_1, v_2, \cdots v_n\}$ represents the sample space of network nodes.

$$V = \begin{bmatrix} v_1 \\ v_2 \\ \vdots \\ v_n \end{bmatrix} = \begin{bmatrix} v_{11} & v_{12} & \cdots & v_{1m} \\ v_{21} & v_{22} & \cdots & v_{2m} \\ \vdots & \vdots & & \vdots \\ v_{n1} & v_{n2} & & v_{nm} \end{bmatrix} = \{v_{ij}\}^{n \times m}, \tag{1}$$

where where v_{ij} is the j-dimensional value of node i, $i \in \{1,2,\cdots,n\}$, $j \in \{1,2,\cdots,m\}$. Using the K-medoids clustering algorithm, these n vectors $v_i (i = 1,2,\cdots,n)$ are classified into K clusters, and the center of cluster $V_k (k = 1,2,\cdots,K)$ should be c_k, assuming $C = \{V_1,V_2,\cdots,V_K\}$, $c = \{c_1,c_2,\cdots,c_K\}$. In this clustering method, $\forall i \neq j$, $V_i \cap V_j = \Phi$, and $V = \bigcup_{1 \leq k \leq K} V_k$. $|V|$ represents the number of samples i.e., n. $|V_k|$ represents the number of nodes in V_k. In K-medoids clustering algorithm, the high similarity between nodes in a cluster indicates that the sum of the distance between each node and the corresponding cluster center is small. $d(v_i,c_k)$ represents the distance between v_i to c_k, and it is defined as in (2).

$$d(v_i,c_k) = \|v_i - c_k\| = \sqrt{\sum_{j=1}^{m}(v_{ij} - c_{kj})^2}. \tag{2}$$

In order to reflect the affiliation between v_i and V_k, membership function is defined as in (3), therefore, the objective function can be defined as in (4).

$$u_{ik} = \begin{cases} 1 & (d(v_i,c_k) = min\{d(v_i,c_h)\}) \\ 0 & (d(v_i,c_k) \neq min\{d(v_i,c_h)\}). \end{cases} \tag{3}$$
$$1 \leq h \leq K, \quad v_i \notin c$$

$$J(c) = \sum_{k=1}^{K}\sum_{i=1}^{n} u_{ik} d(v_i,c_k) = \sum_{k=1}^{K}\sum_{i=1}^{n} u_{ik} \sqrt{\sum_{j=1}^{m}(v_{ij} - c_{kj})^2}. \tag{4}$$

In summary, the mathematical model of the clustering problem for network nodes is expressed as in (5),

$$\begin{aligned} Min \quad & J(c) \\ s.t. \quad & C = \{V_1,V_2,\cdots,V_K\} \\ & c = \{c_1,c_2,\cdots,c_K\} \\ & \forall i \neq j, V_i \cap V_j = \Phi \\ & V = \bigcup_{1 \leq k \leq K} V_k \end{aligned} \tag{5}$$

3   KCIGA for Network Nodes

3.1   The Basic Idea of KCIGA

As K-medoids clustering algorithm is very easy to fall into local optimum, and is sensitive to the choice of initial cluster centers. To address the problems, the K-medoids clustering based on improved genetic algorithm (KCIGA) is presented in this paper. The improved genetic algorithm that uses self-adaptive genetic operator, dynamically adjusting the crossover rate and mutation rate, can avoid premature and slow convergence phenomenon in SGA. The KCIGA algorithm flow chart is showed in Fig. 1.

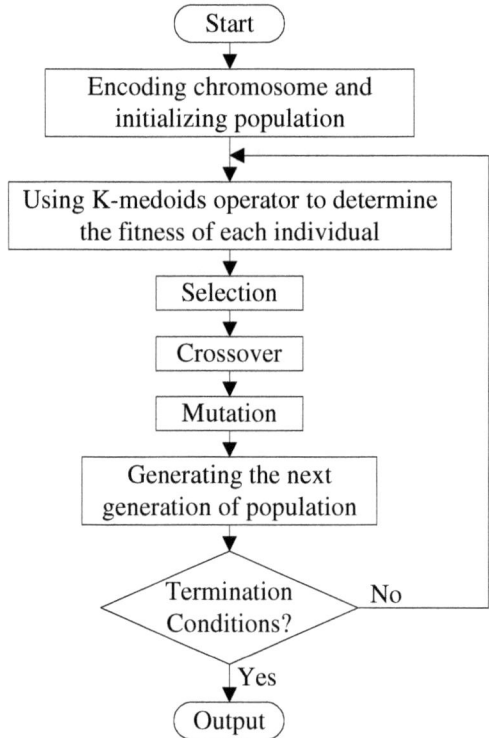

Fig. 1. KCIGA algorithm flow chart

In the proposed algorithm, K nodes are selected randomly as initial cluster centers to form a chromosome, a total of N chromosomes selected, then the fitness of each individual in the initial population is computed. Individuals selected according to their fitnesses, high-fitness individuals are selected for crossover and mutation operation, at the same time, low-fitness individuals are eliminated. Finally, the next generation of population is made of high-fitness individuals. For individuals in the new generation of population, and if there is an individual which stisfies the algorithm termination condition, then the algorithm terminates and outputs the optimal solution of clustering. Otherwise, go to continue. In this way, within each new generation of population, the average fitness is rising, each cluster center is closer to the optimal cluster center, and finally the optimal solution of clustering is gained.

3.2   Description of KCIGA

STEP 1: Encoding chromosome and initializing population. In this paper, we use real-coded, the value of chromosome gene corresponds to cluster center number, the length of the chromosome is the number of cluster, and the specific code form is expressed as $c=\{c_1,c_2,\cdots;c_K\}$, K is the number of cluster center of a chromosome. Select K cluster centers randomly to form a chromosome, if the center randomly selected has already exist in the   same chromosome, then remove the center and reselect until it reaches K centers. In this way, the population A whose size is N can be obtained finally.

STEP 2: Determining the fitness of each individual. This algorithm uses the inverse of objective function $J(c)$ as the fitness function, that is,

$$F(c) = 1/J(c),\qquad(6)$$

Where the smaller $J(c)$ is, the greater fitness will become, so the better clustering effect is. The fitness of each individual is determined by formula (6).

STEP 3: Slection operator. Proportional selection operator is used to select individuals, that is, the probability of an individual being selected is proportional to its fitness. The fitness of individual A_i is $F(A_i)$, then it is selected with probability $p(A_i)$, $p(A_i)$ can be obtained by the following formula.

$$p(A_i) = p(A_i)/\sum_{j=1}^{N} p(A_j).\qquad(7)$$

STEP 4: Crossover and mutation operators. The main goal of crossover and mutation operators is to create diverse and potentially promising new individuals. To address the premature or slow convergence phenomenon in SGA, self-adaptive operator is used to dynamically adjust the crossover rate p_c and mutation p_m. They are calculated as follows.

$$p_c = \begin{cases} p_{c1} - \dfrac{(p_{c1}-p_{c2})(F_{ave}-F')}{F_{max}-F_{ave}}, & F' > F_{ave} \\ p_{c2} & ,F' \le F_{ave} \end{cases}\qquad(8)$$

$$p_m = \begin{cases} p_{m1} - \dfrac{(p_{m1}-p_{m2})(F''-F_{ave})}{F_{max}-F_{ave}}, & F'' \ge F_{ave} \\ p_{m2} & ,F'' < F_{ave} \end{cases}\qquad(9)$$

where F_{ave} is the average fitness of each generation population; F_{max} is the largest individual fitness in the population; F' is the larger fitness of the two parent individuals; F'' is the fitness of mutating individual. According to the experiment, where the coefficient values is as follows, $p_{c1} = 0.95$, $p_{c2} = 0.65$, $p_{m1} = 0.5$, $p_{m2} = 0.1$.

STEP 5: Generating the next generation of population. For one population, to carry out selection, crossover and mutation operations to produce a new population. In this process, the probability of an individual being selected is proportional to its fitness. In order to protect the existing outstanding individuals and make individuals with lower fitness produce new excellent individuals through crossover and mutatation. Those individuals with high fitness have lower crossover rate and mutation rate; but individuals with low fitness have a higher crossover rate and mutation rate. In addition, the elitist strategy is used. If the highest fitness of individual in current population is larger than the fitness of the best individual so far, then the individual with highest fitness in current population is viewed as the new best individual so far. Otherwise, the worst individual in current generation is replaced by the best individual [12].

STEP 6: Termination conditions. There are two conditions for the algorithm terminates. One is that the iteration number exceeds the maximum number of iterations. Two is that the optimal result is obtained in algorithm execution. If any one of the above two conditions is satisfied, the algorithm would terminate and turn STEP 2 to continue.

4   Experiments and Results

4.1   Experiments

In order to simulate the underlying physical network, the hierarchical network topology which contains 6000 nodes is generated by BRITE for the experiments. Network topology is divided into 10 autonomous domains, each domain contains 600 nodes. The delay between nodes of inter-domain and intra-domain are 40ms-100ms and 5ms-30ms respectively. For the purpose of evaluating the effect of clustering algorithm, ACD(average cluster delay) is used as the criteria of evaluation. It is defined as in (10). The smaller the value of ACD is, the better the clustering effect is.

$$ACD = \frac{\sum_{k=1}^{K} \left(\sum_{i,j \in V_k \wedge i \neq j} D(i,j) \Big/ \left(|V_k| (|V_k|-1) \right) \right)}{K} , \tag{10}$$

where $D(i,j)$ is the delay between node i and node j. Meanwhile, AND (average network delay) is defined as in (11), this value can be used to compare with ACD. AND is the average delay of all the network nodes.

$$AND = \frac{\sum_{i,j \in V \wedge i \neq j} D(i,j)}{n(n-1)} . \tag{11}$$

4.2   Results and Analysis

The network coordinates of all the nodes are obtained from the Vivadli, then network nodes are divided into clusters by KCIGA. In this paper, 30 experiments about KCIGA are done, and different initial population is selected in each experiment. The value of ACD in each experiment is recorded and shown in Fig. 2. We can see that ACD is much smaller than AND in the figure, which indicates KCIGA is reliable and effective for network nodes clustering.

In oreder to compare the performance of KCIGA, K-medoids and KGA(K-mediods based on standard genetic algorithm), 30 experiments are separarately done to them, different initial population or initial cluster centers selected in each experiment. The performance comparision is shown in Fig. 3. The clustering result of KCIGA is the best of them, and KCIGA can restrain to a stable optimum value under different initial population. K-medoids is sensitive to the initial cluster centers and very easy to fall into the local extremum. When different initial cluster centers are choosed in each experiment, the fluctuation of ACD value is big. The premature and slow convergence phenomenon sometimes occurs in KGA, therefore, the clustering result of KGA is inferior than that of KCIGA.

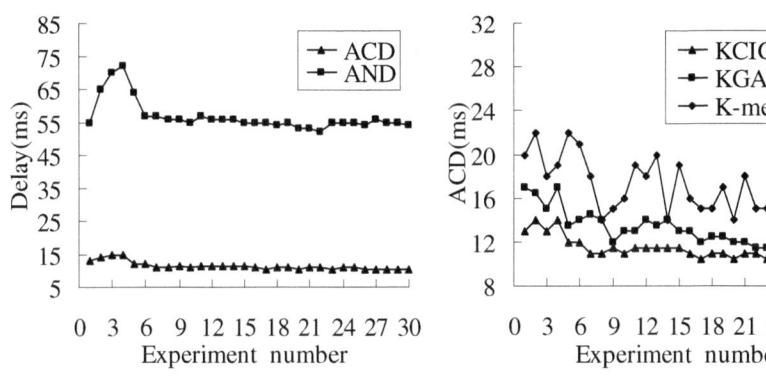

Fig. 2. The effect of KCIGA   Fig. 3. The performance of KCIGA

5   Conclusion

A new network nodes clustering method which is based on the coordinates of nodes is presented in this paper. In this method, Vivaldi is used to generate the coordinates of nodes, then, nodes are clustered by K-medoids. However, K-medoids is sensitive to initial cluster centers and easy to get stuck at local optimal solutions. To address this problem, KCIGA is presented. Experiment results indicate that KCIGA is effective and reliability. Moreover, using Vivaldi to forecast the distance between nodes will not bring big end-to-end burden for measuring when the network scale is large, thus KCIGA has high scalability. In future research work, first, reducing the influence of the error of Vivaldi to clustering results will be studied; second, large-scale distributed application such as P2P, application layer multicast will be studied to reduce network traffic and raise routing efficiency based on network nodes clustering.

References

1. Agrawal, A., Casanova, H.: Clustering Hosts in P2P and Global Computing Platforms. In: Proceedings of the 3rd IEEE/ACM International Symposium on Cluster Computing and the Grid, pp. 367–373. IEEE CS Press, Tokyo (2003)
2. Deng, Z.-j., Song, W., Zheng, X.-f.: P2PKMM: A Hybrid Clustering Algorithm over P2P Network. In: Third International Symposium on Intelligent Information Technology and Security Informatics, pp. 450–454. IEEE CS Press, Jinggangshan (2010)
3. Li, W., Wen, L.: P2P Traffic Control Method Based on Network Distance Measurement and Clustering. Computer Engineering 35(7), 93–95 (2009)
4. Zheng, L.-m., Li, X.-d., Li, X.-y., Sun, W.-d.: Survey on P2P overlay clustering technology research. Application Research of Computers 27(3), 806–810 (2010)
5. Huang, D.-y., Huang, J.-h., Zhuang, L., Li, Z.-p.: A Distributed P2P Network Model Based on Active Network. Journal of Software 15(7), 1081–1089 (2004)
6. Krishnamurthy, B., Wang, J.: On network-aware clustering of web clients. In: ACM SIGCOMM, pp. 97–110. ACM Press, Stockholm (2000)
7. Zhang, X.-y., Zhang, Q., Zhang, Z.-s.: A construction of Locality-Aware Overlay Network: Overlay and Its Performance. IEEE Journal on Selected Areas in Communications 22(1), 18–28 (2004)
8. Jiang, Y., You, J.-y., Shi, H.-b.: Network Coordinates-based Peer-to-Peer Hosts Clustering. In: IFIP International Conference on Network and Parallel Computing, Tokyo, Japan, pp. 205–210 (2006)
9. Yang, L., Zhong, C., Lu, X.-y.: Advances for Distributed Clustering Algorithms Based on P2P Networks. Micro-electronics & Computer 26(8), 83–85 (2009)
10. Yang, T.-f., Zhang, X.-p., Liu, Y.-w.: Spatial clustering algorithm with obstacles constrains by QPSO and K-Medoids. Electronic Design Engineering 19(2), 74–77 (2011)
11. Dabek, F., Cox, R., Kaashoek, F., Morris, R.: Vivaldi: A Decentralized Network Coordinate System. In: ACM SIGCOMM, pp. 631–637. ACM Press, Portland (2004)
12. Jin, Y.-p.: An Intelligent Exam-paper Generating Method Based on Genetic Algorithm. Ms D Thesis. Harbin Engineering University, Harbin, China (2009)

Learning Form Experience: A Bayesian Network Based Reinforcement Learning Approach

Zhao Jin[1], Jian Jin[2], and Jiong Song[3]

[1] Yunnan University, 650091, Kunming, China
[2] Hongta Group Tobacco Limited Corporation, 653100, Yuxi, China
[3] Yunnan Jiao Tong Vocational and Technical College, 650101, Kunming, China

Abstract. Agent completely depends on trail-and-error to learn the optimal policy is the major reason to make reinforcement learning being slow and time consuming. Excepting for trail-and-error, human can also take advantage of prior learned experience to plan and accelerate subsequent learning. We propose an approach to model agent's learning experience by Bayesian Network, which can be used to shape agent for bias exploration towards the most promising regions of state space and thereby reduces exploration and accelerate learning. The experiment results on Grid-World problem show our approach can significantly improve agent's performance and shorten learning time. More importantly, our approach makes agent can take advantage of its learning experience to plan and accelerate learning.

Keywords: Reinforcement learning, Learning experience, Bayesian network, State transition.

1 Introduction

Because the learning mechanism of trial-and-error, Reinforcement Learning (RL) makes agent can autonomously learn without human's assistance [1], which brings the major advantage of RL. But the trial-and-error mechanism is also the reason to cause RL being slow and time-consuming, because the trial-and-error mechanism essentially is exhaustive search [2].

Shaping (guiding) agent's exploration to avoid complete trial-and-error (exhaustive search) is an effective way to accelerate RL [3]. Wang [4] calculates Bayesian posteriors to help agent improving action selection quality for reaching the goal state faster. Amizadeh [5] uses Bayesian approach to measure the uncertainty of reaching state and receiving rewards, and help agent selecting actions for reaching the states with more rewards. Doshi [6] uses Bayesian method to incorporate prior knowledge into the decision making of state transition. Joseph [7] introduces how use Bayesian method to balance exploration and exploitation in the learning process, and he thinks more exploitation would lead less exploration.

These above works show the power and fitness of Bayesian method in RL. First, Bayesian method is naturally fit to represent prior knowledge, especially the knowledge acquired from prior learning; second, it is very powerful to deal with the uncertainty of action selection and the state transition. But we think the function of

B. Liu and C. Chai (Eds.): ICICA 2011, LNCS 7030, pp. 407–414, 2011.

Bayesian method in RL is still under-estimated. In most application, Bayesian method only be used to deal with local optimal, for example, to help agent deciding to select which action and reach which next state[4][5], or provide scattered and cluttered prior knowledge[6][7][8][9]. The global and overall knowledge representation and inference capability of Bayesian method are not developed in depth.

In this paper, we propose an approach to model the state transition that lead agent achieving the goal state by Bayesian Network, to shape (guide) agent's exploration in whole course. The Bayesian Network has two function: on the one hand to shape agent following more effective state transitions for reducing exploration; on the other hand, it can be used to accumulated the knowledge that agent learned in training episode, and make agent taking advantage of knowledge more thoroughly for more exploitation but not exploration[7].

During the learning process, we record the state trajectory agent passed in every training episode, and eliminate all state loops exist in the state trajectory to get acyclic state trajectory. In the acyclic state trajectory, each state would have a fixed distance to the goal state, which is measured by the number of steps that agent move from this state to the goal state. From these accumulated acyclic state trajectory, we would find the shortest distance of each state to the goal state. Then these states would be the node of Bayesian Network, the shortest distance of each state to the goal state would establish the structure of Bayesian Network, and the state transition probability would be the parameters of Bayesian Network. So the Bayesian Network is made up for shaping agent to follow state transitions which have the shortest distance to the goal state.

The experiment results on Grid-World problem show our approach can significantly improve agent's performance and shorten learning time. More importantly, our approach provides a kind of way to make agent can take advantage of its own experience to accelerate learning.

2   Creating the Structure of State Transitions Bayesian Network

The knowledge agent learned from training episode are uncertain, accumulative, and changing along with the learning process, because the trial-and-error learning mechanism. Therefore it is necessary to find some fit knowledge frame to represent such stochastic knowledge. In order to satisfy the requirement, it is natural to select Bayesian Network as the knowledge representation frame.

Bayesian Networks is a kind of graph model to present the stochastic dependency relationship between variable. In Bayesian Networks, the node is the stochastic variable, and the directed edge is the relationship between variables, the degree of dependency is measure condition probability [10]. In our approach, Bayesian Network is used to model the states transitions that lead agent achieving the goal state. The node is the state, the structure is established by the shortest distance of the state to the goal state, which is acquired by accumulated acyclic state trajectory, so to select the shortest state transition path would shape agent reaching the goal state faster and reducing exploration.

We have already given the method to eliminate state loops in original state trajectory to get the acyclic state trajectory in [11]. Here we just give an example on Grid-World problem to show the original state trajectory and the acyclic state trajectory. Fig.1 gives the original state trajectory agent wandered in the state space of Grid-World   problem, and Fig.2 gives the acyclic state trajectory.

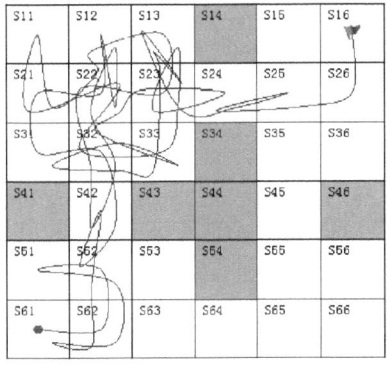
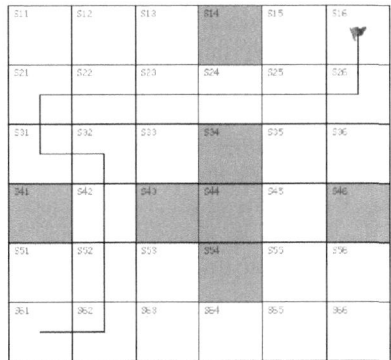

Fig. 1. Original state trajectory   **Fig. 2.** Acyclic state trajectory

Algorithm 1. Creating the structure of sate transitions Bayesian Network
Input
sg: the goal state
total_ast: the accumulated acyclic state trajectories
Output
L_bp: the structure of sate transitions Bayesian Network
Step 1. Initialization:
Create a null list set: L_bp
Create a null list l1, and add sg into l1
Add l1 into L_bp, and let k←1
Step 2. While k < the length of tota_ast Do
Create the kth state layer lk, and add lk into L_bp
For each acyclic state trajectory t in tota_ast Do
    Find the index i of sg in t
    If t [i-k] is not null
        If t [i- k] not exist in any layer between L_bp [2] and L_bp [k - 1]
            Add t [i- k] into lk
        End if
        For each state layer lm between L_bp [k + 1] and L_bp [k + m] Do
            If t [i- k] exist in lm
                Remove t [i- k] from lm
            End if
        End for
    End If
End For
k←‖⇓1

End While
Step 3. Return L_bp
End Algorithm 1

From these accumulated acyclic state trajectory, we would find the shortest distance of each state to the goal state. Then these states would be the node of Bayesian Network, the shortest distance of each state to the goal state would establish the structure of Bayesian Network, and the state transition probability would be the parameters of Bayesian Network. We give the method to create the structure of Bayesian Network in Algorithm 1.

Fig.3 gives the acyclic state trajectories after finishing six times training episode and Fig.4 is the structure of Shaping Bayesian network created by Algorithm1 from these acyclic state trajectories.

S61 -> S51 -> S52 -> S42 -> S32 -> S22 -> S23 -> S24 -> S25 -> S15 -> S16

S61 -> S62 -> S63 -> S64 -> S65 -> S55 -> S45 -> S35 -> S36 -> S26 -> S16

S61 -> S51 -> S52 -> S62 -> S63 -> S64 -> S65 -> S66 -> S56 -> S55 -> S45 -> S35 -> S36 -> S26 -> S16

S61 -> S62 -> S52 -> S53 -> S63 -> S64 -> S65 -> S55 -> S45 -> S35 -> S36 -> S26 -> S16

S61 -> S62 -> S52 -> S53 -> S63 -> S64 -> S65 -> S55 -> S45 -> S35 -> S25 -> S26 -> S16

S61 -> S51 -> S52 -> S42 -> S32 -> S31 -> S21 -> S22 -> S23 -> S24 -> S25 -> S15 -> S16

Fig. 3. Accumulated acyclic state trajectories after 6 times training episode

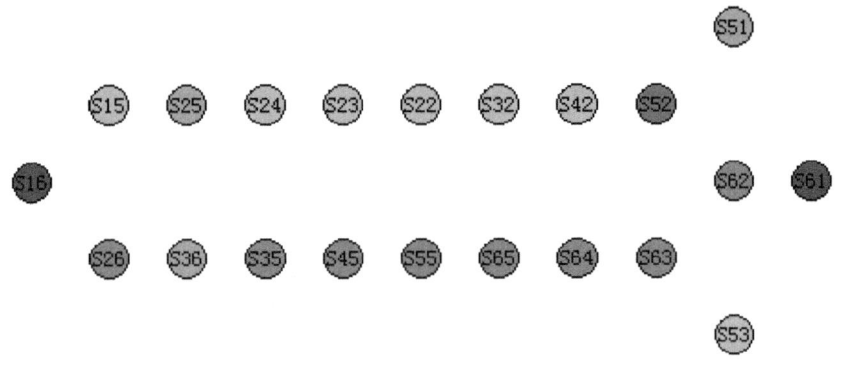

Fig. 4. The structure of Bayesian Network

3   Computing the Parameters of Bayesian Network

The Bayesian Network can not only represent the stochastic dependency relationship between variable, but also compute quantitatively the dependency degree, which can provide clear information for agent's decision. From these accumulated acyclic state trajectories, we can count the times that every state was passed by agent, and also count the path numbers that from this state to the goal state, so we can get the state transition probability (parameters) according to these two counted number. The formal procedure to compute the parameters of Bayesian Network is given in Algorithm 2.

Algorithm 2. Computing the parameters of Bayesian Network
Input:
total_ast: the accumulated acyclic state trajectories
Layer_CS: the structure of Bayesian Network
Output:
   Trans_p: the state transition probability of Bayesian Network
Step 1. Initialization:
Define the structure Trans_p as:
Trans_p
{
   Start_state
   End_state
   Trans_times
   Passed_times
   t_p
}
Create null list Trans_ps to store the state transition probability between states
Step 2. Counting the state path number Trans_times:
i←0; j←0; k←0
For each shortest path l in total_ast do
   Create new structure Trans_p
   Trans_p. Start_state← Layer_CS [i+1][k]
   Trans_p. End_state← Layer_CS[i][j]
   Trans_p. Trans_times++
   If Layer_CS[i][j] and Layer_CS[i+1][k] are adjacent directly then
        Trans_p.t_p←1.0
   End if
   Add Trans_p to Trans_ps
End for
Step 3. Counting the Passed_times of Start_state
For each Trans_p in Trans_ps do
    Trans_p. Passed_times←the times that Trans_p. Start_state appeared in total_ast
End for
Step 4. Computing the state transition probability in each Trans_p of Trans_ps
   For each Trans_p in Trans_ps do
        Tans_p.t_p← Trans_p. Trans_times / Trans_p. Passed_times
   End for
End Algorithm 2

   By Algorithm 2, according to the accumulated acyclic state trajectories and the structure of Bayesian Network, we can compute the state transition probability, which are the parameters of Bayesian Network. By now, we get the whole Bayesian Network, and can be used to shape agent's exploration from the start state to the goal state, which gives agent the whole course guide.

Fig.5 gives the state transition probability of the structure of Bayesian Network, and Fig.6 gives the complete Bayesian Network, which parameters can be check in Fig.5.

start_state	end_state	transition_probability
S15	S16	1
S26	S16	1
S25	S15	1
S25	S26	1
S36	S26	1
S24	S25	1
S35	S25	1
S24	S36	0.3333333
S35	S36	1
S23	S24	1
S23	S35	0.3333333
S45	S35	1
S33	S23	1
S55	S45	1
S32	S33	1
S65	S55	1
S42	S32	1
S64	S65	1
S52	S42	1
S63	S42	0.25
S52	S64	0.25
S63	S64	1
S53	S52	1
S82	S52	0.3333333
S51	S52	1
S53	S63	1
S82	S63	1
S51	S63	0.3333333
S81	S53	0.3333333
S81	S82	1
S81	S51	1

Fig. 5. The state transition probability of Bayesian Network

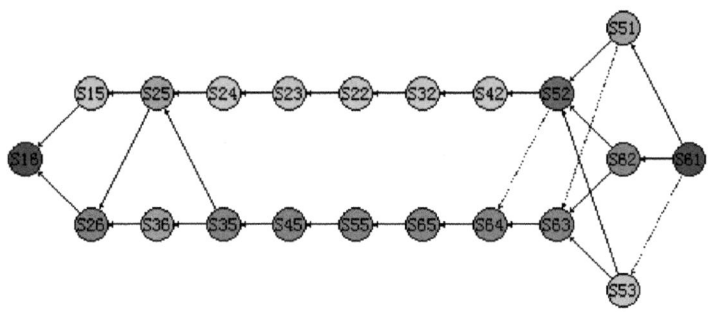

Fig. 6. The State Transitions Bayesian Network

4   Experiments

In a 10×10 Grid-World problem, we compared the time to reach the goal state, and the time to find the optimal policy between traditional reinforcement learning and reinforcement learning with Bayesian Network in different 20 times learning. The comparison results are shown in Fig.7 and Fig.8.

From Fig.4, it is obvious that shaping can significantly reduce agent's exploration for ineffective state space area, and make it reach the goal state faster. From Fig.5, it is clear that shaping can significantly reduce the learning time. The experiment results show our approach is very effective.

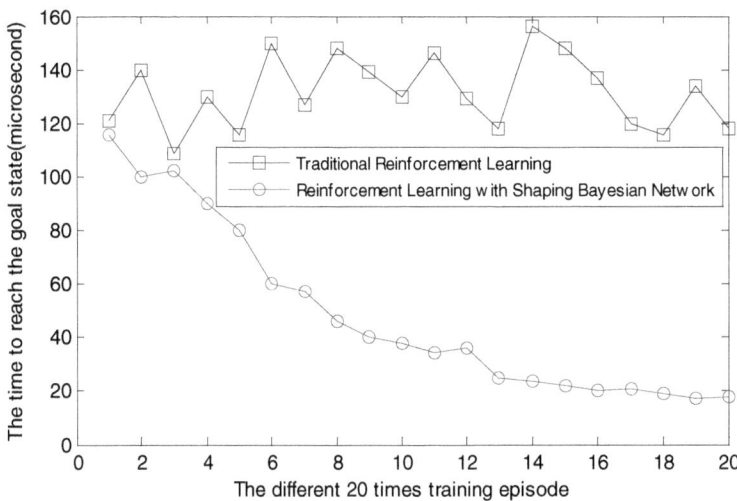

Fig. 7. The time comparison to reach the goal state

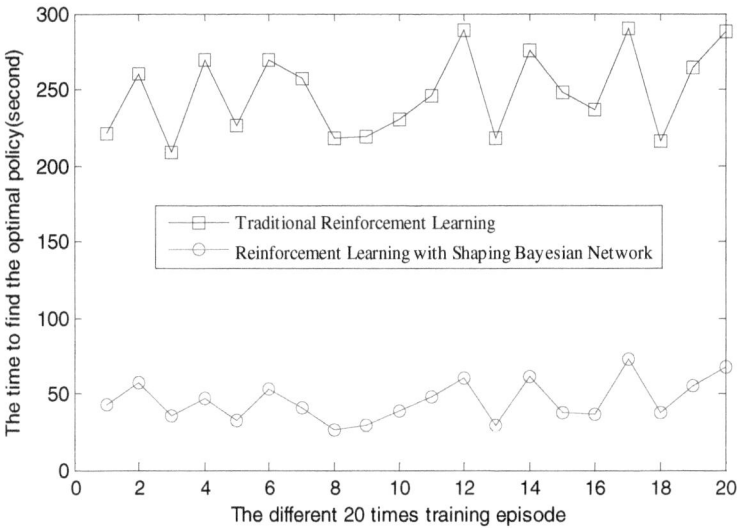

Fig. 8. The time comparison to find the optimal policy

5   Conclusion

We propose a Bayesian Network based reinforcement learning approach, which can significantly reduce the state space that agent has to explore. By representing and inferring knowledge that agent learned from training on the Bayesian Network, agent can do more from exploitation but not exploration, which can greatly accelerate learning and shorten the learning process. The experiment results show the applicability and effectiveness of our approach.

References

1. Sutton, R.S., Barto, A.G.: Reinforcement Learning: An Introduction. MIT Press (1998)
2. Sertan, G., Faruk, P., Reda, A.: Improving reinforcement learning by using sequence trees. Machine Learning 81(3), 283–331 (2010)
3. Grzes, M., Kudenko, D.: Online learning of shaping rewards in reinforcement learning. Neural Networks 23(4), 541–550 (2010)
4. Wang, t., Daniel, L.: Bayesian Sparse Sampling for On-line Reward Optimization. In: Proceedings of the 22nd International Conference on Machine Learning, pp. 956–963 (2005)
5. Amizadeh, S., Ahmadabadi, M.: A Bayesian Approach to Conceptualization Using Reinforcement Learning. In: 2007 International Conference on Advanced Intelligent Mechatronics, pp. 1–7 (2007)
6. Doshi, F., Pineau, J.: Reinforcement Learning with Limited Reinforcement: Using Bayes Risk for Active Learning in POMDPs, pp. 256–263 (2008)
7. Joseph, R., Peter, S.: Online Kernel Selection for Bayesian Reinforcement Learning. In: Proceedings of the 25th International Conference on Machine Learning, pp. 816–823 (2008)
8. Bob, P., Craig, B.: A Bayesian Approach to Imitation in Reinforcement Learning. In: Proceedings of IJCAI 2003, Proceedings of the Eighteenth International Joint Conference on Artificial Intelligence, pp. 712–720 (2003)
9. Firouzi, H., Ahmadabadi, M.N.: A Probabilistic Reinforcement-Based Approach to Conceptualization. International Journal of Intelligent Systems and Technologies 3, 48–55 (2008)
10. Pearl, J.: Probabilistic reasoning in intelligent systems: networks of plausible inference, pp. 79–119. Morgan Kaufmann, San Mateo (1988)
11. Jin, Z., Jin, J., Liu, W.: Autonomous Discovery of Subgoals Using Acyclic State Trajectories. In: Zhu, R., Zhang, Y., Liu, B., Liu, C. (eds.) ICICA 2010. LNCS, vol. 6377, pp. 49–56. Springer, Heidelberg (2010)

The Simulation System of Cytokine Network Based IMMUNE-B

Xianbao Cheng and Bao Chen

Guang zhou Technology & Business Collage, Department of Computer
{luyu1233,chenbao2010}@163.com

Abstract. Introduces two kinds of cytokine network model that already exist, and describes some of problems in the two models. This paper presents a new model system: IMMUNE-B, This model is created using B method. B method is one of the most popular international practical formal method currently, people use it to write software system specification, system design and description, It has many advantages. And the model uses computer program to simulate the process of cytokine network of the immune system. The simulation results are basically consistent with the prediction, it illustrate the model have it's real research value. It provides a new way for research of the cytokine network.

Keywords: cytokine network, B method, IMMUNE-B model, computer program, simulate.

Cytokine network is the development direction of Immunology, and also is a hot research currently, but it is also a difficulty. On the one hand, although there are many people in this area explore vigorously, and made a lot of research results. From the study result we can see many problems still exist, although these results found a number of cytokines, and research the role of cytokines in-depth, but still no clear what the role of cytokines in the process of its role in the end, Some only know the biological role of cytokines or know the results their role, which is not enough for the network of factors[1]. On the other hand, previous studies often used more traditional methods, these traditional methods have significant limitations. With the development of computer technology, computer studies in various fields has become a useful auxiliary tool, why do not we think of ways to use computers to simulate the role of cytokines in order to find the regular pattern from it?[2].

In previous studies, there have been many immunological models: such as: IMMSIM model, Multi-Agent immunization model, these models, both aim at simulation some of the features of the immune system, but not focus on the processes immune that occur. For the inadequacies of these models, we used method B to create a IMMUNE-B model. The model simulate the role of the process about the cytokine network of the immune system with a computer program, the system can be controled and viewed, and other models can not match advantage in the time control.

B method is one of the most popular international practical formal method currently, people use it to write software system specification, system design and description. B method is produced based on VDM and Z, because some of its significant advantages:

B. Liu and C. Chai (Eds.): ICICA 2011, LNCS 7030, pp. 415–423, 2011.
© Springer-Verlag Berlin Heidelberg 2011

such as the symbols and methods of B method used support most of the software process: from requirements analysis, specification, software design through system implementation; Layered software structure along with the identification and validation is the guiding principle of B method[3]. The advantages coupled with the strong support of software tools, B has been successfully applied in many fields in Europe and the United States, including real-time, simulation, information processing and engineering. The B method in biological engineering, artificial life areas has a special application, also it has a large number of software tools to support, so descript such complex network such as cytokine network, B is the best option. We use it to describe a cytokine of the network model and then use computer programs to implement, equivalent to a network of immune factors move to an abstract from the real world. In the abstract world of their simulation, we not only know the final results, but also can be observed the dynamic process in the simulation. we can control on time, so that we may find out the regular pattern of the role of cytokine in the network, so as to establish the correct cytokine network model provides a strong basis, then verify the factor from the role of this model, I believe it will bring great convenience cytokine network in the future. IMMUNE-B model is designed primarily for the process of the factor activity cell about cytokine network can be simulated by computer, the process of cytokines and antigen affect and the relationships between them.

1   The Two Cytokine Network Models Describes

As the network complex and large, there is a long road about cytokines research, people to explore factors in every possible way to study it. In the existing research achievements, basically has the following kinds of model:
IMMSIM model

(1) Basic characteristics
IMMSIM basic mathematical mechanism is a strengthening of the cellular automata. The so-called strengthening of IMMSIM model is used some special agreement in automatic machine, mainly contains:
The implementation of the rules is the probability of the event, achieved through the introduction of random number.
Each grid location contains a number of individuals, and a collection of individuals close to the location only contains one other individual.
Individuals can move from one location to another location.
Model Based IMMSIM software has been developed, such as CIMMSIM, IMMSIM3 etc..

(2) Model components
① Grid
② Definition of individual
③ Definition of status
④ Instruction expression and affinity expression

⑤ Interaction rules

⑥ Clonal selection

(3) Immune response simulation

A typical simulation injected antigen twice, thus observing the system changes. First, each position in the grid has been injected the amount of immune cells and molecules, they are in the initial state. The same location in one step all the interactions are random, then some of the cells and molecules will die, some will be born. The molecules of cells can spread to other locations. the process again and again. Part of the B cells secrete antibodies, and participate in high-frequency variations, in order to obtain a greater degree of recognition[4]. Up to a certain degree of affinity, B cells become memory cells. The same antigen can then be injected to observe the secondary immune response, because the memory cells can quickly identify the pattern of this antigen, typically antigen will be quickly eliminated.

Multi-Agent-based model of immune

Multi-Agent Systems: Managing a certain number and type of agent to accomplish specific objectives. The agent, in general, are some of the entities: they have pre-set goals, have a norm of conduct, know the information about the surrounding environment, and to judge according to environmental information, communication, cooperation, learning and doing the decision-making.

Each agent can stand alone (autonomous), can also exist in a system be composed by a lot of agent, each agent has a set of individual goals, different agent have different goals. Meanwhile, all agents Cooperation to complete the overall goal of a system. Each agent is to maximize the realization of their goals[5].

Multi-Agent System and the immune system of comparison:

They are both consisted by many autonomous entities, the immune cells of immune system and the agent in Multi-Agent System both has autonomy.

They have individual goals and overall objectives. Individual target immune cells is to survive,while the global goal is to eliminate antigens.

They both have the ability to learn. Such as immune regulation and immune system memory, Multi-Agent Systems learning algorithms.

They both have the ability of adjustable. They can control their own activities according to changes in the environment. System have communication skills and competitiveness.

Two systems have some mechanism to maintain the whole system work. For example, the clonal selection algorithm in the immune system, , the learning algorithm and decision-making processes in Multi-Agent system.

I. IMMUNE-B model creation

(Fig. 1) is based on several common factors in immune cells and immune function and their interactions, and ultimately arrive at a factor of network control diagram[6].

Figure is consisted of three parts: cells, cytokines, antigens. Interaction between them, the figure the source of the cytokines, the role of the object and its role in functions described in Table A[7].

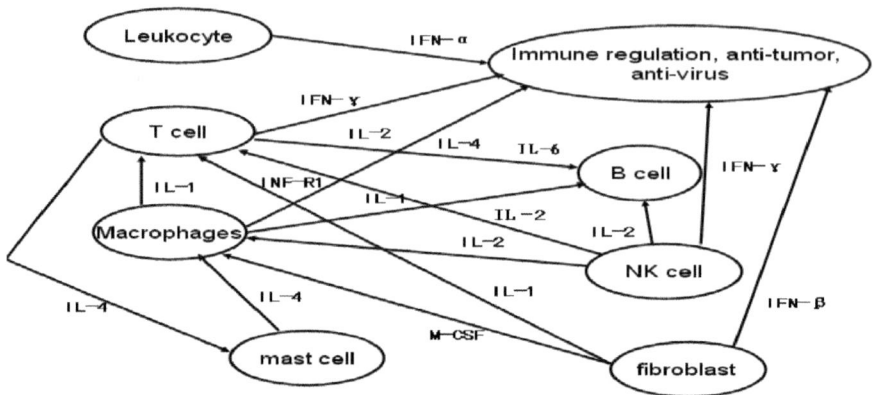

Fig. 1. The cytokine network regulation graph

Table 1.

Cytokines	Secretion of source	The role of the object	Main function
IL-1	Macrophages Fibroblasts	T cells B cells	Activated T cells,B cells
IL-2	T cells	T cells, B cells, Macrophages	Promoting T cells, B cells, Stimulation of NK cells, to enhance its destruction
IL-4	T cells Mast cells	B cells, Mast cells	Promoting T cells, B cells, giant cellsActivated macrophages
IL-6	T cells	B cells	Stimulation of NK cells, enhanc its destruction
IFN-α	Leukocyte	Antigen	Anti-tumor, inhibition of viral replication, inhibition of cell growt
IFN-β	Fibroblasts	Antigen	Anti-tumor, inhibition of viral replication, inhibition of cell growth
IFN-γ	T cells	Antigen	Anti-tumor, inhibition of viral replication, inhibition of cell growth
TNF-R1	Macrophages	Antigen	Anti-tumor, inhibition of viral replication, inhibition of cell growth
M-CSF	Fibroblasts	Macrophages	Activated macrophages

2   IMMUNE-B Model Description

IMMSIM model focuses on the immune response came from the function of the immune system simulation, only depicts the immune response process of the immune system, and Multi-Agent Based Model cannot be immune from the time of the regulation on the network. For the inadequacies of the above model, we use Method B to establish a IMMUNE-B model, and then use computer programs to implement[8]. In the simulation process, we know the final result not only, but also can observed the dynamic process, control them in time, so that We may find out   factor rules of the cytokine network, so as to provides a strong basis and more further factor in establishing the correct network model, Then use this model to validate the factors of cytokine, I believe the future will bring great convenience for cytokine network research. IMMUNE-B model is designed primarily to be able to simulate the process of cells activity by computer, process of the role about cytokines and antigen, and the relationships between them[9]. Here are IMMUNE-B model specified:

1. Process of implementation is probability event, the computer program achieve by a random number; An image is an individual, it can move from one location to another location; Individuals of interaction have been agreed to perform in the system[10].

2. Model constitutes

(1) Image

Use an image to represent the individual of cytokine network.

(2) Define the individual

Individuals including cells, cytokines, antibodies and antigens. There are seven cells of individual: Leukocyte (WC), T cells (T), B cells (B), NK cells (NK), Macrophages (GM), Mast cells (MC), Fibroblasts (FC). There are nine individual cytokines: IL-1、 IL-2、 IL-4、 IL-6、 IFN-α、 IFN-β、 IFN-γ、 TNF-R1、 M-CSF. And another antigen Ag and antibody Ab.

(3) State and action definition

The model set the corresponding state of the individual cells, only when they Taking effect with the outside world, we can decide whether to change the system state according to system definition. The implementation of the action can only occur when certain conditions are met, these rules are defined in systems, such as: cell growth, activation, division, death, and cytokine secretion and the interaction between individuals.

3   IMMUNE-B Model Formal

Users can according to the system define, selectively inject the cell types and antigens into the system initialization state. Then, observe the results of the program in a limited time, After the program over, the system can generate time and antibodies, antigens and cell function curve. According to the results of multiple runs, we can find some of the rules[11].

Model of B formal achieve:
MACHINE
    C ell
SETS
    CELL;
    STATUS={ unmature,mature,split,dead }
    CATEGORY;
CONSTANTS
MaxSize,Maxage,DivAge
PROPERTIES
    MaxAge∈CATEGORY→NAT
    MaxSize∈ CATEGORY→NAT
DivAge∈CATEGORY→NAT
VARIABLES
cell,category ,dx,dy,color,size,reage,status
INVARIANT
cell   CELL
category∈cell→CATEGORY
    dx∈NAT
    dy∈NAT
    color∈CATEGORY→(0…255)
    size∈CATEGORY→NAT
reage∈CATEGORY→NAT
status∈cell→STATUS
INITIALIZATION
cell,status,category:= Ø,Ø, Ø, dx:∈NAT,dy:∈NAT||
color:∈CATEGORY→(0…255)
size:=0...MaxSize
reage:=0..MaxAge
OPERATION
i ←   Creat_cell(cate)
PRE
  CELL-cell≠Ø
  cate∈CATEGOTY
THEN
ANY j WHERE
j∈CELL-cell
THEN
  i:=j
 cell:=cell∪{j}
 categoty(j):=cate
 END
END;
i←Get_cell( )
PRE
cell≠Ø TEHN i: ∈cell
END;

System has three constants: MaxSize, Maxage and DivAge, Represent the corresponding cell types of cell size, cell life, cell division Age in the system.

```
    MACHINE
    Cytokine
USES
    Cell
SETS
    CYTOKINE;
    CATEGORY;
VARIABLES
    cytokine,category,actobject
INVARIANT
    cytokine ⊆ CYTOKINE
    category∈CATEGORY
    actobject∈(cytokine,cell)
INITIALIZATION
    cytokine,category,actobject:= Ø, Ø, Ø
OPERATIONS
k ←   Creat_cytokine(c)
PRE
    CYTOKINE-cytokine≠Ø
    c∈CATEGOTY
THEN
    ANY j WHERE
        j∈CYTOKINE-cytokine
    THEN
    k:=j
categoty(j):=c
cytokine:=cytokine∪(j)
    END
END;
k←Get_cytokine( )
PRE
cytokine≠Ø TEHN k: ∈cell
END;
```

The machine has three variables, cytokine, category, actobject (factor role of the object), which actobject has a relationship with the cell collection.

4   System Simulation

According to the description of language B, we know that a necessary condition for system operation (T cells, B cells, antigen involved), to design a minimal system and initializ the system.

After Initialized the system into simulation program, during the program runs time, cell growth, division and death; cells secrete cytokines; antigen stimulated cells, cell activation; antibodies kill antigen are based on the biological instructions, thus ensuring the authenticity of the simulation. Here is a screenshot of the system is running (Fig. 2).

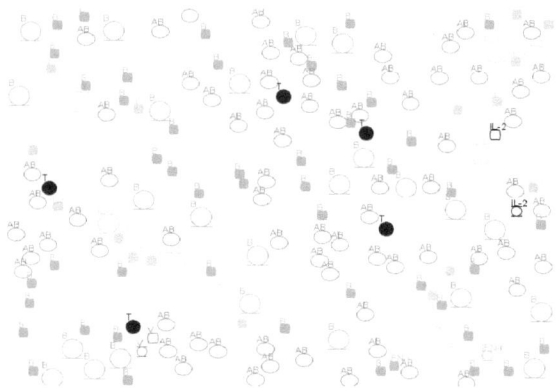

Fig. 2. System running interface

During the simulation, we mainly be regulated from the time, in waiting time, from each results, we record the number of cells, cytokines, antigen and antibody according to the system curve (see Figure 4), to see how much difference the results of each run. In addition, we can zero the number of antigens to observe changes in the number of other entities. However, due to the immune system of immune biology is cyclical, so we mainly use the first approach, in order to verify the biological immune rules or discover rules, so wen can promote the study of cytokine network.

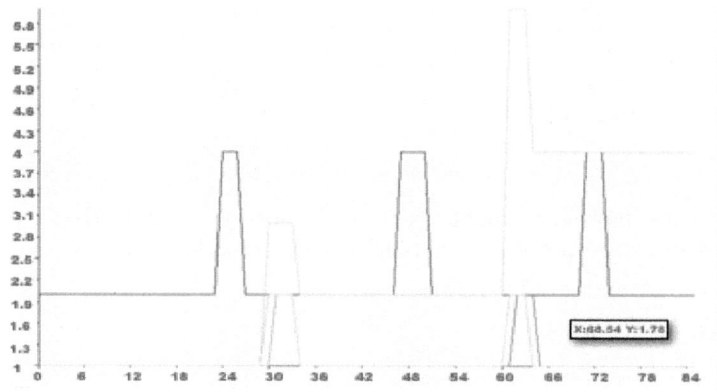

Fig. 3. Curve of real-time

5   Concluding Remarks

We use B method to establish formal model of a cytokine network, and based on this model, we achieve the immune cytokine network system by computer program. We compared the data with the experimental results obtained from the Animal Science and University of Medical. The results showed that: the program's result is almost the same with our forecast, this shows that the model we defined has some validity, the model has practical research value. This model will great help the cytokine network research, and provides a new approach for cytokine networks research[12]. In the course of the entire system, we have not consider affect outside the network environment, such as temperature, concentration and other individual factors. In order to represent the cytokine network more accurately, we must take full account of these factors.

References

1. Qiu, Z.: Translation, B method. Electronics Industry Press (June 2004)
2. Lano, K.: The B Language and Method: A Guide to Practice Formal Development. Springer, Heidelberg (1996)
3. Liu, Z.-Q., Ding, Y.-S., Zhang, X.-F.: Gio-entity inspired from immuneTSTS model for distributed intrusion detection system. Dy—nBxnics of Continuous, Discrete and Impulsive Systems-Series B: Applications & Algorithms 13(8), 46–50 (2009)
4. Lu, D. (ed.): Modern Immunology. Shanghai Science and Technology Press (December 1995)
5. Patti, J.M., Mia, K., Hon, S.I.: Virology 391(1), 64 (2009)
6. Chavez, L., Bais, A.S., Vingron, M., Lehrach, H., Adjaye, J., Herwig, R.: In silico identification of a core regulatory network of OCT4 in human embryonic stem cells using an integrated approach. BMC Genetics (2009)
7. Zou, S., Teng, T., Yang, X.-p.: Based on the B method of authentication technology. Modern Electronic Technology (October 2007)
8. Welch, D.R., Genoni, T.C., Cohen, S.A., Glasser, A.H.: Particle-in-Cell Modeling of Field Reversed Configuration Formation by Odd-parity Rotating Magnetic Fields. Journal of Fusion Energy 29(6) (2010)
9. Zou, S.: Formally Specifying T Cell Cytokine Networks with B Method. In: Zhang, J., He, J.-H., Fu, Y. (eds.) CIS 2004. LNCS, vol. 3314, pp. 385–390. Springer, Heidelberg (2004)
10. Zou, S.R.: Formally Specifying T Cell Cytokine Networks with B Method. In: Zhang, J., He, J.-H., Fu, Y. (eds.) CIS 2004. LNCS, vol. 3314, pp. 385–390. Springer, Heidelberg (2004)
11. Yun, H., Zhao, Y., Wang, J.: Modeling and simulation of fuel cell hybrid vehicles. International Journal of Automotive Technology 11(2) (2010)
12. Zou, S.: Modeling T cell Activation with B Mothod. In: SCBA Internation Symposium, p. 369 (2004)

Fuzzy H-inf Control of Flexible Joint Robot

Feng Wang and Xiaoping Liu

Automation School,
Beijing University of Posts and Telecommunications, Beijing, China
Wangfeng098252@yahoo.cn

Abstract. In this paper, a fuzzy H-inf control approach for flexible joint robot is proposed. First, the Takagi and Sugeno(T-S) fuzzy model is applied to approximate the flexible joint robot. Next, a fuzzy controller is developed based on parallel distributed compensation principle(PDC), and H-inf performance is introduced to restrain the influence of the bounded external disturbance. The sufficient conditions for the stability of the flexible joint robot control system are proposed by using Lyapunov function combined with the decay speed and linear matrix inequality(LMI). Finally, the simulation example is given to demonstrate the performance and robust of the proposed approach.

Keywords: Flexible joint robot, T-S fuzzy model, LMI, PDC.

1 Introduction

The control of flexible joint robot[1] is a difficult and complicated problem, since the flexible joint robot links are longer and lighter causing the end effector to undergo oscillations with inherent uncertainty. To handle these uncertainties, the Takagi and Sugeno(T-S) fuzzy model-based control approach[2-4] has been employed for its advantage and efficiency in nonlinear and uncertainty control system.

In [5], the stable control approach via T-S fuzzy model based on Lie algebras is introduced for single-link flexible joint robot. In [6], sliding mode control approach is employed for two-degree rigid robot. Stable control method based T-S fuzzy model of the flexible joint robot system both network-induced delay and data packet dropout is employed in [7]. Tracking control approach based T-S fuzzy model of the mobile robot with actuator saturation is applied in [8]. In [9], adaptive fuzzy neural network control via T-S fuzzy model is applied for two-link robot manipulator including actuator dynamics.

When the flexible joint robot is manipulated in the practical control process, the external disturbance is inevitable, which would damage the control performance of flexible joint robot system. H-inf performance could be applied to limit the influence of external disturbances.

In this paper, the T-S fuzzy model is applied to approximate the flexible joint robot. Next, a fuzzy H-inf controller is proposed based on parallel distributed compensation principle, the approximation error is also considered, and H-inf performance is introduced to restrain the influence of external disturbance under the given decay speed. The sufficient conditions and design procedures for the stability of

B. Liu and C. Chai (Eds.): ICICA 2011, LNCS 7030, pp. 424–431, 2011.
© Springer-Verlag Berlin Heidelberg 2011

the flexible joint robot control system are proposed by using Lyapunov function combined with linear matrix inequality. Finally, the simulation example is given to demonstrate the performance and robust of the proposed approach.

2   T-S Fuzzy Model for Flexible Joint Robot

In [5], it's given the mathematic model of flexible joint robot as follows:

$$I_1\ddot{\theta}_1 + mgl\sin(\theta_1) + k(\theta_1 - \theta_2) = 0$$
$$I_2\ddot{\theta}_2 + k(\theta_2 - \theta_1) = u \tag{1}$$

where u is the torque input. I_1 is the link inertia. I_2 is the motor inertia. m is the mass. g is the gravity. l is the link length. k is the stiffness. θ_1 and θ_2 are angular positions of first and second joints respectively.

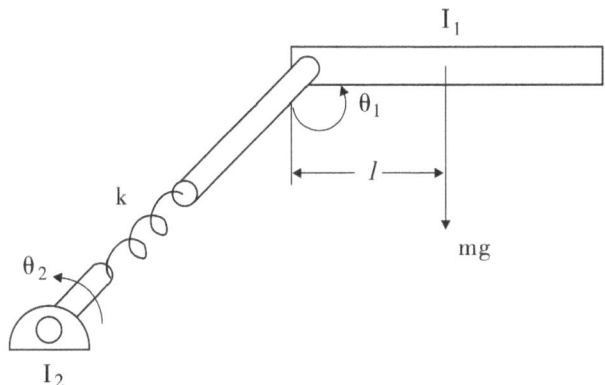

Fig. 1. Flexible joint robot system

Let $x_1 = \theta_1$. $x_2 = \dot{\theta}_1$. $x_3 = \theta_2$. $x_4 = \dot{\theta}_2$. External disturbance could be considered, so (1) can be rewritten as (2)

$$\dot{x}(t) = f(x(t)) + g(x(t))u(t) + w(t) \tag{2}$$

where $x = \begin{bmatrix} x_1 & x_2 & x_3 & x_4 \end{bmatrix}^T$. $w(t)$ is the bounded external disturbance.

$$g(x(t)) = \begin{bmatrix} 0 & 0 & 0 & \dfrac{1}{I_2} \end{bmatrix}^T . \quad f(x(t)) = \begin{bmatrix} 0 & 1 & 0 & 0 \\ -\dfrac{mgl\sin(x_1)}{I_1 x_1} - \dfrac{k}{I_1} & 0 & \dfrac{k}{I_1} & 0 \\ 0 & 0 & 0 & 1 \\ \dfrac{k}{I_2} & 0 & -\dfrac{k}{I_2} & 0 \end{bmatrix} x .$$

For the nonlinear part in the model, then it is considered to approximate the flexible joint robot with T-S fuzzy model.

Define the ith fuzzy rule of FJR: IF the angular position of FJR link tip $x_1(t)$ is F_i, THEN

$$\dot{x}(t) = A_i x(t) + B_i u(t) + w(t)$$
$$y(t) = C_i x(t) , \quad i = 1, 2, \cdots, r.$$

where, F_i are fuzzy sets. r is the number of IF-THEN rules. $x(t)$ is state vector. $x_1(t)$ is the premise variable of fuzzy inference rules. $y(t)$ is the angular position of FJR link tip. Then the final fuzzy system is inferred as follows

$$\dot{x}(t) = \sum_{i=1}^{r} h_i(x_1(t))(A_i x(t) + B_i u(t)) + w(t) \tag{3}$$

$$y(t) = \sum_{i=1}^{r} h_i(x_1(t)) C_i x(t)$$

3   Fuzzy H-inf Controller Design

According to the PDC principle, the fuzzy controller is given as follow:

The fuzzy rule i: IF the angular position of FJR link tip $x_1(t)$ is F_i, THEN $u(t) = K_i x(t)$, where K_i are control gains.

The overall fuzzy control law is represented by

$$u(t) = \sum_{i=1}^{r} h_i(x_1(t)) K_i x(t) \qquad i = 1, 2, \cdots, r \tag{4}$$

By respectively substituting (4) into (3), and for convenience, let $h_i = h_i(x_1(t))$, we obtain

$$\dot{x}(t) = f(x(t)) + g(x(t))u(t) + w(t) = \sum_{i=1}^{r}\sum_{j=1}^{r} h_i h_j (A_i + B_i K_j) x(t)$$

$$+ [f(x(t)) - \sum_{i=1}^{r} h_i A_i x(t)] + [g(x(t)) - \sum_{i=1}^{r} h_i B_i] u(t) + w(t) \tag{5}$$

$$= \sum_{i=1}^{r}\sum_{j=1}^{r} h_i h_j (A_i + B_i K_j) x(t) + \Delta f(x(t)) + \Delta g(x(t)) + w(t)$$

where

$$\Delta f(x(t)) = f(x(t)) - \sum_{i=1}^{r} h_i A_i x(t)$$

$$\Delta g(x(t)) = \sum_{i=1}^{r}\sum_{j=1}^{r} h_i h_j (g(x(t)) - B_i) K_j x(t)$$

If $\left\| \Delta f\left(x(t)\right) \right\| \le \left\| \sum_{i=1}^{r} h_i \Delta A_i x(t) \right\| , \left\| \Delta g\left(x(t)\right) \right\| \le \left\| \sum_{i=1}^{r} \sum_{j=1}^{r} h_i h_j \Delta B_i K_j x(t) \right\|$

and defined $\left[\Delta A_i \quad \Delta B_i \right] = \left[\delta_i A_p \quad \eta_i B_p \right]$

where $\left\| \delta_i \right\| \le 1 , \left\| \eta_i \right\| \le 1 , i = 1, 2, \cdots r$.

Then

$$(\Delta f\left(x(t)\right))^T (\Delta f\left(x(t)\right)) \le (A_p x(t))^T (A_p x(t))$$

$$(\Delta g\left(x(t)\right))^T (\Delta g\left(x(t)\right)) \le (\sum_{i=1}^{r} h_i B_p K_j x(t))^T (\sum_{i=1}^{r} h_i B_p K_j x(t))$$

The approximation error in the closed-loop flexible joint robot control system is bounded by the specified structured bounding matrices A_p and B_p .

For the existence of external disturbance $w(t)$ would affect the control performance, H-inf performance is introduced to restrain the influence of $w(t)$ to control system. If considered initial condition,

$$\int_0^{t_f} x^T (t)Qx(t)dt \le x^T (0)Px(0) + \rho^2 \int_0^{t_f} w^T (t)w(t)dt \qquad (6)$$

where t_f is terminal time of control. ρ is prescribed attenuation level, for restraining the influence of $w(t)$. ρ is less than 1. P and Q are positive definite weighting matrix.

4   Design of Flexible Joint Robot Control System

Theorem. Considered flexible joint robot control system (5), and given the decay speed β satisfied $\dot{V}(x(t)) \le -2\beta V(x(t))$, the theorem in [10] is also applied. If there exists positive definite matrix P , Q and a prescribed $\rho^2 > 0$, such that the following matrix inequalities are satisfied:

$$A_i^T P + A_i P + PB_i K_j + K_j^T B_i^T P + A_p^T A_p$$
$$+(B_p K_j)^T (B_p K_j) + (2 + 1/\rho^2)PP + Q + 2\beta P < 0 \ i, j = 1, 2, \cdots, r \qquad (7)$$

Then the closed-loop flexible joint robot control system (5) is stable, and the H-inf performance (6) is guaranteed.

Design Procedures:

Step 1) Select fuzzy rules and fuzzy membership function for flexible joint robot system (3).

Step 2) Select the decay speed β . Weighting matrix Q . Bounding matrices $\Delta A_i (= \delta_i A_p)$ and $\Delta B_i = (\eta_i B_p)$.

Step 3) Select the level ρ^2 and solve theorem corresponding the LMI to obtain P and K_i.

Step 4) Decrease ρ^2 and repeat Step 3 until P cannot be found.

Step 5) Check the assumptions of

$$\|\Delta f(x(t))\| \leq \left\|\sum_{i=1}^{r} h_i \Delta A_i x(t)\right\| \cdot \|\Delta g(x(t))\| \leq \left\|\sum_{i=1}^{r} \sum_{j=1}^{r} h_i h_j \Delta B_i K_j x(t)\right\|.$$

If they are not satisfied, the bounds for all elements in ΔA_i and ΔB_i should be adjusted (expanded), then repeat Step 3-4.

Step 6) Fuzzy control law in (4) is obtained.

5   Simulation Example

In this simulation example, the parameters[13] of flexible joint robot are $k = 100 Nm/rad$. $g = 9.8 m/s^2$. $m = 1kg$. $I_1 = I_2 = 1kgm^2$. $l = 1m$. $w = \sin(t)$. The T-S fuzzy model is given as follows:

Rule 1: IF the angular position of FJR link tip $x_1(t)$ is $-\pi/2$, THEN

$$\dot{x}(t) = A_1 x(t) + B_1 u(t) + w(t)$$
$$y(t) = C_1 x(t)$$

Rule 2: IF the angular position of FJR link tip $x_1(t)$ is 0, THEN

$$\dot{x}(t) = A_2 x(t) + B_2 u(t) + w(t)$$
$$y(t) = C_2 x(t)$$

Rule 3: IF the angular position of FJR link tip $x_1(t)$ is $\pi/2$, THEN

$$\dot{x}(t) = A_3 x(t) + B_3 u(t) + w(t)$$
$$y(t) = C_3 x(t)$$

where

$$A_1 = \begin{bmatrix} 0 & 1 & 0 & 0 \\ -106.2389 & 0 & 100 & 0 \\ 0 & 0 & 0 & 1 \\ 100 & 0 & -100 & 0 \end{bmatrix}, \quad A_2 = \begin{bmatrix} 0 & 1 & 0 & 0 \\ -109.8 & 0 & 100 & 0 \\ 0 & 0 & 0 & 1 \\ 100 & 0 & -100 & 0 \end{bmatrix}, \quad A_3 = A_1,$$

$$B_1 = B_2 = B_3 = \begin{bmatrix} 0 & 0 & 0 & 1 \end{bmatrix}^T, \quad C_1 = C_2 = C_3 = \begin{bmatrix} 1 & 0 & 0 & 0 \end{bmatrix}.$$

$$A_p = \begin{bmatrix} 0.1 & 0.5 & 0 & 0 \\ 2.3 & 0.1 & 0 & 0 \\ 0 & 0 & 0.1 & 0 \\ 0 & 0.2 & 0 & 0.17 \end{bmatrix} . B_p = \begin{bmatrix} 0 & 0 & 0 & 0.002 \end{bmatrix}^T .$$

And the initial values $x_0 = [1.2 \quad 0 \quad 0 \quad 0]^T$. $\beta = 6$. The fuzzy membership functions are shown in Fig. 2

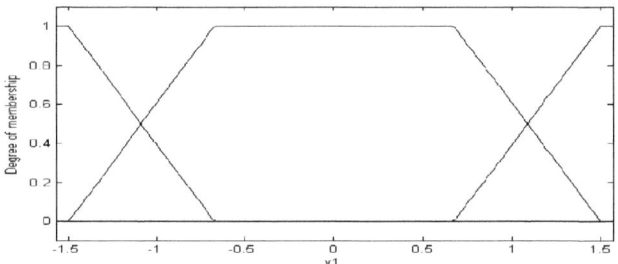

Fig. 2. Fuzzy membership functions

We apply the theorem and solve the correspondent LMI with LMI toolbox of MATLAB. Then we obtain $t_{min} = -0.048302$, so the closed-loop control system is stable. Finally we get the positive definite matrix

$$P = \begin{bmatrix} 7.024 & 1.2243 & 5.1285 & 0.0919 \\ 1.2243 & 0.4979 & 2.6165 & 0.0513 \\ 5.1285 & 2.6165 & 15.2963 & 0.3017 \\ 0.0919 & 0.0513 & 0.3017 & 0.0072 \end{bmatrix}$$

The control gain matrixes are

$$K_1 = K_2 = K_3 = [-3.0318e03, -1.802e03, -1.0822e04, -255.9066] .$$

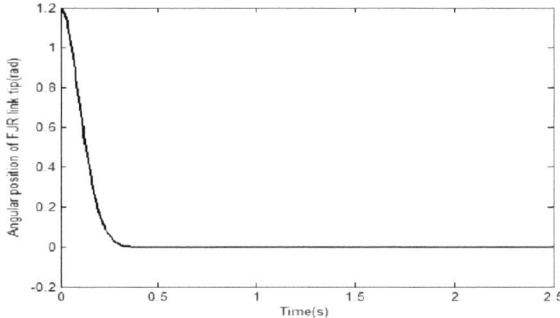

Fig. 3. The control effect $y(t)$

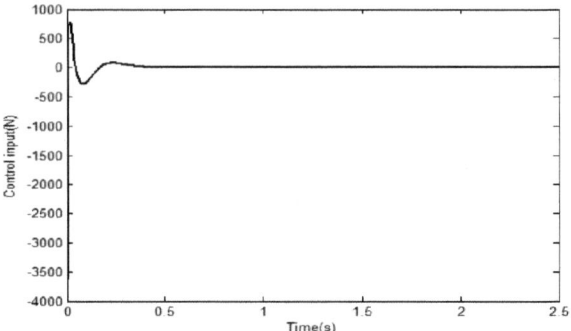

Fig. 4. The control input $u(t)$

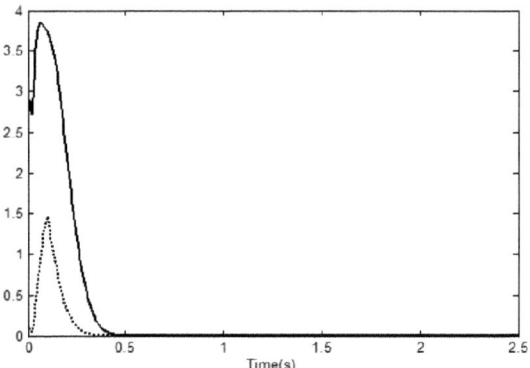

Fig. 5. The plots of $\left\| f(x(t)) - \sum_{i=1}^{r} h_i A_i x(t) \right\|$ (dashed line) and $\left\| \sum_{i=1}^{r} h_i \Delta A_i x(t) \right\|$ (solid line)

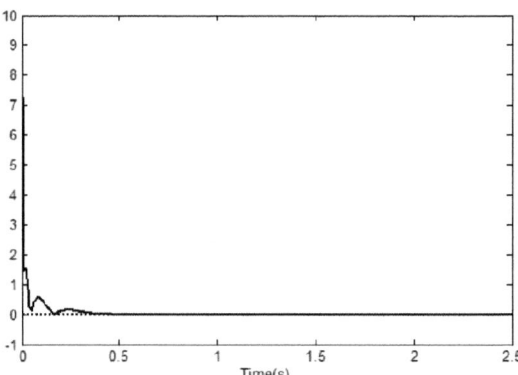

Fig. 6. The plots of $\left\| \sum_{i=1}^{r} \sum_{j=1}^{r} h_i h_j (g(x(t)) - B_i) K_j x(t) \right\|$ (dashed line) and $\left\| \sum_{i=1}^{r} \sum_{j=1}^{r} h_i h_j \Delta B_i K_j x(t) \right\|$ (solid line)

Figs. 3-6 present the simulation results for angular position of flexible joint robot link tip with fuzzy H-inf controller. It can be seen that the control performance of flexible joint robot system with fuzzy H-inf controller is good, and H-inf performance restrain the influence of external disturbance $w(t)$ to control system.

6   Conclusions

In this paper, a fuzzy H-inf control approach for flexible joint robot is proposed, and H-inf performance is applied to reduce the influence of external disturbance. The simulation example is given to illustrate that the fuzzy H-inf controller designed is effective and good robust for flexible joint robot system.

References

1. Ozgoli, S., Taghirad, H.D.: A Survey on the Control of Flexible Joint Robots. Asian Journal of Control 8, 1–15 (2006)
2. Takagi, T., Sugeno, M.: Fuzzy Identification of Systems and Its Applications to Modeling And Control. IEEE Transactions System, Man, and Cybernetics SMC-15, 116–132 (1985)
3. Precup, R.E., Hellendoorn, H.: A Survey on Industrial Applications of Fuzzy Control. Computers in Industry 62, 213–226 (2011)
4. Feng, G.: A Survey on Analysis And Design of Model-based Fuzzy Control Systems. IEEE Transactions on Fuzzy Systems 14, 676–695 (2006)
5. Banks, S.P., Gurkan, E., Erkmen, I.: Stable Controller Design for T-S Fuzzy Systems Based on Lie Algebras. Fuzzy Sets and Systems 156, 226–248 (2005)
6. Liang, Y.W., Xu, S.D., Liaw, D.C., Chen, C.C.: A Study of T–S Model-Based SMC Scheme With Application to Robot Control. IEEE Transactions on Industrial Electronics 5, 3964–3971 (2008)
7. Jiang, X.F., Han, Q.L.: On Designing Fuzzy Controllers for a Class of Nonlinear Networked Control Systems. IEEE Transactions on Fuzzy Systems 16, 1050–1060 (2008)
8. Chen, H., Ma, M.M., Wang, H., Liu, Z.Y., Cai, Z.X.: Moving Horizon H $^{\infty}$ tracking Control of Wheeled Mobile Robots With Actuator Saturation. IEEE Transactions on Control Systems Technology 17, 449–457 (2009)
9. Wai, R.J., Yang, Z.W.: Adaptive Fuzzy Neural Network Control Design via A T-S Fuzzy Model for A Robot Manipulator Including Actuator Dynamics. IEEE Transactions System, Man, and Cybernetics Part B-Cybernetics 38, 1326–1346 (2008)
10. Chen, B.S., Tseng, C.S., Uang, H.J.: Robustness Design of Nonlinear Dynamic System via Fuzzy Linear Control. IEEE Transactions on Fuzzy Systems 7, 571–585 (1999)
11. Tseng, C.S., Chen, B.S., Uang, H.J.: Fuzzy Tracking Control Design for Nonlinear Dynamic Systems via T-S Fuzzy Model. IEEE Transactions on Fuzzy Systems 9, 381–392 (2001)
12. Tong, S.C., Wang, T., Li, H.X.: Fuzzy Robust Tracking Control for Uncertainty Nonlinear Systems. International Journal of Approximate Reasoning 30, 73–90 (2002)
13. Spong, M.W.: Modeling And Control of Elastic Joint Robots. Transactions of the ASME 109, 310–319 (1987)

The Application of Different RBF Neural Network in Approximation

Jincai Chang[*], Long Zhao, and Qianli Yang

College of Science, Hebei United University, Tangshan Hebei 063009, China
jincai@heut.edu.cn, zhaolong1985@126.com

Abstract. The value algorithms of classical function approximation theory have a common drawback: the compute-intensive, poor adaptability, high model and data demanding and the limitation in practical applications. Neural network can calculate the complex relationship between input and output, therefore, neural network has a strong function approximation capability. This paper describes the application of RBFNN in function approximation and interpolation of scattered data. RBF neural network uses Gaussian function as transfer function widespreadly. Using it to train data set, it needs to determine the extension of radial basis function constant SPEAD. SPEAD setting is too small, there will be an over eligibility for function approximation, while SPREAD is too large, there will be no eligibility for function approximation. This paper examines the usage of different radial functions as transferinf functions to design the neural network, and analyzes their numerical applications. Simulations show that, for the same data set, Gaussian radial basis function may not be the best.

Keywords: Radial basis function, neural network, tight pillar, numerical approximation.

1 Introduction

Numerical calculations used in engineering and scientific research applications are widespread, and new numerical methods develop constantly. However, there are common disadvantages of the of classical function approximation theory: compute-intensive, poor adaptability, high models and data requirements and imitations in practical applications. Now, value algorithms studies by using the neural network is popular at home and abroad. In the practical application of neural networks, the majority are BP neural network or its variations. Based on radial basis function network (RBFNN), this paper studies the numerical calculation. The key of improving the radial basis function neural network performance is the hidden layer neurons, hidden layer selection strategy and the output layer connection weights update method [1]. Usually making Gaussian function as transfer function, this paper adopts compactly supported positive definite radial basis function (Wu function) to design a new neural network.

---

[*] Project supported by National Natural Science Foundation of China (No.61170317) and Natural Science Foundation of Hebei Province of China (No. A2009000735, A2010000908).

B. Liu and C. Chai (Eds.): ICICA 2011, LNCS 7030, pp. 432–439, 2011.

2   Questions

Radial basis function (RBF) neural network [2] (referred to as RBF network) is a neural network architecture proposed by J. Moody and C. Darken in the late 1980s, which is a single hidden layer of three-layer feed-forward network. Present evidences show that radial basis function networks can approximate any continuous function with arbitrary accuracy. Nowadays, RBF neural network uses Gaussian function widespreadly as transfer function. Using it to train the network, firstly, it needs to determine the extension of radial basis function constant SPREAD. SPREAD setting is too small, it means a lot of hidden dollars should be needed to adapt the changes in the function, and it will lead the large number of neurons, which will not only slow down the network training speed, but also make the process of approximation not smooth, and there will be an over eligibility for function approximation. On the contrary, the larger the SPRED is, the wider influence sphere for hidden layer neurons of the input vector is, and the better the smoothness of hidden layer neurons is. But too large SPEAD means there being a lot of overlap in neuron input area, there will be no eligibility for function approximation, which can not make the network produce different responses, even for network design and training.

Example: Suppose the following input / output samples, the input vector to [-1,1] interval vector compose of the upper interval P, the corresponding expected vector T.

P=-1:0.1:1;T=[-0.96 -0.577 -0.0729 0.377 0.641 0.66   0.461 0.1336 -0.201 -0.434 -0.5 -0.393 -0.1647 0.0988 0.3072 0.396 0.3449 0.1816 -0.0312 -0.2187 -0.3201];

Take the enter variable as the abscissa, the expected value as the vertical axis, draw the training sample data points.

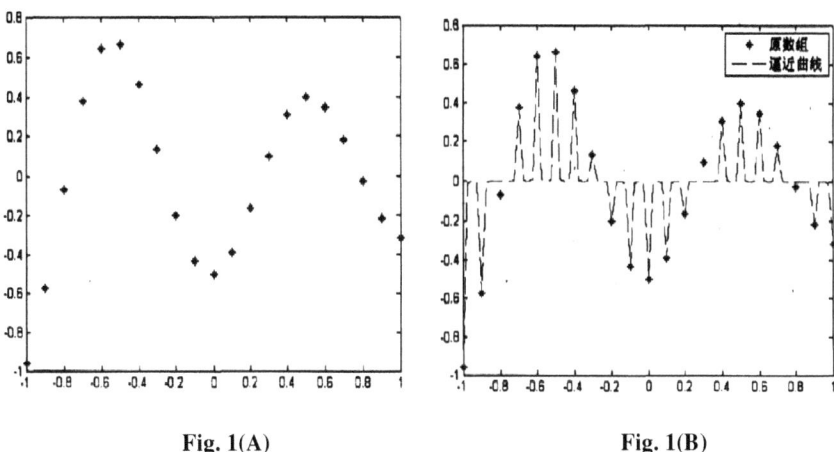

Fig. 1(A)   Fig. 1(B)

Fig. 1. (A) shows the input samples from the plot; (B) shows the small setting Spread from the plot; (C) the large setting Spread from the plot

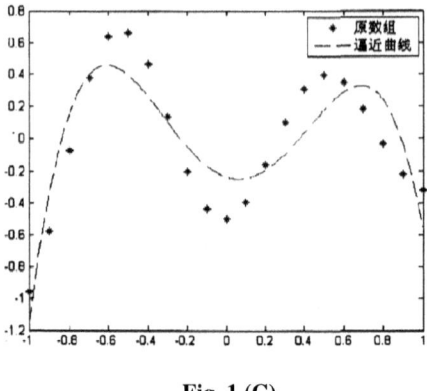

Fig. 1 (C)

Fig. 1. (*Continued*)

When a radial basis function network is established, according to different training set and accuracy requirements, many different SPREAD values should be selected to compared repeatedly network output value and target value in oder to determine the better approximation, and this brings certain difficulties and inconveniences to the actual conditions. Based on the above analysis, the study uses a compact support radial basis function as activation function to design neural network and analyzes their function approximation applications.

3   Compact Support Radial Basis Function

3.1   Single Variable Function of Compact Support

An expression of function [4]:

$$f(x) = \begin{cases} \alpha \exp(\dfrac{1}{x-\beta+2-\gamma^2}) & |x-\beta| < \gamma \\ 0 & |x-\beta| \geq \gamma \end{cases}$$

Where $\alpha > 0$ determines the role height of the function; β is radial symmetry; $\gamma > 0$ is radial symmetry of function radius. It can be verified that the function has a positive definite, radial symmetry, compact support; without loss of generality. Take $\alpha = 1, \beta = 0, \gamma = 1$, then the function expression is:

$$f(x) = \begin{cases} \exp(\dfrac{1}{x^2-1}) & |x| < 1 \\ 0 & |x| \geq 1 \end{cases}$$

The function of the first derivative expression is:

$$f(x) = \begin{cases} \dfrac{-2x}{(x^2-1)^2}\exp(\dfrac{1}{x^2-1}) & |x| < 1 \\ 0 & |x| \geq 1 \end{cases}$$

The second derivative of the function expression is:

$$f(x) = \begin{cases} \dfrac{6x^4-2}{(x^2-1)^4}\exp(\dfrac{1}{x^2-1}) & |x| < 1 \\ 0 & |x| \geq 1 \end{cases}$$

Function 2 [6]: $f(x) = (1-x)_+^5(1+5x+9x^2+5x^3+x^4)$.

3.2   Multi-variable Function of Compact Support

Multivariate radial basis function uses the expression of product of a single variable generalized radial basis function[4]:

$$F(x) = \begin{cases} \prod_{i=1}^{n}\alpha_i\exp(\sum_{i=1}^{n}\dfrac{1}{(x_i-1)^2-\gamma_i^2}) & |x_i-\beta_i| < \gamma_i, i = 1,\cdots,n \\ 0 & |x_i-\beta_i| \geq \gamma_i, i = 1,\cdots,n \end{cases}$$

4   Design Compactly Supported Radial Basis Neural Network

The weighting function of hidden units uses MATLAB neural network toolbox DIST, the input function is NETPROD, activation function takes its own written CSPDRBF, the output layer neuron activation function is the pure linear function PURELIN, weighting function is DOTPROD, and output function is NETSUM.

The detailed algorithms are:

(1) to simulate the network with all of the input samples;

(2) to find an input sample with the maximum error;

(3) to add a RBF neuron whose weight equals to the sample input vector transpose; threshold $b = \sqrt{-\log(0.5)}/2$.

(4) to make the output of RBF neural network as a linear dot product of input neurons, re-design linear network layer, making the error smallest;

(5) When the mean square error does not reach the required error performance indicators, and the number of neurons does not meet the required limit, repeat the above steps until the network mean square error achieves the specified error performance index, or number of neurons gets the specified limit required.

5   Simulation

Scattered   data   interpolation   problem: Given   n   points   on   the   plane $(x_i, y_i)(i = 1, 2, \cdots, N)$ and its corresponding function value $z_i (i = 1, 2, \cdots, N)$, require construction binary function $f = f(x, y)$ to satisfy the interpolation conditions.

$$f(x_i, y_i) = z_i (i = 1, 2, \cdots, N)$$

For this problem, the RBFNN, the input layer selects two neurons, representing (x_i, y_i) , select an output layer neurons, representing z_i .

To the surface $f(x, y) = \exp(-81((x-0.5)^2 + (y-0.5)^2)/4)/3$, do the interpolation experiment. We select a set of points in $[0,1]$ as a test, the comparison between original function iamge and the interpolated image (Fig.2 and Fig.3).

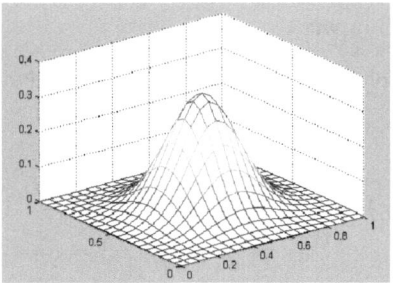

Fig. 2. The original function image   **Fig. 3.** The interpolation function image

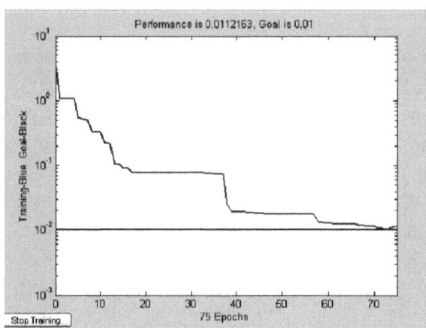

Fig. 4. Network performance error curve

Error curve during training: after a 75-step training, mean square error 0.007 can be up to our pre-specified error requirement 0.01(Fig.4).

Binary function

$$z = f(x, y) = 3(1-x)^2 e^{-x^2-(y+1)^2} - 10(\frac{x}{5} - x^3 - y^5)e^{-x^2-y^2} - \frac{1}{3}e^{-(x+1)^2-y^2}$$

can produce a convex surface, containing three local maximum points and three local minimum p ✕ oints, called the peak function peaks in MATlAB. Use it to do numerical experiments, take 25 * 25 points as the set of test points from [3,3] to test designing approximation results of different pillars tight radial basis neural network.

1) When Gaussian function $f(x) = \exp(-x^2)$ is the transfer function, take 25×25 points as the set of test points network training from [3,3].

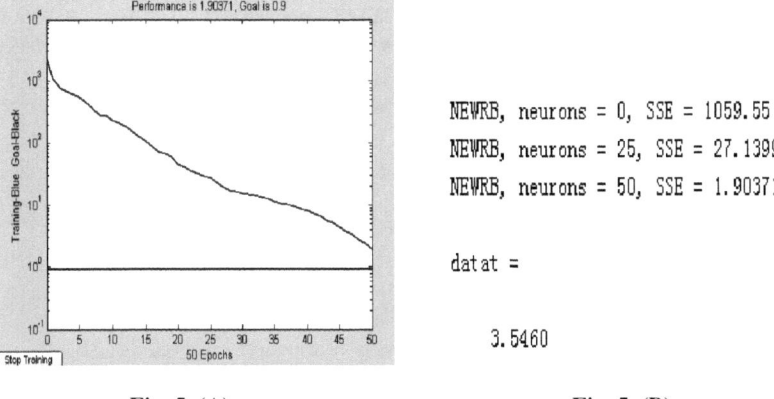

NEWRB, neurons = 0, SSE = 1059.55
NEWRB, neurons = 25, SSE = 27.1399
NEWRB, neurons = 50, SSE = 1.90371

dat at =

3.5460

Fig. 5. (A)   Fig. 5. (B)

Fig. 5. (A) Gaussian error curve network performance; (B) Gaussian function as transfer function of the network training time

2) Compactly supported positive definite function 1:
    When

$$f(x) = \begin{cases} \exp(\dfrac{1}{x^2-1}) & |x| \le 1 \\ 0 & |x| > 1 \end{cases}$$

the transfer function, take 25×25 points as the set of test points network training from [3,3].

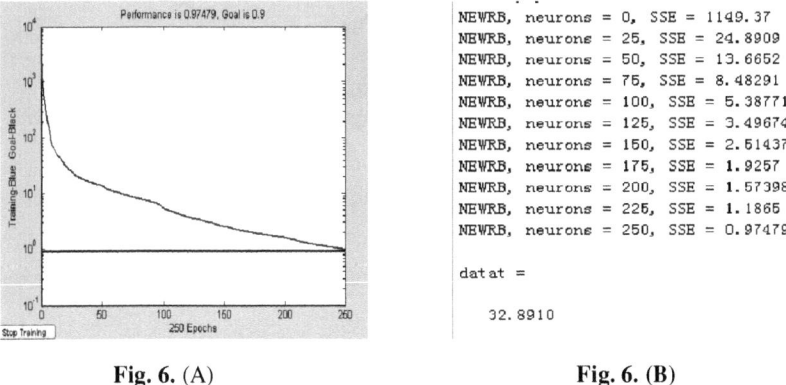

NEWRB, neurons = 0, SSE = 1149.37
NEWRB, neurons = 25, SSE = 24.8909
NEWRB, neurons = 50, SSE = 13.6652
NEWRB, neurons = 75, SSE = 8.48291
NEWRB, neurons = 100, SSE = 5.38771
NEWRB, neurons = 125, SSE = 3.49674
NEWRB, neurons = 150, SSE = 2.51437
NEWRB, neurons = 175, SSE = 1.9257
NEWRB, neurons = 200, SSE = 1.57398
NEWRB, neurons = 225, SSE = 1.1865
NEWRB, neurons = 250, SSE = 0.97479

dat at =

32.8910

Fig. 6. (A)   Fig. 6. (B)

Fig. 6. (A) Compactly supported function a network performance error curve; (B) is compactly supported function a transfer function of the network training time

Compactly supported positive definite function 2: F for the transfer function, take 25×25 points as the set of test points network training from [3,3].

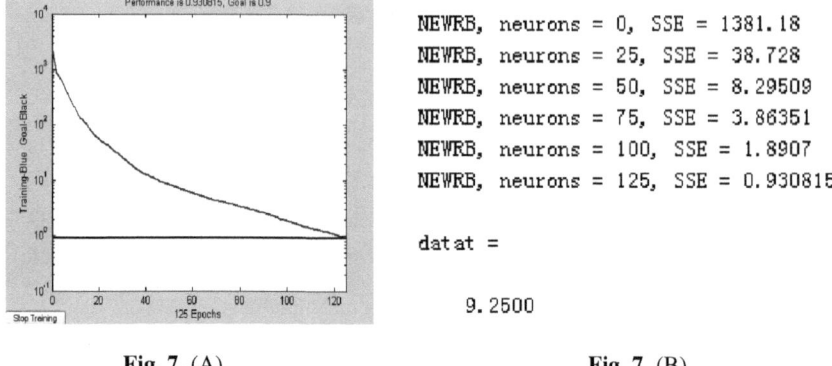

NEWRB, neurons = 0, SSE = 1381.18
NEWRB, neurons = 25, SSE = 38.728
NEWRB, neurons = 50, SSE = 8.29509
NEWRB, neurons = 75, SSE = 3.86351
NEWRB, neurons = 100, SSE = 1.8907
NEWRB, neurons = 125, SSE = 0.930815

datat =

9.2500

Fig. 7. (A)   **Fig. 7.** (B)

Fig. 7. (A) Network performance compact support function error curve; (B) Compactly supported functions the network training time transfer function

Analyzing from error curve, the Gaussian function stops training without reaching the target error. Compactly supported functions compactly supported functions 1 and 2 basically reach the target error.

Analyzing from network training time, the training time of compactly supported function1 is shorter than that of compactly supported function2, the less the number of neurons is, the higher the efficiency is.

In summary, the compact support function2 is chosen to do peaks function approximation.

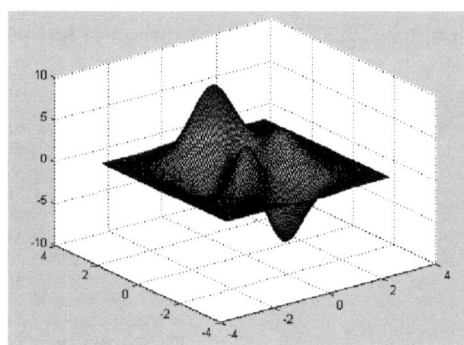

Fig. 8. Compactly supported function 2 as the transfer function of the interpolated image

6   Collusion and Outlook

Through the above example analysis, it can be seen that the Gaussian function is not better than compactly supported functions in interpolation. When a compactly supported function is choosed, once transfer function is determined, and then the expansion constants Spread is also determined. It needs not train repeatedly to determine the constant extend Spread. Compactly supported functions2 is more effective than compactly supported function1. The reason is that polynomial functions are more rapid than the exponential functions in the calculation, leading to a small number of neurons and high efficiency. Neural networks can also handle the data points with noise and missing data point set. It hopes to use compactly supported functions in the above in next step.

References

1. Liao, B., Chen, F.: Radial basis function networks in scattered data interpolation in the application. China University of Science and Technology, Beijing (2001)
2. Wen, X.: MATLAB Neural Network Design. Science Press, Beijing (2001)
3. Ren, Y.: Numerical Analysis by Matlab. Higher Education Press, Beijing (2007)
4. Li, J., Song, G.: A kind of compact support radial basis functions. Shenyang University of Technology 27(2), 226–228 (2005)
5. Zhou, J., Yao, S., Wang, F.-L.: Radial basis function (RBF) network of research and implementation. Mining 10(4), 71–75 (2001)
6. Wu, Z.: Scattered data fitting model. Method and theory. Science Press, Beijing (2007)
7. Sen, C., Su, J.: The application of orthogonal polynomials. Guangxi Agricultural University Press (1997)
8. Chen, T., Chen, H.: Approximation capability to function of several variablea, nonlinear functionals and operalors by radial basis function neaural networke. IEEE Trans. on Neural Networks 6, 904–910 (1995)
9. Gu, P., Yan, X.: Neural network approach to the reconstruction of free form surfaces for reverse engineering

Travel Demand Prediction in Tangshan City of China Based on Rough Set

Juan Li[1], Li Feng[2], and Guanchen Zhou[2]

[1] Hebei United University (Hebei Institute of Technology),
Department of academic Affairs
[2] Hebei United University, Add: No. 46 Xinhua West Road,
Tangshan Postal Code: 063009
lijuanzw@126.com

Abstract. Travel demand prediction plays a scientific guidance role in tourism industry planning and future development. In this paper, the rough set theory is applied to analyze and predict the tourism of Tangshan City in the future based on the sample data of the quantity of tourists of Tangshan City from 2001 to 2010 and the growth range of quantity of tourists was obtained. The result shows that the tourism market of Tangshan City has both a huge development potential and a bright development prospect.

Keywords: Rough set, Travel demand, Quantity of tourists.

1 Introduction

From 21st century, Tangshan City has kept a high growth rate of economy. In 2001, the gross domestic product of Tangshan exceeded RMB 100 billion Yuan for the first time; in 2005, it exceeded RMB 200 billion Yuan; and in 2008, it exceeded RMB 300 billion Yuan and in 2009 it exceeded RMB 378.144 billion Yuan. Great changes have taken place in the consuming behaviors of consumers with the increase of the income of the resident, accounting for the rapid development of the tourism industry as a consumption terminal industry. In 2010, Tangshan City received a total of 1,538,000 domestic and oversea tourists with a growth of 23.3% than that of 2009 and realized the total tourism revenue of RMB 9.51 billion Yuan with a growth of 48.8% than that of 2009.

The rapid development of tourism industry promotes the development of urban construction and propaganda of city image. However, how to establish a scientific and operable travel demand prediction model and grasp the tourism market development tendency accurately, with the purpose to prepare the tourism regulation and control policies applicable to the local economic characteristic of Tangshan City, have a great strategic significance in the sustainable and healthy development of tourism industry of Tangshan City and even Hebei Province.

Traditional travel demand prediction methods include time series model prediction method, causal model prediction method as well as regression model prediction method [1-3]. Although the kinds of model prediction methods have a good prediction results

B. Liu and C. Chai (Eds.): ICICA 2011, LNCS 7030, pp. 440–446, 2011.
© Springer-Verlag Berlin Heidelberg 2011

for the tourism markets of stable development, the tourism market are often subjected to the influence and restriction of many uncertain factors, so the prediction results by means of such kinds of methods are not quite ideal. However, rough set is a kind of mathematical tool describing the disintegrity and uncertainty and may make an effective analysis of the inaccuracy,inconformity and disintegrity and other incomplete information to discover the tacit knowledge from them[4]. At present, the rough set method is widely applied in such fields as artificial intelligence, mode identification and intelligent information processing and so on and has achieved many good achievements. In this paper, the rough set theory is applied to predict and analyze the travel demand of Tangshan City and the results shows that the method is quite effective to solve the nonlinear problems.

The rough set theory was proposed by Poland scholar Pawlak in 1982. The core idea of rough set theory is to describe any subset X through a pair of lower and upper approximation operators under the equivalent relation. It has as strong qualitative analysis ability and is an effective method for expressing the inaccurate and uncertain knowledge and has the advantage of acquisition of knowledge from a large amount of data and shows a strong life force from uncertainty reasoning and incomplete system. Therefore, it is widely applied in such fields as rule generation, intelligent control, decision analysis, knowledge acquisition and machine learning, etc.

2   Basic Theory of Rough Set

The set $U \neq \Phi$ composed by definite research objects is called as universe of discourse in rough set theory. Any subset in the universe of discourse is sometimes called as concept and may form division. Φ is regarded as a special concept.

Division $\delta : \delta = \{X_1, X_2, \cdots, X_n\}$, $X_i \subseteq U$, $X_i \neq \Phi$, and $X_i \cap X_j = \Phi$, as for $i \neq j$, $i, j = 1, 2, \cdots, n$.

2.1   Approximate Set

Set a knowledge base $K = (U, R)$ and define two subsets for each subset $X \subseteq U$ and an equivalent relation $R \in ind(K)$:

$$\underline{R}X = \bigcup\{Y \in U / R \mid Y \subseteq X\},$$

$$\overline{R}X = \bigcup\{Y \in U / R \mid Y \cap X \neq \Phi\}$$

Call them as R lower approximate set and R upper approximate set of X.

The upper approximate set and lower approximate set may also be expressed by the following equation:

$$\underline{R}X = \{x \in U \mid [x]_R \subseteq X\},$$
$$\overline{R}X = \{x \in U \mid [x]_R \cap X \neq \Phi\}$$

Set $bn_R(X) = \overline{R}X - \underline{R}X$ is called as R boundary region; $pos_R(X) = \underline{R}X$ is called as R positive field; $neg_R(X) = U - \overline{R}X$ is called as R negative field.

As shown in the Fig. 1 below.

2.2   Knowledge Reduction

The so-called knowledge reduction is to delete the irrelevant or unimportant knowledge under the condition that the classification capability of knowledge base is unchanged. Two basic concepts in knowledge reduction are—reduction and core [5].

Set R is a family of equivalent relation, $R \in$ R, and if ind (R)= ind (R- $\{R\}$), it calls R as unnecessary in R; Otherwise, it calls R as necessary. If each $R \in$ R is necessary in R, it calls R as independent; otherwise, it calls R dependent [6].

Set $Q \subseteq P$, if Q is independent and $ind(Q) = ind(P)$,it calls Q as one reduction of P. Obviously, P may have many kinds of reduction and the sets composed of all the necessary relations in P are core of P and are recorded as $core(P)$.

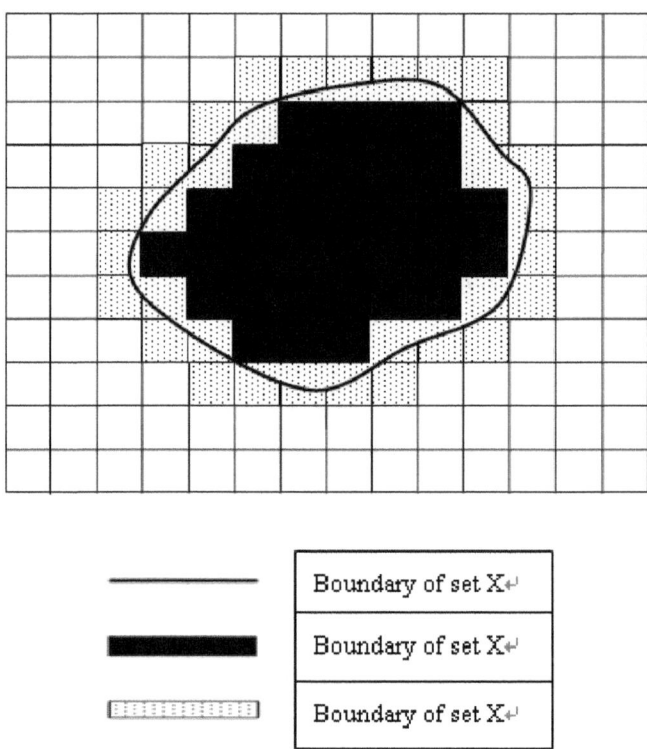

————————	Boundary of set X
■■■■■■■■■	Boundary of set X
▦▦▦▦▦▦▦	Boundary of set X

Fig. 1. Rough Approximation

2.3 Information System

Quaternion $S = (U, A, V, f)$ is a knowledge representation system in form, where:

U :object nonempty finite subset called as universe of discourse;

A :attribute nonempty finite subset;

$V = \bigcup_{a \in A} V_a, V_a$ is range of attribute a;

$f : U \times A \to V$ is an information function and endows each attribute of each object with a information value, namely, $\forall a \in A$, $x \in U$ and $f(x, a) \in V_a$

The knowledge representation system is called as information system and $S = (U, A)$ is used to replace $S = (U, A, V, f)$. Where $A = C \cup D$, $C \cap D = \Phi$, C is called as condition attributes set and D is called as decision attribute set.

2.4 Decision Table

One information system may be expressed by one information table and if attribute $A = C \cup D$ and $C \cap D = \Phi$ it will call information system (U, A) as a decision table, and it will call C as condition attribute and call D as decision attribute.

3 Prediction in Quantity of Tourists in Tangshan City of China Based on Rough Set

The normalization processing is applied to discrete the attribute and remove the redundant individual in the decision table and the attribute reduction is used to obtain the necessary attribute and core attribute and ultimately get the briefest decision rule, so as to determine the growth value[7] by means of scenario analysis.

3.1 Factors Affecting Travel Demand

Factors affecting travel demand are of multiple kinds and the major following factors shall be considered through collection and analysis of data: tourism resources; a. tourism environment; b. traffic, c. service quality; d. tourism expense; e. population in tourism-generating region; f. quantity of tourists. The growth rates of the factors above are selected as the attribute in decision table and the growth rate quantity of tourists is the decision attribute. The growth rate of index is defined as:

Growth rate =(actual value of certain an index in t+1 year-(actual value of certain an index in t year)/ (actual value of certain an index in t year) *100%.

3.2   Establishment of Decision Table

It shall apply the rough set related theories to discrete the original data and apply the supervised static decision table continual attribute discretization methods to make the breakpoint tend to position at the areas of intensive attribute values and have less interval divisions. After investigation and survey on the growth rate of quantity of tourists of the latest ten years, we divide the interval into $(-\infty,0),(0,2),(2,5)$.

The quantity of tourists from 2001 to 2010 is selected for research purpose to obtain the decision table 1 after discretization.

Table 1. Decision Table of Tourism Resources of Tangshan City

Attribute / Object	Condition attribute						Decision attribute
	a	b	c	d	e	f	g
1	2	2	2	2	1	2	3
2	3	3	2	3	2	3	3
3	3	3	2	3	3	3	3
4	2	3	2	3	2	2	3
5	2	2	1	2	2	1	3
6	1	1	1	1	1	1	1
7	1	1	1	1	1	1	1
8	1	1	1	1	1	1	2
9	1	1	3	1	1	1	3
10	1	1	1	1	1	1	2

According to the table above, no consideration will be made for data from 2007 to 2010. Attribute reduction is made for the following decision tables:

$$U/A=U/A-\{a\}=U/A-\{b\}=U/A-\{d\}=U/A-\{f\}\neq U/A-\{c\}\neq U/A-\{e\}$$

Therefore, attributes c, e shall not be subjected to reduction and the traffic factor and tourism expense factor are indispensable in the development of tourism industry. Attribute set after reduction is $\{c, e, f\}$.

3.3   Determination of Decision Rule

After reduction of attribute set, the following definitive decision rule sets are obtained:

$$r_1 : (c=2 \wedge e=1) \vee (c=2 \wedge e=2) \vee (c=2 \wedge e=31) \vee (c=1 \wedge e=1) \rightarrow g=3$$
$$r_2 : (c=1 \wedge e=1) \vee (c=3 \wedge e=1) \rightarrow g=2$$
$$r_3 : (c=1) \vee (e=2) \rightarrow g=1$$

The tourism industry development tendency is optimistic driven by the rapid economic development and the growth rate of the number of tourists is expected to be at the level of 2%-5% under the condition that the growth rates of tourism expense is not high and the traffic conditions are convenient.

4   Conclusion

The analysis above shows that the traffic condition and tourism sites in Tangshan City influence the development speed of the tourism industry in the city directly, which is in accordance with the development status of current tourism market. Based on the consideration above, the travel agency has the vehicles capable of reaching the tourism destination directly and the expense of group tourism is more economic than that of individual tourism, so the new service industry has a broad development scope. At the same time, the travel agency assists the tourism industry for propaganda and they supplement each other to promote the development of tourism industry of Tangshan City. In this paper, the rough set method is applied to predict the travel demand and realize an accurate and dynamic analysis of the tourism income and the number of tourists, and the analysis results show that application of rough set method in prediction of tourism is practical and effective.

References

1. Wang, P.-y., Wang, Y.-j.: On Tourism Economic Analysis and Forecast in the Tourism Planning. Tourism Tribune (4), 47–50 (2000)
2. Wu, J., Liu, X.p.: Model of Time Series Predication for International Tourist Foreign Exchange Earnings in China. Basic Sciences Journal of Textile Universities 14(1), 51–55 (2001)

3. Liu, S.: Dynamic Prediction of Passenger Source of Tourism Destination. Journal of Qufu Normal University: Natural Science (4), 107–110 (2003)
4. Han, Z., et al.: A Survey on Rough Set Theory and Its Application. Control Theory & Applications 16(2), 153–157 (1999)
5. Mai, F., Lu, L., Liang, Y.: Study on Tourism Destination Value Based on Rough Sets. Ecological Economy (11), 98–100 (2005)
6. Zhang, Q.-g.: Introduction to Artificial Neural Networks, pp. 97–107. China Waterpower Press, Beijing (2004)
7. Sun, Y., Zhang, L., Lu, R.: Tourist Quantity Forecast by Using Neural Network. Human Geography 17(6), 50–52 (2002)

WSPOM: An Ontology Model for Web Service Planning

Liming Liu and Hongming Cai

School of Software,
Shanghai Jiao Tong University,
Shanghai, P.R. China
andyliuliming@gmail.com,
cai-hm@cs.sjtu.edu.cn

Abstract. With the growth of web services' quantity and complexity, the problem that whether the existing web services could satisfy the requirements of business has become more complex. Based on the OWL, WSPOM (Web Service Planning Ontology Model) is proposed, WSPOM consists of General Ontology, Task Ontology and Web Service Ontology. First, the General Ontology is built up to describe the state of the world and the changes each Operation could make to the world. The changes are described in graph difference vector. Then one satisfactory relation all over the ontology model is proposed. Based on the satisfactory relation, whether the web service could satisfy the business demand will be validated, then the Web Service Grounding would be generated in WSDL or code stub. Then one framework is developed based on WSPOM, the result shows that the framework proposed could resolve the problem efficiently.

Keywords: Ontology, WSPOM, Semantic Web, Web Service.

1 Introduction

The advent of Web services is a proof that nowadays the need for communication among loosely coupled distributed systems is bigger than ever. Web services offer a well-defined interface through which other programs may interact by sending messages based on Internet protocols and Web standards. They may also be combined in order to achieve a complex service whose functionality cannot be achieved by a single one, a procedure that is called service composition.

The term "Semantic Web" is often used more specifically to refer to the formats and technologies that enable it. These technologies include the Resource Description Framework (RDF [1]), a variety of data interchange formats (e.g. RDF/XML, N3, Turtle, N-Triples), and notations such as RDF Schema (RDFS [2]) and the Web Ontology Language (OWL [3]), all of which are intended to provide a formal description of concepts, terms, and relationships within a given knowledge domain. Increasingly, OWL has become one of the most popular languages for representing ontology in the Semantic Web.

In this paper, we present WSPOM (Web Service Planning Ontology Model). In WSPOM, we first divide the domain ontology of business into general ontology, task

B. Liu and C. Chai (Eds.): ICICA 2011, LNCS 7030, pp. 447–454, 2011.

ontology and web service ontology, and describe them in OWL, then describe one method to validate the web service whether could satisfy the business demand based on WSPOM. Furthermore, as a web service planning model, WSPOM can guide the business analysts to figure out what kind of the services their organization lack of, and provide the service grounding based on the result figured out.

The rest of this paper is organized as follows. Section 2 summarizes the researches worked in recent years related to our studies. In section 3, we introduce a new ontology model called WSPOM and present every part of WSPOM detailed. The web service planning framework based on this ontology model is introduced in section 4. Section 5 describes an example based on our approach. Last section (Section 6) concludes this paper.

2   Related Work

The current research about web service mainly focuses on using semantic to allow automatic discovery, matching and composition of web service. Semantic Web initiative tries to solve such problems related to knowledge representation by suggesting standards, tools and languages for information annotation. Thus, the combination of Semantic Web techniques with Web services seems the best way to give the appropriate semantic notion to Web services in order to achieve the desirable level of automation. Towards this need, many languages and frameworks have been proposed, such as WSMO [4], WSDL-S [5] or OWL-S [6]. In this progress, the ontology language plays more and more important role in semantic web.

Ongoing web service research is not limited to the automation matching, composition and invocation [7]. The problem that whether web service in one organization could satisfy all the functional and non-functional requirements of business tasks, and how to optimize the web services to save the cost is also important. T. G. J. Schepers[8] present a lifecycle approach to SOA governance. Min Luo[9] give out one approach to practical SOA, it includes three perspectives: Service Modeling, Enterprise Service Bus and Governance.

Eric Bouillet[10] give us a domain-independent, general purpose knowledge engineering and planning framework that supports the construction of planning domains and problems based on OWL ontology, and the integration of the planning process with description logic (DL) reasoning.

All these researches concerned the Web Service Ontology and Business Domain Ontology separately, and made great achievements. But the problem that whether the existing web services could satisfy the requirements of business has become more complex. For resolving this problem, we present an ontology model to resolve the problem through validating the web service whether could satisfy the business demand and how to optimize the existing web services.

3   Web Service Planning Ontology Model (WSPOM)

The WSPOM focuses on the concepts and operations ontology, and it consists of general ontology, task ontology and web service ontology. As shown in figure 1.

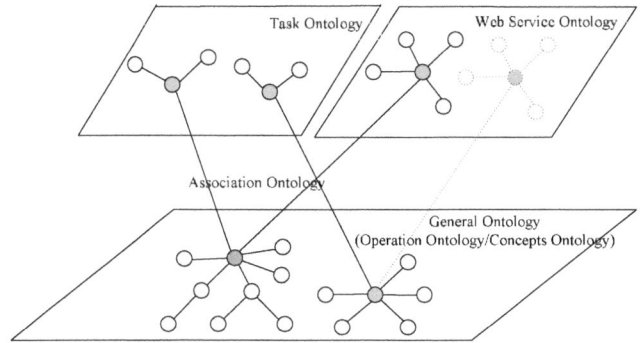

Fig. 1. The structure of Web Service Planning Ontology Model (WSPOM)

The most important concept in General Ontology is the Operation, and it could be described by one OWL graph. For simplicity, we depict it in the grey node in Fig1. The satisfactory relation is most based on the graph homomorphism or equality. Based on this model, we could decide what business task could be satisfied already, and what kinds of web service the organization should provide. In this section, we formally describe how to describe general ontology, task ontology, and web service ontology in OWL.

A) General Ontology

The general ontology is used to describe the most basic elements in business domain such as the concept ontology and the operation ontology. Concept ontology describes the concepts, properties and individuals. In WSPOM, we call them OWL Resource. The operation ontology describes changes to the concept ontology itself causes. Every operation has its preconditions and effects described using the OWL graph. The first definition in WSPOM is OWL Model consists of Classes, Individuals and Property, where Individuals are instances of classes, and properties may be used to relate one individual to another. A class defines a group of individuals that belong together because they share some properties. Properties can be used to state relationships between individuals or from individuals to data values. OWL Triple is a member of the set Classes ×Property ×Classes. A Triple Pattern is a member of the set (Classes∪Individuals) ×Property × (Classes∪Individuals). An OWL Graph is one set of OWL Triples; furthermore, the OWL Graph Pattern consists of the Triple Patterns. Another concept in General Ontology is the Operation. An operation can cause the state of the world to change, and from out of the ontology domain, the only thing we can notice is the changes to the OWL Graph, and so the operation can be described in OWL Graph Delta.   The Operation is described formally as:

Definition 1: OP=<P, E>

OP stands for Operation. P is the precondition of this operation which could be described in OWL Graph Pattern, and E is the effects of this operation which can be described as OWL Graph Delta. Let us show how to get the OWL Graph Delta. First let the current state is G, after applying the OP to Ontology, O, and some changes will happen in the ontology. These changes could be one process right transfer from one person to another or other changes alike. Then the state of the Ontology O would come into G'.The ontology could be described in OWL Graph, so if we define the graph subtracting operation of OWL Graph, we can get the OWL Graph Delta through subtracting G by G'.

Definition 2: E=Sub<G', G>=<NT, PT, CHG>

Sub is the symbol for subtracting operation, NT is the negative OWL Triples generated by this subtracting operation, PT is the Positive OWL Triples, CHG stands for the changes to the nodes of the triples.

B) Task Ontology

The task ontology is built upon general ontology, and the task definition is the most important in it. There are three kinds of tasks: atomic, composite and simple task. Atomic tasks correspond to those can perform by engaging it in a single Operation.

Operation, User does not constrain to people, the computer or other sub-system can also be modeled as User.

Definition 3: Atom Task=<OP, C, PreT>

OP reference to the Operation concept in General Ontology, and the C means the candidates who take charge of this task. The tasks in business environment may not be independent to others; the most common scenario is that one task depends on some tasks. PreT is the precondition task.

Definition 4: Task= {Atom Task and Composite Task}

According to the complication, the task could be divided into Atom Task and Composite Task. Atom Task means those tasks could be described by one OP in definition 1 and one candidate to finish it. Composite tasks are decomposable into other (non-composite or composite) processes; their decomposition can be specified by using control.In another word, the task ontology provides one view to the whole ontology with greater granularity comparing to the General Ontology.

C) Web Service Ontology

The Semantic Web should enable greater access not only to content but also to services on the Web. Users and software agents should be able to discover, invoke, compose, and monitor Web resources offering particular services and having particular properties, and should be able to do so with a high degree of automation if desired. In WSPOM, we put more concentration on the abilities web service possesses. Such abilities are described by OP in General Ontology. Other dimensionalities of Web Service are presented in Input, Output and PreWS.

Definition 5: Atom Web Service=<OP, I, O, PreWS>

A Web Service can have two sorts of purpose. First, it can generate and return some new information based on information it is given and the world state. Information production is described by the inputs and outputs of the process. Second, it can produce a change in the world. This transition is described by the preconditions and effects of the process.

Definition 6: Web Service= {Atom Web Service and Composite Web Service}

Just like the Task in definition 4, we use Web Services as the model for process decomposition and assembly, which means one Web Service, could be assembled by some other Atom Web Services.

D) Satisfactory Relation

In WSPOM we put more concentration on the whole tasks whether could be satisfied by the whole set of web services. We say there is one Satisfactory Relation between one Task and Web Service if and only if the OPs of one Task in definition 5 could be covered by OPs of one web service or one series of web services in definition 3. Let X and Y are two Operations defined in General Ontology, if the $X.P \subseteq Y.P$ and $X.E==Y.E$, then we say the X could be covered by Y. $X.E==Y$. E holds not only when they have the exactly OWL graph, if the two OWL graph describing X.E and Y.E has the Equality relation in OWL or could be reasoned by OWL reasoned, they would be covered by each other.

Definition 7: $ST=\{T|\exists WSi \in WSO :(\forall OP \in T.OP , \exists OPi \in WSi .OP : OPi$ Cover OP)\}

ST means the Tasks has been satisfied by the existing Web Services. WSO means the whole set of web services in Web Service Ontology. Based on definition 8, we can give out the algorithm which could figure out what kind of web service we lack of but should provide.

4   Overall Framework and Planning Algorithm

The main goal of our WSPOM is to help the business analysts and service providers to find out what web services we lack of but should provide. Figure 2 shows the overall WSPOM framework, and it consists of Ontology Access and Reasoning

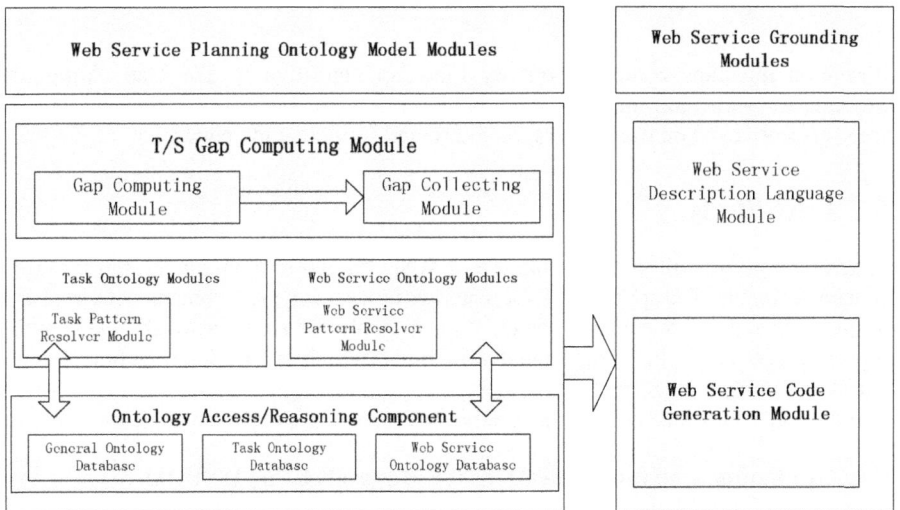

Fig. 2. Web Service Planning Ontology Model Framework

Component, Task Ontology Module, Web Service Ontology Modules, Task/Service Gap Computing Module, and Web Service Grounding Module, and one Web Service Code Generation Module throughout the entire system.

The framework depends on some preparations including the ontology modeling of WSPOM. And Ontology Access/Reasoning Component would provide one access API to the ontology model for the Task Pattern and Web Service Pattern Resolver Modules. At the meantime, the reasoning is also applied to the origin Ontology Model to fetch the truth not been depicted straightly. But for getting the efficiency and the consistence result, the reasoning of the Ontology Model would not run parallel to the Computing actions. Once the Operation in the Task Ontology and the Web Service Ontology are ready, then the Task/Service Gap Computing Module will compute out the gap between the two based on the Satisfactory Relation described in WSPOM which originates from the Operation and its precondition, effects described in changes, which is in the form of OWL Graph. The FindGap algorithm below shows the concrete steps.

The Algorithm for Finding the Gap between Task Ontology and Web Service Ontology

```
Program FindGap (GO,TO,WSO)
  GO+=Reason(GO);//Use the Reasoner Engine for OWL
  TO+=Reason(TO);//To find the entailed Relation
  WSO+=Reason(WSO);
  Begin
    for(T : TO)
          TOperations=
            (Find Operations In WSO Mapping to T)
          For(WS : WSO)
              If(WS Satisfies T)
              Tag T with satisfied
    Output Ts in TO without the satisfied Tag
  end.
```

Based on the gaps computed out by FindGap algorithm in T/S Gap Computing Module, we will figure out that what business task could not be satisfied by existing web services based on the task graph pattern and service graph pattern.

5   Case Study

In this section, we will introduce the use of WSPOM and matching algorithm through a concrete example. The applications for materials process is typical in manufacturing based business operations. When one person want to apply for some materials, like printing papers or steer plates, he or she should write one materials application note first, then the manager would examine the note to decide approve this application or not. If approve, then the applicant would contact the warehouse keeper to pick up the materials.

5.1   An Example about Applications for Materials Using WSPOM

Based on the description, There should be three Operations: Approve Operation, Reject Operation, and Withdraw Operation, three task: Approve Task, Reject Task, and Withdraw Task. We imagine there only have the service which has two Operation Approve Operation and Reject Operation. Now the WSPOM for this process could be presented in figure 3.

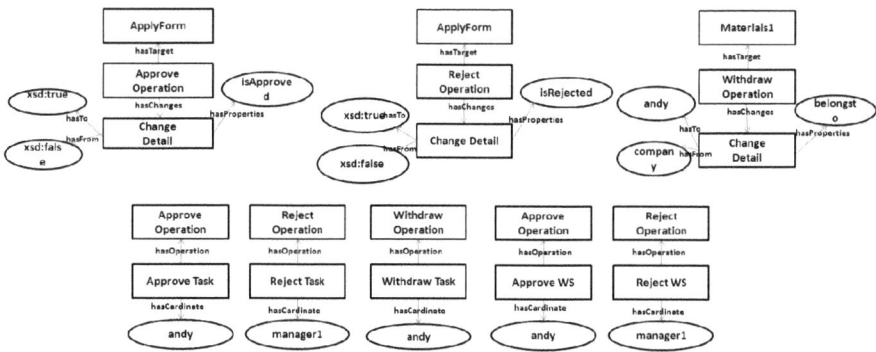

Fig. 3. WSPOM for Applications For Materials Process

Let us take the Approve Task as an example. The Withdraw Task could be described formally as Withdraw Task= {Withdraw OP, Employee, Approve Task}. The Withdraw OP has {Precondition for W, Effects for W}. Both the Precondition and Effects could be depicted by one OWL Graph. Employee is the candidate in charge of finishing the task, and Approve Task is the pre-task which also could be described by our WSPOM formally. Finally our T/S Gap computing module computes based on the Satisfactory Relation and some other metrics, but failed to find one web service to satisfy the Withdraw Task, then the result would be transferred to the Web Service Grounding Module to generate the web service description.

```
<?xml version="1.0" encoding="UTF-8" standalone="no"?>
<wsdl:definitions xmlns:soap="http://schemas.xmlsoap.org/wsdl/soap/"
    xmlns:tns="http://ist.sjtu.edu.cn/WithdrawWebService/"
    xmlns:wsdl="http://schemas.xmlsoap.org/wsdl/"
    xmlns:xsd="http://www.w3.org/2001/XMLSchema"
    name="WithdrawWebService" targetNamespace="http://ist.sjtu.edu.cn/WithdrawWebService/">
  <wsdl:types>
    <xsd:schema targetNamespace="http://ist.sjtu.edu.cn/WithdrawWebService/">
      <xsd:element name="WithDrawMaterials">
        <xsd:complexType>
          <xsd:sequence>
            <xsd:element name="candidate" type="xsd:string" />
            <xsd:element name="materials" type="xsd:string"></xsd:element>
            <xsd:element name="quantity" type="xsd:string"></xsd:element>
          </xsd:sequence>
        </xsd:complexType>
      </xsd:element>
      <xsd:element name="WithDrawMaterialsResponse">
        <xsd:complexType>
          <xsd:sequence>
            <xsd:element name="out" type="xsd:string"/>
          </xsd:sequence>
        </xsd:complexType>
      </xsd:element>
    </xsd:schema>
  </wsdl:types>
  <wsdl:message name="WithDrawMaterialsRequest">
    <wsdl:part element="tns:WithDrawMaterials" name="parameters"/>
  </wsdl:message>
  <wsdl:message name="WithDrawMaterialsResponse">
    <wsdl:part element="tns:WithDrawMaterialsResponse" name="parameters"/>
  </wsdl:message>
  <wsdl:portType name="WithdrawWebService">
    <wsdl:operation name="WithDrawMaterials">
      <wsdl:input message="tns:WithDrawMaterialsRequest"/>
      <wsdl:output message="tns:WithDrawMaterialsResponse"/>
    </wsdl:operation>
  </wsdl:portType>
  ...
</wsdl:definitions>
```

Fig. 4. WSDL generated by Web Service Description Language Module

5.2   Discussion

By defining the Operations and Changed each Operation makes to the state of the world, our framework could compute the gap between the Task Ontology and the Web Service Ontology describing the ability the existing web services could provide. Then the Web Service Description Language Module could generate the WSDL for the needed web service, moreover, once we get the WSDL file, we could generate the code stub for the web service automatically. The framework depends on the pre-built ontology, so the ontology building is also important for the result.

6   Conclusion

In this paper, we present The WSPOM which is especially suitable to the REST service. It is inspired by the existing problem that when the number and complexity of web service increase, it's hard to figure out what web service the organization need to provide or develop. The WSPOM consists of General Ontology, Task Ontology and Web Service Ontology. So the business analysts and web service provider could describe them in the same base. It's designed to resolve the web service automatic planning problem. In the future, further research work will be carried out in the management of ontology and automatic building of the General Ontology, Task Ontology and Web Service Ontology through data mining or some other theory. Furthermore the computation algorithm should also be optimized.

Acknowledgments. This research is supported by the National Natural Science Foundation of China under No.70871078, the National High Technology Research and Development Program of China ("863" Program) under No.2008AA04Z126, and Shanghai Science and Technology Projects 09DZ1121500.

References

1. Manola, F., Miller, E.: RDF Primer, W3C Recommendation, February 10 (2004)
2. Brickley, D., Guha, R.V.: RDF Vocabulary Description Language 1.0: RDF Schema W3C Recommendation, February 10 (2004)
3. Hitzler, P.: OWL 2 Web Ontology Language Primer, W3C Recommendation, October 27 (2009)
4. WSMO: Web Service Modeling Ontology D2v1.3, http://www.wsmo.org/TR/d2/v1.3/
5. Akkiraju, R., Farrell, J.: Web Service Semantics - WSDL-S, W3C Member Submission, November 7 (2005)
6. OWL-S: Semantic Markup for Web Services, http://www.w3.org/Submission/OWL-S/
7. Li, N., Cai, H.: Functionality semantic indexing and matching method for RESTful Web Services based on resource state descriptions. In: 2nd International Workshop on Computer Science and Engineering (2009)
8. Schepers, T.G.J.: A lifecycle approach to SOA governance. In: SAC 2008 (2008)
9. Luo, M., Zhang, L.-j.: Practical SOA: Service Modeling. In: Enterprise Service Bus and Governance in ICWS (2008)
10. Bouillet, E., Feblowitz, M., Liu, Z., Ranganathan, A., Riabov, A.: A Knowledge Engineering and Planning Framework based on OWL Ontologies. In: ICKEPS 2007 (2007)

A Requirement Group Based Web Service Scheduling Model

Xinyong Wu, Lihong Jiang, Fenglin Bu, and Hongming Cai

School of Software, Shanghai Jiao Tong University, Shanghai, P.R. China
sinron.wu@gmail.com, {jiang-lh,bu-fl}@cs.sjtu.edu.cn,
hmcai@sjtu.edu.cn

Abstract. The growth of concurrent requests from increasing applications based on web service aggravates the load of web service, and may result in low performance or even breakdown of the invoked web service. However, some requests are similar or related, and can be integrated into a slightly complex one, which can be resolved by the service provider in a unified way. Based on this fact, a requirement group-based web service scheduling model (RGBWSSM) is proposed. It concentrates on the integration of similar or related requests and the capability matching between the integrated request and web services, both based on OWL-S. Moreover, a case study is implemented and the results show that this model is effective and useful in reducing web service load in the environment of high concurrent service requests.

Keywords: Web Service Load, Similar Request, Request Group, OWL-S.

1   Introduction

Web Services are getting more and more popular due to its good features, which include platform independence, program language independence, loose coupled, easy interaction, and so on. As a result, applications based on web service are growing rapidly, making some web services in certain domains, such as E-commerce, invoked by an increasing number of requesters. Too many requests put forward the new challenge to web service load capacity, and may result in overload or even breakdown of the invoked web service from the service providers' perspective in one side, and increased response time or failed invocation from the service consumers' perspective in the other side.

Although the distributed feature of web service allows it to scale in response to new resource demands by introducing new servers hosting the particular, needed component, flash traffic patterns can drive a web service into overload. This leads to poor performance as the system is unable to keep up with the demands placed on it, and users see increased response time for their requests [1].

In order to effectively avoid overload of web service, in this paper, we propose a requirement group-based web service scheduling model (RGBWSSM). It's based on the fact that some requests are often similar or related, and can be grouped into a request group that can be resolved by the service provider in a unified way.

B. Liu and C. Chai (Eds.): ICICA 2011, LNCS 7030, pp. 455–462, 2011.

2   Related Work

OWL-S [2] enable users and software agents to automatically discover, invoke, compose, and monitor web resources offering services. With its capacity of semantically representation OWL-S makes it possible to search web services that satisfy the requester's requirement and find similar requests without human assistance.

In [3], capability matching is described as the match between IOPE of advertisements service and IOPE of requests service. The matching process is just the comparing between two ontology concepts substantially.

In order to enhance the query efficiency and supporting the run-time substitutions of the temporally unavailable services, the conception of service group is proposed in [4]. It refers to a web service collection in which each member may implement the same function.

Response time is a significant quality of QoS (Quality of Service) [5]. It's a function of load intensity, which can be measured in terms of arrival rates or number of concurrent requests. To avoid long response time and improve the web service performance, many researches about load balancing for web services are made. For example, in [6], three models are described, which are client-oriented, broker-oriented, and server-oriented models, to achieve suboptimal request assignment. A model is investigated that not only based on QoS specification but also load balancing to support dynamic web service selection in [7]. In [8], a scheduling mechanism was studied that helps the BPEL engine to select between different web service providers according to their current workload to improve the throughput of a BPEL engine. And in [9], an enhancement to the conventional Round Robin DNS load balancing technique is proposed by taking consideration of the status of the servers and distributing the services requests based on the performance matrix of the servers in the cluster.

However, these researches all focus on load balancing in the server side by distributing service requests to different web service server. Service providers often have to expand their servers for the increasing requests, which will increase the cost.

3   Architecture of the Model

The architecture of the model is based on SOA (Service-Oriented Architecture) [10]. As shown in figure 1, besides the OWL-S Request Pool, there are four modules and two objects in this model. Details of each module will be given below.

Grouping Module: This module takes in requests presented by OWL-S as input. With the semantic support of Domain Ontology Base, it compares the requests and integrates the similar ones to construct a request group.

Matching Module: This module finds a service group to meet the requirements of a request group.

Service Invocation Module: This module is in charge of invoking all the services of the service group to get results routed to every member of the request group.

Result Routing Module: This module takes charge of routing the results produced by service invocation module to each member correctly.

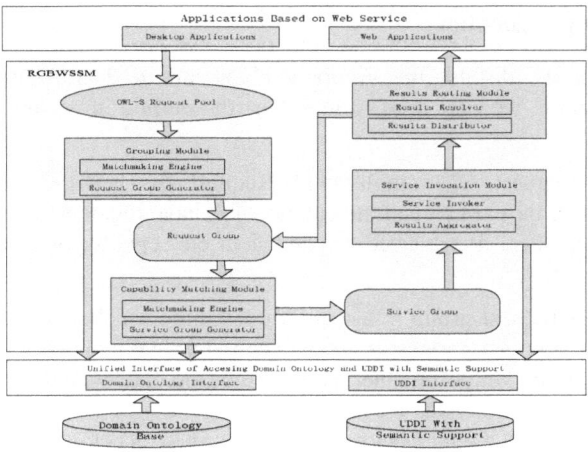

Fig. 1. Architecture of RGBWSSM

The definitions of the two objects named request group and service group will be given in section 4 and section 5.

4   Grouping Requests

4.1   Request Description with OWL-S

According to OWL-S, web service ontology consists of three parts, which are ServiceProfile, ServiceModel and ServiceGrounding. The upper ontology for ServiceProfile is logically divided into three parts: contact information referring to the entity that provides the service, functional description, and description of a host of properties that are used to describe features of the service. Specifically, the functional description specifies the inputs required by the service and the outputs generated; furthermore, the profile describes the preconditions required by the service and the expected effects that result from the execution of the service.

According to the above description, we extend the IOPE and model a request as below:

$$\text{Request} = <I, O, P, E, A>$$

I = Inputs, describing the objects that will be taken by web service as its inputs.

O = Outputs, which describe the expected results.

P = Preconditions, which present conditions that should be true for the expected results obtained by the requester.

E = Effects, which depict the changes after the request is implemented.

A = Address, which tells where the request is from.

4.2   Grouping Algorithm

In this section we discuss the grouping algorithm in detail. To understand the grouping algorithm we first need to introduce the definition of similar requests and request group.

Definition 1. Request A is similar with Request B when the inputs of A have relationships with those of B, and the outputs of A have relationships with those of B. The relationships here are referred to equivalent concepts to, or sub-concepts of, or super-concepts of.

Definition 2. A request group is a request containing a collection of other requests and the information that pertains to them. The individual requests represented within the group are similar with each other. A request group may contain only one request if there are no other requests that are similar with it. The inputs and outputs of a request group should cover all the member requests', described as below:

for all inputs of request group $Input_{Rg}$

$$Input_{Rg} = Max_{concept}\{Input_R \mid R{\in}Rg\} . \tag{1}$$

The method $Max_{concept}$ is to get the top concept of a certain input of the requests according to ontology concept. $Input_R$ is an input of request R which is a member of the request group Rg. $Input_R$ corresponds with $Input_{Rg}$. So do all the outputs.

The main idea of grouping algorithm is that for all the requests in the OWL-S Request Pool, if one Request can be added to a RequestGroup in the RequestGroupList, add it to that RequestGroup, and if not, instantiate a RequestGroup object, and assign the Inputs and Outputs of the Request to the RequestGroup's, then add this new RequestGroup to RequestGroupList. When all the requests are added to a corresponding RequestGroup, return the RequestGroupList. The main control of the grouping algorithm is shown as below.

The main control of the grouping algorithm

```
GroupRequests(RequestList){
  RequestGroupList = empty list;
  for all Request in RequestList do{
    if (AddToGroup(Request, RequestGroupList) is false){
      Instantiate a RequestGroup object as TempGroup;
      TempGroup.Input = Request.Input;
      TempGroup.Output = Request.Output;
      Add TempGroup to RequestGroupList;}}
  return RequestGroupList;}
```

To judge whether a Request can be added to a RequestGroup or not, we compare the Inputs and Outputs of the Request with the RequestGroup's according to definition 1. As shown in the pseudo code below, in the function BecomeMemberOf, we compare a Request with a RequestGroup. If the Inputs and Outputs of the Request hasRelations--- equivalent concepts to, or sub-concepts of, or super-concepts of, with the RequestGroup's, we assert that this Request can become a member of the RequestGroup. And if the "hasRelations" is super-concepts, in other word, the Inputs or Outputs of the Request subsume those of RequestGroup, the Inputs or Outputs of the Request will be assigned to the RequestGroup's so as to guarantee that the Inputs and Outputs of the RequestGroup cover all the members' Inputs and Outputs.

The details of adding a request to a request group

```
AddToGroup(Request, RequestGroupList){
  for all RequestGroup in RequestGroupList do{
    if(BecomeMemberOf(Request, RequestGroup) is true){
      return true;}
    else continue;}
  return false; }
BecomeMemberOf(Request, RequestGroup){
  If(Request.Input hasRelations with RequestGroup.Input
      and Request.Output hasRelations with
      RequestGroup.Output){
    If(Request.Input subsumes RequestGroup.Input){
      RequestGroup.Input = Request.Input;}
    If(Request.Output subsumes RequestGroup.Output){
      RequestGroup.Output = Request.Output;}
    Add Request to RequestGroup;
    return true;}
  return false; }
```

Once request groups are generated, web services which are expected to implement the requirements should be found and invoked. In next section, the capability matching between request group and web services will be described.

5   Matching between Request Group and Web Services

In order to find appropriate web services that can implement all the requirements of a request group, the concept of service group is proposed.

Definition 3. A service group is a service that contains a collection of other web services and the information that pertains to them. The individual services within the group are executed cooperatively or independently to implement the requirements of a request group. When the services within the group are executed cooperatively, the service group becomes a composite service.

For example, there is a buyer who wants to get a wine list in which the price of each wine is low than 50 dollars. And there are two sellers, of which one sells white wine, and the other sells red wine. In this case, the two sellers' services should be collected together to be executed independently because white wine and red wine both are wine. However, when a buyer wants to buy a red wine which price is low than 50 dollars, it will need two services to be executed cooperatively. One takes charge of verifying if there is any red wine of which the price is low than 50 dollars, and the other is responsible for finishing this transaction if the expected red wine exists.

Since web service composition is a hot topic in recent years, and there's much research about it, we will not discuss it in this paper. We just consider the situation where web services in service group are executed independently.

The matching between request group and web service capability is also based on OWL-S. M Paolucci brought forward four matching levels in [11], which are exact matching, plug in matching, subsumes matching and fail. Exact matches are of course preferable to any another; plugIn matches are the next best level; subsumes is the third best level; fail is the lower level and it represents an unacceptable result.

Besides the outputs matching between advertisement and request, the inputs matching should be implemented too because no other than that the inputs of request meet the inputs of service can the service be invoked successfully.

In order to compare the matching accuracies, we quantize the four matching levels with matching score: Exact = 1, Plug in = 0.8, Subsumes = 0.5, and Fail = 0.

Definition 4. The matching score between a request group Rg and a service group Sg, sim(Rg, Sg), is a function of summing matching scores between the member services in the service group and the request group:

$$sim(Rg\ Sg) = \sum\{sim(Rg\ S)\mid S \in Sg\}. \tag{2}$$

In the function above, sim(Rg, S) is the matching score between service S which is a member of the service group Sg and the request group Rg. As we mentioned above, capability matching consists of input matching and output matching, so sim(Rg, S) should consider both the input matching score and the output matching score:

$$sim(Rg\ S) = sim_{in}(Rg\ S) + sim_{out}(Rg\ S). \tag{3}$$

We find a service group that has the highest matching score to provide service for a request group according to definition 4.

The results from the invocation of the service group will be routed to each member request of the request group according to their requirements and addresses.

6   Case Study

A wide application of web service is in E-commerce, where information searching is the most common operation. In this section, we will study a familiar case grounding on our model by extending the example given in the section 5.

6.1   An Example about Searching Wine using RGBWSSM

There are several requests with the purpose of searching a certain kind of wine which price is in a price range. The input of these requests may be two price instances, of which one specifies the minimum price, and the other specifies the maximum price. The output may be a certain kind of wine list. And there are two web services in the UDDI with Semantic Support. Both of them accept a price range as input, and return a red wine or white wine list. We identify them as service R and service W. The requests are depicted in a nutshell in table 1.

Table 1. The description of the requests

Request	Input		Output	Address
	MinPrice(dollars)	MaxPrice(dollars)		
A	0	50	A Wine List	Aa
B	20	50	A Red Wine List	Ab
C	0	30	A Red Wine List	Ac
D	10	50	A White Wine List	Ad
E	0	40	A White Wine List	Ae

According to the grouping algorithm presented in section 4, these five requests will be grouped to one request group, and the price range of the request group should be from 0 to 50 dollars. Through the capability matching by the matching module, the two services R and W will be searched and grouped to a service group. These two services are invoked by the service invocation module independently and return a list of composite results, which will be routed to each member request of the request group according to their requirements and addresses, as shown in figure 2.

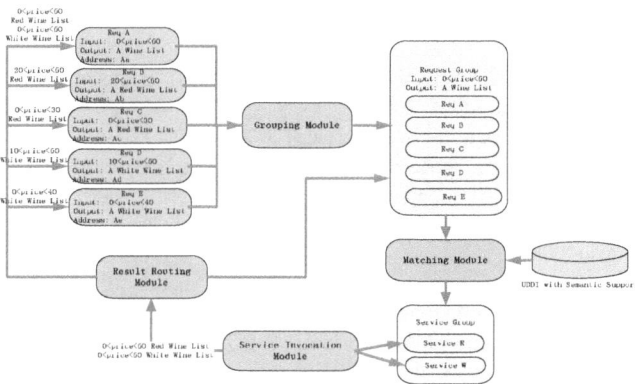

Fig. 2. The process of the case

We send the five requests described above at the same time to test the model. And we get the same return results for each request using RGBWSSM model as that not using RGBWSSM model costing almost the same time averagely. However, we find that the invocation times of the two services using RGBWSSM (just 1 time) are largely lower than that not using RGBWSSM (3 times). In other words, the loads of the two services are reduced largely by using RGBWSSM.

6.2   Discussion

Since the load of a web service can be measured in terms of the invocation times during a unit of time, we assert that it can reduce the load and improve the performance of a web service effectively by decreasing the requests, especially in the environment of high concurrent service requests. And we can surmise that by grouping more requests to a request group the more effectiveness will be acquired.

7   Conclusion and Future Work

The load of web service boosts rapidly because of the rapid increasing of applications based on web service, resulting in overload or even breakdown of web service. In this paper, we propose a requirement group-based web service scheduling model (RGBWSSM) to reduce the load of web service. We discuss the grouping requests and matching between request group with web services in detail. In particular, a case study is given and the results show that our model is useful and effective in reducing the load of web service. In the future, we will extend criterion according to which a request is determined to be similar with another request to get more complete request groups so as to reduce the load of web service to greater extent.

Acknowledgments. This research is supported by the National Natural Science Foundation of China under No. 70871078, the National High Technology Research and Development Program of China ("863" Program) under No. 2008AA04Z126, and Shanghai Science and Technology Projects 09DZ1121500.

References

1. Porter, G., Katz, R.H.: Effective Web Service Load Balancing Through Statistical. Communications of the ACM (2006)
2. OWL-S: Semantic Markup for Web Service (2006),
   http://www.ai.sri.com/daml/services/owl-s/1.2/overview/
3. Guo, R., Le, J., Xia, X.L.: Capability Matching of Web Services Based on OWL-S. In: Proceedings of the 16th International Workshop on Database and Expert Systems Applications (2005)
4. He, L., Liu, L., Wu, C.: A revised service group model for web service availability mangement. In: Fifth International Joint Conference on INC, IMS and IDC. IEEE (2009)
5. Menasce, D.A.: QoS Issues in Web Services. IEEE Internet Computing (2002)
6. Badidi, E., Serhani, M.A., Esmahi, L., Elkoutbi, M.: Practical Load Distribution Models for Web Services based environments. In: 2008 International MCETECH Conference on e-Technologies (2008)
7. Zhao, Q., Tan, Y.: A Load Balancing Based Model for Dynamic Web Service Selection. In: 2009 Second International Symposium on Computational Intelligence and Design (2009)
8. Ferber, M., Hunold, S., Rauber, T.: Load Balancing Concurrent BPEL Processes by Dynamic Selection of Web Service Endpoints. In: 2009 International Conference on Parallel Processing Workshops (2009)
9. Chin, M.L., Tan, C.E., Bandan, M.I.: Efficient Load Balancing for Bursty Demand in Web based Application Services via Domain Name Services. In: 2010 8th Asia-Pacific Symposium on Information and Telecommunication Technologies, APSITT (2010)
10. Gottschalk, K., Graham, S., Kreger, H., Snell, J.: Introduction to Web services architecture. IBM Systems Journal 41(2) (2002)
11. Paolucci, M., Kawamura, T., Payne, T.R., Sycara, K.: Semantic Matching of Web Services Capabilities. In: Horrocks, I., Hendler, J. (eds.) ISWC 2002. LNCS, vol. 2342, p. 333. Springer, Heidelberg (2002)

An Approach to Enhance Safe Large Data Exchange for Web Services

Chen Wang[1], Qingjie Liu[2], Haoming Guo[3], and Yunzhen Liu[3]

[1] Institute of Geophysics, China Earthquake Administration
[2] Institute Of Disaster Prevention Science And Technology
[3] Beihang University Computer School, New Main Building G1134, Xueyuan street 37#,
Haidian district, Beijing, China
guohm@nlsde.buaa.edu.cn

Abstract. As a stateless distributed component, web service could not transfer large data simply and safely under confine of network communication. This problem affects service centric system's performance for data concentrated applications. To address the issue, LDT4WS is introduced to enhance web service to transfer large data. In LDT4WS, large data is converted into binary and divided into smaller packs by sender. Receiver creates context for packs' relationship by which packs could be organized orderly and integrated. In data transferring, the network failure would be checked and only the failed packs would be transferred. Upon LDT4WS, a data exchange platform is built for CAE's SPON. In the system, large data are transferred safely by web service.

Keywords: SOA, Web Service, Data Transfer.

1   Introduction

In recent years, web service has been widely accepted by IT industry for its simplicity and is viewed as one de-facto model for web based applications[1]. It enhances mutual message exchange between components of any OS that supports net communication. In web service's architecture, SOAP is the core element that is a light-weight and XML based protocol. It's designed to facilitate structured and prefabricated message exchange for web applications. SOAP can be incorporated with many internet data transfer protocol, such as HTTP. For the convince, service centric webs are built upon web service through which accomplish works for SOAP request over HTTP[2].

The SOAP is designed extendable for applications[3][4]. It may contain any data in its body. However, while being implemented with specific network communication protocol, its capacity is about to be limited. The problem seriously affects performance of web service in data concentrated applications[5]. For example, in CEA(China Earthquake Administration), sensors are deployed all over the country. Every sensor works to collect data of surrounding environment. The data are aggregated in provincial center. To manage the data, a system, named Seismological Precursors Observation Net(SPON), is built upon standard web service. In SPON those provincial centers are responsible to transfer the daily aggregated data to national center at specific time. The

B. Liu and C. Chai (Eds.): ICICA 2011, LNCS 7030, pp. 463–471, 2011.
© Springer-Verlag Berlin Heidelberg 2011

average size of data that should be transferred may be 1G bytes. As result, the SOAP message of transferred data is huge.

The example above shows issue about large data transfer of web service. In many data concentrated applications, message exchange requirement is different from simple request-response approach of traditional web service. As in SPON's case, data sender creates connection with receiver as long as system allowed. To accomplish large data transferring, the data is about to be divided into smaller packs. One by one, the data pack is sent to receiver. The receiver receives data packs and their order. Once all data are transferred successfully, the receiver recreates whole data by the data packs[6][7]. The whole mechanism of lager data transferring for web service is called Large Data Transfer for Web Service(LDT4WS). In the mechanism, data sender and receiver are created to accomplish large data transfer work. The protocol, named Large Data Transfer Protocol for WS is core for both to organize data transfer work and guarantee integrity of data. In transferring, one data transfer work will be divided into a serial of smaller tasks. With information of the work, data sender creates task context with receiver initially. The context is utilized to organize order of data pack and maintain transferring process while network failure takes place. Through the approach, large data could be safely transferred in service centric web system.

2   Related Works

In conventional web applications, large data transfer between sender and receiver is realized by long and static connection. As discussed above, in many environments, the connections are not allowed for safety and reliability reasons[8][9].

In the world of web services, SOAP is the core for messaging. However it only supports structured data messaging. For the problem, SwA[10] is introduced in 2000. SwA had simple goals of providing a solution for packaging binary data along with SOAP envelope using existing SOAP1.1 and MIME technologies. In a nutshell, SwA defines an extension to SOAP1.1 transport binding mechanism. The extension allows a SOAP1.1 message to be carried within a MIME multipart/related package without having to change the SOAP1.1 processing rules. Through SwA, large data could be converted into binary and transferred. However, the solution offered by SwA only focuses on how to transfer binary data through web service. While dealing with large, it can not address confine of network communication. Furthermore, by SwA, transfer failure could not be repaired locally. Once a data transfer task fails, both the sender and receiver can not acknowledge how many data has been received rightly. The whole task will be start over again.

The problem could be addressed by dividing the large data into smaller packs. However, as a stateless distributed component, web service can't create session on both sides to build context of data packs. As result, the data packs could not be organized orderly and the large data can't be integrated without information of organization. In 2004, WSRF was introduced. The purpose of the OASIS WSRF TC[11] is to define a generic and open framework for modeling and accessing stateful resources using Web services. It includes mechanisms to describe views on the state, to support management of the state through properties associated with the Web service, and to describe how these mechanisms are extensible to groups of Web services. Through WSRF, sender and receiver could build session for transfer works. However, WSRF does not provide support of transaction and revocation. If one pack's transfer fails, the information about it can not be refreshed on both sides synchronously. The order of packs may be confused.

The confine of web service has slow down it's pace in development. In most data concentrated application, as SPON's example, large data needs to be transferred, managed and processed. As web service's inability to realize safe and reliable large data transfer, whole environment's performance will be affected. Consequently, service centric systems needs a mechanism to enable web service to transfer large data safely.

3   LDT4WS

The objective of LDT4WS is to enhance web service for large data transfer in data concentrated applications. As introduced above, network communication mechanism won't allow long and static connection between both sides. As result, In LDT4WS, the data will be converted into binary and transfer work is divided into a serial of smaller packs. At the beginning, sender and receiver create cache as local task context that works to maintain orders of data packs and integrity of whole data. Following the creation of task context, packs are to be transferred. After all packs are transferred successfully, the data will be re-constructed by the cached packs. In the process, if transfer failed, the sender needs to re-send current pack utile the receiver receives it successfully. The graphic below shows whole process.

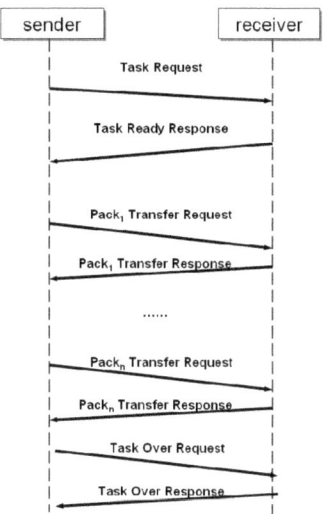

Fig. 1. Process of LDT4WS

3.1   The Message Protocol of LDT4WS

LDT4WS enable web services to transfer larger data by SOAP over HTTP in data concentrated applications. For long and safe connection between sender and receiver is not allowed in most system, LDT4WS addresses the issue by dividing large data into smaller packs. As result, the data transfer task includes steps of starting task, transferring packs and ending task. Accordingly, the protocol of LDT4WS consists of three parts: task request, pack transfer request and task over request.

The task request is to instruct receiver to create task context for oncoming large data transfer task. It's definition is shown as below:

tskReq = {senderID, dataSize, packSize, packNum, time}

senderID is to specify sender's identity. by the identity, receiver distinguish data's origin. dataSize is to specify oncoming data'size. By the information, receiver creates cache accordingly. packSize is to specify the data pack's size. packNum is to specified the total number of data packs. By the information, receiver determines whether a task could be ended. Time is to specify the starting time of the task. Once a receiver receives a task request, it prepares data cache and return task ready response to sender. Task ready response message's definition is shown as below:

tskResp = {taskID, receiverID, readyStatus}

in Task ready response's definition, taskID is to specify current task's identity. the identity is generated by receiver. If sender continues to send data pack, the taskID will be attached into data pack by sender and used as index in receiver's local data cache. receiverID is to specify identity of receiver. readyStatus is to specify whether receiver is ready to work. In the response, if receiver is not ready for the task, readyStatus will be "error" and sender won't start current task, or readyStatus will be "ok" and sender will continue following works.

As sender begin to transfer data packs, it send pack transfer request to receiver. the message 's definition is shown as below:

pckTransReq={taskID, senderID, packIdx, startIdx, packSize, time, value}

In pack transfer request's definition, taskID is to specify current task's identity. The identity is obtained in Task ready response message. senderID is to specify identity of sender. packIdx is to specify current data pack's index in serial of data packs of whole data. By the information, receiver reorganize data pack's order in local cache and integrate whole original data. startIdx it to specify current pack's head data's position in whole data. packSize is to specify current data pack's size. time is to specify pack transferring time. value is to carry real data content of current data pack.

In data pack transferring, receiver return pack transfer response message to sender after it receives data pack and cache successfully. The pack transfer response message's definition is shown as below:

pckTransResp={taskID, receiverID, packIdx, packSize, status, time }

In pack transfer response message, taskID is to specify current data transfer task's identity. receiverID is to specify receiver's identity. packIdx is to specify the data pack's index which has been received and cached locally. packSzie is to specify the received data pack's size. status is to specify whether current data pack transfer work has been implemented successfully. If it failed the statue will be "error" or will be "ok". By the information, sender decides to continue transferring following data packs or re-transferring the failed data pack. time is to specify when the data pack is transferred from sender.

After all data pack is transferred to receiver successfully, sender sends task over request to receiver. By the information, receiver reorganizes data packs and integrates whole data. The definition task over request is shown as below:

tskOverReq = {taskID, senderID, dataSize, packNum, time}

in task over request's definition, taskID is to specify identity of task which is to be ended. senderID is to specify sender's identity. dataSize is to specify size of whole data. By the information, receiver checks whether all data are received and cached rightly. packNum is to specify the number of packs that data is divided into. time is to specify when the task is to be ended.

Once receiver receivers task over request, it reorganize data packs by order in local cache and integrate whole data. After checking integrity of data, it return task over task over response message. The message's definition is shown as below:

tskOverResp={taskID, receiverID, receivedSize, time}

In the definition, taskID is to specify current task's identity. receiverID is to specify receiver's identity. receivedSize is to specify received data's size. time is to specify when the task is to be ended.

3.2   Sender and Receiver in LDT4WS

In LDT4WS, sender divides large data into smaller packs to transfer. Sender' composition is shown as below:

sender = {ID, taskID, dataValue, packsize, packIdx }

in sender's definition, ID is to specify sender's identity. taskID is to specify current data transfer task's identity. As introduced above, taskID is generated by receiver. In data pack transferring, sender needs to attach taskID into data pack to enable receiver to distinguish current task. dataValue is real data which is to be transferred. In task, the data will be divided into packs by packszie and tranfered. In transferring, packIdx is to mark current data pack's index by which receiver organizes data pack's order and integrates data.

Receiver is to receive data pack and integrates data when task is over. For web service is stateless. Receiver needs to create local caches to maintain data packs while transferring. The composition of receiver is shown as below:

receiver ={ID, packCache}

ID is to specify receiver's identity. packCache is to cache data pack locally. Because receiver may receives data from multi senders simultaneously, it's composition is shown as below:

packCache= {cachei| i=1,2,....n}

cache is created for one task. In transferring one cache is to reserve all data packs of related task in order. It's composition is shown as below:

cache = {taskID, tskReq, {pack1, pack2,......packn}};

taskID is current task's identity. by its reference, receiver locate specific cache for task. tskReq is task request message sent by sender. By the message receiver compares the received data's size with original size information and decide whether the task is implemented successfully. packi is array to cache data packs it received.

3.3   Transferring Large Data

The sender's sending process is shown as below:

```
SendData: sender × dataValue × receiver → Boolean
SendData(s, data , r) =
                    packIdx = 0 ;
                    let rid= r(ID, - )
                    let sid = s(ID,-,-,-,-) ;
                    let pszie = s(-,-,-,packsize,-);
                    let num = countPack(sizeof(data) , pszie) ;
                    let startTime = currentTime ;
                    let tskReq = (sid , sizeof(data) , pszie , num, startTime)
                    let response = TaskStart (tskReq, r)
                    let statue = response(-,-, rStatus);
                    if(statue = fasle)
                        return fasle ;
                    let tID = response(taskID,-, -);
                    While(packIdx < num)
                        { let pack = getPackFromData(data, packIdx, packSize);
        let pckReq= (tID, id, packIdx, getPosion(packIdx, packSize),packSize,
currentTime, pack)
                            let return = TransferPack (pckReq, r)
                            let result = return(-, -, -, -, status, -) ;
                            if(result = error)
                            {do() { let return = TransferPack (pckReq, r) ;
                                    let re = return(-, -, -, -, status, -) ;
                                    if(re = ok) break ;
                                    if (outOfTime(currentTime, startTime) =
true)
                                        return fasle ;}
                            packIdx ++ ;
                            packIdx → s(-,-,-,-,idx) ; } }
                    let  overReq  =  tskOverReq(tID,  sid,  sizeof(data),  num,
currentTime)
                    let endResult = EndReq(overReq, rid) ;
                    let res = endResult (-, -, -, -, status, -) ;
                    return res ;
```

At the beginning of task, the receiver receive request and prepare for transferring. The process is as below:

TaskStart: tskReq × receiver →tskResp
TaskStart (treq ,   r) = let tid = createTaskID() ;
                    let rid = r(ID, -)
                    let cacheSize = treq(-, dataSize,-, -,-)
                    let c = cache (tid, treq, -) ;
                    addCache(r, c) ;
                    let resp = { tid , rid, true} ;
                    return resp ;

In transferring, receiver receives data packs and organize them in local cache. The procesds is as below:

    TransferPack: pckTransReq× receiver →pckTransResp
    TransferPack(req , r) = let tid = req(taskID, -, -, -, -, -, -) ;
                        let pack = req(-, -, -, -, -, -, value) ;
                        let c = getCache(r, tid) ;
                        addPack(c, pack) ;
        let res = pckTransResp{tid, r(ID,-), req(-,-,idx,-,-,-), req(-,-,-,size,-,-), true, req(-,-,-,-,-,t)};
                        return res ;

while all data packs are received, receiver reconstructs data by packs. The process is shown as below:

EndReq: tskOverReq× receiver →tskOverResp
EndReq(req, r) = let tid = req(taskID, -, -, -, -) ;
                    let c = getCache(r, tid) ;
                    let data = appendAll(c) ;
                    let size = req(-, -, dsize, -, -) ;
                    let res = tskOverResp(tid , r(ID,-) , size , currentTime) ;
                    return res ;

The graphic below show activity of sender and receiver in transferring

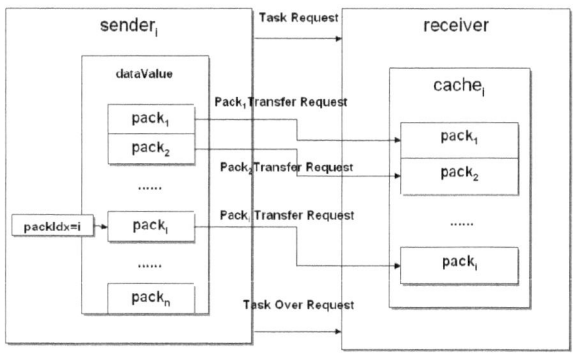

Fig. 2. Activities of Sender and Receiver

4   Application and Test

LDT4WS is to address issue of transferring large data with web service for data concentrated applications. As introduced above, CAE'S Seismological Precursors Observation Net (SPON) is a typical environment of data concentrated application. In SPON, the data need to be aggregated into national center from provincial center daily for further research. By web services, SPON provide data sharing, data management and other services. For the daily data transferred between national center and provincial centers, a data exchange platform is built for CAE's SPON upon LDT4WS. In test, the network band is set to be 1M bytes/s for data transfer. The data is about 1G bytes. In different implementations, the pack sizes are to be 1k, 4k, 16k, 256k, 512k and 1024k. The data is transferred by a single implementation as comparison. The graphic below shows result of the test.

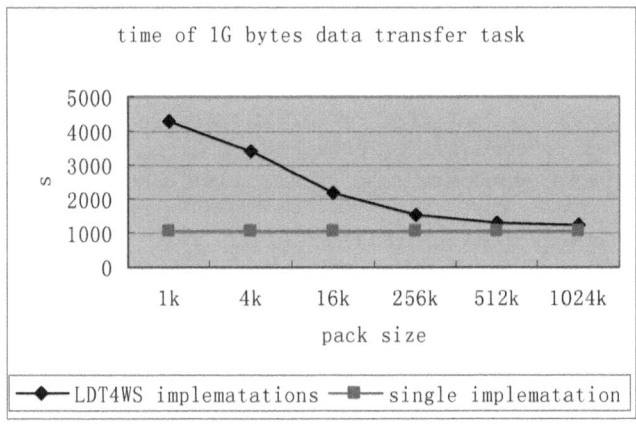

Fig. 3. Implementation of Large Data Transfer

As the data is divided into smaller packs and transferred continuously. The invocations may take more time than the single implementation. As pack size are set bigger, the implementations take shorter time. As shown by the graphic, when the pack size is set to be 1024k, the time is close to the single implementation.

The test above shows effectiveness of LDT4WS for data concentrated applications. Through LDT4WS, large data could be transferred in service centric system safely.

5   Conclusion

In data concentrated environments, large data needs to be transferred by web services. However, due to confine of network communication and web service's stateless feature. The large data could not be simply and safely transferred. This paper introduced an approached, called large data transfer for web service(LDT4WS), to address the issue. Upon LDT4WS, a data exchange platform is built for CAE'S SPON to enhance reliable large data exchange for China's earthquake research.

As network condition may change, the LDT4WS may adjust automatically pack size to adapt the situation. The future work will be focused on this development to achieve better speed and performance

Acknowledgement. This research work was supported by China Earthquake Administration's program for Seism-Scientific Research "Research in Online Processing Technologies for Seismological Precursory Network Dynamic Monitoring and Products" (NO. 201008002)

References

1. Jensen, F.: Introduction to Computational Chemistry. Wiley (2006)
2. Newcomer, E.: Understanding SOA with Web Services (Independent Technology Guides). Addison-Wesley (2004)
3. Schelp, J., Winter, R.: Business application design and enterprise service design: a comparison. International Journal of Services Sciences 1, 206–224 (2008)
4. Chaouchi, H.: The Internet of Things: Connecting Objects. Wiley-ISTE (2010)
5. Demirkan, H., Kauffman, R.J., Vayghan, J.A., Fill, H.-G., Karagiannis, D., Maglio, P.P.: Service-oriented technology and management: Perspectives on research and practice for the coming decade. Electronic Commerce Research and Applications 7(4), 356–376 (2008)
6. Yan, L., Zhang, Y., Yang, L.T., Ning, H.: The Internet of Things: From RFID to the Next-Generation Pervasive Networked Systems (Wireless Networks and Mobile Communications). Auerbach Publications (2008)
7. Hu, W., Bulusu, N., Chou, C.T., Jha, S., Taylor, A., Tran, V.N.: Design and evaluation of a hybrid sensor network for cane toad monitoring. ACM Trans. Sen. Netw. 5, 1–28 (2009)
8. Schelp, J., Winter, R.: Business application design and enterprise service design: a comparison. International Journal of Services Sciences 1, 206–224 (2008)
9. Papazoglou, M.P., Van Den Heuvel, W.-J.: Service-oriented design and development methodology. International Journal of Web Engineering and Technology 2(4), 412–442 (2006)
10. Lawrence, K.: Web Services Security SOAP with Attachments (SwA) Profile 1.1. In: 2004 OASIS Web Services Resource Framework (WSRF) TC (2004)

Collect Environmental Field Data
through Hybrid Network Technologies

Jiuyuan Huo[1,2] and Yaonan Zhang[1,*]

[1] Cold and Arid Regions Environmental and Engineering Research Institute,
Chinese Academy of Sciences, Lanzhou 730000, China
[2] Information Center, Lanzhou Jiaotong University, Lanzhou 730070, China
yaonan@lzb.ac.cn

Abstract. Although manual field data collection in the environmental research are simple, it is extremely difficult for geoscientists to get environmental data in areas of high altitude and harsh climate frequently. To assist scientists to transmit data and control equipments remotely, moreover minimizing human workload and time delays, we presented a data collection model by harnessing hybrid network technologies for acquiring field data in this paper. This model integrates multiple modern network technologies to achieve the remote control of equipments and data transfer between the field station and data center in research institute. Several different network technologies are deployed and tested at Mafengou catchment of Heihe River Basin and Cryosphere Research Station on Qinghai-Xizang Plateau. These experiments demonstrate a noticeable improvement in efficiency and precision has been achieved.

Keywords: Field Data Collection, Environmental Monitoring, Wireless Communications, Hybrid Networking.

1 Introduction

Cold and Arid Regions Environmental and Engineering Research Institute (CAREERI) have about 16 experiment and research stations, such as the Tianshan Glacier Observation Station, Cryosphere Research Station on Qinghai-Xizang Plateau, Shapotou Desert Experiment and Research Station and so on. The observation point of these field stations are in possession of a wide variety of environmental monitoring devices, and ground temperature, active layer, eddy correlation, and other environmental factors were observed. Although many studies use dataloggers to record these environmental factors, the analysis are usually performed after the field data is transferred to the data center of research institute.

Nowadays, field data collection in environmental research still mainly depends upon papers or a laptop. Although these methods are simple, researchers have to come to the field to collect data. Due to the monitoring points were distributed in a large area, it is extremely difficult for geoscientists to get environmental data in the areas of high altitude and harsh climate. This kind of manual practice is considered time

---

* Corresponding author.

B. Liu and C. Chai (Eds.): ICICA 2011, LNCS 7030, pp. 472–479, 2011.

consuming and labour intensive, and resulted in time delay of field data collection, transmission and processing. Considering disadvantages of these methods, a more efficient mechanism is needed. Rapid developments in network communications and computer fields could help the geoscientists to transmit environmental data and control equipments remotely, moreover minimizing human workload and time delays [1], [2],[3].

In this paper, we presented a data collection model that simplifies the collection process by harnessing hybrid network technologies for acquiring environmental field data. The model integrates multiple modern network technologies to achieve the remote control of equipments and data transfer between field stations and data center in research institute. Several different network technologies are deployed and tested at Mafengou catchment of Heihe River Basin, Cryosphere Research Station on Qinghai-Xizang Plateau. The experiments demonstrate a noticeable improvement in efficiency and precision achieved with the using of these hybrid network technologies.

This study focuses on the collection of geospatial and environmental data through multiple hybrid network technologies. To this end, in Section 2 we introduced related work of the technologies of environmental data collection. In Section 3, we discussed in detail several hybrid network technologies which could be used to support our data collection model. In Section 4, we presented system architecture of environmental field data collection model through hybrid network technologies. In Section 5, we described several different network technologies deployed and tested at research stations and discussed the results. Finally, in Section 6 we concluded the paper and described the future work.

2   Related Work

Due to the research fields are often far away in remote regions, the collection of environmental field data is a typical time-consuming and lab-intensive tasks for the geoscientists. In the last two decades, network technologies have been one of the most exciting and fastest developing technologies in the world. A lot of researches focus on various network and computer technologies to help the scientists to acquire the data from field in the literature [3],[4],[5].

Tseng utilized global system or mobile communication (Global System for Mobile Communications, GSM) and short message service (SMS) to conduct field data acquisition and proceed to investigate the corresponding feasibility. They developed a field data collection prototype system that is composed of field monitoring and host control platforms. Data transmission, communication, and control of these two platforms are accomplished using GSM-SMS technology [6].

Vivoni presented an integrated system developed for environmental and geolocation data acquisition that is intended to streamline the collection process. The system consists of software applications and hardware components that enable wireless, mobile and Internet computing during field campaigns. A prototype system has been tested in field trials in Cambridge, Massachusetts, USA and Newcastle, New South Wales, Australia [7].

Tsai and Yang have developed a synchronous system integrated with wireless and speech technologies to enhance the cooperation between construction workers and application devices. System tests and efficiency evaluation in a material management case study demonstrate that this system increased productivity, reduced operation time and simplified subprocesses for activity completion [8].

Wang and Zhang in our research team discussed the wireless transmission method used in the experiment for transmitting character data and video data from weather stations and interested area in real time. A system also have been developed which can browse wireless transmission character data and video data by Web and store them into database automatically [9].

The above studies have accomplished lots of work in data collection and transmission. But these researches just related to a part of network technology, and do not establish a system framework by a variety of hybrid network technologies. In this paper, we propose a hybrid network framework using a variety of techniques to help scientists to collect field data, monitor status of equipments and control equipments.

3   Hybrid Network Technologies

Nowadays, a large number of wired, wireless, satellite networking technologies have been developed and studied. Applying advanced and matured networking technologies could facilitate to build a collection and transmission network of field observation data. The most common and associated wireless network transmission technology were introduced as follows.

3.1   GSM-SMS Network Technology

Global System for Mobile Communications (GSM) is the world's most popular standard for mobile telephony systems. The GSM Association estimates that 80% of the global mobile market uses the standard. GSM differs from its predecessor technologies in that both signaling and speech channels are digital, and thus GSM is considered a second generation (2G) mobile phone system [10]. GSM also pioneered low-cost implementation of the short message service (SMS) which called text messaging.

3.2   GPRS/CDMA/3G Network Technology

General packet radio service (GPRS) is a packet oriented mobile data service on the 2G and 3G cellular communication systems global system for mobile communications (GSM). It is a best-effort service, as opposed to circuit switching, where a certain quality of service (QoS) is guaranteed during the connection. In 2G systems, GPRS provides data rates of 56-114 kbit/s [11]. GPRS coverage of China Mobile has reached almost 100% in the central city and 80% or so in remote areas.

Code division multiples access (CDMA) is a channel access method used by various radio communication technologies. CDMA employs spread-spectrum technology and a special coding scheme to allow multiple users to be multiplexed over the same physical channel. CDMA is a form of spread-spectrum signaling, since the modulated coded signal has a much higher data bandwidth than the data being communicated [12].

3G is a generation of standards for mobile phones and mobile telecommunications services fulfilling specifications by the International Telecommunication Union [13]. International Telecommunication Union (ITU) had identified W-CDMA, CDMA2000 and TD-SCDMA as the three mainstream 3G standards. A 3G system must allow simultaneous use of speech and data services, and provide peak data rates of at least 200 kbit/s. The main advantage of 3G technology can greatly increase the system capacity, improve communication quality and data transfer rate.

3.3   Satellite Network Technology

Satellite communication use artificial earth satellites as relay stations to forward the radio waves to complete the communication between two or more earth station. Satellite communication system is consisting of satellite communication satellite and earth station. Compared with other communication means, satellite communication has many advantages such as covered a wide range areas, larger telecommunication capacity, and better transmission quality.

Tibet earthquake observation network which established by Satellite Network, observing systems along the Qinghai-Tibet Railway which implemented by GMS/GPRS data transmission technologies and application of the MiWAVE system in the real-time monitoring in emergency handling of Tangjiashan barrier lake have verified and showed the importance and assistance of multiple networking technology to field observing systems.

4   System Architecture of Environmental Field Data Collection Model

According to the discussion above, this paper will use hybrid network technologies to facilitate field data collection for environmental research. To attain this goal, system architecture of environmental field data collection model by multiple hybrid network technologies was designed and several preliminary experiments have been done to gain practical experience. A data collection system for these purposes should:

1) Make use of any available network technologies to achieve data transmission task;
2) Efficiently collect and store environmental field data into database;
3) Aggregate field data from multiple observing stations;
4) Provide convenient data access and data visualization to geoscientists;
5) Get the status of the data logger;
6) Remote control the data logger.

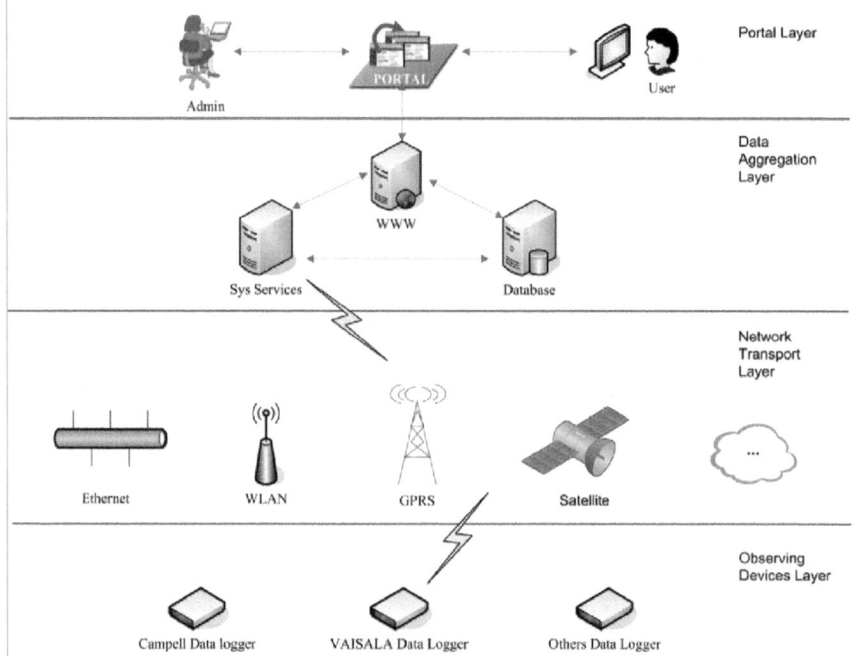

Fig. 1. Conceptual Diagram of Environmental Field Data Collection Model

To understand the model more clearly, a conceptual diagram of environmental field data collection model was shown in Fig.1. System architecture could be divided into four layers: Observing Devices Layer, Network Transport Layer, Data Aggregation Layer, Portal layer.

1) Observing Devices Layer: A collection of data logger devices which records multiple observing data, such as Cambpell CR1000, CR3000, CR23X data logger, VAISALA Milos520 device and other data loggers. These data loggers were deployed in the region of field and make a connection to Network Transport Layer to perform the data transmission by its network modular or a data transmission device.

2) Network Transport Layer: Perform the data transmission through Ethernet cable, Wireless LAN, GPRS/CDMA/3G, satellite and other hybrid network transmission technologies to transmit the observing data back to data center.

3) Data Aggregation Layer: Receive observing data from Network Transport Layer and distinguish the data source to store them into the appropriate database; and response to the task request from system portal to provide following system services: user management and system management, data access, data visualization, device status view and device remote control.

4) Portal Layer: Receive the interaction command from users to manage and maintain the system, download data, and show visualization result and so on.

5   Case Studies

Several different network technologies are deployed and tested at Mafengou catchment of Heihe River Basin and Cryosphere Research Station on Qinghai-Xizang Plateau. We discussed these preliminary experiments in detail as follows:

5.1   Mafengou Catchment of Heihe River Basin

A small catchment named Mafengou catchment which is located in Heihe upper reaches and the coordinate is 99°52' 5'' E, 38°15' 5''N was chose as our experiment area. There are four sets of weather station in the catchment, and their elevation are 2972m, 3402m, 3710 and 4163m respectively. Four weather stations are lies away kilometers from the base station in the catchment, and none of them has electricity power support or cell phone signal coverage. The base station has electricity power support and ADSL support. Besides, only two of the stations can unobstructed see the base station, the rest two stations have obstacles on their view [9]. Thus we adopted ZigBee wireless protocol to achieve data transmission to base station. ZigBee is a specification for a suite of high level communication protocols using small, low-power digital radios based on the IEEE 802.15.4 standard [14]. A set of ZigBee wireless transmission module produced by Shanghai Shunzhou and a set of microwave aperture plate antenna have been assembled to transmit observing data and video data from four weather stations to the base station. In base station, we have deployed a data server which was used to receive and store data to database.

5.2   Cryosphere Research Station on Qinghai-Xizang Plateau

Cryosphere Research Station on Qinghai-Xizang Plateau (CRS) of Chinese Academy of Sciences is located in Golmud city of Qinghai Province. Most observation stations of CRS are located Qinghai-Tibet Plateau and distributed from Xidatan to Naqu along Qinghai-Tibet Highway with a range of more than 700 km. Since monitoring instruments of CRS were deployed in scattered locations and wide areas, and most locations are in harsh environment for researchers to reach. We found the observed data is small, thus we adopted GPRS/CDMA/3G wireless network technologies to achieve data transmission.

As shown in Fig. 2, the structure of the experiment was consist of three layers, the most front-end is GPRS Modem connected to a data logger, the middle layer is the GPRS network   provided by China Mobile, and the upper layer is the data center in our institute. By using a WaveCom's GPRS modem to connect a data logger with a standard RS232 data line to data logger's serial interface, and then modem made a constant connection with data center through China Mobile's GPRS network. Users could use the software for data loggers to acquire field data, get the status of data logger and configure parameters of data logger remotely.

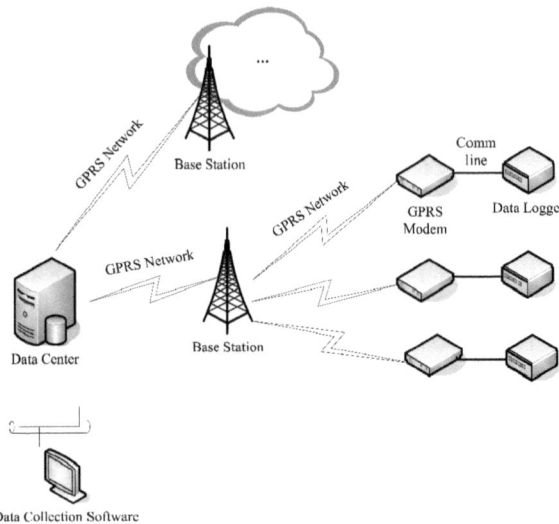

Fig. 2. Structure of the Experiment in CRS

We have deployed two sets of WaveCom GPRS Modems in the Cambpell CR3000 data loggers located in Xidatan comprehensive observing field and Wudaoliang comprehensive observing field of CRS, and realized the observation data transmission from field station to data center.

6   Conclusion and Future Work

Manual collection of environmental field data is simple, but the researchers have to spend lots of time for data collection and transmission. It is a tough work for those observing stations which located at a high altitude and harsh natural environment. Geoscientists could make use of advanced network technologies to facilitate them to transmit environmental data and control equipments remotely. By harnessing hybrid network technologies, we presented a data collection model for acquiring environmental field data and remote control of equipments. We tested and verified several different network technologies in different field stations. The experiments demonstrate a noticeable improvement in efficiency and precision achieved with the use of the hybrid network technologies. It also enhances collaborative research of geoscientists to solve problems and improve their research environment.

We have just presented a general framework and the related technologies for the environmental field data collection model. In the future, we will implement this model to help geoscientists to transmit environmental data and control equipments in an easy way.

Acknowledgments. This work is supported by Informationization Foundation of Chinese Academy of Sciences (CAS), "The E-Science Environment for Ecological and Hydrological Model Research in Heihe River Basin" (Grant number: 29O920C61);

Project for Incubation of Specialists in Glaciology and Geocryology of National Natural Science Foundation of China (Grant number: J0930003/ J0109); and Second Phase of the CAS Action-Plan for West Development (Grant number: KZCX2-XB2-09-03).

References

1. Tamelen, P.G.V.: A comparison of obtaining field data using electronic and written methods. Fisheries Research 69, 123–130 (2004)
2. Findley, D.J., Cunningham, C., Hummer, J.E.: Comparison of mobile and manual data collection for roadway components 19(3), 521–540 (2011)
3. Shu, L.L., Nan, Z.T.: A Data Acquistion System for Near Real-Time Field Observation Based on Twitter-like Services and GSM/SMS Network. Journal of Glaciology and Geocryology 32(5), 976–981 (2010)
4. Gu, Q., Lu, C., Guo, J.P., Jing, S.G.: Dynamic management system of ore blending in an open pit mine based on GIS/GPS/GPRS. Mining Science and Technology (China) 20(1), 132–137 (2010)
5. Mills, J.W., Curtis, A., Kennedy, B., Kennedy, S.W., Edwards, J.D.: Geospatial video for field data collection. Applied Geography 30, 533–547 (2010)
6. Tseng, C., Jiang, J., Lee, R., Lu, F., Ouyang, C., Chen, Y., Chang, C.: Feasibility study on application of GSM-SMS technology to field data acquisition. Comput. Electron. Agr. 53, 45–59 (2006)
7. Vivoni, E.R., Camilli, R.: Real-time streaming of environmental field data. Computers & Geosciences 29, 457–468 (2003)
8. Tsai, M., Yang, J., Lin, C.: Integrating wireless and speech technologies for synchronous on-site data collection. Automation in Construction 16, 378–391 (2007)
9. Wang, Y., Zhang, Y.N., Luo, L.H., Zhao, G.H.: An e-Science Environment Study Using Wireless Transmission Technique. In: 2010 6th International Conference on Wireless Communications Networking and Mobile Computing (WiCOM), pp. 1–4. IEEE Press, China (2010)
10. GSM Technical Data, http://www.cellular.co.za/gsmtechdata.htm
11. General packet radio service from Qkport, http://about.qkport.com/g/general_packet_radio_service
12. Viterbi, A.J.: CDMA: Principles of Spread Spectrum Communication. Addison-Wesley (1995)
13. Xia, J.: The third-generation-mobile (3G) policy and deployment in China: Current status, challenges, and prospects. Telecommun Policy 35, 51–63 (2011)
14. ZigBee Alliance, http://www.zigbee.org/

A New Model of Web Service Composition Framework

Zhang Yanwang

Department of Computer and Information Science, Hechi University,
Yizhou, Guangxi 546300, China
zyw0719@tom.com

Abstract. With the rapid development of web service as well as its enlargement in the quantity, a single web service can no longer satisfy the needs of customers. It has been a hot research topic that how to combine the multiple web services together to form a composite service for meeting the customers' increasing needs. To the problems existed in the current web service frames, particularly the insufficiencies at web service composition, this paper implemented the algorithm through its own designed agent middleware Proxy, and combined Proxy with the current web service frame, finally designed a new frame model of web service composition. Applying heuristic service composition system can better complete logical service composition of business and quality optimization of web service.

Keywords: Web, service composition, frame, model.

1 Introduction

With the emerging e-commerce, the web-based application models achieved a rapid growth. Web application has already developed from localization into globalization, from B2C into B2B, from the localized into the distributed, web service has already been an effective solution to e-commerce. Web service is a new distributed computing mode as well as an effective mechanism for integrating data and information on web. The new framework and efficient execution mode of web service, its combination with other mature technologies and integration with other web services are all important technologies for solving the problems in practical application.

Web service is a software system supporting for the interaction between network-based machines, it can be created and executed under different platforms. However, see from the current development situation of web service, there are still some defects existed and many problems need to be solved for achieving a mature web service framework though many codes and standards have already been established. The industry gradually realized the philosophy that it can't fully support the service oriented architecture (SOA) just through creating these standards, because all these standards essentially are the standardization and unification of the existing technologies, which is beneficial to the solving of interoperating problem at different levels but can do nothing to the deficiencies of technology itself. Secondly, as a distributed computing system based on internet, this service oriented collaborative application system has great demands on the aspects of adaptability to the open

B. Liu and C. Chai (Eds.): ICICA 2011, LNCS 7030, pp. 480–486, 2011.
© Springer-Verlag Berlin Heidelberg 2011

dynamic network computing environment as well as the constantly variable personalized needs from users; while the current existed software developing technique and run-time supporting technology still can't completely meet this kind of demand, especially the aspects of evolution on dynamic allocation, relocation and self-adapting still need a deep research. Therefore, the current usage of web service is relatively monotonous, mainly for encapsulating the corporation's original information system and releasing it through a series of criterions, which will then be used by users via network.

2   The Web Service Composition Framework

Along with the rapid development of web service and its corresponding technologies, the enterprise-level service integration also transitioned into web service-based service oriented system architecture. The main study problem of web service composition is how to combine those existing, self-governing web services and efficiently build an enterprise service application. At present, the service composition mainly has two kinds of classified modes, one mode is classifying according to the dynamics of service composition, and the other mode is according to the execution route. Following are some kinds of service composition classified according to their dynamics:

Static service composition, which means to know users' demands for service before designing a new service, the service composition supplier to choose services according to some certain strategies, while the service is no longer changeable after being provided to users. Semi-dynamic web service composition is a process or flow, which implement the given target through interaction with service integration. The stability (static) of flow and variability (dynamic) of flow executed services symbolize the characteristics of semi-dynamics. Dynamic web service composition, namely automatically generate composite flow according to the demands occurred during the operation as well as automatically search and combine web services so as to supply service, which symbolize the dynamics of the whole service lifecycle. Currently, the web service composition mode applied by the industry is mostly the static or the semi-dynamic. For example, the BPELAWL technology launched by IBM, Software and other companies as well as the EFlow technique proposed by HP are all based on manually generated workflow patterns. Dynamic web service composition takes advantage of Semantic web service technology formed by the combination of Semantic web technique and web services technology [2'] to dynamically combine web services and generate new business according to users' demands, which provides more flexibility and gradually becomes a hot research topic.

Classifying according to execution route, service composition can be divide into single path routing service composition and complex path routing service composition. There is only one execution route existed in the single path routing service composition, which means there is only a logical relationship of sequential control between services. Web service composition derives from the software reuse, and its fundamental idea is to create a new or better-quality web service for meeting users' demands through combinations in some certain order or changes among combine sequences. There are different design styles for web service composition framework designing, and one simple design framework will be introduced in the following. The typical web service composition framework is consisted by two user roles (service requester and service provider) and five components (translator, composition manager, execution engine, service matcher and service library), and the selectable components ontology base among provides ontology definition and reasoning support for service description, as shown in figure 1.

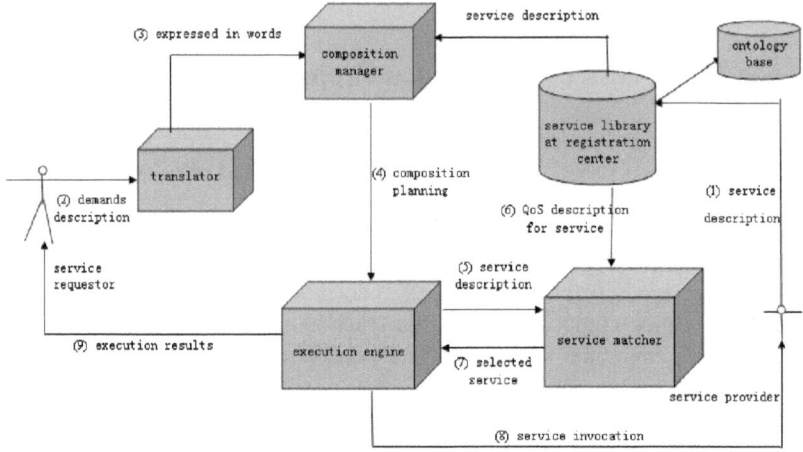

Fig. 1. Web service composition framework

The flow of the service composition framework is shown as following:

(1). Service requesters send requested service information to the service base at registration center through service registry.

(2-3). Processed by translator, the demand submitted by service requesters is transformed from natural language into meaningful demand description and then be passed on to composition manager.

(4). Composition manager, according to the demand description and the service description from service base, generates the composition plan satisfied for the service demand and then passes it to execution engine.

(5-7). Execution engine passes the composition plan to service matcher, who then chooses the web service most suitable to the service description and return its handle to execution engine.

(8). Execution engine calls and monitors the web service execution according to the composition plan and web service handle.

(9). Finally, pass the execution results on to service requesters.

3   The Design of Heuristic Service Composition System

For the rapid development of web service, the previous web service framework combined only by three modules of Registry, Provider and Requester has started to expose its defects. More and more service providers begin to register large amount of node web service from Registry, and meanwhile service web service demanded by requesters can't be satisfied by the single service at Registry. Therefore, a model with the function of intelligently combine web services is needed. This paper proposed a improved version of web service framework, which includes an extra design of Proxy agent function model besides those three basic modules. Proxy, who is the central control module of the framework, is used for analyzing users' service requests and according to which choosing suitable web services for composition, then return the solving plan to users. And users can implement communication with Provider through this solving plan and then bind the corresponding web service.

According to the requests sent from Requester, Proxy module chooses the suitable service in Registry. If all the services in Registry can't meet Requester's demands individually, then there need to combine services according to certain strategies as well as carry out routing optimization in accordance with certain algorithms. After that, Proxy return the solving plan, specific information of web service composition, to Requester, while Requester binds the specific Provider according to the information and call the particular web service for completing request.

3.1   Designs for Demand Analysis Function Module

According to users' request, specifically the input parameters, output parameters provided by users, this function module automatically establish the text information for request message in a certain format. The format is shown in figure 2.

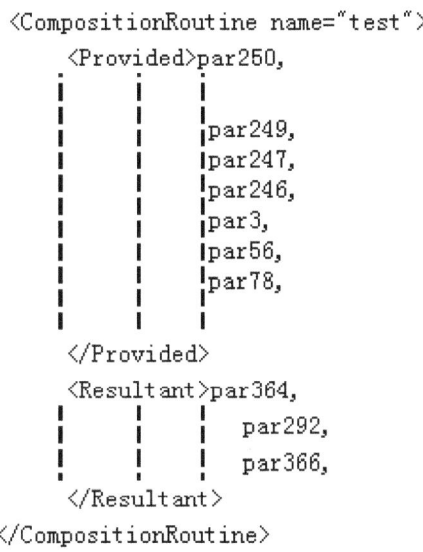

```
<CompositionRoutine name="test">
    <Provided>par250,
              par249,
              par247,
              par246,
              par3,
              par56,
              par78,

    </Provided>
    <Resultant>par364,
               par292,
               par366,
    </Resultant>
</CompositionRoutine>
```

Fig. 2. The format of request message

Because web service has characteristics of self-descriptiveness and self-distribution as well as their existed heterogeneity in semanteme, it is necessary to use ontology for matching so as to improve the service matching efficiency. Therefore, to analyze the semanteme of service request information, it can be implemented through establishing an ontology interpreter. The basic method of using ontology interpreter is to compute the approximation degree between the semanteme of parameter from request information and the recorded semanteme in history database through using of ontology and judge whether this approximation degree satisfies a certain threshold value. This approximation degree is determined by the distance of concept semanteme calculated through ontology. The greater distance between the two concepts, the lower approximation degree exists. Instead, the smaller distance between them, the larger approximation degree occurs. If the approximation degree meet the assumed threshold value, then only need to call out the history record from database and return the record content to users; if not, then need to proceed to the next step.

3.2   Design for Workflow Function Module

Workflow design module: if in requirement analysis module, there is no corresponding record found when searching the history database, then it is needed to design for workflow. The main function of this module is to design workflow according to the request message. Service composition designer firstly search for the corresponding service in Registry (service is described by WSDL), then automatically design the workflow according to a certain strategy which is a function-based service composition. In the paper, this module generates heuristic workflow by using the SCF algorithm introduced in the third chapter above (of course, here considers the expandability and variability when designing this module if different algorithms are allowed for choosing strategies generated from workflow), and the workflow generated is also represented through the particular XML format. Each node in the generated workflow represents an abstract service, e.g., <service name="ws983">.

As in the workflow generating process, an abstract service may correspond to multiple concrete services, in other words, an abstract service can be replaced by multiple concrete services in same function but different QoS attributes. So it is needed to collect these concrete services with same function but different QoS attributes together. The function of workflow service integration constitutor is to collect these concrete services and form a candidate service list for abstract service according to functional attributes. In the list, each abstract service corresponds to certain amount of concrete services in same function but different QoS attributes. This abstract service list is to facilitate the service composition among those non-functional attributes services in the next step.

3.3   Design for QoS Evaluation Function Module

QoS evaluator: after generating corresponding workflow in workflow function module, as the related routing of workflow generated at this moment is based on functional attributes without considering the QoS attribute of service, while QoS data of the whole related routing are needed for conducting non-functional attributes service composition. The function of QoS evaluator is to calculate the QoS of concrete services according to certain web service QoS computing model. QoS evaluator, firstly, need to create a QoS attribute list for recording QoS data, and in the list here includes service name, cost (uniform unit as $), time (uniform unit ms), reliability, reputation. This list needs to be updated constantly in the future execution process, for example: users at service providing end may adjust price, so the corresponding cost in QoS attribute list also needs to be adjusted. Execution time is an average value of multiple execution time values gained from many times of service invocation by users, therefore, after each successful invocation of service, the corresponding time queue also requires data updating according to information returned by users. Similarly, Reliability and Reputation also need updating.

Table 1. QoS attribute list

Service Name	Cost ($)	Time (ms)	Reliability	Reputation
S1	15	40	0.8	0.8
S2	25	30	0.9	0.9
S3	......	......	......	......

Another main function of QoS evaluator is to calculate service composition according to QoS computing model, and then gain the QoS values for the whole composition. Among, the computing model can be evaluated according to the QoS computing model for web service researched.

3.4 Design for Service Composition Function Module

Service composition function module is the most important module in Proxy module structure, while the function of the rest modules is to provide data and data analysis for it. The final solving plan for the whole service composition is achieved through this module.

After workflow function module finishing the design for workflow, the workflow achieved is only a composition of abstract services, while the main task of service composition module is to implement a global optimum concrete service composition based on workflow according to service QoS attribute. Considering the expandability and low coupling of the module, the global optimum algorithm is different optional optimization algorithms. After completing the concrete service composition through optimization algorithm based service composition module, the solving plan achieved needs to be stored into the solving plan database for the next inquiry by users.

4 Conclusion

With the rapid development of web service as well as its enlargement in the quantity, a single web service can no longer satisfy the needs of customers. It has been a hot research topic that how to combine the multiple web services together to form a composite service for meeting the customers' increasing needs. To the problems existed in the current web service frames, particularly the insufficiencies at web service composition, this paper implemented the algorithm through its own designed agent middleware Proxy, and combined Proxy with the current web service frame, finally designed a new frame model of web service composition. Applying heuristic service composition system can better complete logical service composition of business and quality optimization of web service.

References

1. Ling, X.: SOA Overview. Computer Applications and Software 24(10), 121–125 (2007)
2. SOAP. Simple Object Access Protocol 1.2 (2008), http://www.w3.org/TR/soap
3. WSDL. Web Service Description Language 1.1 (2009),
   http://www.w3.org/TR/wsdl
4. Clement, L., Hately, A.: UDDI version 3.0.3 (October 2008),
   http://uddi.org/pubs/uddi_v3.htm
5. Mao, X.: SOA Principle, Method and Practice. Publishing House of Electronics Industry, Beijing (2007)
6. Benatallah, B., Dumas, M.: The Self-service Environment for Web Service Composition. IEEE Internet Computing 7(1), 40–48 (2009)

Multi-receiver Identity-Based Signcryption Scheme in the Standard Model

Yang Ming[1], Xiangmo Zhao[1], and Yumin Wang[2]

[1] School of Information Engineering, Chang'an University,
Xi'an, Shaanxi 710064, China
[2] State Key Laboratory of Integrated Service Networks Xidian University,
Xi'an, Shaanxi 710071, China
{yangming,xmzhao}@chd.edu.cn, ymwang@xidian.edu.cn

Abstract. Signcryption is a novel cryptographic primitive that simultaneously provides the authentication and encryption in a single logic step. To adapt multi-receiver setting, motivated by Waters' identity based encryption scheme, the first multi-receiver identity-based signcryption scheme in the standard model was proposed. Then we prove its semantic security under the assumption of Decisional Bilinear Diffie-Hellman and its unforgeability under the hardness of Computational Diffie-Hellman Problem.

Keywords: Signcryption, Identity-based, Standard model, Bilinear pairing.

1 Introduction

In 1984, Shamir [1] first proposed the idea of identity-based (simply ID-based) public key cryptography (ID-PKC). The main idea of ID-PKC is that the user's public key can be calculated directly from his/her identity such as email addresses. Private keys are generated for the users by a trusted third party, called Key Generation Center (KGC) using some master key related to the global parameters for the system. The direct derivation of public keys in ID-PKC eliminates the need for certificates and some of the problems associated with them.

Confidentiality, integrity, non-repudiation and authentication are the important requirements for many cryptographic applications. A traditional approach to achieve these requirements is to sign-then-crypt the message. The concept of signcryption was firstly proposed by Zheng [2]. The idea of this kind of primitive is to perform signature and encryption simultaneously in order to reduce the computational costs and communication overheads. Since then, there are many signcryption schemes proposed. By combining the ID-based cryptography and signcryption, Malone-Lee [3] proposed the first ID-based signcryption scheme along with a security model. But Libert and Quisquater [4] pointed out that scheme in [3] is not semantically security. Chow et al. [5] given an ID-based signcryption scheme that provides both public verifiability and forward security. In 2003, Boyen [6] proposed a secure ID-based signcryption scheme in the random

B. Liu and C. Chai (Eds.): ICICA 2011, LNCS 7030, pp. 487–494, 2011.

oracle model. In [7] Chen and Malone-Lee improved Boyen's scheme [6] in efficiency. In [8] Barreto et al. proposed an efficient ID-based signcryption scheme. In 2009, based Waters scheme [9], Yu et al. [10] proposed the first identity based signcryption scheme without random oracles. However, Jin et al. [11] pointed out that Yu et al.'s scheme [10] does not satisfy semantic security and present an improved scheme in the standard model.

To adapt multi-receiver setting, Duan and Cao [12] defined the security notion for multi-receiver ID-based signcryption (MIBSC) in 2006. In this security model, the adversary commits ahead of time to multiple identities it will attack (called selective multi-identity attack). Duan and Cao also proposed a MIBSC scheme. Recently, Tan [13] showed that Duan and Cao's scheme [12] is not secure against adaptive chosen ciphertext attacks under their defined security model. In 2007, Yu et al. [14] proposed a new MIBSC scheme. However, Selvi et al. [15] showed that Yu et al.'s scheme [14] does not satisfy the unforgeability and proposed an improved scheme. But, Li et al. [16] pointed out that Yu et al.'s scheme [14] and Selvi et al.'s scheme [15] does not satisfy the confidentiality and proposed an efficient MIBSC scheme. In 2009, Selvi et al. [17] presented a new scheme which is efficient in transmission length, storage of keys and computation at both ends.

Provably security is the basic requirement for MIBSC schemes. The security of all the schemes [12-17] described above was only proven secure in the random oracle model. The random oracle model is a formal model in analyzing cryptographic schemes, where a hash function is considered as a black-box that contains a random function. Although the model is efficient and useful, it has received a lot of criticism that the proofs in the random oracle model are not proofs. Therefore, to design a provable secure MIBSC scheme in the standard model remains an open and interesting research problem.

In this paper, we proposed the first multi-receiver ID-based signcryption (MIBSC) scheme in the standard model. Using the Waters' scheme [9], we give a concrete secure scheme. We prove its semantic security under the hardness of Decisional Bilinear Diffie-Hellamn problem and its unforgeability under the computational Diffie-Hellman assumption.

2   Preliminaries

2.1   Bilinear Pairings

Let $\mathbb{G}_1$ and $\mathbb{G}_2$ be two multiplicative cyclic groups of prime order q and let g be a generator of $\mathbb{G}_1$. The map e is said to be an admissible bilinear pairing with the following properties:

- **Bilinearity:** For all $u, v \in \mathbb{G}_1$, and $a, b \in \mathbb{Z}_q$, $e(u^a, v^b) = e(u, v)^{ab}$.
- **Non-degeneracy:** $e(g, g) \neq 1$.
- **Computability:** There exists an efficient algorithm to compute $e(u, v)$ for all $u, v \in \mathbb{G}_1$.

2.2   Complexity Assumptions

Decisional Bilinear Diffie-Hellman (DBDH) Problem. Given $g, g^a, g^b, g^c \in \mathbb{G}_1$, for unknown $a, b, c \in \mathbb{Z}_q^*$ and $Z \in \mathbb{G}_2$, decide whether $Z = e(g,g)^{abc}$.

Defined the advantage of a polynomial algorithm $\mathcal{A}$ against the DBDH problem is $|\Pr[\mathcal{A}(g, g^a, g^b, g^c, e(g,g)^{abc}) = 1] - \Pr[\mathcal{A}(g, g^a, g^b, g^c, Z) = 1]| \geq \varepsilon$.

Definition 1. *The (t, ε) DBDH assumption holds if no t-time adversary has at least ε advantage in solving DBDH problem.*

Computational Diffie-Hellman (CDH) Problem. Given $g, g^a, g^b \in \mathbb{G}_1$, for unknown $a, b \in \mathbb{Z}_q^*$, compute g^{ab}.

The success probability of a polynomial algorithm $\mathcal{A}$ in solving CDH problem is denoted as $Succ_{\mathcal{A}}^{CDH} = \Pr[\mathcal{A}(g, g^a, g^b) = g^{ab}] \geq \varepsilon$.

Definition 2. *The (t, ε) CDH assumption holds if no t-time adversary has at least ε in solving CDH problem.*

3   Formal Model of Multi-receiver ID-Based Signcryption Schemes

A multi-receiver identity-based signcryption scheme consists of four algorithms (**Setup, Extract, Signcrypt, Unsigncrypt**) [12, 14, 16]. According to [16], there are two security requirements that multi-receiver ID-based signcryption scheme need to satisfy. **(1) Confidentiality**. This property is semantical security (i.e. the ciphertext is indistinguishability against adaptive chosen multi-ID and ciphertext attack. **(2) Unforgeability**. The property is existential unforgeability against adaptive chosen messages attacks.

4   The Proposed Scheme

Setup: Given a security parameter k, the PKG chooses groups $\mathbb{G}_1$ and $\mathbb{G}_2$ of prime order q, a generator g of $\mathbb{G}_1$, a admissible bilinear pairing $e : \mathbb{G}_1 \times \mathbb{G}_1 \to \mathbb{G}_2$, and hash functions $H : \{0,1\}^* \to \{0,1\}^l$ and $H_m : \{0,1\}^* \to \{0,1\}^{n_m}$. The PKG chooses a random value $\alpha \in Z_q^*$, computes $g_1 = g^\alpha$ and selects $g_2 \in \mathbb{G}_1$. Furthermore, the PKG picks $u', m' \in \mathbb{G}_1$ and vectors $\mathrm{u} = \{u_i\}$, $\mathrm{m} = \{m_i\}$ of length n_u and n_m, respectively, whose entries are random elements from $\mathbb{G}_1$. The system parameters are $params = \{\mathbb{G}_1, \mathbb{G}_2, e, p, g, g_1, g_2, H, H_m, u', m', \mathrm{u}, \mathrm{m}\}$ and the master secret key g_2^α.

Extract: Let u be a bit string of length n_u, representing an identity and let $u[i]$ be the i-th bit of u. Define $\mathrm{U} \subset \{1, 2, \cdots, n_u\}$ to be the set of indices i such that $u[i] = 1$. A private key d_u for identity u is generated as follows. Pick $r_u \in Z_q^*$ and compute $d_u = (d_{u1}, d_{u2}) = (g_2^\alpha(u' \prod_{i \in \mathrm{U}} u_i)^{r_u}, g^{r_u})$.

Signcrypt: To send a message $M \in \{0,1\}^l$ to receivers, the signcrypter picks $r \in Z_q^*$ randomly and does the following:

1. Compute $\sigma_1 = g^r$ and $\sigma_2 = M \oplus H(e(g_1, g_2)^r)$.
2. Compute $\sigma_3 = d_{A2} = g^{r_A}$ and $T_i = (u' \prod_{i \in U_{B_i}} u_i)^r$ for $i = 1, \cdots, n$.
3. Compute $h = H_m(M, \sigma_1, \sigma_2, \sigma_3, T_1, \cdots, T_n, L)$. Here h is an n_m bit string and $h[j]$ denotes the j-th bit of h, and $M \subset \{1, \cdots, n_m\}$ denotes the set of i for which $h[j] = 1$. Compute $\sigma_4 = d_{A1} \cdot (m' \prod_{j \in M} m_j)^r$.

The ciphertext is $\sigma = (\sigma_1, \sigma_2, \sigma_3, T_1, \cdots, T_n, \sigma_4, L)$, where L is a label that contains information about how T_i is associated with each receiver.

Unsigncrypt: When receiving σ, the receiver u_{B_i} follows the steps below.

1. Find appropriate T_i by L and compute $M = \sigma_2 \oplus H(e(d_{B_i1}, \sigma_1) \cdot e(d_{B_i2}, T_i)^{-1})$.
2. Compute $h = H_m(M, \sigma_1, \sigma_2, \sigma_3, T_1, \cdots, T_n, L)$ and generate the corresponding set M, the set of all j for which $h[j] = 1$. Accepted the message if and only if the following equality holds:

$$e(\sigma_4, g) = e(g_1, g_2)e(u' \prod_{i \in U_{B_i}} u_i, \sigma_3)e(m' \prod_{j \in M} m_j, \sigma_1)$$

5   Analysis of the Proposed Scheme

5.1   Security

Theorem 1. *(Confidentiality) Assume there is an adversary $\mathcal{A}$ that is able to distinguish two valid ciphertexts during the defined in confidentiality game with an advantage ε when running in a time t and making at most q_e private key extraction queries, q_s signcryption queries and q_u unsigncryption queries. Then there exists a distinguisher $\mathcal{D}$ that can solve an instance of the DBDH problem with an advantage $\varepsilon' = \frac{\varepsilon}{2^{n+2}(q_e+q_s+q_u)^n(n_u+1)^nq_s(n_m+1)}$ in a time $t' = t + (9q_e + nq_s)t_e + 6q_ut_p$, where n denotes the number of receivers, t_e denotes the time of an exponentiation in $\mathbb{G}_1$ and t_p denotes the time of a pairing in $(\mathbb{G}_1, \mathbb{G}_2)$.*

Proof. Assume that there is an polynomially bounded adversary $\mathcal{A}$ that is able to break the semantic security of our scheme, then there exists a distinguisher $\mathcal{D}$ that can decide whether $Z = e(g, g)^{abc}$ or not with a non-negligible advantage when receiving a random instance (g, g^a, g^b, g^c, Z). $\mathcal{D}$ runs $\mathcal{A}$ as subroutine and acts as challenger and interacts with $\mathcal{A}$ as described below.

Initial. The distinguisher $\mathcal{D}$ chooses randomly the following elements:

- two integers $0 \leq l_u \leq q$ and $0 \leq l_m \leq q$.
- two integers $0 \leq k_u \leq n_u$ and $0 \leq k_m \leq n_m$ ($l_u(n_u+1) < q, l_m(n_m+1) < q$).
- an integer $x' \in Z_{l_u}$ and n_u-dimensional vector $(x_1, \cdots, x_{n_u}) \in Z_{l_u}$.
- an integer $y' \in Z_{l_m}$ and n_m-dimensional vector $(y_1, \cdots, y_{n_m}) \in Z_{l_m}$.
- an integer $z' \in Z_q$ and n_u-dimensional vector $(z_1, \cdots, z_{n_u}) \in Z_q$.
- an integer $\omega' \in Z_q$ and n_m-dimensional vector $(\omega_1, \cdots, \omega_{n_m}) \in Z_q$.

To make the notation easy to follow, we define four functions:

$$F(U) = x' + \sum_{i \in U} x_i - l_u k_u \text{ and } J(U) = z' + \sum_{i \in U} z_i$$
$$K(M) = y' + \sum_{i \in M} y_i - l_m k_m \text{ and } L(M) = \omega' + \sum_{i \in M} \omega_i$$

$\mathcal{D}$ sets system parameters as follows:

- $g_1 = g^a$ and $g_2 = g^b$.
- $u' = g_2^{-l_u k_u + x'} g^{z'}$ and $u_i = g_2^{x_i} g^{z_i} (1 \le i \le n_u)$, which means that, for any identity U, we have $u' \prod_{i \in U} u_i = g_2^{F(U)} g^{J(U)}$.
- $m' = g_2^{-l_m k_m + y'} g^{\omega'}$ and $m_i = g_2^{y_i} g^{\omega_i} (1 \le i \le n_m)$, which means that, for any message M, we have $m' \prod_{i \in M} m_i = g_2^{K(M)} g^{L(M)}$.

Finally, $\mathcal{D}$ returns all parameters to $\mathcal{A}$.

Phase 1. $\mathcal{D}$ answers the private key extract queries, signcryption queries and unsigncryption queries as follows.

- **Private key extraction queries:** When the adversary $\mathcal{A}$ issues a private key extraction query on an identity u, $\mathcal{D}$ acts as follows:
  (1) If $F(u) = 0 \bmod l_u$, $\mathcal{D}$ aborts and reports failure.
  (2) If $F(u) \ne 0 \bmod l_u$, $\mathcal{D}$ picks a random $r_u \in Z_q^*$ and computs $d_u = (d_{u1},$
  $$d_{u2}) = (g_1^{-\frac{J(u)}{F(u)}} (g_2^{F(u)} g^{J(u)})^{r_u}, g_1^{-\frac{1}{F(u)}} g^{r_u}).$$
- **Signcryption queries:** At any time, the adversary $\mathcal{A}$ can perform a signcryption query on a plaintext M for the signcrypter identity u_A and a receiver list $u_{B_1}, \cdots, u_{B_n}$, $\mathcal{D}$ acts as follows:
  (1) If $F(u_A) = 0 \bmod l_u$, $\mathcal{D}$ aborts and reports failure.
  (2) If $F(u_A) \ne 0 \bmod l_u$, $\mathcal{D}$ first obtains the private key for u_A as he does in response to the private key extraction query, and then runs **Signcrypt**$(M, d_A, u_{B_1}, \cdots, u_{B_n})$ to answer $\mathcal{A}$'s query.
- **Unsigncryption queries:** At any time, the adversary $\mathcal{A}$ can perform a unsigncryption query on a ciphertext σ for the sender identity u_A and a receiver list $u_{B_1}, \cdots, u_{B_n}$, $\mathcal{D}$ acts as follows:
  (1) If $F(u_{B_i}) = 0 \bmod l_u$, $\mathcal{D}$ aborts and reports failure.
  (2) If $F(u_{B_i}) \ne 0 \bmod l_u$, $\mathcal{D}$ first obtains the private key for u_{B_i} as he does in response to the private key extraction query, and then runs **Unsigncrypt**(M, u_A, d_{B_i}) to answer $\mathcal{A}$'s query.

Challenge. After a polynomially bounded number of queries, the adversary $\mathcal{A}$ chooses identities $u_A^*, u_{B_1}^*, \cdots, u_{B_n}^*$ on which he wishes to be challenged. Note that $\mathcal{D}$ fails if $\mathcal{A}$ has makes a private key extraction query on $u_{B_i}^* (1 \le i \le n)$ during the Phase 1. Then $\mathcal{A}$ submits two messages $m_0, m_1 \in \{0, 1\}^l$ and $u_A^*, u_{B_1}^*, \cdots, u_{B_n}^*$ to $\mathcal{D}$. $\mathcal{D}$ will abort if $F(u_{B_i}^*) \ne 0 \bmod l_u (1 \le i \le n)$. Otherwise, $\mathcal{D}$ flips a fair binary coin $\gamma \in \{0, 1\}$ and constructs ciphertext of m_γ as follows.
   $\mathcal{D}$ randomly chooses a number $r^* \in Z_q^*$ and computes

$$h_\gamma^* = H(m_\gamma, g^c, m_\gamma \oplus H(Z), g_1^{-\frac{1}{F(u_A^*)}} g^{r_u^*}, (g^c)^{J(u_{B_1}^*)}, \cdots, (g^c)^{J(u_{B_n}^*)}, L)$$

M_γ^* denoted the set of 1 for which $h_\gamma^*[j] = 1$. If $K(M_\gamma^*) \neq 0 \bmod q$, $\mathcal{D}$ aborts. Otherwise, $\mathcal{D}$ sets the ciphertext as

$$\sigma^* = \begin{pmatrix} g^c, m_\gamma \oplus H(Z), g_1^{-\frac{1}{F(u_A^*)}} g^{r_u^*}, (g^c)^{J(u_{B_1}^*)}, \cdots, \\ (g^c)^{J(u_{B_n}^*)}, g_1^{-\frac{J(u_A^*)}{F(u_A^*)}} (g_2^{F(u_A^*)} g^{J(u_A^*)})^{r^*} (g^c)^{L(M_\gamma^*)} \end{pmatrix}$$

Phase 2. The adversary $\mathcal{A}$ then performs a second series of queries which are treated in the same as the Phase 1. It is not allowed to make the private key extraction query on $u_{B_i}^* (1 \leq i \leq n)$ and it is not allowed to make an unsigncryption query on under $u_{B_i}^* (1 \leq i \leq n)$.

Guess. At the end of the simulations, the adversary $\mathcal{A}$ outputs a guess γ'. If $\gamma' = \gamma$, $\mathcal{D}$ answers 1 indicating that $Z = e(g, g)^{abc}$, otherwise, $\mathcal{D}$ answers 0 to the DBDH problem.

This completes the description of simulation. It remains only to analyze the success probability and running time of $\mathcal{D}$. Analogy to [9], we can obtain the success probability of $\mathcal{D}$ is $\varepsilon' = \frac{\varepsilon}{2^{n+2}(q_e+q_s+q_u)^n(n_u+1)^n q_s(n_m+1)}$ and the total running time is at most $t + (9q_e + nq_s)t_e + 6q_u t_p$. Thus, the theorem follows.

Theorem 2. *(Unforgeability) Assume there is an adversary $\mathcal{A}$ that is able to break our proposed scheme during the defined in unforgeability game with an advantage ε when running in a time t and making at most q_e private key extraction queries, q_s signcryption queries and q_u unsigncryption queries. Then there exists an algorithm $\mathcal{B}$ that can solve an instance of the CDH problem with an advantage $\varepsilon' = \frac{\varepsilon}{8(q_e+q_s+q_u)q_s(n_u+1)(n_m+1)}$ in a time $t' = t + (9q_e + nq_s)t_e + 6q_u t_p$, where n denotes the number of receivers, t_e denotes the time of an exponentiation in $\mathbb{G}_1$ and t_p denotes the time of a pairing in $(\mathbb{G}_1, \mathbb{G}_2)$.*

The proof of Theorem 2 found in the full version of the paper due to the space limitation.

5.2   Efficiency

We compare our proposed scheme with existing five schemes in Table 1. In the following table, we denote by E an exponentiation, by P a pairing, M a scalar multiplication, RO a random oracle model and SM a standard model. We assume that the bit length of element in $\mathbb{G}_1$ is $|\mathbb{G}_1|$. We consider the pre-computation here and do not take hash function evaluations into account.

From Table 1, we know that our proposed scheme has slightly higher computation cost than schemes in [12,14-17] in Unsigncryption algorithm. We note that the computation of the pairing is the most-consuming. In Signcryption algorithm, the computation cost of our scheme is less than the schemes in [12, 14, 15] and more than schemes in [16, 17]. To the best of our knowledge, it is the first scheme which is proven secure in the standard model.

Table 1. Comparison of Efficiency with Existing Schemes

Schemes	Signcryption cost	Unsigncryption cost	Ciphertext cost	Security Model
[12]	(n+3)M+1E+1P	4P+1M	$|m| + (n+3)|\mathbb{G}_1| + |L|$	RO
[14]	(n+3)M+1E+1P	3P+1M	$|m| + (n+3)|\mathbb{G}_1| + |L|$	RO
[15]	(n+2)M+1E+1P	4P+1M	$|m| + (n+3)|\mathbb{G}_1| + |L|$	RO
[16]	(n+1)M+1E	2P+1M+1E	$|m| + (n+1)|\mathbb{G}_1| + |L|$	RO
[17]	(n+2)M+1E	3P+3M+2E	$|m| + 3|\mathbb{G}_1| + |L|$	RO
Our	(n+3)M+1E	5P	$|m| + (n+3)|\mathbb{G}_1| + |L|$	SM

6   Conclusions

In this paper, we proposed a concrete multi-receiver ID-based signcryption scheme based on Waters' identity-based encryption scheme. To our best knowledge, this is the first multi-receiver ID-based signcryption scheme that can be proven secure in the standard model. The scheme is proved semantic security and existential unforgeability under the assumption of decisional bilinear Diffie-Hellman and computational Diffie-Hellman respectively.

Acknowledgments. This work is supported by Natural Science Foundation of Shaanxi Province (NO: 2010JQ8017), the Special Fund for Basic Scientific Research of Central Colleges, Chang'an University (NO: CHD2009JC099).

References

1. Shamir, A.: Identity-based cryptosystems and signature schemes. In: Blakely, G.R., Chaum, D. (eds.) CRYPTO 1984. LNCS, vol. 196, pp. 47–53. Springer, Heidelberg (1985)
2. Zheng, Y.: Digital Signcryption or How to Achieve Cost (Signature & Encryption) << Cost(Signature) + Cost(Encryption). In: Kaliski Jr., B.S. (ed.) CRYPTO 1997. LNCS, vol. 1294, pp. 165–179. Springer, Heidelberg (1997)
3. Malone-Lee, J.: Identity based signcryption., http://eprint.iacr.org/2002/098
4. Libert, B., Quisquator, J.J.: A new identity based signcryption scheme from pairings. In: Proceedings of ITW 2003, pp. 155–158. Elsevier, Paris (2003)
5. Chow, S.S.M., Yiu, S.M., Hui, L.C.K., et al.: Efficient forward and provably secure ID-based signcryption scheme with public verifiability and public ciphertext authenticity. In: Lim, J.-I., Lee, D.-H. (eds.) ICISC 2003. LNCS, vol. 2971, pp. 352–369. Springer, Heidelberg (2004)
6. Boyen, X.: Multipurpose Identity-Based Signcryption. In: Boneh, D. (ed.) CRYPTO 2003. LNCS, vol. 2729, pp. 383–399. Springer, Heidelberg (2003)
7. Chen, L., Malone-Lee, J.: Improved identity-based signcryption. In: Vaudenay, S. (ed.) PKC 2005. LNCS, vol. 3386, pp. 362–379. Springer, Heidelberg (2005)

8. Barreto, P.S.L.M., Libert, B., McCullagh, N., Quisquater, J.-J.: Efficient and provably-secure identity-based signatures and signcryption from bilinear maps. In: Roy, B. (ed.) ASIACRYPT 2005. LNCS, vol. 3788, pp. 515–532. Springer, Heidelberg (2005)
9. Waters, B.: Efficient identity-based encryption without random oracles. In: Cramer, R. (ed.) EUROCRYPT 2005. LNCS, vol. 3494, pp. 114–127. Springer, Heidelberg (2005)
10. Yu, Y., Yang, B., Sun, Y., et al.: Identity based signcryption scheme without random oracles. Computer Standards and Interfaces 31, 56–62 (2009)
11. Jin, Z., Wen, Q., Du, H.: An improved semantically-secure identity-based signcryption scheme in the standard model. Computers and Electrical Engineering 36, 545–552 (2010)
12. Duan, S., Cao, Z.: Efficient and provably secure multi-receiver identity-based signcryption. In: Batten, L.M., Safavi-Naini, R. (eds.) ACISP 2006. LNCS, vol. 4058, pp. 195–206. Springer, Heidelberg (2006)
13. Tan, C.H.: On the security of provably secure multi-receiver ID-based signcryption scheme. IEICE Transactions on Fundamentals of Electronics, Communications and Computer Sciences E91-A(7), 1836–1838 (2008)
14. Yu, Y., Yang, B., Huang, X., Zhang, M.: Efficient identity-based signcryption scheme for multiple receivers. In: Xiao, B., Yang, L.T., Ma, J., Muller-Schloer, C., Hua, Y. (eds.) ATC 2007. LNCS, vol. 4610, pp. 13–21. Springer, Heidelberg (2007)
15. Sharmila Deva Selvi, S., Sree Vivek, S., Gopalakrishnan, R., et al.: Cryptanalysis of ID-based signcryption scheme for multiple receivers, http://eprint.iacr.org/2009/238
16. Li, F., Hu, X., Nie, X.: A new multi-receiver ID-based signcryption scheme for group communications. In: Proceedings of International Conference on Communications, Circuits and Systems, pp. 296–300. IEEE CS (2009)
17. Sharmila Deva Selvi, S., Sree Vivek, S., Srinivasan, R., Pandu Rangan, C.: An Efficient Identity-Based Signcryption Scheme for Multiple Receivers. In: Takagi, T., Mambo, M. (eds.) IWSEC 2009. LNCS, vol. 5824, pp. 71–88. Springer, Heidelberg (2009)

Hypergraph Based Network Model and Architechture for Deep Space Exploration

Xiaobo Wang[1], Junde Song[2], and Xianwei Zhou[1]

[1] Department of Communication Engineering,
School of Computer&Communication Engineering,
University of Science and Technology Beijing, Beijing 100083, China
[2] School of computer, Beijing University of Posts and Telecommunications,
Beijing 100876, China
wang_xiao_bo@163.com, jdsong@butp.edu.cn, xwzhouli@sina.com

Abstract. Deep space communication plays a key role in deep space exploration, it has some special requirments in deep space environment, in this paper, we focus on the network model and architecture for deep space exploration. Because the multiple access technique was widely used in deep space exploration communication, a scenario of deep space exploration come up in this paper, then a network model was proposed which based on hyperpraph theory to address the problem of connectivity, frequency spectrum resource scarcity, mutual interference etc., and how to construct a hyperedge was described in detail. Taking into account the specificity of deep space, combined with the concept of DTN, a novel network architecture was proposed which applies to the deep space exploration.

Keywords: Deep space exploration, hypergraph theory, network model, protocol architecture.

1   Introduction

As soon as human beings could see the neighboring planets, we are eagering to reach them. Also it is the human nature to explore, imagine and probe the unknown world they can see and reach. Nowadays, we are closing to the dream of exploration and utilization of the deep space through our efforts, highly sophisticated spacecrafts were launched into the lolar system and sent back variety of information about our neighboring planets. In 1957, USSR launched the first artificial satellit Sputnik-1, and in 1959 Lunar 1 was launched, it was the first lunar flyby which discovered the solar wind, this can be viewed as the start of human deep space exploration. In the following about half a centry, many countries carried out a lot of work about deep space exploration. Some of which were representative as following table shows[1]:

B. Liu and C. Chai (Eds.): ICICA 2011, LNCS 7030, pp. 495–504, 2011.

Table 1. Some past representative deep space exploration missions

Launch time	mission name	Mission purpose	overview
September 12, 1959	Luna 2 - USSR	Lunar Hard Lander	The first spacecraft to impact the surface of the moon.
April 30, 1966	Surveyor 1 - USA	Lunar Soft Lander	The first American soft landing on the lunar surface.
June 12, 1967	Venera 4 - USSR	Venus Atmospheric Probe	The first probe to be placed directly into the atmosphere and to return atmospheric data.
July 16-24, 1969	Apollo 11 - USA	Lunar Manned Lander	the first manned lunar lander
August 17, 1970	Venera 7 - USSR	Venus Lander	The first successful landing of a spacecraft on another planet.
May 28, 1971	Mars 3 - USSR	Mars Orbiter/ Soft Lander	The lander was released and became the first successful landing on Mars.
May 30, 1971	Mariner 9 - USA	Mars Orbiter	The first US spacecraft to enter an orbit around a planet other than the Moon.
May 26, 1973	Skylab - USA	Space Station	America's first space station.
7 April 2001	2001 Mars Odyssey - USA	Mars Orbiter	with the objective of conducting a detailed mineralogical analysis of the planet's surface.

China also accelerated the process of deep space exploration. On 24 October 2007, the Chang'e-1 was launched, its scientific objectives include analysis of the amount and distribution of elements on the Moon's surface. After the success of Chang'e-1, Chang'e-2 was launched on 1 October 2010 which is the predecessor of Chang'e-3. Then China will launch the Tiangong 1 target spacecraft and an unpiloted Shenzhou VIII spacecraft in the near future. And China's own deep space network will take shape in the next three to five years to promote its exploration of the solar system.

Communication plays a key role in deep space exploration. The communication system is responsible for telemetry&control and information transmission as well as transmission of image, video, scientific data, etc. But the deep space communication has its differences with the normal terrestrial communication and satellite communication, and will face the following challenges which will affect networking and communication links[2-3]:

Heterogeneous network interconnection and node access. In a deep space exploration network, there may be several kinds of coexist networks, such as planet surface network, planet satellite network. Therefore, how these networks interconnect must be considered, also the handover and access problem of the nodes must be solved.

Very Long and varying Propagation Delay Time. Due to the long distance between planets, even the speed of light imposes serious limitations on interplanetary travel[4]. The movement of the planets and spacecrafts also made the delay time variable. The end-to-end round trip time for the Mars-Earth communication network varies from 9 minutes to 50 minutes[5].

Bandwith asymmetry between forward link and reverse link. Data rate asymmetry of forward link and reverse link is typically on the order of 1000:1 in deep space communication. Most data are delivered from deep spacecraft to Earth in spacecraft-Earth communication[5].

Energy constraint. Almostly all the current artificial satellites and planet rovers are depend on solar energy to run their system. And transmission across long distance need very high transmission power, so an energy-efficient transmission power control algorithm is important.

Blackouts or Intermittent Connectivity[6]. In deep space communication, periodic link outrages may occur for the reason of planet/spacecraft movement and harsh environment in deep space.

Lack of Fixed communications infrastructure. In the scenario of deep space exploration, both on the planet surface and between the planets there is hardly any fixed infrastructure to be helpful to the inter-spacecrafts communication.

The frequency spectrum has becoming a restricted resource. The bands 2025-2110 and 2200-2290 MHz which atracted most interests has more than 400 spacecrafts registered. For the popurse of using spectrum effectively, new recommendation on deep space spectrum utilization was proposed and accepted at Spectrum Frequency Consultation Group(SFCG) 23.

In order to deal with the above problems, people has developed the concept of IPN[7].and DSIN[8], which has their particular structure adapted to the environment of deep space communication. A new technique which called delay&disruption-tolerant networking(DTN) has been proposed[9]. A method called store and forward was used in DTN, it does not assume a continuous end-to-end connection, in such sase, the TCP/IP is not suitable. Moreover, with the increase in the number of spacecraft, more spacecrafts may be accomodated within a particular band allocation. To achieve this goal, the multiple access technique was widely used to improve the efficiency of spectrum resource use.

In order to deal with the problems which will affect the network handover and node access in heterogeneous networks, we focus on the network infrastructure and the protocol architechture of deep space planet exploration networks. Based on the multiple access and self organized technique, we use hypergraph theory to model the network and proposed the concept of hyperedges, then explained how to form a hyperedge in the network. Then we proposed an architechture based on improved DTN.

2   Overview of Related Works

In this section, we will introduce some inspiring works related to the network infrustructure and protocol architechture in deep space planet exploration.

2.1   Multiple Access Techniques Used in Deep Space Exploration

NASA's Space Communication and Navigation(SCaN)office has been designing an agency-wide space communication and navigation architecture to surpport NASA space exploration and science missions out to 2030[10]. The study team developed multiple access scenarios for combinations of links for near earth relay, lunar relay, Mars relay,lunar Direct-To-Earth(DTE)/ Direct-From-Earth(DFE), and Mars DTE/DFE links. The Multiple Access subteam recommendated to continue to use existing techniques for ground network and space network[11], and they did a lot of analysis to further develop and analyze both CDMA and FDMA.

In the field of space exploraion, multiple access is not a comparatively new technique, Code Division Multiple Access (CDMA) was used in the NASA's Space Network(SN) program as a multiple access mode in the start, and the Deep Space Network(DSN) utilized Frequency Division Multiple Access (FDMA) system in Mars exploration. And in the future exploration of moon and Mars, we will need for greater multiple access abilities[12].

2.2   Delay Tolerant Networks

With the development of communication technologies in deep space exploration, a new architechture must be proposed to meet the requirments in deep space communication. NASA and JPL started the project of "Interplanetary Internet (IPN)" in 1998[13], they described a scenario of IPN[12], it consists of the backbone network and planetary network. In order to address the problems in the IPN communication which mentioned above, Kevin Fall et al. proposed the concept of DTN(delay/desruption tolerant network, DTN) to describe the architechture used in IPN[13]. And DTN was designed with the characteristics which made it can operates in the high-propagation delay environment, such as near-planetary, planetary surface environments or deep space communication.

The core idea of DTN is added bundle layer into the network architechture, the bundle layer is an intermediate layer between the transport layer which below it and the application layer which above it. The bundle layer provides "store-carrying-forward" services to the application layer, to deal with intermittent connectivity, it store and forwards "bundles", bundles are the variable-sized protocol data unit(PDU) of the "bundle protocol"[14].

Researchs usually concentrate on congestion control, access control and routing based on DTN architechture. In 2009, NASA tested the first deep-space communications network, engineers from the Jet Propulsion Laboratory, used software called Disruption-Tolerant Networking (DTN) to transmit dozens of space images to and from a NASA science spacecraft located more than 20 million miles from Earth[15].

3   Network Model Based on Hyperpraph Theory

3.1   Scenario of Deep Space Planet Exploration

In the future, whether what kind of deep space communication network was constructed, there must be more and more spacecrafts in the space especially in the scenario of planet exploration, so the multiple access(MA) is considered to be a method to support the simultaneous communication. Particularly, the need for MA capability will increase when several deep space nodes(vehicles, Astronauts or satellites) appear in a same relay nods's antenna beam.

For the exporation of deep space planets, we propose a 4-layer structure network model as figure 1 shows.

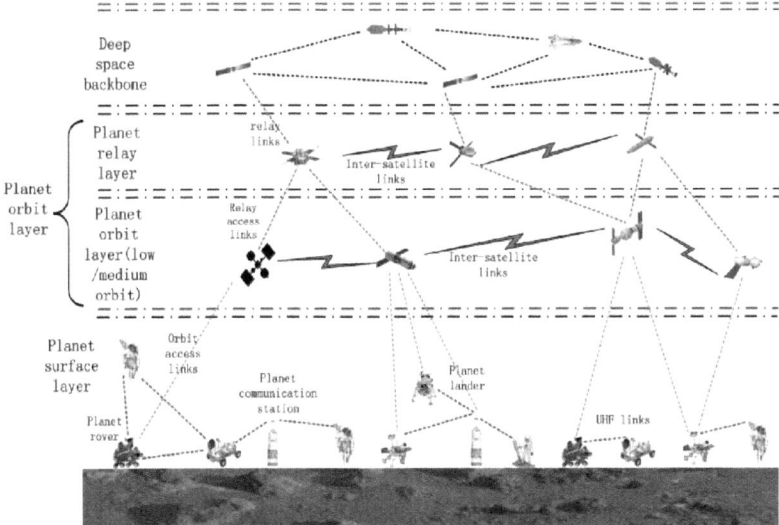

Fig. 1. Scenarios of deep space planet exploration

The first layer is planet surface layer, it support multiple access and autonomous networking of surface nodes, and the surface nodes can communicate to planet orbit layer through planet communication station(PCS) or the nodes which has the access ability to orbit layer. The responsibility of orbit layer is provide extra communication links for the isolated surface nodes or as the forwarding layer betweeen surface layer and planet relay layer, and it also responsible for planet monitoring and data collecting. The planet relay layer is in charge of provide communication between deep space backbone[16] and orbit layer, the orbit layer can communicate with relay layer through multiple access technique.

So in the above scenario of deep space planet exploration, we must take the following consideration into the design of network architechture:

(1) *Self-adaptive.* Because of the uncertainty of deep space or planet surface environment, the network need the ability of perceive, response to the surroundings.

(2) *Self-healing.* The deep space is a complex environment, it's hard to ensure the connectivity of the network, so the self-healing ability is important.

(3) *energy efficient.* Deep space nodes usually depend on solar energy to work, it is power constrained, so improve energy efficient and increase the network lifetime is a design goal.

(4) *Heterogeneous network.* In the deep space exploration, we need numerous nodes, and these nodes can be different space crafts, so the network is a heterogeneous network.

(5) *Random access.* In the deep space planet exploration, multiple access technique is widely used, and the planet surface nodes usually move under a stochastic mobile model, then a node may move from one access point to another, so the problem of handoff and its randomness must be considered.

3.2   Hypergraph Based Network Model

3.2.1   Introduction and Preliminaries

In 1970 C.Berge proposed the concept of hypergraph for the first time, and described hypergraph theory in detail[17]. Hypergraph theory is constructed on the base of graph theory and set theory.

A binary relation $H = (V, E)$ is a hypergraph[18], in which, $e_i \neq \emptyset (i = 1, 2, \cdots, m)$ $\bigcup_{i=1}^{m} e_i = V$, and $V = \{v_1, v_2, \cdots, v_n\}$ is a finite set. $v_1, v_2, \cdots, v_n$ are vertex of hypergraph, $E = \{e_1, e_2, \cdots, e_m\}$ is edge set of the hypergraph, and $e_i = \{v_{i_1}, v_{i_2}, \cdots, v_{i_j}\}$ $(i = 1, 2, \cdots, m)$ is called edge of hypergraph. And a hypergraph $H = (V, E)$ differs from an ordinary graph in the edge, hyperedges are arbitary sets of nodes, it means a hyperedge connects multiple nodes, but an ordinary edge connects only two nodes.

3.2.2   Hypergraph Based Network Model

In order to improve the network performance in connectivity, power efficiency etc., we use hypergraph to model the network. In the hypergraph modeled network $H = (V, E)$, V denotes the nodes in the network, E denotes the hyperedge of the network. And in the deep space exploration network, the hyperedge E is constitute of the nodes of the same access point. As described above, multiple access technique is widely used in deep space communication, and the nodes belong to the same access point will have better communication performance.

Based on the above description, a scenario of hypergraph based deep space exploration network is shown in Fig.2, the oval denotes hyperedges, and the dotted line denotes wireless link between nodes that can communicate with each other. And if E_i and E_j are adjacent hyperedges, and there is a node $n_{i1} \in E_i$, if also there exists a n_{i1}'s neighbor $n_{j1} \in E_j$, then n_{i1} and n_{j1} form a new hyperedge E_{ij} called connecting hyperedge. Two adjacent hyperedges can transfer data through connecting hyperedges between them, and in this way, a hyper-route can be setup.

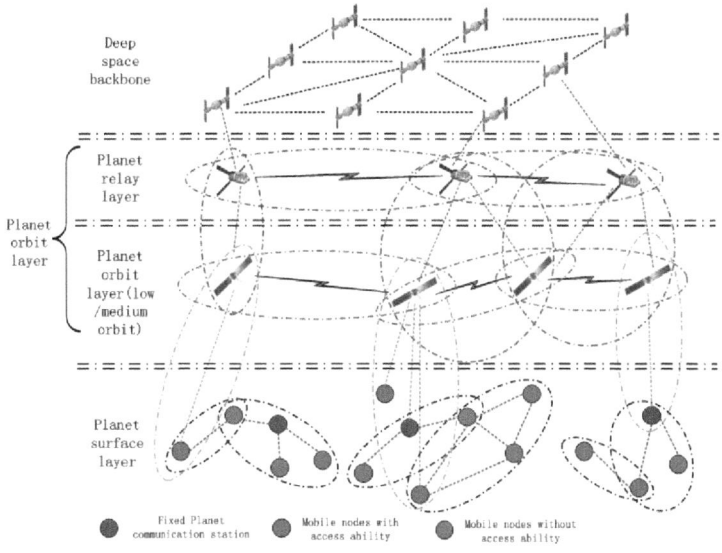

Fig. 2. The concept of hypergraph based network model

Although the network capacity can be increased by techniques, capacity also suffers due to the mutual interference of the channel. And the network model which based on hypergraph theory, provide a way to implement various kinds of interference cancellation methodes and also networking and routing.

4   Using DTN and Hypergraph within an Architechture

According to the characteristic of deep space exploration, for the aim of environment self-adaptive, energy efficient, etc, we proposed an architecture to face the requirments in deep space which based on some existing research results[19][20], as shown in Fig.3.

The three-dimensional protocol stack shows a 7 layer structure, includes application layer, bundle layer, transport layer, hypernetwork layer, topology control layer, data hyperlink layer, and physical layer. In the right side of the stack, there are three modules: the security mechanism module provide function of electromagnetic interference suppression in the physical layer, identify the malicious nodes and provide data encryption; the access management module responsible for access point selection and handover; the energy saving module has the function of perceive the remaining energy of the node, adjust the transmission power according to the environment.

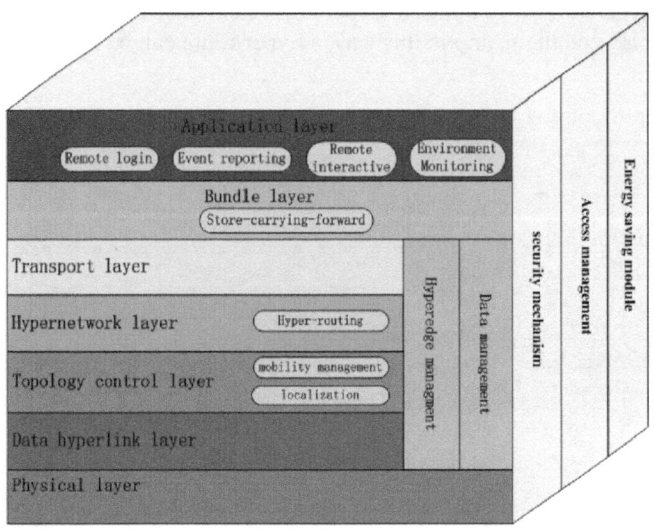

Fig. 3. The hyper-network architecture of deep space exploration

In our layered protocol stack, the bundle layer, transport layer and physical layer are basically the same as conventional DTN architecture. We primarily emphasize on the application layer, hypernetwork layer, topology control layer and data hyperlink layer:

(1) The application layer in our protocol stack constitute of software systems which face to various deep space environment, it is a high-level human interface. It provide services include remote login, event reporting, remote interactive and environment monitoring.

(2) The main popurse of hypernetwork layer is to realize low power consumption, muti-hop hyper-routing. The hypernetwork layer can determine and filter routes according to the information provided by bottom layers and the three modules in the right side of the stack.

(3) The topology control layer is responsible for adjust the network topology, mobility management and localization. According to the environmental parameters, node's remaining energy, the topology control layer will be able to adjust the topology through transmission power control and hyperedge mangement.

(4) The data hyperlink layer is responsible to form hyperframes, reduce the collision of wireless media access within hyperedges and perform error control.

5   Conclusions and Future Work

The deep space exploration network is different with terrestrial networks in networking, resource constraints, environmental parameter, etc. So we must construct new network architecture which is suitable for deep space environment. In this paper, first we described the typical deep space exploration scenario, established the network topology model based on hypergraph theory, and analysed the special needs in deep space environment. Then proposed a hyper-network architecture for deep space exploration. Our network topology model and protocol architecture provide a framework for future research of each protocol layer in deep space exploration environment, and we expect our work can be a useful guidance.

Based on the idea of this paper, the next research work include efficient hyperedge establish schemes and routing in the hypergraph theory modeled network.

Acknowledgments. This work is supported by the National High Technology Research and Development Program of P. R. China (no. 2009AA062705), the National Natural Science Foundation of China under Grant No.60903004 and the National Research Foundation for the Doctoral Program of Higher Education of China under Grant (No. 20090006110014).

References

1. http://www.solarviews.com/eng/craft1.htm
2. Chen, C., Chen, Z.: Towards a routing framework in ad hoc space networks. International Journal of Ad Hoc and Ubiquitous Computing 5(1), 44–55 (2010)
3. Zhou, X., Yin, Z., Wang, J., Liu, T., Wang, C.: Deep Space Communications. National Defense Indrustrial Press, Beijing (2009)
4. http://www.spaceacademy.net.au/spacelink/commdly.htm
5. http://www.ece.gatech.edu/research/labs/bwn/deepspace/transport.html
6. Gajurel, S.: Space Communication and dynamic routing solutions (2006), http://vorlon.case.edu/~sxg125/Projects/
7. Akyildiz, I.F., Akan, O.B., Chen, C., Fang, J., Su, W.: InterPlaNetary Internet: state-of-the-art and research challenges. Computer Networks 43(2), 75–112 (2003)
8. Xianwei, Z., Long, Z., Zhimi, C., Huan, H., Jianping, W., Yueyun, C.: Hypernetwork model and architecture for deep space information networks. In: Proceedings of the 2010 IEEE International Conference on Future Information Technology (ICFIT 2010), Changsha, China, pp. 448–452 (2010)
9. Fall, K.: A delay-tolerant network architecture for challenged internets. In: Proc. ACM SIGCOMM 2003, Karlsruhe, Germany, pp. 27–34 (Augest 2003)
10. Deutsch, L., Stocklin, F., et al.: Selecting Codes,Modulations, Multiple Access Schemes, and Link Protocols for future NASA Missions. In: IEEE Aerospace Conference, Big Sky (March 2008)
11. Akyildiz, I.F., Akan, O.B., Chen, C., et al.: InterPlaNetary Internet: state of the art and research challenges. Computer Networks 43(2), 75–112 (2003)
12. Bhasin, K., Hayden, J.L.: Space Internet architecture and technologies for NASA enterprises. International Journal of Satellite Communications 20(5), 311–332 (2002)
13. Fall, K.: A delay-tolerant network architecture for challenged internets. In: Proceedings of the 2003 Conference on Applications, Technologies, Architectures, and Protocols for Computer Communications, pp. 27–34. ACM Press (2003)
14. Paul, S., Pan, J., Jain, R.: Architectures for the future networks and the next generation Internet: A survey. In: IEEE 18th Annual Symposium on High Performance Interconnects (HOTI) 2010, Mountain View, CA, August 18-20, p. 123 (2010)
15. http://www.techbriefs.com/component/content/article/3552

16. Akyildiz, I.F., Akan, O.B., Chen, C., Fang, J., Su, W.: InterPlaNetary Internet: state-of-the-art and research challenges. Computer Networks 43(2), 75–112 (2003)
17. Berge, C.: Graph and Hypergraph. North Holland, Amsterdam (1973)
18. Berge, C.: Graphs and hypergraphs. Elsevier, New York (1973)
19. Cerf, V., et al.: Delay-Tolerant Networking Architecture, RFC 4838, Internet Engineering, Task Force (April 2007)
20. Scott, K., Burleigh, S.: Bundle Protocol Specification, RFC 5050, Internet Engineering Task Force (November 2007)

Cost-Benefit Analysis for Adaptive Web Service Substitution with Compensation

Ying Yin, Xizhe Zhang, and Bin Zhang*

College of Information Science and Engineering, Northeastern University
Shengyang, 110004, China
yy_00000000@163.com

Abstract. Nowadays, plenty of enterprises are willing to outsource their internal business processes as services and make them accessible via the Web. Toward some business process, Web service based application require more transactional support beyond traditional transactions, In this situation, direct substitution may violate the atomicity character of transaction. This will lead to the inconsistency of transaction Web service. Moreover, executed data were stored in the memory and could not be released which consequently result in the serious results such as "missing" of data. However, most of previous works on service replacement algorithms leak the cost analysis with compensation. In this paper, we extend the substitution framework by supporting compensation. A correlation analysis and behavior matching can be executed to obtain the range of cascading compensation and decide the range of substitution. Further more, we analyze the single node substitution cost and path substitution cost in details. After that, this paper presents a cost-benefit function which considers customer's preference and selects the optimal substitution strategy by regulating parameters at will.

Keywords: Substitution, Cost-Benefit Analysis, Compensation.

1   Introduction

With service-oriented computing architecture (Service Oriented architecture, SOA) to further the promotion and application, Nowadays, plenty of enterprises are willing to outsource their internal business processes as services and make them accessible via the Web[1,2]. In addition, they can dynamically combine individual services to provide new value-added services. However, in practical applications, business process is always in a dynamic, distributed environment[3]. Service quantity with the same function (such as response time, security, reliability etc) and service implementation strategy also changing.

Exception or failure induced by dynamic changes in the service space are often happen[4,5]. How to realize reliable, stable Web service business process in the real-time change's dynamic network environment has already become hot spot and key in the present distributed computing. To achieve this, transaction

---

* Corresponding author.

B. Liu and C. Chai (Eds.): ICICA 2011, LNCS 7030, pp. 505–512, 2011.
© Springer-Verlag Berlin Heidelberg 2011

support has become a serious issue for Web service technology, since it is critical to ensure a correct integration and a reliable execution of the integrated business process[6,7].

It's well known to all, a variety of abnormal behaviors may lead to the entire workflow interruption or termination. The traditional solution of facing unusual or failure is to build a Web service substitution model. When the combination of flow in a composite service fails to meet the user needs, we need to find a candidate service or a set of services for replacement[4,5,8,9,10]. For example, tao proposed a substitution method based on a combination of services from the current service to the interrupted services and built a backup path. As long as a web service fails, it automatically turned back up services. Other researchers implement it by other technologies, such as Petri Nets, process algebra[11,12] to complete the substitution. However, it is not feasible toward business process due to lacking the transaction support. It is necessary to design a substitution with compensation framework. More important, cost-benefit analysis is the vital step for customer to select the optimal substitution strategy.

With all these problems above, we present a flexible substitution method with compensation to maintain the stability and credibility. The main contributions of this paper are as follow: (1) This paper considers transaction-based gained for substitution to maintain the atomicity; (2) a substitution framework with compensation was proposed to support transactional Web service substitution better; (3) based on mined behavior pattern previously, the system complete the substitution as quickly as possible; (4) a benefit-cost function which considers customer's preference was presented.

The remainder of this paper is organized as follows: Section 2 states an adaptive substitution framework with compensation, the basic definitions related and an efficient cost-benefit function. Section 3 presents related work for comparison with our approach. We summarize our research and discuss some future work directions in Section 4.

2   Adaptive Substitution Framework with Compensation

In real application, dynamic Web services composition is the first step, it is more important for automatic business process which have the capability to adjust to environment variations(self-reconfiguring). That is if one component service node becomes unavailable, a mechanism is need to ensure that the business process is not interrupted and the failed service is quickly and efficiently replaced. In this paper, we propose an adaptive substitution framework which considers compensation-support during the execution process. Figure 2 shows the overall substitution framework with compensation. Based the framework, we illustrate the process of substitution with compensation. After a long time of a set of web services were initiated and implemented, we obtain plenty of original service log. Further, by extracting service path and logic relation between services using data mining method, we get preprocessed service log which can be used to mine service behavior patterns. The substitution will be trigger as long as the system monitors the failure during the execution process. The cost-benefit analysis model with compensation will start-up with matching the service behavior pattern, and finally, selecting the optimal strategy to replacement.

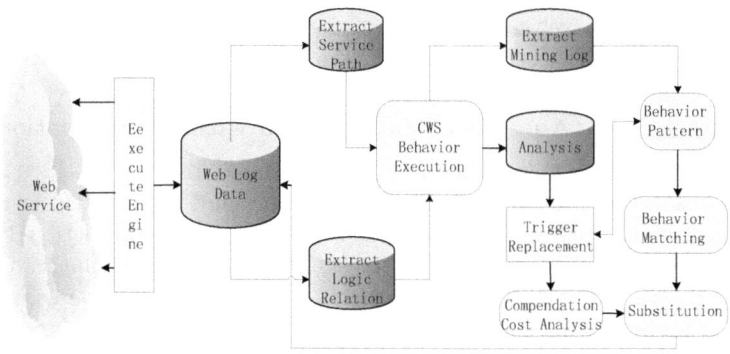

Fig. 1. The framework of Substitution with Compensation

2.1   Basic Definition

There are two strategies: node substitution and subsequence(subgraph) substitution. Let give some definitions related to failure services.

Definition 1. *Interrupt Web Service (IWS).* *Given composite Web service flow* $\mathcal{S}=\{s_1, s_2, ..., s_n\}$, *During the execution, the phenomenon of s_i's failure or unavailable which make the whole flow interrupt, we say service s_i (i ∈ {1,2,...,n}) is the interrupt Web service node.*

Definition 2. *Successor Service Set (SSS).* *Given composite Web service flow* $\mathcal{S}=\{s_1, s_2, ..., s_n\}$, *During the execution, if service s_i is failure or unavailable, we say the service sequence set from node s_i to final point $\mathcal{S}_f$ **Successor Service Set**.*

Definition 3. *Ancestor Service Set (ASS).* *Given composite Web service flow* $\mathcal{S}=\{s_1, s_2, ..., s_n\}$, *During the execution, if service s_i is failure or available, we say the service sequence set from beginning point $\mathcal{S}_b$ to node s_i **Ancestor Service Set**.*

For example, the abstract composite Web service $\mathcal{S}$, $\mathcal{S}=\{s_1, s_2, s_3, s_4, s_5, s_6\}$. For one of the composite Web service sets $S_i=\{s_{11}, s_{21}, s_{41}, s_{52}, s_{72}, s_{81}\}$, during the execution, if service s_{41} is failure, the interrupt Web service IWS= s_{41}, corresponding Successor Service Set and Ancestor Service Set are **SSS**$(s_{41})=\{s_{41}, s_{52}, s_{72}, s_{81}\}$ and **ASS**$(s_{41})=\{s_{11}, s_{21}\}$ respectively.

Adaptive capability is refers to during the execution of a composite Web service, if a failure service happened, a mechanism is need to ensure that the running process is not interrupted and the failed service is quickly and efficiently replaced. At first, the system need to mark the failure nodes and edges connected with it and switch to substitution with compensation model, then select the right strategy to replace. If composite web service with transactional property fails at a point, well-executed services of this structure must undo [7] and rollback to starting state.

2.2   Cost-Benefit Analysis of Substitution with Compensation

As the pre-mention, a transaction may consists of several services. If considering transactions, for business process that has executed half way, a failed service may prevent it from completing the execution. According to the transaction it belongs to, well-executed services of this structure must undo [7] and rollback to starting state of the transaction. Meanwhile, the system should have the capability to compensate, eliminate the effect of the rollback transaction and replace the failed transaction by matching a new service or a set of services which have the same behavior with failure transaction

The more rollback services, the more cost of substitution with compensation. The best strategy is backward rollback with the minimal services length. At first, the atomic transaction which containing failed service must be rollback. Second, the behavior of replacement services need consistent with completed service, that is the pre-condition of replacement services should contains the post-condition of terminate services. Last but not least, select an optimal candidate services with the following two points: (1) the range of rollback services is as short as possible; (2) the behavior of replacement services will compiled with rollback services.

Based on the analysis, we can deduce the following formulas of substitution cost. Formula (1) and formula (2) are the substitution cost without considering compensation. Formula (3) and formula (4) are the substitution cost with considering compensation.

(1) For node substitution:

$$SC(NS) = Q_r(s_i')$$

(2) For single forward extended subgraph substitution:

$$SC(FES) = \sum_{i=1}^{l} Q_r(SSS(g'))$$

(3) For single backward extended subgraph substitution:

$$SC(RES) = \sum_{i=1}^{rl} Q_{Undo}(ASS(g)) + \sum_{i=1}^{rl} Q_r(ASS(g'))$$

(4) Bi-dirential (forward-backward) extended subgraph substitution Cost is:

$$SC(DES) = \sum_{i=1}^{rl} Q_{Undo}(ASS(g)) + \sum_{j=1}^{rl+x} Q_r(SSS(g'))$$

where rl is the roll back length of subgraph in ASS, $rl + x$ is the length of subgraph for replacement. g represents the subgraph need to be replaced and g' represents the new subgraph of replacement.

From the above formulas, we can see substitution cost not only depends on the numbers of Web services need to be compensated in the original graph(Q_{Undo}), but also the cost of substitution.

Let illustrate the range of substitution using a figure. Figure 2 shows the relationship between compensation range and substitution range. Based on the previous work, we can deduce the transactional granularity(i.e. the transaction which it belongs to). A transaction may cover some operations of one service or some services. Based on the operations or services covered by transaction, we can obtain the compensation range. How to select the optimal path or subgraph to substitution is vital for automatic composite business. During the substitution process, in order to obtain the minimal value of SC, we require the minimal steps of rollback and "maximal" forward steps. For the reason that minimal steps of rollback means the minimal compensation and maximal forward steps means increase the chance for choosing the minimal cost from more service nodes of forward graph.

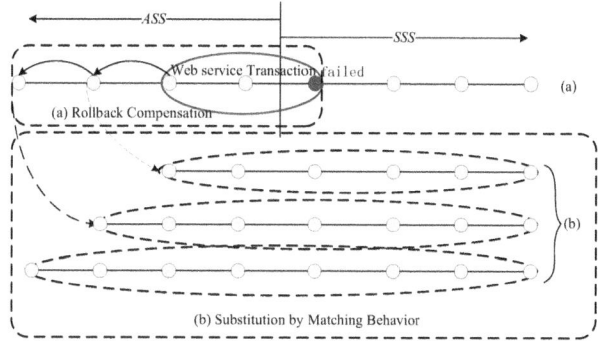

Fig. 2. range

2.3 Cost-Benefit Analysis of Service Substitution

based on the previous analysis, we can get the cost-benefit of substitution with compensation which shows as the formula below.

Given a composite services $cp_i=\{s_1, s_2,..., s_n\}$, if Web service s_i $i \in \{1,2,...,n\}$ was failed, the length of failure service to the final service is l, $l \leq n$, the length of rollback is x, then substitution cost with compensation (SC) is computed by the following formula:

$$SC = \sum_{i=1}^{rl} Q_{Undo}(g) + \sum_{j=1}^{rl+x} Q'_r(g')$$

where Q'_r is the cost of new replacement composite services and Q_{Undo} is the compensation cost.

Further more, the user may have different personal requirement due to various customer's preference. This paper describes the characters and gives a personal QoS^p driven service substitution model with compensation. Our QoS^p driven model select the compensation services and substitution service by his/her personal preferences and satisfy the users SLA. The QoS^p driven service substitution model shows as formula:

$$SC^p = \sum_{i=1}^{rl} Q_{Undo}(g) + \sum_{j=1}^{rl+x} Q_r(g')$$

$$= w_1(\sum_{i=1}^{rl} Q^t_{Undo}(g) + \sum_{j=1}^{rl+x} Q^t_r(g')) + w_2(\sum_{i=1}^{rl} Q^p_{Undo}(g) + \sum_{j=1}^{rl+x} Q^p_r(g'))$$

where, SC^p denotes the personal substitution service cost-benefit QoS function. For simplicity, we only consider two parameters: time and price. Q^t_{Undo} and Q^t_r denotes the compensation time and substitution time respectively, Q^p_{Undo} and Q^p_r denotes the compensation price and substitution price respectively, w_1 and w_2 denotes the weight of time and weight of price appointed by user separately. In this work, we select the path with the maximal score according to the proposed substitution algorithm. If there are several paths with the maximal score one of them is selected randomly.

3   Related Work

Substitution is one of the important mechanisms guarantee the system reliability. There are two class of substitution mechanisms at present. one substitution strategy is replacement oriented service function[4,5,8,9,10]. The key to this approach is to establish a set of redundant services for each component service. Then, if one component service fails, the service can be replaced with an alternative member of the same function. Another substitution strategy is replacement oriented quality(i.e.QoS)[4,5]. In Ref[4], all the replacement composite services are backed up before the execution of the composite service. Because of the dynamic nature of Web services, such two approaches do not consider the QoS in the execution of the composite service. In ref[13], the author proposed composite service replacement algorithm for global optimization. the method focus on reselecting the unexecuted services when the failure was triggered and ensure the global QoS as soon as quickly. However, they only analysis the QoS requirement for replacement without ensuring the overall system consistency. All in all, those replacement algorithms ignoring transaction support will fail even when satisfying the requirements from function or semantic. In this circumstance, the system will be interrupt and the application was limited.

4   Conclusion

In this paper, we discuss transaction-support substitution method which are important to maintain service reliability, and extend the substitution framework by supporting compensation. A correlation analysis and behavior matching can be execute to obtain the range of cascading compensation and decide the range of substitution. Further more, we analysis the single node substitution cost and path substitution cost in details. After that, we give a benefit-cost function which consider customer's preference and select the optimal strategy by regulating parameters at will. In the future work, we will discuss and add more parameters to the substitution function in order to satisfy more user's requirements.

Acknowledgments. National Natural Science Foundation of China under grants (No. 61100028, 61073062,60903009,60803026); This work was supported by China Postdoctoral Science Foundation (No. 20100481204); The Fundamental Research Funds for the Central Universities under grants (No. N090304006); Industry tackle key problem plan of Liaoning Province(No.2010216005); Natural Science Foundation of Liaoning Province(No. 20102061).

References

1. Riegen, M.V., Husemann, M., Fink, S., Ritter, N.: Rule-Based Coordination of Distributed Web Service Transactions. IEEE T. Services Computing (TSC) 3(1), 60–72 (2010)
2. Liu, A., Li, Q., Huang, L., Xiao, M.: FACTS: A Framework for Fault-Tolerant Composition of Transactional Web Services. IEEE T. Services Computing (TSC) 3(1), 46–59 (2010)
3. Yu, Q., Liu, X.M., Bouguettaya, A., Medjahed, B.: Deploying and managing Web services: issues, solutions, and directions. The VLDB Journal 17, 537–572 (2008)
4. Yu, T., Zhang, Y., Lin, K.J.: Efficient Algorithms for Web Services Selection with End-to-End QoS Constraints. ACM Trans. Web 1(1), 1–25 (2007)
5. Yu, T., Lin, K.J.: Adaptive algorithms for finding replacement services in autonomic distributed business processes. In: The 7th International Symposium on Autonomous Decentralized Systems, Chengdu, China, pp. 427–434 (2005)
6. Alrifai, M., Dolog, P., Balke, W.T., Nejdl, W.: Distributed Management of Concurrent Web Service Transactions. IEEE Transactions on Services Computing 2(4), 289–302 (2009)
7. Michael, S., Peter, D., Wolfgang, N.: An environment for flexible advanced compensations of Web service transactions. ACM Transactions on the Web 2(2), 1–36 (2008)
8. Jorge, S., Francisco, P., Marta, P., Ricardo, J.: Ws-replication: A framework for highly available web services. In: Proc. of the 15th International Conference on World Wide Web, pp. 357–366. ACM press, Edinburgh (2006)
9. Taher, Y., Benslimane, D., Fauvet, M.C.: Towards an approach for web services substitution. In: Proc. of 10th International Database Engineering and Applications Symposium, pp. 166–173. IEEE CS Press, Delhi (2006)

10. Yin, Y., Zhang, X., Zhang, B.: An efficient service substitution algorithm based on temporal composite behavior graph. In: Proc. of the 6th Web Information Systems and Applications Conference, pp. 126–131. IEEE Press, Jiangsu (2009)
11. Pathak, J., Basu, S., Honavar, V.: On Context-Specific Substitutability of Web Services. In: Proc. of IEEE International Conference on Web Services, pp. 192–199. IEEE Press, Salt Lake City (2007)
12. Kuang, L., Xia, Y.J., Deng, S.G., Wu, J.: Analyzing Behavioral Substitution of Web Services Based on Pi-calculus. In: Proc. of IEEE International Conference on Web Services, pp. 441–448. IEEE Press, Miami (2010)
13. Gerardo, C., Massimiliano, D.P., Raffaele, E.: Qos-aware replanning of composite web services. In: Proc. of the IEEE International Conference on Web Services, pp. 121–129. IEEE Press, Orlando (2005)

Study of WSNs Trust Valuation Based on Node Behavior

Qi Zhu

Department of Information Engineering, Jilin Business and Technology College
zhuqi5678@126.com

Abstract. In WSNs, security system based on password system from within the network attacks and to identify malicious nodes there is no effective treatment methods, trust-based security mechanisms have come into being, This paper discusses in detail the specific steps of trust node, proposed the node behavior evaluation mechanism based on the node trust, The paper has used ANP and the F-ANP method successively for the node behavior's trust appraisal, and has carried on the contrast, finally obtains F-ANP more superior performance.

Keywords: Wireless Sensor Networks, Behavior Trust, ANP, F-ANP.

1 Introduction

With the ability increased in all aspects of sensor nodes, the supporting technologies of WSNs are gradually mature, and the application background of WSNs rapidly expands. So the demand on safety and reliability of WSNs is growing. In WSNs, the security system which is Password-based, it does not have effective method for responding to the attacks from inside of networks and identifying the malicious node.

This article discussed trust evaluation is aimed at assessing the behavior trust of ordinary nodes; it is a dynamic assessment method. All nodes trusted create and update are direct or indirect establish variety of act evidence the foundation. Behavior evidence is node and node in the interactive process, obtained directly from the software and hardware used to detect quantitative assessment of the overall foundation of trust value node, it is objective, and subjective trust itself does not have features, therefore, the behavior evidence is calculates the node behavior trust the foundation.

2 WSNs Network

Wireless Sensor Networks (WSNs) be composed massive inexpensive sensor node in the monitors region by the deployment, through the wireless communication way forms jumps organization's network system, its goal is the cooperation sensation, gathering and the processing network covers in the region the sensation object information, gives the viewer concurrently [1], the architecture shown in Figure 1.

B. Liu and C. Chai (Eds.): ICICA 2011, LNCS 7030, pp. 513–520, 2011.

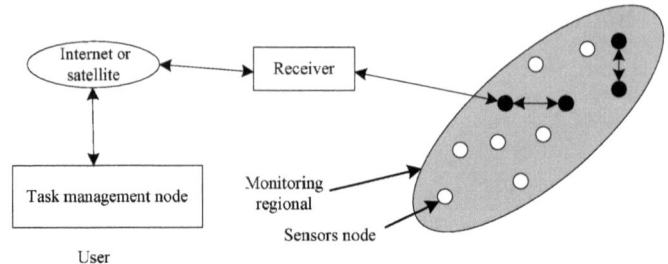

Fig. 1. Wireless sensor networks architecture

3   WSNs Trust Management

The trust itself takes the concept in a sociological, has generally, fuzzy, uncertainty. If carries on it the quantification, then needs to choose the appropriate model. Any trust model should include the following 4 steps, as shown in Figure2 :(1) collects the evidence or the collection information, namely collection appraisal object original evidence.   (2)Appraises the score, namely to appraises the object based on the collection evidence the behavior to carry on the allocation. (3) Appraises the service degree of satisfaction, namely basis score judgment service degree of satisfaction. (4) Serves feedback, namely according to the degree of satisfaction carries on the feedback, including reward, penalty.

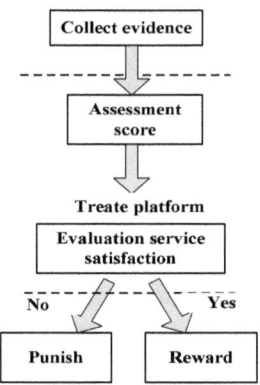

Fig. 2. Generic Trust Model

At present regarding the wireless sensor network trust management's research mainly concentrates on   to the node carries on the confidence level appraisal, through strengthens wireless sensor aspects and so on network security, toughness to the confidence level appraisal.   The trust which proposed based on the node behavior evidence's trust management system management system measures and appraises, its essence is uses one relative method to the node security, the performance information carries on the measure and the appraisal, can under the good response distributed environment WSNs polytropy and the uncertainty, and this method suits the trust collection of information appraisal the automation to realize.

4   Node Trust Appraisal Based on ANP Method

4.1   Node Trust Evaluation

The trust appraisal is aims at the ordinary node the behavior trust appraisal, is one dynamic appraisal method. All node's trust establishes and renews is direct or the indirect establishment above each behavior evidence foundation. The behavior evidence is the node and the node in the interactive process, obtains directly according to the software and hardware examination uses for the quantitative assessment node overall trust the foundation value, it has the objectivity, itself does not have the trust subjective characteristic. Therefore, the behavior evidence is calculates the node behavior trust the foundation.

1) The Establishment of ANP Model
This article establishes wireless sensor node trust appraisal's model like Figure 3, ANP divides into system's element two parts, the key-course and the network level. The first part is the key-course the policy-making criterion which and carries on including the question goal, namely Figure 3 first. The second part is the network level, namely Figure 3 second, third, fourth, In the element group (for example performance attribute B1, security attribute B2) between as well as between various daughter elements' relations are not mutually independent, has the interdependence to relate and to feed back the relations, therefore has formed network architecture [2].

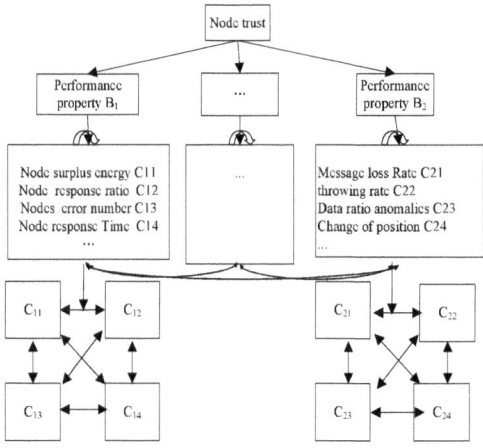

Fig. 3. ANP model of trust evaluation for node

2) The Establishment of Super-Matrix
Accordance with the above method to calculate the other matrix, and denoted as:

$$W_{ij}, \quad w_{ij} = \begin{pmatrix} w_{i1}^{(j1)} & \cdots & w_{i1}^{(jn_j)} \\ \vdots & \ddots & \vdots \\ w_{in_i}^{(j1)} & \cdots & w_{in_i}^{(jn_j)} \end{pmatrix}$$

n_j is j first element group containing the number of elements, which w_{ij} of Column vector is B_i the element of group elements B_j on the degree of $C_{j1}, C_{j2}, \cdots, C_{jn_j}$ $(j = 1, \cdots N)$ influence of the elements in sorted vector. If the elements group of elements B_j from the impact of the elements B_i, then $w_{ij} = 0$, may obtain exceed matrix

$$W = \begin{pmatrix} w_{11} & \cdots & w_{1N} \\ \vdots & \ddots & \vdots \\ w_{N1} & \cdots & w_{NN} \end{pmatrix}$$

5   Node Trust Appraisal Based on F-ANP Method

5.1   Establishment Models

F-ANP and the ANP method's difference lies in the two judgment matrices the constitution, the F-ANP use triangle fuzzy number replaces the statty scale, and the F-ANP method improves in the ANP method foundation comes, therefore its model and ANP method's same, as shown in Figure 3. Model various part of expressions and ANP are the same.

5.2   Construction of Fuzzy Judgment Matrix

Definition 1. If the fuzzy number, where, M is called triangular fuzzy number; its membership function can be expressed as [4]:

$$\mu_M(x) \begin{cases} \dfrac{x-l}{m-l}, l \le x \le m \\ \dfrac{x-u}{m-u}, m \le x \le u \\ 0, other \end{cases} \tag{1}$$

In Equation 1, l and u are upper and lower bounds of M, m is the median Set $M_1 = (l_1, m_1, u_1), M_2 = (l_2, m_2, u_2)$, then the following algorithms:

(1) $M_1 \oplus M_2 = (l_1, m_1, u_1) \oplus (l_2, m_2, u_2) = (l_1 + l_2, m_1 + m_2, u_1 + u_2)$

(2) $M_1 \otimes M_2 = (l_1, m_1, u_1) \otimes (l_2, m_2, u_2) = (l_1 l_2, m_1 m_2, u_1 u_2)$

(3) $1 / M_1 = (l_1, m_1, u_1)^{-1} \approx (1 / u_1, 1 / m_1, 1 / l_1)$

In the traditional ANP method, after obtaining the judgment matrix, must carry on the relative weight the computation, must obtain the relative weight in here, must first use in the formula (2) computing matrix each element to synthesize Cheng Duzhi who compares with other elements, records is S_i

$$S_i = \sum_{j=1}^{m} M_{E_i}^j \otimes [\sum_{i=1}^{n} \sum_{j=1}^{m} M_{E_i}^j]^{-1} \tag{2}$$

And M_{E_i} for the fuzzy judgment matrix's in fuzzy number, S_i is the triangle fuzzy number, m is the rectangular array number, n is a matrix line of number.

After obtaining n triangle fuzzy number, must calculate each triangle fuzzy number to be bigger than other fuzzy numbers the possible degree. From formula (3) calculate:

$$V(M_1, M_2) = \begin{cases} 1 & m_1 \geq m_2 \\ \dfrac{l_2 - u_1}{(m_1 - u_1) - (m_2 - l_2)} & m_1 < m_2, u_1 \geq l_2 \\ 0 & \text{other} \end{cases} \tag{3}$$

V (M1, M2) expressed that M1 is bigger than M2 the possible degree.

5.3   The Establishment of Super-Matrix

Build up super matrix method is homology to ANP method, first what to arrive is, then get super matrix W according to the method in ANP, and add power matrix A.

5.4   Example Analysis

1) Construct Misty Judgment Matrix. Take function as a time the standard carry on function and safety more while trusting the standard that the valuation value maximizes and list compare the two matrixes, such as table 2. Table 2 medium of the data use triangle the faintness count to describe opposite importance.

Table 2. each cluster with criteria A compared to the Importance of B_1

	Performance (B1)	Security (B2)
Performance (B1)	(1,1,1)	(1,1,2)
Security (B2)	(1/2,1,1)	(1,1,1)

According to the formula (2) Computation synthesis important Degree Value:
$S_{B1} = (.4, 0.5, 0.87); S_{B2} = (0.3, 0.5, 0.58)$

According to the formula (3), Computation each group of elements important possibility than other elements (SB1)=0.5; d(SB2)=0.5,After process normalization, obtains the weight vector is $W_{B1}=(0.5,0.5)^T$, Similarly we may also establish in the trust evaluation value are biggest, under safety criterion element level fuzzy judgment matrix. Obtains weight vector $W_{B1}=(0.5,0.5)^T$. From this we may obtain a weighting matrix,

$$A = \begin{pmatrix} 0.5 & 0.5 \\ 0.5 & 0.5 \end{pmatrix}$$

Table 3. Pair-wise comparison matrix of performance cluster with the number of position change

C24	C11	C12	C13	C14
C11	(1,1,1)	(1/3,1/2,1)	(1,2,4)	(1/4,1/3,1/2)
C12	(1,2,3)	(1,1,1)	(1/6,1/4,1/2)	(1,1,1)
C13	(1/4,1/2,1)	(2,4,6)	(1,1,1)	(1/5,1/4,1/3)
C14	(2,3,4)	(1,1,1)	(3,4,5)	(1,1,1)

In the model, there are two elements in the network layer, the performance property and security properties. in assessing the value of the greatest trust, elements of the element of time, establish 2 compare matrix, as in a position to change frequently (c24)certain standards, performance property sets the elements of comparison between 2 compare matrix, as shown in table 3. the relative weight that is super matrix part. Get the normalized weight vector $W=(0.17818,0.145503,0.27190,0.40589)^T$.

2) The Establishment of ANP Structure of the Super-Matrix
Likewise, extracts other matrices with the similar method, and obtains the corresponding relative weight. Obtains all relative weights is possible to construct the ANP matrix, as shown in Table 4, in this article application the ultra matrix which is composed of the interaction element's 16 22 comparison relative weights.

Table 4. Super-matrix

	C11	C12	C13	C14	C21	C22	C23	C24
C11	0.39788	0.28608	0.29874	0.26542	0.26175	0.35202	0.44109	0.17818
C12	0.19783	0.21743	0.29129	0.26855	0.22437	0.18629	0.02549	0.14503
C13	0.18027	0.27505	0.23574	0.26855	0.21418	0.14881	0.20392	0.27190
C14	0.22402	0.18143	0.17423	0.19749	0.29970	0.31288	0.32950	0.40589
C21	0.17343	0.17248	0.13227	0.29118	0.26216	0.22304	0.14778	0.20249
C22	0.30242	0.30112	0.30526	0.13688	0.39211	0.32648	0.29784	0.25402
C23	0.25855	0.25981	0.30967	0.19145	0.26162	0.19092	0.25090	0.29483
C24	0.26560	0.26659	0.25280	0.38049	0.08410	0.25956	0.30348	0.24866

In ultra matrix's each sub block is the normalization, but the ultra matrix is in itself not the normalization. Therefore, must to the ultra matrix element weighting, obtain a weighting ultra matrix. Afterward, is the ANP core work, namely the solution ultra matrix, the weighting matrix from the multiplication, until obtains a product restraining, long-term stability up to limit matrix, as shown in Table 5.

Table 5. Limited super-matrix

	C11	C12	C13	C14	C21	C22	C23	C24
C11	0.15605	0.15605	0.15605	0.15605	0.15605	0.15605	0.15605	0.15605
C12	0.09532	0.09532	0.09532	0.09532	0.09532	0.09532	0.09532	0.09532
C13	0.10984	0.10984	0.10984	0.10984	0.10984	0.10984	0.10984	0.10984
C14	0.13316	0.13316	0.13316	0.13316	0.13316	0.13316	0.13316	0.13316
C21	0.09988	0.09988	0.09988	0.09988	0.09988	0.09988	0.09988	0.09988
C22	0.14187	0.14187	0.14187	0.14187	0.14187	0.14187	0.14187	0.14187
C23	0.12394	0.12394	0.12394	0.12394	0.12394	0.12394	0.12394	0.12394
C24	0.13054	0.13054	0.13054	0.13054	0.13054	0.13054	0.13054	0.13054

Because this is a very complex computational process, the manual operation is nearly impossible to complete. May using tool software and so on MATLAB solve, limit ultra matrix each line of non-zero values which obtains is the same. Each line of correspondence's value is opposite for various elements in the goal stable weight. Takes limit ultra matrix any row to obtain the final ordering vector.

5.5   F-ANP and the ANP Analyses Compare

We may see from the above example: Lies in the following two aspects based on the F-ANP computation regarding the wireless sensor node trust appraisal's merit: First, compares the AHP appraisal method, the F-ANP method has the ANP method merit, may respond that the wireless sensor network element group (for example performance attribute B1, security attribute B2) between as well as between various daughter elements' incident cross-correlation characteristic, Uses some node times the behavior evidence, uses F-ANP and the ANP method appraisal node trust value separately, and compares two methods the results, as shown in Figure 4. Two methods fully have both reflected the trust characteristic: "falls quickly, rises slowly". But comparatively speaking, F-ANP method curve change even smoother, stable.

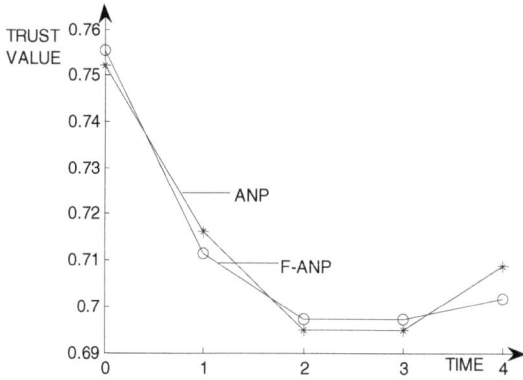

Fig. 4. Evaluation effects comparison between the ANP and F-ANP

6   Summary

This article proposes wireless sensor node trust appraisal based on the network analytic method (ANP) and the F-ANP. Using the ANP method analysis node each characteristic's between dependence and the feedback relations, establish the ANP model, uses two methods to establish the ultra matrix, through the example explained how two methods do apply in the wireless sensor network application, finally compares two methods the appraisal effects, proves the F-ANP method the superiority.

References

1. Ren, F., Huang, H., Lin, C.: Wireless sensor network. Journal of Software 14(7), 1282–1291 (2003)
2. Wang, L.: Network analytic method (ANP) theory and algorithm. Systems Engineering Theory and Practice 3, 45–50 (2001)
3. Sun, H., Tian, P.: Network-level analysis and scientific decision-making. Decision science theories and methods, pp. 3–8. Ocean Press (2001)
4. Van Laarhoven, P.J.M., Pedrycz, W.A.: Fuzzy extension of Statty's Priority theory. Fuzzy Sets and Systems 11, 229–241 (1983)
5. Guoxing, Z., Weisong, S., Julia, D.: A resilient trust model for WSNs. In: Proceedings of the 7th International Conference on Embedded Networked Sensor Systems, Berkeley, California, USA, pp. 411–412 (November 2009)
6. Lin, C., Wang, Y., Tian, L.: Development of Trusted Network and Chanllenges It Faces. ZTE Communications 6(1), 13–17 (2008)
7. Tian, L.Q., Ni, Y., Lin, C.: Node Behavior Trust Evaluation Based on Behavior Evidence in WSNs. ICFCC V1, 312–317 (2010)
8. Boukerch, A., Xu, L., EL-Khatib, K.: Trust-based security for wireless ad hoc and sensor networks. Computer Communications 30, 2413–2427 (2007)
9. Saaty, T.L.: Decision making with the analytic hierarchy process. Services Sciences 1(1), 83–96 (2008)
10. Tian, L., Lin, C., Ji, T.: A Kind of Quantitative Evaluation of User Behaviour Trust Using AHP. Journal of Computational Information Systems 3, 412–420 (2007)

Optimal Design of Weigh for Networks Based on Rough Sets

Baoxiang Liu[*] and Shasha Hao

College of Science, Hebei United University Tangshan Hebei United University
Liubx5888@126.com, shashaheut@163.com

Abstract. When the traditional rough neural network is structured, The selection of initial weights are random values between (0,1).This article address this issue, proposed an application of rough set theory attribute importance, replaced with the attribute importance method of initial weights. Finally, with instance validation, compared to the traditional rough neural network,This method is not only to accelerate the network convergence rate, but also enhances the adaptability of BP neural network.

Keywords: rough set, neural network, initial weights, attribute importance.

1 Introduction

Rough set theory is a novel, efficient method of soft science calculating, since scholars Pawlak who is from Poland in 1982 proposed the theory, it has become a powerful tool to describe the incomplete, inaccurate and noisy data. Rough set method [1] does not require the additional information given in advance,You can remove the redundant input, Simplify the input space of expression. The Algorithm is simple and easy to operate. Because of its powerful capabilities, it has been widely used in machine learning, pattern recognition, signal processing and other fields. Attribute reduction is one of the core, That is maintained without changing the classification ability, delete redundant attribute. Currently, the rough set has become a powerful tool of how to describe the incomplete dates, inaccurate and noisy data.

Neural network is a more mature artificial intelligence approach. With the deepening of research, the applications are more and more widely. However, the design of the network architecture, especially the determine of the initial weights is a bottlenecks problem that is encountered. It impact on its application in practice directly. Current neural networks have been widely used function approximation, pattern recognition, classification, prediction, data compression and other areas. But for the neural network to determine the initial weights, it has not been clearly defined theoretical. In this paper, shows the neural network method for determining the initial weights.

---

[*] Project supported by National Natural Science Foundation of China(No. 61170317) and Natural Science Foundation of Hebei Province of China(No. A2011209046).

2   About Rough Set and Neural Network

2.1   Rough Set Theory

Mathematician Z. Pawlak in Polish proposed rough set theory[6], it use Decision-making system to describe the problem, rough set methods and models based on a very intuitive decision   table based   on two-dimensional.   In   such a   two-dimensional decision table, Column means the property, line means the object. Each line represents a message of the object.

Definition 1. Quaternion $s = (U, A, V, F)$ is a knowledge representation system,

U : Non-empty finite set of objects $U = \{x_1, x_2, x_3, ..., x_n\}$ called the domain.

A : The non-empty finite set of properties $A = C \cup D$, $C \cap D = \varnothing$, C called the condition attribute, D called the decision attribute.

V : $V = \bigcup_{a \in A} V_a$, V_a is the range of property a;

f : $U \times A \rightarrow V$ is a information functions,Each of its object's property give information a value, that $\forall a \in A$, $x \in U$, $f(x, a) \in V_a$.

The data of knowledge representation system uses relational table to express. The row of the relational table correspond to the object of study, column correspond to the object's properties. The information of the object express its value through specifying the property value.

Definition 2. If $P \subseteq A$,Define can not distinguish relationship of attribute set P :

$$ind(P) = \{(x, y) \in U \times U \mid \forall a \in P, f(x, a) = f(y, a)\} \tag{1}$$

If $(x, y) \in ind(P)$, so that claimed x and y is indistinguishable of P.

Definition 3. R is a equivalence relation, $R \in$ **R,** if

$$ind \ (\text{R}) = ind \ (\text{R} - \{R\}), \tag{2}$$

call R is unnecessary in R; Otherwise call R is necessary in R. The so-called knowledge reduction, in the same conditions to maintain the ability of the knowledge of the classification. Delete irrelevant or unimportant knowledge.

Definition 4. In the decision-making system, the dependence of condition attributes relative decision attribute can defined as follows

$$\gamma_c(D) = \sum_{X \in U/D} \frac{card(\underline{C}(X))}{card(U)} \tag{3}$$

$card(X)$ means the cardinal number of the set X, $\gamma_c(D)$ measure the knowledge of the description of property, the objects of domain can correctly classified to the ratio of the corresponding decision class.

Definition 5. C   and   D   are   condition   attribute set   and decision   attribute set respectively, The importance of a subset of attributes $C' \in C$ about D defined as:

$$\sigma_{CD}(C') = \gamma_C(D) - \gamma_{C-C'}(D) \tag{4}$$

Specially when $C' = \{a\}$, The importance of property $a \in C$ about D defined as:

$$\sigma_{CD}(a) = \gamma_C(D) - \gamma_{C-a}(D) \tag{5}$$

2.2 BP Neural Network

BP algorithm is a supervised learning algorithm[7]. The main idea is: for q Input learning samples: $P^1, P^2, P^3, ..., P^q$, their corresponding output samples are $T^1, T^2, T^3, ..., T^q$,The purpose of learning is to use Network's actual output $A^1, A^2, A^3, ..., A^q$ to compare with the Target vector $T^1, T^2, T^3, ..., T^q$ to modify its weight, make A and T are as close as possible.

Specific training of BP network algorithm is as follows:
Initialization
Before starting training, all of the network values must be initialized. Usually set small initial random number.
Select the training samples to provide to the network

Randomly selected sample of a training x_j, O_k to provided to the network
The output of the hidden layer is:

$$y_i = f(net_i) = f(\sum_j w_{ij} x_j - \theta_i) \tag{6}$$

The calculated output of output node is as follows:

$$O_l = f(\sum_i T_{li} y_i - \theta_l) = f(net_l) \tag{7}$$

The error formula of output node :

$$E = \frac{1}{2}\sum_l (t_l - O_l)^2 = \frac{1}{2}\sum_l (t_l - f(\sum_i T_{li} y_i - \theta_l))^2$$
$$= \frac{1}{2}\sum_l (t_l - f(\sum_i T_{li} f(\sum_j w_{ij} x_j - \theta i) - \theta_l))^2 \tag{8}$$

Adjustment between the input layer to hidden layer weights and hidden layer of threshold units.
Adjustments weights and threshold between the hidden layer and the output layer.
Update the Learning samples.
Select the next training sample, return 3), continue to train the network, until all training samples are over.
Training cycle to train the network.

2.3   Traditional Rough Set and Neural Network Coupling Method

The traditional rough set and neural network[3] coupling method is rough set neural network as a front-end processor. Based on the ability to ensure the same classification, attribute reduction compressed space dimension, to achieve feature selection, simplify the training set, stream line the training set to build neural network and training the neural network. It can guarantee a good generalization ability and a relatively high accuracy.

Rough set analysis's reduction operation can remove the excess property. Reduce the size of the dataset. Without losing the original data set contains useful information, on one hand it can improve the representative of the data. Reduce the interference of the noise, the trained neural network is not easy to occurred the phenomenon. The other hand, it reduces the training data, to reduce the training time, and to Improve efficiency.

The united of the rough set and neural network, General steps are as follows:

Decision   table constructed from   the source   data,   using rough   set analysis to find the minimum reduction and core.

Reduce the size of decision tables.

Build the neural network on the data after reduction.

Using a variety of learning algorithm of the neural network to study the network.

3   The Determine of the Initial Weights in Neural Network Based on Rough Set

Because of the Learning process is to modify the network weights [9], and make the output error to achieve the required. Because the system is nonlinear, whether the initial value of learning is the local minimum, whether can weaken. If the initial value is too large, makes the weighted input activation out of the saturated zone. Result its derivative $f'(x)$ is very small. Because of δ proportional to the $f'(x)$, when $f'(x) \to 0$, there are $\delta \to 0$, so that $\omega \to 0$. Making the adjustment process almost to   a   halt,   so,   generally always want   the   initial weighted value of each neuron's output to close to zero. This can ensure the weight of every neuron to Adjust in their S-type activation function. So Initial weights is always Random number in [-1,1].

The criteria of learning [10] is that: if the net do the wrong decision, through the learning of the network, should reduce the likelihood of the network making the same mistakes. Such as: If the output of the network is correct, increased the connection weights, so that Network face the same input once again, can make right judgment. If the output is wrong, then adjust the network connection weights to the direction of the Reduction of the input weights. The aim is to reduce the possibility of making the same mistake next time.

Based on the above ideas, in this paper it gives the method of the initial weights between input layer and hidden layer.

1) Training sample data preprocessing;

2) Group the training samples, packet number and the number of the hidden layer neurons are the same.

3) Work out the attribute importance;

4) shine upon the attribute importance to the $(0, \sqrt[j]{S_1})$, as the initial weights between input layer and hidden layer of the neural network;

According to the above statement, based on the tradition rough set and neural network coupled, set the weights, proposed an algorithm that optimize BP neural network based on rough set.

Step1: Get data, the data is divided into two parts, one part as the training sample, the other as test samples;

Step2: Data pre-processing with rough set, data normalization;

Step3: Use matlab[2] as a training tool, setting all the parameters, train the training samples. Find the best number of the neurons, get the trained neural network;

Step4: Group the data, the number of the group is equal to the number of the hidden layer neurons, calculation the attribute importance degree of each set of data, as a initial weights of the neural network;

Step5: Re-train the neural network.

4   Experimental Verification

The Data of a place in southwest China's seismic data is as a sample source, achieve earthquake prediction based on neural network. Based on these seismic data, 7 predict Factor and actual magnitude as the input and target vectors. The predictor:

a: Cumulative frequency of earthquakes $M \geq 3$ in six months;

b: Release of energy accumulated value within six months;

c: b value;

d: Number of unusual earthquake swarm;

e: the number of seismic bands;

f: Weather it is in the active region;

g: Earthquake magnitude of the related earthquake zone.

17 data samples after process by rough set, randomly selected 10 samples as a training samples, another 7 as the test samples.

Table 1. Training samples

	a	b	c	d	e	f	g	h
X1	0	0	0.62	0	0	0	0	0
X2	0.3915	0.4741	0.77	0.5	0.5	1	0.3158	0.5313
X3	0.2835	0.5402	0.68	1	0.5	1	0.3518	0.5938
X4	0.6210	1.0000	0.63	1	0.5	1	1.0000	0.9375
X5	0.4185	0.4183	0.67	0.5	0	1	0.7368	0.4375
X6	0.2160	0.4948	0.71	0	0	1	0.2632	0.5000
X7	0.9990	0.0383	0.75	0	0	1	0.9474	1.0000
X8	0.5805	0.4925	0.71	0	0	0	0.3684	0.3750
X9	0.0810	0.0629	0.76	0	0	0	0.0526	0.3125
X10	0.3915	0.1230	0.98	0.5	0	0	0.8974	0.6563

Table 2. Test samples

	a	b	c	d	e	f	g	h
X11	0.0270	0.0742	0.62	0	0	0	0.2105	0.1875
X12	0.1755	0.3667	0.77	0	0.5	1	0.7368	0.4062
X13	0.4320	0.3790	0.68	0.5	0	1	0.2632	0.4375
X14	0.4995	0.4347	0.63	0	0	1	0.6842	0.5938
X15	0.6885	0.5842	0.67	0.5	0.5	1	0.4211	0.6250
X16	0.5400	0.8038	0.71	0.5	0.5	1	0.5789	0.7187
X17	0.1620	0.2565	0.75	0	0	1	0.4737	0.3750

The 1-7 is input factor of learning samples, actual magnitude is the output factor. Network trained with the data in the table. All the data has been deal.

Use the single hidden layer BP network to do the earthquake prediction. Because of that the input samples have 7-dimensional input vector, input layer has 7 neurons, for empirical formula:

$$m = \begin{cases} r+0.618(r-m), & (r \geq m) \\ m-0.618(m-r), & (r < m) \end{cases} \tag{9}$$

The hidden layer has ten neurons is the best. The network has only one output data, so the output layer has only one neuron. The network should be $7 \times 10 \times 1$.

Use the following code to create a BP network:

threshold=[0 1;0 1;0 1;0 1;0 1;0 1;0 1];
net=newff(threshold,[10,1],{'tansig','logsig'},'traingdx');
The enactment of the Training parameters is below:
The training code is below:
net.trainParam.epochs=1000;net.trainParam.goal=0.001;net=train(net,P,T);

Variable P and T are input vector and target vector, after 351 trainings, the target error of the network meet the requirements.

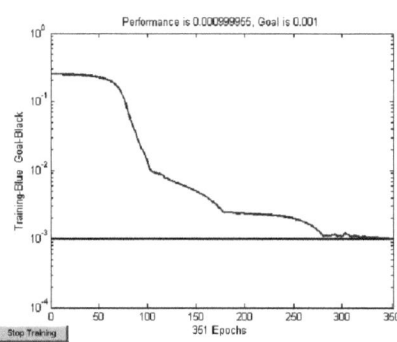

Fig. 1. The target error of the network meet the requirements

When the Network training is over, use another data to test the network:
Y=sim(net,P_test);
P_test is called the training data.
Output:
    Y= 0.2360 0.3865 0.3571   0.5433 0.6591 0.6807 0.3518
After anti-normalization, we get the magnitude prediction, compared with the actual magnitude, get the prediction error:

Table 3. The enactment of the Training parameters

actual magnitude	magnitude prediction	prediction error
4.4	4.5550	0.1550
5.1	5.0369	0.0631
5.2	4.9626	0.2374
5.7	5.5387	0.1613
5.8	5.9393	0.1393
6.1	5.9782	0.0218
5.0	4.9258	0.0742

Put the attribute importance into the network, The Step is the same as the above, after 275 Training, Training achieves the requirements error:

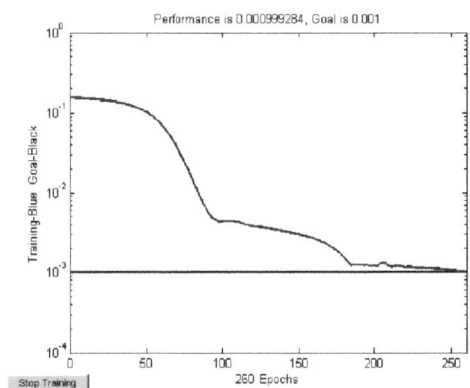

Fig. 2. Training achieves the requirements error

Using the same test data to test, get the output:

Table 4. Prediction error

actual magnitude	magnitude prediction	prediction error
4.4	4.4540	0.0540
5.1	5.0810	0.0190
5.2	5.0026	0.1974
5.7	5.6438	0.0562
5.8	5.9747	0.1747
6.1	6.0431	0.0569
5.0	4.9283	0.0717

Experimental results show that: Compared to the traditional rough neural network, the method of given weight can improve the training speed.

Rough set theory provides a new approach for the optimization of neural network weights. The importance of the condition attributes takes the place of the initial weights of the neural network. On the basis of how to improve the training speed of neural network, and improve the prediction accuracy.

References

1. Hai, J.: Neural network theory, pp. 191–209. Mechanism Industry Press, Beijing (2009)
2. Zhou, K., Kang, Y.: Neural network model and Matlab programmer, pp. 140–148. TsinghuaUniversity Press, Beijing (2004)
3. Wu, Y., Zhang, L.: The generalization ability of neural networks and structural optimization algorithm. Computer Science 19(6), 21–25 (2002)
4. Zhao, W., Huang, D., Ge, Y.: Application of genetic algorithms to the structure optimization of radial basis probabilistic neural networks. IEEE Transactions on Neural Networks 12(1), 1243–1246 (2002)
5. Pawlak, Z.: Rough Set-Theoretical Aspects of Reasoning about Data, pp. 9–30. Kluwer Academic Publishers, Dordrecht (1991)
6. Chen, W.C., Chang, N.B., Chen, J.C.: Rough set-based hybrid fuzzy-neural controner design for inudsrtial wastewater treatment. Water Resaerch 37(1), 95–107 (2003)
7. Wanfu, Z., Shijun, Z.: The Structural Optimization for Neural Network Based on Rough Sets. Computer Engineering and Design (9), 4210–4212 (2009)
8. Wang, H.: A Study on the Fusion of Granular Computing Based on Rough Set and Neural Networks. Computer Engineering and Design (8) (2010)
9. Liu, B.: The Research of Rough Set on Data Preprocessing; Baldonado, M., Chang, C.-C.K., Gravano, L., Paepcke, A., Liu, B.: The Research of Rough Set on Data Preprocessing. The Stanford Digital Library Metadata Architecture. Int. J. Digit. Libr. 1, 108–121 (1997)
10. Zhou, C.: Attribute reduction based on rough set. Hefei University of Technology (2008)

Campus Type Network Virtual Community System

Gong Dianxuan[1], Wei Chuanan[*,2], Wang Ling[1], and Peng Yamian[1]

[1] College of Sciences, Hebei United Universtiy, Tangshan, 063009, China
[2] Department of Information Technology, Hainan Medical College, Haikou 571101, China
dxgong@heuu.edu.cn

Abstract. Along with the development of network technology, people can get unlimited number of network resources. To effectively use and manage these resources, many kinds of so-called "virtual community" internet platform appear.This paper introduces a campus type network virtual community system, mainly discusses the design and the most important part of its realization. With the platform of application system based on web, using the at present the most popular Java programming language, the structure is build on the J2EE platform.

Keywords: J2EE platform; design structure, MVC model, AJAX asynchronous request.

1 Introduction

With the development of network technology, people can get unlimited number of network resources. However, how to most effectively use and manage these resources becomes a big problem. Many kinds of network platform structures, mostly with the so-called "virtual community" forms arise at the historic moment. In addition to web portal, the professional BBS and wikis platform, government agencies, education institutions and unit, the company and freedom group, have set up their own information collection and management system, the purpose is to the more easily control the latest internal information and dynamic in time, accumulate precious experience, disclose the innovation, facilitate the improvement of members and obtain the maximization of decision support.

The school is the most appropriate to set up virtual community units, can be refined to discipline. Because the school has stable user groups, against commercial influence; and the school is the most suitable for access to the latest technology reform and development of the technology for network, and the adapt cycle is short; The teacher and students is of high dependency and research interests to network technology. So building practical, friendly type campus network virtual community is very popular. However, unfortunately, a lot of schools have not achieved a similar environment. In fact, there are a lot of knowledge wealth has lost, although

---

* Corresponding author.

B. Liu and C. Chai (Eds.): ICICA 2011, LNCS 7030, pp. 529–536, 2011.

they originally belong to oneself. The most advantage of open network platform is it can accumulation all the strengths of people, make study easy to get help, lead the organization more positive. Members can be very convenient to record and management their learning achievement, can communicate fast on the network.

This paper is organized as follows: in section 2 is the system uses and technical background; the system design and code realization is in section 3; the last section is conclusion.

2   Technical Background and System Structure

2.1   Technical Background

The platform implement the most popular Java programming language based on web application system. The structure is built in Java language on the basis of the J2EE platform. The J2EE platform is an excellent open source technology to achieve this kind of application system. Short for Java 2 Platform Enterprise Edition, J2EE is a platform-independent, Java-centric environment from Sun for developing, building and deploying Web-based enterprise applications online. The J2EE platform consists of a set of services, APIs, and protocols that provide the functionality for developing multitier, Web-based applications. For short, J2EE is a system structure based on Java 2 platform to simplify enterprise solution and development, deployment and management related complex problems. The foundation of J2EE technology is the core of Java platform or Java 2 platform of the standard edition, it not only consolidates many advantages from the standard version, but also gives overall support for EJB (Enterprise JavaBeans), Java Servlets API, JSP(Java Server Pages) and XML technology.

At present the J2EE is very popular, and is widely used from Banks, securities system, to the enterprise information platform, and even some small company. In the beginning, the J2EE is very noble product, at that time EJB is used as the core of the J2EE , development and Deployment costs is high, the cost is high, the learning curve of the developers is steep. Nowadays, micro edition J2EE technology make itself become popular. Micro edition J2EE technology is simplified form classic J2EE technology, it keeps the structure, good scalability and the maintainability of the J2EE technology; and simplified the development of J2EE technology and reduced the deployment cost of J2EE technology.

Standard edition J2EE technology with rigorous architecture has many advantages and excellent distributed architecture. It can be called a programming art. But it is too luxurious that limits more people use it. Micro edition J2EE technology made the J2EE platform at a very fast speed occupied the market of various kinds of information platform such as electronic commerce, the electronic government affairs, etc.

This system is built on three frameworks: struts, hibernate and spring.

2.2   Overview of System Design

System Module Chart

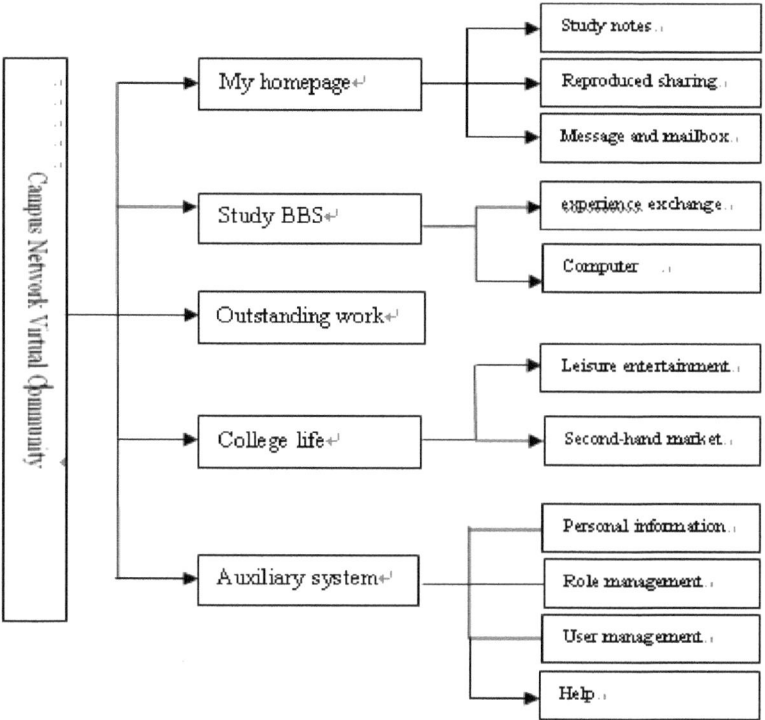

Fig. 1. System Module Chart

Explain of Module Chart

The system is divided into five modules, nether "My home page", "Study BBS", "Outstanding works", "College life" and "Auxiliary system".

"My home page" module has three more function point, respectively is "study notes", "Reproduced sharing" and "message and mailbox". Here, students can easily summarize the records on important content of course or experience, check the past records, create and management course, can search interesting or excellent works from other members, also can send or receive text messages.

"Study BBS" provide a public communication platform, students can exchange learning experience, launch questions, help answer questions, sharing and seek literature etc.

"Outstanding works" accumulation excellent resources and classify them according to different subject.

"College life" pays attention to the students' learning and life demand. "Leisure entertainment" with many interesting things such as puzzle games and problems could provides students a good place to relax. Second-hand market serves as an information platform helping students realize second-hand business.

"Auxiliary system" responses for users and administrators to manage their information and members, such as password revision, permission setting and user information change. Help student and teachers to make full use of this system.

Of course, all modules have administrators.

3   System Design and Code Realization

This section mainly explains the basic design method of functional module. Take "my homepage" as the example, introduce the process.

3.1   Database Design

The purpose of the application system based on web is to store and display public information, so the database is the most important basis, and its design is critical especially in big and frequently visited system.

The design of the structure of this system is not complicated, database structure and the description of the relationship is as follows:

Database: Mysql 5.0.

Database name: community.

Database table: article, bbsarticle, book, classtype, menu, message, mydoc, person, role, rolemenu.

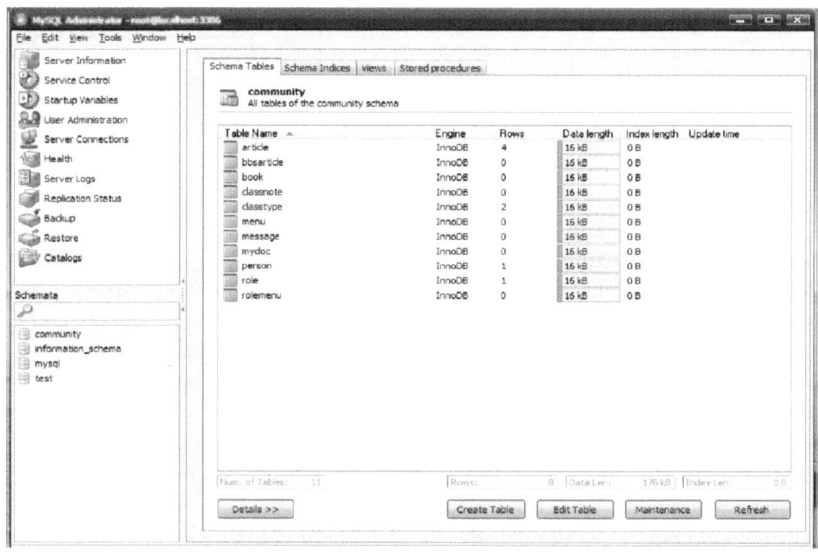

Fig. 2. The Database

Now we show two database table:then have the mapping relation one to more,namely an article belong to a category, a category includes several articles.

Table 1. Database table: article

Column Name	Datatype	NOT NULL	AUTO INC	Flags	Default Value	Comment
id	VARCHAR(255)	Y		BINARY		ID
authName	VARCHAR(255)	Y		BINARY		Auther name
title	VARCHAR(255)	Y		BINARY		Title
content	VARCHAR(255)	Y		BINARY		Content
sContent	VARCHAR(255)	N		BINARY	null	Abstract
isDeleted	VARCHAR(255)	Y		BINARY		Delete sign
createdBy	VARCHAR(255)	Y		BINARY		founder
updateBy	VARCHAR(255)	Y		BINARY		modifier
createDate	DATE	Y				Creation date
updateDate	DATE	Y				Updata date
modify	VARCHAR(255)	Y		BINARY		Number of change
isOpen	VARCHAR(255)	Y		BINARY		Open sign
classType	VARCHAR(255)	Y		BINARY		category

Table 2. Database table: classtype

Column Name	Datatype	NOT NULL	AUTO INC	Flags	Default Value	Comment
id	VARCHAR(255)	Y		BINARY		ID
classTypeName	VARCHAR(255)	Y		BINARY		Category
classTypeNm	VARCHAR(255)	Y		BINARY		Category number
isDeleted	VARCHAR(255)	Y		BINARY		Delete sign
createdBy	VARCHAR(255)	Y		BINARY		founder
updateBy	VARCHAR(255)	Y		BINARY		modifier
createDate	DATE	Y				Creation time
updateDate	DATE	Y				Update time
modify	VARCHAR(255)	Y		BINARY		Namber of change

3.2 Framework Design

This system is realized by joint the three framework struts, hibernate and spring.

Struts was responsible for the display of the data collection, submit layer, return to show the client. It realized the MVC structure and is widely used. The details is shown in Fig. 3.

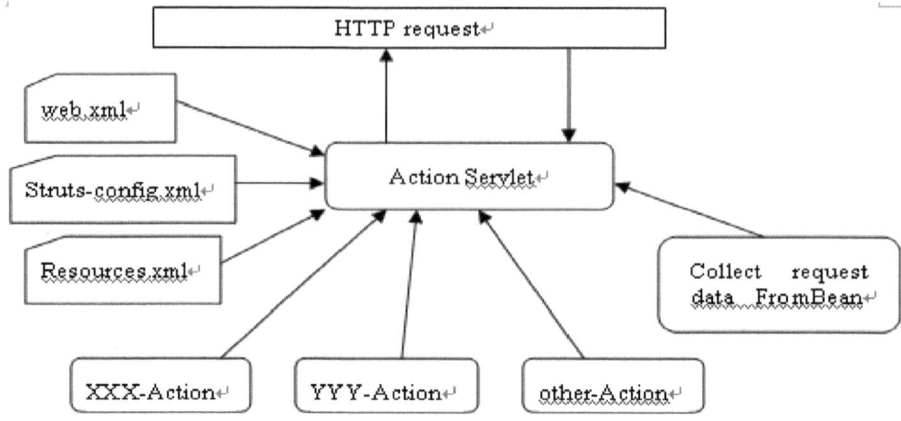

Fig. 3. Chart for struts

Hibernate encapsulating the operations related to database make programmers can control the objects directly and greatly simplified the operation of SQL. Spring manages the relationship between objects.

3.3   Code Design

The default start page is "my home page" when a user enter the system, and users need to input the correct user name and password, the so-called identity authentication. Different user can access different permissions of resources, and must let system accurately know if users are still in the login state, a process to realize this can be show in following Fig. 4.

In the code, the user relevant information Person class is saved in Session class. It is convenient for system to get the user information at anytime and log records. Session class is managed by web container so that a single user will only produce one example which is the single mode.

Code design from request received to page update: First, index_live.js loads the default page and defines some other JavaScript response function of events. Function listNote is in file myMainPage.js takes asynchronous interactive way AJAX to update page as other system request. Detail process can be seen in Fig. 5.

4   Conclusion

In this paper, we mostly use the chart to show the design idea of structure, request processing and code. We believe the design is the most important and it determines the difficulty of maintenance and extension.

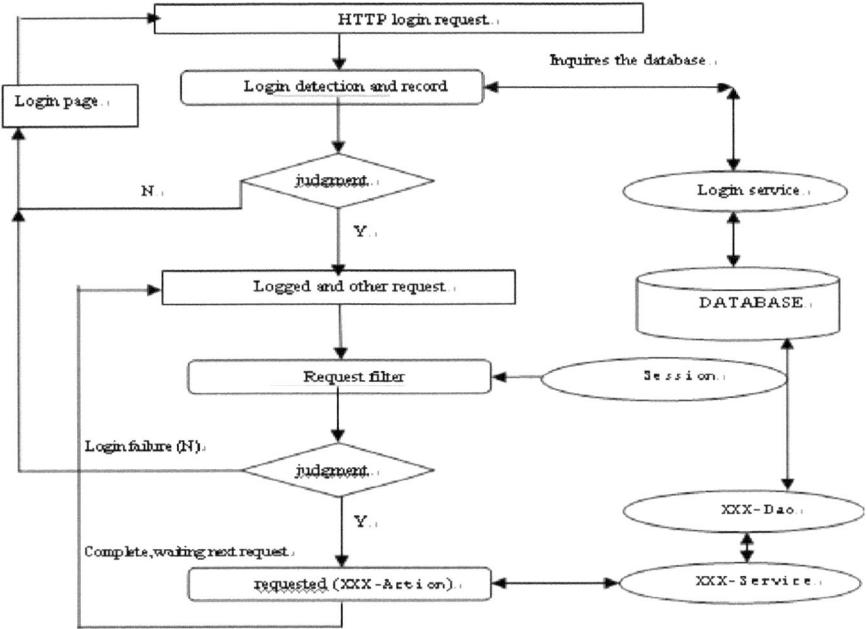

Fig. 4. Chart for Code Design 1

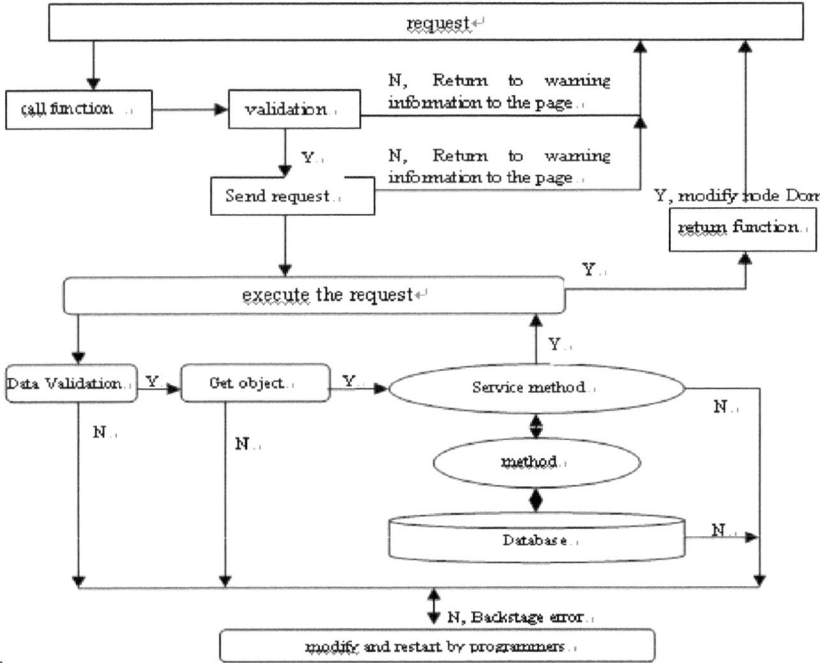

Fig. 5. Chart for Code Design 2

Acknowledgments. This job is supported by the Educational Commission of Hebei Province of China (Grant No. Z2010260).

References

1. Duan, X.G.: Design and Implement the Information Management System of State Key Laboratory Based on Java Web. Master's degree thesis. Hunan Universtiy, Hunan (2005)
2. Zhang, Y.L., Wang, Z.D.: The Survey of Network Performance Test Tools. Computer Engineering and Applications 21 (2002)
3. Cassioe, G., Elia, G., Gotta, D., et al.: Web Performance Testing and Measurement: a com approach, http://www.stickyminds.com/
4. Lin, S.J., Lin, K.S.: Handbook of JSP.2.0 Technical. Publishing House of Electronics Industry, Beijing (2006)
5. Lin, S.L., Wang, K.R., Meng, H.L.: Java optimization programming. Publishing House of Electronics Industry, Beijing (2007)
6. Li, J.: Discuss on design idea by Java web. China Computer & Communication 10 (2010)

The Establishment and Research of a Neural Network-Based Evaluation Model for Chinese Universities' Online PE Courses

Yunzhi Peng

Physical Education Institute, Hunan University of science and technology,
411201 Xiangtan, China
doctor2000@yeah.net

Abstract. Purpose: Applying principles of BP neural networks to the evaluation of the teaching quality of online PE courses at Chinese universities, proposes a neural network-based model for evaluating the online PE teaching quality. Method: To integrate the theory concerning online PE course development with the practice of Hunan Normal University's online PE courses teaching to make a empirical analysis. Result: It also brings forward a step size algorithm that can be implemented within the MATLAB environment. Conclusion: The application of artificial neural networks to online PE teaching quality evaluation not only avoids the elements of subjectivity from experts in the evaluation, but also brings about satisfactory results. Thus, this approach has wide applicability.

Keywords: Establishment, Neural Network-based, Evaluation Model, Online PE Courses.

1   Introduction

For the past few years, the Chinese government has paid more and more attention to the important role of educational technology in the promotion of educational modernization. During the 9th Five-Year Plan period, the Central Committee of CPC and the State Council of China held the third National Education Work Meeting, issued Decision on Deepening Educational Reform and Promoting Quality Education and made special elaboration on the educational technology and the application of information technology in education. The Ministry of Education stressed at national education conferences that the reform of teaching content, curricular system, teaching methods should rely on the platform of educational technology and also pointed out that we should be fully aware of the role of educational technology. "Construction Project" was then officially launched.[1] On April 29, 2001, Joint Research and Development Center for Modern Distance Education was formally founded. The center was preparing to provide a series of systematic online courses which covers middle school, high school, undergraduate and postgraduate education.[2] As can be seen from the above-discussed, the issue of these policies and technical standards has brought opportunities as well as challenges to the development of online PE courses.

B. Liu and C. Chai (Eds.): ICICA 2011, LNCS 7030, pp. 537–544, 2011.

Developing online PE courses implies a challenge to the traditional physical education. It also means the temporal shrinkage and the spatial extension of physical education and brings about a trend that the legitimately unified development of physical education is becoming pluralistic. Nowadays, with rapid increases in the number and scale of online PE courses, quality has become the main constraint for the further development of online PE courses, This issue has gained increasing attention. [3] The basic features of artificial neural networks include nonlinear mapping, classification of learning and real-time optimization, and thus artificial neural networks have opened up a new way for research into pattern recognition and nonlinear classification.[4] Compared with traditional teaching methods, the application of online course technology has greatly improved teaching effects. And a scientific evaluation system should not only effectively promote courseware development but also contribute to the application of courseware and the improvement of teaching qulity. The paper adopts the theory and methods of artificial neural network to achieve the overall evaluation of the online PE courses at colleges and universities, aiming at using the quality management theory to standardize online PE course development and establishing a new system for evaluating and managing online PE course quality.

BP(Back Propagation): neural network, also called multilayer feedforward neural network, has many characteristics. For example, there are only connections between neurons and those on the adjacent layers, there are no connections among neurons on the same layer, there are no feedback connections among neurons, and a highly nonlinear mapping relationship exists between the input and the output. If the number of input nodes is n and that of output nodes is m, then the BP network is the mapping from n-dimensional Euclidean space to m-dimensional Euclidean space.[5]Therefore, the normalized indices of the online PE teaching quality evaluation are used as input vectors of the BP network model, and the evaluation result is used as the output of the BP network model. Then enough samples are used to train the network to gain experience, knowledge, subjective judgments and inclination towards the importance of the indices from evaluation experts. Thus, the weight coefficients that the BP network model possesses are the correct internal knowledge representation obtained through adaptive learning. Based on the attribute value of each index of the online PE teaching quality evaluation, we can obtain the results of the online PE teaching quality evaluation. [6] In accordance with experts' experience, knowledge, subjective judgments and their inclination towards the importance of the indices, we can achieve the effective combination of qualitative analysis with quantitative analysis and guarantee the consistency and objectivity of the evaluation.

2   Experimentation

2.1   The Philosophy and Guiding Principle in Online PE Courses Design

Online course design is a process by which the perfect and harmonious integration of education, technology and art is achieved. [7] When designing online courses, we need to integrate educational theory, artistic forms, and information technology means together, taking into consideration the permeation of educational theory into all phases of online course teaching design. [8] In addition, we also need to ensure that the online courses' interface will demonstrate artistic features and that the advantages of technical means will bring about increases in teaching effectiveness and efficiency. We can invent a virtual space model for online course design(see Figure 1), a model in which education represents the X-coordinate, technology the Y-coordinate, and art the Z-coordinate. In such a three-dimensional model for online course design, when deciding which aspect(s) should be highlighted, we need to pay attention not only to teaching objectives, but also to actual conditions such as funding and capacities of different developers. [9]

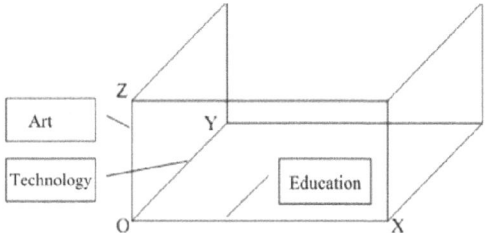

Fig. 1. The 3-Dimensional graph of education, technology and art

Since in PE courses there are a number of events having artistic qualities, many people take part in these events for the pursuit of beauty. This is why we should pay more attention to the artistic quality of online PE courses, with particular attention to the integration of art with the interface design of online courses. [10] We should present the teaching content and resources in an artistic way in order to stimulate and maintain students' interest and confidence in PE learning. Artistic expression involves the overall layout of the interface, the use of point, line and plane, font using, color matching, sound quality, etc. The use of multimedia technology has greatly improved the expressive power of the artistic design of online courses. For example, designers should pay attention to the sizes of elements like images and abstract graphics, animations, three-dimensional images, and texts and their locations on the screen. Other elements, like the pace of alternate switching between images and animations or video clips, the brightness and cold and warm colors of images and the background, and shapes, sizes, positions of links, buttons and hotspots, should also be taken into account. Designers should produce vivid, harmonious, and aesthetic interfaces in accordance with different subjects, and ensure that the courses can demonstrate both changes and consistency in terms of rhythm and rhyme, so that learners can have a relaxing and pleasant aesthetic experience in the world of online learning, and acquire much knowledge and information within a relatively short period of time. [11]

Network technology, mainly as an objective factor, serves as a means and a support environment of educational communication. Artistic originality mainly functions as a subjective factor. Course designers should take the initiative to utilize existing theories of network technology, and pay attention to the tight combination of technology and art. Only in this way, can we enjoy the benefits of technology, realize all kinds of artistic imagination, and meet learners' needs for high-quality online courses. To motivate students' enthusiasm for learning, we must adhere to the people-oriented principle and maintain the attractiveness of online teaching. [12] All things, including the interface layout, knowledge formulating, questions and exercises design, should be given full consideration so that we can create a challenging but delightful learning environment for students.

2.2   The Establishment of a Neural Network-Based Index System

The quality evaluation of online PE courses is a process by which people analyze those factors that influence online PE teaching quality and rank them in accordance with the extent of each factor. Therefore, from the overall perspective of online PE teaching, we can establish the following index system. (see Figure 2)

Figure 2 Overall perspective of online PE teaching index system

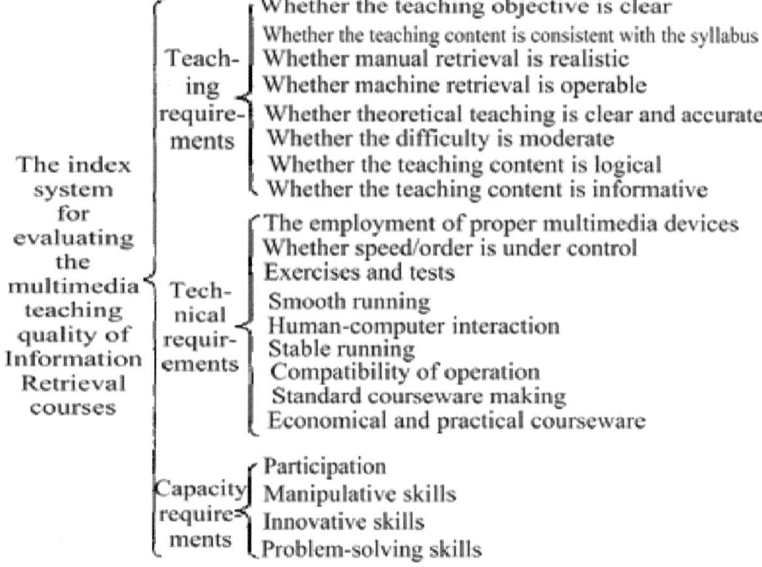

Fig. 2. Overall perspective of online PE teaching index system

The online version of the volume will be available in LNCS Online. Members of institutes subscribing to the Lecture Notes in Computer Science series have access to all the pdfs of all the online publications. Non-subscribers can only read as far as the abstracts. If they try to go beyond this point, they are automatically asked, whether they would like to order the pdf, and are given instructions as to how to do so.

Please note that, if your email address is given in your paper, it will also be included in the meta data of the online version.

3   BP Neural Network-Based Model for Managing Online PE Course Quality

In the course of PE teaching quality evaluation model establishment, the choice of network model structure is a very important job. A good network model structure can reduce the network training times and improve the accuracy of online learning.

a. Determine the number of neurons of the input layer. According to the index system, there are 21 indices that influence the multimedia teaching quality of Information Retrieval courses. Therefore, the number of neurons on the input layer is 21, namely n=21.

b. Determine the number of neurons of the output layer. The evaluation result is used as the output of the network, and therefore the number of neurons of the output layer is 1, namely m=1.

c. Determine the number of hidden layers. The more hidden layers there are, the slower the learning speed of the neural network is. According to Kolmogorov's theorem, under a reasonable structure and proper weight conditions, a three-layer BF network's approximation of any continuous functions can be realized. Therefore, we choose a three-layer BF network that has a relatively simple structure.

d. Determine the number of neurons of the hidden layer. Generally speaking, the number of neurons of the hidden layer is determined in accordance with the network's convergence performance. After summarizing a large number of network structures, we obtain an empirical formula: s=0.43nm+0.12m2+2.54n+0. 77m+0. 35+0. 510. According to the above formula, the number of neurons on the hidden layer is 8, namely s=8.

e. Determine the transfer function of neurons. The transfer function of network neurons goes like

$$f(x) = \frac{1}{1 + e^{-x}}$$

f.   Determine the model structure. BP neural network model structure, as is shown in Figure 3, is determined according to the above results.

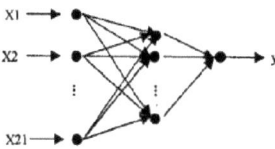

Fig. 3. BP Neural Network Model Structure

3.1   BP Neural Network Learning Algorithm

a. Initialize network and learning parameters, add a random number in θ[-2/n,2/n] to connection weights wij, wj and threshold θj, i=1,2,...,n; j=1,2,...,s.

b. Choose the training data of a sample pair xp=[x1,x2,...,x21] as input vectors, yp as desired output.

c. Calculate the output of neurons of each hidden layer with the input vector xp, the connection weight wij and the threshold θj.

$$y_j = \cfrac{1}{1 + \exp\left|-\left|\sum_{i=1}^{n} w_{ij} x_i - \theta_j\right|\right|}$$

i=1,2,...,n; j=1,2,...,s

d. Calculate the output of neurons of the output layer with the hidden layer output yi, the connection weight wj and the threshold θ.

$$y = \cfrac{1}{1 + \exp\left|-\left|\sum_{j=1}^{s} w_j y_j - \theta\right|\right|}$$

e. Calculate the alignment error of the output layer with the desired output yp and the actual output of the network y.

$$\sigma = (y_p - y)y(1 - y)$$

f. Calculate the alignment error of the hidden layer with wj, σ, yj

$$\sigma_j = y_j(1 - y_j)\sigma w_j$$

g. With wj, σ, yj and θ, calculate the new connection weight between the next hidden layer and output layer and the threshold of output neurons.

$$\theta(t+1) = \theta(t) + \eta(t)\sigma + \alpha[\theta(t) - \theta(t-1)]$$

$$\theta(t+1) = \theta(t) + \eta(t)\sigma + \alpha[\theta(t) - \theta(t-1)]$$

$$\eta(t) = \eta\left|1 - \frac{t}{T+M}\right|$$

In the above formulas, η(t) represents the step size, α the momentum coefficient which values from 0 to 1 and normally 0.9; η represents the initial step size; t represents learning times; T represents iteration times; M is any positive number.

h. With σj, xi, wij and θ, calculate the new connection weight between the next output layer and hidden layer, that between hidden layers, and the neuron threshold of the hidden layer.

$$w_{ij}(t+1) = w_{ij}(t) + \eta(t)\sigma_j x_i + \alpha[w_{ij}(t) - w_{ij}(t-1)]$$

$$\theta_j(t+1) = \theta_j(t) + \eta(t)\sigma_j + \alpha[\theta_j(t) - \theta_j(t-1)]$$

i. Obtain allowable errors after iterative calculations. Choose another sample pair for training and repeat the above algorithm until all the samples have received training and the neural network evaluation model has been established.

3.2 The Realization of a Neural Network-Based Model for Evaluating Online PE Teaching Quality

We use MATLAB, a high-performance visual software environment for numerical computation, to build a three-layer BP neural network. Within this network, there are

21 neurons of the input layer, 1 neuron of the output layer, 8 neurons of the hidden layer, 1000 learning times and 600 iteration times. Also, the initial step size is 0.9, M values 400, the momentum coefficient is 0.9, and the allowable error is 0.001. We choose 10 pairs of sample data and 10 pairs of test data, and adopt the MATLAB neural network for learning training. Finally, we establish a neural network-based evaluation model. After verifying the test data, we obtain some satisfactory results. The comparison of these test results with expert evaluation results is shown in Chart 1.

No.	Test value	Expert eva-luation value	Class
1	0.406	0.4	pass
2	0.411	0.4	pass
3	0.607	0.6	fair
4	0.213	0.2	poor
5	0.810	0.8	good
6	0.591	0.6	fair
7	0.789	0.8	good
8	0.611	0.6	fair
9	0.991	1.0	excellent
10	0.588	0.6	fair

Chart. 1. The Comparison of Test Results with Expert Evaluation Results

4 Conclusion

Results Through discussing and analyzing this topic, we can draw following conclusions as to how to establish a system for managing online PE course quality.

Given sharp increases in the number and scale of online PE courses, the quality management of online PE courses should not simply rely on universities' administrative and summative assessment.

The application of neural networks to online PE teaching quality evaluation avoids the elements of subjectivity from experts, and thus is widely applicable. This approach, by employing a variable step size algorithm, has dramatically improved the performance of network convergence and the training efficiency, and thereby can bring about satisfactory evaluation results.

Online PE course development must follow a certain pattern under which the participation of the school staff and close collaboration among relevant departments are indispensable. Universities should strive to develop a student-oriented quality management philosophy.

The quality of online PE courses depends on a university's efforts to control the whole process of course development. To control the course development process, we employ practices used in the project management process, which includes project management, planning management, stage management and project delivery.

We must ensure that functional departments have assumed their responsibility on all stages of the project. At the same time, we need to make sure that major parts, including responsibilities, resources, curriculum implementation, analysis and improvement, and much smaller parts within these major parts are under control.

References

1. Zhang, B.: Study on the Development Strategies of Education. Tianjin People's Publishing House, Tianjin (2003)
2. The Secretariat of National Quality Management and Quality Assurance Standardization Technical Committee (editing), Comprehension and Implementation of the National Standard of Quality Management System(2000 Version). Standards Press of China (2002)
3. Xie, Y., Ke, Q.: Development and Application of Network Courses. Electronic Industry Press, Beijing (2005)
4. Dong, Y., Huang, R.: Establishment and Revise of the Accreditation Standards for the Courseware Quality of Network Courses. J. Electronic Education Research 6, 65 (2003)
5. Chen Jiang, F., Yao, S.: Application of Virtual Reality Technology in the Technical Simulation of Sports. J. Exercise and Sports Sciences 9, 36 (2006)
6. Peng, G., Shen, W., Tang, Y.: Study on the Establishment of the Evaluation System and Standards for the Multimedia CAI Courseware of the Sports Technology Courses. J. Exercise and Sports Sciences. 10, 51–53 (2004)
7. Zeng, Q., Deng, S., Lu, Y.: Network Course Design of Health Education. Journal of Beijing Sport University 28, 100–110 (2005)
8. Zhang, S., He, J.: Research on Optimized Algorithm for BP Neural Networks. Computer and Modernization 161, 73–75 (2009)
9. He, Y.-h., Zhou, H.: Research of Multi-source Information Data Fusion Technology Based on Artificial Neural Network. Compter Knowledge and Technology 5, 149–150 (2009)
10. Liang, L., Huang, Y.-q., Zhang, X.-q.: Computer Network Performance Evaluation Based on Neural Network. Computer Engineering 36, 105–106 (2010)
11. Zhang, H., Wu, Y.: Research of Multimedia Communications Control Mechanism Based on Neural Network. Journal of Computer Applications 29, 16–18 (2009)
12. Ba, J., Yu, H.: Optimization of Neural Network with Fixed-point Number Weights and Its Application. Journal of Computer Applications 29, 230–233 (2009)

Verification Tool of Software Requirement for Network Software

Tao He[1], Liping Li[2], and Huazhong Li[1]

[1] Software Engineering Department, Shenzhen Institute of Information Technology,
Shenzhen, China
[2] Computer and Information Institute, Shanghai Second Polytechnic University,
Shanghai, China
he_tao@foxmail.com, llping2000@yahoo.com.cn, lihz@sziit.com.cn

Abstract. A model checking tool OWLSVerifyTool is proposed, designed and developed to verify Web service composition model of software requirement in this paper. It can convert OWL-S documents into Petri nets document described in PNML, then analysis and verify it with Petri nets with engine in dynamic context. Compositing the DL reasoning engine Pellet and F-logic-based reasoning engine Flora-2, it can play their respective advantages to reason and verify static model in static context of software requirement. The automated validation tool can effectively verify software requirement meta-model based on Web service described with OWL-S.

Keywords: Web Service, Software Requirement, Verification Tool.

1 Introduction

The network software disposed in network environment is a kind of special ultra-large service-oriented computation of complex software system. Requirement engineering of Network software are facing many problems at present [1, 2], because of its dynamic topology, uncertainty of users, and its continuous increasing requirements. However, present software requirement modeling and verification technique lack enough support to service-oriented computation, and are unable to gather the Web service resources in the network effectively, provide high dependable Web service resources to enhance development of information system efficiency [3].

Requirement verification is an important process in software requirement engineering. If without it, the project may lead to be unsuccessful. The design and development of verification tool is very important, for its enhancing efficiency of verification, improving software development process, and guarantee software quality. Semantic Web service language has carried on the clear description to Metamodel various levels of software requirement, and carries on the formalized modeling by Petri net and F-logic, and verifies the uniformity of Web service semantic restraint.

Verification of Web service combination has two kinds of research methods at present: verification based on work flow BPEL4WS and based on semantic OWL-S. Now regarding the latter there are few research achievements. In literature [4] for

B. Liu and C. Chai (Eds.): ICICA 2011, LNCS 7030, pp. 545–552, 2011.

controls flow and the data flow on the modeling, it transforms directly the OWL-S process model into the simpler Promela modeling, and verifies it with SPIN. However, its data flow modeling is too simple to verify whether the input /output type match.

This paper has designed OWL-S model examination prototype tool OWLSVerifyTool. It takes OWL-S storage documents (*.owl or *.xml) as the input, simultaneously carries on dynamic model and static model verification. The dynamic model, transforms with the document format switch to the PNML document, then directs or transforms the PNML document to corresponding form, then input to Petri net verification DiNAMiCS and Tina engine to verify it. The static model, combining description logic DL inference and Flora-2 rule, takes OWL-S storage documents(*.owl or *.xml) as input, simultaneously unifies DL inference (Pellet) and the Flora-2 rule to carry on inference alternately, carries on the analysis verification of the static model, and output the results.

2   General Organization of Verification of Verification Tools

OWL-S verification tool OWLSVerifyTool is mainly composed of dynamic model verification and static model verification modules. The designing structure of verification tool is shown in Fig. 1.

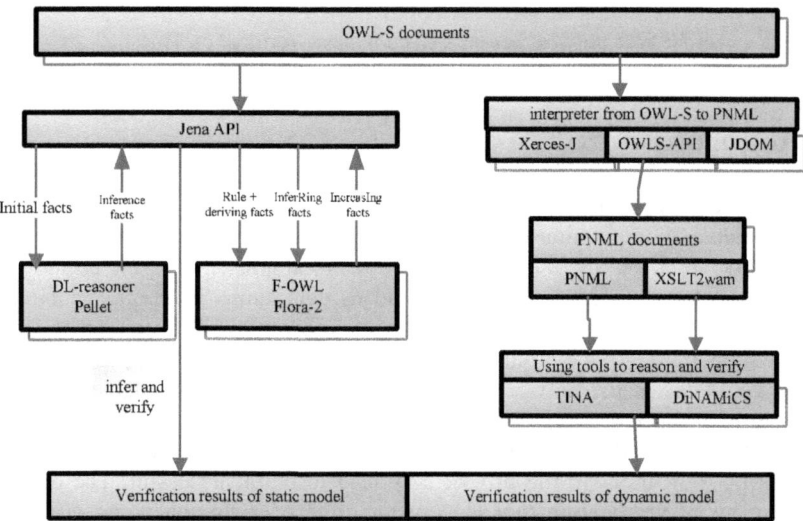

Fig. 1. General architecture diagram of validation tool

OWL-S is the Web service frame described in OWL language, but OWL represents ontology by class, attribute, value, and concept relations and so on. When describing

RGPS software requirement, it often uses essential factor and relational of standard ontology, a series of axioms as well as the formal restraint semantic to build its model. This paper calls it "static model". A static model, often not evolving, can only express invariant and special condition. formalism methods such as Z language, DL, VDM, F-logic, which are based on set theory and the first-order predicate calculus, may use for modeling it. This paper uses DL and F-logic. To verification of static model, its realization of definition standard, accuracy, uniformity and completeness of inference must be considered.

In the static model verification module it transforms the OWL-S documents through Jena API. First, inputs it with rules to the Pellet engine to infer for obtaining the new fact, then combine the new fact and the original fact, transforms it to the form that the engine Flora-2 can accepts. Eventually, inputs it with rules to engine Flora-2 to infer and verify. The output is the verification result of static model.

Uses the Petri net in the dynamic model aspect to take the verification model, and carries on the verification with DiNAMiCS and Tina engine; mainly verify its accuracy (activeness, boundedness), Reachability, final state, security, and so on.

As shown in Fig. 1, first inputs OWL-S documents in unified user interface. Transforms the OWL-S documents into the PNML documents in the dynamic model verification module with interpreter respectively, simultaneously carries on the PNML documents verification. Because the input form of DiNAMiCS engine is different from PNML slightly, which is the wam form, first transforms it into wam form by XSLT, and then inputs it to DiNAMiCS to carry on inference and model checking. Tina may directly input PNML form documents to carry on it. After verification by two engines, merge the results to obtain dynamic model verification result of the OWL-S documents.

3   Petri Nets Verification Module

3.1   Interpreter of the Transforming from OWL-S to PNML

This interpreter can carry on analysis of OWL-S service and transform it into PNML form. For the transformation from OWL-S service to Petri net can use reuse many methods, algorithms, and reuse tools to inspect the equivalence of Petri net (for example literature [5,6]). Narayanant and McIlraith have first defined the Petri net semantics of DAML-S in [7] (the OWL-S preceding edition). However, their semantics is not the combinatorial property for it is unable to process any-order control structure. DaGen tool [8] transforms Petri net semantics of DAML-S description in [7] to referring Petri net. DaGen inserts to a Reference Net Workshop (Renew), and causes the Petri net simulator execution of Reference Network as well as the graph draw. However, DaGen has not had the intermediate Petri net file of transform. The interpreter described in this paper can transform OWL-S process model expressing service behavior to Petri network described by PNML format.

When execute the reasons, how to share the input/output data in different process? In fact, when input and output data are shared by many processes, the OWL-S process model may be defined by the input/output binding mechanism. [9, 10] The interpreter deals with this problem by carrying out a suitable analysis sentence of OWL-S process model.

This paper comes through the XML resolver to transform OWL-S documents to the PNML form. The interpreter from OWL-S to the PNML is a Java Servlet. Input a URL pointing to OWL-S service description (or file system path) from Web client of the interpreter, and sends it to the serve, analyzing by the Servlet in the background conversion, and returning the Petri net describing in PNML of OWL-S service.

3.2   PNML Verification and XSLT Transformation

1)   PNML Correctness Verification module: To change the default, adjust the template as follows.

Through to the OWL-S documents' transformation, the Web service which the OWL-S documents describe definitely may use the Petri net simulation and indicate by the PNML document that like this may verify the Web service operation flow which using the Petri net's correlation analysis method and the tool the OWL-S documents describe whether to have in the flowage structure design question. Therefore the next stage is transforms, but results in the PNML document hands over by PNML Correctness the Verification module processes. PNML Correctness Verification module construction is like chart 2.

PNML Correctness the Verification module contains the PNML resolver, the accurate verification, the security, the Reachability, the deadbolt lock, finally the shape, and the durable verification and so on. PNML Parser is responsible for the PNML document which analyzes transmits. The accurate verification, the security, the Reachability, the deadbolt lock, the shape, the durability through the execution coverage diagram, may reach analysis methods separately finally and so on chart, incidence matrixes, condition agenda, migration matrix to carry on the verification. But the PNML document's proving program is first starts by the accuracy and the secure nature, next is the Reachability nature, and finally is the deadbolt lock, the final state and the durable nature. So long as this verification step has an item will be unable through to transmit makes a mistake harms the information and stops the entire proving program.

PNML Correctness Verification module contains:

*PNML decoder
PNML Parser is responsible to receive the PNML document after OWL-S file conversion. First analyzes the Web service operation flow which describes in the PNML document, again storehouse which describes the PNML document, migration and arc by array way storage. PNML Parser through analyzes the PNML document and separately the storehouse, the migration and the arc by the array form storage, will then transmit these three objects by the parameter way for the nature proving program. The nature proving program after receiving the Petri net model the image parameter the basis itself uses again the analysis method, constructs by the parameter in information may reach the tree, the incidence matrix and so on mathematics type.

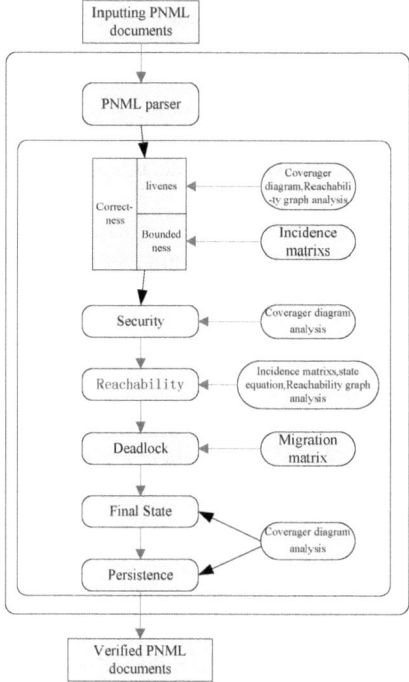

Fig. 2. PNML Correctness Verification module structure

* The secure verification

For Web service composition, the tool will use cover tree analysis to verify the Petri-Net model of the security property. Therefore realizes the cover tree and the coverage diagram method application procedure analysis in the PNML accuracy verification proxy service combine the Petri-Net model after PNML the resolver analysis Web.

* Reachability verification

The Petri-Net model regarding the Web services the Reachability nature to use the incidence matrix, the equation of state or may reach the chart the analysis method verification. Therefore is by realizes the incidence matrix and the equation of state method application procedure analysis in the PNML accuracy verification proxy service combines the Petri-Net model after PNML the resolver analysis Web.

3.3   Calculation Method Verification Tool DaNAMiCS and Tina

DaNAMiCS is a modeling verification tool, can use for to analyze the Petri net and to color the Petri net, and can reduce the modeling grid complexity. DaNAMiCS includes suppressing the arc the support to help a system model the foundation. The DaNAMiCS important merit is that it supports some analysis tool and the method, for instance matrix invariant and migration matrix, structure analysis, as well as some simple and advanced performance analysis. Because DaNAMiCS has compared to other OWL-S verification tool more analysis tools, we use XSLT to transform the PNML documents are the wam forms, like this can induct them to DaNAMiCS.

Tina (time Petri net analyzer) is a tool that analyzes the Petri net and a time Petri net. It may construct and reach the chart and can carry on the analysis to the Petri net's structure:

The accurate analysis confirmed that a system's integrity is maintained, it including analyzes net's activeness and the boundedness. We will use the accuracy to represent the net to be live and have, and with the accuracy explained that this model net expressed the correct system.

Besides these essential attributes, we must confirm some other attributes, including net whether to contain the final state, net whether safe (security), whether it is lasting (durability) and so on.

1) Analysis tools and methods:

Coverage diagram
We must inspect the construction algorithm of the coverage diagram. It will need to renew, so that processes increases newly suppresses the arc and has the capacity storehouse institute order of complexity. Carry out the algorithm that must be correct and the most superior movement.

* Analyze the correctness with invariant
DaNAMiCS will be able to calculate the incidence matrix. It will be able to determine P- and the T- invariant from the incidence matrix. This may use for surveying the boundedness and the deadbolt lock that does not exist.

* Analyze the correctness with coverage analysis
The user will be able to choose the option in DaNAMiCS, which domain through assigns the functional analysis to need to investigate. The coverage diagram will be produced for the accuracy and tests.

Coverage analysis
In many situations, the invariant analysis cannot produce about the Petri net model accurate conclusion. It needs to carry on the spreadability analysis. It is a two stepped process. The first stage is the coverage diagram construction. The coverage diagram is all may reach marking the set. The second section is this stage analysis. The first stage is called the coverage diagram production, has the greatest time consumption and the complexity; the second section analyzes the section, it will process afterward.

2) The nature analysis of Coverage diagram:
After discussion chart product, now it will analyze the coverage diagram to appraise the following nature, such as activeness, boundedness, final state, security and durable process and so on.

    a) Activity
A Petri net most important nature is an activeness, this relates judges some system whether to collapse or whether systematic some part infinite loop. A Petri net's live performance through the following two rules, but determined from the coverage diagram:
    * If a net is live and has, then it has the strong connection coverage diagram.
    * A net has, the net is lives, and when only all migration in the coverage diagram connects in the module the demonstration is a label finally at least. A strong connection chart is the random point may arrive in the chart from the chart through a

series of ways other all point charts. A chart's final module is a series of points. A strong connection chart has a correct final module. Loads the above two principles to carry on the spreadability analysis to a useful form, we may say, if each migration can cause a coverage diagram all final module's marking to enable, then the net is active. Thus, the definite active question became finds connects a module's question finally, this was the algorithm question which easy to understand. Abbreviate this algorithm specific code here.

b)   Boundary analysis

In DaNAMiCS, permits an expansion network's class use. Not only it is not always possibly uses these expansions to determine a Petri net's accuracy, This paper therefore the mark storehouse institute and does not have, moreover, if this storehouse were considered that possibly does not have, uses a different mark to mark. If a storehouse institute possibly does not have, we send a letter the number to the user to point out that corresponding storehouse institute possibly does not have. From this start, the user will be able not but to use the heuristic method to determine whether that storehouse institute accurate doesn't have.

c)   Terminal analysis

The final state is that in Petri net chart a marking may arrive from each other. This is very important to the software, namely, regardless of the current condition is anything, can always arrive at the ultimate objective. The final state existence is easy to calculate. If final, strong connects module's quantity to be equal to 1, then this Petri net contains the final state.

d)   Security analysis

If a Petri net does not have the storehouse to be able in the net the packet of energy including an above request to be willing, then calls it safely. This is the Petri net very important attribute; net's storehouse in the system is the condition mark. If the storehouse contains a request to be willing, then the condition is effective, otherwise the condition is untenable. A storehouse contains is more than a request to be willing not to have the logical significance in these net's type, usually the expression somewhere has a mistake in the design. As mentioned above, if in a Petri net's storehouse institute the request is willing quantity are most, then it is safe. The security may be determined by all mark of linear search chart simply. If has not met has the storehouse contains an above request to be willing a marking, then this net is safe.

e)   Durability

If a Petri net enables the migration to random two, an initiation's migration ever does not forbid other migration to enable, then calls it lastingly. If a lasting net's migration enables, it will maintain enables to initiate until it.

4   Conclusion

This paper has realized the integration static model and the dynamic model inspection is a body's automated verification tool prototype; In the static model aspect, gives the method which DL description logic reasoning (Pellet) and the F-logic rule (Flora-2) unifies, has used two kind of system forward reasoning fully and latter to the inference merit, combined based on DL inference engine Pellet and based on F-logic inference engine Flora-2, displays its respective superiority to carry on the inference and the verification. An OWL DL subset may transform is F-logic, and also provides the reverse support.

Should automate the verification tool prototype to be able service to provide the effective verification support to OWL-S the description Web, have the important theory and the practical significance, and has the widespread application prospect.

Acknowledgments. This work was supported in part by a grant from Basic Research Program of Shenzhen Science and Technology Research and Development under grants JC201006020791A and JC201006020820A, Natural Science Foundation of Guangdong Province, China, under grant S2011040000672, Guangdong Provincial Education and science project of the 11th "five-year plan", Research Fund of Shenzhen Institute of Information Technology.

References

1. He, K., Liang, P., Li, B., et al.: Meta-modeling of Requirement for Networked Software, An Open Hierarchical & Cooperative Unified Requirement Framework URF. DCDIS2B 14(S6), 293–298 (2007)
2. Wang, J., He, K., Li, B., et al.: Meta-models of Domain Modeling Framework for Networked Software. In: Proceedings of The Sixth International Conference on Grid and Cooperative Computing, GCC 2007, Urumchi, China, pp. 878–885 (August 2007)
3. Matthias, K., Benedikt, F., Katia, S.: OWLS-MX: A hybrid Semantic Web service matchmaker for OWL-S services. Web Semantics: Science, Services and Agents on the World Wide Web 7(2), 121–133 (2009)
4. Morimoto, S.: A Survey of Formal Verification for Business Process Modeling. In: Bubak, M., van Albada, G.D., Dongarra, J., Sloot, P.M.A. (eds.) ICCS 2008, Part II. LNCS, vol. 5102, pp. 514–522. Springer, Heidelberg (2008)
5. Qi, G., Tianshi, C., Haihua, S., Yunji, C., Weiwu, H.: On-the-Fly Reduction of Stimuli for Functional Verification, ats. In: 2010 19th IEEE Asian Test Symposium, pp. 448–454 (2010)
6. He, F., Le, J.: Hierarchical Petri-Nets Model for the Design of E-Learning System. In: Hui, K.-c., Pan, Z., Chung, R.C.-k., Wang, C.C.L., Jin, X., Göbel, S., Li, E.C.-L. (eds.) EDUTAINMENT 2007. LNCS, vol. 4469, pp. 283–292. Springer, Heidelberg (2007)
7. Hallé, S., Hughes, G., Bultan, T., Alkhalaf, M.: Generating Interface Grammars from WSDL for Automated Verification of Web Services. In: Baresi, L., Chi, C.-H., Suzuki, J. (eds.) ICSOC-ServiceWave 2009. LNCS, vol. 5900, pp. 516–530. Springer, Heidelberg (2009)
8. Moldt, D., Ortmann, J.: DaGen: A Tool for Automatic Translation from DAML-S to High-Level Petri Nets. In: Wermelinger, M., Margaria-Steffen, T. (eds.) FASE 2004. LNCS, vol. 2984, pp. 209–213. Springer, Heidelberg (2004); Kornack, D., Rakic, P.: Cell Proliferation without Neurogenesis in Adult Primate Neocortex. Science 294, 2127–2130 (2001)
9. Staats, M., Heimdahl, M.P.E.: Partial Translation Verification for Untrusted Code-Generators. In: Liu, S., Araki, K. (eds.) ICFEM 2008. LNCS, vol. 5256, pp. 226–237. Springer, Heidelberg (2008)
10. Emilia, O., Tomi, J.: A Translation-based Approach to the Verification of Modular Equivalence. Journal of Logic and Computation 19(4), 591–613 (2008)

Synchronization Dynamics of Complex Network Models with Impulsive Control

Yanhui Gao

Department of Basic Courses,
North China Institute of Science and Technology,
East Yanjiao, Beijing, China, 101601
yanhuigao@gmail.com

Abstract. In this paper, synchronization dynamics of complex network models with impulsive control is investigated. Based on impulsive control theory on dynamical systems, Two sufficient impulsive consensus protocol for such networks is proposed.The theoretical results are applied to chaos synchronization of a small-world networks model composing of the representative Duffing oscillators, the Numerical simulation also demonstrate the effectiveness of the proposed control techniques.

Keywords: Complex dynamical networks, impulsive control techniques, chaos synchronization, scale-free networks.

1 Introduction

Over the past decade,complex networks have been intensively studied in various disciplines, such as social, biological, mathematical, and engineering sciences [1-8]. Many researches have fund the real-world network is neither regular nor completely random. To interpolate between these two extremes, Watts and Strogatz [1998] introduced the interesting concept of small-world networks. The so-called small-world networks have intermediate connectivity properties but exhibit a high degree of clustering as in the regular networks and a small average distance between vertices as in the random networks. Another significant recent discovery in the field of complex networks is the observation that a number of large-scale and complex networks are scale-free, that is, their connectivity distributions have the power-law form [5,8,9]. A scale-free network is inhomogeneous in nature:most nodes have very few connections and a few nodes have many connections.

The research on complex networks has been not only focused on their topological structure but on their dynamical processes.Recently, the synchronization in small-world and scale-free networks has been addressed [9-16]. Some synchronization criteria have been obtained in the literature[15]. In [16], Wang and Chen studied the synchronization by applying local feedback injections to a fractions of network nodes.

In the present work we are trying to apply impulsive control to the nodes, because impulsive control is more efficient and useful in a great number of real

B. Liu and C. Chai (Eds.): ICICA 2011, LNCS 7030, pp. 553–560, 2011.

life applications. It can give a efficient method to deal with system which cannot endure continuous disturbance.We also investigate that for the scale-free network, the different pining of impulsive control: to pin impulsive control to specially selected nodes and to pin the control to the randomly selected nodes has different effect. That is because its inhomogeneous topology in nature:most nodes have very few connections and a few nodes have many connections.

2   A Coupled Complex Dynamic Model with Impulsive Control

Now suppose that a complex network consisting if N identical linearly and diffusively coupled nodes, with each node being an n-dimensional dynamical system. The state equations of this dynamical network are given by

$$\dot{x}_i = f(x_i, t) + c \sum_{j=1, j\neq i}^{N} a_{ij}\Gamma(x_j - x_i), i = 1, 2, \cdots. \tag{1}$$

where $x_i = (x_{i1}, x_{i2}, \cdots, x_{in})^T \in R^n$ are the state variables of node i,the constants $c_{ij} > 0$ represents the coupling strength between node i and node j, $\Gamma = (\gamma_{ij}) \in R^{n\times n}$ is a matrix linking coupled variables, and if some pair $(i, j), 1 \leq i, j \leq n$,with $\gamma_{i,j} \neq 0$, then it means two coupled nodes are linked through their ith and jth state variables, respectively.

In the network (5) the coupling matrix $A = (a_{ij}) \in R^{N\times N}$ representing the coupling configuration of the network, In this paper we assume that A is symmetric and diffusive coupling .furthmore if there is a link between node i and j $(i \neq j)$, then $a_{ij} = a_{ji} = 1$; otherwise $a_{ij} = a_{ji} = 0 (i \neq j)$. If the degree k_i of the node i is defined to be the number of connections, then

$$\sum_{j=1, j\neq i}^{N} a_{ij} = \sum_{j=1, j\neq i}^{N} a_{ji} = k_i, i = 1, 2, \cdots, N.$$

Let the diagonal elements be $a_{ii} = -k_i, i = 1, 2, \cdots, N$. then, the coupling matrix is symmetric and one eigenvalue of A is zero, with multiplicity 1, and all the other eigenvalues of A are strictly negative.
the the equation can be written as

$$\dot{x}_i = f(x_i, t) + c \sum_{j=1}^{N} a_{ij}\Gamma x_j, i = 1, 2, \cdots, N. \tag{2}$$

Suppose that we want to stabilize network(4) on a homogeneous state $s(t)$,$s(t)$ can correspond to an equilibrium point,a periodic orbit,or a chaotic orbit. In the following we consider $s(t)$ as an arbitrary orbit satisfy $\dot{s}(t) = f(s(t))$. We want to achieve the goal of control by applying impulsive controller. Consider at the time $t = t_k$ add impulsive control to the system (4),then we get the complex dynamic network as follows:

$$\begin{cases} \dot{x}_i(t) = f(x_i(t), t) + c \sum_{j=1}^{N} a_{ij} \Gamma x_j(t), t \neq t_k, \\ x_i(t_k^+) = s(t_k) + b_{ik} \Gamma'(x_i(t_k) - s(t_k)), t = t_k, \qquad i = 1, 2, \cdots, N. \qquad (3) \\ x_i(t_0^+) = x_{i0}. \end{cases}$$

Here $\Gamma' = diag\{\gamma_1', \gamma_2', \cdots, \gamma_n'\}$, if the ith node's jth vector is added impulsive control ,then $\gamma_j' = 1$, else $\gamma_j' = 0, (j = 1, 2, \cdots, n)$. We also assume that $t_{k-1} < t_k$, and $\lim\limits_{k \to \infty} = +\infty$

Now we want to find some sufficient conditions make the system(5) synchronize to the sate $s(t)$, that is for all solutions $x_i(t)(i = 1, 2, \cdots, N)$ of system(5)satisfy

$$\lim_{t \to \infty} \|x_i(t) - s(t)\| = 0, i = 1, 2, \cdots, N$$

3 Synchronization Analysis

To simplify our discussion ,we assume $\Gamma' = I_n$, let $X = [x_1, x_2, \cdots, x_N]^T$,

$S(t) = \overbrace{[s(t), s(t), \cdots, s(t)]}^{N}{}^T, \eta = [\eta_1, \eta_2, \cdots, \eta_N]^T$.Then we make a translation $\eta(t) = X(t) - S(t), \eta_i = x_i - s(t)$,the synchronization of system (5) is changed to the stability of η at 0,that is

$$\lim_{t \to \infty} \|x_i(t) - s(t)\| = \lim_{t \to \infty} \|\eta_i\| = 0, i = 1, 2, \cdots, N$$

To investigate the local stability we linearize the above equation at $s(t)$,get the following

$$\dot{\eta}_i = \Sigma(t)\eta_i + c \sum_{j=1}^{N} a_{ij} \Gamma \eta_j$$

$\Sigma(t) = (\sigma_{ij}(t)) = Df(s(t), t)$ is the Jacobi matrix.
So the system(5)can be transformed to the following:

$$\begin{cases} \dot{\eta}_i(t) = Df(s(t), t) + c \sum_{j=1}^{N} a_{ij} \Gamma \eta_j(t), t \neq t_k, \\ \eta_i(t_k^+) = b_{ik} \Gamma' \eta_i(t_k), t = t_k, \qquad i = 1, 2, \cdots, N. \qquad (4) \\ \eta_i(t_0^+) = \eta_{i0}. \end{cases}$$

Let $\lambda_1 = 0 \geq \lambda_2 \geq \lambda_3 \geq \cdots \geq \lambda_N$ be the eigenvalues of matrix A, and $\Phi = [\Phi_1, \Phi_2, \cdots, \Phi_N] \in R^{N \times N}$ be the corresponding eigenvector basis satisfying

$$A\Phi_k = \lambda_k \Phi_k, k = 1, 2, \cdots, N$$

As A is a symmetric matrix, let $\Phi^{-1} = \Phi^T$,Then we make another translation

$$\eta = \Phi\nu, \text{so } \nu = \Phi^{-1}\eta, \nu = [\nu_1, \nu_2, \cdots, \nu_N],$$

When $t \neq t_k$

$$\dot{\nu} = \Phi^{-1}\dot{\eta} = \Phi^{-1}DF(\bar{X},t)\Phi\nu + c\Phi^{-1}A\Phi\Gamma\nu = Df(\bar{x},t)\nu + c\Lambda\Gamma\nu$$

When $t = t_k$

$$\nu_i(t_k^+) = \Phi_i\eta(t_k^+) = \sum_{s=1}^{N}\Phi_{si}\eta_s(t_k^+) = \sum_{s=1}^{N}\Phi_{si}b_{sk}\eta_s$$

As $\eta = \Phi\nu, \eta_s = \sum_{j=1}^{N}\Phi_{sj}\nu_j$, substitute η with ν get

$$\nu_i(t_k^+) = \sum_{j=1}^{N}(\sum_{s=1}^{N}\Phi_{si}b_{sk}\Phi_{sj})\nu_j,$$

let $C_{ij}(k) = \sum_{s=1}^{N}\Phi_{si}b_{sk}\Phi_{sj}$,so the original system (5)has been changed to the following system

$$\begin{cases} \dot{\nu}_i = Df(\bar{x},t)\nu_i + c\lambda_i\Gamma\nu_i, & t \neq t_k, \\ \nu_i(t_k^+) = \sum_{j=1}^{N}C_{ij}(k)\nu_j & t = t_k, \quad i = 1,2,3,\cdots,N \\ \nu_i(t_0^+) = \nu_{i0}. \end{cases} \quad (5)$$

Theorem 1. If there exist a symmetric positive definite matrix $P \in R^{n \times n}$ and constants $\alpha, \beta_k (k = 1, 2, \cdots, +\infty)$ for any $t \in (t_{k-1}, t_k)$ satisfy

$$(PDf(s(t))^T)^s \leq \alpha I,$$

$$\frac{\rho_2}{\rho_1}\sum_{i=1}^{N}\sum_{j=1}^{N}C_{ij}(k)^2 \leq \beta_k.$$

Here ρ_1 and ρ_2 are the smallest and biggest eigenvalues of matrix$P$. If $\beta_k exp(\frac{\alpha}{\rho_1}(t_k - t_{k-1})) \leq 1$,then system(6)stable at 0, that is system (5)synchronize to $s(t)$.

If $\beta_k exp(\frac{\alpha}{\rho_1}(t_k - t_{k-1})) < 1$,then system(6)asymptotically stable at 0, that is system (5) asymptotically synchronize to $s(t)$.

Proof. Define a Lyapunov candidate $V = \sum_{i=1}^{N}\nu_i^T P\nu_i$, then

$$\rho_1\sum_{i=1}^{N}\nu_i^T\nu_i \leq \sum_{i=1}^{N}\nu_i^T P\nu_i \leq \rho_2\sum_{i=1}^{N}\nu_i^T\nu_i.$$

When $t \neq t_k$

$$\dot{V} = \sum_{i=1}^{N}(\dot{\nu}_i^T P\nu_i + \nu_i^T P\dot{\nu}_i) \leq \sum_{i=1}^{N}\nu_i^T(Df(\bar{x},t)^T P)^s\nu_i,$$

As the the assumption in the theorem1 $(PDf(s(t))^T)^s \leq \alpha I$, So we get

$$\dot{V} \leq \sum_{i=1}^{N} \nu_i^T \alpha I \nu_i \leq \frac{\alpha}{\rho_1} \sum_{i=1}^{N} \nu_i^T P \nu_i = \frac{\alpha}{\rho_1} V.$$

When $t = t_k$

$$V(t_k^+) = \sum_{i=1}^{N} \nu_i(t_k^+)^T P \nu_i(t_k^+) \leq \rho_2 \sum_{i=1}^{N} \nu_i(t_k^+)^T \nu_i(t_k^+) = \rho_2 \sum_{i=1}^{N} \sum_{s=1}^{n} (\sum_{j=1}^{N} C_{ij}(k)\nu_{js})^2,$$

Using cauchy inequality, Changing the sum sequence, we get

$$V(t_k^+) \leq \rho_2 \sum_{i=1}^{N} (\sum_{j=1}^{N} C_{ij}(k)^2)(\sum_{j=1}^{N} \sum_{s=1}^{n} \nu_{js}^2),$$

$$(\sum_{j=1}^{N} \sum_{s=1}^{n} \nu_{js}^2) = \sum_{i=1}^{N} \nu_i^T \nu_i,$$

so

$$V(t_k^+) \leq \frac{\rho_2}{\rho_1} (\sum_{i=1}^{N} \sum_{j=1}^{N} C_{ij}(k)^2) V.$$

As the the assumption in the theorem1

$\frac{\rho_2}{\rho_1} \sum_{i=1}^{N} \sum_{j=1}^{N} C_{ij}(k)^2 \leq \beta_k$, so $V(t_k^+) \leq \beta_k V$,

Using the impulsive theorem in[22] If $\beta_k exp(\frac{\alpha}{\rho_1}(t_k - t_{k-1})) \leq 1$, then system(6) stable at 0, that is system (5)synchronize to $s(t)$.

If $\beta_k exp(\frac{\alpha}{\rho_1}(t_k - t_{k-1})) < 1$, then system(6)asymptotically stable at 0, that is system (5) asymptotically synchronize to $s(t)$.

Theorem 2. If there exist n positive constants $p_1, p_2, \cdots, p_n$, and $\alpha, \beta_k (k = 1, 2, \cdots, +\infty)$ for any $t \in (t_{k-1}, t_k)$ satisfy

$$p_s \sigma_{ss}(t) + \sum_{j=1, j \neq s}^{n} p_j |\sigma_{js}(t)| \leq \alpha, \qquad\qquad s = 1, 2, \cdots, n.$$

$$\frac{p_2}{p_1} \sum_{i=1}^{N} |C_{ij}(k)| \leq \beta_k, \qquad\qquad j = 1, \cdots, N$$

Here p_1 and p_2 are the smallest and biggest eigenvalues of n positive constants $p_1, p_2, \cdots, p_n$.If $\beta_k exp(\frac{\alpha}{\rho_1}(t_k - t_{k-1})) \leq 1$,then system(6)stable at 0, that is system (5)synchronize to $s(t)$.

If $\beta_k exp(\frac{\alpha}{p_1}(t_k - t_{k-1})) < 1$, then system(6)asymptotically stable at 0, that is system (5) asymptotically synchronize to $s(t)$.

Remarks. 1Theorem1-2presents a sufficient condition for the synchronization of a complex dynamic system(6) with impulsive control. Theorem1-2 shows that the synchronization of system(6)is determined by impulsive intervals, impulsive strength and coupling strength. The new results can be applied to the synchronization of any given dynamical network,such as small-word network and sale-free network.

Specially when $t = t_k$,the impulsive strength adding to any node are the same, that is $b_{1k} = b_{2k} = \cdots = b_{Nk} = b_k$ $\eta_i(t_k^+) = b_k I\eta_i, \nu_i(t_k^+) = b_k I\nu_i$then the origin system(5) has the following form

$$\begin{cases} \dot{\nu}_i = [Df(s(t),t) + c\lambda_i\Gamma]\nu_i, t \neq t_k \\ \nu_i(t_k^+) = b_k I\nu_i, t = t_k, \qquad i = 2, \cdots, N \\ \nu_i(t_0^+) = \nu_{i0} \end{cases} \qquad (6)$$

Corollary1. Let γ_k be the largest eigenvalue of $((Df(\overline{x},t) + (Df(s(t),t))^T)$, $t \in R^+ t \in (t_{k-1}, t_k], (k = 1, 2, \cdots)$,if there exist a constant $\xi > 1$ satisfy

$$ln(\xi b_k^2) + \gamma_k\tau_k \leq 0, \qquad k = 1, 2, \cdots \qquad (7)$$

then system(7)stable at 0, that is system (5)synchronize to $s(t)$. $s(t), 0 < \tau_k = t_k - t_{k-1} < \infty(k = 1, 2, \cdots)$ is impulsive interval.

Corollary2. If there exist β_k and $\xi > 1$ satisfy

$$\sum_{j=1,j\neq s}^{n} |\sigma_{js}| + \sigma_{ss} \leq \beta_k, 1 \leq s \leq n$$

$$ln(\xi b_k) + \beta_k\tau_k \leq 0, \qquad k = 1, 2, \cdots \qquad (8)$$

then system(7)stable at 0, that is system (5)synchronize to $s(t)$. $0 < \tau_k = t_k - t_{k-1} < \infty(k = 1, 2, \cdots)$ is impulsive interval.

4   A Simulation Example

This section present an example to show the effectiveness of the above synchronization criteria and give a figure without impulsive control to show the effectiveness of the impulsive control.

Consider a dynamical network consisting of 50 identical Duffing oscillators. Here dynamics is described by

$$\begin{pmatrix} \dot{x}_{i1} \\ \dot{x}_{i2} \end{pmatrix} = \begin{pmatrix} x_{i2} \\ -p_1 x_{i1} - x_{i1}^3 - px_{i2} + qcos(\omega t) \end{pmatrix}$$

Where
$p_1 = -1.1, p = 0.4, q = 2.1, \omega = 1.8$ and $1 \le i \le 50$ And the network system is small-world ,its algorithm is generated as section 2,and its parameters are chosen as:
$N = 50, k = 2, p = 0.1, c = 0.1$,Where the constant c is the coupling strength,p is the rewiring probability,k is the number of the linked nearest neighbor.

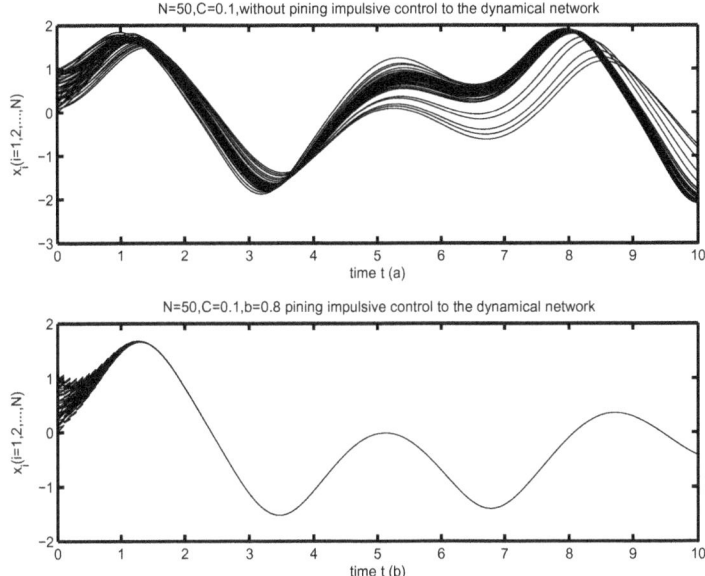

Fig. 1. a: The asynchronous figure of a dynamical system consisting of 50 nodes without impulsive control. b: The synchronous figure of a dynamical system consisting of 50 nodes without a impulsive control. Which $b_{ik} = 0.8; \Gamma' = I; t_k - t_{k-1} = 0.1.$

5 Conclusions

In this work,we add the impulsive control to the dynamical system,and analyzed the synchronization of the new model with impulsive control.Several synchronization criteria have showed that the model's synchronization is determined by means of single node's dynamical property ,the eigenvalues of the coupled configuration matrix and the impulsive intervals and impulsive strength.A smaller couple strength with impulsive control can make the model synchronize while on the same condition the system can not achieve synchronization without impulsive control.

References

1. Barabsi, A.L., Albert, R.: Emergence of scaling in random networks. Science 286, 509–512 (1999)

2. Strogatz, S.H.: Exploring complex networks. Nature 410, 268–276 (2001)
3. Newman, M.E.J., Watts, D.J.: The structure and function of complex networks. SIAM Rev. 45, 167–256 (2003)
4. Watts, D.J., Strogatz, S.H.: Collective dynamics of small-world. Nature 393, 440–442 (1998)
5. Barabsi, A.L., Albert, R., Jeong, H., Bianconi, G.: Power-law distribution of the world wide web. Science 287, 2115a (2000)
6. Barabsi, A.L., Albert, R.: Emergence of scaling in random networks. Science 285, 509–512 (1999)
7. Albert, R., Jeong, H., Barabsi, A.L.: Error and attack tolerance of complex networks. Nature 406, 378–382 (2000)
8. Albert, R., Barabsi, A.L.: Statistical mechanics of complex networks. Rev. Mod. Phys. 74, 47–97 (2002)
9. Wang, X.F.: Complex networks:Topology,dynamics and synchronization. Int. J. Bifurc. Chaos 12(5), 885–916 (2002)
10. Barahona, M., Pecora, L.M.: Synchronization in small-world system. Phys.Rev.Lett. 89(5) (2002)
11. Almaas, E., Kulkarni, R.V., Stoud, D.: Characterizing the structure of small-world networks. Phys.Rev.Lett. 88(9) (5002)
12. Wang, X., Chen, G.: Synchronization in small-world dynamical networks. Int.J.Bifurc.Chaos 12(1), 187–192 (2002)
13. LiX, X., Wang, F., Chen, G.R.: Pinning a complex dynamical network to its equilibrium. IEEE Trans. Circuits and Systems-I 51(10), 2074–2087 (2004)
14. Erdös, P., Rényi, A.: On the evolution of random graphs. Publ. Math. Inst. Hung. Acad. Sci. 5, 17–60
15. Lü, J., Yu, X., Chen, G.: Chaos synchronization of general complex dynamical networks. Phys. A 334, 281–302 (2004)
16. Wang, X., Chen, G.: Pinning control of scale-free dynamical networks. Phys. A. 310, 521–531 (2002)
17. Chen, G.R., Zhou, J., Liu, Z.: Global synchronization of coupled delayed neural networks and applications to chaotic CNN models. International Journal of Bifurcation and Chaos 14(7), 2229–2240 (2004)
18. Zhou, J., Chen, T.P., Xiang, L.: Robust synchronization of delayed neural networks based on adaptive control and parameters identification. Chaos, Solitons, Fractals 27(4), 905–913 (2006)
19. Lin, P., Jia, Y.M.: Average Consensus in Networks of Multi-Agents with both Switching Topology and Coupling Time-Delay. Physica. A 387, 303–313 (2008)
20. Zhou, J., Xiang, L., Liu, Z.R.: Synchronization in Complex Delayed Dynamical Networks with Impulsive Effects. Phys. A 384, 684–692 (2007)
21. Cai, S.M., Zhou, J., Xiang, L., Liu, Z.R.: Robust Impulsive Synchronization of Complex Delayed Dynamical Networks. Phys. Lett. A 372, 4990–4995 (2008)
22. Zhou, J., Wu, Q.J.: Exponential Stability of Impulsive Delayed Linear Differential Equations. IEEE Trans. Circuit Syst. II 56, 744–748 (2009)

Design and Test of Nodes for Field Information Acquisition Based on WUSN

Xiaoqing Yu[1,3], Pute Wu[1,2,3], and Zenglin Zhang[1,3]

[1] Northwest Agriculture and Forestry University, Shaanxi,Yangling, 712100, China
[2] National Engineering Research Center for Water Saving Irrigation at Yangling, Institute of Soil and Water Conservation of Chinese Academy of Sciences, Shaanxi, Yangling, 712100, China
[3] Research Institute of Water-saving Agriculture of Arid Regions of China, Shaanxi, Yangling, 712100, China

Abstract. The wireless sensor network technology was researched. Some wireless underground sensor network nodes and a sink node based on embedded technology and RF technology were designed innovatively. WUSN node consists of sensor, the processor, wireless communication module and power module, including processor using MSP430 microcontroller, RF modules adopting nRF905 communication module which having 433/868/915 MHz 3 ISM channel, the sink node is made up RF transceiver module, the core control circuit, information processing, data storage, LCD module and power supply. The nodes which acquired soil parameters information were regularly distributed in the monitoring area. The sink node collected the information of nodes that were sent in way of a single jumping or multiple hops and implemented fusion, analysis, processing, storage and display of information. For 50% sands, 35%silt, and 15% clay, a bulk density of 1.5 g/cm3 and a specific density of 2.6 / cm3, test for different soil moisture (5%, 10%, 15%, 20% and 25%) in three different frequencies, result shows that radio signal path loss is the minimum in the low frequency and low moisture. Moreover, the changes of node deployed depth (0.2 m, 0.4 m, 0.6 m,0.8 m, 1m, 1.2 m, 1.4 m, 1.6m, 1.8 m and 2m) affected signal attenuation under 433MHz, it is concluded that the best WUSN node buried depth for effective transmission.

Keywords: WUSN, MSP430, Sink node, Frequency, Depth, Field information acquisition.

1 Introduction

Farmland information is the important and the deciding factor in agricultural production. Collection method is one of the primary technical problems in farmland information research field and realization if modern agricultural production, which can collect variable information of crop growth environment many-side, accurately, rapidly and effectively. It is also the key and decisive factor of the modern efficient agricultural production. Perception, processing, management decision-making and

B. Liu and C. Chai (Eds.): ICICA 2011, LNCS 7030, pp. 561–568, 2011.

information integration control of farmland information has become focus in the field of contemporary international agricultural science and technology research [1].

The wireless sensor network technology has been applied in farmland information monitoring field [2-4], it has achieved good scientific research achievements. At present application domain, underground sensing systems require data loggers or motes deployed at the surface with wiring leading to underground sensors in order to avoid the challenge of wireless communication in the underground. All of these existing solutions require sensor devices to be deployed at the surface and wired to a buried sensor. While the usefulness of these applications of sensor network technology is clear, there remain shortcomings that can impede new and more varied uses. These equipment exposed on the ground not only influence farming, wireless transmitting functions of wireless node also be affected because of geography, meteorology and natural factors. Based on these disadvantages, the Wireless Underground Sensor networks (WUSN) provide a new method for underground monitoring. WUSN has also becoming a new research direction in the agricultural industry. Sensors equipments with wireless receiving and send module have been completely deployed in certain depth of soil, induction module sending data in the way of wireless when it perceives data. Many sensor nodes formed into sensor network, which complete automatically the whole process of perception and collect data. WUSN have several remarkable merits, such as concealment, ease of deployment, timeliness of data, reliability and coverage density [5, 6]. Besides monitoring soil ingredients in underground, wireless underground sensor network can also be used for monitoring soil motion, forecasting landslide, debris, underground ice motion and volcanic eruptions [7, 8], it has higher value for study.

WSN relevant theoretical research and practical application has existed in the international. J.A.Lopez has applied wireless sensor network in precise viticulture [9], O.Green has used wireless sensor network to monitor temperature change of feed storehouse [10], Lili applied wireless sensor network in greenhouse environment monitoring [11], Zhang rongbiao realized greenhouse wireless communications of wireless sensor network based on ZigBee [12], Cai yihua designed farmland information acquisition node based on WSN [13], Feng youbing has applied wireless sensor network in water-saving irrigation [14].

Reality application or imagine about the various sensors monitor system in underground environment has a part writings. A.Sheth proposed tension induction module attached to wireless sensor may forecast landslide [15], K.Martinez introduces a sensor that test ice parameter system [16], G.W.Allen proposes the use of sensor network in monitoring volcanic activity [8], these systems do not really construct wireless underground sensor network. This paper introduces the design of wireless sensor nodes and sink node of agricultural application based on wireless underground network. Furthermore, Signal attenuation experiment is conducted under different frequency and different moisture content. This work will lead to characteristics relationship among frequency, burial depth and path loss in farmland environment.

The rest of the paper is organized as follows. Section 2 presents the process of node design. Section 3 describes the design of sink node. Section 4 specifies the test and analysis. Finally, the paper is concluded in Section 5.

2   Design of WUSN Node

Wireless sensor node is designed through the modularizing design method. The entire node system structure consists of sensor module, processor module, wireless communication module, and energy supply module. In farmland information monitoring applications, soil parameter information is the major factor. According to the application requirements, wireless underground sensor node for soil moisture acquisition is designed. The block diagram is shown in Fig.1.

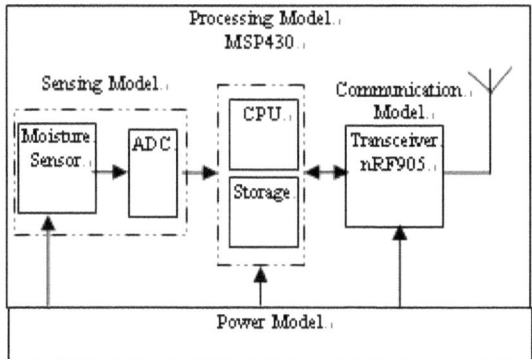

Fig. 1. Architecture of wireless underground sensor network node

2.1   Design of Sensor Model

Node collects farmland necessary information through the sensor module. Processor module retained interfaces for various sensors. Here, it takes moisture sensor XR61-TDR2 for example. XR61-TDR2 has some advantages, such as high stability, high sealing, strong resistant to squeeze ability, good shielding effect and anti-interference, long transmission distance, less influence on soil qualities, volume miniaturization and low price. It can obtain the precise soil moisture content information for processor processing. Table 1 shows sensor voltage signal and measurement for system.

Table 1. Sensor voltage and measurement

Sensor category	Sensor type	Output voltage/V	Measuring range
Soil moisture	XR61-TDR2	0~2.5	0~100%(m3/m3)
Soil temperature	DS18B20	0~5	-55~125℃
Soil salt	NT18-TYC-2	--	0.02~0.15N 0.15~0.3N

2.2   Design of Processing Model

The 16 bit series MSP430 microcontroller is launched by TI Company and adopted as the main control chip. It has a unique advantage in low power applications. MSP430 has very high levels of integration, a single chip usually integrated 12 bit A/D comparator, multiple timer, USART, watch dog, oscillator, a large number of I/O port and high-capacity memory. This microcontroller has cost-effective, many function, strong anti-interference ability,serial programming is very convenient. Circuit principle diagram is shown in Fig.2.

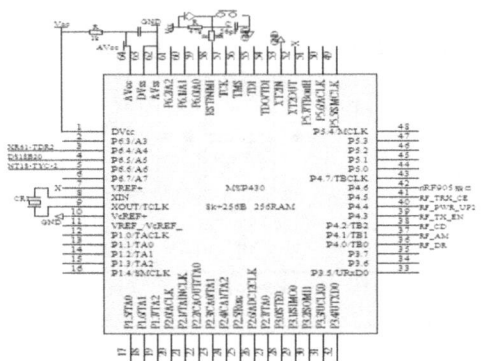

Fig. 2. Diagram of processor block

2.3   Design of Communication RF Model

Wireless communication RF module implements the communication between the nodes. The Norway Nordic company nRF905 monolithic RF transceiver chip is adopted. It has two kinds of working mode and energy saving mode, they are power lost pattern, the standby mode, Shock Burst TM receiving mode and Shock Burst TM

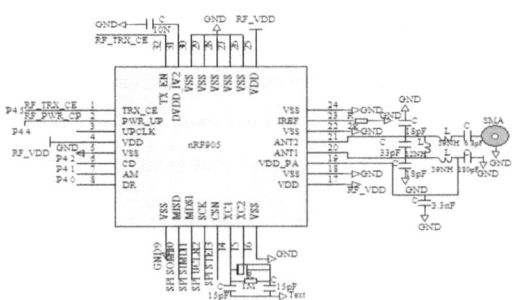

Fig. 3. Control diagram of nRF905 block

send mode separately. The chip can work freely in 433/868/915 MHz 3 ISM channel, communication with micro-controller by using SPI interfaces, configuration is very convenient, power consumption is very low. Fig.3 illustrates communication with the processor.

3   Design of SINK Node

3.1   Hardware Design

Wireless underground sensor network node collect data, sink node is responsible for data gathering, analysis, processing, storage and display. Hardware structure of sink node can be divided into RF transceiver module, the core control circuit, information processing, data storage, LCD module and power module, the structure is shown in Fig.4.

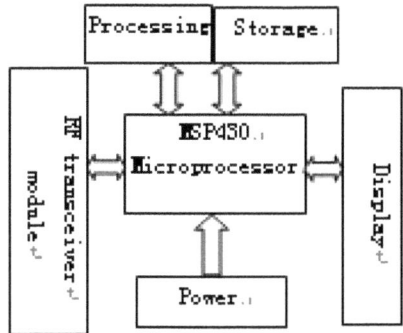

Fig. 4. Structure of sink node

After RF transceiver module receiving signal, the signal is processed and converted by regulate circuit, then is stored into MSP430 microprocessor. In addition, the wireless underground sensor node composed nets. Sink node send data queries, WUSN node can will query data sent to sink node. Specially, relevant information can be displayed in LCD screen according to demand of customers.

3.2   Software Design

For ease of management and scheduling, the function achieved by node is defined as the events of processing, every event completes corresponding function. The events are string connected in a certain relationship, which can achieve function. Sink node workflow is shown in Fig.5.

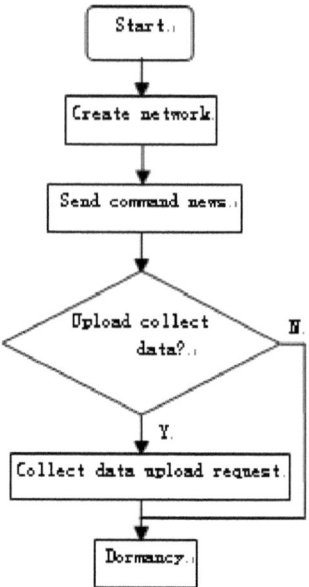

Fig. 5. Flow chart of sink node

4   Experiments and Analysis

4.1   Test Method

When wireless underground sensor network nodes get farmland information, reflection, scattering and diffraction may exist simultaneously in the process of wireless electromagnetic wave transmission in soil and interface between soil and air.

RF frequency is the core of electromagnetic signals is influenced. In addition, agricultural environment dynamic changes constantly as seasonal variation, soil water content will cause great path loss to the radio signal propagation. In order to path loss minimum when underground sensor nodes are deployed, it is necessary to find a proper depth, which guarantee deployment of sensor nodes is the most economic, signal paths loss is minimum, can transmission effectively.

In the trial, we assume the sand particle percent as 50%, the silt percent as 35%, the clay percent as 15%, the bulk density as 1.5 g/cm3, and the solid soil particle density as 2.6 / cm3 unless otherwise noted. According to three different frequencies of RF modules nRF905, the attenuation of signal strength is measured in different soil volumetric water content VWC (VWC were taken as 5%, 10%, 15%, 20% and 25%). Meanwhile, the path loss is measured in different depth h WUSN node deployed (h were taken as 0.2 m, 0.4 m, 0.6 to 0.8 m, 1-m, 1.2 m, 1.4 m, m, 1.6m, 1.8 m and 2m) is under the same frequency.

4.2   Test Analysis

Path loss is the difference in value between real received signal strength and source signal strength level, namely the signal attenuation extent, it reflects directly efficiency of wireless electromagnetic signal transmission. Fig.6 describes the path loss of the wireless signals is caused by soil volumetric water content change in different frequencies. Fig.7reflects the relationship between WUSN nodes deployed depth and the path loss under 433 MHz RF frequencies.

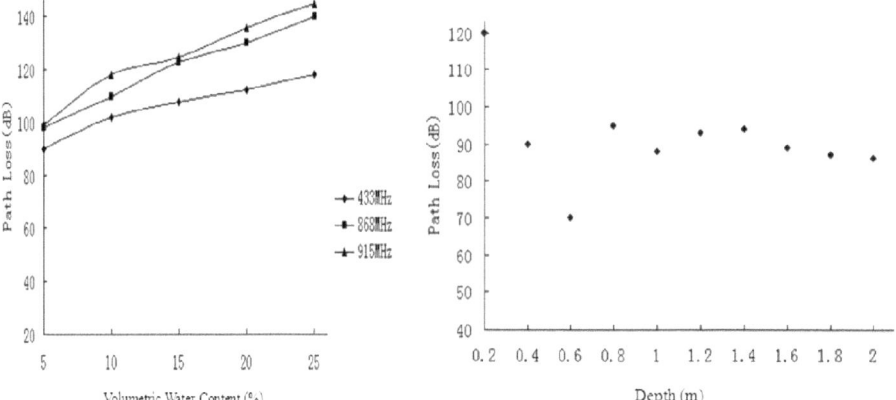

Fig. 6. The relationship among path loss, operating frequency and volumetric water content

Fig. 7. The relationship between path loss and depth

5   Conclusions

(1) According to the application requirement of farmland environmental information, wireless underground sensor network technology is researched. Underground sensor node and sink node are developed combined with embedded processors, which realized real-time dynamic collection, transmission, store and display for the farmland soil information parameter. Nodes satisfy requirements of low power consumption, low cost, high real-time, high reliability for farmland information collection.

(2) For three different frequencies of wireless RF modules, path loss of radio signal is analyzed in different volumetric water content through changing of soil volumetric water content. Experiment shows that soil attenuation is the minimum in the low frequency RF and low volumetric water content.

(3) In 433MHz RF frequency, path loss is influenced by different depth WUSN node deployed in the soil. The results indicate that signal attenuation is the minimum when node is deployed in suitable depth.

Acknowledgements. The authors wish to thank the National Engineering Research Center for Water Saving Irrigation, which partially supported this work through the project "National 863 plan"(2006AA100217), "National science and technology support plan"(2007BAD88B10)and "National Natural Sciences Foundation project"(40701092). The authors are also grateful to the anonymous reviewers for their valuable feedback.

References

1. Sawant, H., Tan, J., Yang, Q., Wang, Q.: Using Bluetooth and Sensor Networks for Intelligent Transportation Systems. In: Proceedings of IEEE Intelligent Trans. Portation Systems Conference, Washington, D.C., USA, pp. 767–772 (2004)
2. Boulis, A., Srivastava, M.B.: A Framework for Effieient and Programmable Sensor Networks. In: Proceedings of OPENARCHZ 2002, New York, pp. 117–128 (June 2002)
3. Stojmenovic, I.: Handbook of sensor networks. CRC Press (2005)
4. Chen, J., Cao, X., Cheng, P., Xiao, Y., Sun, Y.: Distributed Collaborative Control for Industrial Automation with Wireless Sensor and Actuator Networks. In: IEEE Transactions on Industrial Electronics (2010)
5. Akyildiz, I.F., Su, W., Sankarasubramaniam, Y., Cayirci, E.: Wireless Sensor Networks: A Survey. Computer Networks, 393–422 (2002)
6. Akyildiz, I.F., Stuntebeck, E.P.: Wireless underground sensor networks: Research challenges. Ad Hoc Networks, 669–686 (2006)
7. Akyildiz, I.F., Vuran, M.C., Sun, Z.: Channel modeling for Wireless Underground Communication in Soil. Physical Communication (2009)
8. Allen, G.W., Lorincz, K., Welsh, M., Marcillo, O., et al.: DePloying A Wireless Sensor Network on An Active Volcano. IEEE Internet Computing, 18–25 (2006)
9. Lopez Riquelme, J.A., Soto, F., Suardiaz, J., et al.: Wireless sensor networks for precision horticulture in Southern Spain. Computers and Electronics in Agriculture, 25–35 (2009)
10. Green, O., Nadimi, E.S., Blanes, V., et al.: Monitoring and modeling temperature variations inside silage stacks using novel wireless sensor networks. Computers and Electronics in Agriculture, 149–157 (2009)
11. Li, L., Li, H., Liu, H.: Greenhouse Environment Monitoring System Based on Wireless Sensor Network. Transactions of the Chinese Society for Agricultural Machinery, 228–231
12. Zhang, R., Gu, G., Feng, Y., et al.: Realization of Communication in Wireless Monitoring System in Greenhouse Based on IEEE802.15.4. Transactions of the Chinese Society for Agricultural Machinery, 119–122 (2008)
13. Cai, Y., Liu, G., Li, L., et al.: Design and test of nodes for farmland data acquisition based on wireless sensor network. Chinese Society of Agricultural Engineering, 176–178 (2009)
14. Feng, Y., Zhang, R., Gu, G.: Application of Wireless Sensor Network in Water-Saving Irrigation. China Rural Water and Hydropower, 24–26 (2007)
15. Sheth, A., Tejaswi, K., Mehta, P., et al.: Senslide: A Sensor Network Based Landslide Prediction System. In: Proceedings of Sensys 2005-The 3rd International Conference on Embedded Networked Sensor Systems, pp. 280–281 (2005)
16. Martinez, K., Ong, R., Har, J.: A Sensor Network for Hostile Environments. IEEE Secon, 81–87 (2004)

Application of Model Driven Architecture
to Development Real-Time System
Based on Aspect-Oriented[*]

Wei Qiu[1] and Li-Chen Zhang[2]

[1] School of Computer Science, JiaYing University MeiZhou City GuangDong Province,
514015, China
[2] Faculty of Computer Science, Guangdong University of Technology, GuangZhou City
Guang Dong Province, 514015, China
qiuwei@jyu.edu.cn, lchzhang@gdut.edu.cn

Abstract. A way to specify Aspect-based software architectures for real-time systems is introduced. Component models are specified taking the Model Driven Architecture (MDA) approach, and employing UML notations. First, the principle of the developing process based on the Aspect-Oriented approach and the new concepts of UML-specified component architectures are addressed. Then, the conceptual framework architecture for the design of embedded real-time systems is presented, in which platform-independent component models are built. AOP is a new software development paradigm, which could attain a higher level of separation of concerns in both functional and non-functional matters by introducing aspect, for the implementation of crosscutting concerns. Different aspects can be designed separately, and woven into systems. This article introduces the technology of MDA, aspect-oriented, real-time systems. The paper takes the Aspect-oriented to the MDA modeling by the UML extension mechanisms, and presents a method, which is Aspect-Oriented MDA. In this article, UML profile is utilized to construct the meta-modal specifications respectively for common Aspect-Oriented and AspectJ. So the core business logic and the crosscutting aspects can be modeled as separate, modular Aspect-Oriented PIM's and PSM's.

Keywords: Real-time System, Aspect-Oriented, Model-Driven Architecture, Platform Independent Model, Platform Specific Model.

1   Introduction

In developing software for embedded systems one has to consider non-functional and resource constraints besides software quality aspects such as re-usability, because the

---

[*] Wei QIU , born in 1974, received M.S. degree in Software Engineering from Guangdong University of Technology of China in 2004. He is an Associate Professor at School of Computer Science and JiaYing University. His main research interests Software engineering, data mining, e-commerce technology, Real-time Systems. The work supported by the Major Program of National Natural Science Foundation of China under Grant No.90818008.

B. Liu and C. Chai (Eds.): ICICA 2011, LNCS 7030, pp. 569–576, 2011.

correct operation of such a system is not only dependent on the correct functional working of its components, but also dependent on its non-functional properties. Embedded systems often have limited resources such as processing power, storage capacity and network bandwidth [1]. A developer has to cope with these constraints and make sure that the software will be able to run on the constrained system. In addition, embedded systems also have timing constraints on their computations. Missing a time constraint can be catastrophic or annoying. In order to cope with these special requirements, it is indispensable to improve and enhance the component technologies to design software for embedded systems. This paper summaries lessons from several projects where a model-driven approach and component based development was implemented to address productivity problems. The important issues are surprisingly non-technical and require paying careful attention to team structure and project organization. In the paper, the author analysis non-functional requirements of the real-time systems at first, and then apply aspect-oriented MDA modeling to develop an example of real-time systems, and propose how to model aspects of timer and real-time constraints. Finally, the author use extended sequence diagram to dynamically show how aspect impact real-time systems.

2   MDA and Real-Time Component Model

Model-Driven Software Development provides a set of techniques that enable the principles of agile software development (www.agilealliance.org) to be applied to large-scale, industrialized software development [2]. MDA envisages systems being comprised of many small, manageable models rather than one gigantic monolithic model, and allows systems to be designed independently of the technologies they will eventually be deployed on. The approach is based on two essential concepts, viz., the Platform Independent Model (PIM) and the Platform Specific Model (PSM), which separates the specification of system functionality from the specification of the implementation of that functionality on a specific technology platform and provides a set of guidelines for structuring specifications expressed as models [3]. Both PIM and PSM can be expressed in extended UML [4]. This article introduces the technology of MDA, aspect-oriented, real-time systems and UML[5]. This article takes the AO to the MDA modeling by the UML extension mechanisms, and presents a method, which is Aspect-Oriented MDA. In this article, UML profile is utilized to construct the meta-modal specifications respectively for common Aspect-Oriented and AspectJ [6]. So the core business logic and the crosscutting aspects can be modeled as separate, modular Aspect-Oriented PIM's and PSM's.

2.1   Modeling Aspect-Oriented PIM for Real-Time System

Aspect-oriented platform-independent PIM needs to be expressed in the form's structure and behavior characteristics, such as crosscutting aspects of the operation, the selected connection point of the structure of crosscutting relations [7]. If the direct use of object-oriented model or a simple extension that will lead to the case in the absence of the model specification appears ambiguous. In the MDA model transformation in the end lead to failure. Therefore need to use PIM meta-model to define [8]. In addition, cross-platform support of model exchange and reuse, aspects of the PIM meta-model level should also be defined in order to have good features and a unified platform-independent specifications [9]. On the left hand side of Fig. 1 a usage of the MDA approach is shown. A system can horizontally be split into multiple functions, each of which has a model of its subsystems. These models can be considered to be views on an overall system PIM. The PIM can be converted into a PSM running on a specific platform. The converting technology is transformations such as in Fig. 1: transformation T1 integrates some requirements of a system for defining a PIM, transformation T2 converts the PIM into PSMs for each deployment platform.

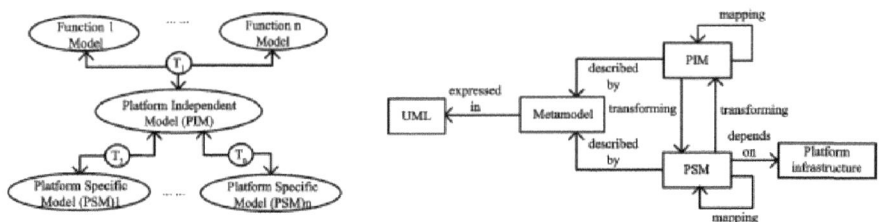

Fig. 1. Meta-model of Aspect-oriented MDA approach

According to the basic concepts of AOP, the aspect PIM meta-model describes the modeling aspect-oriented crosscutting model elements required for the establishment of aspect-oriented specification provides the definition of PIM, the aim is to develop the framework of MDA can First, the platform independent model, namely the establishment of aspect-oriented PIM. Meta-model of the various aspect-oriented platform has some common characteristics of abstract, semantically covers a total of aspect-oriented nature and behavior, ignoring the details of the specific platform that can guide the establishment of aspect-oriented PIM [10].A platform-independent real-time component model is constructed by defining a set of stereotypes, a Contract class which governs some functional, non-functional constraints, and assigning a Component Meta-mode. Here some kinds of components are defined meeting the requirements in the embedded real-time domain, such as Active Component, Passive Component, and Event Component [11]. Each of them has own attributes (such as scheduling, resources and initialization). We now detail each of these stereotypes.

2.2   Modeling Aspect-Oriented PSM for Real-Time System

From the component developer's point of view, a component has a unique identifier and set of properties. It encapsulates a well-defined piece of the overall application functionality. An application is assembled from collaborating components accessing each other through a well-defined component interface [12]. The external view on a component is a set of provided and required interfaces, which may be exposed via ports. A component may also have an internal view in the form of a realization, which is a set of class instances or smaller components that collaborate to implement the services exposed by the component's provided interfaces while relying on the services of its required interfaces.

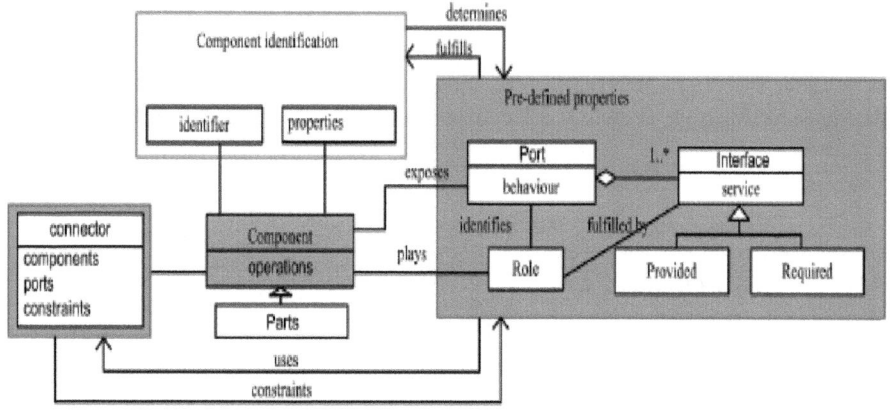

Fig. 2. Component meta-model of Aspect-oriented

The concept can be used to specify both logical and physical components. Now we can define a component meta-model as shown in Fig. 2, which illustrates the component concepts and reflects both user's and developer's view. The component architectures are structured by subcomponents, connectors, ports and interfaces. A component owns a unique identifier and set of properties, and defines a set of communication ports of which provide interfaces. Components can exchange data with each other through ports and connectors, only. A component can be a composite one containing other component.

3   The Aspects of PSM to Development Real-Time System

The principle of this development process is to establish, first, a PIM, then, to refine it according to a specific situation and, next, to derive one or more PSM's from this PIM. These PSM's are the results of mappings to several platform models. Both PIM and PSM's are developed according to the rules defined within a meta-model of which can be described in UML. If necessary, the PIM and PSM's can be optimized to reduce their complexity. The PSM produced last is used to generate executable code, preferably in an automatic way [8]. The transformation from one PIM to several PSM's is the core of MDA, while the one from a PSM to executable code is straightforward and nowadays supported by a number of commercial tools.

3.1 Requirement Analysis of Real-Time Aspects for the Washer

In order to more clearly understand how to complete the MDA in the aspect-oriented modeling, especially in real-time system. Following through the example of a real-time system is discussed. The system simulates the operation of self-automatic washer process.

Overview of washer system functions:

l) Users before washing clothes texture and weight according to the control panel, select the amount of water, washing time and washing methods. Then there is the remaining amount will include the laundry card into the card slot.

2) When you press the Start button, less the cost of laundry card, while washer started to work in accordance with user requirements and tips ringing sound. After running, the motor starting time decreasing the cycle time. When you press the Cancel button to stop working.

3) When the washer work will be suspended when the outer door laundry work, stop the timer, close the outer door, then continue to work while continuing to timing.

4) The time taken to reach zero when the laundry is completed, run the washer motor power and lights will automatically turn off prompt, suggesting that the bell will ring three times to prompt the laundry work is done.

5) The motor record time and with each washing had been running all the time before adding, in the control panel shows the total hours of work.

6) Require the user card, the controller displays in the three seconds less than the balance of the card.

7) Require the user presses the start button a second charge.

From the above system (show figure 3) can be summarized in the basic classes, as shown, which includes the user, the washer controller (selection panel), motor, door sensors, indicators, and they may have various properties, such as the controller of the water, laundry format and more.

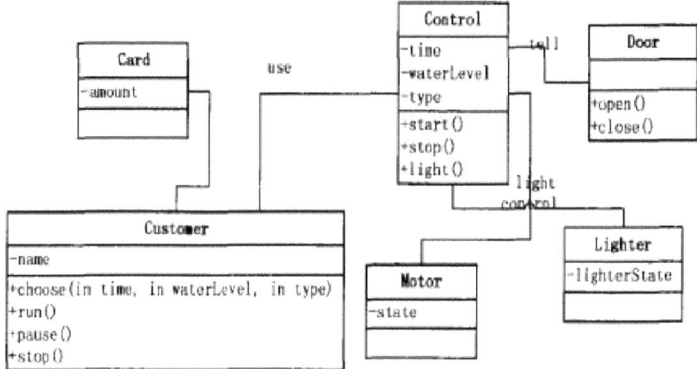

Fig. 3. Core Class of Washer

The washer system has the typical characteristics of real-time systems. Mentioned earlier, the biggest feature of real-time system is required for the system within a certain time limit to respond. In this system, after the display card charge back fees, and laundry are two functions need to be completed within a certain time interval. In addition, after the start of washing laundry in the countdown and begin recording the motor to run this time. These four mechanisms function with time can be divided into two categories; the first two and the time constraints related to the last two and timing related. Since this is a very simplified system, the system in practical applications there will be many areas with similar time constraints or timer functions. Should therefore be both time-related aspects in the model.

3.2   Aspect Time for PIM and PSM

Timer to time constraints and modeling, the literature OMG's modeling time model can be used as a case. PIM to establish aspect-oriented, by analyzing the demand for shows, you should create a time dimension, which it has to count down the operation of the motor cycle time and total time. Processes for each washing into a timer set for each user laundry countdown, but also to motor run time, the motor controller should calculate the total time to retain a variable to be updated after each washing.

In order to run the motor all the time, time, Time to introduce a totalTime variable, in the laundry after each time the user running the second time to the total running time of the motor. Can also be seen from the figure are structural cross-section of the process.Cut into the time constraints the timer before running the method, while the value of the specified time from the constraint starts to count down. Countdown to zero if the methods have not received the news of the end, it means that did not meet the time constraints, go to the appropriate error handling. In the end of the method (after) compare method is called, first stop the countdown, then the time spent compared with minTime and maxTime obtained meets all of the time constraints.

3.3   Added the Core Aspect to PIM and PSM

Washing machine is in fact the basic class diagram has been described in its core function as the core PIM can be transformed. After transformation,   the core of PSM. A clearer response to cross-cutting impact in this figure only those associated with the timing of the core classes that are modeled with 00 ideas, and because the specific implementation technologies through the combination of model transformation, which expressed as ordinary Java classes. The area has been by the time was converted to, Figure 4 show a time-based AspectJ PSM. Aspects and core classes for the connection to the following figure to express their relationship with the crosscut.

The principle of this development process is to establish, first, a PIM, then, to refine it according to a specific situation and, next, to derive one or more PSMs from this PIM. These PSMs are the results of mappings to several platform models. Both PIM and PSMs are developed according to the rules defined within a meta-model which can be described in UML. If necessary, the PIM and PSMs can be optimized to reduce their complexity. The PSM produced last is used to generate executable code, preferably in an automatic way. The transformation from one PIM to several PSMs is the core of MDA, while the one from a PSM to executable code is straightforward and nowadays supported by a number of commercial tools.

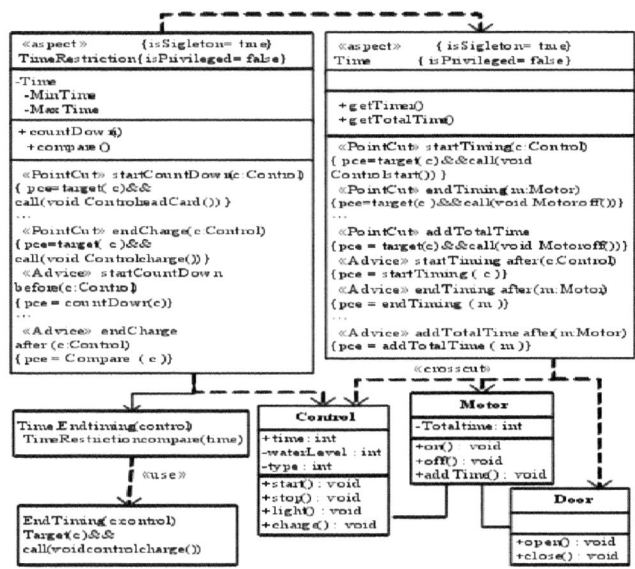

Fig. 4. The Core Aspects of PSM by Adding Timing Aspects of PIM and Time Constraints PSM

4   Conclusion

Real-time systems have many non-functional requirements, which crosscut entire system and are difficult to be handled. This paper presents a real-time system development method based on aspect-oriented Model-Driven Architecture (MDA). It separates the handing of non-functional requirements from the functional ones in design phase, which reduces the complexity of real-time systems development and improves reusability and maintainability of real-time systems, and modularization of crosscutting concerns. Application case proves that the method is effective. A developer has to cope with these constraints and make sure that the software will be able to run on the constrained system. In addition, embedded systems also have timing constraints on their computations. Missing a time constraint can be catastrophic or annoying. In order to cope with these special requirements, it is indispensable to improve and enhance the component technologies to design software for embedded systems.

Acknowledgments. The first author wish to thank Li-Chen ZHANG. This work is supported by the Major Program of National Natural Science Foundation of China under Grant No.90818008,   and by the National Natural Science Foundation of China under Grant No.60774095, and by the Natural Science Foundation of Guangdong Province of China under Grant No.07001774.

References

1. Bjoerkander, M., Kobryn, C.: Architecting Systems with UML 2.0. IEEE Software 20(4), 57–61 (2003)
2. Boulet, P., Dekeyser, J.L., Dumoulin, C., Marquet, P.: MDA for SoC Embedded Systems Design, Intensive Signal Processing Experiment. In: Proc. SIVOES-MDA, Workshop at UML 2003, San Francisco, pp. 166–174 (2003)
3. Monperrus, M., Jaozafy, F., Marchalot, G., Champeau, J., Hoeltzener, B., Jézéquel, J.-M.: Model-driven simulation of a maritime surveillance system. In: Proceedings of the 4th European Conference on Model Driven Architecture Foundations and Applications, ECMDA 2008, pp. 246–267 (2008)
4. Douglass, B.P.: Real-Time UML – Developing Efficient Objects for Embedded Systems, 2nd edn., pp. 332–358. Addison-Wesley (2002)
5. Monperrus, M., Jézéquel, J.-M., Champeau, J., Hoeltzener, B.: Measuring models. In: Jörg, R., Christian, B. (eds.) Model-Driven Software Development: Integrating Quality Assurance, pp. 576–592. IDEA Group (2008)
6. Elrad, T., Filman, R.E.: Aspect-Oriented Programming. Communication of ACM 44(10), 29–32 (2010)
7. Monperrus, M., Long, B., Champeau, J., Hoeltzener, B., Marchalot, G., Jézéquel, J.-M.: Model-driven architecture of a maritime surveillance system simulator. System Engineer 13, 1030–1045 (2010)
8. Papapetrou, O., et al.: Aspect Oriented Programming for a component-based real life application: A case Study. In: 2004 ACM Symposium on Applied Computing, pp. 1555–1558 (2004)
9. Staron, M., Meding, W., Nilsson, C.: A framework for developing measurement systems and its industrial evaluation. Inf. Softw.Technol. 51, 721–737 (2009)
10. Zhang, L., Liu, R.: Aspect-Oriented Real-time System Modeling Method Based on UML. In: Proc.of RTETA 2005, pp. 546–551. IEEE Press (2005)
11. Kniesel, H.Y.: Towards Visual AspectJ by a Meta Model and Modeling Notation. In: Proc. of the 6th International Workshop on Aspect-Oriented Modeling, Chicago, Illinois,USA, pp. 1134–1146 (2005)
12. Monperrus, M., Jaozafy, F., Marchalot, G., Champeau, J., Hoeltzener, B., Jézéquel, J.-M.: Regular Paper: Model-driven generative development of measurement software. In: Software System Model, pp. 1–16 (2010); Published online: (June 13, 2010)

Application Integration Patterns Based on Open Resource-Based Integrated Process Platform

Yan Zheng, Hongming Cai, and Lihong Jiang

School of Software, Shanghai JiaoTong University, Shanghai, China
zhengyan.claire@gmail.com, {cai-hm,Jiang-lh}@cs.sjtu.edu.cn

Abstract. The push toward more interactive and flexible applications, motivated by huge demand in terms of cost saving, shorter development cycle, faster adjustment, and more reliable execution, has generated the need for integrating the different applications and attract more business people involved in. New approaches and tools for enterprise lightweight collaboration and composition, such as Mashup and Web service, bring various fresh and significant elements into present system architecture. In this work, we firstly indentify and classify the integration architecture framework systematically. Subsequently, five typical patterns for enterprise integration are identified, characterized, and evaluated with focus on the practical applicable scenarios. Within the current research project ORIPS, a resource-based platform that allows for different options to realize resources reorganization and application integration, verifies the correctness and feasibility of integration framework and patterns, and creates promising solution in rapid system development.

Keywords: Integration, Resource, Web Service.

1 Introduction and Motivation

Along with the extensive use and fast update of enterprise software, the company generates an urgent need for integrating different applications in terms of efficient communication and reliable interaction. To respond to rapidly changing business requirements, the company is suggested to develop new applications by reusing their existing resources including applications, web services and data assets, for lower cost, higher quality and shorter development cycle. Integration has led to a large body of research and development in several areas such as data integration, service integration, and recently light weight composition named Enterprise Mashup.

The purpose of the enterprise integration is to realize intra- and inter- enterprise function invocation and information exchange in two main scenarios: integrating internal systems of each enterprise, referred to as enterprise application integration (EAI), and integration with external entities, referred to as business-to-business integration (B2B). In the internal, there is a need to integrate data applications relocated to various systems, while the expectation of the external interactions is that services of each enterprise can communicate seamlessly with those of its partners [1].

In Fact, there are many feasible proposals and technologies advanced for software integration issues after several decades' deep research such as data integration.

B. Liu and C. Chai (Eds.): ICICA 2011, LNCS 7030, pp. 577–584, 2011.

However, they cannot meet the growing needs. Moreover, rapid changes and huge demands in business desire new integration approaches for situational and ad-hoc usage that allow for End User Development (EUD) first introduced by Martin[2], but most of today's software lack in providing its users intuitive ways to modify or recompose the original applications. Most existing approaches insufficiently support EUD for infrequent, situational, and ad-hoc integration and collaborations [3]. Nowadays, enterprise lightweight composition approaches and tools turn to be fairly popular within the concept such as software as a service (SaaS) and Mashup since they are promising solutions to unleash the huge potential of integrating the mass of users into development.

Service-oriented Architectures (SOAs) provide an architectural paradigm and abstractions that allow simplifying integration. There are a number of technologies available to realize SOA. Among them XML-based Web service is a step in the right direction, as the best-known enabler supporting SOA [4]. Web service as a programmable Web application with standard interface descriptions that provides universal accessibility through standard communication protocols, provide a holistic set of XML-based, ad-hoc, industry-standard languages and protocols, using Web Service Description Language (WSDL) for description, Universal Description Discovery and Integration (UDDI) for publication and discovery, and Simple Object Access Protocol (SOAP) for transportation. The service built under the Representation State Transfer (REST) architecture, referred to RESTful service, is another step forward on the process side [5]. In REST architecture, applications in the network are modeled as a set of resources, which are uniquely addressable using a URI (Uniform Resource Identifier). REST promotes a client-server, stateless and layered architecture.

We believe it is useful to study the integration in terms of layers, which address various parts of problem at different level of abstractions. At a conceptual level, information systems are designed around three layers: presentation, application logic and resource management. Presentation layer plays the role in communicating with external entities, human users or other computers. Application logic layer involves programs that implement the actual operation requested by the client through the presentation layer. Resources management layer encompasses the data that resides in database, file systems, or other information repositories [6].

Based on the current research project ORIPS (Open Resource-based Integrated Process Platform), a resource-based platform that allows for different options to realize resources reorganization and enterprise integration in system development. The integration framework and integration patterns are proposed to support multi-layer enterprise application integration.

2   Enterprise Integration Patterns

2.1   Enterprise Resources Integration Platform for Layer Perspectives

From the discussion above, we should know it is necessary to deeply understand the constituent components of the relatively representative system that needs to be integrated due to the various integration perspectives. In this paper, the architecture is separated into five layers horizontally, which goes from lower lever layer to higher lever layer that mostly build on top on the lower ones and may or may not be need depending on the application[7].

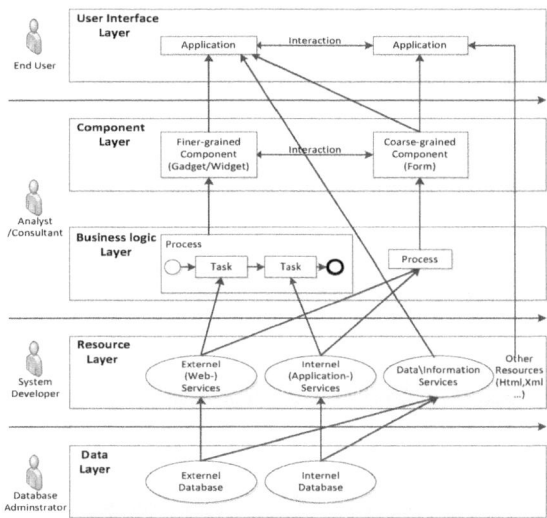

Fig. 1. Enterprise resources integration platform

Figure 1 summarizes conceptualization with regard to the involved actors and their roles in the development and allocation process. The lowest data layer solves the core problem of system integration, data interaction. Data service as a major advance in the data-level integration should be taken into consideration [8]. On the resources layer, the external and internal resources are located. Standardized interfaces abstract these resources from their technical implementation and facilitate the loose coupling of different resources fulfilling a central requirement of service-oriented architecture [9]. These resources are provided by internal developers or offered by external vendors. The business logic layer refers to business process modeling, business process execution language for Web Services (BPEL4WS) and BPEL engine. On the highest level of abstraction, knowledge workers create, adopt, use and share applications. They facilitate the adding and removing of pre-built components as well as accessible services and other resources, e.g., by linking well-defined input and output ports with a graphical development tool and therefore personalizing their work environments to fulfill individual, situational business needs [10].

2.2 Enterprise Resources Integration Patterns

There are several ways to achieve the integration and collaboration from heterogeneous systems. In this paper we present five patterns, including UI integration, component integration, business logic integration, resource integration and data integration, each of them describing one option for realizing the integration. The five integration patterns in different system layers describe how to integrate resources from disparate systems. We are exploiting the opportunity created by recent rapid emergence of SOA and its derivatives for enterprise application integration. As provided by previous examples, we will discuss the characteristics for each integration pattern from the perspective of the enterprise in accordance with the implementation from the simplest to the hardest way. Each business scenario could find its corresponding integration patterns and vice versa. However, all patterns have certain advantages and disadvantages, shown in table 1.

Table 1. Five integration patterns' comparison

Pattern	Advantage	Disadvantage	Participates
UI Integration	Few integration operation, easy to implement by non-IT stuff	No automated data exchange, functionality limited	End user
Component Integration	Fine-grained integration with more flexibility, easy to implement by non-IT stuff	Gadget needs to be available, pre-defined and limited functionality, homogeneous building platform	Analyst /Consultant
Business Logic Integration	Reusing exiting business logic, easy to implement by non-IT stuff	Task and Web service need to be available, complex to execute	Analyst /Consultant
Resource Integration	Reorganize service resources with great flexibility	Web service needs to be available, low touch by the end user	System Developer
Data Integration	Data service makes low complexity on data access and control.	Data service needs to be available, low touch by the end user	System Maintainer

When you choose a pattern for integration, it always tends to suffer the trade-off between reach and richness. UI integration has high reach due to the fact that it has nearly no technical requirements for most participants in the collaborations. On the other hand, the richness would be quite low, because of the limitation of the pre-defined applications which makes the extensible integration difficult or not possible. Data integration on the contrary has a very low reach, as it strictly depends on the code or even hard-wired connections to be established. Still, it allows a high richness, as the completely interacted systems can be configured to handle any data exchanges. Beside the corporate development strategy, the trade-off between reach and richness should also be taken into consideration when company chooses a proper pattern.

3   Enterprise Integration Implementation Based on ORIPs

Enterprise Integration mostly begins with the existing resource, the employees' information and inner logic, or specific function in legacy system that ran well before. Then an efficient business process should be designed and described in standardized modeling language that improves the comprehension and execution of the process. Moreover, the existing resources are expected to operate following the pre-defined sequence and decision logic. Ideally, all the functions are running as we assumed.

ORIPs as a powerful integrating and developing platform can satisfy the demand of enterprise system integration. Services management, business process visualized design, simulation and optimization are realized in the ORIPs.

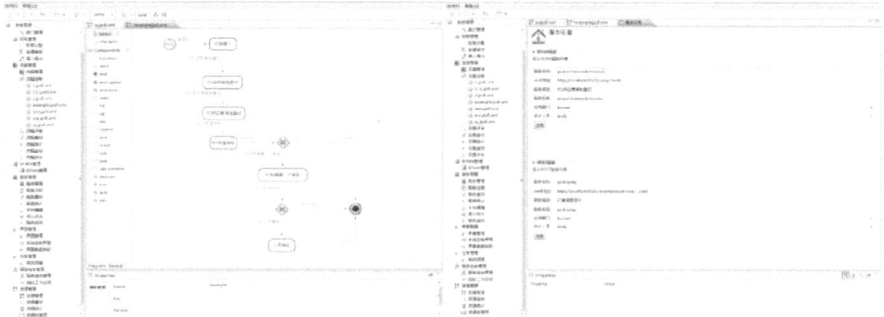

Fig. 2. Mockups of ORIPs for services registration and business process modelling

In this session, we will explain and verify the integration pattern through the operation in ORIPs step by step. A simple scenario is given as follows: The procurement company A is in charge of goods purchasing. The sales company B, as a supplier, provides the goods for the purchaser. A typical purchasing request between the two companies would like this: an employee at the company A creates Request for Quotation (RFQ) specifying the type and quantity. An employee at company B can approve or decline this request depending whether they have enough materials to fulfil the order by checking their inventory IT system. Ideally, supplier can accept this request and give a respond to the purchaser including the price and delivery time, etc. After comparison among several proposals from each supplier, the procurement department chooses one company as their supplier and creates the purchase order (PO) to enter the procurement process.

As depicted in figure 3. Consultants in company A design reasonable and efficient workflow, named process here, consisting of various tasks in a configurable sequence. Company B could package the quotation function into an independent task as a node in their original process. Thereby company A could add the quotation task into the procurement process. Furthermore, if the quotation function refers to more than one task, we could package them into process and we could see that tasks and processes are flexible enough for invocation and reorganization. Business logic integration mainly refers to the process and task supporting by both the internal service and external service. The integration in this layer targets on the reusing of the existing business logic and external business function. The consultant could arrange tasks in optimized executing sequence in terms of efficiency and lower cost by defined them into process unit using the Business Process Execution Language (BPEL) for the sake of future reusing, and the process and task concept enable the workflow invocation from other systems. Process as a set of tasks with specified functions could be defined in different granularities, which is a very important in system analysis and design. Each task binds the specific services downwards and presents on the components upwards, typically related to one Form.

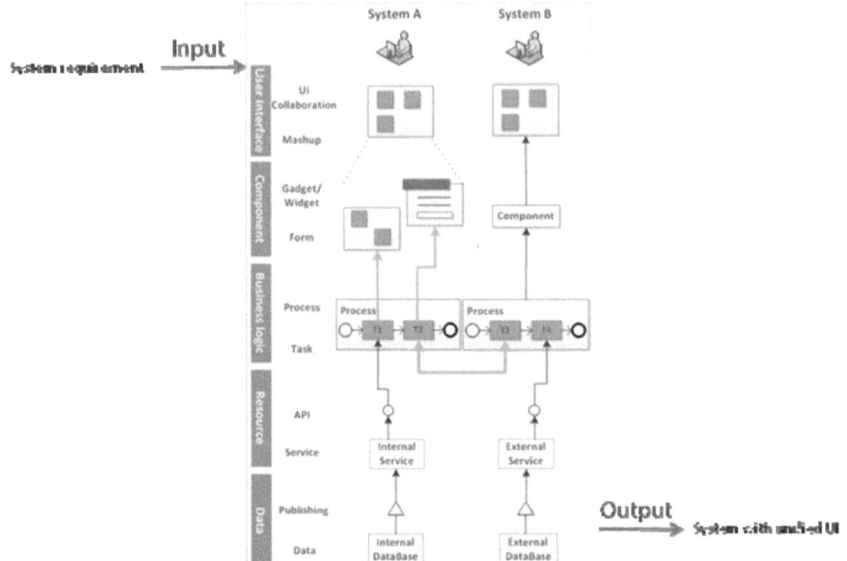

Fig. 3. Business logic integration of an sample scenario

Step 1: Discovery existing resources and register in the system service library

First of all, we should distinguish the necessary resources and make sure they are available. Then, publish and register them into ORIPs. Some of the services could be drawn in the old system, e.g., creating RFQ and PO, and services in other systems should be invocated with authority, e.g., evaluating the quotation.

There are three possibilities at the beginning of the integration. First and the simplest, all necessary services are standby and you just register them in to our system service library. The register course analyzes the service information and extracts the operations and input and output parameters in the WSDL. Secondly, users want to find services that match the functions they specified. They could query their target services via key words since services have been named following specified principles, or discovery services adopting the semantics. System allow for services management and auto expression as a service tree, considering the services' semantic, which makes it easier to find the proper services. The query results are registered in the library as well. In the third case, the ideal services are not available. So, users have to add new services with the help of IT staff. The frequency of third case depends on the capacity of the existing resource library. The services accumulation play important role in the integration in term of efficiency and automation.

Step 2: Analyze and design the business process using embedded modeling tools

Secondly, figure out the reasonable and efficient business logic. Developers will draw the procurement process in the business modeling interface by dragging the shapes provided. Each task in this process is corresponding to a step in the procurement process.

System users, especially business people, prefer to draw the business process in a visualization way via drag-and-drop. Here, we define the unit in a "process" as a "task", component to drag. Referred to previous scenario, PO submission is a task of the procurement process. The relation among tasks, can be added in the process, such as "and", "or", "not" or even more complicated logic.

ORIPs adopt Business Process Modeling Notation (BPMN) as the default business modeling language while various kinds of language are supported at the same time, such as Event Process Chain (EPC), used in ARIS, SAP prominent modeling tool. In addition, our system allows to export the process or input the process that created in other modeling tools due to the advantage of the standardization, and the smooth conversion among several modeling languages, such as from the BPMN to JPDL (jBPM Process Definition Language) and vice versa.

It's an essential and common method to describe the business process using workflow diagram following some standardized principles, BPMN 2.0 here. In this way, we could define the complex relation, execution sequence and the decision logic among tasks. Moreover, it could be easily executed through the process execution language and workflow engine, BPEL and JBPM are adopted in this system.

Step 3: Draw the UI corresponding to each task in the process.

In this step, users could design their own UI corresponding to specified task by drag-and-drop. For PO submission task, you could add textbox on a Form to present the PO number and another textbox to let the purchaser input the price and quantity. Each task corresponds to one Form. Moreover, the better UI design the more user experience improving. Other web pages developing tool can involve in through the input method in ORIPs that support standard XHTML.

Step 4: Bidding UI to services

After the personalized UI have finished, we need to attach the executable function to the UI component. Each component, e.g., textbox, label, grid, get a unique id during UI editing. In this step, we need mapping operations and parameters of the specified service to the component id in order to realize the service function. The mapping allows to two kind of service invocation, SOAP and REST, which adopt different invocation path. After four steps have been set up correctly, the customized system is accomplished.

A complete system integration building is start from the requirement and end in the new system with unified UI. The definition of system integration in not limited to coordinate several existing systems. The better way to realize system integration is to collect the function chips and arrange them into a new system by the business requirement.

The integration method saves efforts on creating a new system than other traditional developing way. From the perspective of cost, it greatly shortens the system development cycle, from several months to few days. Also, integrating development leads to retrench IT stuff. Ideally, end users could develop a complete system without the participation of the IT stuff, while more attention could be paid on the business logic to reduce the mistaken understanding. Furthermore, once the system is created and authorities granted, end users could execute the functions in a process with specified role. In our scenario, for example, manager in procurement department could execute the purchasing workflow and verified PO submitted by the employee. In addition, the established business process is flexible enough to change the sequence of the tasks or add new tasks in. Moreover, ORIPs also support the B/S structure. Users managed by work group, could design and execute the new workflow on the website.

4   Conclusion

The identified patterns, which describe ways how to integrate applications in different architecture layers, including UI layer, component layer, business logic layer, resources layer and data layer, show the possibility and systematicity in the integration architecture involving four participated roles in the course of integration to construct new applications that fulfill their individual and heterogeneous needs. In addition, we line up an efficient way to construct executable business process by ORIPs and verify the feasibility and efficiency of the integration method.

We believe the end user is the focus of next wave of research and development work in the various approaches in SOA. In addition, we should consider the influence brought by the introduction of cloud environment.

Acknowledgments. This research is supported by the National Natural Science Foundation of China under No.70871078, the National High Technology Research and Development Program of China ("863" Program) under No.2008AA04Z126, and Shanghai Science and Technology Projects 09DZ1121500.

References

1. Boualem Benatallah, L., Nezhad, H.R.M.: Service oriented architecture: Overview and directions, pp. 116–130 (2008)
2. Martin, J.: An Information Systems Manifesto. Prentice Hall, New Jersey (1984)
3. Spahn, M., Dörner, C., Wulf, V.: End User Development: Approaches Towards a Flexible Software Design. In: ECIS 2008: 16th European Conference on Information Systems, Galway, Ireland (June 2008)
4. Alonso, G., Casati, F., Kuno, H., Machiraju, V.: Web Services - Concepts, Architectures and Applications. Springer, Berlin (2004)
5. Richardson, L., Ruby, S.: The Resource-Oriented Architecture. In: Restful Web Services. O'Reilly, USA (2007)
6. Sneed, H.M.: Integrating Legacy Software into a Service Oriented Architecture. In: CSMR 2006, pp. 3–14. IEEE CSP (2006)
7. Wirtsch, D., Beck, R.: Enterprise Mashup Systems as Platform for Situational Applications. Business & Information Systems Engineering 2, 305–315 (2010)
8. Carey, M.: Data delivery in a service-oriented world: the bea aqualogic data services platform. In: SIGMOD (2006)
9. High, R., Krishnan, G., Sanchez, M.: Creating and maintaining coherency in loosely coupled systems. IBM Syst. J. 47(3), 357–376 (2008)
10. Simmen, D., Altinel, M., Markl, V., Padmanabhan, S., Singh, A.: Damia: data mashups for intranet applications. In: Proceedings of the 14th International Conference on Management of Data, Vancouver (2008)

Towards a Verbal Decision Analysis on the Selecting Practices of Framework SCRUM

Thais Cristina Sampaio Machado, Plácido Rogério Pinheiro,
Marcelo Marcony Leal de Lima, and Henrique Farias Landim

University of Fortaleza (UNIFOR) – Graduate Course in Applied Informatics Av.
Washington Soares, 1321 - Bl J Sl 30 - 60.811-341 - Fortaleza – Brasil
thais.sampaio@edu.unifor.br, placido@unifor.br,
marcelomarcony@hotmail.com, hflandim@gmail.com

Abstract. Considering that agile methodologies, in focus Framework SCRUM, are always more popular for Development Software Companies, and noticing that the mentioned companies cannot always apply every characteristics of the framework, this paper presents an application of methodologies from the Verbal Decision Analysis (VDA) framework to generate a rank of the SCRUM characteristics to be applied in a company. The paper consists on an application of a questionnaire for a group of experienced ScrumMasters and considers the elicitation of preferences of a decision maker for creating the final rank of alternative. The methodology from VDA framework selected to be applied for ranking the alternatives was ZAPROS-LM. The final rank of alternatives indicates a list of SCRUM practices, from the more preferable to less preferable one, according to the decision maker responses.

Keywords: Agile, Framework, SCRUM, Elicitation of Preferences, Verbal Decision Analysis, ZAPROS-LM.

1 Introduction

Development Software Organizations used to focus on process definition for generating high quality products. However many plans and documentations became hard to maintain because of wrong estimative and projects deployed latter. So, companies which used to adopt maturity models, like Capability Maturity Model (CMMi), for defining and improving their processes, became interested in manager their projects applying agile methods [2].

The use of agile methodologies for managing projects became more popular between Development Software Companies, aiming to create high quality products in less time and spending less documentation.

The paper selected a specific agile methodology for studying: framework SCRUM. The framework is applicable for managing the development of software, group management, feedback for the team and correction of impediments. SCRUM is composed by steps and practices to apply.

B. Liu and C. Chai (Eds.): ICICA 2011, LNCS 7030, pp. 585–594, 2011.

The problem is that, usually, the organizations are not capable of implementing every characteristics of SCRUM. Hence, which would be the best practices of it to be implemented by the organization?

First, experienced ScrumMasters were interviewed through a questionnaire. Thus, it was possible to characterize the SCRUM practices, according to the experience of many professionals.

The SCRUM practices can be described qualitatively, based on a set of multiple criteria. Therefore, the paper studies an area called Multicriteria, which is an approach to support the process of decision making [6]. The characteristics were evaluated qualitatively, applying verbal decision analysis. The method [6] ZAPROS-LM, which belong to the Verbal Decision Analysis (VDA) framework, was used [4] for solving problems that has qualitative nature and are difficult to be formalized, called unstructured [19].

The mentioned method, ZAPROS-LM, aims to rank a group of alternatives from the best to the inferior one. The final ranking will be valuable because the organization which wants to apply SCRUM practices could choose as many practices as its necessity.

2   Framework SCRUM

Considered recent, the "agile" term for software development was created in 2001, as a response for the traditional models of software development.

The bigger concept for agile is "Agile Manifest" [3], which defines some important characterizations:

"We are uncovering better ways of developing software by doing it and helping others do it. Through this work we have come to value:

• Individuals and interactions over processes and tools
• Working software over comprehensive documentation
• Customer collaboration over contract negotiation
• Responding to change over following a plan

That is, while there is value in the items on the right, we value the items on the left more."

Framework SCRUM is an agile method different from the others for focusing on project management, not development. It was developed by Ken Schwaber and Jeff Sutherland to help organizations to carry complex projects [5].

SCRUM assumes that the software development is unpredictable to be planed completely initially, so it must guarantee visibility, inspection and fast adaptation, as can be seen in its pillars [5].

The framework is based on some practices, like: short iterations (from 1 to 4 weeks), close relation with the product owner, planning meetings, daily monitoring, visible charts of activities, and so on.

3   Verbal Decision Analysis

Decision making is a special kind of human activity aimed at the conclusion of an objective for people and for organizations. In the human world, emotions and reasons become hard to separate. In personal decisions or when the consequences reach them, the emotions often influences the decision making process [1].

Generally, multi-criteria decision support methods are based on well structured mathematical models. Even if the description of the problems is initially defined in a qualitative way, later they are transformed into the required quantitative form, in accordance with the model established for the corresponding method [16].

According to [16] in the majority of multi-criteria problems, exists a set of alternatives, which can be evaluated against the same set of characteristics (called criteria or attributes). These multi-criteria (or multi-attribute) descriptions of alternatives will be used to define the necessary solution.

The Verbal Decision Analysis (VDA) framework is structured on the assurance that most decision making problems can be qualitatively described. The Verbal Decision Analysis supports the decision making process by the verbal representation of problems [7][8] [10][11][12][13][14][15][20][21][23][24].

The methodologies of decision making support allow evaluating the alternatives considering the multiple criteria and the decision maker's preferences, which become responsible for the decisions. As a multi-criteria decision support approach, the process does not aim to show a solution for the decision maker, but it aims to help the decision making process [6].

The decision maker's ability to choose is very dependent on the occasion and the stakeholders interest's, although the methods of the decision making are universal.

According to [6], the methods of verbal decision analysis are: ZAPRO-III, ZAPROS-LM, PACOM and ORCLASS. The first three have the goal to establish a ranking of the alternatives from some order of preference. The last is the only methodology for classification from the VDA framework.

4   Interview

Aiming to collect information and opinion with a high number of ScrumMasters, a questionnaire was developed and applied with 6 experienced professionals with the framework SCRUM.

As the first part of the questionnaire, we qualified the professionals. There were selected the answers of ScrumMasters which has similar experiences in projects. The answers considered for the research were the ones which the professional has the following characteristics:

Leading SCRUM projects until 3 years of experience;
Leading until 6 projects applying SCRUM;
Has ever led team with no experience and moderate experience with the framework SCRUM;

The questionnaire aims to characterize the alternatives, which are some practices of the framework SCRUM. All the interviewed analyzed the SCRUM practices according to a group of criteria and criteria values. In the end of each response, it was possible to create a table expressing the interviewee's choices as opinion about the relation between the SCRUM practice and the criteria values.

The answers obtained with the questionnaires were analyzed and created a new table with the responses. For each SCRUM practice, the final table was composed by the answer with major quantity of interviewee's choices. Table 3 presents the final result of the questionnaires.

5   Methodology ZAPROS-LM

5.1   Overview and Structure

As long as the other methods that belong to the Verbal Decision Analysis framework, methodology ZAPROS-LM is also applied to solve problems described qualitatively and supports the decision making process [6].

ZAPROS-LM was created aiming to establish a rank of alternatives from an initial set. For applying the mentioned methodology, there are formulated hypothetical alternatives, which will be compared to real ones.

Dependency between the criteria and criteria values can occasionally occur and must be eliminated through new questions to the decision maker.

The final scale is constructed and its ordering will be shown to real and hypothetical alternatives.

For applying the methodology, a pair of criteria was selected from the list to be compared. A simple rule is formulated to generate questions and conclude the comparison between the selected criteria [6]:

Two middle criteria values are being compared (example X2 and Y2): one becomes more preferable than the other (Y2);

The less preferable one (X2) is compared with the inferior value on the scale of the second criterion (Y3): one becomes more preferable than the other (X2);

The less preferable criterion value (Y3) in the second comparison is compared to the inferior value of the second criterion (X3). Analogous, one becomes more preferable than the other;

Thus, according to the decision maker's answers, the scale of criteria is created [6]. And latter, after answering the analogous comparison between all criteria, a final scale will be developed. To construct the joint ordinal scale, it's necessary to compare all possible pairs of values upon all criteria.

6   Application of ZAPROS-LM

6.1   Criteria Definition and Alternatives

As the first step to apply ZAPROS-LM, the criteria and criteria values, which the alternatives are going to be evaluated against, were defined and associated. For each criterion, there is a scale of values associated [8][9][18]. The criteria and criteria values are stated in Table 1:

Table 1. Criteria and associated values

Criteria	Values of Criteria
A – Difficult degree for implementation	A1. Low: It's implementation doesn't require experience with the framework SCRUM. A2. Medium: It's implementation requires a little experience with the framework SCRUM or can be learned on the job. A3. High: It's implementation requires experience (maturity) about framework SCRUM.
B – Time consumption	B1. Gain: The consumption of time in the project for executing the activity is less than the previous model. B2. Not changed: There is no extra time in project for executing the activity than the previous model. B3. Lose: There is extra time in project for executing the activity comparing to the previous model.
C – Cost for the project	C1. Gain: The new activities are able to provide to the project an economy of cost. C2. Not changed: The new activities do not change the cost of the project. C3. High cost: The new activities are able to increase to the project new costs.

The alternatives for the application will be the practices of SCRUM. Notice that the alternatives are practices from the framework SCRUM that can be assigned to any project (the organization may have a process defined in maturity model or not).

The practices were described in Table 2.

Table 2. Board of Alternatives Identification

ID	Alternatives
Prac1	Sprints (or iterations) with 1 to 4 weeks
Prac2	A product backlog and a sprint backlog creation and prioritization
Prac3	Planning meeting – part 1
Prac4	Planning meeting – part 2
Prac5	Daily Meeting
Prac6	Burn down chart and visible activities board
Prac7	Sprint Review
Prac8	Sprint Retrospective
Prac9	Release Planning

Analyzing each questionnaire, a final board was created with the summary of the answers. Table 3 presents the sum of the interview's answers and finally the characterization of each alternative according to each criterion values (identified in Table 1), described as Final Vector.

Table 3. Characterization of Alternatives

Criteria/ Alternatives	Difficult degree for implementation			Time consumption			Cost for the project			Final Vector
	A1	A2	A3	B1	B2	B3	C1	C2	C3	
Prac1	1	5	0	5	1	0	4	2	0	A2B1C1
Prac2	2	1	3	3	1	2	1	2	3	A3B1C3
Prac3	3	1	1	1	3	2	1	2	3	A1B2C3
Prac4	2	1	3	2	1	3	1	2	3	A3B3C3
Prac5	4	2	0	5	1	0	5	1	0	A1B1C1
Prac6	1	3	2	1	5	0	0	6	0	A2B2C2
Prac7	2	3	1	0	4	2	0	5	1	A2B2C2
Prac8	3	2	1	3	1	2	2	1	3	A1B1C3
Prac9	1	3	1	1	3	1	1	4	0	A2B2C2

7   Computational Results of ZAPROS-LM

The methodology application follows its explanation described in [6][13].

The comparison is made for hypothetical alternatives, next to the best possible alternative (A1B1C1). The vector composed by the hypothetical alternatives is:

V = (211, 311, 121, 131, 112, 113)

For the first comparison, the criteria A and B were chosen. Next, the pair of criteria A and C was compared and latter the pair B and C was chosen. The ordering for the comparison followed the formulated rule described in section 5.1.

As long as the comparisons were done, a matrix could be filled with the decision maker responses. The matrix must be filled with the following values:

0 – the elements were not compared;
1 – element in the row is MORE preferable than element in the column;
2 – element in the row is EQUALLY preferable than element in the column;
3 – element in the row is LESS preferable than element in the column;

Table 6 presents the mentioned matrix filled with the final decision maker answers.

Table 4. Matrix of comparison

		A2	A3	B2	B3	C2	C3
		211	311	121	131	112	113
A2	211	2	1	1	1	1	1
A3	311		2	2	1	2	1
B2	121			2	1	2	1
B3	131				2	2	2
C2	112					2	1
C3	113						2

After all paired comparison, according to the matrix of responses, Table 5 presents the summary of partial scale of preferences for all the comparison:

Table 5. Summary of partial scale of preferences

Criteria compared	Order of preferences
A x B	A2 $\prec$ B2 $\prec$ A3 $\prec$ B3
A x C	A2 $\prec$ C2 $\prec$ A3 $\prec$ C3
B x C	C2 $\prec$ B2 $\prec$ C3 $\prec$ B3

Concluded the comparisons and created the partial order of preferences, then the joint ordinal scale can be generated:

A1B1C1 $\prec$ A2 $\prec$ C2 $\prec$ B2 $\prec$ A3 $\prec$ C3 $\prec$ B3

For all the vectors of hypothetical alternatives, one rank value is assigned. Using this ordered scale, Table 6 presents the relation between the rank values and the hypothetical alternatives:

Table 6. Relation between Vectors and Rank value

Vector	Rank value
111	1
211	2
112	3
121	4
311	5
113	6
131	7

Each real alternative is composed by the vectors of hypothetical alternatives, for example: the SCRUM practice (real alternative) Prac1, which vector is 211, is composed by the hypothetical alternatives 211 (for criterion A), 111 (for criterion B) and 111 (for criterion C). The vectors that formulate each real alternatives is described in Table 3.

Analogous, Table 7 shows the real alternatives and the vectors divided according to the hypothetical vectors. For each one, the table shows the related rank value, followed by the ordered final rank value of the alternative:

Table 7. Rank values for the real alternatives

Prac1	Vector	211	111	111	Ordered rank
Rank value	2	1	1		112
Prac2	Vector	311	111	113	Ordered rank
Rank value	5	1	6		156
Prac3	Vector	111	121	113	Ordered rank
Rank value	1	4	6		146
Prac4	Vector	311	131	113	Ordered rank
Rank value	5	7	6		567
Prac5	Vector	111	111	111	Ordered rank
Rank value	1	1	1		111
Prac6	Vector	211	121	112	Ordered rank
Rank value	2	4	3		234
Prac7	Vector	211	121	112	Ordered rank
Rank value	2	4	3		234
Prac8	Vector	111	111	113	Ordered rank
Rank value	1	1	6		116
Prac9	Vector	211	121	112	Ordered rank
Rank value	2	4	3		234

The next step for applying the methodology is a pair comparison between the rank values of real alternatives. As result, the final scale of preferences with the real alternatives is:

Prac5 $\prec$ Prac1 $\prec$ Prac8 $\prec$ Prac3 $\prec$ Prac2 $\prec$ Prac6 and Prac7 and Prac9 $\prec$ Prac4

8   Conclusions

The framework SCRUM is an agile model for managing the development software process which has been discussed very much.

It is composed by practices that can be described qualitatively, based on a set of multiple criteria. Therefore, the paper studies an area called Multicriteria, which is an approach to support the process for decision making [6]. The characteristics were evaluated qualitatively, applying verbal decision analysis.

The ZAPROS-LM method, from the Verbal Decision Analysis framework, was applied to rank the alternatives existent in a rank from the most preferable to the less preferable one.

This paper presents SCRUM practices which were analyzed verbally and ordered by applying ZAPROS-LM, aiming to be selected for a Software Development Company that cannot implement all of them. Thus, the Organization may choose and apply as many practices as it needs, with knowledge about the more preferable ones.

The paper contribution is to prove that verbal decision analysis methodologies can be applied in real problems of elicitation of preferences and decision making, helping Software Development Companies that would like to implant part of SCRUM practices.

9   Future Works

As future works, more research can be done applying other methodologies for classification [22][26], or considering another criteria to evaluate the alternatives, or applying hybrid methodologies for solving the problem [1].

More research will be done when the use of selected practices applied before the methodology in a real software development organization, to study the results of the SCRUM practices for projects.

Acknowledgment. The first author is thankful to the Organization FUNCAP and the second author is thankful to the National Counsel of Technological and Scientific Development (CNPq) for the support received on this project.

References

1. Machado, T.C.S., Menezes, A.C., Tamanini, I., Pinheiro, P.R.: A Hybrid Model in the Selection of Prototypes for educational Tools: An Applicability In Verbal Decision Analysis. In: IEEE Symposium Series on Computational Intelligence SSCI (to appear, 2011)
2. Marcal, A.S.C., Freitas, B., Furtado, M.E.S., Soares, F.S.F., Belchior, A.D., Maciel, T.M.: Blending Scrum Practices and CMMI Project Management Process Areas. Innovations in Systems and Software Engineering 4, 23–35 (2008)
3. Beck, K., et al.: Manifesto for Agile Software Development. Disponível em (2001), 1.http://agilemanifesto.org/
4. Larichev, O.: Ranking Multicriteria Alternatives: The Method ZAPROS III. European Journal of Operational Research 131 (2001)
5. Schwaber, K.: Agile Project Management With Scrum, Microsoft (2004)
6. Larichev, O.I., Moshkovich, H.M.: Verbal decision analysis for unstructured problems. Kluwer Academic Publishers, The Netherlands (1997)
7. Tamanini, I., Pinheiro, P.R.: Challenging the Incomparability Problem: An Approach Methodology Based on ZAPROS. In: Modeling, Computation and Optimization in Information Systems and Management Sciences. CCIS, vol. 14(1), pp. 344–353. Springer, Heidelberg (2008), doi:10.1007/978-3-540-87477-5
8. Tamanini, I., Machado, T.C.S., Mendes, M.S., Carvalho, A.L., Furtado, M.E.S., Pinheiro, P.R.: A Model for Mobile Television Applications Based on Verbal Decision Analysis. In: Sobh, T. (org.) Advances in Computer Innovations in Informations Sciences and Engineering, vol. 1, pp. 399–404. Springer, Heidelberg (2008), doi:10.1007/978-1-4020-8741-772
9. Machado, T.C.S., Menezes, A.C., Pinheiro, L.F.R., Tamanini, I., Pinheiro, P.R.: Toward The Selection of Prototypes For Educational Tools: An Applicability In Verbal Decision Analysis. In: 2010 IEEE International Joint Conferences on Computer, Information, and Systems Sciences, and Engineering, CISSE (2010)
10. Tamanini, I., Carvalho, A.L., Castro, A.K.A., Pinheiro, P.R.: A Novel Multicriteria Model Applied to Cashew Chestnut Industrialization Process. Advances in Soft Computing 58(1), 243–252 (2009) doi:10.1007/978-3-540-89619-7 24

11. Tamanini, I., Castro, A.K.A., Pinheiro, P.R., Pinheiro, M.C.D.: Towards an Applied Multicriteria Model to the Diagnosis of Alzheimer's Disease: A Neuroimaging Study Case. In: Proceedings of 2009 IEEE International Conference on Intelligent Computing and Intelligent Systems, Shanghai, China, vol. 3, pp. 652–656. IEEE Press, Beijing (2009), doi:10.1109/ICICISYS.2009.5358087

12. Larichev, O., Brown, R.: Numerical and verbal decision analysis: comparison on pratical cases. Journal of Multicriteria Decision Analysis 9(6), 263–273 (2000)

13. Moshkovich, H., Larichev, O.: ZAPROS-LM– A method and system for ordering multiattribute alternatives. European Journal of Operational Research 82, 503–521 (1995)

14. Larichev, O.I.: Method ZAPROS for Multicriteria Alternatives Ranking and the Problem of Incomparability. Informatica 12, 89–100 (2001)

15. Tamanini, I., Pinheiro, P.R., Carvalho, A.L.: Aranau Software: A New Tool of the Verbal Decision Analysis, Technical Report. University of Fortaleza (2007)

16. Gomes, L.F.A., Moshkovich, H., Torres, A.: Marketing decisions in small businesses: how verbal decision analysis can help. Int. J. Management and Decision Making 11(1), 19–36 (2010)

17. Tamanini, I.: Improving the ZAPROS Method Considering the Incomparability Cases. Master Thesis I Graduate Program in Applied Computer Sciences. University of Fortaleza (2010)

18. Machado, T.C.S., Menezes, A.C., Pinheiro, L.F.R., Tamanini, I., Pinheiro, P.R.: Applying Verbal Decision Analysis in Selecting Prototypes for Educational Tools. In: 2010 IEEE International Conference on Intelligent Computing and Intelligent Systems (ICIS), Shanghai, China (2010)

19. Simon, H., Newell, A.: Heuristic Problem Solving: The Next Advance in Operations Research. Oper. Res. 6, 4–10 (1958)

20. Dimitriadi, G.G., Larichev, O.I.: Decision support system and the ZAPROS-III method for ranking the multiattribute alternatives with verbal quality estimates. European Journal of Operational Research (December 2002)

21. Tamanini, I., Pinheiro, P.R.: Applying a New Approach Methodology with ZAPROS. In: XL Simpósio Brasileiro de Pesquisa Operacional, 2008, João Pessoa, Brazil. XL Simpósio Brasileiro de Pesquisa Operacional, pp. 914–925 (2008)

22. Brasil Filho, A.T., Pinheiro, P.R., Coelho, A.L.V.: The Impact of Prototype Selection on a Multicriteria Decision Aid Classification Algorithm. In: Sobh, T. (org.) Innovations and Advanced Techniques in Computing Sciences and Software Engineering: SpringerLink, vol. 1, pp. 379–382 (2010)

23. Mendes, M.S., Carvalho, A.L., Furtado, E., Pinheiro, P.R.: Towards for Analyzing Alternatives of Interaction Design Based on Verbal Decision Analysis of User Experience. Internacional Journal of Interactive Mobile Technologies (IJIM) 4, 17–23 (2010)

24. Mendes, M., Carvalho, A.L., Furtado, E., Pinheiro, P.R.: A co-evolutionary interaction design of digital TV applications based on verbal decision analysis of user experiences. International Journal of Digital Culture and Electronic Tourism 1, 312–324 (2009)

25. Pinheiro, P.R., Furtado, M.E.S., Mendes, M.S., Carvalho, A.L.: Analysis of the interaction design for mobile TV applications based on multi-criteria. IFIP, pp. 389–394 (2007)

26. Brasil Filho, A.T., Pinheiro, P.R., Coelho, A.L.V.: Towards the Early Diagnosis of Alzheimer's Disease via a Multicriteria Classification Model. In: Ehrgott, M., Fonseca, C.M., Gandibleux, X., Hao, J.-K., Sevaux, M. (eds.) EMO 2009. LNCS, vol. 5467, pp. 393–406. Springer, Heidelberg (2009)

Development of Portable Electrocardiogram Signal Generator

Ai-ju Chen[1] and Yuan-juan Huang[2]

[1] School of Electronic and Electrical Engineering, Wuhan Textile University,
Wuhan, China 430073
[2] Department of Gynecology and ObstetricsQianjiang Municipal Hospital,
Qianjiang City, Hubei Province, China 433100
aijuchen@126.com

Abstract. To meet the rising popular demand on the treatment of diseases and health security, a portable medical electrocardiogram (ECG) signal generator is developed. With the integrated circuit design, microcomputer control technology, the communication technology and the corresponding software technology, the design utilizes Samsung's ARM chip S3C2410 as the whole system master chip, including micro controller, Ethernet communication interface, serial communication interface, USB communication interface, power supply circuit, DA conversion circuit and signal processing circuit. It not only realizes the functions of normal ECG signal generator, but also filters out actual human signal in ECG data. It can also compare different manufacturers of ECG products' performance through standard of database of ECG data.

Keywords: ECG signal generator, ARM, micro controller, microcomputer control technology, communication technology.

1 Introduction

Human electrocardiosignal need in displayed on ECG products, can observe the human heart function whether in good condition. Early in 1887 home famous French electrophysiological Waller apply Lippman Alzheimer describe out first electrocardiogram in the history Pioneered the electrocardiogram record precedent. 1911 British electrical engineers William du Bois Duddell according to paper of William Einthoven design the electrocardiogram apparatus first face to market. Then, human through ECG product acquisition ECG signals and shows waveform, the ECG signals to the human body, and judging human heart function is in good condition by observing waveform

Electrocardiogram instrument is the most basic and necessary way to check heart attack, high quality means of the electrocardiogram (ECG) is the right-hand man for doctor diagnosed heart disease. The performance must meet people requirements in the process of ECG products used, and ensure the accuracy of measuring parameters and efficient and normal use. Therefore, before ECG products was put to use, need testing to its performance and other parameters. ECG simulation signal generator is a kind of testing tool which used in ECG product development.

B. Liu and C. Chai (Eds.): ICICA 2011, LNCS 7030, pp. 595–601, 2011.

Current medical use of ECG signal generator, most is imported, expensive. Domestic production product by division components, they have large size, no less function and without its core technology, can't download lesions ECG waveform. In order to test for ECG products of performance and the contrast similar products, This topic design of ECG signal simulation generator, through digital signal reduction by ECG original analog signals to simulate human body, or send other analog signal to ECG products, achieve the purpose of test ECG products. The whole system realizes several different format of ECG signal data of transmission through serial and net, storage and analog output and polarization voltage function, and can also produce sine wave and square-wave and normal ECG wave can be used waveform, such as amplitude, frequency of regulation. This ECG signal generator not only with the function of normal ECG signal generator, but also can reduction actual human signal which put in ECG data, can comparison different manufacturer of ECG product performance according to ECG data in the standard of ECG database , predictable it with prospect broad market.

2   System Introduction

ECG simulation signal generator forms by the hardware and software of upper and lower level computer, the task of PC is sending ECG data down to machine work, the task of lower machine is responsible for receiving and storage ECG data. PC is based on Visual C++ integrated environment of software system. through Visual C++ complete program design. Lower place machine is based on embedded system by the ARM9, ARM9 core board, serial ports, Ethernet, DAC, LCD, input/output, and other plug-ins interface component.p

2.1   System Design Brief

ECG simulation signal generator including micro controller, Ethernet communication interface, serial communication interface, USB communication interface, power supply circuit, DA conversion circuit and signal processing circuit. This design using Samsung's ARM chips as the whole system master S3C2410 chips. The whole system hardware diagram as shown in figure 1 shows:

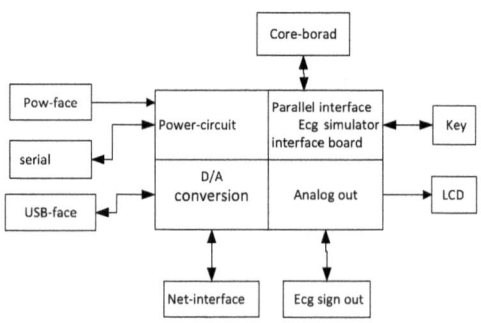

Fig. 1. Hardware diagram of the whole system

This system integrated circuit design, microcomputer control technology, the communication technology and the corresponding software technology etc., all research contents include: 1, S3C2410 lower level computer control program development; 2, PC user interface development, including read ECG data on PC, and display of serial ECG data files Ethernet portal transmission; 3 the development the firmware of ECG simulation signal generator system.

2.2   System Software Design

1) total program flow chart

This design software of mainly divided into the PC and lower place machine , PC based on Visual C++ to implementing software system integration environment, using Visual C++ language programming complete design. Lower level computer software design applications ARM development environment ADSv1.2, simulators ARM v2.2 Mutli - ICE, based on embedded operating system mu C/OS - □, realize the data storage and receive data from PC and in response to the keyboard's button action .and display the operation interface on narrow LCD screen . Figure 2 for its total program flow chart.

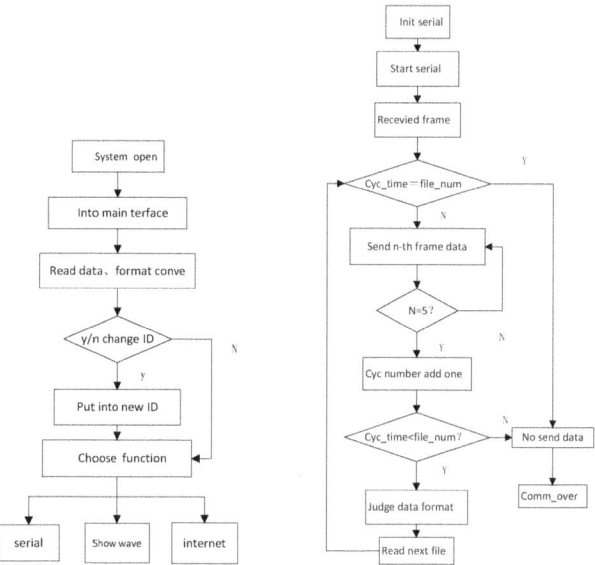

Fig. 2. Total program flow chart   **Fig. 3.** Serial communication flow chart

2) serial communication module

Lower place machine request five frames, which PC send fourth frame to under machine is actual ECG data, under mainly introduced the overall process (FIG.3) and the fourth frames (figure 4) delivery program (including batch send).

Realize : in advance to stored the path name of send documents, first documents send finished, use string functions Mid () to obtain the next file path name, also count on cycling times add 1, then carries on the data read, display and send. When cycling time is no shorter than the file number (file path is empty), no longer response at the request frame of the lower level computer.

Sending fifth frame means a communications over, said the judge relationship between send cycle times and the number of reading files, we can know whether send documents, if cycling times less than files number means need to send documents, then can covering a file through the path name, then the communication according to first time method.

The fourth frame delivery process figure 4 shows.

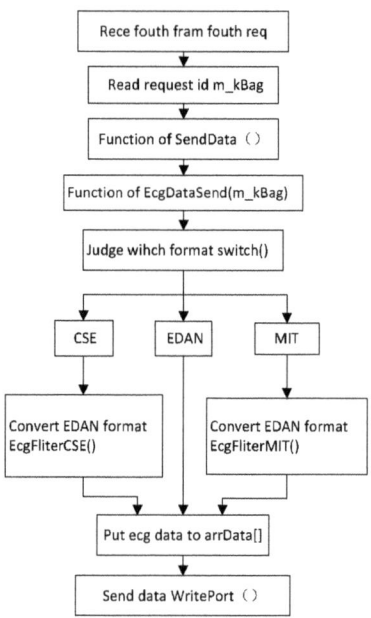

Fig. 4. Serial communication sending fourth frame process

SendData (m): serial send function according to what call, is frame through the function of the parameters to convey information. If m = 1, so is sending the first frame data, and so on.

Switch (m): what kind of data for judging open format, if m = 1 () function of CSE format is expressed open, m = 2 says MIT format, m = 3 says EDAN format, m = 4 says ECG workstation format.

ArrData []: deposit ECG data.

WritePort (): wrote the serial port function in CSerialPort, PC is to use this function send data to the next place machine, such as: WritePort (arrData, 521).

3) Ethernet communication module

Lower place machine request four frames, among them the third frame for actual ECG data frame, only use one frame to sent 16,000 * N a bytes of ECG data complete. The total flow chart is similar with serial interface communication. the main difference is in the third frame, through function of send in the C Socket, each function Send 1,000 bytes, complete use 16 * N times. And different with serial interface communication is, in serial communication, lower place machine sent to up machine   16000N / 1000 (for the integer subcontractor when several add 1) times bag of the fourth frame request, and in Ethernet communications, lower place machine only request once, PC using circular statement can send all data . Batch sent is similar. With serial interface communication The third frame delivery process as shown in figure 5 shows.

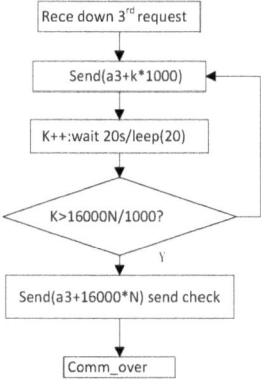

Fig. 5. Send third frame flow chart

3   Function Test

3.1   Waveform Display

Four formats of ECG data are normal display, for example the EDAN format data show,. EDAN format: eight guide league 10s of data, sampling rate for 1000Hz. EDAN format without filtering, ECG waveform ECG data for the original figure 6 the twelve guide league waveform figure.

3.2   Filtering Effect Testing

CSE data format: the original sample rate for 500 Hz, after conversion is 1000Hz.

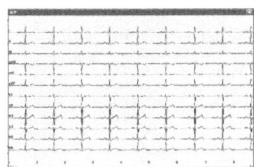   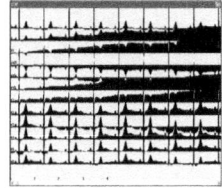   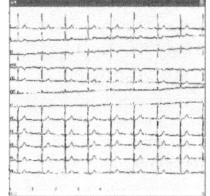

Fig. 6. EDAN waveform   **Fig. 7.** CSE data filtering former waveform   **Fig. 8.** CSE after data filtering waveform

Figure 7, figure 8 MIT data filtering the contrast between the before and after waveform can be seen filtering, aliasing composition removed, to display clear ECG waveform.

The MIT data formats: sample rate for 250Hz, after converting is 1000Hz. The filtering effect is same as CSE data filtering, here no longer, for example.

3.3   Serial Communication Test

In order to check data sent from PC to down place machine, using a serial port debug assistant to examine send data whether is right, using a serial port debug assistant to simulate a machine up send request to under a machine. Batch download need continuously to send several files, and serial debugging assistant without function of continuous send frame , when batch download test by lower level computer.

PC should not only sends each frame data down, but also achieve the effect of continuous send, so one-time sent complete five frame data. With lower place machine (responsible for receiving data frames), can see effect of the PC send, as shown in figure 9 below.

Fig. 9. Serial send finished

RECEIVEING1 means received the first frame, after received 3th frame, it then received subcontractor data of the fourth frame. It displays RECEIVEING4 until after received bag 405, Display STORE SUCCESSFUL FILE after collect all .if Lower place machine show receiving success, also suggests that PC data sending finished, communication is successful.

3.4   Ethernet Communications Test

Due to effect of realize each frame same as the serial communication, here no longer be detailed introduction.

4   Closing

This topic research of ECG simulation generator basically achieve the expected effect, PC can clear display three different formats length of data of original ECG waveform, and can batch and continue sent to down machine by serial ports and Ethernet, the under machine can receive correctly and stored it. In output it will reduce to analog sign, the output waveform without distortion and analog signal amplitude accurate, and can choice poling voltage by any standard key, buttons can control the size of positive polarization voltage. Basically achieve the expected effect.

References

1. Xiang, k., Wang, c.: Design of Suppositional ECG Signal Generator Based on ECG Database,China Medical Equipment CNKI:SUN:YLSX.0 (May 8, 2009)
2. Tan, h.q.: Construction of the C Language Programme. Tsing Hua University Press, Beijing (2005)
3. Mu, l.s.: The Development and Application of Visual C++6.0. Sing Hua University Press, Beijing (2003)
4. Qi, j., Yang, w.: S3C2410-based Communication Programme Research Under the Embedded Linux. Nuclear Electronics & Detection Technology,CNKI:SUN:HERE.0 (December 20, 2010)
5. He, l.s.: Design of 12 lead ECG signal generator based on DDS arithmetic. Chinese Journal of Scientific Instrument CNKI:SUN:YQXB.0 (Febrary 9, 2010)
6. Yu, x.f.: Principle and design of medical electronic equipment. South China University of Technology Press, Guang zhou (2003)
7. Jian, x.z., Xu, l.: The embedded high-speed synchronous acquisition device based on ARM9 and FPGA. Microcomputer Information CNKI:SUN:WJSJ.0 (Febrary 2, 2011)
8. Hua, c.y.,Tong, s.b.: Analog electronics technology. Higher Education Press (2006-05)
9. Wan, L., Zhang, Y.: ECG Data Processing Mechanism of Remote Wireless ECG Monitor. Computer Engineering (2010)
10. Cao, H.: Develop of ECG system on-chip based on the fution mixed-mode FPGA. Microcomputer & Its Applications (2010)

Research on Random Collision Detection Algorithm Based on Improved PSO

Ting-dong Hu

Shenzhen Kunyiziyuan Electronic Ltd, Shenzhen, China
Td.hu@kunyiziyuan.com

Abstract. In order to improve the real-time of collision detection algorithm, this paper introduces particle swarm optimization (PSO), PSO simple and easy to operate, and search capability and convergence speed have a greater advantage. To reduce the random collision detection algorithm missed some of the interfering elements and to improve the accuracy of collision detection, using the OBB bounding box surrounding the basic geometric elements instead of the basic geometric elements characterized as a random sampling point collision detection method. The complex three-dimensional models of the collision problem are transformed into simple two-dimensional discrete space optimization problems, and improve the algorithm in real time.

Keywords: Random collision detection, PSO algorithm, OBB, particles, collision detection.

1   Introduction

With the development of collision detection technology, emergence of random collision detection algorithm. It is based on accuracy for speed of collision detection algorithms, it can be used in relatively high real-time interactive systems, it become collision detection field a new of research in recent years [1]. Particle swarm optimization (PSO) is a recently developed an optimization algorithm based on swarm intelligence, it have wide applications in the neural network optimization and fuzzy control. PSO simple and easy to operate, and search capability and convergence speed have a greater advantage. Random collision detection algorithm operating the basic geometric elements of point, triangle, tetrahedron, etc., and these basic elements of the space is discrete. PSO algorithm can quickly search in the discrete space, this paper combine random collision detection algorithm with the quick optimization of PSO algorithm the advantages, and formation of a rapid, real-time and efficient collision detection algorithm.

2   Particle Swarm Optimization

PSO algorithm [2] is Propose by James Kennedy and Russell Eberhart in 1995; it is an evolutionary computing technology, the algorithm is simulated foraging behavior of birds flying, through the collaboration between individuals to search for the

B. Liu and C. Chai (Eds.): ICICA 2011, LNCS 7030, pp. 602–609, 2011.

optimal solution. Set,in N dimensional search space, there are m particles, in which the i particle position vector is $X_i = (x_{i1}, x_{i2},..., x_{iN})(i = 1,2,...,m)$, Particle velocity vector is $V_i = (v_{i1}, v_{i2},..., v_{iN})$. the i-th particle best position (optimal solution) is P_{best} ., PSO the best position denoted by G_{best} ,PSO algorithm formula:

$$V_{id}(t+1) = v_{id}(t) + c_1 * Rand_1 *(P_{best} - x_{id}(t)) + c_2 * Rand_2 *(G_{best} - x_{id}(t)) \qquad (1)$$

$$X_{id}(t+1) = x_{id}(t) + v_{id}(t+1) \qquad (2)$$

Which, $Rand_1, Rand_2$ is [0, 1] random number; c_1, c_2 is acceleration factor, the particle in each dimension has a maximum speed limit V_{max} $(V_{max} > 0)$, if one dimension exceeds set speed V_{max} , the speed is limited to V_{max} . The PSO basic algorithm schematic as Fig. 1.

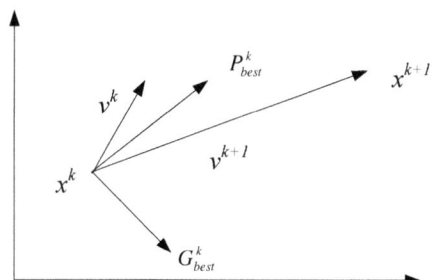

Fig. 1. The basic principle schematic drawing of PSO Algorithm

3   Random Collision Detection Algorithms

3.1   Algorithm Overview

Random collision detection algorithm currently is divided into two categories: the first Average-Case algorithm is proposed by Klein and Zachmann. Algorithm idea is based on the the region bounding the intersection of two objects number of features as the the test standard, and each node of the feature vectors do not do precise collision detection, to achieve the purpose of controlling the rate of detection [3].

The second algorithm is proposed by Guy and Debunne [4], Raghupathi [5] a surface such as random sampling, assuming random sampling of the collision, and then step through the sampling of the local minimum to reach the accurate detection algorithm. The movement of objects is very small time difference between time slots, typically a few hundredths of a second, so the location of the object only minor differences, the relative displacement between the characteristics of teams is small, so the movement of objects with spatial and temporal correlation [6, 7].

3.1   The Algorithm Improve

In this paper use random for collision detection algorithm   based   on feature
sampling , due to random collision detection algorithm   random collision of two
objects in question into the search feature on the distance between the issues.
Therefore, the characteristics of the distance between should be able to be calculated.
Algorithm real is to calculate the distance between the basic geometric elements; the
algorithm requires the user to set the distance threshold. If the threshold is set large,
easy to miss has occurred the geometric elements to interfere with and affect the
algorithm accuracy. In contrast, a smaller set of values will increase the geometric
elements on without interference, increasing the complexity of the algorithm.

Point-based particle swarm algorithm robustness is poor; to determine the spatial
relationship between adjacent points is more complex. To overcome these
shortcomings, this paper, the basic geometric elements surrounded by the bounding
box as a sampling of feature objects. This paper selects the OBB bounding box. OBB
bounding box making the model more elements intersection test simple. Before
making accurate collision detection, first determine whether the OBB bounding box
intersection, excluding most of the disjoint interference Yes, there are likely to
intersect the basic elements of the basic details of the interference detection. Solve the
point-based particle swarm search algorithm is easy to miss elements of interference
problems. Enhance the robustness of the particle search. To surround the basic
geometric elements of the center of OBB bounding box collision detection algorithm
as the characteristics sampling, calculate the distance from the center of bounding
box. Bounding box of the center distance as shown Fig. 2.

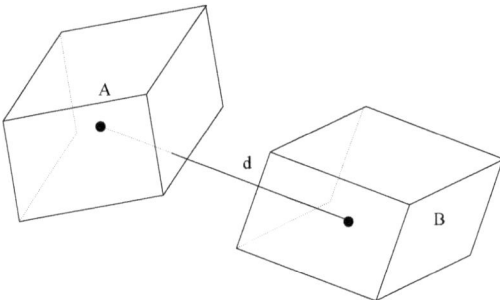

Fig. 2. The center distance of Oriented Bounding Box

4   The Application of PSO

Feature samples consisting of two-dimensional space is discrete, and PSO algorithm
for discrete optimization of space. random Collision detection algorithm based on
improved PSO main solve problem is the PSO search space, the particle's position and
velocity update and algorithms to terminate the flight conditions.

4.1   The Search Space of Particles

A, B is a sampling of feature points of the two collections, two objects at least one pair between the characteristics $P(a_i, b_j)$, which $a_j \in A(0 < i < N_a)$, $b_j \in B(0 < j < N_b)$ so $L(P) \leq \delta$, L is the distance function, δ is the collision threshold. Search space as shown in Fig 3 and Fig 4, the search space in the location and the optimal solution will be movement and change with the object of the fitness, so the search space is dynamic [8].

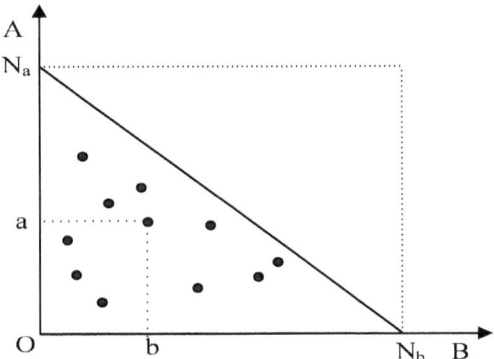

Fig. 3. The discrete space of Two-dimensional

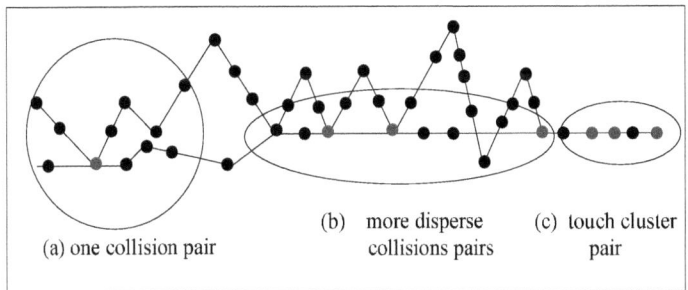

Fig. 4. The two zigzag Model Collision

4.2   Calculate and Update the Particle Position, Velocity

This calculation is the distance between two bounding box center. Fitness function for the Euclidean distance between the centers of two, used to reduce the amount of square of the distance calculation[9].

$$L(X_a, X_b) = (x_a - x_b)^2 + (y_a - y_b)^2 + (z_a - z_b)^2 . \qquad (3)$$

A particle on behalf of Two objects sampling characteristics of a combination, each feature has of three-dimensional objects in space location. Optimization goal is to find a combination of minimum distance between features. Initial position of particles:

$$P_k = (x_i, x_j) \tag{4}$$

Which, $i \in N_a, j \in N_b$

$$V_k = (v_i, v_j) . \tag{5}$$

Which, $i \in N_a, j \in N_b$

(1) Update the particle position. Calculate the fitness value of each particle, and and global, individual optimal values. Fitness function coordinates of the particle in the search space, returns the current position assigned to the fitness value, if the value is less than at this time to adapt to the individual particles or the global optimum, then update the global best particle position of the individual or, otherwise not be updated.

(2) Renewal particle velocity. Particle velocity's control is the entire optimized core. The particle velocity changes along with individual and the overall situation most superior position's change, toward sufficiency night-watch superior direction acceleration.   This article according to the formula (4), (5) renews each granule the speed and the position. And, on granule in search space each unvaried some maximum speed V_{max} , uses for to the granule speed to carry on the limit, causes the speed control in the scope $[-V_{max}, V_{max}]$.

4.3   Condition of Algorithm Termination

Because based on the grain of subgroup's collision examination belongs to the multi-peak value optimization question. Therefore, in time internal, must seek repeatedly satisfies the collision condition the solution. If the obtained result satisfies the terminal condition, then the stop iterates and records the result. Here, the terminal condition has two kinds:

1, find first interference later to terminate the granule the evolved advancement, but if in the evolved advancement, had not found the interference spot, when then the granule evolved algebra achieved the biggest evolved algebra, the termination moves;

2, if wants to seek for all impact points between the geometric model , the terminal condition may suppose to achieve the biggest evolved algebra[10].

5   Process of Collision Detection

The algorithm mainly divides into two stages: First, pretreatment stage; Second, examines the stage in detail. The pretreatment stage according to the object expressed that the model and the accuracy requirement carries on the characteristic sampling. Here surrounds the basic geometric element the OBB surrounding box central point to carry on the sampling as the characteristic. The detailed examination divides into the initial survey and the precise examination, first used a grain of subgroup to optimize the algorithm the intelligent search characteristic to find the interference fast the OBB surrounding box. In goal space stochastic initialization certain amount granule, has the initial point and the speed. The granule in the iteration, carries on the position and the speed renewal. Returns all has the interference OBB surrounding box, further carries on again asks the junction precisely. What returns has collision's OBB surrounding box, has not been able to determine surrounds the basic geometric element whether have the collision. Needs to carry on the intersection test of basic geometric elements. Uses between the triangles the intersection test[11].

Algorithm process as follows:

Step1 with OBB surrounds box algorithm to construct the OBB surrounding box for the input geometric model, calculates OBB the surrounding box the central place.

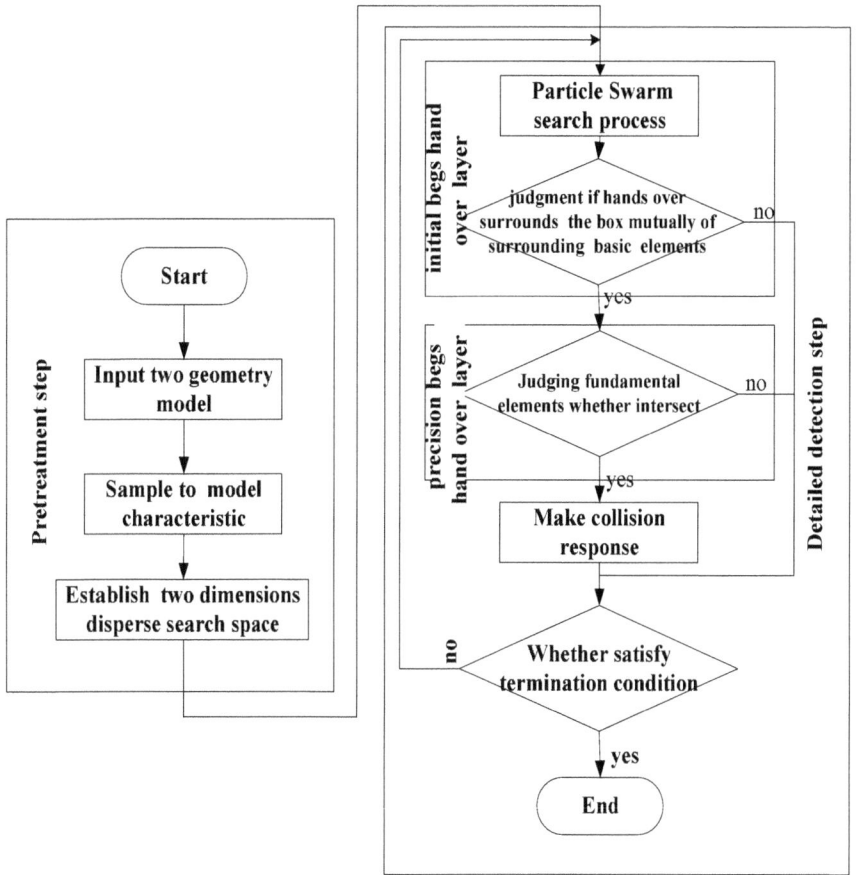

Fig. 5. The Flow Chart of PSO Optimized the Random Collision Detection Algorithm

Step2 establishes two-dimensional search space D, carries on the characteristic sampling. Coexisting enters in the corresponding array.

Step3 grain of subgroup carries on the search in two-dimensional space D, the judgment surrounding basic geometric element OBB surrounding box whether to have the interference, does not interfere, and transfers Step6.

Step4 If the surrounding box has the interference, then judgment triangle whether to intersect, does not intersect, then transfers Step7.

Step5 if the triangle is intersection, after making collision's response.

Step6 algorithm whether satisfy the terminal condition, satisfies, then withdraws from the grain of subgroup searching algorithm, otherwise, transfers Step3.

Step7 algorithm whether to satisfy the terminal condition, satisfies, then withdrawal procedure, otherwise, transfers Step4.

Algorithm flow chart shown in Fig 5:

6   Summary

This article first introduces the PSO algorithm the basic principle and the corrective method. Next introduces two kind of random collisions detection algorithm. And the feature-based random sample of collision detection algorithm has been improved, with the basic elements of the bounding box instead of the basic characteristics of sampling points. Then from the granule search space, the granule position and the speed renewal, and so on several conveniences introduced the PSO algorithm application. Finally gives the algorithm the flow.

References

1. Kimmerle, S.: Collison detection and post-processing for physical cloth simulation: [Dissertation], Tübingen, pp. 28–31 (2005)
2. Kennedy, J., Eberhart, R.C.: Particle swarm optimization. In: Proceeding of 1995 IEEE International Conference on Neural Networks, pp. 1942–1948. IEEE, New York (1995)
3. Klen, J., Zachmann, G.: Adb-trees: Controlling the error of time-cntical collision detection. In: 8th International Fall Workshop Vision, Modeling, and Visualization (VMV), Germany, pp. 19–21 (2003)
4. Guy, S., Debunne, G.: Monte-carlo collision detection. Technical Report RR-5136, INRIA, pp. 564–567 (2004)
5. Raghupathi, L., Grisoni, L., Faure, F., Marchal, D., Cani, M.-P., Chaillou, C.: An intestione surgery simulator: Real-time collision processing and visualization. IEEE Transactions on Visualization and Computer Graphics, 708–718 (2004)
6. Cohen, J.D., Lin, M.C., Manocha, D., Ponamgi, M.: I-COLLIDE: An interactive and exact collision detection system for large-scale environments. In: Symposium on Interactive 3D Graphics, pp. 189–196 (1995)
7. Guibas, L.J., Hsu, D., Zhang, L.: H-walk: Hierarchical distance computation for moving convex bodies. In: Oz, W.V., Yannakakis, M. (eds.) Proceedings of ACM Symposium on Computational Geometry, pp. 265–273 (1999)

8. Kimmerle, S., Nesme, M., Faure, F.: Hierarchy accelerated stochastic collision detection. In: Proceeding of Vision, Modeling, Visualization, pp. 307–314 (2004)
9. Klen, J., Zachmann, G.: Adb-trees: Controlling the error of time-cntical collision detection. In: 8th International Fall Workshop Vision, Modeling, and Visualization (VMV), Germany, pp. 19–21 (2003)
10. Wang, T., Li, W., Wang, Y., et al.: Adaptive stochastic collision detection between deformable objects using particle swarm optimization. In: EvoWorkshops: EvolASP, pp. 450–459 (2006)
11. Guy, S., Debunne, G.: Monte-carlo collision detection. Technical Report RR-5136, INRIA, pp. 564–567 (2004)

Active Disturbance Rejection Decoupling Control Method for Gas Mixing Butterfly Valve Group

Zhikun Chen, Ruicheng Zhang, and Xu Wu

College of Electrical Engineering, Hebei United Universtiy, Hebei Tangshan 063009, China
rchzhang@yahoo.com.cn

Abstract. This paper proposed an active disturbance rejection control (ADRC)method for gas mixing butterfly valve group. According to the serious decoupling, uncertainty, many disturbances and nonlinear etc of the gas mixing butterfly valve group, the model of the plant is established, proposed the ADRC based on static decoupling and ESO dynamic decoupling in order to eliminate the coupling between the loops. Simulation results show that the designed controller not only has good decoupling pedormance, but also ensures good robustness and adaptability under modeling uncertainty and external disturbance.

Keywords: gas mixing, active disturbance rejection control(ADRC), extended state observer(ESO), decoupling.

1   Introduction

Gas is an important raw material in steel and nonferrous metal smelting enterprises. The gas from coke oven and blast furnace through two butterfly valves and mixed, then pressured by the pressure machine as the production of mixed gas. When the gas mixing, butterfly valve opening both affect flow also affect pressure.There is serious coupled between the flow control and pressure regulator, at the same time the regulation of the four butterfly valves are also interaction. This implies that there is a relatively strong coupling in this system.At present, most of domestic compression station mainly by manual control. There are three common methods of gas mixture pressure control in automatic control of the compression stations: pressure and calorific value dual-loop controller, feedforward compensation decoupling and intelligent decoupling. Pressure and calorific value of the dual-loop adjustment method using the variable matching circuit, by appropriate matching the amount to be transferred and adjust to eliminate the coupling, this method is not accurate, the ability to inhibit the disturbance is not strong. Feedforward compensation decoupling method using blast furnace gas pipeline pressure, coke oven gas pipeline pressure, mixed gas pressure front of the pressure machine as the perturbation, control the system by static decoupling manner. However, due to the limitations of control methods, control effect is poor[1-4]. With the development of intelligent control, intelligent decoupling control strategy is proposed by Ren Hailong. They through analysis the law and the physical structure of the gas mixture pressure process, they divided the control loop into two: calorific value and the pressure decoupling control

B. Liu and C. Chai (Eds.): ICICA 2011, LNCS 7030, pp. 610–617, 2011.

loop, pressure control loop. In this paper, they design the intelligent decoupling control strategy for the calorific value and the pressure decoupling control loop. However, due to the two valves Series on the blast furnace gas pipes and the two valves series on the coke oven gas pipes is very close. The literature one pointed that :this is a typical strong coupling system. Document 5-7 was not performed detailed design. Although, the literature 1 has been decoupled the two valve system based on feedforward compensation. but the system has the blast furnace gas pressure changes, coke oven gas pressure change and changes in user traffic disturbance. So, ability of this system that inhibition disturbance is not strong.

In this paper, a novel disturbance rejection based decoupling control approach is proposed. The method was applied it to the gas mixing process, to remove the coupling between systems.

2   The Model of the Gas Mixing Butterfly Valve Group

In the blast furnace, coke oven gas pressure during the mixing, when the controller to calculate the control values of coke oven and blast furnace valve group, the butterfly valve controller according to the series gain matrix to decision that how to adjust the two valves of corresponding valve group in the end.

Due to the basic idea of the pressure distribution essentially the same of blast furnace values and coke oven values, following, for example by coke oven butterfly group. Butterfly valve series system specifically for the following:

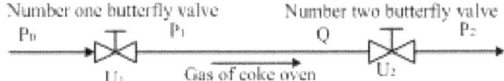

Fig. 1. Pipe system of pressure-flow

In the fig 1, P_0 is coke oven gas pressure befer the number one butterfly valve. P_1 is coke oven gas pressure befer the number two butterfly valve. P_2 is coke oven gas pressure before the number two butterfly valve. u_1, u_2 are the opening of number one and number two butterfly valves. Q is the gas flow of coke oven pipeline.

According to the literature12, pressure-flow process can be described as

$$Q = u_1(P_0 - P_1) = u_2(P_1 - P_2) = \frac{u_1 u_2}{u_1 + u_2}(P_0 - P_2) \tag{1}$$

When two loops are in open loop,the first amplification factor is

$$\left.\frac{\partial Q}{\partial u_1}\right|_{u_2=const} = \left(\frac{u_2}{u_1 + u_2}\right)^2 (P_0 - P_2) \tag{2}$$

Closed loop pressure, the first amplification factor is

$$\frac{\partial Q}{\partial u_1}\bigg|_{P_1=const} = P_0 - P_1 = \frac{u_2}{u_1+u_2}(P_0-P_2) \tag{3}$$

According to the definition of a relative gain

$$\lambda_{11} = \frac{\dfrac{\partial Q}{\partial u_1}\bigg|_{u_2=const}}{\dfrac{\partial Q}{\partial u_1}\bigg|_{P_1=const}} = \frac{u_2}{u_1+u_2} \tag{4}$$

From equation (1) and solve for u_1 and u_2 into equation (4).can be used pressure to represent the gainm, is

$$\lambda_{11} = \frac{P_0-P_1}{P_0-P_2} \tag{5}$$

can also find the relative gain of the u_2 and flow Q, is λ_{12}

$$\lambda_{12} = \frac{P_1-P_2}{P_0-P_2} \tag{6}$$

If use P_1 to describe the pressure-flow system, is

$$P_1 = P_0 - \frac{Q}{u_1} = P_2 + \frac{Q}{u_2} = \frac{u_1P_0+u_2P_2}{u_1+u_2} \tag{7}$$

Another gain can be determined,to strike a partial derivative for equation (7), can be, respectively derived the relative gain of u_1, u_2 for two channels of the P_1.

Then, pressure—flow system of input-output relationship can be expressed by relative gain matrix as

$$\begin{bmatrix} Q \\ P_1 \end{bmatrix} = \Lambda \begin{bmatrix} u_1 \\ u_2 \end{bmatrix} \tag{8}$$

Among

$$\Lambda = \begin{bmatrix} \lambda_{11} & \lambda_{12} \\ \lambda_{21} & \lambda_{22} \end{bmatrix} = \begin{bmatrix} \dfrac{P_0-P_1}{P_0-P_2} & \dfrac{P_1-P_2}{P_0-P_2} \\ \dfrac{P_1-P_2}{P_0-P_2} & \dfrac{P_0-P_1}{P_0-P_2} \end{bmatrix} \tag{9}$$

If in the system P_1 close to P_2, Λ is very close to the unit matrix, instructions to control the flow with a valve 1, with a valve 2 to control pressure P_1 is appropriate. If P_1 close to P_0, use valve 2 to control flow and with a valve 1 to control pressure P_1 is appropriate.if P_1 close to the midpoint of (P_0-P_2), while adjusting the two valves are the best.

Known, a system of coke oven gas pipeline pressure is P_0=6.2kPa, P_1=5.5kPa , P_2=5.0kPa, The gain matrix is

$$\Lambda = \begin{bmatrix} \lambda_{11} & \lambda_{12} \\ \lambda_{21} & \lambda_{22} \end{bmatrix} = \begin{bmatrix} 0.58 & 0.42 \\ 0.42 & 0.58 \end{bmatrix}$$

$\lambda_{i,j}$ very close to 0.5, its shows that there are serious coupling in this system. From equation (8), we could be find the mathematical model of this control system can be expressed as

$$\begin{bmatrix} y_1(s) \\ y_2(s) \end{bmatrix} = \bar{G}(s)\Lambda \begin{bmatrix} u_1(s) \\ u_2(s) \end{bmatrix} \tag{10}$$

Among, $y_1(s) = Q(s)$ and $y_2(s) = P(s)$ $\overline{G}(s) = diag\{G(s), G(s)\}$ is transfer function of controlled object.

In debugging the process of trial and error

$G_1(s) = \dfrac{K}{T_0 s + 1}$, In the equation, $K = 2, T_0 = 1.5$.

3   ADRC Decoupling Control Design

The ADRC is based on the idea that in order to formulate a robust control strategy. Although the linear model makes it feasible for us to use powerful classical control techniques such as frequency response based analysis and design methods, it also limits our options to linear algorithms and makes us overly dependent on the mathematical model of the plant. Instead of following the traditional design path of modeling and linearization and then designing a linear controller, the ADRC approach seeks to actively compensate for the unknown dynamics and disturbances in the time domain. This is achieved by using an extended state observer (ESO) to estimate the total dynamics, lumping the internal nonlinear dynamics and the external disturbances. Once the external disturbance is estimated, the control signal is then used to actively compensate for its effect, then the system becomes a relatively simple control problem. More details of this novel control concept and associated algorithms can be found in [8]-[10]. A brief introduction is given below.

In many practices, the performances of the controlled system are limited by how to pick out the differential signal of the non-continuous measured signal with stochastic noise. In practice, the differential signal (velocity) is usually obtained by the backward difference of given signal, which is very noisy and limits the overall performance. Han [8] developed a nonlinear tracking-differentiator (TD) to solve this problem effectively. It is described as

$$\begin{cases} e_0 = r_{1i} - r_i^* \\ \dot{r}_{1i} = -R \cdot fal(e_0, \alpha_0, \delta_0) \end{cases} \tag{11}$$

R —tracking parameter, δ_0 —control parameter, ω_1 tracking ω^*

The function of *fal* as following

$$fal(e,\alpha,\delta) = \begin{cases} |e|^{\alpha} \, \mathrm{sgn}(e) & |e| > \delta \\ \dfrac{e}{\delta^{1-\alpha}} & |e| \le \delta \end{cases} \tag{12}$$

The extended state observer (ESO) proposed by Prof. Han [8] is a unique nonlinear observer designed to estimate $f(\cdot)$:

$$\varepsilon_i = z_{1i} - y_i \tag{13}$$

$$\begin{cases} \dot{z}_{1i} = z_{2i} - \beta_{1i} \, fal(e,\alpha,\delta) + b_{0i} v_i \\ \dot{z}_{2i} = -\beta_{2i} \, fal(e,\alpha,\delta) \end{cases} \tag{14}$$

where $\beta_{01}, \beta_{02}, \beta_{03}$ and β_{04} are observer gains, b_0 is normal value of b.

Once the design of TD and ESO is accomplished, the general error and its change between the reference and the estimated states can be defined as e_1, e_2 and e_3. The nonlinear proportional derivative (N-PD) law is used to synthesis the preliminary control action, which can be described as

$$\begin{cases} e_i = r_i - z_{1i} \\ \dot{z}_{3i} = r_i - z_{1i} \\ e_{0i} = z_{3i} \\ v_{0i} = k_{pi} \, fal(e_i,\alpha_i,\delta_1) + k_{di} \, fal(e_1,\alpha_1,\delta_1) \end{cases} \tag{15}$$

Which reduces the plant to a double integrator, which in turn is controlled by the nonlinear PD controller. Where k_{pi}, k_{di} are the gains of PD controller.

Plus the control action to cancel out the external disturbance, then the total control action of ADRC can be determined as follows

$$v_i = (-z_{2i} + v_{0i}) / b_{0i} \tag{16}$$

The active disturbance rejection decoupling control system is given as Figure 2.

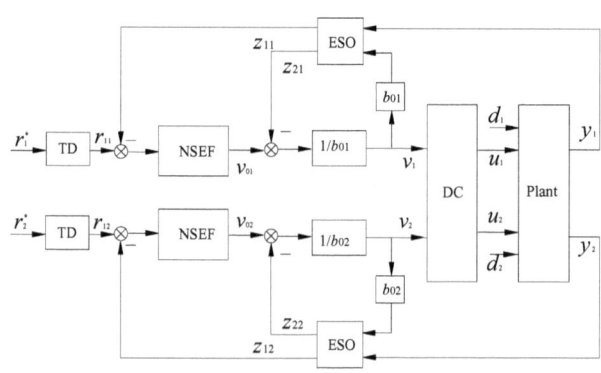

Fig. 2. Architecture of ADRC for gas mixing butterfly valve group

4   Simulation

To verify the feasibility of the control program, figure 2 shows the block diagram of the simulation test.

For the nominal state, two channels each with unit step value were given, the simulation results shown in figure 3. Figure (a) is response curve of flow channel for a given value of the step change. Curve 1 is the curves of flow, curve 2 is curves of the pressure; figure (b) is response curve of pressure channel for a given value of the step change. Curve 1 is the curves of pressure, curve 2 is curves of a flow. Visible, Although the coupling of two valve series system is more serious, but decoupling through decoupling method based on growth hormone two-way adjustment mechanism, when the flow rate Q changes, pressure remained almost unchanged, and the output is no static error, decoupling works well, and vice versa.

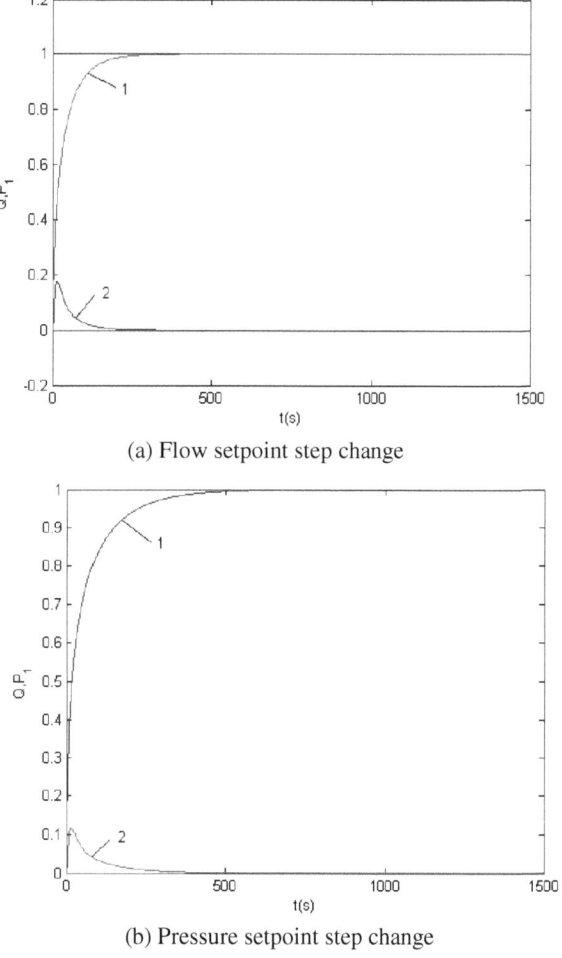

(a) Flow setpoint step change

(b) Pressure setpoint step change

Fig. 3. Responses of gas mixing butterfly valve group

When t=500s, increase amplitude of 0.5 load interference signal in the-channel. The simulation results shown in figure 4. Figure (a) is response curve of the flow channel plus interference signal, curve 1 is the curves of flow, curve 2 is curves of the pressure; figure (b) is response curve of the Pressure channel plus interference signal.

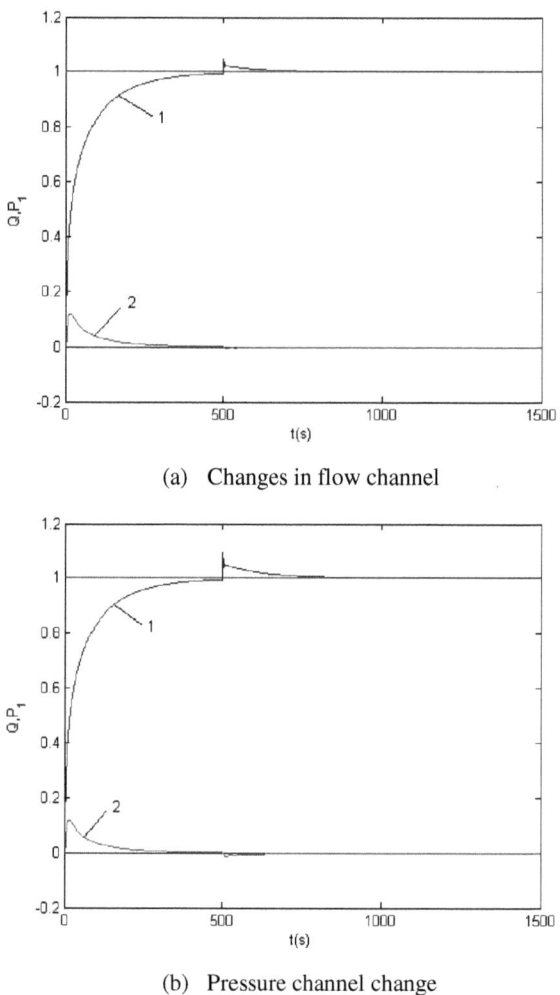

(a)   Changes in flow channel

(b)   Pressure channel change

Fig. 4. Responses of disturbance rejection

Curve 1 is curves of the pressure, curve 2 is the curves of flow. Visible, butterfly valve series system has little effect of interference,so it is can negligible.

5   Conclusions

This paper use decoupling method based on ADRC to remove gas mixing system coupling. Simulation results show that the control system not only has better tracking performance, anti-disturbance ability, and the decoupling works well, to solve the gas mixture system valve in series strong coupling, time-varying, confounding factors such as more adverse impact on the system.So,the control method has good prospects.

Acknowledgments. This work was supported by the Natural Science Foundation of Hebei Province (grant number:F2009000796).

References

1. Qin, l.: Decoupling and Smith compensator with a mixture of gas heating value control. Control Engineering, 73–75 (2002)
2. Meng, c., Xia, c.: SMAR intelligent regulator and control of mixed gas calorific value. Hebei Metallurgical, 53–55 (2002)
3. Wu, x.f.,Sun, h.f.,Wei, c.y.: Automatic control of furnace gas pressure regulator. Metallurgical Automation, 64–65 (2002)
4. Huang, j.m.,Cao, j.m.: Computer-controlled system of gas mixing. Automation and Instrumentation, 28–30 (2000)
5. Ren, h.l., Li, s.y.: Mixture gas pressure and calorific value of the fuzzy decoupling control. Automation Instrumentation, 65–67 (2004)
6. Cao, w.h., Wu, m., Hou, s.y.: Method and application of gas mixture pressurized process intelligent decoupling control, pp. 780–785. Zhong nan University News (2006)
7. Guo, j.: Intelligent control of blast furnace,coke oven gas pressure and calorific value of mixture. Engineering Edition, pp. 231–235. Xinjiang University News (2001)
8. Han, j.q.: ADRC control technology - an estimated uncertainty compensation control technology. National Defence Industry Press, Biejing (2008)
9. Li, S.L., Yang, X., Yang, D.: Active disturbance rejection control for high pointing accuracy and rotation speed. Automatica,1854–1860 (2009)
10. Qing, Z.: On active disturbance rejection control:stability analysis and applications in disturbance decoupling control. National University of Singapore (2003)
11. Jin, y.h., Fang, c.z.: Process Control. Tsinghua University Press, Biejing (1995)

An LED Assembly Angle Detection System

Li Chunming, Shi Xinna, Xu Yunfeng, and Wu Shaoguang

College of Information Science and Engineering,
Hebei University of Science and Technology, Yuxiang Street 26, Shijiazhuang, China
shixinna1985@163.com

Abstract. In view of the present problem of the LED display brightness not uniformity, the text puts forward a kind of LED light assembly angle automatic detection system. First, make use of stepping motor drive the LED display module rotation. In the process of LED rotation, use CCD camera get each image of LED display module with Angles of each step; And then by using the template matching principle pinpoint positions of leds in LED display module. Finally, detect the brightness of the led, and according to the step motor's stepping Angles to calculate the deflection Angles of leds.

Keywords: LED light emitting diode, template matching, Euclidean distance, stepping motor.

1 Introduction

Along with the development of the display technology [1], The LED display screen with its characteristics of energy saving, environmental protection and high brightness has got welcome from people. But at present, the LED display screen brightness uniform problem is the focus of attention. The LED display screen is mainly composed by many LED light emitting diodes [2]. The problems of affect the LED display screen brightness not uniformity are so many. For instance, the LEDs exist differences in photoelectric characteristic aspects, in human visual characteristic aspects and so on. In order to improve the screens' brightness display effect, Predecessors are mainly aim to adjustment the aspects of LED light-emitting diode brightness and color, but this adjustment method also can lead the LED display screens brightness nonuniform. Therefore, the paper from the piont of LED emitting diode assembly angle and test to the LED display screens brightness uniformity.

LED light-emitting diode assembly Angle detection system is mainly aimed to the LED display assembly process,Because of the LED light-emitting diode deflection Angle difference lead to the LED display screens' brightness not uniformity and inspect it. The systems mainly include the location of the LED lights, the lightspot detection of LED lights and the design of the stepping motor driver circuit. Below introduce respectively.

B. Liu and C. Chai (Eds.): ICICA 2011, LNCS 7030, pp. 618–624, 2011.

2   The Location of the LED Lights

The location of the LED lights is mainly use of the template matching method. In modern image processing filed,the template matching technology is the most important technology.And in the more filed there will be more and more widely application. Template matching technology refers to first give a known source image, also known as template image. And then, use a smaller detected image, namely matching images and template images are compared,used to decide in the template images if exist a same or similar area in this matching image.If exist such an area, still can determine the position of the area[3].

2.1   The Classification of the Template Matching Algorithm

Template matching algorithms are mainly consist the matching algorithm based on the characteristics, based on the gray level related matching algorithm, based on the model's matching algorithm and based on transform domain matching algorithm[4].

The image matching algorithms based on the characteristics are refers to according to the geometric relations of the two images owned the same characteristics and calculate theirs registration parameters. The characteristics of the images owned mainly consist point, edge and surface features etc, due to the surface characteristics' calculate is very complex, so mainly use images' point characteristics and edge character to match. The matching algorithm mainly used in those images which own obvious edges and corners.Use a certain feature matching algorithm can find these feature points soon[5].

The image matching algorithms based on the characteristics mainly consist SUSAN algorithm, Harris algorithm and SIFT algorithm. Below is use SIFT algorithm detection get the results[6].

The first image shows an example of matching produced by the demo software. Features are extracted from each of the two images, and lines are drawn between features that have close matches. In this matching example, many matches are found and only a small fraction are incorrect.

As shown in figure 1 is the original image,

Fig. 1. The original image

As shown in figure 2 is the SIFT Corner detection result.

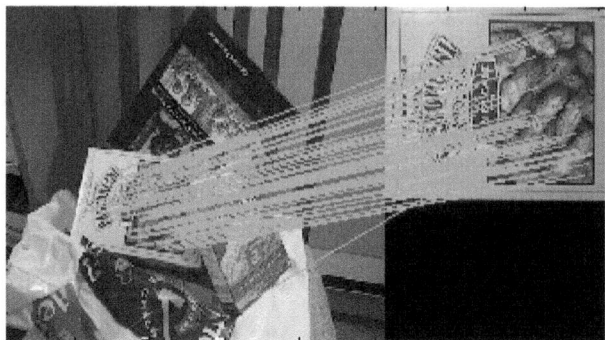

Fig. 2. SIFT Corner detection result

Because of the image's pixel gray values contain image's all information, it does not need to extract image's feature but matching by all useful image information of the image contains. Generally use the whole image gray-scale information directly and set up the similarity measure between two images, and then use a certain search method for a similarity measure maximum or minimum value to the transformation model's parameter values.

Commonly similarity measures consists gray level square difference's sum of two images、Sequential similarity detection and mutual correlation etc. The paper is mainly used the two images' gray square sum to the template matching.

2.2   The Template Matching Algorithms Based on the Gray Level

Based on the gray level template matching algorithm refers to Per-pixel between a certain size of the real-time image window gray matrix and the reference image of all possible Windows gray scale matrix ,according to some similarity measure methods to search and compare. The performance of this kind algorithm mainly depends on similarity measure search strategy and the size of the match window's choice. In this method the size of the Match window's choice is the most important question method ,the big window for the situation of existing keep out in the scenery or image not smooth will appear of mismatch problem,the small window can't cover the enough strength changes.So it can adapt to the size of the match area to achieve a good match results.This kind of matching algorithm is mainly applied to those matching simple rigid body or radiation transform scene[7].

Because of this text deal the image with no obvious characteristics, the structure is simple,So the paper mainly introduce based on gray's template matching algorithm.

This method is mainly in the template image find out the location of the matching image, the matching process is mainly through matching image in the template image in translation and done.In here use $S(t)$ said the template image,use $S_2(t)$ said the image matching, use European distance between two images as the standard of comparison. When the little picture in the large image along horizontal and vertical direction translation (m, n), theirs' error squares sum can be expressed as [8]:

$$D(m,n) = \sum_{x,y} [S_1(x,y) - S_2(x-m,y-n)]^2$$

By the above equation unfold can be that:

$$D(m,n) = \sum_{x,y} (S_1^2(x,y) - 2S_1(x,y)S_2(x-m,y-n) + S_2^2(x,y))$$

Because of the $S_2(t)$ is the matching image, therefore $\sum_{x,y} S_2^2(x,y)$ is a constant. Because of $\sum_{x,y} S_1^2(x,y)$ can be approximately as a constant, $V(m,n) = 2S_1(x,y)S_2(x-m,y-n)$ shows the cross-correlation degree between matching images and template images. So when cross-correlation degree between two images is the maximum, the error square sum between two images is the least. Namely the matching is the best. Consequently when $D(m,n)$ getting a minimum value, namely when $V(m,n) = 2S_1(x,y)S_2(x-m,y-n)$ getting a maximum vale, The Cross-correlation degree between of them is the biggest.

As shown in figure3 is the Template image $S_2(t)$, as shown in figure4 is the $S_1(t)$.Through the above matching algorithm conclude the matching result as shown in figure 5.

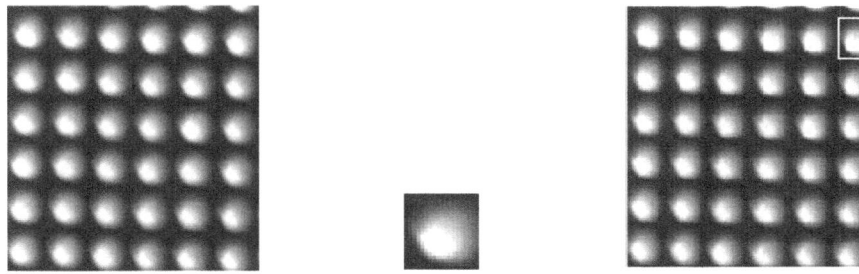

Fig. 3. Template image $S_2(t)$ Fig. 4. Matching image $S_2(t)$ Fig. 5. Matching results

3   The Lightspot Detection of LED Lights

According to the last step find out the position of the LED in the LED display module. After find out the position of the LED, There will be aim at the brightness of the light doing a concrete analysis. To the lightspot detection of the LED is mainly used to analysis the lamp deflect direction[9]. Come to the part of the brightness of the light of detection the text used to the method of Euclidean distance testing. Namely come to the brightness of one lamp in one Angle and with given template doing a template matching, the test results will come out. Namely output the brightest parts of the coordinates and the brightest region of the time and the distance of the template.

We first select a most bright template of 8 * 8, as shown in figure 6 shows, namely it's gray values are all 255.And then through the template with the picture of figure7 used of Euclidean distance doing a lightspot detection. And then output the picture of the finally matching result. As shown in figure8.

   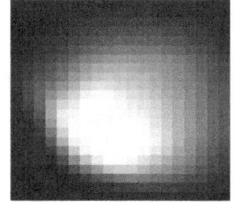   

Fig. 6. 8*8 template   **Fig. 7.** Single lamp image   **Fig. 8.** Matching results

And then the stepping motor drives the LED display screens rotate, in each angle of the step repeat the steps of the lightspot detection, and also compared the output distances. When the distance have a smallest value output the angle of the light. Calculate the step angle of the time namely calculate the light deflection angle.

4   Stepping Motor Driver Circuit

Stepping motor is an electrical machinery executive component of convert electrical pulses signal to mechanical displacement. Whenever a pulse signal applied to the motor control winding, the shaft is turned a fixed Angle, through send out the pulse continuously, and then the motor shaft will step by step to turn [10].

In this paper, the designed stepping motor driver circuit is divided into four parts, the first part is the pulse signal circuit. The article chooses the Single-chip microcomputer is AT89S51, through the single-chip microcomputer generates to the pulse signals control the step motor rotation. The frequency of the input pulse signal determines the speed of the step motor, the number of the input pulses decided the Angle of the stepping motor turned. The second part is the pulse distributor, the article's choose is PMM8713, its main function is add the pulse signal by certain order take turns up to the each phase winding and control the power way of the windings. Through control the sequence of the every phase windings' power to control the turning of the step motor. However, because of the pulse distributor's output current is little, stepping motor's rotation can not drive very well, need to connect an external power drive circuit. So in the third part design a power amplifier circuit, because of the output signal from the circular distributor is digital level, and the need of stepping motor is high power voltage signal. So the article used a photoelectric coupling circuit make the step motor's high voltage signal and the microcomputer's weak current signal isolation and level conversion. The fourth part is stepping motor circuits. The choice of the step motor is mainly consideration its step distance horns, that is through the accuracy requirement of the load, make the load's minimum resolution ratio convert to the motor shaft and calculated the step motor's step distance horns.

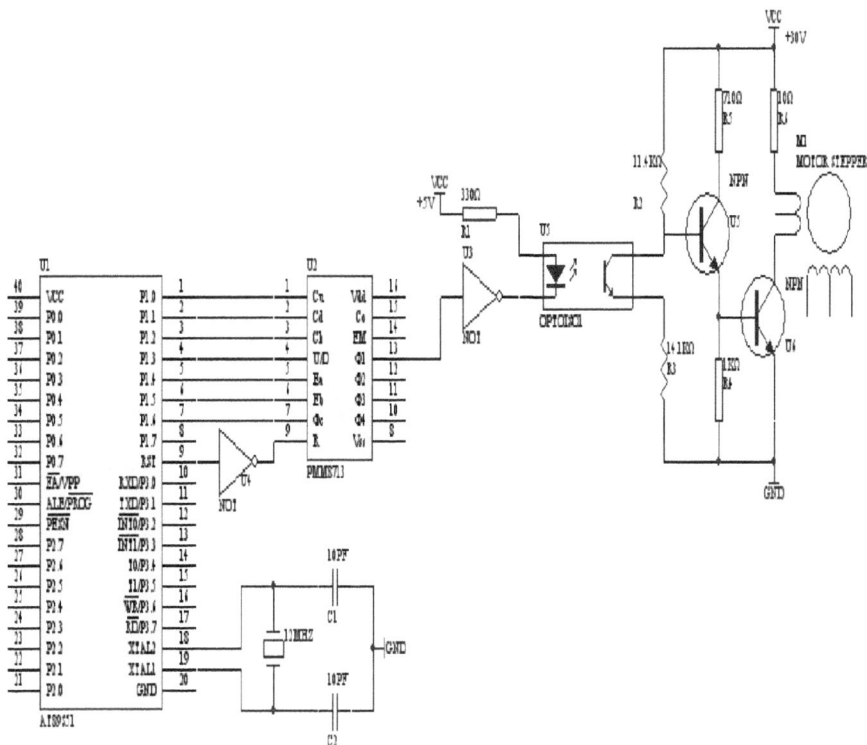

When choose stepping motors, we also can consider its static torque, the step motor's static torque is mainly determined by the load. And also considering the three-phase step motor's step distance Angle, noise and vibration are also very moderate, price is not expensive, therefore this paper choose a stepping motor of model number is 75BF003.

5   Conclusion

Use the detection system of the LED assembly Angle to research LED display module's brightness uniformity is the new direction of the present study. The detection system of the paper introduce can be detect the LED display module's brightness uniformity very well, make a great contribution to the LED display screen's development and application.

Acknowledgments. The paper is partially supported by the Hebei Administration of Science and Technology Foundation (10213570), Hebei Province Department of Education Foundation(2010239), and Research Foundation of Hebei University of Science and Technology.

References

1. Ma, X.: The research and implementation of the LED display screen's γcorrection. Lightning and Control, 92–96 (2010)
2. Xu, J.: The adjustment technology application with the LED display screen's brightness and chromaticity. Modern Display, 146–147 (2010)
3. Du, J.: Two kinds of Fast Image Registration Algorithms based on the gray level. Master's degree thesis (2007)
4. Mount, D.M., Netanyahu, N.S., Moijne, J.: Efficient algorithms for robust feature matching. Pattern Recognition 32, 17–38 (1999)
5. Zhang, H., Zhang, K., Li, Y.: Image matching research progress. The Computer Engineering and Application, 42–45 (2004)
6. Yu, P., Yuan, H., Zhao, Z., et al.: The interest points in the image recognition matching method research. The Computer Engineering and Application, 132–135 (2010)
7. Rao, J.: The research of the image matching method based on the gray level. Master's degree thesis (2005)
8. Wei, N.: Study on Fast Image Registration Algorithms in Pattern Recognition. Doctoral Dissertation (2009)
9. Chang, Y., et al.: A kind of fast detection method for the full-color LED display screen's brightness uniformity. Lamps and Lighting, 33–35 (2009)
10. Huang, L.: Stepping motor control based on AT89C51 Single-chip micro computer. Electronic Production, 19–22 (2006)

Research on Simplified Decoding Algorithm of RA Code

Yanchun Shen[1], Yinpu Zhang[1], and Jinhu Zhang[2]

[1] College of Tangshan,Tangshan, Heibei, China
[2] Tangshan Kaiyuan Autowelding System Co., Ltd., Tangshan, Heibei, China
cnsyc@126.com

Abstract. RA code, using BP iterative decoding algorithm for decoding in AWGN channel, is linear time encoding and decoding algorithm and its performance is closer to the Shannon limit. In order to improve decoding speed and reduce decoding complexity, this paper studies the simplified decoding algorithm of RA code, puts forward the minimal sum algorithm of RA code, and proposes the poly-line decoding algorithm based on BP algorithm. The simulation results show that error code rate of the system reduces and its performance closes to channel capacity with increasing message length. When the number of iterations increases, the error code rate drops with better system performance.

Keywords: RA code, BP iterative decoding algorithm, minimal sum algorithm, poly-line decoding algorithm.

1   Decoding Algorithm of RA Code

The regular RA code expressed by Tanner Fig. can use message passing algorithm to achieve decoding. This algorithm is known as Belief propagation (BP) or BP decoding algorithm. [7]

In addition, decoding algorithms similar to Turbo code can be considered to be used for decoding. First is to decode the accumulator and repeated codes, and then use the loop decoding method of outside information. It can be seen from BP algorithm passing message by the side and posterior probability density of relevant bits of the variable nodes.

Probability density of bit is composed of the non-negative real numbers pair p_1 and p_0 satisfying the relation $p_0 + p_1 = 1$ and is represented by the log likelihood ratios (LLR) $\ln \dfrac{p_0}{p_1}$. There are four different types of information in BP decoding algorithm:

(1) Passing information $m[c, u]$ from the check node to information node;

(2) passing information $m[u, c]$ from the information node to check node;

(3) passing information $m[c, y]$ from the check node to parity node;

(4) passing information $m[y, c]$ from the parity node to check node.

B. Liu and C. Chai (Eds.): ICICA 2011, LNCS 7030, pp. 625–632, 2011.

The initial value of all received bits of code is $B(y) = \ln \dfrac{p(x=0|y_i)}{p(x=1|y_i)}$, where x

refers to the value of bit node and y_i the received value of the channel.

Belief decoding algorithm of RA code:

(1) Initialize $m[c,u], m[u,c], m[c,y], m[y,c]$ and set all values to 0;

(2) Update $m[y,c]$:

$$m[y,c] = \begin{cases} B(y) & y = y_{qn} \\ \\ B(y) + m[c',y] & c' \neq c \ (c',y) \in E \end{cases}$$
(1)

(3) Update $m[u,c]$

$$m[u,c] = \sum_{c'} m[c',u] \, c' \neq c, (u,c') \in E$$
(2)

(4) Update check nodes $m[c,y]$ and $m[c,u]$:

$$m[c,y] = \begin{cases} m[u,c] & c = c_1 \\ \\ 2 artanh\left(\tanh\left(\dfrac{m[u,c]}{2}\right) \bullet \tanh\left(\dfrac{m[y',c]}{2}\right)\right) & (u,c),(y',c) \in E, \ y \neq y' \end{cases}$$
(3)

$$m[c,u] = \begin{cases} m[y,c] & c = c_1 \\ \\ 2 artanh\left(\tanh\left(\dfrac{m[y,c]}{2}\right) \bullet \tanh\left(\dfrac{m[y',c]}{2}\right)\right) & (y,c),(y',c) \in E, \ y \neq y' \end{cases}$$
(4)

In the formula $\tanh(x)$ is hyperbolic tangent function and $\tanh\left(\dfrac{x}{2}\right) = \dfrac{e^x - 1}{e^x + 1}$.

(5) Repeat step (2), (3), and (4) for K times.
(6) Sum for each node information.

$$s(u) = \sum_c m[u,c] \quad u \in U$$
(5)

If the result shows $s[u] > 0$, then decode the current node into 0, otherwise 1.

The above algorithms can be operated on computer, and in order to reduce the distribution volume, the first step (4) can be considered to change multiplication into addition by taking logarithms from both sides. [1]

2   Minimal Sum Algorithm of RA Code

BP decoding algorithm of RA code decoding algorithm is a loop decoding technology. BP algorithm can achieve linear time decoding to obtain superior performance, but the update of check node requires massive using of hyperbolic tangent and inverse hyperbolic tangent functions and other complex computing which are hard to be realized by hardware and impacting their application in next-generation communication system.

In the study on LDPC codes, considering various factors many scholars put forward the sub-optimal decoding algorithms, including minimal sum decoding algorithm and improved algorithm based on minimal sum algorithm: normalization algorithm and migration algorithm. [2] As the minimal sum algorithm drops lots of data and the lifted performance of the improved minimal sum algorithm is very limited, the poly-line approximation algorithm is further proposed based on BP algorithm, [3] of which the decoding complexity and performance has obtained good compromise.

3   Proposing of the Minimal Sum Algorithm

The minimal sum algorithm is a simplified decoding algorithm proposed by Fossorier, etc. in the study of LDPC codes, which simplifies the update operation of check nodes. The minimal sum algorithm of BP algorithm is: the check node information update of degree d_c can be estimated by formula (6):

$$m[c,u] \approx \prod_{i=1}^{d_c} \mathrm{sgn}\big(m[u_i,c]\big) \times \min\big(\big|m[u_1-1]\big|,\big|m[u_2-1]\big|,\cdots,\big|m[u_{d_c-1},c]\big|\big) \qquad (6)$$

Prove: there is information update relationship between check nodes of degree:

$$\tanh\frac{m[c,u]}{2} = \prod_{i=1}^{d_c-1} \tanh\frac{m[u_i,c]}{2} \qquad (7)$$

In the above formula u and u_i are variable nodes, while c is check node. $u_i \in N(c)\setminus u$ refers to the adjacent nodes of c with the exception of u.

$\tanh(x)$ and $\tanh^{-1}(x)$ is the monotonically increasing odd functions which have the property:

$$\tanh(x) = \mathrm{sgn}(x) \bullet \tanh(|x|) \qquad (8)$$

$$ar \tanh(x) = sgn(x) \bullet ar \tanh(|x|) \tag{9}$$

From formula (7) it can be derived directly:

$$sgn(m[c,u]) = sgn(m[u_1,c])\,sgn(m[u_2,c]) \cdots sgn(m[u_{d_c-1},c])$$

$$= \prod_{i=1}^{d_c-1} sgn(m[u_i,c]) \tag{10}$$

And because $0 \le \tanh(|x|) \le 1$

$$\prod_{i=1}^{d_c-1} \tanh\frac{|m[u_i,c]|}{2} \le \min(\tanh\frac{|m[u_1,c]|}{2},\cdots,\tanh\frac{|m[u_{d_c-1},c]|}{2})$$

$$= \tanh(\min(\frac{|m[u_1,c]|}{2},\cdots,\frac{|m[u_{d_c-1},c]|}{2})) \tag{11}$$

Therefore:

$$|m[c,u]| \approx \min(|m[u_1,c]|,|m[u_2,c]|,\cdots,|m[u_{d_c-1},c]|) \tag{12}$$

Formula (6) can be obtained from (10) and (12).

From formula (6) it can be seen that the information update of check node only needs minimal sum and symbols multiplication operations. Because the minimal sum operation is achieved by comparing which is equal to addition operation, and the symbol multiplication can be operated as modulo 2. Therefore, compared with hyperbolic tangent and inverse hyperbolic tangent computing of BP algorithm, the minimal sum algorithm is simpler. Whenever updating the Information of check node, the absolute value outputted is decided by the smallest and the second smallest value regardless of node degree, and retaining these two values is enough.

In AWGN channel, the minimum algorithm makes it easier for hardware to achieve, and estimation of channel parameters is not required. As y can be used directly as node to input values for computing, this decoding algorithm has been widely applied.

4   Minimal Sum Algorithm of RA Code

The information update of check node in BP algorithm of RA code is in line with the formula (7). Thus, RA code can apply the minimal sum algorithm.

Apply the formula (6) , we get:

$$m[c,y] \approx sgn(m[u,c])\,sgn(m[y',c])\min(|m[u,c]|,|m[y',c]|) \tag{13}$$

The same can be obtained by formula (1):

$$m[c,u] \approx \mathrm{sgn}(m[y,c])\,\mathrm{sgn}(m[y',c])\,\min(|m[y,c]|,|m[y',c]|)$$ (14)

It can be seen from the above two formulas that decoding according to these two formulas is simpler than BP algorithm, needing only addition and comparison operations on check nodes.

The following Fig. shows that although the minimal sum algorithm has dramatically simplified the operation, the system performance of minimal sum algorithm is worse than BP algorithm.

5   Improved Algorithm of RA Code Algorithm

The minimal algorithm mentioned previously retains only the smallest and second smallest values of check node and the other ignored information does not get to use, and therefore the lost information makes the decoding performance compromised. This section describes feature of using check node function to replace curve approximation by straight line, which does not only retain the information of check node but also simplify operation.

5.1   Poly-line Algorithm

Poly-line algorithm is a method using piecewise linear function to represent nonlinear function, which is first put forward by Chua and Kang. Poly-line algorithm has a quite simple structure which uses a small number of parameters to estimate the continuous function. The algorithm is widely used in the field of modeling, analysis and estimation of large number of nonlinear dynamic systems. [4]

Poly-line function is defined in a sequence of intervals, and the simplest poly-line function is absolute value function:

$$|x| = \begin{cases} -x & x < 0 \\ 0 & 0 \\ x & x > 0 \end{cases}$$ (15)

5.2   Poly-line Approximation Decoding Algorithm

Updated equation according to IRA code check node:

$$m[c,v] = \gamma^{-1}\left(\prod_{i=1}^{d_c-1}\gamma(m[v_i,c])\right)$$ (16)

Obtain the following theorem:

Theorem: output information L of check node of IRA code degree $a+2$ is

$$L = \prod_{i=1}^{a+1} \mathrm{sgn}(x_i) f\left(\sum_{i=1}^{a+1} f(|x_i|) \right) \tag{17}$$

In this formula $x_i (i = 1, \cdots, a, a+1)$ is input information from variable node to check node and L is output information.

Variable node includes node information and parity node, and the expression of $f(x)$ is

$$f(x) = \ln \frac{(e^x + 1)}{(e^x - 1)} \tag{18}$$

And it can be inferred form formula (17) that:

Inference: update relation of regular RA code check node information is

$$L = \mathrm{sgn}(x_1) \bullet \mathrm{sgn}(x_2) f (f(|x_1|) + f(|x_2|)) \tag{19}$$

It can be seen from (17) that the rest are addition operation except $f(x)$, and therefore t it is only needed to transform $f(x)$ to simplify computing. The $f(x)$ in (18) contains exponent, logarithm and addition operation, and $f(x)$ is first considered to be linearized. [5]

3 piecewise linear function

$$f_1(x) = \begin{cases} -4.67x + 5 & x \le 0.8841 \\ -0.214 + 1.07 & 0.8841 < x \le 5 \\ 0 & x > 5 \end{cases} \tag{20}$$

5 piecewise linear function

$$f_2(x) = \begin{cases} -10.2854x + 5 & x < 0.301 \\ -1.7434x + 2.42 & 0.301 \le x < 0.8841 \\ -0.5736x + 1.3912 & 0.8841 \le x < 1.9 \\ -0.0972x + 0.486 & 1.9 \le x < 5 \\ 0 & x \ge 5 \end{cases} \tag{21}$$

Replacing functions $f_1(x)$ and $f_2(x)$ with $f(x)$ can avoid logarithmic and exponential operations, and only addition and multiplication can greatly simplify the complexity of algorithm.

Following is the curve chart of $f_1(x)$, $f_2(x)$ and $f(x)$:

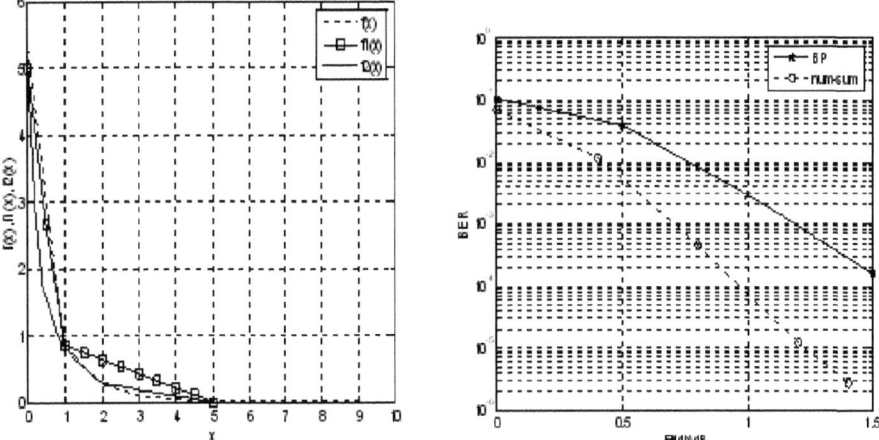

Fig. 1. performance simulation chart of minimal sum algorithm

Fig. 2. the curve chart of $f_1(x)$, $f_2(x)$ and $f(x)$

Fig. 2 shows the simulation results of BP iterative decoding algorithm of RA code and of its poly-line algorithm. In the Fig. BP represents BP algorithm, LINE is poly-line algorithm. It can be seen that when the SNR of BP algorithm is about 0.9dB, the corresponding code error rate is 10^{-5}; when the SNR of poly-line algorithm is about 0.9dB, its corresponding code error rate is $10^{-4.66}$. Therefore, their simulation performance is close. As the poly-line decoding algorithm do approximate to the original function, it avoids logarithmic and exponential operations and needs only addition and multiplication. Therefore, it greatly reduces the algorithm complexity and optimizes BP algorithm.

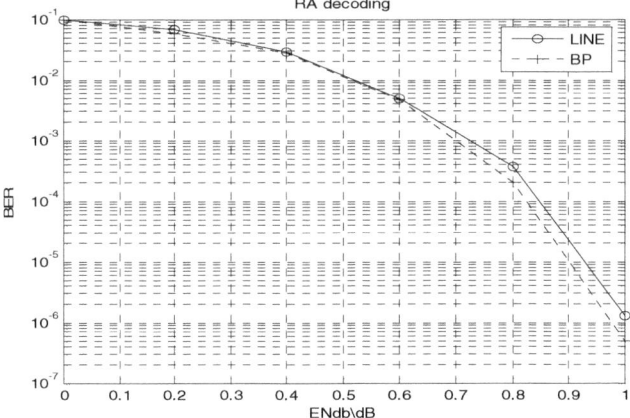

Fig. 3. Simulation chart of RA code poly-line algorithm

6   Complexity Comparisons of Various Algorithms

Table 1 lists calculate amount needed for one information update on a check node of regular RA codes in BP algorithm, minimal sum algorithm, 5 piecewise linear approximation algorithm, and 3 piecewise linear approximation algorithm. Table 1 shows the order of the four algorithms' complexity from simple to complex is: minimal sum algorithm, 3 piecewise linear approximation algorithm, 5 piecewise linear approximation algorithm and BP algorithm. As the simplified algorithm does not require the hyperbolic tangent and arctangent and other complex calculations, their complexity is far less than that of BP algorithm.

Table 1. Update complexity of three algorithms on RA code check node

Algorithm	$\tanh$	$\tanh^{-1}$	Multiplication	Addition
BP algorithm	3	3	12	-
Minimal Sum algorithm	-	-	-	3
3 piecewise linear approximation	-	-	6	15
5 piecewise linear approximation	-	-	6	18

7   Conclusion

This chapter applies minimal algorithm in RA code decoding algorithm and poly-line approximation algorithm based on BP algorithm; and it also studies a variety of decoding algorithms of RA codes, and finally compares the complexity of various algorithms. The results show that: Relative to BP iterative decoding algorithm of RA code, the complexity of minimal sum is significantly reduced. But for RA code, the performance of the system has reduced significantly; The performance of poly-line approximation algorithm is closer to that of BP algorithm, and the complexity is also reduced, so this algorithm has superior performance for RA code. Simplified decoding algorithm of RA code obtained good compromise between complexity and performance, showing high practical value.

References

1. Gao, H.: RA codes encoding and decoding algorithms. Journal of Luoyang Normal University 2, 69–73 (2004)
2. Fossorier, M.P.C., Mihaljevic, T., Dholakia, A.: Reduced complexity iterative decoding of lowdensityparity check codes based on belief propagation. IEEE Trans. Commun. 47(5), 673–680 (1999)
3. Chen, J.: Reduced complexity decoding algorithms for low-density parity check codes and turbo codes. PhD Thesis. University of Hawaii (2003)
4. Gao, H.: RA code decoding simplifies the algorithm. Journal of Sichuan University (Engineering Science Edition) 36(1), 107–110 (2004)
5. Gao, H., Xu, Z., Wu, Y.: IRA code decoding algorithm to simplify. Journal of University of Electronic Science and Technology of China 34(1), 40–43 (2005)

Comparative Study on the Field of Time-Frequency Measurement and Control Based on LabVIEW

Hao Chen, Ying Liu, and Yan Shan

Natural Science College of Hebei United University,
063009 Tangshan, China
chen_2@tom.com

Abstract. LabVIEW(Laboratory Virtual Instrument Engineering Workbench) is a graphical programming language development environment, it is widely used in industry, academia and research laboratories. It is seen as a standard data acquisition and instrument control software. This paper studied on high-precision frequency measurement methods based on a series of new principles and new methods, and then proposed two algorithms based on measuring the frequency of LabVIEW. Combined with modern measurement theory, the three-point method and energy rectangular balance, this algorithm improved signal processing capabilities LabVIEW. The experiment verified the correctness and feasibility of two methods, which can both effectively reduce the frequency measurement error and noise; and can be used in different fields.

Keywords: LabVIEW,three-point method,energy rectangle balance method, frequency measurement, Virtual Instrument.

1 Introduction

LabVIEW——Laboratory Virtual Instrument Engineering Workbench is a innovative software product of NATIONAL INSTRUMENTS, referred to as NI. It is currently the most widely used and fastest growing, most powerful graphics software integrated development environment. [1] It creates the basic conditions for the users to easily and quickly build their own instrument system needed in actual production with its intuitive and easy program manner, a number of source-level device drivers, and a wide variety of analysis and expression function supports. After twenty years of development, LabVIEW has been fully proved to be a very powerful and best software system for automated test, measurement engineering design, development, analysis and simulation experiment. Now it has been widely used in automotive, electronics, chemical, biological and life sciences, aviation, aerospace, and many other are as. In the data acquisition environment, the signal frequency measurement and signal frequency tracking is a basic and important issue. Only after the frequency of the signal measured, it is possible to realize the sampling of its whole cycle, creating favorable conditions for the next digital signal processing. Frequency measurement is also a period measurement, commonly used methods can be divided into time domain and frequency domain methods. [2] Generally speaking, the common measurement data are mostly presented by time-domain approach, they are

B. Liu and C. Chai (Eds.): ICICA 2011, LNCS 7030, pp. 633–640, 2011.

reflected in the rectangular coordinate system, the horizontal axis represents time variable and the vertical axis represents the physical quantities that change over time. The main advantage of time-domain method is direct, relatively simple and easy to estimate error. Let us start from two aspects, supplemented by examples, to explore the basic method of frequency measurement.

2   Principles and Procedures of Time-Domain

Three-point method is a data fitting method based on trigonometric functions transformation. Assuming the measured function is the sine function, on condition of interval sampling (including non-integral period sampling), we can use three adjacent data samples to derive linear equations solving the signal frequency, and then fit the coefficients of the equation, the find the frequency finally.[3]

Derived as follows(hereinto, U_m is the amplitude of the carrier, ω is a sine wave frequency, φ is the phase angle):

To establish the signal:

$$\mu(t) = U_m \sin(\omega t + \varphi)$$

If

$$\omega t + \varphi = \alpha$$

So

$$\mu(t) = U_m \sin \alpha$$

As if

$$\theta = \frac{\omega}{F_S} = 2\pi \frac{f}{F_S}$$

Hereinto F_s is sampling frequency, so

$$f = \frac{\theta F_S}{2\pi}$$

The three Border upon data swatch can be showed:

$$\mu_i = U_m \sin \alpha_i$$
$$\mu_{i+1} = U_m \sin(\alpha_i + \theta)$$
$$\mu_{i+2} = U_m \sin(\alpha_i + 2\theta)$$

Using triangle transform,

$$\mu_i + \mu_{i+2} = U_m[\sin \alpha_i + \sin(\alpha_i + 2\theta)]$$
$$= 2U_m \sin(\alpha_i + \theta) \cdot \cos \theta$$
$$= 2\mu_{i+1} \cdot \cos \theta$$

So, order

$$x(n) = 2\mu_{i+1}$$

Just

$$y(n) = \mu_i + \mu_{i+2}$$

So

$$y(n) = \cos\theta \cdot x(n) \qquad (1)$$

$$f = \frac{Fs}{2\pi}\arccos\theta \qquad (2)$$

Formula (1) is the linear equation needed. Here are n equations and only one unknown number $\cos\theta$, with least square method we can get a more accurate slope $\cos\theta$, and then use formula (2) to find the frequency.

The general form of linear fitting is $F = mX + b$, where F is the series representing best fitting, m is the slope; b is intercept.

2.1 Experimental Test

Start LabVIEW, select the option to open a new panel, Then use the controls on the control object template (controls) and display objects (indications) to create a graphical user interface (ie front panel). [4]In LabVIEW, the linear fitting function is Linear Fit Coefficients; its function is to find the slope and intercept of a linear equation, so that the line determined by them can describe a set of input data best. [5]

The icon of this function is shown in Figure 1:

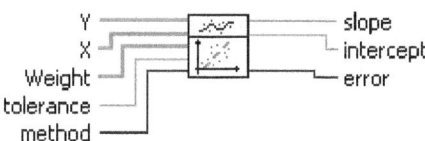

Fig. 1. Linear Fit Coefficients Function

Thereinto, Y Values and X Values contain at least two points, namely, n ≥ 2; slope is the slope of the linear fitting made , intercept is the intercept of linear fitting done.

The equation given in formula (1) is relatively simple, the intercept is 0.

LabVIEW gives the function Pulse Measurements, which can be directly used to calculate the cycle of waveform. As shown, the program will use it to do the calculation and comparison with the above method.[6]

In order to make the process close to the actual collection process, random noise is superimposed on the signal source. The signal array generated through the loop produce the desired array $y(n), x(n)$, and then find $\cos\theta$ through linear fitting; finally obtain "measurement frequency of 1". Meanwhile, as a comparison, the program also uses a function of Pulse Measurements computing cycles to make "measured frequency 2." Both of the two measurements give relative error.

Table 1. Program front panel

Sampling frequency	Frequency setting	Measuring frequency 1	Relatively error 1.
513.001	50.23	50.2311	0
Swatch numbers	Yawp value	Measuring frequency 2.	Relatively error 2.
120	0	50.25	-0.05

Sampling frequency	Frequency setting	Measuring frequency 1	Relatively error 1.
513.001	50.23	50.237	-0.02
Swatch numbers	Yawp value	Measuring frequency 2.	Relatively error 2.
120	0.01	50.3	-0.14

Table 1 shows the program front panel. Front panel displays the results of two experiments when the noise is 0 and 0.01. The entire experiment is conducted under the condition of non-full-cycle sampling, about 10 samples per cycle and 10 cycles of data collection. The results indicate that in the case of noise-free, the relative error of "Measuring frequency 1" is 0, the relative error "Measuring frequency 2" is 0.05%; When 1% noise is superimposed, the relative error of "measuring frequency 1" is 0.02%, the relative error of "Measuring frequency 2" is 0.14%. Thus, the effect of three-point method is good.

3   Principles and Procedures of the Frequency Domain

Frequency domain method is based on FFT. Fast Fourier transform (FFT) algorithm is simple and fast, and the use of multiple characteristic quantities does not increase its calculated amount, these advantages of the Fourier transform make the system more suitable for online application and analysis. [7] Signals in the frequency domain are usually described by the rectangular coordinate system in which the horizontal axis is frequency and vertical axis is amplitude or phase. On the frequency of measurements, a basic problem is how much is the minimum number of the horizontal axis scale, that is the frequency resolution ratio problem. If Fs is the sampling rate, $\#s$ is the sample size, the frequency resolution ratio is :

$$\Delta f = \frac{F_s}{\#s}$$

In the frequency domain, the thinking of dealing with the problem is to refine the smallest distinguished unit further by interpolation method, which is to refine Δf . [8]

First, we analyze the spectrum, commonly we use the energy barycenter of discrete window spectrum function to get gravity method correction frequency formula.[2] p_i indicating the amplitude of number i spectral line, x_i the abscissa of p_i , borrowing the concept of mechanics, we assume the number i spectral line forming a torque on the origin(called the energy moment), which is $p_i x_i$, for all N spectral lines, the total energy moment is $\sum_{i=1}^{N} p_i x_i$ assume that there is a center of gravity in the x-axis at x_0, the energy applied to the entire signal from opposite direction is p_0. Without considering the case of leakage, make the energy moment on x-axis balanced, i.e. $p_0 x_0 = \sum_{i=1}^{n} p_i x_i$, the diagram shown in Figure 2.

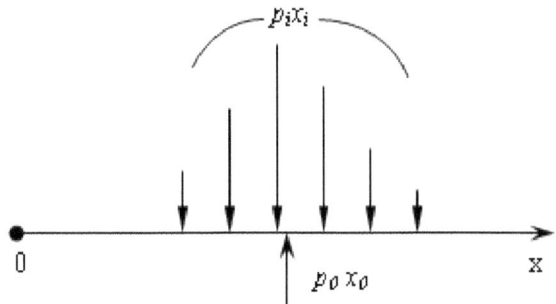

Fig. 2. .Sketch map of energy rectangle balance method

Owing to p_0 can be showd: $p_0 = \sum_{i=1}^{N} p_i$, so that $x_0 \sum_{i=1}^{N} p_i = \sum_{i=1}^{N} p_i x_i$, thereby educed:

$$x_0 = \frac{\sum_{i=1}^{N} p_i x_i}{\sum_{i=1}^{N} p_i}$$

Finally, abscissa multiply $\Delta f = \dfrac{Fs}{N}$, gained frequency:

$$f = \frac{\sum_{i=1}^{N} p_i x_i}{\sum_{i=1}^{N} p_i} \frac{F_s}{N} \tag{3}$$

In formula (3), F_s is the sampling rate, N is the number of samples.

3.1   Experimental Test

We conducted a simulation test with LabVIEW test.

In the experiment, we generated a simulate sinusoidal signal, took its non-full-cycle samples, and then calculate its rate of work, and then calculated the signal frequency according to formula (3). The results showed that, if we change the signal type to other waveforms such as square wave, you will find that the error is very large, or simply incorrect. The reason for this result is that there are multiple signal frequency components square wave, which means there are more than one energy moment in the same direction. At this point if it is to balance with an energy moment, the calculated frequency is not the fundamental frequency.

Based on this, we intercepted the first 100 points on part of the spectrum line of square wave signal power spectrum and designed the improved process shown in Figure 3, as in the actual testing process, the signal will be inevitably mixed with noise, the noise signal superimposed on the signal, which made the wave zero-crossing point shifted. Therefore, the simulate improvement program takes harmonics, DC component and noise and other factors into consideration.

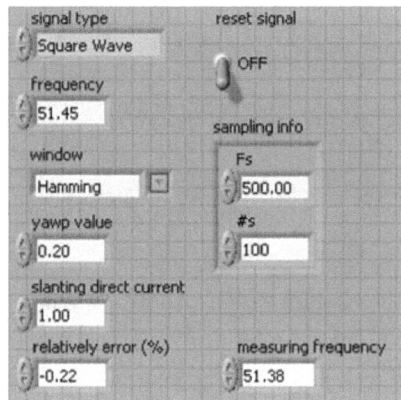

Fig. 3. Improvement program front panel

First, the simulated signal source has been modified, adding the function of DC bias and noise superposition. In the data processing section, first, use of Basic Averaged DC-RMS function to measure DC component, and then subtract it from the signal waveform data. Experiments show that it is clearer and more effective to do so than to divide Dc spectral lines after getting the power spectrum. Then use improved algorithm to get the power spectrum, which uses Array Max & Min function to find the index M of the largest element, Then take a 2M-long sub-array from the first element of original fundamental frequency, we can assume that this array contains all the Function spectrum lines of fundamental frequency. Since ultimately we do not care about the harmonic,so when setting the sampling parameters (sample rate and sample size),we only considered the fundamental wave. The following is the simulated data comparison got from the experiment.

Experiment 1, a typical wave and windows adopted, no noise, the signal amplitude is 1, the signal frequency is 51.45Hz, $F_s = 500, \#s = 100$. Simulated data shown in Table 2.

Table 2. The relative error of frequency

Window	Sine wave (the basic algorithm)/%	Sine wave /%	Triangular wave /%	Sawtooth wave/%	Square wave/%
Rectangular window	-0.89	1.16	1.06	-3.09	2.65
Hanning	0.00	0.04	0.03	-0.71	0.17
Hanning	-0.01	0.05	0.05	-0.58	0.21
Blackman	0.00	0.04	0.04	-1.21	0.48

Experiment 2 was based on Experiment 1, noise increased by 0.1, the original signal amplitude is 1. simulated data shown in Table 3.

Table 3. The frequency relative error after noise superimposed

Window	Sine wave /%	Triangular wave /%	Sawtooth wave /%	Square wave /%
Rectangular window	1.52	1.50	-2.08	2.83
Hanning	-0.04	-0.05	-0.62	-0.01
Hanning	-0.01	-0.02	-0.46	0.05
Blackman	-0.04	-0.05	-1.14	0.30

Experiment 3 is based on the experiment 1, let # s = 500. Simulated data shown in Table 4.

Table 4. The frequency relative error after changing sample number

Window	Sine wave (the basic algorithm)/%	Sine wave /%	Triangular wave /%	Sawtooth wave/%	Square wave/%
Rectangular window	0.1763	0.40	0.38	-0.30	0.89
Hanning	0.0000	0.01	0.01	-0.24	0.14
Hanning	0.0028	0.01	0.01	-0.24	0.15
Blackman	0.0000	0.01	0.01	-0.39	0.13

We can see from the above experiments:

The method is suitable for a variety of additional windows, but different windows correspond to different precision.

That algorithm has better inhibitory effect on noise.

Increasing the number of samples can significantly improve the precision.

4   Conclusion

LabVIEW language program is easy to program and modify, it has good man-machine interface and powerful data analysis function library.[9] In this paper, two frequency measurement techniques based on LabVIEW software virtual instrument are proposed, the methods have their respective advantages of easy frequency calculation and small measurement error. Both methods have advantages for different areas. The main problem of three-point method is that it requires the signal can only be a sine wave. In some literatures, the use of the method is also discussed in the case of harmonics and DC component. In this paper improve the original energy rectangular balance algorithm, simplify coordinate system and formula, the frequency measurement technique based on this has the advantages of simple computation, anti-noise and high precision.

Two methods are implemented on NI's LabVIEW, they make full use of computer software and hardware resources and processing power, also have the features of high cost performance, easy operation and flexibility.[10] In recent years, the frequency measurement technology is covering more and more areas, its accuracy is increasing and it is more and more closely linked to different disciplines. Extending the measurement methods described in this article to communications, instrumentation, measurement technology, electronics, physics, and other domains closely linked to frequency measurement technology has a certain practical significance.

References

1. Wang, L., Tao, M.: Proficiency in Labview 8.0. Publishing House of Electronics Industry, Beijing (2007)
2. Yao, J., Jie, H.-y.: Design Of Frequency Measurement Software System Based On Virtual Instl-Ument. Foreign Electronic Measurement Technology 26(11) (2007)
3. Hou, G., Wang, K.: Labview7.1 Program And Virtual Instrument Design. Tsinghua University Press, Beijing (2005)
4. Chai, J.-z., Han, F.-c., Su, X.-l.: Application Based on Virtual Instrument in Frequency Measurement of Power System. Journal of Electric Power 21(1) (2006)
5. Ding, K., Wang, L.-q.: Spectrum Of The Discrete Energy Centrobaric Method. Vibration Engineering Journal (14) (2001)
6. Chen, H., Cao, F.-k.: Frequency Measurement Based on Labview 3-Point Method. Modern Scientific Instruments (3) (2007)
7. Xue, H., Zhang, L.: Overview Of Nonuniform Fast Fourier Transformation. Computerized Tomography Theory And Applications 19(3) (2010)
8. Chen, H., Shan, Y.: Research and Improvement of Frequency Measurement Based on Energy Rectangular Balance Method. Noise And Vibration Control 28(3) (2008)
9. Wu, C.-d., Sun, Q.-y.: Labview Virtual Instrument Design And Application Procedures. People's Posts And Telecommunications Publishing House, Beijing (2008)
10. Zhang, T.-y., Jiang, H.: Virtual Frequency Characteristics of the Design of the System Test. Electrical Measurement Technology and Watch (9) (2004)

Protocol to Compute Polygon Intersection in STC Model

Yifei Yao[1], Miaomiao Tian[2], Shurong Ning[1], and Wei Yang[2]

[1] School of Computer and Communication Engineering,
University of Science and Technology Beijing, Beijing, 100083, P.R. China
[2] Department of Computer Science and Technology,
University of Science and Technology of China, Hefei, 230027, P.R. China
yaoyifei@mail.ustc.edu.cn,fancyning@163.com,
{tianmiaomiao,qubit}@ustc.edu.cn

Abstract. Intersection and union of convex polygons are basic issues in computational geometry, they can settle lots of matters such as economy and military affairs. To solve the problem that traditional method of making the polygons public could not satisfy the requirements of personal privacy, the protocol to compute intersection and union of convex polygons in secure two-party computation(STC) model was investigated. Along with the scan line algorithm and secret comparison protocol, the protocol completes the calculation without leaking so much information. The security and complexity analysis of the protocol are also given.

Keywords: Secret comparison, intersection, union, STC.

1 Introduction

Along with the issues of privacy-preserving turning more and more attractive, secure computation of basic algorithms become popular questions. Polygonal intersection and union are basis of computational geometry and computer graphics, they are of significance both in theory and practice [1]. Many problems need polygonal intersection such as removing hide line, pattern recognition, component position, linearity programming and so on [2,3].

In former applications, people always collect the polygons' information together and solve it by a trust third party (TTP). But the demand of privacy makes it hard to find such an agency trusted by both partners. In this paper, we study how to calculate polygonal intersection and union in STC model. This solution does help in economy and military affairs. For example, a new company hopes to build a shopping mall, it must review if there is another company working at the same area. Both of them want to know weather their orbits meets or not without leaking their own border information. Meanwhile, military affairs also refer to the intersection question often. Fortunately, SMC technique [4] can help to achieve the goal.

In this paper, we devise a protocol to compute intersection and union of convex polygons approximately, and then analyze its security, complexity and applicability. The paper is organized as follows. In section 2 we describe preliminaries.

B. Liu and C. Chai (Eds.): ICICA 2011, LNCS 7030, pp. 641–648, 2011.

The basic comparison protocol is introduced in section 3 and the STC protocol is presented in section 4. Then in section 5 we discuss the protocol's complexity and security. At last we conclude the paper in section 6.

2   Preliminaries

2.1   Secure Multi-Party Computation

In a multi-agents network, SMC helps two or more parties complete the synergic calculation without leaking private information. Generally speaking, SMC is a distributed cooperation. In this work, each party hold a secret as input, and they want to implement the cooperative computation while knowing nothing about other's data except the final result [5,6].

After [4] in 1982, the technology of SMC has already come into more and more domains such as data mining, statistical analysis, scientific computation [7], electronic commerce, private information retrieval (PIR), privacy-preserving computation geometry (PPCG) [8,9], quantum oblivious transfer and so on[10].

Previous methods work on a third-party who is trusted by all parties. A TTP can get enough information to complete the calculation and broadcasts the result. But the hypothesis itself is insecure and unpractical. Therefore, an executable protocol which can preserve participants' privacy becomes more and more dramatically. It is known that any secure computation problem can be solved by a circuit protocol, but the size of the corresponding circuit is always too large to realize. So investigators choose to design special protocol for special use.

2.2   Secret Comparison Protocol

In 1982, Yao A.C. brought forward the famous millionaire's problem: two millionaires, say Alice and Bob, want to know which is richer, without revealing their respective wealth [4].In 2004, Qin brings forward a method for two parties comparing if $a = b$ privately. There are three parties taking part in the protocol: A, B, and an oblivious third party C who helps A and B to check if their private value a and b are equivalent or not. The method validates its security by computational indistinguishability through homomorphism encryption and Φ-hiding assumption. It returns which one is greater or equal to the other, while Yao's method couldn't returns the equality message. This protocol is complex in computation and safe to resist decoding, it reduces the communication of random data perturbation techniques in Yao's method.

2.3   Models

Computational Model. In this paper, we study the problem under a semi-honest model, in which each semi-honest party follows the protocol with exception that he keeps a record of all its intermediate computations, and he will never try to intermit or disturb with dummy data.

Security Model. We name I_A and I_B as the input instance of Alice and Bob, and name O_A and O_B as the corresponding output. C represents the computation executed by the two partners, then $(O_A, O_B) = C(I_A, I_B)$ [2]. A protocol for executing C is secure when it satisfies two conditions as follow:

1. There is an infinity set $D_A = \{(IA_i, OA_i)|i = 1, 2, \ldots\}$ that $(O'_A, O_B) = C(I'_A, I_B)$ for $\forall (I'_A, O'_A) \in D_A$.

2. There is an infinity set $D_B = \{(IB_i, OB_i)|i = 1, 2, \ldots\}$ that $(O_A, O'_B) = C(I_A, I'_B)$ for $\forall (I'_B, O'_B) \in D_B$.

In the execution, it's inevitable to leak some message. The adversary can obtain something through security analysis, but it's not enough for him to get the certain value. Although the protocol is not zero-knowledge, it is a desirable way to achieve high efficiency.

2.4   Related Algorithms in Computational Geometry

Firstly, we find out the maximal and minimal x-coordinate values, noted as a and b, then divide into k equidistant bands perpendicularly between a and b. The k bands form a memory serial, and we distribute the n points of set S into the memory serial. At last, we pick the maximal and minimal y-coordinate values of each band and save them as set S^*. S^* has $2k + 4$ points at most, we construct its convex to form the approximate outline. It will be found in $O(1)$. Then, a scanning work in linear time can set up these fragments. At last, we pick them up and move out the void peaks at the margin.

Meanwhile, a protocol computing the generating set of intersection and union of polygons can be modified at the last stage. The outermost point with the same ordinate is outputted instead of primary maximum point. An example is showed as Fig. 1

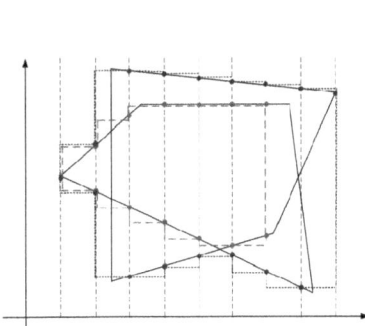 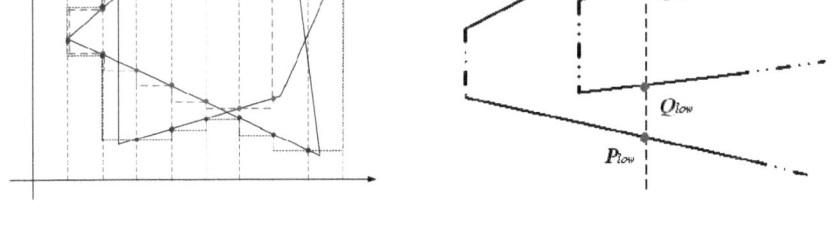

Fig. 1. Example of proportionate partition   **Fig. 2.**   Basic comparison protocol: Result 1

3 Building Blocks

In this section, we introduce the secure building blocks. It performs comparison on one scanning bean. It's a basic tool for the latter protocols.

We assume that P_{high} and P_{low} belong to Alice, Q_{high} and Q_{low} belong to Bob. They are on the same scanning bean, and are ranked by their y-coordinate.

There will be four instances appearing on each bean as follows.

Result 1. $P_{high} > Q_{high}$ and $P_{low} < Q_{low}$: Then Q_{high} and Q_{low} belong to polygonal intersection, P_{high} and P_{low} belong to polygonal union (Fig. 2).

Result 2. $P_{high} < Q_{high}$ and $P_{low} < Q_{low}$:

Result 2.1: if $P_{high} > Q_{low}$ then P_{high} and Q_{low} belong to polygonal intersection, Q_{high} and P_{low} belong to polygonal union (Fig. 3).

Result 2.2: if $P_{high} < Q_{low}$ then no one on this scanning bean belongs to the intersection, the four points all belong to the union (Fig. 4).

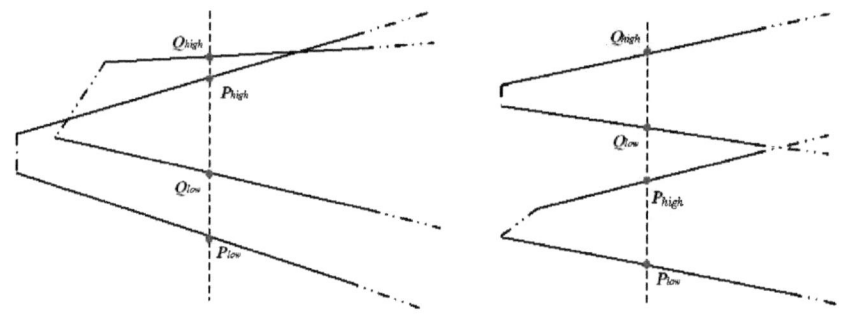

Fig. 3. Result 2.1   **Fig. 4.** Result 2.2

Result 3. $P_{high} > Q_{high}$ and $P_{low} > Q_{low}$:

Result 3.1: if $Q_{high} > P_{low}$ then Q_{high} and P_{low} belong to the polygonal intersection, P_{high} and Q_{low} belong to the polygonal union (Fig. 5).

Result 3.2: if $Q_{high} < P_{low}$ then no one on this scanning bean belongs to the intersection, the four points all belong to the union (Fig. 6).

Result 4. $P_{high} < Q_{high}$ and $P_{low} > Q_{low}$: Then P_{high} and P_{low} belong to polygonal intersection, Q_{high} and Q_{low} belong to polygonal union (Fig. 7).

We summarize the basic comparison protocol as below:

Protocol 1. Basic Comparison Protocol

Input: Alice has P_{high} and P_{low}, while Bob has Q_{high} and Q_{low} at each bargained scanning bean.

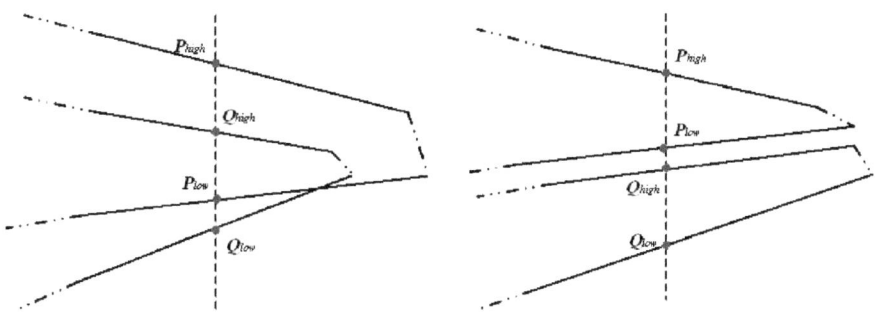

Fig. 5. Result 3.1   **Fig. 6.** Result 3.2

Output: Both Alice and Bob know which of his point is on the polygonal borderline with no information leaking to the other.

Alice cooperates with Bob to compare (P_{high}, Q_{high}) and (P_{low}, Q_{low}) using the secret comparison protocol in 2.2.

Case 1: if $P_{high} > Q_{high}$ and $P_{low} < Q_{low}$ then we get result 1 and terminate.

Case 2: if $P_{high} < Q_{high}$ and $P_{low} < Q_{low}$ then we continue to compare P_{high} and Q_{low}: if $P_{high} > Q_{low}$ then we get result 2.1 , else we get result 2.2, and terminate.

Case 3: if $P_{high} > Q_{high}$ and $P_{low} > Q_{low}$ then we continue to compare Q_{high} and P_{low}: if $Q_{high} > P_{low}$ then we get result 3.1 , else we get result 3.2, and terminate.

Case 4: if $P_{high} < Q_{high}$ and $P_{low} > Q_{low}$ then we get result 4 and terminate.

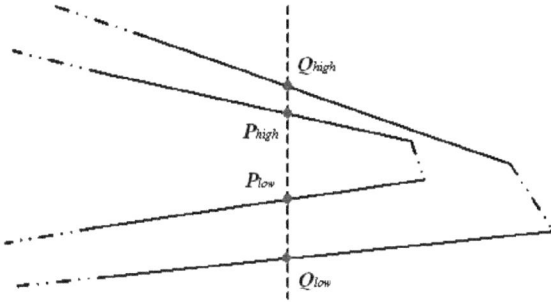

Fig. 7. Basic comparison protocol: Result 4

Complexity and security analysis. Protocol 1 has the complexity $O(1)$ times of secret comparison protocol, and it could complete the comparison privately.

4   Protocol to Compute Intersection and Union of Convex Polygons Approximately in STC

4.1   Secure Protocol for Approximate Intersection of Convex Polygons in STC

In this section, we discuss the secure protocols for approximate intersection of two polygons in unproportionate and proportionate partition.

Protocol 2. Secure Two-Party Protocol for Approximate Intersection of Two Polygons in Unproportionate Partition.

Input: Alice's and Bob's private convex polygons

Output: the approximate intersection of the two polygons

Step1: Alice and Bob announce to each other the x-coordinate of each peak or selected x-coordinates to form the unproportionate partition scanning beans.

Step2: On each scanning bean, Alice has P_{high} and P_{low} Bob has Q_{high} and Q_{low}. They invoke the Basic Comparison Protocol in section 3 to know which point is on their approximate intersection without leaking any other information.

Step3: They repeat step 2 until all the scanning beans are finished.

Protocol in Unproportionate Partition. Protocol under proportionate partition is similar to that of unproportionate partition, the only difference is in step 1, and we modify it as below.

Step1': Alice and Bob choose their own greatest and least x-coordinate to compare securely for announcing the maximum and minimum value, and negotiate about the number of regions. Then, they carve up $n + 1$ scanning beans proportionately between the two values.

Thus it can be seen that the complexity of this protocol is correlative to the partition number n. Although reducing partition number will preserve parties' privacy better, it descends precision meanwhile.

Protocol of Generating Set. The protocol of generating set is similar to that of subset, the difference is in step2. We get it when change Basic Comparison Protocol into Basic Generating Set protocol in 2.4.3.

4.2   Secure Protocol for Approximate Union of Polygons in STC

Protocol 3. Secure Two-Party Protocol for Approximate Union of Two Polygons in Unproportionate Partition.

Input: Alice's and Bob's private convex polygons

Output: the approximate union of the two polygons

Step1: Alice and Bob announce their x-coordinate of each peak or selected x-coordinates to form the unproportionate partition scanning bean.

Step2: On each scanning bean, Alice has P_{high} and P_{low}, and Bob has Q_{high} and Q_{low}. They invoke the Basic Comparison Protocol in section 3 to know which point is on their approximate union without leaking any other information. Alice or Bob only knows if her/his point is the maximum or minimum value but nothing else.

Step3: They carry out step 2 repeatedly until all the scanning beans are finished.

5 Analysis

In this section, we analyze the complexity and security of the protocols.

Complexity analysis. Secure two-party protocol to compute intersection or union of convex polygons in unproportionate partition has time and communication complexity $O(m + n)$ times of Basic Comparison Protocol. The corresponding protocols in proportionate partition has time and communication complexity $O(l)$ times of Basic Comparison Protocol, where l is the number of regions the both bargained on. Protocol for generating set likes the foretype.

Security Analysis. Protocol 1 (Protocol 2, Protocol 3) can execute securely without leaking privacy.

Now, we analyze the message leaked at each case in Basic Comparison Protocol. On each scanning bean, Alice gets to know if her P_{high} and P_{low} are on the outline. In case 1, Alice sees Q_{high} and Q_{low} of Bob's are between P_{high} and P_{low}, Bob gets that his Q_{high} and Q_{low} are not on the outline and $P_{high} > Q_{high}$, $P_{low} < Q_{low}$ thereby. In case 2, if 2.1 happens, they see P_{high} is seated between Q_{high} and Q_{low}, and Q_{low} is greater than P_{high}. The rest may be deduced by analogy. So, Alice or Bob only knows the relative position of her/his point and the other's but not the value.

When we consider this problem, some message is predetermined to leak out. If anyone knows his point is on the outline, he immediately sees the other's corresponding point is not on the outline. The acceptance or rejection indeed discloses some information about big or small on the same scanning bean, but it is inescapable. Our method can not guarantee this kind of message but only prevent from leaking any needless information.

Because the four points is independence and there is no rule between them, neither can analyze to know the other's information through the secure intersection or union protocol. This does preserve the parties' privacy.

6 Summary

The intersection and union of convex polygons are basic issues in computational geometry, and the demand of privacy-preserving calls on secure protocols for special fields. We have proposed protocols to compute approximate intersection and union of convex polygons in STC model. Detailed analysis about security and complexity are also presented. By the help of secret comparison, the protocols use Basic Compare Protocol as sub-protocol and gain in privacy and efficiency at the price of precision appropriately. Along with the development of SMC, our future work would like to settle the problem in more complex settings, such as multi-party model and malicious behavior.

References

1. Du, W., Mikhail, J.A.: Privacy-preserving Cooperative Scientific Computation. In: 14th IEEE Computer Security Foundations Workshop, Nova Scotia, Canada, pp. 273–282 (2001)
2. Yao, Y., Huang, L., Luo, Y.: Privacy-preserving Matrix Rank Computation and Its Applications. Chinese Journal of Electronics 17(3), 481–486 (2008)
3. Li, S., Dai, Y.: Secure Two-Party Computational Geometry. Journal of Computer Science and Technology 20(2), 259–263 (2005)
4. Yao, A.C.: Protocol for Secure Computations (extended abstract). In: 21st Annual IEEE Symposium on the Foundations of Computer Science, pp. 160–164. IEEE Press, New York (1982)
5. Yehuda, L., Benny, P.: An Efficient Protocol for Secure Two-Party Computation in the Presence of Malicious Adversaries. In: 26th Annual International Conference on Advances in Cryptology, Barcelona, Spain, pp. 52–78 (2007)
6. Hawashin, B., Fotouhi, F., Truta, T.M.: A Privacy Preserving Efficient Protocol for Semantic Similarity Join Using Long String Attributes. In: Proceedings of the 4th International Workshop on Privacy and Anonymity in the Information Society, PAIS 2011, article 6. ACM, New York (2011)
7. Dowsley, R., van de Graaf, J., Marques, D., Nascimento, C.A.: A Two-Party Protocol with Trusted Initializer for Computing the Inner Product. In: Chung, Y., Yung, M. (eds.) WISA 2010. LNCS, vol. 6513, pp. 337–350. Springer, Heidelberg (2011)
8. Hardt, M., Talwar, K.: On the Geometry of Differential Privacy. In: Proceedings of the 42nd ACM Symposium on Theory of Computing, STOC 2010, pp. 705–714. ACM, New York (2010)
9. Eppstein, D., Goodrich, M.T., Tamassia, R.: Privacy-preserving Data-oblivious Geometric Algorithms for Geographic Data. In: Proceedings of the 18th SIGSPATIAL International Conference on Advances in Geographic Information Systems, GIS 2010, pp. 13–22. ACM, New York (2010)
10. Dachman-Soled, D., Malkin, T., Raykova, M., Yung, M.: Efficient Robust Private Set Intersection. In: Abdalla, M., Pointcheval, D., Fouque, P.-A., Vergnaud, D. (eds.) ACNS 2009. LNCS, vol. 5536, pp. 125–142. Springer, Heidelberg (2009)

Adaptive Pole-Assignment Speed Control
of Ultrasonic Motor

Lv Fangfang[1], Shi Jingzhuo[1], and Zhang Yu[2]

[1] HenanUniversity of Science and Technology, 471003 Luoyang, China
[2] Southeast University, 210000 Nanjing, China
lvfangfang000@126.com, sjznew@163.com,
hit19991008@yahoo.com

Abstract. The obviously time-varying nonlinearity of ultrasonic motor makes the good performance of speed control difficult to be obtained. Aiming at the nonlinear problem of ultrasonic motor's speed control, a kind of adaptive pole assignment speed control strategy is proposed in this paper. The parameters of motor system are identified online, and then the controller is designed online according to the identified parameters of motor. Novel method for determining initial value of parameters of online identification algorithm is investigated, that realized efficient self-tuning control using a small amount of data. The proposed strategy is robust and relatively simple, and the excellent performance of the proposed control scheme is examined through experimental results.

Keywords: Ultrasonic motor, speed control, adaptive, online identification.

1   Introduction

Because of containing nonlinear processes such as piezoelectric electromechanical energy conversion and mechanical energy friction transmission during the operation, ultrasonic motor (USM) shows significant time-varying nonlinearity [1-3]. In order to improve the control performance of ultrasonic motor control system, we must consider how to deal with the essential nonlinear effect of ultrasonic motor system [4-6]. Real-time perceiving characteristics changes of motor system, and then adjusting control rule of controller online, therefore, realizing the "servo" action between the change of motor characteristic and adjustment of controller, it's an effective way to overcome time-varying nonlinear effect. The control strategy such as neural network or adaptive control can be used to achieve the "servo" action. The online self-learning neural network control has won many applications in ultrasonic motor control field [7-9]. However, because of the complicated nonlinear features and fast time-varying speed of ultrasonic motor, we often need to adopt more complex network structure and learning algorithm. This significantly increased the amount of on-line calculation of control algorithm.

Aiming at the problems of USM speed control, this paper tries to give a relatively simple self-tuning control strategy for ultrasonic motor speed control. This strategy adopts online identification algorithms to estimate real-time change of motor parameters,

B. Liu and C. Chai (Eds.): ICICA 2011, LNCS 7030, pp. 649–656, 2011.

real-time adjustment controller is designed to adapt electrical system changes and improve the robustness of control. Because body temperature raise rapidly caused by friction, the continuous operation time of ultrasonic motor should be short enough. Pole assignment control strategy is used to design the speed controller. Aiming at the short-time working characteristics of ultrasonic motor, this paper put forward the identification method which only needs a small amount of data to acquire reasonable identification parameter estimation and so improve the control effect. Experiments show that the proposed control strategy can significantly improve the control effect.

2   Self-Tuning Pole Assignment Speed Control Strategy

In order to cope with the nonlinear time-varying characteristics of ultrasonic motor system, Pole assignment self-tuning speed control strategy is designed to improve dynamic control performance. The block diagram of designed ultrasonic motor speed control system is shown in Fig. 1. An opto-electric encoder, E, is used to measure the rotating speed of ultrasonic motor. And the voltage amplitude controllers of phase A and B are designed to make the voltage amplitude of two phase equal to the given value by adjusting the duty ratio of PWM signals. In the figure, *Nref* and *Uref* are the given value of rotating speed and amplitude of driving voltage, respectively. In the dotted line box, there is the designed pole assignment self-tuning speed controller. Real-time parameters can be obtained through the part "online identification of motor system model parameters", then parameters of speed controller are adjusted to adapt to the time-variable motor system. The output controlled variable of controller is the frequency of driving voltage.

Phase-shift H-bridge driving circuit based on DSP is used in this paper. DSP is used to implement the control strategies shown in Fig.1, which is type DSP56F801. The traveling ultrasonic motor used here is the commercial used ultrasonic motor, Shinsei USR60. The principal specification of USR60 motor is shown in Table 1.

Table 1. Specifications of USR60 motor

Item	Value
Driving voltage	130Vrms
Rated torque	0.5Nm
Rated output power	5.0W
Rated rotating speed	100r/min
Max. torque	1.0Nm
Temperature range	-10 ~ +55°C
Weight	260g

In Fig. 1, the self-tuning speed controller includes three parts: model parameters identification, pole assignment controller parameters calculation, and the controller.

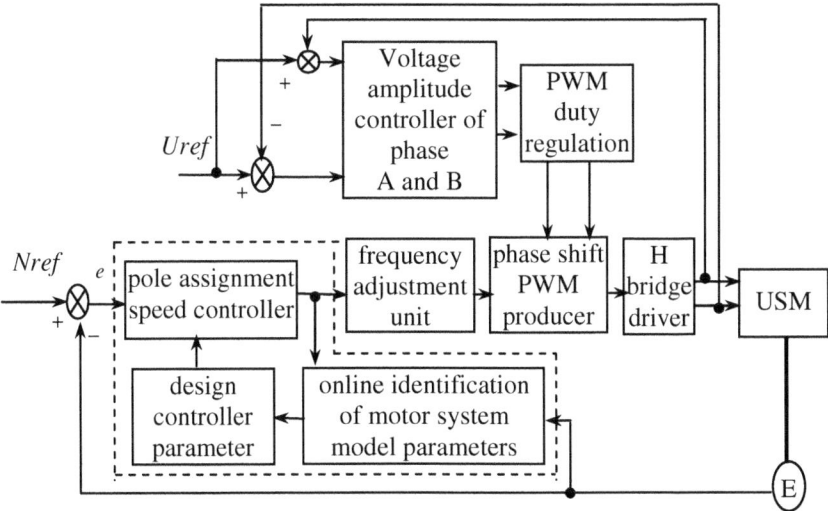

Fig. 1. Structure of self-tuning pole assignment speed control system

2.1   Identification of Model Parameters

Mainly due to the high temperature of motor body caused by the friction operation process, ultrasonic motor usually adopts short-time continuous working, and expects high control performance during the short-time working. To keep good speed control performance, it requires the identification algorithm should tend towards time-varying parameters rapidly and realize accurately track as far as possible in the condition of insufficient data. So the model parameters online identification of ultrasonic motor speed control faced harsh environment different to general. It also needs careful selection and design of the identification algorithm to realize truly self-tuning control.

By simulation, different identification algorithms are compared and analyzed. The used USM model used in simulation is identified offline based on measured data. Parameters' identification performance and speed control effect comparison show that, the identification algorithm contains forgetting factor is significantly better than it excluding forgetting factor. Model parameters' tracking error of extended and generalized least squares algorithm is small, while the control effect which using extended least-square algorithm is superior to generalized least squares algorithm. Thus, in the self-tuning speed control of ultrasonic motor, using extended least-square recursion algorithm that contains forgetting factor is relatively appropriate.

The recursive calculating formula for extended least squares algorithm is

$$\begin{cases} \hat{\theta}(k) = \hat{\theta}(k-1) + K(k)[y(k) - \varphi^T(k)\hat{\theta}(k-1)] \\ K(k) = P(k-1)\varphi(k)[\lambda + \varphi^T(k)P(k-1)\varphi(k)]^{-1} \\ P(k) = [I - K(k)\varphi^T(k)]P(k-1)/\lambda \end{cases} \tag{1}$$

The difference model of ultrasonic motor system is

$$A(z^{-1})y(k) = z^{-d}B(z^{-1})u(k) + C(z^{-1})\xi(k) \qquad (2)$$

Here

$$A(z^{-1}) = 1 + a_{11}z^{-1} + a_{22}z^{-2} \qquad (3)$$

$$B(z^{-1}) = b_{00} + b_{11}z^{-1} \qquad (4)$$

$$C(z^{-1}) = 1 + c_1 z^{-1} \qquad (5)$$

In Equ.(1), $\phi^T(k) = [-y(k-1), -y(k-2), u(k-1), u(k-2), \xi(k-1)]$, $y(k)$ and $u(k)$ are the present value of speed and frequency control words respectively. The vector of estimated parameters, $\theta^T(k)$, is $[a_{11}, a_{22}, b_{00}, b_{11}, c_1]$. λ is the forgetting factor, K is gain vector, P is estimated error covariance matrix. Normally, K is in direct proportion to P. The greater the difference between the real value and estimated value, the greater the gain vector K will also be produced, and the correction of model parameters is bigger.

$\hat{\theta}(k)$ can be obtained online according to the above recursive equations, namely the present value of estimated parameters, $\theta^T(k) = [\hat{a}_{11}, \hat{a}_{22}, \hat{b}_{00}, \hat{b}_{11}, \hat{c}_1]$. These identified parameters can be made use for calculation of controller parameters in next step.

2.2   Parameters Calculation of Pole Assignment Controller

The pole assignment speed controller [10] is

$$F(z^{-1})u(k) = G(z^{-1})[y_r(k) - y(k)] \qquad (6)$$

Here

$$\overline{F}(z^{-1}) = 1 - (1 + f_1)z^{-1} + f_1 z^{-1} \qquad (7)$$

$$G(z^{-1}) = g_0 + g_1 z^{-1} + g_2 z^{-2} \qquad (8)$$

$$A_c(z^{-1}) = 1 - d_1 z^{-1} \qquad (9)$$

Among them, f_1, g_0, g_1 and g_2 are undetermined parameters. $A_c(z^{-1})$ is the expected closed-loop characteristic polynomial, reflecting the control performance requirements.

When the parameters of ultrasonic motor system model in Equ. (2) have gotten through identification, the undetermined parameters can be obtained by Diophantine equation

$$A(z^{-1})\overline{F}(z^{-1}) + z^{-d}B(z^{-1})G(z^{-1}) = A_c(z^{-1}) \tag{10}$$

Substituting Equ. (7), (8) and (9) for the symbols in Equ. (10), then make the left and right sides corresponding coefficients equal, the following equations can be obtained

$$
\begin{cases}
a_{11} + b_{00}g_0 - f_1 - 1 = d_1 \\
a_{22} - a_{11} - a_{11}f_1 + f_1 + b_{11}g_0 + b_{00}g_1 = 0 \\
-a_{22} - a_{22}f_1 + a_{11}f_1 + b_{11}g_1 + b_{00}g_2 = 0 \\
a_{22}f_1 + b_{11}g_2 = 0
\end{cases}
\tag{11}
$$

and

$$
\begin{bmatrix}
-1 & b_{00} & 0 & 0 \\
1-a_{11} & b_{11} & b_{00} & 0 \\
a_{11}-a_{22} & 0 & b_{11} & b_{00} \\
a_{22} & 0 & b_{00} & b_{11}
\end{bmatrix}
\begin{bmatrix}
f_1 \\
g_0 \\
g_1 \\
g_2
\end{bmatrix}
=
\begin{bmatrix}
1-a_{11}+d_1 \\
a_{11}-a_{22} \\
a_{22} \\
0
\end{bmatrix}
\tag{12}
$$

According to Equ. (12), online calculation of speed controller parameters can be realized. After the new parameters of controller are calculated online, the output of pole assignment speed controller, namely controlled variable, can be calculated using Equ. (6)

$$
\begin{aligned}
u(k) = (1+f_1)u(k-1) - f_1u(k-2) \\
+ g_0(y_r(k) - y(k)) + g_1(y_r(k-1) - y(k-1)) \\
+ g_2(y_r(k-2) - y(k-2))
\end{aligned}
\tag{13}
$$

Here, $u(k)$ is the present controlled variable, $y_r(k)$ is the present given value of speed and $y(k)$ is the present value of speed.

3   Experiments of Pole Assignment Self-Tuning Speed Control

According to the above algorithms, DSP software is programmed to realize pole assignment self-tuning speed control of ultrasonic motor.

3.1   Initial Values of The Parameters of Identification Algorithm

In order to make identification algorithm can be calculated online, we need to determine the initial value $\varphi(0)$, $\theta(0)$ and $P(0)$ of matrix $\varphi(k)$, $\theta(k)$ and $P(k)$ in Equ.1. Initial values of estimation matrix of model parameters, $\theta(0)$, can be set according to ultrasonic motor model parameters values which are identified offline with the measured data. The initial values of system state variables matrix, $\varphi(0)$, can be set according to actual state of motor when it starts from zero speed, namely makes $\varphi(0)=[0,0,u(0),u(-1),0]$. Among them, $u(-1)=u(0)$ is the value of initial frequency control character of the actual system.

State estimation error covariance matrix P is directly related to model parameters correction function, usually there are two methods to determine its initial value. One is taking $P(0) = \alpha I$, which I is the unit matrix, α is a constant that is large enough, generally taking 10^6-10^{10}. The other is offline identification using enough measured data. And then take the terminal value of P as the initial value. But the experiments show that when taking the offline calculated P as the initial value of online identification, performance of speed control is poor and sometimes even can cause the motor suddenly stopped. There are two reasons. On the one hand, the off-line and on-line identification are realized in computer and DSP respectively. Differences of running environments bring different rounding errors. On the other hand, ultrasonic motor system has obviously time-varying nonlinearity. In order to reflect the present motor and control program's actual situation more accurately, this paper adopts online identification to obtain $P(0)$. Because only a few data can be tested during one operation process, multiple online identifications are made during multiple times of speed control processes, and then take the final value of P as $P(0)$. An initial value $P(0)$ is still needed to realize the multiple online identifications, it is set as the offline calculated P.

Specifically, the system as shown in Fig. 1 is used to do the experiments, adopting P value from offline identification as the initial value of online identification. Then, speed step-response experiments are done to realize online identification, but the identified parameters are not used to modify controller parameters. Namely, remove the part "design controller parameters" in Fig. 1 temporarily. The final value of P after step-response of speed is used as the initial value $P(0)$ of next experiment of speed step-response. To accumulate enough data points, multiple speed step-response experiments are done continuously, then take the final P value as the initial value of online identification.

So far, initial value of $\varphi(0)$, $\theta(0)$ and $P(0)$ for identification algorithm have been obtained.

3.2   Experimental Results of Self-tuning Control

To validate the control performance of the proposed speed controller, the measured speed step-responses for different given values are shown in Fig.3 and Fig.4 respectively. For convenient comparison, the control responses of other two control methods are also shown in the figures simultaneously. One is pole assignment control with fixed parameters which without self-tuning. Another one is a kind of model gain self-tuning control proposed in [10], the control parameters are adjusted according to the gain of offline identified models. Some indexes of control performance are given in Table 2 to compare the effects of these three control methods, the data are corresponding to the curves shown in Fig.2 and Fig.3.

According to Fig.2, Fig.3 and Table 2, by using the pole assignment control strategy that is proposed in this paper, its overshoot maintain zero and adjusting time of speed step-response control is significantly reduced. Meanwhile, its maximum error absolute value of steady speed is close to the situation of the other two control strategies. The control performances and indexes under different conditions effectively improved system response speed, control steadiness and robustness.

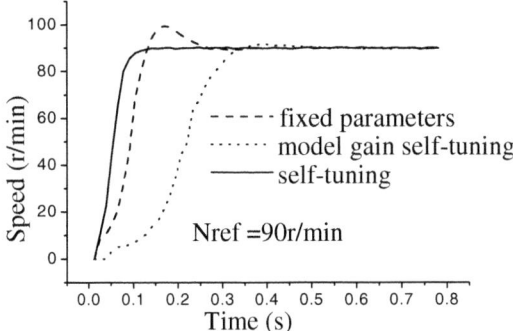

Fig. 2. Speed control response comparison (90r/min)

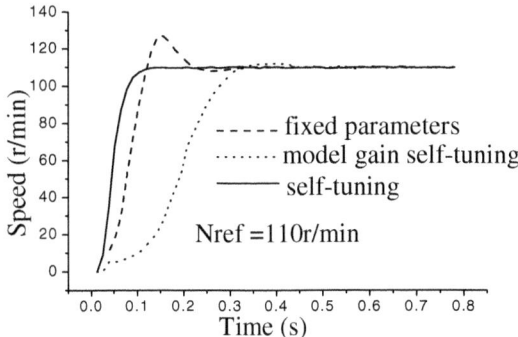

Fig. 3. Speed control response comparison (110r/min)

Table 2. Performance comparison between three pole assignment controllers

given speed (r/min)	adjusting time (s)						overshoot (%)		
	I		II		III		I	II	III
	e<5%	e<1%	e<5%	e<1%	e<5%	e<1%			
120	0.195	0.273	0.299	0.442	0.078	0.117	18.2	2.3	0
110	0.208	0.299	0.299	0.442	0.091	0.130	15.3	2.0	0
100	0.208	0.247	0.325	0.442	0.104	0.143	11.6	1.3	0
90	0.221	0.260	0.312	0.455	0.104	0.130	10.6	1.7	0

I: pole-assignment control with fixed parameters, II: model gain self-tuning control,

III: self-tuning control proposed in this paper.

4 Conclusions

Aiming at the problems of USM speed control, this paper presents a robust USM speed control strategy. The strategy can track the motor characteristic variations through online identification, and then tune the controller parameters to adopt the varied motor characteristic. To reduce the algorithm complexity, this paper designs a relatively simple pole assignment control strategy and extended least square online identification algorithm, meanwhile the offline controller parameters adjustment formula is deduced for online calculation. Aiming at the characteristics of ultrasonic motor short-time working, we propose a method of determining initial value of parameters of online identification algorithm, which realized efficient adaptive control using a small amount of data. The experimental results show that the proposed control algorithm is effective. Furthermore, the work of this paper demonstrates that proper online adaptive control strategy can significantly improve ultrasonic motor speed control effects.

References

1. Radi, B., Hami, A.E.: The Study of The Dynamic Contact In Ultrasonic Motor. Applied Mathematical Modelling 20, 3767–3777 (2010)
2. Chen, W.S., Shi, S.J., Liu, Y.X., Li, P.: A New Traveling Wave Ultrasonic Motor Using Thick Ring Stator With Nested PZT Excitation. IEEE Transactions on Ultrasonics, Ferroelectrics and Frequency Control 57, 1160–1168 (2010)
3. Hua, Z., Zhirong, L., Chunsheng, Z.: An efficient approach to optimize the vibration mode of bar-type ultrasonic motors. Ultrasonics 50, 491–495 (2010)
4. Bal, G., Bekiroglu, E.: A Highly Effective Load Adaptive Servo Drive System for Speed Control of Travelling-Wave Ultrasonic Motor. IEEE Transactions on Power Electronics 20, 1143–1149 (2005)
5. Puu-An, J., Ching-Chih, T.: Equivalent circuit modeling of an asymmetric disc-type ultrasonic motor. IEEE Transactions on Instrument and Measurement 58, 2351–2357 (2009)
6. Frédéic, G., Betty, L.-S., Julien, A.: Stability analysis of an ultrasonic motor for a new wave amplitude control. IEEE Transactions on Industrial Application 45, 1343–1350 (2009)
7. Lin, F.J., Hung, Y.C., Chen, S.Y.: FPGA-Based Computed Force Control System Using Elman Neural Network for Linear Ultrasonic Motor. IEEE Transactions on Industrial Electronics 56, 1238–1253 (2009)
8. Tien-Chi, C., Chih-Hsien, Y.: Generalized regression neural- network-based modeling approach for traveling-wave ultrasonic motors. Electric Power Components and Systems 37, 645–657 (2009)
9. Faa-Jeng, L., Syuan-Yi, C., Po-Huan, C., et al.: Interval type-2 fuzzy neural network control for X-Y-Theta motion control stage using linear ultrasonic motors. Neurocomputing 72, 1138–1151 (2009)
10. Shi, J.Z., Liu, B.: Self-Tuning Pole Assignment Speed Control of Traveling Wave Ultrasonic Motor. In: Proceedings of the CSEE, vol. 30, pp. 215–219 (2010)

Research on Algorithm of Image Processing of Butt Weld

Li Jun[1,2], Huo Ping[1], Li Xiangyang[1], and Shi Ying[3]

[1] College of Mechanical Engineering, Hebei United University, Tangshan, China
[2] Tangshan Kailuan Orient Power Co. Ltd., Tangshan, China
[3] Baoding Yindingzhuang Wastewater Treatment Plant, Baoding, China
Huoping009@163.com

Abstract. In accordance with the characteristics of the butt weld image, this article designs the image processing to extract the size of the weld with according to the butt weld .Firstly, the algorithm of gray closing and mathematical morphology in the method is taken to smooth the interferences of surface scratches and glistens effectively. Secondly, the processed image is detected the image edge with Canny operator, and using the researching method to extract the image character, record the edge coordinate of welding line and achieve the image model to model transformation. Finally, the date information of slope gaps is obtained by the least-squares algorithm to do the edge fit to derive the linear equation. The method in this paper can denoise and protect images has good practicality for its relatively high measurement precision, which are of great value towards the engineering practice.

Keywords: Slope gaps, welding seam, algorithm of Image processing, mathematical morphology.

1   Introduction

Vision sensing technology is consistently being used in the Tracing Process of the weld seams. Being the core and foundation of the whole vision tracing system, the arithmetic process for images plays a crucial role in the result of weld seam detection. Thus, it becomes especially important to develop a set of arithmetic process for images which suits the seam's characteristics and is of stronger counter-interference ability. With the development of the technology of processing images, the mathematical morphology has become a crucial theory for computers to handle the figures. The morphology analyses figures from collective angle. And as a theory of processing and analyzing nonlinear images plays an important part in the pre- and post processing of images [1]. This passage designs the process of editing images in accordance with the feature of butt-welded, decreases and smoothes the interference by utilizing mathematical morphology during the process of the arithmetical processing of images.

2   Gray-Scale Morphological Filtering

Mathematical morphology image processing adopts the structure elements with certain structures and characteristics (i.e. patterns) to measure the patterns of the

B. Liu and C. Chai (Eds.): ICICA 2011, LNCS 7030, pp. 657–662, 2011.

image, whose aim is to find the specific geometrical structure of the original collection which is embodied in the collection after its transformation. This transformation is realized by assembling the characteristics of structure elements. Thus the selection of structure elements greatly affects the processing effect. Dilation and erosion are the most fundamental logical calculation in the image processing of the mathematical morphology.

2.1   Erosion and Dilation

Dilation.   Dilation is an action of lengthening or widening the Binary images. Its way and degree are controlled bay collection of a structure element, as in (1).

$$A \oplus B = \left\{ x \mid (\hat{B})_x \cap A \neq \Phi \right\}$$ (1)

The process of A's dilation though B is the process of B's original point and points on and around X to correspond.

    If point on B is belonging to the scale of X, we can regard the original point of B as the point on X under the dilation.

    That is the result of "or" calculation between structure element and the Binary images which has been overlapped by the element is 0, then the peel of that image is 0, or 1. Thus Dilation is often used for filling the holes among the edges, which forms after the cutting of materials and items.

Erosion.   To erode is to contract or thin the Binary images, as the process of dilation, degree and form of eroding images are controlled by a structure element. The erosion of A through B is $A \Theta B$, as in (2).

$$A \Theta B = \left\{ x \mid (B)_x \subseteq A \right\}$$ (2)

The erosion of A through B is completely corresponding the original point of B to the points on X. If all the points in B can be found in X, then original point of B in X is preserved, or is deleted if not been found. If the results of making an "and" calculation between the structure elements and the Binary images which it overlaps are all 1, then the Pixel of that image is 1. Or is 0 if not. Therefore, it's of great importance to get rid of the useless small dots which accounts for certain space.

2.2   Opening and Closing Operation of Morphology

In the practical application of image processing, it is more to use expansion and corrosion in the form of various combinations. The most commonly used is a combination of corrosion and expansion: opening operation and closing operation. With the effectively combination, Opening and Closing operations can be made a more superior performance of the filter [2]. Gray-scale morphological is the natural expansion from binary morphology to grayscale space. The biggest different is that the binary morphological operations are a collection of objects, while grayscale morphological operations are the object of image functions.

The grayscale image opening operation. In the grayscale opening operation, input image: f, structure element: b, as in (3).

$$f \circ b = (f \Theta b) \oplus b \qquad (3)$$

From the perspective of image, opening removes the smaller bright details compared with the structural elements. While maintain the overall gray value image and the large bright areas not affected [3].

The grayscale image closing operation. The grayscale image closing operation, as in (4).

$$f \bullet b = (f \oplus b) \Theta b \qquad (4)$$

From the perspective of image, opening removes the smaller dark details compared with the structural elements. While maintain the overall gray value image and the large bright areas not affected.

3   The Image Arithmetic Processing for Butt Weld

This paper uses the grayscale image that filmed by industrial CCD as the research object, as is shown in Figure 1. Firstly we preprocess the weld image, to eliminate noise, realize weld image edge accurate testing.

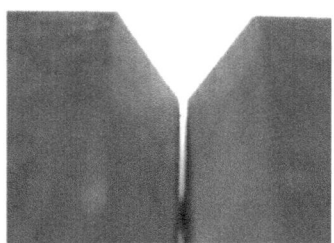

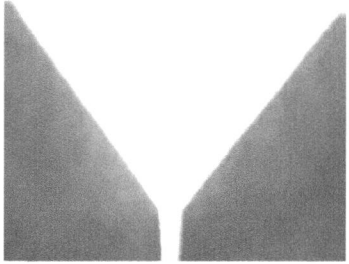

Fig. 1. Weld image

Fig. 2. Gray-scale morphological closed operations

3.1   Image Preprocessing

The main purpose of the image preprocessing is to eliminate the irrelevant information in image. Image preprocessing includes grayscale processing, image cropping, filtering and closing gray-scale image. In the image there exists noise such as the surface reflective, scratches, small protrusions, which will inevitably give further difficult to subsequent handling. In order to improve the efficiency of the image processing algorithm, this paper makes use of gray scale mathematical morphology closing to solve this difficult problem. As is shown in Figure 2, small dark details (surface scratches, edges dark protrusions) were filtered, bright part affected lesser, weld edge basic was not affected.

3.2   Image Postprocessing

The human eyes can easily identify the edges of objects from the pretreated images of welds. But it is necessary for the computer to make further threshold segmentation and edge detection of the images to identify the information of the welds effectively. Therefore post processing of the images is also necessary. These include: binarization, edge detection, feature extraction.

The binarization segmentation of gray images.   When we do the analysis and recognition of the images, we should first segmentate the effective parts from the image. The threshold segmentation of the images is a widely-used technique of image segmentation. which mainly use the image of the target to extract the gray background characteristics of their differences, by choosing a suitable threshold, the image with different gray levels two regions (object and background) to distinguish between the combination, if the inputted gray value of pixel is bigger than the given threshold then the output value of pixel assigns to 1, otherwise assigns to 0. So it results the corresponding binary image [4], [5] This paper uses the global threshold value .Using the best threshold function of MATLAB in selecting threshold; we can get the results shown in Figure 3. The binarization of image makes the features more salient, but there are many small teeth in the edge of the welds.

   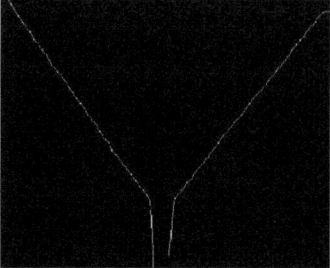

Fig. 3. The results after binarization   **Fig. 4.** Canny edge detection

Edge Detection of images. It is necessary to detect the edge of welds after the weld images have been binarized. The more appropriate the edge detection operator is, the more closely the got weld center to the actual weld center and the higher the precision is. After several experiments the final selection in this paper is canny edge detection operator. Through edge detection of the weld image, the result shown in Figure 4.

Image Feature Extraction. By preprocessing, segmentation and edge detection processing, we can get a relatively clear outline of the weld edge image. But in order to extract the inflection point coordinates of the edge points on both sides and find the pixel width of the weld, it is also necessary to extract the features of the weld images. In this paper, the records of the edge coordinates are achieved by the searching algorithm.

Obtained by the search method is a two-dimensional array which uses image coordinates of edge points as the elements. The digital image is a matrix and the location information of each pixel is based on the order that the row ranks first and the list last. So the order of the two-dimensional array's columns must been exchanged. Because the vertical axis of the image makes downward positive, the second column should be made negative in order to keep the image the same. This completes the extraction of the edge point's coordinates in the image as shown in Figure 5.

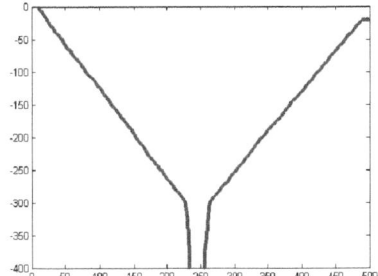
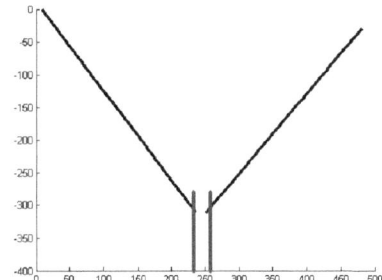

Fig. 5. Weld seam edge image in coordinate   **Fig. 6.** The results of least square linear regression

lower left, upper right, lower right and they are fitted by the least squares in this paper. It's shown in Figure 6.

This is the principle of the fit by the least squares: fit the n pairs of data (x_k, y_k), $(k = 1, 2, ...n)$ using $\varphi(x)$, and make the error of the sum of the squares $\sum_{k=1}^{n}[y_k - \phi(x_k)]^2$ the smallest so we can get the method of obtaining $\varphi(x)$.

Fitting by a straight line: $y = \varphi(x) = a_0 + a_1 x$, as in (5):

$$\begin{cases} na_0 + (\sum_{k=1}^{n} x_k)a_1 = \sum_{k=1}^{n} y_k \\ (\sum_{k=1}^{n} x_k)a_0 + (\sum_{k=1}^{n} x_k^2)a_1 = \sum_{k=1}^{n} x_k y_k \end{cases} \quad (5)$$

That a_0, a_1 are the results of Linear Equations [6], [7].

The fitting equations obtained by using MATLAB platform are as follows:

The fitting equation of upper left edge is:

$$y = -1.3719x + 12.1244 \quad (6)$$

The fitting equation of lower left edge is:

$$x = 232 \tag{7}$$

The fitting equation of upper right edge is:

$$y = 1.2314x - 621.3757 \tag{8}$$

The fitting equation of lower right edge is:
So we can get the coordinates of the two turning points are: (232, -306.1564), (258, -303.6745).

$$x = 258 \tag{9}$$

The pixel distance between the two turning points of the image weld: ΔX is 26; ΔY is 2.5.

4   Conclusion

In the process of the weld image's processing, we adopted mathematical morphology to eliminate the work piece's scratches, the surface reflection, untrue edge and other interferences, finally got a clear, smooth image of the weld edge. And in the feature extraction, width information is obtained by using searching method and the least square method. Experiments show that the image processing algorithm is simple and efficient and it can meet the engineering practical requirements.

References

1. Lei, T.: A Method of Removing Fake Edge Based on Morphological Filtering. Journal of Lanzhou Jiaotong University, 104–106 (2008) (in Chinese)
2. Cohen, J.: A Coefficient of Agreement for Nominal Scales. Educational and Psychological Measurement, 37–46 (1996)
3. Lin, H.: Method of Image Segmentation on High-resolution Image and Classification for Land Covers. In: Fourth International Conference on Natural Computation (ICNC 2008), pp. 563–565. IEEE Computer Society (2008), doi:10.1109/ICNC.2008.870
4. Vincent, L.: Morphological Grayscale Reconstruction: Definition. Efficient Algorithms and Applications in Image Analysis, 633–635 (1992)
5. Wen, H.: Research of Image Processing Algorithm Based on Mathematic Morphology. Harbin Engineering University, 11–12 (2007) (in Chinese)
6. Yorozu, Y., Hirano, M., Oka, K., Tagawa, Y.: Adaptive algorithm of edge detection based on mathematical morphology. Journal of Computer Applications 9, 997–1000 (2009)
7. Dai, Q.Y., Yun, Y.L.: Application development of mathematical morphology in image processing. Control Theory and its Application 4, 13–16 (2001)

Simulation of Microburst Escape
with Probabilistic Pilot Model

Zhenxing Gao[*] and Hongbin Gu

Research Center of Flight Simulation and Advanced Training,
Nanjing University of Aeronautics and Astronautics, P.R. China
z.x.gao@nuaa.edu.cn

Abstract. Simulation of large aircraft approach and landing in microburst wind shear was studied for flight safety research. A real-time flight dynamics model with wind shear effects was built based on Boeing747 modeling data. A parameterized three-dimensional microburst model was formulated by vortex ring and Rankine vortex principle. Further more, a parameterized human pilot model was developed to simulate pilots' control behavior during microburst encountering. A pilot-aircraft-microburst environment model was constructed for further study. Since pilots would have variable control behavior, a group of pilot was modeled by treating the characteristic parameters as random variables. To study the safety of pitch guidance strategy recommended by FAA, the Monte Carlo Simulation was adopted to obtain a numerical approximation of the probability density function of the minimum altitude. The results indicate that the 3-D microburst model can generate wind vectors with high fidelity. The dynamics model with wind shear effects is reasonable and valid. Credible and valuable conclusions of escape strategy and safety can be acquired by Monte Carlo simulation.

Keywords: Microburst, wind shear, pilot model, flight simulation, Monte Carlo.

1 Introduction

Low altitude wind shear, in particular the microburst, is the variation of wind direction and velocity. The microburst is induced by strong downdraft from thunderstorm and diverges near the ground to yield severe wind shear [1]. If the microburst do take place near aerodrome, it would do harm to takeoff and landing. During approach and landing, headwind was encountered firstly, lift increased and the aircraft flied above glide slope. Pilot may decrease the throttle to maintain on the slope. However, strong downdraft and tailwind came soon. With inadequate thrust and low airspeed, severe consequence would be suffered. The FAA has recommended three kinds of strategies for microburst escape [2].

In order to examine the feasibility and safety of escape strategies, a microburst model should be developed together with a flight dynamics model (FDM) with wind

---

[*] A Lecturer in Nanjing University of Aeronautics and Astronautics, P.R. China.

B. Liu and C. Chai (Eds.): ICICA 2011, LNCS 7030, pp. 663–670, 2011.

effects. Since pilots would make different reactions in flight, research cannot be exerted on single or several pilots. An effective method is to use probabilistic model which contains parameters describing pilots' characteristics such as skills, experience, emotion, etc. By choosing suitable probability density function (PDF) for each parameter, the variation of pilots could be included in simulation.

This paper firstly proposes a parameterized three-dimensional (3D) microburst model to support flight simulation. The parameters can be stochastic. Secondly, a Boeing747 FDM with wind shear effects is inferred. Thirdly, a probabilistic pilot model is designed whose parameters can subject to specific probabilistic distribution. With the combination of pilot, aircraft and microburst model, the Monte Carlo method is used to judge the pitch guidance strategy. Based on the simulation, some results and discussion are explained in the last part of this paper.

2   Modeling of Microburst Wind Field

2.1   Modeling with Vortex Ring

In Fig. 1, a primary ring with the strength of Γ is placed above the ground while an imaginary ring with the strength of $-\Gamma$ is placed below the ground symmetrically.

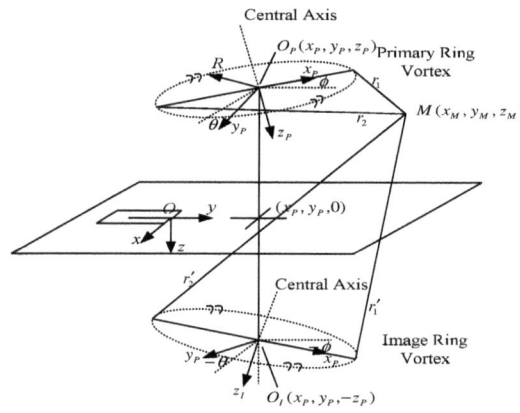

Fig. 1. Microburst Modeling with Vortex Ring

As far as the reference point $M(x_M, y_M, z_M)$ is concerned, the stream function of point M is:

$$\psi = \psi_P + \psi_I = -\frac{\Gamma}{2\pi}[\frac{0.788k^2(r_1+r_2)}{0.25+0.75\sqrt{1-k^2}} - \frac{0.788k'^2(r_1'+r_2')}{0.25+0.75\sqrt{1-k'^2}}] \tag{1}$$

In Eq. (1), ψ_P is the Stokes' stream function of the primary ring, while ψ_I is the imaginary ring's. The induced velocities $[W_x, W_y, W_z]^T$ of point M can be acquired by deriving Eq. (1).

2.2 Induced Velocity in Vortex Core

A realistic vortex ring has a core because of viscosity. In the core, the velocity reduces to zero gradually. Combined with Rankine vortex, the core is regarded as a cylinder, and the vorticity distributes uniformly to make the vortex filament velocity be zero[4]. The fluid field outside the core still conforms to stream function. The velocity vector in the vortex core can be acquired by spatial analytic geometry.

2.3 Parameterized Microburst Model

According to the modeling procedure above, a microburst model can be described by $\vec{\vartheta} = [x_P, y_P, z_P, \Gamma, R, r]$. In flight simulation, some of the parameters can be set as random variables which subject to specific probabilistic distribution.

When a microburst model was designed with $z_P = -2000\,ft$, $R = 6000\,ft$, $r = 1500\,ft$, $\Gamma = 30\,ft/s$, the wind vectors on the cross section of central axis show in Fig. 2.

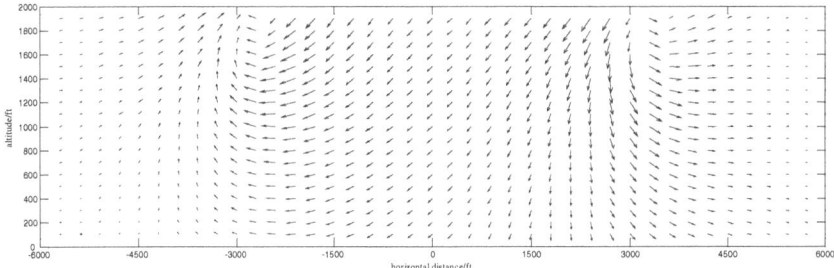

Fig. 2. Microburst Wind filed Simulation

3 Flight Dynamics Modeling with Wind Effects

3.1 Dynamics Equations in Static Atmosphere

The dynamics equations in static atmosphere are quoted from [5]. The equations contain three force equations, three moment equations, three kinematic equations and three navigation equations.

3.2 Dynamics Equations with Wind Effects

If wind shear exists, the ground speed is not equal to airspeed due wind effects. For large aircraft, the fuselage and wing can be comparable with microburst wind field. So the wind vector and its gradients should be considered.

The wind vector in ground coordinate $[W_{xE}, W_{yE}, W_{zE}]^T$ should be transformed to body frame by Eq. (2), in which L_{BE} is transition matrix.

$$[W_{xB} \quad W_{yB} \quad W_{zB}]^T = L_{BE}[W_{xE} \quad W_{yE} \quad W_{zE}]^T \qquad (2)$$

Derive Eq. (2),

$$\begin{bmatrix} \dot{W}_x \\ \dot{W}_y \\ \dot{W}_z \end{bmatrix}_B = L_{BE} \begin{bmatrix} \dot{W}_x \\ \dot{W}_y \\ \dot{W}_z \end{bmatrix}_E - \begin{bmatrix} 0 & -r & q \\ r & 0 & -p \\ -q & p & 0 \end{bmatrix}_B L_{BE} \begin{bmatrix} W_x \\ W_y \\ W_z \end{bmatrix}_E \tag{3}$$

In this paper, the Boeing747 modeling data is adopted[6][7]. An integrated Boeing747 FDM with high fidelity has been built for further study.

4   Probabilistic Pilot Modeling

Compared to pilot model in frequency domain, the time domain model considers the actual human processing characteristics and could simulate more precisely[8]. So the pilot model used in this paper is in time domain. The effect of pilot model is applied to only one control input namely the elevator deflection. As far as pitch guidance escape is concerned, the pilot needs to do nothing on aileron and rudder deflections. Maximum throttle settings are also prerequisite.

4.1   Single Pilot Modeling

As shown in Fig. 3, the pilot model consists of four components [9]:

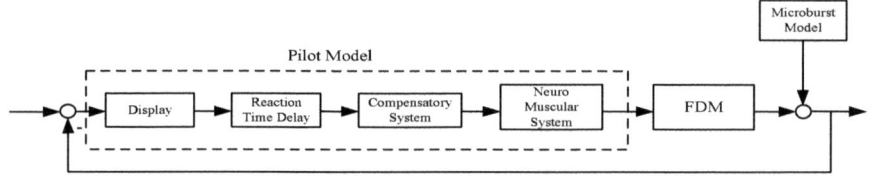

Fig. 3. Structure of Pilot Model

The Display block forms the medium by which the pilot observes the instantaneous state of the aircraft. Pilot's decision is based on the combination of his experience and the readings of all kinds of flight instruments. There are various sources of error in this procedure, such as observation error, instruction error, etc. An observation noise needs to be included in it.

Reaction time delay denotes pilot's observation delay, processing and decision making delay, etc. Evidently, the overall time delay depends on the pilot's training, experience and knowledge.

Compensatory system, which forms the core of pilot model, is analogous to a controller. There are two kinds of control tasks in pitch guidance strategies:

(1) Pitch angle control

According to the pitch guidance strategy, advised by FAA, when detecting wind shear, the aircraft's throttle should push to full with pitch angle $\theta_c = 15^o$. An elevator PID controller is designed:

$$\delta_e = K_{Pe}(\theta - \theta_c) + K_{De}(\dot{\theta} - \dot{\theta}_c) + K_{Ie}\int_0^t(\theta - \theta_c)d\tau \tag{4}$$

(2) Stall prevention control

This controller is used to represent the behavior of the pilot by pitching down the airplane to reduce the angle of attack and increase the airspeed in order to prevent the aircraft from stalling.

Neuromuscular dynamics system characterizes the personality, skill and experience of each pilot. Besides, a kind of motor noise is caused due to natural factors such as physical stability, mental stress and intensity of concentration.

4.2   Probabilistic Pilot Modeling

A single pilot may not be sufficient to make decisions in case of research study that might involve pilots with a wide variation in characters. To model a group of pilots, the variations can be expressed as random variables with suitable probability density functions.

According to single pilot model, the parameters that will vary depending on the level of pilot's training, experience, mental and emotional state, are as follows:

(1) The time constant τ, in the reaction time delay.

(2) The time constant τ_N, in the neuromuscular dynamics system.

(3) The noise power in the observation noise.

(4) The noise power in the motor noise.

Even in case of a pilot with the best performance, the parameters listed above will not be zero. These parameters in a group of pilots are modeled using Rayleigh probability density function. In general, a Rayleigh random variable is a good model for measurement error in a number of different physical situations [10].

With regard to τ, based on the prior experience with the behavior of most pilots, it is known that varying the parameter τ from 0.15s to 0.3s models the entire range of pilots encountered in practice[11]. The range of values for τ_N is 0.1s to 0.2s [12].

It is assumed that the maximum observation error does not exceed 5% of the corresponding command values. When it comes to pitch-guidance strategy, the noise power for the pitch angle measurements are $0.0017\,rad^2$.

The limits of the motor noise depend on the training, experience, and the emotional stability. In our study, the value of noise power for elevator deflections, is assumed to be equal to that of corresponding observation noise.

5   Monte Carlo Simulation

Based on microburst model and Boeing747 FDM, together with the probabilistic pilot model, the pitch guidance strategy is examined in this paper. Take the same coordinate as Fig. 1, the microburst locates at 7000m south of runway. By uncontrolled flight simulation, the flight states change acutely when encountering the wind. In the end, the "aircraft" crashes at 114.8s. The trimmed approaching flight state can be shown in Fig. 4 with real lines.

5.1   Microburst Escape Simulation

If the pitch guidance escape begins at 80s, the simulation results show in Fig. 4. Although the flight state changes acutely, the "aircraft" flight apart microburst safely. Simulation results also show that if escape time is late, the aircraft would inevitably crash.

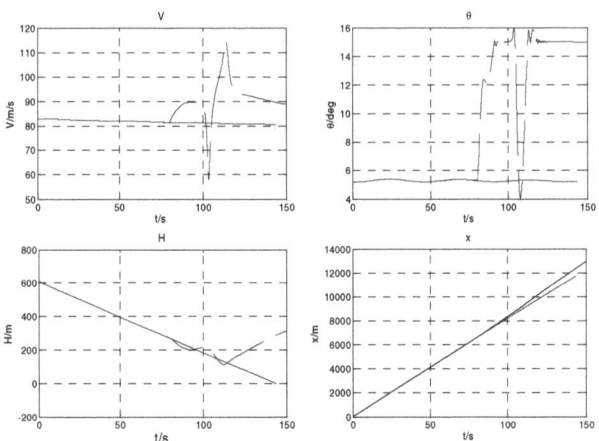

Fig. 4. Pitch Guidance Escape

5.2   Monte Carlo Simulation

Among all kinds of escape strategies, the altitude loss is the most important factor to be considered. The escape safety can be qualified by the minimum altitude the aircraft flies during escape maneuver. h_{min} becomes a random variable since both the microburst and pilot parameters are random variables. In order to compare the escape strategies, the probability density function or at least some statistical properties of h_{min} should be determined. Obviously, it is difficult to formulate h_{min} to be an algebraic function of the microburst and pilot parameters. Therefore, the Monte Carlo simulation is adopted to obtain a numerical approximation of the probability density function of h_{min}.

h_{min} depends on the initial condition at which the aircraft enters the microburst, the parameters of the microburst, and the guidance strategies. Hence,

$$h_{min} = h_{min}(\bar{h}_0, \bar{\vartheta}, \bar{K}) \tag{5}$$

In this paper, $\bar{h}_0 = \left[V_x, V_y, V_z, p, q, r, \phi, \theta, \psi, x_E, y_E, z_E\right]^T$ describes the flight state. The microburst parameters can be expressed as $\bar{\vartheta} = [x_P, y_P, z_P, \Gamma, R, r]^T$. The pilot parameters are $\bar{K} = \left[\tau, \tau_N, n_o, n_m\right]^T$.

The probability distribution function of $h_{\min}$ can be defined as:

$$F_{h_{\min}} = Probability(h_{\min} \le h) \tag{6}$$

Note that $F_{h_{\min}}(0) = \Pr(h_{\min} \le 0)$ is the probability of crash. In this study, the microburst intensity, Γ is assumed to be Rayleigh distributed because it is utilized to fit the statistical data from the JAWS[4]. Other parameters subject to uniform probability density functions. The pilot parameters used in this paper are:

$$[\tau, \tau_N, n_o, n_m]^T = [0.225, 0.15, 0.007, 3]^T$$

Based on thousands of experiments, the probabilistic density function of $h_{\min}$ can be acquired as shown in Fig. 5:

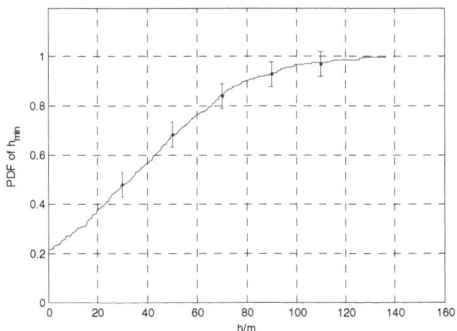

Fig. 5. Probability Density Function of $h_{\min}$

6   Conclusions

The safety of microburst escape strategies can be affected by the following factors:

(1) Pilot characteristics, including skills, experiences, knowledge, emotion, etc;
(2) Microburst conditions, including its outburst dimensions, the intensity, etc;
(3) High-frequency atmospheric turbulence may have adverse effects;

Besides, there are many kinds of unknown factors affecting the escape performance. By Monte Carlo Simulation, the numerical approximation of some key factors could be acquired instead of complicated reasoning.

In the future, based on the pilot-aircraft-microburst environment research platform, the dive guidance and altitude guidance strategies can be studied deeply. Furthermore, the safety of lateral escape can also be validated.

Acknowledgements. This work was supported by postdoctoral foundation (1001013C) and National Natural Science foundation (61039002).

References

1. Gao, Z., Gu, H.: Research on Modeling of Microburst for Real Time Flight Simulation. Journal of System Simulation 20(23), 6524–6528 (2008) (in Chinese)
2. Dogan, A.: Guidance Strategies for Microburst Escape. ProQuest Dissertations and Theses, AAI9977147 (2000)
3. Dogan, A., Kaewchay, K.: Probabilistic Human Pilot Approach: Application to Microburst Escape Maneuver. Journal of Guidance, Control, and Dynamics 30(2), 357–369 (2007)
4. Gao, Z., Gu, H., Hui, L.: Real-Time Simulation of Large Aircraft Flight through Microburst Wind Field. Chinese Journal of Aeronautics 22(5), 459–466 (2009)
5. Brian, L.S., Frank, L.L.: Aircraft Control and Simulation. John Wiley and Sons, Canada (2003)
6. Rodney Hanke, C.: The Simulation of a Large Jet Transport Aircraft Volume I: Mathematical Model. NASA CR-1756 (1971)
7. Rodney Hanke, C., Nordwall, D.R.: The Simulation of a Jumbo Jet Transport Aircraft Volume II: Modeling Data. NASA CR-114494 (1970)
8. Pool, D.M., Zaal, P.M.T., van Paassen, M.M.: Identification of Multimodal Pilot Models using Ramp Target and Multisine Distance Signals. Jounral of Guidance, Control, and Dynamics 34(1), 86–97 (2011)
9. Kaewchay, K., Dogan, A.: Design of a Probabilistic Human Pilot: Application to Microburst Escape Maneuver. In: AIAA Atmosphere Flight Mechanics Conference and Exhibit (2005)
10. Gestwa, M., Bauschat, J.M.: Development of a Fuzzy-Controller with a State Machine as a Cognitive Pilot Model for an ILS Approach. In: AIAA Modeling and Simulation Conference and Exhibit (2007)
11. Schroeder, J.A., Grant, P.R.: Pilot Behavioral Observations in Motion Flight Simulation. In: AIAA Modeling and Simulation Conference and Exhibit (2010)
12. Beukers, J.T., Stroosma, O., Pool, D.M.: Investigation into Pilot Perception and Control During Decrab Maneuvers in Simulated Flight. Journal of Guidance, Control, and Dynamics 33(4), 1048–1063 (2010)

A Method for Space Tracking and Positioning of Surgical Instruments in Virtual Surgery Simulation

Zhaoliang Duan, Zhiyong Yuan[*], Weixin Si,
Xiangyun Liao, and Jianhui Zhao

School of Computer, Wuhan University,
Wuhan 430072, China
{dzlwhu,wxsics,xyunliao}@gmail.com,
{zhiyongyuan,jianhuizhao}@whu.edu.cn

Abstract. As the unique interface for users to communicate with virtual environment, space tracking and positioning apparatus of surgical instruments is an indispensable part of virtual surgery simulation system. A method based on stereoscopic vision is proposed to construct a suit of space tracking and positioning apparatus of surgical instruments. It is able to capture spatial movements of simulated surgical instrument in real time, and provide corresponding six degree of freedom information with the absolute error of less than 1 mm. In order to verify its feasibility, this method is integrated into soft tissue deformation simulation in virtual surgery, and the experimental results show that the developed apparatus is highly accurate, easily operated, and inexpensive.

Keywords: Virtual surgery simulation, Stereoscopic vision, Surgical instruments, Space tracking and positioning.

1   Introduction

As the cutting-edge interdisciplinary research field of information and medical sciences, research on virtual surgery simulation system has significant application value for reducing surgery risks, cutting training cost and protecting human health [1]. In order to simulate the interaction between surgical instrument and virtual organ tissue vividly in virtual surgery simulation, the surgical instrument must be tracked and located accurately in real time.

Currently, there have been some available three dimensional trackers in the filed of virtual reality. According to their physical properties, they are roughly classified into five subcategories: mechanical tracker [2], magnetic tracker [3-4], ultrasonic tracker [5], optical tracker [6-7] and hybrid tracker [8]. Some of them can provide high positioning accuracy, such as [6], and have been used in some medical applications. However, these existing devices are very expensive, therefore can only be popularized in a limited number of medical centers and research institutes. A 3D surgical instrument tracking and positioning method with a high performance-price ratio has been highly desirable for computer-based virtual surgery simulation systems.

---

[*] Corresponding author.

B. Liu and C. Chai (Eds.): ICICA 2011, LNCS 7030, pp. 671–679, 2011.
© Springer-Verlag Berlin Heidelberg 2011

In order to make this goal come true, we present a method based on stereoscopic vision for space tracking and positioning of surgical instruments. This method employs three cameras to capture the motion images of simulated surgical instrument in real time. After a series of computer processing, including camera calibration, reconstruction of 3D coordinates of markers on simulated surgical instruments and so on, we can obtain the six degree of freedom information of simulated surgical instruments, thereby positioning the instrument. At the end of this paper, we apply the presented method to accomplish interactive virtual organ tissue deformation simulation. The experimental results show that it is feasible and effective in virtual surgery simulation systems.

The rest of the paper is organized as follows. Section 2 describes the methodology for tracking and positioning surgical instruments. Section 3 gives the implementation of the presented method in details. Section 4 provides experimental results, and then section 5 concludes this paper.

2   Methodology for Space Tracking and Positioning

2.1   Basic Principle

Based on stereoscopic vision [9], the basic principle of the proposed method is to intimate people's eyes to perceive 3D objects, i.e., people's eyes can observe the objects in the surrounding environment from two different viewpoints, and further recover 3D information of objects.

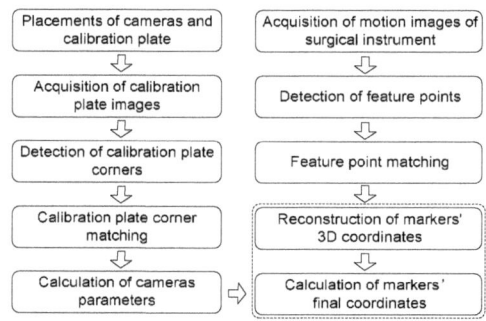

Fig. 1. Flowchart of our presented method

In turn, our presented method is to utilize cameras to recover 3D coordinates of two markers on the simulated surgical instruments. Aiming at this goal, each marker must be covered by at least two cameras. If we use two cameras to track simulated surgical instrument, we can detect four feature points, the corresponding image regions of markers within the motion image of simulated surgical instrument, at a time, then we classify these four feature points into two pairs of identical points, and calculate their image coordinates respectively. Along with the camera parameters, 3D coordinates of two markers can be obtained through least square method [10].

Considering the virtual surgery simulation system is extremely strict with precision, we employ three cameras to capture the movement of simulated surgical instrument for the purpose of minimizing the system error introduced by image acquisition. Three cameras construct three pairs of cameras groups, and each camera group includes two cameras. As for each marker, three pairs of cameras groups obtain three groups of 3D coordinates. By calculating their average values, we get the final and more accurate 3D coordinates of markers. Figure 1 shows the flowchart of the presented method.

2.2   System Construction

The space tracking and positioning apparatus of surgical instrument based on our proposed method consists of a simulated surgical instrument, three cameras and a computer.

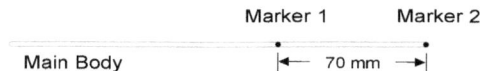

Fig. 2. Abstraction of surgery instrument

Two markers with 70 mm space distance are deployed on the simulated surgical instrument, and specific distribution of markers has been shown in Figure 2. As for the actual simulated surgical instrument, its main body is white, while two markers are black.

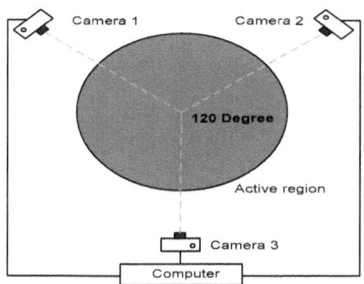

Fig. 3. Hardware distribution of the space tracking and positioning apparatus

Figure 3 illustrates the hardware distribution of our developed space tracking and positioning apparatus. The gray circular area is active region of simulated surgical instrument. The degree of included angle constructed by any two cameras and the center of gray circular area is 120.

3   Implementation

3.1   Camera Calibration

Camera calibration is the most basic step in stereoscopic vision [9]. Its purpose is to obtain camera parameters, once we obtain camera parameters, we do not need to calculate them again until camera is moved. Generally speaking, the detailed calibration process includes the following five steps.

(1) **Step 1.** Generation of planar calibration plate: We adopt a regular 7×7 black-and-white checkerboard as the pattern on the calibration plate, the size of each checker is 30×30 mm.

(2) **Step 2.** Acquisition of calibration plate images: We utilize multithread and soft trigger techniques to synchronously capture calibration plate images. Figure 4 gives two calibration plate images.

(a)   (b)

Fig. 4. Captured calibration plate images

(3) **Step 3.** Corner detection: In this paper, we employ the function, cv Find ChessboardCorners() to detect corners of calibration plate image [11]. After that, cvFindCornerSubPix() is used to get more accurate image coordinates of corners. Figure 5 shows the corner detection results of calibration plate images.

(a)   (b)

Fig. 5. Corner detection results

(4) **Step 4.** Corner matching: Actually, the distributions of corners in different rectangular arrays are not one-to-one corresponding, so corner matching is indispensable. In our paper, we design two basic transformation functions performing on the rectangular corner array: clockwise rotation function and horizontal flip function. The different combination of these two functions can achieve all transformation of the rectangular array needed in experiment.

(5) **Step 5.** Calculation of camera parameters: When it comes to calculation of camera parameters, the two relatively popular algorithms are Tsai two-step method [12] and Zhang's algorithm [13], and they are both highly accurate and robust. Compared with Tsai two-step method, Zhang's algorithm expects camera is supposed to capture

calibration plate from different viewpoints, but the camera and calibration plate should be fixed all the time and can not be moved in our application. In this situation, we choose Tsai two-step method to calculate the camera parameters.

3.2   Reconstruction of 3D Coordinates of Markers

In order to recover 3D coordinates of two markers on simulated surgical instrument, we first need to extract corresponding feature points, and then match feature points from three cameras to form identical points. After these, least square method is used to calculate the 3D coordinates of markers.

Owing to effect of illumination, it is unavoidable to generate shadows of simulated surgical instrument and hands of operator in motion images of simulated surgical instrument. Besides, the noises are also inevitable to be introduced during camera imaging. Therefore, image preprocessing is quite necessary.

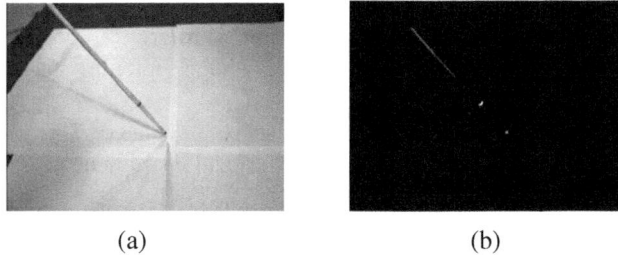

(a)   (b)

Fig. 6. The motion image of simulated surgical instrument and image preprocessing result

Image preprocessing includes the following steps:

Step1. Convert the RGB image into gray image,Subtract the gray image of corresponding background image to remove background;

Step 2. Apply median filter to the residual image for noise removal.

Step 3. Utilize threshold segmentation method based on statistics to complete initial shadow removal. The threshold value is set as 101 through statistics.

The purpose of feature point extraction is to further remove redundant areas, and finally obtain two dimensional (2D) image coordinates of marker by calculating the barycentre of corresponding feature point, so their grey histograms will appear Gaussian peaks [14].

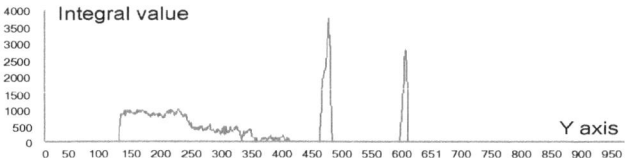

Fig. 7. The integral projection of Figure 6.b on Y axis

As shown in Figure 7, the integral projection of Figure 6.b on Y axis forms a blue curve. Two highest peaks represent the projections of two feature points on Y axis, others areas correspond to the projections of redundant areas. If we set a proper threshold value, it is quite easy to isolate projections of feature points on Y axis. In the same way, we also can get their projections on X axis. Through experiments, we find that the threshold value is 2000. In this way, we can determine the feature points. At last, we take the coordinates of barycenter of each feature point as its 2D image coordinates.

It is worth notice that the Gaussian peaks would overlap if the simulated surgical instrument is vertical in active region. Therefore, we should calculate the integral projection of image on Y axis first. After getting Y coordinates of feature points, we divide the original image into two parts using average value of obtained Y coordinates. Then, we determine X coordinates of feature points by calculating the integral projection of two sub-images on X axis.

During the actual surgery, marker 1, as shown in Figure 2, always moves above marker 2. As for any image, the feature point with larger Y coordinate therefore is image region of marker 1, and another one is image region of marker 2. In this way, the extracted feature points are simply matched.

The final 3D coordinates of maker 1 can be calculated. Thus, we complete the space tracking and positioning of simulated surgical instrument.

4   Experimental Results

The detailed configuration of our experimental platform is as follows:

- Computer: Intel Core Duo CPU @2.66GHz, 2GB memory;
- Basler acA1300-30gc Camera: 1092×962 resolution, 30FPS
- Software: Microsoft Visual C++.net 2005

4.1   Error Analysis

As we mentioned above, the actual distance between the two makers on the simulated surgical instrument is fixed and it measures 70 mm. The table 1 provides the distances calculated by the 3D coordinates of the two makers. As can be seen, the absolute error of our developed apparatus is less than 1mm, which totally satisfies the precision requirement of current virtual surgery simulation systems.

There mainly exist three types of errors in our developed apparatus. The first one is the algorithm error which comes from the implementation of algorithms. The second error is the generation and placement of planar calibration plate and the third one is the generation of simulated surgical instrument manually. Overall, our developed apparatus is precise, and the major reason for the errors is the fabrication errors, which can be decreased by using professorial calibration plate and machined simulated surgical instrument.

Table 1. The distance between two markers computed by their coordinates

	Image coordinates			Space coordinates	Distance (mm)	Error (mm)
	Camera C_1	Camera C_2	Camera C_3			
1	510.01, 226.23	285.76, 407.24	952.65, 337.45	(43.79, -38.61, 92.68)	70.21	-0.21
	477.09, 380.54	346.31, 595.17	945.31, 474.26	(64.08, -29.52, 26.08)		
2	663.15, 226.67	258.64, 336.43	827.08, 417.54	(-19.3, -27.34, 92.39)	69.37	0.63
	625.61, 384.06	317.18, 508.61	830.26, 555.04	(-1.14, -16.55, 26.32)		
3	674.41, 170.26	259.24, 262.90	804.00, 342.18	(-24.80, -22.31, 117.62)	69.11	0.88
	634.01, 333.78	321.10, 438.32	995.31, 164.45	(-5.91, -11.83, 51.98)		
4	456.09, 104.61	273.32, 277.64	997.67, 301.41	(60.87, -42.92, 144.09)	68.90	1.10
	401.13, 258.51	350.13, 488.40	731.05, 129.39	(89.87, -34.88, 82.11)		
5	429.32, 190.17	556.72, 239.15	757.01, 257.53	(63.21, 52.81, 140.14)	68.97	1.03
	366.14, 366.27	621.82, 436.19	647.91, 239.28	(92.44, 61.47, 78.27)		
6	553.67, 261.04	524.30, 268.42	952.08, 337.12	(21.52, 61.07, 114.55)	68.78	1.22
	507.17, 438.30	580.19, 447.20	945.31, 474.16	(43.50, 71.32, 50.19)		

4.2 Application

In order to verify the feasibility of our presented method in virtual surgery simulation systems, we apply it to accomplish interactive organ tissue deformation simulation based on Force Asynchronous Diffusion Model [15]. Using our method, we can achieve real time interaction. Figure 9 shows deformation results virtual organ tissue at different times.

Fig. 9. Interactive organ tissue deformation simulation

5   Conclusion

In this paper, we present a method based on stereoscopic vision to construct a suit of space tracking and positioning apparatus of surgical instruments. It consists of a simulated surgical instrument, three cameras and a computer. Three cameras are used to capture the motion images of simulated surgical instrument in real time. After a series of computer processing, we can obtain the six degree of freedom information of simulated surgical instruments with the absolute error of less than 1 mm, thereby positioning the instrument. Then, we analyze the sources of error, and integrate the developed apparatus into soft tissue deformation simulation in virtual surgery. The experimental results show that the proposed method is highly accurate and easily operated, even if it is inexpensive.

Acknoledgement. This work was fully supported by a grant from the National Natural Science Foundation of China (Grant No. 61070079).

References

1. Basdogn, C., Sedef, M., Harders, M., et al.: VR-Based Simulators for Training in Minimally Invasive Surgical. IEEE Computer Graphics and Applications 27(2), 54–66 (2007)
2. Mead Jr., R.C., McDowall, I., Bolas, M.: Gimbal-mounted Virtual Reality Display System. United States Patent 6774870 (August 2004)
3. Trejos, A.L., Patel, R.V., Naish, M.D., Schlachta, C.M.: Design of a Sensorized Instrument for Skills Assessment and Training in Minimally Invasive Surgery. In: Proceedings of the 2nd Biennial IEEE/RAS-EMBS International Conference on Biomedical Robotics and Biomechatronics, Scottsdale, Arizona, October 19-22, pp. 965–970 (2008)
4. Yamaguchi, S., Yoshida, D., Kenmotsu, H., Yasunaga, T., Konishi, K., Ieiri, S., Nakashima, H., Tanoue, K., Hashizume, M.: Objective Assessment of Laparoscopic Suturing Skills Using a Motion-tracking System. Surg. Endosc., 771–775 (2010)
5. Stoll, J., Novotny, P., Dupont, P., Howe, R.: Real-time 3D Ultrasound-based Servoing of a Surgical Instrument. In: Proceedings of the 2006 IEEE International Conference on Robotics and Automation, Orlando, FL (2006)
6. Welch, G., Bishop, G., Vicci, L., et al.: High-Performance Wide-Area Optical Tracking. Presence 10(1), 1–21 (2001)
7. Halic, T., Kockara, S., Bayrak, C., Rowe, R.: Mixed Reality Simulation of Rasping Procedure in Artificial Cervical Disc Replacement (ACDR) Surgery. BMC Bioinformatics 11(suppl. 6), S11 (2010)
8. Foxlin, E.: Motion Tracking Requirements and Technologies. In: Stanney, K. (ed.) Handbook of Virtual Environments, pp. 163–210. Erlbaun, Mahwah (2002)
9. Kaufmann, M.: Machine Vision: Theory, Algorithms, Practicalities, 3rd edn. Morgan Kaufmann Publishing Co. Inc. (2004)
10. Burden, R.L., Faires, J.D.: Numerical Analysis, 009th edn., August 9. Brooks Cole (2010)
11. http://sourceforge.net/projects/opencvlibrary/

12. Tsai, R.Y.: A Versatile Camera Calibration Technique for High-accuracy 3D Machine Vision Metrology using Off-shelf TV Cameras and Lenses. IEEE Journal of Robotics and Automation 3(4), 323–344 (1987)
13. Zhang, Z.: A Flexible New Technique for Camera Calibration. IEEE Transactions on Pattern Analysis and Machine Intelligence 22(11), 1330–1334 (2000)
14. Yang, X., Pei, J.H., Yang, W.H.: Real-time Detection and Tracking of Light Point. Journal of Infrared and Millimeter Waves 20(4), 279–282 (2001)
15. Si, W., Yuan, Z., Liao, X., Duan, Z., Ding, Y., Zhao, J.: 3D Soft Tissue Warping Dynamics Simulation Based on Force Asynchronous Diffusion Model. Journal of Computer Animation and Virtual Worlds (CASA 2011 Special Issue of JCVAW) 22, 251–259 (2011)

The Design of Antenna Control System Used in Flight Inspection System

Zhang Yachao, Zhang Jun, and Shi Xiaofeng

Flight Inspection Laboratory, Vision tower room 1208,
No 39 Xueyuan Road, Beijing, China
Zhangyachao0909@gmail.com,
buaazhangjun@vip.sina.com,
shixiaofeng@buaa.edu.cn

Abstract. Antenna control system (ACS)[1] is a very critical sub-system of the flight inspection system(FIS)[2][3], which controls the RF routes switching in the front of the FIS system. In this paper, an ACS system used in FIS system is introduced. The ACS system use the ACS controller, which is designed based on freescale 16-bit microcontroller MC9S12XDP512, to receive the control command the host computer through RS232 serial interface, and then it controls the microwave switches installed in the switch matrix to finally switch the RF routes. Functions like logic interlocks, operation protection are also performed by the ACS controller on different flight inspection operations to make sure no damage would occur in the inspection operations.

Keywords: Antenna control system(ACS), flight inspection system(FIS), freescale 16-bit microcontroller, MC9S12XDP512.

1   Introduction

Flight inspection is an evaluation process, using properly equipped aircraft, regarding continuity, integrity and accuracy of significant parameters from radio navigation aids and procedures, aiming their calibration with international standards. VOR, ILS, DME, MARKER[4][5] and some other devices are different navigation aids and need periodic flight inspection.

Flight inspection system (FIS) is a system specially integrated for flight inspection. Usually the FIS systems are installed on small size business planes like Cessna, Gulfstream, or Bombardier planes. The FIS system majorly incorporates antenna control system, high-precision navigation receivers, calibration sources, analyzer, and operating console.

The antenna control system (ACS) is a critical sub-system of the FIS system, which controls the distribution and commutation of the radio signals received. The ACS system in this paper majorly consists of three blocks: antenna block, microwave switch matrix block and ACS controller block. It covers all the VHF and UHF bands used in the flight inspection. The basic function of the ACS system is to distribute the signals received by antenna block to the 12 FIS receivers, which are the major

B. Liu and C. Chai (Eds.): ICICA 2011, LNCS 7030, pp. 680–687, 2011.
© Springer-Verlag Berlin Heidelberg 2011

function of the FIS system. The ACS system also has to distribute the signals received by the antennas to some auxiliary equipment like analyzer or commute the signals generated by calibration sources to the FIS receivers.

2   Antenna Block

The antennas of FIS system means added antennas to the airplane for flight inspection, which majorly includes VOR/LOC antenna, MARKER antenna, VHF COMM antenna, UHF DATALINK antenna, DME antennas, and some monitoring used antennas like VHF monitoring antenna and L-BAND monitoring antenna.

These antennas cover almost all the flight inspection used bands within VHF and UHF.

3   Microwave Switch Matrix

Microwave switch matrix realizes the physical RF links of the antennas and receivers or other equipment like analyzer, calibration sources.

3.1   Microwave Component Analysis

Major components used in the ACS system are microwave switches and power dividers.

In this paper, both the mechanical and diode microwave switches are used. There are majorly 3 types of switches:

> 2P2T failsafe coaxial mechanical switch.
> 1P7T failsafe multi-position coaxial mechanical switch
> PIN diode switch.

In this design, the DOWKEY mechanical switches like 571 series multi-position switches, 411c series transfer switches, and 412 series transfer switches are chosen. And a JFW 50S pin switch is also used in this design.

Table 1. Microwave switches specification

	Frequency /Hz	Insert loss / dB	VSRW	Isolation /dB	Max. input RF power	Power /DC v
411C	DC-18G	<0.1	<1.10	>85	300W	28
412	DC-12.4G	<0.15	<1.15	>85	400W	28
571	DC-18G	<0.2	<1.20	>80	400W	28
50S	20M-2000M	<1.5	<1.40	>60	+10dBm	5

Note: The specification of 411C and 412 are within 0-1GHz; and 571's is within 0-4GHz.

In this design 3dB power dividers are also used, to split one signal into two evenly.

3.2   RF Route Design

The RF routes of the FIS system can be grouped into 3 types as follows by operating modes:

Normal flight inspection mode:Routing the 75MHz to 1.1GHz signals received from the front 11 antennas to corresponding receivers.;

Space signal analyze mode:Routing the space signals received from the front 7 antennas to the analyzer;

Auto-calibration mode:Routing the aeronautical standard signal generator to the corresponding receivers.

As the FIS system mostly works in the normal flight inspection mode, so the default connections of the switches are used to link the antennas and receivers to minimize the operations. The matrix is in normal flight inspection mode after reset.

When space signal analyze is needed, changing the connection of multi-position switch K11 and corresponding transfer switches will meet the need.

When auto-calibration is required, changing the connection of multi-position switch K12 and corresponding transfer switch will change the matrix into auto-calibration mode.

The overall RF routes is shown as figure 1:

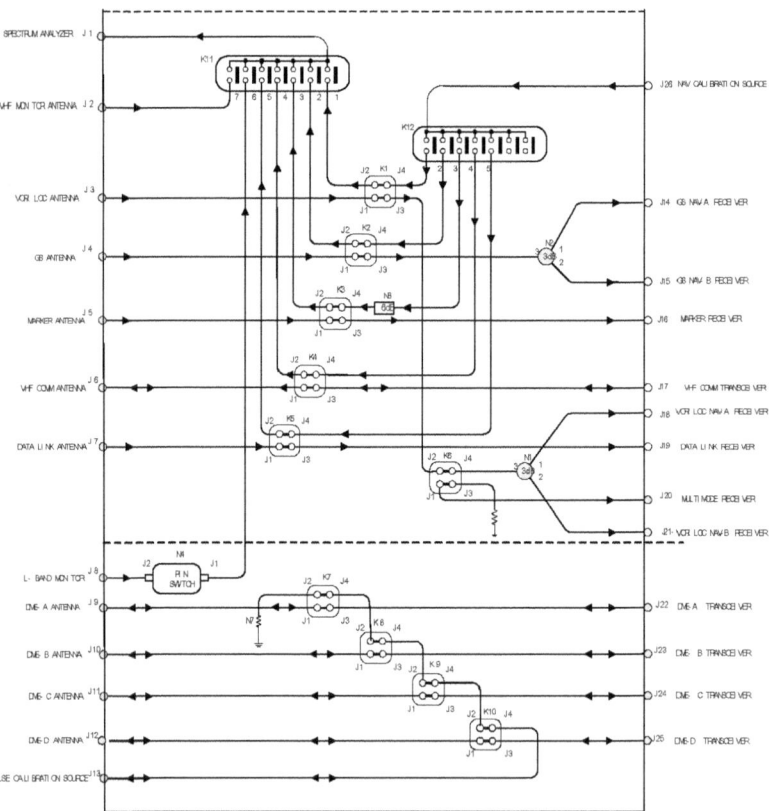

Fig. 1. Switch matrix RF routes

The DOWKEY 571 series multi-position microwave switches are used to distribute the signals from the FIS antennas to the analyzer, and to distribute the signal from the signal generator to the receivers. The DOWKEY 411c and 412 series transfer switches are used to link the FIS antennas to corresponding FIS receivers. When switched, the 411c transfer switches can cooperate with the multi-position switches to realize the analyze mode and auto-calibration mode.

PIN switch is applied to the L-BAND monitor route to realize the protection of the analyzer when doing the L-BAND monitoring.

3   ACS Controller

The ACS controller is used to realize the control of the RF route and communicate with the host FIS computer.

3.1   Protection Interlocks

When doing the flight inspection, in some inspection operation modes, certain operations would do harm to the equipment if no protection is applied. The necessary protections are designed as follows:

VHF monitor mode: If the pilot or flight inspector is using the VHF COMM communication, there would be significant interference as the antennas of VHF monitor and VHF COMM are maybe very adjacent to each other[6]. So in this situation, when the pilot or flight inspector is using the VHF COMM communication, the VHF monitor operation will be interrupted.

VHF COMM analyze mode: The VHF COMM antenna would be occupied when doing the VHF COMM analyze operation. So the request of VHF COMM communication flight inspector would be suppressed. As the same reason with VHF monitor item, when pilot is using the VHF COMM communication, the VHF COMM analyze operation should be interrupt to make sure both the analyze is not interfered and the pilot communication is normal.

VHF COMM calibration mode: The ACS should link the calibration source to the VHF COMM transceiver. For the VHF COMM transceiver is receiving and transmitting when working. The transmission would do great harm to the calibration source. So the transmitting function should be stopped when doing the VHF COMM calibration.

L-BAND monitor mode: The ACS should route the L-BAND antenna to the analyzer. Considering there is a lot of L-BAND equipment on the airplane[7], including equipment of the airplane itself and added FIS equipment. And some of the equipment like DME would transmit at very high power level. The ACS should silent all the FIS L-BAND equipment and cut the L-BAND monitor route with the PIN switch to protect the analyzer, when the airplane L-BAND equipment is transmitting.

3.2   Hardware Design of the ACS Controller

The ACS controller can be divided into microcontroller, power, RS232 communication module, RF switch controlling module and displaying module. The BDM module is used for debugging. The function module diagram is shown as figure 2.

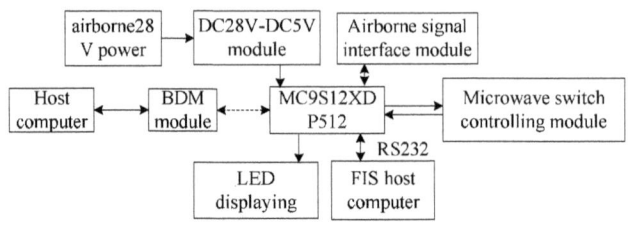

Fig. 2. ACS controller diagram

3.2.1   Microcontroller
In this paper, freescale 16-bit microcontroller MCS12XDP512[8] is used for its rich resource.

The MC9S12XD family is composed of standard on-chip peripherals including up to 512 Kbytes of Flash EEPROM, 32 Kbytes of RAM, 4 Kbytes of EEPROM, six asynchronous serial communications interfaces(SCI), three serial peripheral interfaces (SPI), an 8-channel IC/OC enhanced capture timer, an 8-channel, 10-bit analog-to-digital converter, a 16-channel, 10-bit analog-to-digital converter, an 8-channel pulse-width modulator (PWM), five CAN 2.0 A, B software compatible modules (MSCAN12), two inter-IC bus blocks, and a periodic interrupt timer. The MC9S12XD family has full 16-bit data paths throughout.

3.2.2   Power Module
The airborne DC28V power is chosen as the ACS power supply. The DC28v power is used to power the mechanical switches directly. A DC-DC module with LM2596S-5.0 chip is designed to transform the DC28V to DC5V to the control board of the ACS controller..The circuit is shown as figure 3.

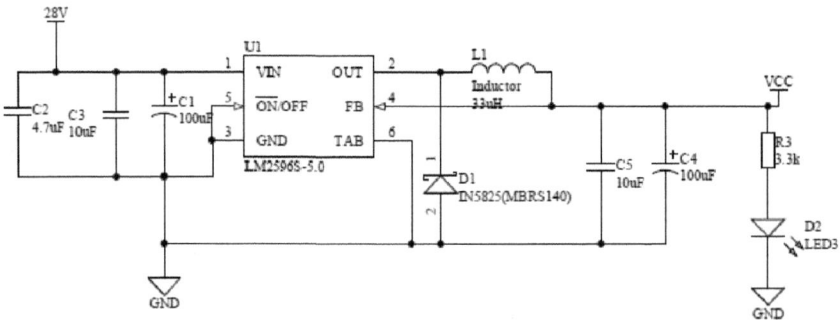

Fig. 3. DC28V-DC5V circuit

3.2.3   Airborne Signal Interface

The voltage level of airborne signals is 28V, while the voltage level of the microcontroller is TTL 5V, so an interface circuit is introduced to do the level switching. The TTL-28v signal level switching circuit is like figure 4.

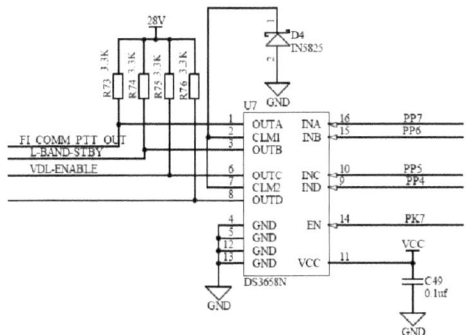

Fig. 4. TTL-28V signal switching circuit

3.2.4   Microwave Switches Controlling Module

The controlling circuit is decided by the type of the microwave switches chosen. In this paper, the controlling circuit has two parts:

1: 1P7T multi-position switches controlling circuit:

When the control pin outputs high level, corresponding position is connected, and corresponding indicator end is connected to the COM end. Part of the controlling circuit is shown as figure 5.

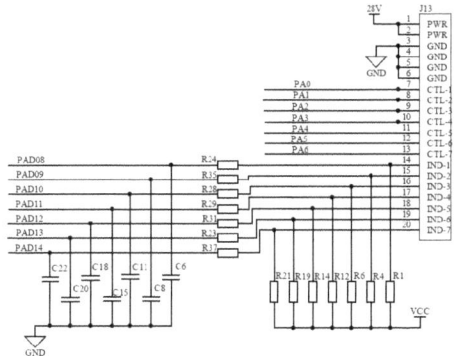

Fig. 5. 1P7T multi-position controlling circuit

2: 2P2T transfer switches controlling circuit:

Normally, the control pin is in low voltage level, the status of the transfer switch is in default status. When it comes to high level, the transfer switch changes into the other status.

The controlling circuit is like figure 6.

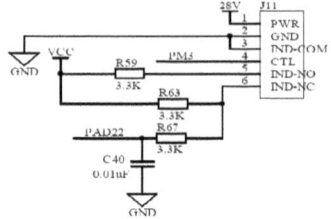

Fig. 6. 2P2T transfer switch controlling circuit

3.2.5   RS232 Communication Module

The RS232 communication module is used to communicate between the ACS controller and the host FIS computer. The ACS controller receives commands, and operates according the commands received.

The MAX232ESE chip is used to realize the RS232 interfacing circuit, see figure 7.

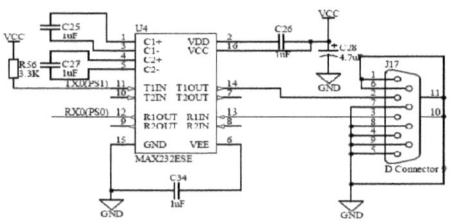

Fig. 7. RS232 communication circuit

3.3   The Software Design for the ACS Controller

The software flow diagram is like figure 8:

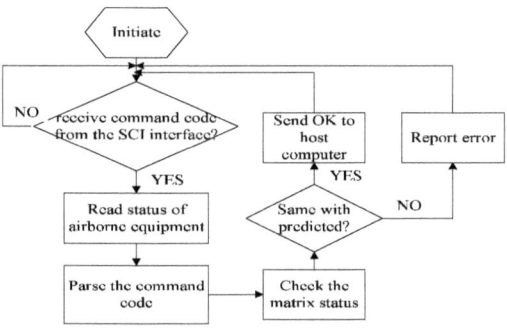

Fig. 8. Software flow diagram

When the ACS controller powers on, it begins to start the work flow: The ACS controller is always ready to receive the command code from the host FIS computer. When it gets a command code, it firstly check the status of the airborne equipment, then to parse the command code, control the RF routes inside the microwave matrix and display the current operation, with consideration of the airborne equipment status.

After that, the ACS controller will check the RF route inside the microwave matrix with the preset route of the command code. If the current route is identically the same with the preset one, it would report the ACS is working fine to the FIS computer, or it would report error.

4   Conclusion

This paper presented one type of antenna control system used in flight inspection system, which has been tested successfully that it could follow the command code sent by the FIS computer and control the RF routes automatically. During certain operation, the operation interlocks are also effective to protect the equipment.

References

1. Marinkovic, V., Pavic, B.: The Control of Antenna System in the radio system of the special purpose. In: International Symposium on Intelligent Signal Processing and Communications Systems, December 13-16, pp. 709–712 (2005)
2. Gucciardo, R.: Flight Inspection- "The State of the Art". In: AIAA/IEEE. Digital Avionics Systems Conference 13th DASC, Phoenix, AZ, USA, 30 October-3 November, pp. 221–226 (1994)
3. Eskelinen, P.: A Computerized Flight Inspection System. IEEE Aerospace and Electronic Systems Magazine 7(3), 5–11 (1992)
4. McFarland, R.H.: ILS-A Safe Bet for Your Future Landing. IEEE Aerospace and Electronic Systems Magazine 5(5), 12–15 (1990)
5. Moir, I., Knight, S.G.: Civil Avionics Systems. AIAA (December 2002)
6. Hou, Y., Su, D., Chen, W., Liao, Y.: Analyzing and Calculating Isolation between Antennas on Airplane Wireless System. In: International Symposium on Electromagnetic Compatibility, Qingdao, China, October 23-26, pp. 378–381 (2007)
7. Brandes, S., Epple, U., Gligorevic, S., et al.: Physical Layer Specification of the L-band Digital Aeronautical Communications System (L-DACS1). In: Integrated Communications, Navigation and Surveillance Conference, Arlington, VA, May 13-15, pp. 1–12 (2009)
8. MC9S12XDP512 Data Sheet. Rev.2.21, Freescale semiconductor (October 2009)

Modeling and Application of Battlefield Synthetic Natural Environment Based on Ontology

Bo Wang[1], Limin Zhang[1], and Yan Li[2]

[1] Naval Aeronautical and Astronautical University of PLA,
264001 Yantai, China
[2] Dalian Naval Academy of PLA, 116018 Dalian, China
songzywb@hotmail.com,
{iamzlm,aaplaly}@163.com

Abstract. The battlefield synthetic natural environment (BSNE) modeling and simulation (M&S) is one of the directions in research on military M&S. Current researches of M&S for the BSNE are analyzed firstly. Ontology is applied to modeling the BSNE in this paper, and definitions of concepts, relationships, and attributes of the BSNE ontology are given. A processing for the BSNE semantic abstraction is presented, which makes the BSNE ontology compose with two main parts: a Spatial Ontology and a Semantic Ontology. A meta-model of the BSNE Spatial Ontology is built and the relationships between the BSNE Spatial Ontology and the Semantic Ontology are discussed. A prototype system for developing and applying the BSNE ontology is setup finally.

Keywords: battlefield natural environment, Spatial Ontology, Semantic Ontology, computer generated forces.

1 Introduction

The modern battlefield can be divided to four spatial domains: the land, the ocean, the air and the space, and military actions are taken place in them. Corresponding to the four domains, the purpose of the battlefield synthetic natural environment (BSNE) modeling and simulation (M&S) is to build up authoritative, integrated, polymorphic and coherent data descriptions and model representations for the whole natural environment to satisfy the simulation requirements of visualizations, sensors and computer generated forces (CGF) [1]. With the development of advanced distributed simulation forwarding to the direction of dimension, network and intelligence, many problems are coming forth for the new requirements of data description, interoperation, extensibility and real-time ability of the BSNE, including:

(1) The data formats of the BSNE applied to battlefield simulation systems are multiplicate, which affects the performance of interoperations [2].
(2) In CGF reasoning processing, entities in CGF systems should react to the behaviors of their operators and environments automatically or even autonomically. Battlefield simulation systems provide terrain-oriented models and data interfaces [3]. But it is imprecise and limited for these systems to describe the natural environment and to support CGF reasoning.

B. Liu and C. Chai (Eds.): ICICA 2011, LNCS 7030, pp. 688–695, 2011.
© Springer-Verlag Berlin Heidelberg 2011

Department of Defence of US advanced eXtensible Modeling and Simulation Framework (XMSF) in 2000[4] which defines standards and frameworks for M&S based on network, and applies ontology technology to solving these problems[5][6].

Ontology shares conceptions by describing concepts, attributes of one domain of knowledge hierarchically to exchange information, interoperate and understand conception between the human beings and machines or among machines. In this paper we advance to modeling and simulating the BSNE based on ontology. Current researches of M&S for the BSNE are analyzed firstly in part 2, and we give definitions of concepts, relationships, and attributes of the BSNE ontology in part 3. In part 4, we also give a processing for the BSNE semantic abstraction and build a meta-model of the BSNE spatial ontologies, discussing the relation between the BSNE Spatial Ontology and the Semantic Ontology. We also set up a prototype system for developing and applying the BSNE ontology.

2   Related Works

The recent researches on M&S of BSNE include two contents: data-based simulation and GIS-based (Geographic Information System) simulation.

2.1   BSNE Simulation Based on Environment Database

The data format in the BSNE simulation includes specific data and generic data. The former builds some types of terrain data format to support movement and path-planning of entities (soldiers, tanks, vehicles etc.) on land. Such formats include: OpenFlight,DTED,DEM,ESRI GRD, etc. The typical application is the VR-Forces of VT MÄK [3]. The generic data format means M&S based on the Synthetic Data Representation and Interchange Standard (SEDRIS) [7]. The SEDRIS advances a uniform description and a data-exchange standard for geographic information, natural phenomena and force entity models in military domains. [8] [9] [10] applies SERDIS as environment database and data exchanging standard to military simulation systems to support modeling and realization of CGF and visualizations. An atmosphere database based on the SEDRIS to support a missile simulation system is designed in [13]. A SEDRIS ontology is built in [14], which maps classes, relationships and enumerations in the SEDRIS to OWL -based (the Web Ontology Language) structures and forms the SedOnto which implements environment data to be shared and objects in environment reasoning in network.

2.2   GIS-Based Geographic Simulation of Battlefield

Applying GIS to military simulation is another aspect for battlefield simulation because of multiple geographic layers and rich geographic information supplied by GIS. Combining the VR-Forces of VT MÄK with the ArcGIS of ESRI, [2] develops a GIS-enabled simulation prototype system. The system replaced intrinsic terrain database of the VR-Forces by GIS layer data, and wrapped the VR-Forces Terrain Application Program Interfaces (APIs) to the ESRI terrain APIs and the ESRI coordinates APIs.

The system also offers GIS-Link components oriented to the ArcGIS, which can be used to connect the system to HLA/DIS (the High Level Architecture/the Distributed Interaction Simulation) simulation systems, and can be extended to connect to the ArcMap or the ArcGlobe to display battlefield situation.

So, we can reach a conclusion that M&S of the BSNE is still in the state of data level, which supports CGF modeling and battlefield visualization with some data standards and/or formats. The SedOnto environment ontology extends data level radically, and does not give the BSNE semantic definitions and criterions. The GIS-enabled BSNE simulation focused on terrain, geographic info and objects on land, and does not give a unified description and layout. Researches on GIS ontologies which define classes, relationships and axioms in GIS provide referrible methods for M&S of the BSNE [15], [16], [17], [18], [19]. Now, we give the formalization description of the BSNE ontology at first.

3   Formalization of BSNE

In processing of working with ontology, the basic elements are classes/categories, attributes and relationships. Classes are sets, concepts, objects or things, which include individuals and instances. Attributes descript classes in detail (parameters, characters, etc) and the basic relationships of classes are inheritance and implement. Now we give the definition of BSNE [11].

Definition 1: VB_Ontology:=<VB_Concepts, VB_Relations, VB_Functions, VB_Axioms, VB_Instances>. Where, VB_Concepts are sets of concepts in the BSNE, VB_Relations are sets of relationship, VB_Functions are sets of functions, VB_Axioms are sets of axioms, and VB_Instances are sets of instances.

Definition 2: VB_Concepts:={C}. Where, C is concept and attribute of concept in the BSNE.

Definition 3: VB_Relations:={$R(c1,c2)|c1, c2 \in VB\_$ Concepts}, which are sets of duality relationships in the BSNE. The relationships in the BSNE, including spatial relations and semantic relations are shown in Fig. 1 [12].

Definition 4: VB_Axioms:={A},where A is an axiom in military domain.

Definition 5: VB_Instances:={$I|I \in VB\_Concept \cup VB\_Relations$}, which are instances in the BSNE.

4   Modeling the BSNE Base-on Ontology

4.1   BSNE Semantic Abstraction

According to the description of the BSNE, the processing of building ontology models of the BSNE is to abstracting the BSNE semantic and building its concepts, relationships, attributes and axioms. We present the process of abstracting BSNE is shown in Fig. 2[15].

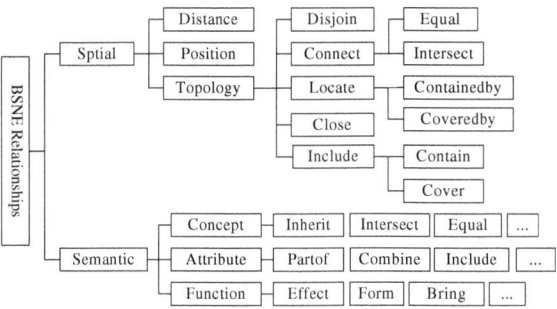

Fig. 1. Relationships in BSNE

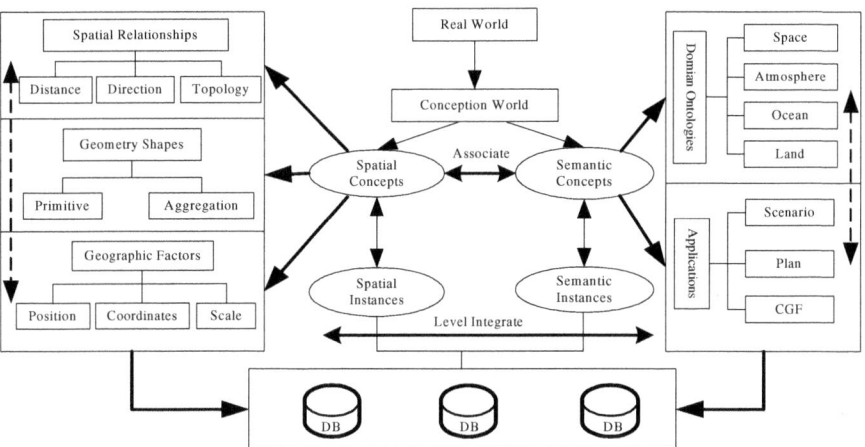

Fig. 2. Processing of Abstraction and Application of the BSNE Ontology

Abstraction of the Real World to the Conception World is the processing to define concepts and relationships in the BSNE, and the Conception World is a formalization and symbolization to the Real World. The Conception World has two main contents: the spatial concepts and the semantic concepts. The spatial concepts orient to the integration based on environment data, which is semantic description to spatial relationships, geometry shapes and geographic factors. The spatial concepts map the environment data elements to ontology elements by establishing the relationships between data elements and ontology elements. The SedOnto is a type of spatial concepts. The semantic concepts are the upper level for the BSNE abstraction. Associations are created between the semantic concepts and the spatial concepts by mapping concepts and relationships to each other reciprocally. So, this unifies the BSNE in data level and semantic level.

4.2 BSNE Spatial Ontology

In this paper, we consider all objects located on the earth, including the land, the ocean, the air or the space, are spatial things, which have spatial attributes. The BSNE spatial things have geographic features, spatial relationships and geometry shapes as cores. The geographic features provide standards and criterions for measuring the spatial relationships. The geometry shapes describe characters making up of the spatial objects, and illuminate the spatial characters of geographic objects. The geometry shapes derive primitive shapes (points, lines, areas, polygons, vectors, etc) and aggregation shapes, provide a bound for the geographic features and the domain things, and also provide basic descriptions for data and types of spatial objects, including points, lines, areas, polygons, vectors, etc. The primitive shapes aggregate the aggregation shapes. The spatial things associate the BSNE objects data with metadata which is defined to abstract metadata here providing description of various domains or instances which can be instanced based on data standard of domains. The meta-model of spatial ontologies is shown in Figure 3.

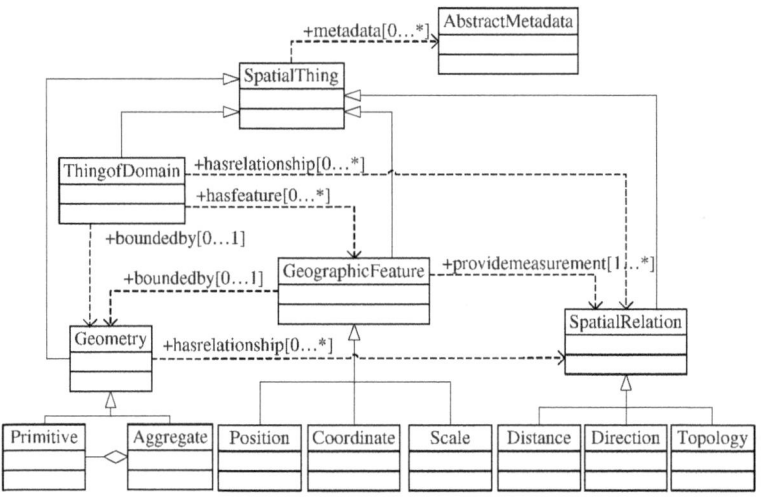

Fig. 3. Meta-model of BSNE Spatial Ontology

The distance relationships in the BSNE are distance measurements, and position relationships are locations (east, south, west, north, up, and down) and their combinations. With scales and coordinates, querying positions and locating targets can be done. The topology relationships should match the model of Region Connection Calculus.

4.3   BSNE Semantic Ontology

The Spatial Ontology is a semantic description oriented to spatial data criterion, but it is oriented to spatial domains and applications. So, it needs to create concepts, relationships, attributes and axioms to describe the BSNE Semantic Ontology. Modeling the BSNE Semantic Ontology needs to take account into the followings:

It needs to provide plain and integrated definitions and classifies for the conceptions and attributes in the BSNE, and also can provide abundant terms and rules for the BSNE.

It can support all of the four domains in the BSNE at the same time, and involve their intersections.

It should complete the associations between the Semantic Ontology and the Spatial Ontology when building the BSNE Semantic Ontology.

For these reasons, there are two ways to build the Semantic Ontology. One is to reuse built ontologies of the earth environment ontologies and the other is to model ontology starting from scratch. Here we adopt the former. That is, we can create associations between the Spatial Ontology and the Semantic Ontology when extending concepts, relationships and attributes of built ontologies and building the Semantic Ontology. The referrible ontologies and terminologies can be SWEET (the Semantic Web for Earth and Environmental Terminology) [20], DAML (the DAPAR Agent Maker Language) [21], the WordNet [22], etc.

5   Application

A prototype system for military simulation based on the BSNE ontology is shown in Figure 4. The bottom of the system is environment database, including geographic information database, terrain database, atmosphere database and ocean database, etc. The data standard level makes all these data to one standard —SEDRIS by tools from SEDRIS. The Semantic Ontology is the SWEET [17] as upper ontology, and is extended to satisfy the military requirements. The SWEET uses keywords form the Global Change Master Directory of NASA as s direction to build ontology terminology and accomplished by the OWL.

When building the BSNE Semantic Ontology, modeling the associations between the two ontologies must be considered which can be worked with conception-collection. The concepts are derived from both the Semantic Ontology and the Spatial

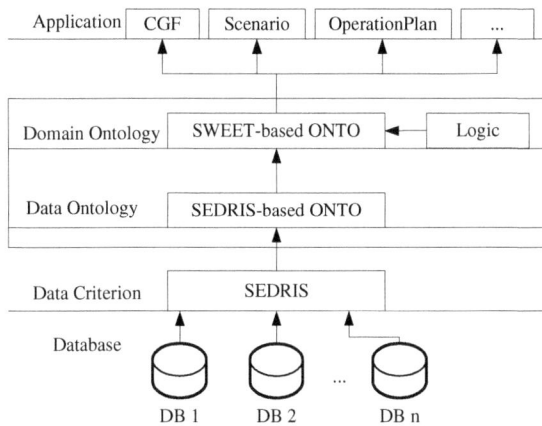

Fig. 4. The Prototype of Simulation System based on the BNSE ontology

Ontology, and modeling the relations must be taken into account at the same time. To accomplish the associations it needs to extend the attributes of its parents including the Semantic Ontology and the Spatial Ontology both, and relationships among concepts need to be reprogrammed. To support ontology reasoning it also needs to import reason logic. Here, we apply the OWL DL (Description Logic) which can satisfy completeness and computability [23].

6   Conclusions and Future Works

In this paper we analyze the current researches on the BSNE, and give definitions of formalization of the BSNE and the processes of building and application BSNE ontology. The meta-model of Spatial Ontology and the methodology of building the Semantic Ontology are advanced. Finally we build a prototype system to model and apply the BSNE ontology. The BSNE ontology can be combined with the ontology for ArcGIS and also can be applied to distributed military simulation system based on HLA/RTI architecture to support geographic reasoning of CGF and interoperation of environment data.

The model in the paper is a static model for the BSNE, which can not involve dynamic evolvements (weather, ocean current) of the environment and the effect of CGF entities to environment (crater and mine field, etc). To these problems, it needs to consider to build the spatial-temporal relationship of the BSNE ontology. And the semantic reasoning based on DL do not support rule sets. It needs to introduce other logics or axioms to support rule-based reasoning, such as SWRL (the Semantic Web Rule Language). These will be researched in-depth in the future.

References

1. Huang, K.-d., Liu, B.-h., Huang, J., et al.: A Survey of Military Simulation Technologies. Journal of System Simulation 16(9), 1887–1895 (2004)
2. Satnzione, T., Johnson, K.: GIS Enabled Modeling and Simulation, GEMS (2007), http://www.dtic.mil/cgibin/GetRTDoc?AD=ADA481213
3. VT MÄK: VR-Forces Developers Guide, Revision VRF-4.0-2-110120 (2011)
4. Brutzman, D., Zyda, M., Pullen, M., et al.: Extensible Modeling and Simulation Framework (XMSF) Challenges for Web-Based Modeling and Simulation. Technical Challenges Workshop – Interim Report of XMSF (2002)
5. Gerber, W.J., Lacy, L.W.: Developing Standard Ontology Behavior Representations to Support Composability. Air Force Research Laboratory of US Final Report for April 2003 to December 2004 (2005)
6. Bhatt, M., Rahayu, W., Sterling, G.: SedOnto: A Web Ontology for Synthetic Environment Representation Based on the SEDRIS Specification. In: Proceeding of the Fall 2004 Simulation Interoperability Workshop, Orlando (2004)
7. An Introduction to SEDRIS, http://www.sedris.org
8. Ma, Q.-l., Lu, X.-x., Xu, H.-x., et al.: Research on Virtual Battle Space Environment and Its Data Representation and Interchange Method. Journal of System Simulation 20(suppl.), 141–146 (2008)

9. Liu, W.-h., Wang, X.-r., Li, N.: Modeling and Simulation of Synthetic Natural Environment. Journal of System Simulation (12), 2631–2635 (2004)
10. Liu, J., Liu, Z., Liu, G.-f.: Research on Synthetic Environment Simulation System of Sea Battlefield Based on HLA. Journal of System Simulation 20(11), 2872–2876 (2008)
11. Perez, A.G., Benjamins, V.R.: Overview of Knowledge Sharing and Reuse Components: Ontologies and Problem-Solving Methods. In: Stockholm, V.R., Benjamins, B., Chandrasekaran, A. (eds.) Proceedings of the IJCAI 1999 Workshop on Ontologies and Problem-Solving Methods (KRR5), pp. 1–15 (1999)
12. Zhao, Z., Huang, Y.-q.: Research on the Representation of Geographical Spatio-temporal Information and Spatio-temporal Reasoning Based on Geo-ontology and SWRL. Journal of Anhui Agri Sci. 37(3), 1375–1379 (2009)
13. Song, X., Xin, L.: Research on Key Technology of Standard Atmosphere Database. In: Proceeding 2008 Asia Simulation Conference/7th International Conference on System Simulation and Scientific Computing (ICSC 2008), pp. 1497–1500 (2008)
14. Bhatt, M., Rahayu, W., Sterling, G.: Synthetic Environment Representational Semantics Using the Web Ontology Language. In: Gallagher, M., Hogan, J.P., Maire, F. (eds.) IDEAL 2005. LNCS, vol. 3578, pp. 9–16. Springer, Heidelberg (2005)
15. Zheng, M., Feng, X., Jiang, Y., et al.: A Formal Approach for Multiple Representations in GIS Based on DL Ontologies. Acta Geodaetica et Cartographica Sinica 35(8), 261–266 (2006)
16. Song, J., Zhu, Y., Wang, J., et al.: A Study on the Model of Spatio-temporal Geo-ontology based on GML. Journal of Geo-information Science 111(14), 442–451 (2009)
17. Raskin, R., Pan, M.: Semantic Web for Earth and Environmental Terminology, http://gcmd.nasa.gov/records/SWEET-ontology.html
18. Visser, U., Stuckenschmidt, H., Schuster, G., et al.: Ontologies for Geographic Information Processing. Computer & Geosciences 28(1) (2002)
19. Fonseca, F.T., Egenhofer, M.J., Agouris, P., et al.: Using Ontology for Integrated Geographic Information System. Transaction in GIS 6(3) (2002)
20. SWEET Ontologies, http://sweet.jpl.nasa.gov/ontology
21. DAML, http://reliant.teknowledge.com/DAML
22. Wordnet, http://wordnet.princeton.edu/wordnet
23. OWL Guide, http://www.w3.org/TR/owl-guide/

Research for Influence of Different Leg Postures of Human Body in Walking on Plantar Pressure

Yi Gao

Criminal Technology Department,
China Criminal Police College, Shenyang, Liaoning, China
gao_yi7666@sina.com

Abstract. This article uses Footscan plantar pressure analysis measuring system to study plantar pressure distribution. After comparing various index parameters of plantar mechanics characteristics of human body's lower limbs in different walking postures, the result indicates that: different walking movements of lower limbs have certain influence on plantar pressure distribution, which results in sharp differences of regional mechanics characteristics.

Keywords: Different Postures of Lower Limbs, Plantar Pressure, Footwork Characteristics.

1 Introduction

Plantar pressure measuring technique is an applied technology using pressure measuring instrument to measure mechanics, geometry and time reference data of plantar pressure while human body is under static or motor process, analyze and study plantar pressure parameters under different conditions, and reveal different plantar pressure distribution characteristics and modes. Our current research is to use Belgian Footscan plantar pressure measuring system to analyze footwork characteristics, measure interaction process and model of action between foot and supported object of human body's lower limbs in different walking postures and the reflection of plantar pressure distribution situation in system, through various comparisons of mechanical parameter, find out regulation characteristics of plantar pressure changes of human body's lower limbs under different walking movements situations and its relationship with walking posture to provide evidence for using footwork characteristics for quantitative inspection to further achieve goals of personal identification.

2 Introduction to Footscan Plantar Pressure Measuring Analysis System

2.1 Formation of the System

The Belgian Footscan USB flat-panel system is a pressure plate procedural testing platform, 2m long, 40cm wide and 2cm thick; its one-step dynamometric flat panel of 0.5m has 4096 sensors with sampling frequency 125·300Hz and resolution ration of 4

B. Liu and C. Chai (Eds.): ICICA 2011, LNCS 7030, pp. 696–703, 2011.

/cm^2, as showed in Fig.1. It is also equipped with a data acquisition box and an external connected laptop. The USB interface is easily connected to computer without additional power. The software could provide plantar pressure and gait analysis function of dynamics, statics, contrast and measurement.

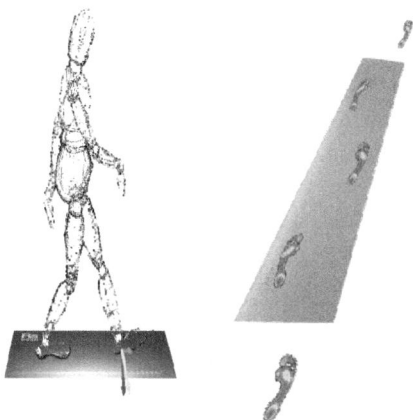

Fig. 1. Graph for plantar pressure measurement system to acquiring human body walking plantar mechanics parameter

2.2   Main Function of Plantar Pressure Measuring System (Footscan)

The standard functions of Footscan system include real-time dynamic display, successive-frame playback, inspection of central pressure, calculation of contract area, evaluation of integrals of pressure and time and graphical analysis. This system could measure movement locus of plantar pressure center and peak pressure of relevant

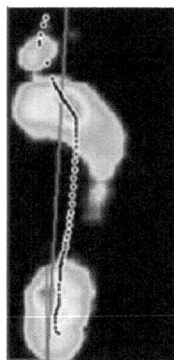

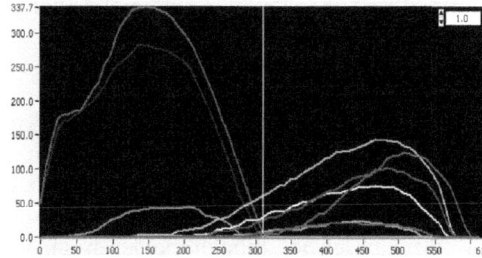

Fig. 2. Graph for plantar pressure distribution

Fig. 3. Resultant action curve of entire interaction process in each area

plantar areas. The software provides plantar pressure and gait analysis function of dynamics, statics, contrast, measurement, etc. The system with clear interface and powerful software functions can provide gait's time, gravity center, pressure, pressure intensity and dynamic analysis, contrasting in left and right, equilibrium analysis, angular measure, static test, body weight distribution, dynamic test, two dimensional analysis, three dimensional analysis, simultaneous analysis for both left and right feet, vector analysis for plantar friction, analysis of plantar pressure and plantar pressure intensity, equilibrium analysis, maximum plantar pressure, average pressure, analysis of impact force, database administration, and output function of information provided for reference, as showed in Fig.2, Fig3 and Table 1.

Table 1. Quantized data of entire interaction process in each area

	Start Time	End Time	% Contact	Max F	Time Max F	Load rate	Impulse	Contact area	Active Contact area	Max peak sensor value in area
	ms	ms	%	N	ms	N/s	Ns	cm	cm	N
Left										
Toe 1	294.1	611.0	51	124.2	507.9	0.83	18.6	9.8	9.8	10.6
Toe 2-5	246.5	570.8	52	19.7	404.7	0.18	3.8	3.8	3.8	7.6
Meta 1	191.0	587.2	64	101.5	484.1	0.47	18.8	5.2	5.2	16.7
Meta 2	119.6	587.2	76	143.9	468.2	0.57	32.1	6.4	6.4	15.1
Meta 3	175.1	579.2	65	75.7	468.2	0.38	15.5	4.1	4.1	10.6
Meta 4	238.6	547.0	50	22.7	436.5	0.16	3.7	5.2	5.2	6.1
Meta 5	7.9	611.1	97	0.0	0.0	0.00	0.0	2.2	2.2	0.0
Midfoot	32.3	301.0	43	45.4	206.3	0.59	6.9	6.4	6.4	7.6
Heel Medial	0.0	332.8	54	337.7	142.8	4.26	69.3	13.1	13.1	19.7
Heel Lateral	0.0	324.8	52	281.7	134.9	4.44	58.9	12.0	12.0	18.2
Right										
Toe 1	254.5	595.1	56	149.9	515.8	0.95	26.2	8.6	8.6	16.7
Toe 2-5	333.8	523.2	31	1.5	341.2	-2.89	0.3	2.6	2.6	1.5
Meta 1	79.6	571.3	80	125.7	436.5	0.42	22.3	9.0	9.0	10.6
Meta 2	71.9	571.3	82	142.4	460.3	0.47	31.7	8.2	8.2	10.6
Meta 3	95.8	563.4	77	116.6	452.3	0.51	29.7	9.0	9.0	10.6
Meta 4	87.6	531.2	73	68.2	230.1	0.96	18.1	10.1	10.1	6.1
Meta 5	87.8	451.8	60	40.9	238.1	0.65	6.5	7.5	7.5	6.1
Midfoot	40.2	301.0	43	77.2	166.7	0.89	9.6	16.5	16.5	4.5
Heel Medial	0.0	229.6	38	192.3	71.4	24.24	26.7	15.0	15.0	10.6
Heel Lateral	0.0	229.6	38	180.2	87.3	22.71	27.0	13.5	13.5	12.1

3   Research Object and Method

3.1   Research Object

Choose a group of homogeneous people of total 30, aging from 20 to 25, heights from 173cm to 178cm and weight from 60kg to 70kg, in good health without any injury of joint of legs or ankles and in normal walking style.

3.2   Test Method

The testees walk barefooted at normal walking speed of 1m/s and rapid walking speed of 2m/s through pressure sensing platform. Then they return and walk through pressure testing platform again to fulfill a gait cycle movement, so that we can obtain a plantar pressure distribution graph, pressure curve graph and relevant data within a full gait cycle through Footscan plantar pressure measuring analysis system, and meanwhile we draw test samples of human body gait and plantar pressure with different leg rising heights and knee joint bend angles randomly for comparative study.

4 Research Result

Collect statistics for test results and make analysis to obtain data of typical value, record plantar pressure characteristics with different lower limb movement characteristics of different leg rising heights and knee joint bend angles, and then draw movement data table, area distribution graph and pressure curve graph of peak pressure intensity characteristic parameters in each plantar area for further description and comparison.

4.1 Comparison of Pressure Intensity Peak in Different Plantar Areas with Different Leg-Lifting Heights

The randomly selected testees are all at the walking speed of approximately 1m/s, and different by almost 30° at their minimum thigh-trunk angle. The testee with greater leg-lifting height is coded as 1, the other as 2. The comparison of two testees' pressure intensity peaks in different plantar areas shows that Testee 1's pressure intensity peaks in Meta 3, 5 are greater than Testee 2's while less than Testee 2's in toes, Meta 1, 2,4 and heel, which indicates that Testee 1's (with greater leg-lifting height) greater pressure intensity is mostly in Meta 3, while Testee 2's (with less leg-lifting height) mostly in Meta 3,4,5, as shown in Fig.4 and 5. Two testees' pressure intensity peaks in different plantar areas are shown in Table 2.

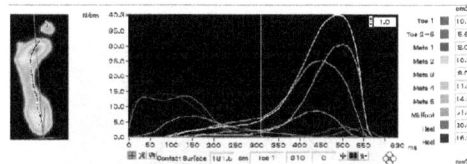

Fig. 4. Curve graph of Testee 1's pressure intensity in different plantar areas

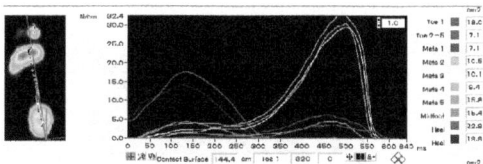

Fig. 5. Curve graph of Testee 2's pressure intensity in different plantar areas

Table 2. Pressure intensity peak in different plantar areas   Unit: N/cm^2

Area Code	Toe 1	Toe 2-5	Meta 1	Meta 2	Meta 3	Meta 4	Meta 5	Heel Medial
1	3.0	0.3	3.4	30.6	40.3	25.4	8.7	13.5
2	3.4	0.1	6.2	32.4	29.5	30.5	4.5	17.6

Through two testees' thigh-trunk angles at similar walking speed, we can conclude a great difference between two testees' thigh-trunk minimum angles, which indicates the great difference between two testees' leg-lifting heights in walking and its effect on angles between footprint centerline and walking direction: the one with greater leg-lifting height has his footprint centerline close to walking direction, while the one with less leg-lifting height has his footprint centerline away from walking direction; Testee 1's (with greater leg-lifting height) greater pressure intensity is mostly in Meta 3, while Testee 2's (with less leg-lifting height) mostly in Meta 3,4,5.

4.2   Comparison of Pressure Intensity Peak in Different Plantar Areas at Different Knee Angles

The two randomly selected testees are of similar height, weight and age, all at the walking speed of approximately 1m/s, and different by almost 30° at the minimum knee angle (angle between thigh and calf). The testee with less minimum knee angle is coded as 3, the other as 4. The comparison of two testees' pressure intensity peaks in different plantar areas shows that Testee 3's pressure intensity peaks in Toe 1, Meta 3 and Heel are greater than Testee 4's while less than Testee 4's in toes, Meta 1, 2,4,5, which indicates that the larger extent the knee bends, the larger the pressure intensity in Toe 1, the larger the pressure intensity in Heel, so footprint pressure is mostly at footprint centerline; on the contrary, when the knee bends to a smaller extent, pressure would at both sides of foot centerline, as shown in Fig.6, 7. Two testees' pressure intensity peaks in different plantar areas are shown in Table 3.

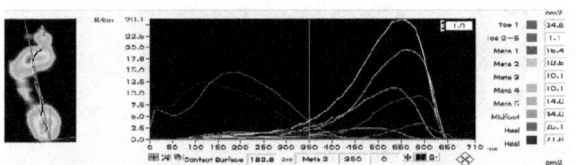

Fig. 6. Curve graph of Testee 3's pressure intensity in different plantar areas

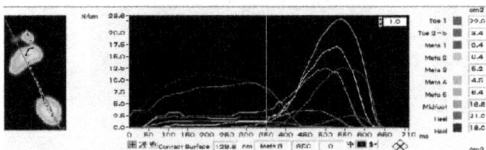

Fig. 7. Curve graph of Testee 4's pressure intensity in different plantar areas

Table 3. Pressure intensity peak in different plantar areas   Unit: N/cm^2.

Area Code	Toe 1	Toe 2-5	Meta 1	Meta 2	Meta 3	Meta 4	Meta 5	Heel Medial
3	3.5	0	9.4	19.5	26.1	11.2	2.5	14.5
4	2.6	0	12.7	22.9	17.1	12.6	3.6	9.4

Through two testees' thigh-trunk angles at similar walking speed, we can conclude that for the one who lifts his calf higher, his knee bends to a large extent, the bending angle is small, footprint centerline is close to walking direction, and accordingly the foot abduction is small; the larger extent the knee bends, the larger the pressure intensity in Toe 1, the larger the pressure intensity in Heel, so footprint pressure is mostly at footprint centerline; on the contrary, when the knee bends to a smaller extent, pressure would at both sides of foot centerline.

4.3 Comparison of Pressure Intensity Peak in Different Plantar Areas at Different Walking Speeds

The experiment has verified that the mechanical characteristics resulted from different swing positions of lower limbs are different among people and the swing posture has some correlation relationship with foot mechanical characteristics. We analyze the corresponding changes in different swing positions of lower limbs and the mechanical characteristics of foot bottom under different walking speeds by the same person, and further analyze the changes in different swing positions of lower limbs and the corresponding changes in mechanical characteristics of the foot under different walking speeds by the people. The experimental data for the same person under the normal walking speed of 1 m/s and brisk walking speed of 2 m/s are selected for comparing and analysis. By comparing the peak pressures of each plantar area in Table4, it shows that the pressure is distributed among Meta 1, Meta 2, Meta 3 and the heel area during fast walking, and among Meta 2, 3, 4 and 5 during normal speed walking. At normal speed the pressure is only present in the Toe 1 area and not significantly present in the other four toe areas; while the pressure in the Toe 1 area is greater than that of the normal walking speed, and the pressure is present in the other four toe areas, suggesting the increasing toe force during fast walking. The pressure of each plantar area is shown in Fig. 8 and Fig. 9.

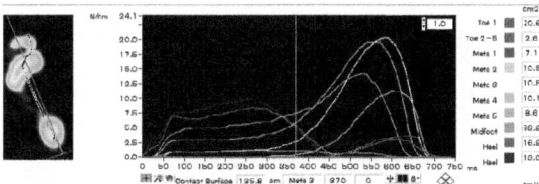

Fig. 8. Pressure intensity in plantar areas at normal speed

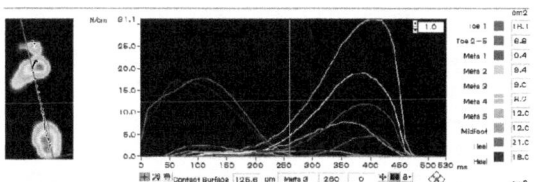

Fig. 9. Pressure intensity in plantar areas at fast speed

Table 4. Pressure intensity peak in different plantar areas   Unit: N/cm^2

Area Code	Toe 1	Toe 2-5	Meta 1	Meta 2	Meta 3	Meta 4	Meta 5	Heel Medial
Normal	3.1	0	4.7	12.5	21.2	21.0	11.9	8.6
Fast	4.7	0.3	11.9	31.1	19.0	7.7	1.3	17.7

After the comparison for the same person at different walking speed, we summarize that when walking accelerates, the maximum acceleration of the knee at X-axis increases significantly, but the minimum angle of the bending knee changes little; the accelerations of ankle in different directions change significantly and the maximum accelerations in different directions increase. At normal speed, the minimum angle of thigh and torso is greater, suggesting lower lifting height of the thigh, while lifting height of the thigh increases during brisk walking. Pressure is mainly distributed among the first, second, third metatarsal areas and the heel area during brisk walking, and among the second, third, fourth and fifth metatarsal areas during normal speed walking. At normal speed the pressure is only present in the Toe 1 area, and not significantly present in the other four toe areas; while the pressure in the Toe 1 area is greater than that of the normal walking speed, and the pressure is present in the other four toe areas, suggesting the increasing toe force during brisk walking.

5   Conclusion analysis and discussion

Through numerous tests and contrast for above parameters, we can conclude:

(1) The thigh-trunk angle is weakly negatively correlated with the toe area and metatarsal area, and positively correlated with the heel area, relatively significantly. By comparing the different thigh-trunk angles of the test subject, it is found that for the higher thigh-lifting height, the central line of footprint is close to the walking direction, while for the lower thigh-lifting height, the central line of footprint deviates far from the walking direction; for the higher thigh-lifting height, the greater pressures are concentrated mainly in the third metatarsal area, while for the lower thigh-lifting height, the greater pressures are concentrated mainly in Meta 3, Meta 4 and Meta 5.

(2) The angle of knee is significantly positively correlated with foot toe areas and metatarsal areas, while significantly negatively correlated with the heel area. By comparing the different knee angles of the test subject, it is found that for the higher lifting height of lower leg, the bending of knee is fairly greater, the central line of footprint deviates little from the walking direction, and accordingly the foot abduction is small while for the lower lifting height of lower leg, the bending of knee is fairly smaller, the central line of footprint deviates far from the walking direction, and accordingly the foot abduction is greater; the greater the bending of knee is, the greater the pressure at the Toe 1 area, and the greater the pressure on the heel area when putting down the foot; footprint pressure is mainly focused on the central line of

footprint, while if the bending of knee is smaller, the pressure will be distributed to both sides of the center line of the footprint.

(3) After the comparison for the same person at different walking speed, we summarize that: at normal speed, the minimum thigh-trunk angle is greater, suggesting lower thigh-lifting height, while thigh-lifting height increases during brisk walking. Pressure is mainly distributed among the first, second, third metatarsal areas and the heel area during brisk walking, and among the second, third, fourth and fifth metatarsal areas during normal speed walking. At normal speed the pressure is only present in the Toe 1 area and not significantly present in the other four toe areas; while the pressure in the Toe 1 area is greater than that of the normal walking speed, and the pressure is present in the other four toe areas, suggesting the increasing toe force during brisk walking.

References

1. Li, S., Shi, F.: Contrastive Analysis for Sports Biomechanics of Different Walking Postures. Journal of Tianjin Institute of Physical Education 22(6), 504–508 (2007)
2. Tang, C.: Primary Exploration for Application of Footscan Gait Analysis Systems in Footprint Examination. Criminal Technique (4), 18–20 (2008)
3. Zhang, Q., Meng, Z.: Contrastive Analysis for Normal Youth's Right and Left Plantar Pressure Distribution Characteristics. Chinese Journal of Chinese Engineering research and Clinical Rehabilitation 11(5), 889–892 (2007)
4. Bertsch, C., Unger, H., Winkelmann, W., et al.: Evaluation of early walking patterns from plantar pressure distribution measurements. Gait and Posture 19, 235–242 (2004)
5. Schmidt, R.: Amarked-based measurement procedure for unconstrained wrist and elbow motions. Journal of Biomechanics 32, 615–621 (1999)

Overall Design and Quality Control
of Equipment Based on Virtual Prototyping

Qingjun Meng and Chengming He

Armored Forces Engineering, 100072 Beijing, China
mengqingjun95@sohu.com

Abstract. The overall design scheme of equipment largely determines the equipment's quality level. The application of Virtual Prototype (VP) technology into the overall design process can find equipment's objection early, and to optimize equipment design, reduce design risk and improve equipment quality. This paper discusses the technical aspects of equipment quality control problems, analyze the feasibility of the use of the virtual prototype in equipment overall design, and established a relatively virtual prototype framework for equipment overall design.

Keywords: equipment, quality control, overall design, virtual prototype.

1   Introduction

The concept of equipment quality consists of not only combat characteristics, such as mobility, firepower, protection and communication, but also requires a good operational performance and operation adaptability[1], including reliability, maintainability, security, testing, safety, survivability, adaptation to military use and other things[2],.As more high-tech applications, and equipment complexity increases, the overall quality of equipment is not a matter of course to increase, and because the application of high technology use and the sharp increase in costs, making the affordability of equipment reduced, extend the development cycle [3].

The key time to the equipment is development phase for Quality control and reduce equipment life cycle cost, because advantages and disadvantages of the overall design of the equipment largely determines whether "eugenics", and the level of economic characteristics. Equipment Quality remains with systems engineering, and it is not only the question of the component units, but also of a "combination" process. The traditional sequential design process has been difficult to adapt to the overall design of today's complex equipment [4].

Virtual prototype is a carrier of parallel design idea and multi-disciplinary engineering design implementation and, based on virtual prototype design method can overcome many problems of traditional design [5], it can be a very short period of time for virtual prototype to "developed and tested "a variety of design, simulation results can be achieved through" test and modify "continuous coordination of the various properties and equipment, optimize the overall design of equipment. It is can

B. Liu and C. Chai (Eds.): ICICA 2011, LNCS 7030, pp. 704–711, 2011.

be said that the introduction of the virtual prototype to equipment design would improve the overall equipment quality and reducing life cycle cost and accelerate the speed of equipment development.

2   Quality Control of Equipment

Quality control of equipment is a complex issue, which not only related to management problems, such as personnel, rules and processes, but also related to technical aspects, for instance methods, tools, methods and so on. This paper, in technical level, discus the use of virtual prototyping equipment in the equipment design stage, and analyses the feasibility of it in quality control.

2.1   The Source of Quality Problems of Equipment

From the equipment design point of view, constitute the source of quality problems, including the following:

(1) Caused by the irrational design of the overall layout, and tactical and technical performance of equipment is difficult to play, such as "dead angle" problem.

(2) The performance of component units of equipment is not high, resulting in combat skills and characteristics of the equipment does not meet the requirements of tactical and technical performance, such as the shooting accuracy problems, communication distance problems, speed problems, and so on.

(3) Because of lack of advisement for using environment, such as vibration, temperature, noise, magnetic field, dust pollution, corrosion, wet conditions, result in failure frequently, and bring on safety issues.

(4) Because of improper maintenance of the design, maintenance, inspection and maintenance difficulties, such as pipe, line interface problems, "open rate", removable interference problems, maintenance and operation of space issues.

(5) Human adaptation problems caused by occupant comfort is not enough to consider operational difficulties, or could easily lead to operational error, prone to fatigue, likely to cause bodily injury and so on.

(6) Inadequate equipment design for the extent of the use of harsh conditions, leads to poor environmental adaptability, for example through the issue of poor, low survivability issues, cylinder pollution problems confined poor.

(7) The irrational support capability caused by the allocation of resources for protection and support system delay, and caused great waste of resources.

2.2   Equipment Quality Improving

Improve the quality of equipment should be from two aspects, one is to improve the quality of its functional units, and bring forward request reasonable for each functional unit, such as functional parameters, performance indicators, technical parameters; the other is to improve the rationality of the overall design and equipment, correct handling of the functional units (various properties) between the intervention, coordination, cooperation and other relations, to optimize the combination of purposes, including equipment layout of the overall structure, comprehensive functions, the basic performance and technical parameters. To achieve these two points, using traditional design methods difficult to do, should be actively used virtual prototyping technology for virtual working and virtual use under the prediction conditions to find bottlenecks restricting the quality of equipment, and the corresponding solution.

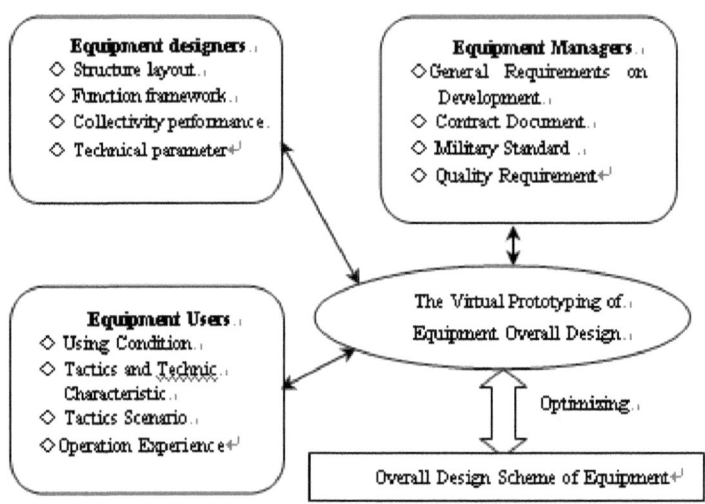

Fig. 1. Process of improvement for equipment quality

In addition, the virtual prototype provides a consultation and communication platform for analysis and discussion of correlative peoples to revise and improve equipment quality. Overall design of equipment can take advantage of the virtual prototype simulation results, and continuously adjust and optimize the overall design; equipment managers can use the Virtual Prototype equipment, keep track of the quality of the intervention and related advice; equipment used may be based on equipment use and operations needs and experience make recommendations to improve equipment. The process of equipment to improve the quality is shown in Figure 1.

3   The Virtual Prototyping of Equipment Overall Design

3.1   The Overall Design of Equipment

Overall design of equipment is on the basis of the information form the component units of equipment, and every component unit has its independence function and certain performance. From the physical composition, equipment has various levels of composition, including systems, subsystems, components, components, parts, but the overall design does not related to the lowest level elements, the components level so far; from the quality point of view, can be divided into general function, specific features, performance, technical parameters, assembly, processes, materials and other aspects of the overall design, and the overall design is concerned with the technical parameters of certain functions and performance of the components[6-8].

From the input and output relations of overall design, their input should be two aspects, one is the features, performance and technical parameters of the component (parts), the other is the relationship between the various components, including conflict, collaboration, interference, and combination, the output should be combined to optimize the overall design of the equipment, specifically including the overall layout of the coordination between the various parts, as well as parts of its technical parameters of the adjustment[9-12].

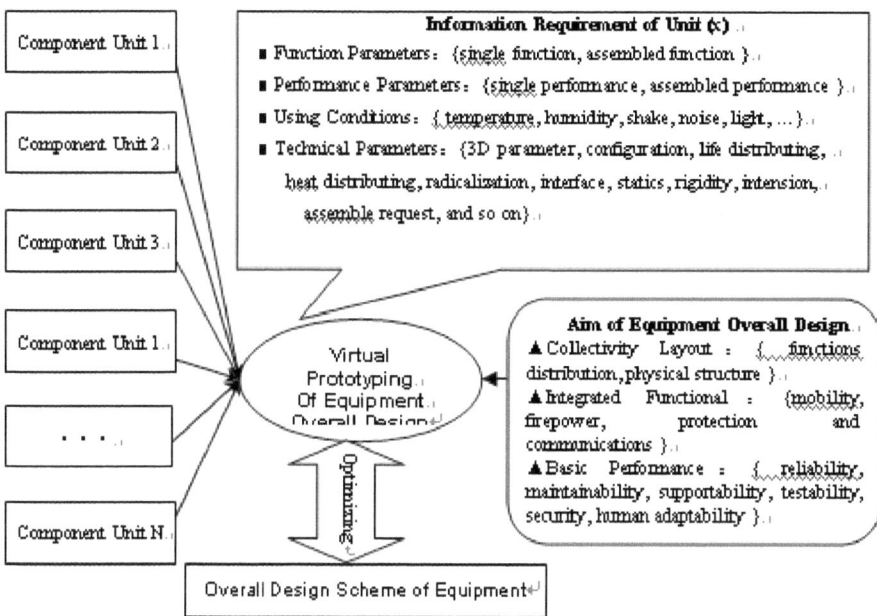

Fig. 2. Overall design of equipment and optimization process

The internal components, structure, working principle, processing technology, materials and other aspects of component units can be shielding like a "black box", so that the data and information processing of the overall design is simple and effective. That overall design process of equipment is not concerned about the internal parts of the information, they focus on parts of the external information, such as component units of functionality, including its own function, and composition with other elements of the common functions; component units of performance, including their own performance common components with other elements of performance; component units of the use of conditions, including temperature, humidity, vibration, noise, light, electromagnetic and other requirements; component units of the technical parameters, including three-dimensional shape and structure parameters, life distribution curve, the work generated when the heat distribution, vibration, noise, radiation, and other component units of the interface force distribution, stiffness, strength, assembly requirements, and other technical parameters.

The general goal of overall design is to affect the level of the integrated elements of equipment quality, such as the overall layout of equipment, integrated functions, including mobility, combat, protection and communication capabilities; basic performance, including reliability, maintainability, supportability, testability, security, human adaptability, support resources and protection systems. Overall design of equipment and optimization process is shown in Figure 2.

3.2   Equipment Virtual Prototyping

Virtual Prototype equipment provides an integration platform for the realization of different aspects of the overall design, which is a simulation systems integration based on collaborative environment for the overall design. Virtual Prototyping framework of equipment overall design is shown in Figure 3.

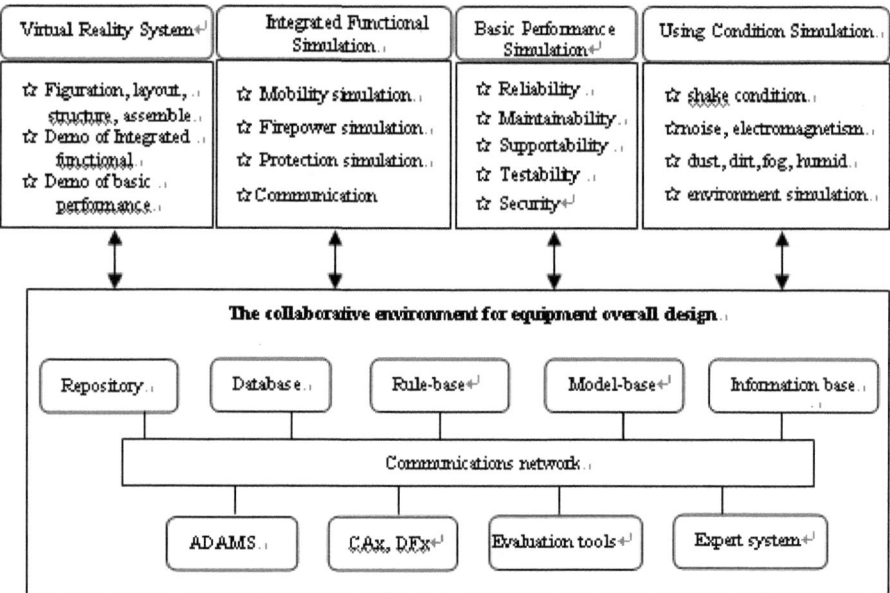

Fig. 3. Virtual Prototyping framework of equipment overall

3.2.1   The Collaborative Environment for Equipment Overall Design

Overall design of equipment relates to personnel in all aspects by staff of equipment design, equipment use, and equipment manage, even in the equipment designers also have their own different division of labor, so the overall design and equipment virtual prototype needed to provide a collaborative platform or environment that allows communication between different professionals.

Collaborative environment of overall design is the of communication networks as channels for information exchange and integration of the various kinds of "base" and "Tools." "Base" category, including knowledge base, database, criteria base, model base and information base, knowledge base which includes experience in the design engineering staff, use of personnel experience in the operation, troops and equipment management experience; database contains various data required for design analysis, expected, distribution, calculation; guidelines base contains a mature design criteria for design of reliability, maintainability, supportability, testability and security; model base contains the equipment three-dimensional model, layout model, functional parameter calculation model, performance model parameters, technical parameters calculation models; information base contains a variety of personnel in the overall design process, the views and suggestions, as well as the overall design and equipment changes and feedback information.

Virtual Prototyping of equipment overall design utilizes a variety of tools, including CAx tools such as CAD, CAE, CAM, etc.; graphics tools; mechanical system dynamic simulation software ADAMS; DFx tools such as performance-oriented (DFP), surface to the assembly (DFA), design for manufacturing (DFM), for analysis (DFF), for test (DFT), for quality (DFQ), for cost (DFC), for service (DFS), etc.; assessment tools, equipment, the overall design of a comprehensive assessment of the merits; expert systems and other smart tools that can analyze the overall design and equipment problems, and optimize the overall design scheme proposed recommendations.

3.2.2   Virtual Reality System

Virtual reality system is equipped with visual design and optimization tools, but also is platform to shape the structure and layout of the display, feature presentation, performance testing. Through the virtual reality system can validate rationalization of the structure and layout of the distribution to reduce space and abnormal "intervention"; can taste a variety of basic functions, such as the operation aiming, driving, launch, there is an intuitive experience; can be descript and testing of equipment performance, such as failure, repair, maintenance, use protection, equipment protection. On this basis, the designer can either design mistakes that can also inspire new design ideas, and can use the party to adopt various parties in charge feedback, eventually forming a practical and constructive design scenario of the overall design for the equipment.

3.2.3   Integrated Functional Simulation

Equipment integrated function mainly refers to the mobility, firepower, protection and communications, 4 major functions, which are comprehensive characteristics reflection of equipment system. For example, the mobility of equipment is a combined effect of dynamic system, transmission system, propulsion system (running system), attitude control system; fire strike capability is combined effect of ammunition systems, weapons systems, fire control system; protection system is a stealth system, active protection system, passive protection systems (armor) the combined effect; communication capability combined effect of early warning systems, detection systems, GPS navigation systems, information processing system, command and control system. The four functions have an interdependent relationship, such protection is closely related to ability and mobility, firepower capability relates to early warning, detection, and location.

Integrated functional simulation of equipment is through simulating model reflects the internal composition of each functional relations, and the linkages between functions, and through the change of conditions and the given input to find functional defects of equipment, find the "short board" of integrated functions.

3.2.4   Basic Performance Simulation

The basic performance of equipment constitutes the basis of quality factors of equipment, including reliability, maintainability, supportability, testability, security, human adaptability, protection of resources and support systems and other aspects, among which the reliability, maintenance , supportability (RMS) is the key.

In equipment use, the incidence of failure is inevitable, to solve the problem of reliability simulation is designed to reduce failure rates by controlling the failure within a reasonable range, and utilizes failure prediction technology to find what result in failure frequently, and makes input for logistical support needs and equipment maintenance requirements.

The task of maintenance simulation is to simulate when the equipment failure (corrective maintenance), or to prevent equipment failure (preventive maintenance), if the various types of maintenance work can be carried out smoothly. According to the structure and equipment, layout, and system structure, it can predict the accessibility of equipment, operational and maintenance difficulty, time required for equipment maintenance work, and so on.

The task of supportability simulation is to simulate the support process and predict whether support resource is reasonable. Equipment charge (electricity), filling (liquid), add (oil), hanging (shells) and other combat preparations largely determines the equipment's combat preparation time, and the usual ease of maintenance. Building virtual prototype simulation model based on can be used to precise support conditions. In addition, the supportability simulation can be used to analysis for support scenario, support resources, and support systems. Also at design stage, according to readiness, life cycle cost, difficulty and other support constraints, reflects the irrational factors of equipment quality and put forward proposals to improve overall design.

4   Conclusion

The overall design of equipment is the key to the quality of equipment, the introduction of the virtual prototype at development phase of equipment for the equipment overall design, largely to reduce the risk of equipment design, and provide a good tool for the quality control of equipment. But it is still a long way to go for the virtual prototype of equipment overall design, not only because of the virtual prototyping technology is still in the development stage, but also because the complexity of the equipment design. All of those need requiring the wisdom of the parties who play for the virtual prototyping of equipment overall design.

References

1. Du, M.-d.: Project architect should be prepared to design the control. Nonferrous Metals Design 27(4), 22–25 (2000)
2. Tang, H.: Application of multidisciplinary and optimization technology in the field of aerospace design. Dual-use Technology and Products (3), 3–5 (2007)
3. Li, S.-l., Yao, S.-w., Yan, Q.-d.: Virtual prototype integration design platform of vehicle transmission system. Computer Integrated Manufacturing Systems 26(7), 245–248 (2009)
4. Yang, G.-l., Chen, Y.-s.: Research on the Application of Artillery Virtual Prototyping framework. Journal of Ballistics 18(1), 51–54 (2006)
5. Yu, L.: Modern product design and concurrent engineering. Jiangxi Nonferrous Metals 13(4), 30–33 (1999)
6. Shirolkar, P.A.: Designing Supportability into Software. Master's Paper, Massachusettts Institute of Technology, 112–116 (2003)
7. Weaks, H.L., James, D., Barrett, A.: Demonstration of the Integrated Supportability Analysis and Cost System. In: Proceedings of the 1997 Winter Simulation Conference, pp. 61–64 (1997)
8. Wang, T.-m.: Simulation-based method of ship equipment development. Ship Science and Technology (1), 9–11 (2001)
9. Li, H.-q., Guo, B.-w., Xu, H.: Virtual Prototype Modeling and Simulation Analysis of Servo Based on ADAMS and SIMULINK 21(21), 6886–6888 (2009)
10. Gao, Q.-r., Sun, H.-y., Li, H.-h., Zhou, X.-l.: Research on machine tool design and simulation based on virtual prototype. Journal of Machine Design 26(7), 16–18 (2009)
11. Zhang, Q.-F., Gu, J.-n., Lv, X.-F., Zhang, P.-F.: Research on virtual prototypes and simulation of a new manipulator clamping device for NC Machine 10, 160–163 (2009)
12. Yang, Y.-l., Zliu, Q.-p.: Kinetics Research and Simulation of CVT Based on Virtual Prototype 36(1), 91–94 (2009)

Simulation System of Air Transport Route Planning

Ming Tong, Hu Liu, and MingHu Wu

School of Aeronautics Science and Engineering,
Beijing University of Aeronautics and Astronautics,
Beijing, China
tm36051207@126.com

Abstract. Air transport is enjoying a rapid development in the modern society due to its advantages on convenience, comfort and speed. However, these advantages are followed by increasingly complex problems: there is higher demand on scientific and timeliness when considering significant designing and decision making in construction and production of air transport. The research on the problem of air transport become possible by virtue of the rapid development of computer simulation, which makes the development of air transportation simulation system become a urgent need nowadays. The most important issue of air transportation simulation is to solve the problem of air transport route planning, which paves roads for the further simulation of the whole process of air transport. Under investigation in this paper is a method in constructing a simulation system for air transport route planning. Though this method, complex tasks of air transport are simplified. Furthermore, domestic air transport system is simulated and its feasibility is verified.

Keywords: Air transport, emulation, route planning.

1 Introduction

Along with economic development and social progress, Air transport is enjoying a rapid development in the modern society due to its advantages on convenience, comfort and speed, and it's development has became one of the factors which have momentous influence on our national economy. However, with the rapid development, the problems of modern air transport are increasingly complex [1]. First, contemporary air transport owns large span in both time and space, the task of long-time and far-distance air transport is growing rapidly, the traffic and flight double in proportion; then in addition to simple passenger / cargo transport mode, the ratio of special tasks takes a larger part in total transport; in the same time, additional temporary, emergency air transport tasks become critical facing kinds of emergencies. Coupled with the gradual opening up of China's low-altitude areas, the rapid development of general aviation is increasing huge pressure on China's aviation environment. The combined effect of these factors will affect the advancement of China's air transport industry, thereby affecting people's lives as well as national development. Therefore, mastering air transport law and improving the efficiency of air transport air are the keys issues to overcome the problems as to promote the development of air transport.

B. Liu and C. Chai (Eds.): ICICA 2011, LNCS 7030, pp. 712–718, 2011.

As the rapid development of science and technology, computer simulation [2] has become the important and indispensable tool of modern theory of production and validation research, program evaluation, decision-making process in support. The production process and transport of current air transport has a variety of state involving the complexity of interrelated factors, the air construction as well as the major design an d decision-making of production demands increasing scientificalness and timeliness. Therefore, the development of air transport simulation systems becomes an urgent need. For the most important issue of the air transport simulation is the route planning, which means reasonable simulating of air transport flight arrangements, extracting and classifying the complex task of air mission, at last forming a simple, regular planning. By reasonable planning, the complex situation of air transport can be simplified; the route planning can provide the further simulate the whole air cushion process.

This article will explore a simulation building of air transport route planning system, simplify the complex air flight planning in a sense, and through simply simulating the domestic air transport system verify the system's theory.

2   Key Technologies

2.1   Route Planning

Flight missions, includes the take-off from the departure airport, the flight in accordance with a predetermined route, while stopping by a series of task demand points or airports, and landing in the airport of destination. The task of route planning is to arrange the flight routes and all the passing by airports into a continuous curve in a sequence, and describe in a series of critical points [3]. Generated trajectory planning path is the premise to test the conflict between flights.

By analyzing the flying through the air transport instance, route planning can be divided into two sections: the flight planning section, ground planning section.

Flight segment is composed by flight airlines planning and flight profiles planning. Flight airlines' planning is used to determine the horizontal flight line program, while flight profiles planning is used to determine the vertical flight line program. According to the abstract treatment of the flight routes, flight airlines planning can be seen as splicing of "Airline" (the airport - the airport based element). Single airline departure, has no additional stops at other airports but the purpose and destination airports (non-stop passing except in the case); Airline information includes a series of key waypoints (such as navigation units, reporting points, etc.) too. Such a connection from the airports, waypoints as the critical points forms the unique horizontal flight path. Depending on whether it's standardized (whether is existed, and could be investigated clearly) flight airlines are divided into standard airline and temporary airline. Standard airline can be directly edited related to civil aviation routes by querying the data, and stored in the database. Temporary airline is added based on the needs of specific tasks, and not stored in the database. After gradually adding flight airlines, it is time to edit the vertical flight profile planning for a single airline. Aircraft mission profile[4], including typical tasks such as taking-off, climbing, descending, accelerating, decelerating, adjusting, airdropping and so on. The system remove the complex cases profile, keep taking-off, climbing, cruising, descending, and landing as the five basic profile sections, and distinctively parameterize each section. The process of setting the mission profile is the free selection and combination of the five basic profiles, while the profile's parameters should ensure a continuous transition between the profile section, no height or speed mutation. Ground planning is simplified to three parts: sliding segment, ground mission segment,. Taking the fact that content of this paper is route planning into account, coupled with the ground plan structure itself is extremely complex as well, the sections are used to quantify by the time, in other words referencing to empirical data, set the default time length of the sliding segment, ground segment, preparing before taking–off segment, at last get the time arrangement of the ground planning.

After the completed the route planning, the system figures out the time arrangements between planes and internal one each aircraft, then get the global air mission time scheduling program according to the whole plan. Scheduling table can be checked by looking at every step, and re-editing the scheduling table makes it possible to adjust the time arrangement and make further changes to the plan.

2.2   Visual Display

During flight simulation, the dynamic display depicts the dynamic behavior of the system, enhances the reality and realism of simulation. The system uses GIS-based, two-dimensional display technology. The display module is mainly used for displaying the visual flight aircraft in the corresponding position according to the state of aircraft. By monitoring the view, it is convenient to detect the busy place and time of the global routes, find where a conflict occurs among the routes' plan. The establishment of the visual display subsystem will help planners to analyze route planning defects, and make it better to propose new method to optimize route planning.

2.3   Database Building

Combined with the organizational characteristics of GIS data in the software development process and considered the variety of data of actual air transportation, basic database sheet should include the following: aircraft data table, airport data table, airline point data table, airline data table and other date information. As the aircraft, airports, routes and other entities involved in a number of parameters, it is necessary to simplify the characteristics of each abstract based on air transport route planning simulation. For example, by simplified airline data table uses name as the primary key, other specific information includes the departure airport, destination airport, flight route, route attributes, and airline point and so on.

3   System Architecture and Running Instance

3.1   System Architecture

Because of the real conditions in air transportation influenced by the airline companies, air traffic control conditions, physical environment and many other factors, the corresponding complex human factors involved between them are uncontrollable and unpredictable, and ignored here. The air transportation airline planning simulation system consists of four parts: Airline input planning module, simulation module, visualization module, and database module. Among them, the simulation module is system's core. Planning module provides the input for the simulation task, the visual display module tricks and display the calculation results of the entire simulation process, basic database module provides model and date support for the simulation system. The architecture is shown in Figure 1.

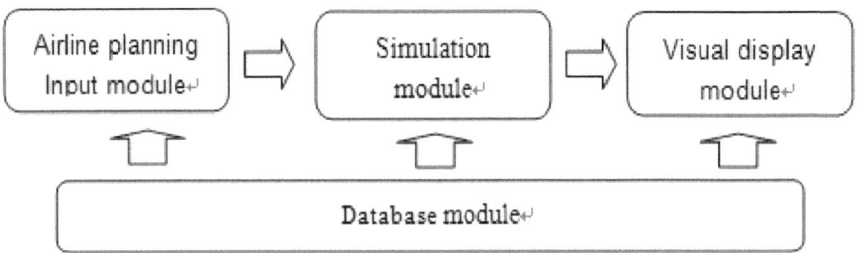

Fig. 1. System Architecture

3.1.1   Airline Planning Input Module

The function of the airline planning input module is to input the task scenario setting of the overall air transport simulation. According to the route planning model ana-lyzed and simplified above, through a series of interface operation, input the setting information of the task-related aircraft and airline planning, and ultimately determine the aircraft flight scheduling.

3.1.2   Simulation Module

Simulation module is the core component of the air transportation airline planning simulation system. It contains the calculate model and the interface between each module.

The module simulation models uses simplified flight equation model (which simpl-ify the flight process as taking-off, climbing, cruising, descending, landing and other parts), and adopts sub-iteration, so the calculation results can ensure sufficient accura-cy. The model calls the relevant physical and mathematical models stored in the data-base module, then calculates with parameters of airline planning input model, and store the calculated data. In the same time, the results are synchronized to the visual display module to achieve the overall simulation monitoring.

3.1.3   Visual Display Module

Visual display module is mainly responsible for visualization simulation of the air transport planning system, mainly implemented on the main map. Main map control uses a layered control method [6], which means that airports, airlines, aircrafts, borders, boundaries and other elements are placed in different layers. In operation, the visual display module functions are:

According to the operation of adding, modifying, deleting of the database, synchronize the map physical layer;

According to real-time information of the aircraft attitude (such as location, weight, height, speed) received from the calculation module, adjust the position of the aircraft layer on the map, and draw the flight- height chart;

Monitoring the simulation data, warn when exception occurs, and if necessary, stop the simulation process.

3.1.4   Database Module

Database module normalizes the query, call, addition, deletion, storage and other operations, provides interface to use data, improves the efficiency and reliability of data using. This module includes basic model database (aircraft model base, airline model base, airline point model base, the airport model base, etc.) and simulation database (airline planning library, the simulation structure libraries, etc.), provide data for the overall system simulation.

3.2   Running Instance

Here use the air transport route planning simulation software based on system mentioned above as an example to show how the system implementation and operation processes.

The software uses VS2010 as the development platform, commercial software SQLITE as the database, and MapX as the map control. Here briefly outline the MapX. MapX [5] is a programmable control based on ActiveX (OCX) technology, uses the consistent data format with MapInfo Professional map, and can achieve most of the MapInfo Professional features. MapX provides software developers with a fast, easy, and powerful map-based component element. Also it supports a variety of high-level language. The system uses MapX, which provides a set of powerful tools for complex business map, data visualization and GIS functionality. MapInfo Professional can be connected by local and server-side database, create maps and charts to reveal the true meaning behind the ranks of the data. MapX also has a strong advantage of data binding features. Through binding data in the database and MapInfo in the connection, MapX makes the map object and relational database data item corresponded to each other. After data binding, it is possible to achieve the query data on the map.

Now simulate a simple Airline planning task. The mission is about flight simulation which involved 30 airplanes and three airlines. The following are the rendering Figures of general system interface, the task parameters setting interface, and the simulation display.

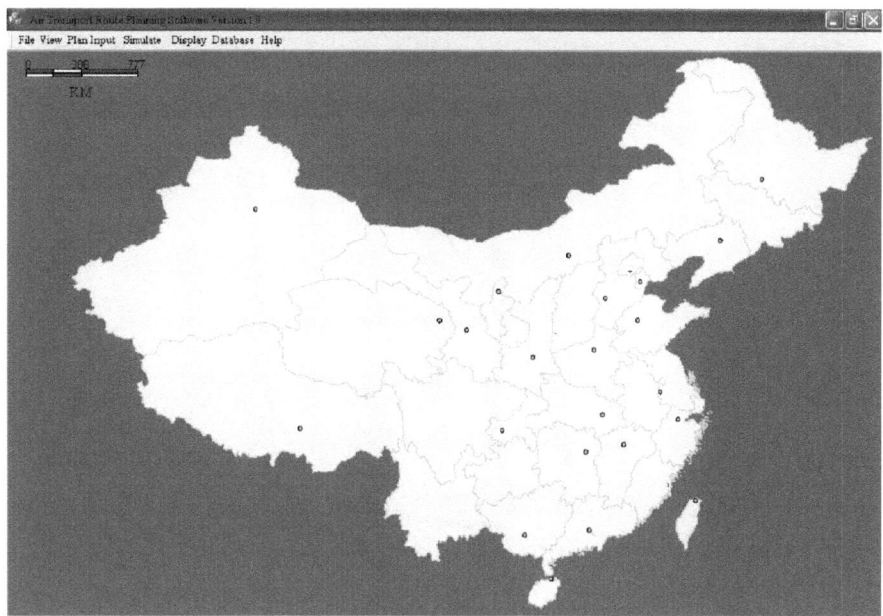

Fig. 2. Software main interface

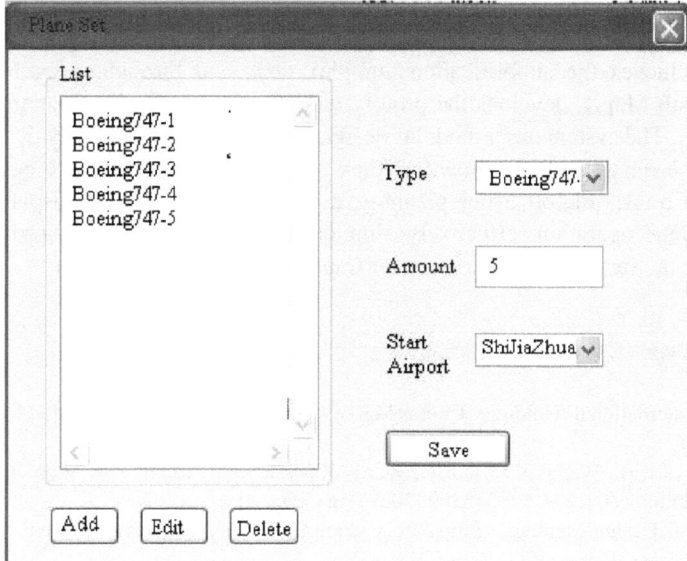

Fig. 3. Task parameters setting interface

Simulation results show that the system successfully enters air transportation planning of multi vehicles and achieve the tracking and display of air transport, initially validates the feasibility of the system.

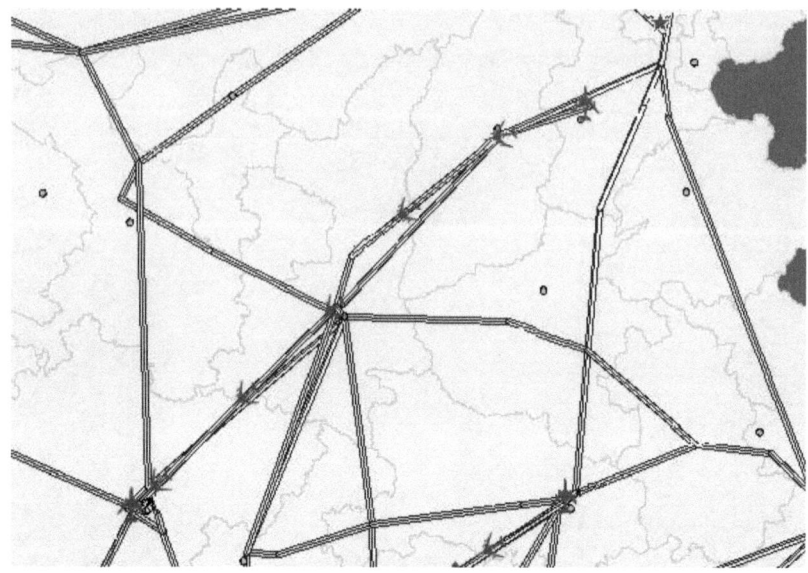

Fig. 4. Simulation display

4   Conclusion

This article explores an air transport airline planning simulation system construction method, achieves the simplification complex task, and through a combination of VS2010 with MapX, develops the prototype of air transport airline planning simulation system. The system has a modular design, which be extended easily.

Prototype running results show that the system can be convenient to complete the input of air transportation airline planning, and achieve a convenient display to monitoring airlines operation, effectively simulate the air transportation, provide some support for the further comprehensive air transport simulation.

References

1. AVIC International Holding: China Market For Civil Aircraft. 2010-1019 (2010) (in Chinese)
2. Chen, Z., Li, B., Wang, X., Xiao, T., Wang, X.: Study on Discipline of Simulation Science and Technology 21(17), 5265–5269 (2009) (in Chinese)
3. 《Aircraft Design Manuals》 Preparatory Committee: Aircraft Design Manuals 7. Aviation Industry Press, Beijing (2000) (in Chinese)
4. Hong, Z., Xu, L.: Domestic Flight Schedule Simulation Using Service Model 19(2), 77–82 (2010) (in Chinese)
5. Yin, X., Zhang, W.: Visual C++ Based MapX Development Technology. Metallurgy Industry Press, Beijing (2009) (in Chinese)
6. Zhai, Y., Chen, H., Wu, C., Duan, X.: MapX-based UAV Navigation System Developing 16(11), 1626–1628 (2008) (in Chinese)

A Corpus Callosum Segmentation
Based on Adaptive Active Contour Model

Jia Tong, Yu Xiao Sheng, Wei Ying, and Wu Cheng Dong

School of Information Science & Engineering,
Northeastern University,Shenyang, China
jiatong@ise.neu.edu.cn

Abstract. Corpus callosum is very important for interhemispheric communication. A new corpus callosum segmentation scheme based on MR image is proposed in this paper. Firstly, the live-wire algorithm is adopted to compute the initial contour of ACM, which have small user interaction time and higher flexibility. Secondly, based on an improve GAC model, this scheme takes Neumann boundary condition as the termination condition of the curve evolution. Through test and comparison, the scheme's high performance for the CC segmentation shows much promise for further clinical applications.

Keywords: corpus callosum segmentation, active contour model, neumann boundary condition, GAC model.

1   Introduction

The corpus callosum (CC) is one of the important brain tissues and plays a special physiological function. Once CC lesions occur, there will be a variety of clinical manifestations which are easily neglected by people. Considering the importance of the CC for interhemispheric communication, one of the goals of clinical research has been to find the size and shape changes taking place in CC [1].

An example image is given in Fig. 1. CC is located in the centre of the image. CC is of interest to medical researchers for a number of reasons. The size and shape of CC have been shown to be correlated to sex [2], age [3] and some diseases (such as

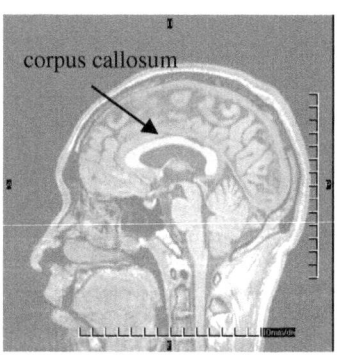

Fig. 1. CC MR image

B. Liu and C. Chai (Eds.): ICICA 2011, LNCS 7030, pp. 719–725, 2011.
© Springer-Verlag Berlin Heidelberg 2011

dyslexia, epilepsy, multiple sclerosis etc) [1,4]. In order to find such correlations in living brains, magnetic resonance (MR) is regarded as the best method to obtain cross-sectional area (and shape) information of CC.

CC is the largest fiber bundle connecting the left and the right cerebral hemispheres in the human brain. Since the higher cognitive functions of the brain are highly affected by the impaired communication between the hemispheres, several studies [5-8] have proposed to analyze the CC. In [5,6], the CC had been traced from the midsagittal MRI slice. Statistical difference analysis was applied to find out which part in the CC contributes significantly to identification of autistic brains. Chung et al. [7] applied a voxel based morphometry approach using a freely available public domain software package to spatially normalize the midsagital MRI slice to a common stereotactic space. Plessen et al. [8] found the midsagittal CC mean shape of both dyslexic and normal brains and noticed that the 2D CC body length can discriminate between the dyslexic and normal subjects.

However, most of these studies suffer from the following problems. Most of these studies are based on either a global shape analysis of the CC, which fails to capture the regional shape deformation, or use a regional representation that is not clinically meaningful.

In this paper we propose an interactive CC segmentation method which consists: (1) because the CC has the higher curvature of the shape, we adopt live-wire algorithm to compute the initial contour of ACM. (2) We compute the edge energy which can make curve evolve along the second order derivative in the image gradient direction. (3) We take the Neumann boundary as the termination conditions of the curve evolution and make it stop at the boundary of CC. Through tests, the method which is proposed in this paper obtains the accuracy segmentation result.

2   GAC Model

Geodesic active contours (GAC) were introduced as a geometric alternative for "snakes" [9]. Snakes are deformable models that are based on minimizing energy along a curve. The curve deforms its shape so as to minimize"internal" and "external" energies along its boundary. The internal part causes the boundary curve to become smooth, while the external part leads the curve toward the edges of the object in the image.

In [10], a geometric alternative for the snake model was introduced. The method works on a fixed grid, usually the image pixels grid, and automatically handles changes in the topology of the evolving contour. The GAC functional is

$$s[c] = \int_0^1 (\alpha + \tilde{g}(c)) |c_p| dp \qquad (1)$$

It may be shown to the equivalent to the arc length parameterized functional.

$$s[c] = \int_0^{L(c)} \tilde{g}(c) ds + \alpha L(c) \qquad (2)$$

Where, $L(c)$ is the total Euclidean length of the curve. One may define

$$g(x, y) = \tilde{g}(x, y) + c, \quad \text{in which case} \quad s[c] = \int_0^{L(c)} g(c) ds$$

One minimization of the modulated arc length $g(c)ds$, The Euler-Lagrange equations as a gradient descent process are

$$\frac{dc}{dt} = (g(c)k - <\nabla g, \vec{N}>)\vec{N} \tag{3}$$

3 CC Segmentation Methods

3.1 Initial Contour Compute

Because CC has the higher curvature of the shape, we adopt live-wire algorithm to compute the initial contour of ACM. The live wire algorithm is a user-steered segmentation method for 2D images based on the calculation of minimal cost paths by dynamic programming and graph search algorithm [11], which can be described as follow.

$$l(p,q) = \omega_z \cdot f_z(q) + \omega_D \cdot f_D(p,q) + \omega_G \cdot f_G(q) \tag{4}$$

Where $l(p,q)$ is the local cost on the directed link from p to q, $f_z(q)$ is the Laplacian Zero Crossing function at q, $f_D(p,q)$ is the gradient direction from p to q, and $f_G(q)$ is the gradient magnitude component function at point q. In (5), each component function is weighted with a weighting factor, such as ω_z, ω_D and ω_G.

The Laplacian Zero-Crossing function is a binary function normally defined as follow.

$$f_z(q) = \begin{cases} 0 & if \ I_L(q) = 0 \\ 1 & if \ I_L(q) \neq 0 \end{cases} \tag{5}$$

Where, I is Laplacian image at point q. The next local cost component function is gradient magnitude function f_G, is defined as follow.

$$f_G = \frac{\max(G) - G}{\max(G)} = 1 - \frac{G}{\max(G)} \tag{6}$$

The gradient magnitude is defined as in (9). Where $I_x = \partial I / \partial x$ and $I_y = \partial I / \partial y$.

$$G_{mag} = \sqrt{I_x^2 + I_y^2} \tag{7}$$

The last local cost function is the gradient direction function, f_D. The gradient direction cost is to penalize sharp changes in boundary direction thereby acting as a smoothness constraint on the boundary, which is defined as in (8).

$$f_D(p,q) = \frac{1}{\pi}\left\{\cos\left[d_p(p,q)\right]^{-1} + a\cos\left[d_q(p,q)\right]^{-1}\right\} \tag{8}$$

Where $d_p(p,q) = \vec{v}_p \cdot L(p,q)$, $d_q(p,q) = L(p,q) \cdot \vec{v}_q$,and

$$L(p,q) = \begin{cases} q-p & \text{if } \vec{v}_p \cdot (q-p) \geq 0 \\ p-q & \text{if } \vec{v}_p \cdot (q-p) < 0 \end{cases} \tag{9}$$

The initial contour compute result based on live-wire algorithm is showed in Fig.2, which have small user interaction time and higher flexibility.

3.2   CC Segmentation Based on ACM

In order to converge to CC edge accurately, edge energy is computed which can make curve evolve along the second order derivative in the image gradient direction. Set original image $I(x,y)$, I_x and I_y to be the first derivative in the horizontal and vertical direction. The image gradient direction vector is defined as:

$$\vec{\xi}(x, y) = \frac{\nabla I}{|\nabla I|} = \frac{\{I_x, I_y\}}{\sqrt{I_x^2 + I_y^2}}$$

The other vertical vector is defined as:

$$\vec{\eta}(x, y) = \frac{\nabla I}{|\nabla I|} = \frac{\{-I_y, I_x\}}{\sqrt{I_x^2 + I_y^2}} , \text{therefore, } \langle \vec{\xi}, \vec{\eta} \rangle = 0 .$$

Let $I_{\varepsilon\varepsilon}$ to be the second derivative in the image gradient direction, $I_{\eta\eta}$ is the other vertical vector. According to $I_{\varepsilon\varepsilon} = I_{\varepsilon\varepsilon} + I_{\eta\eta} - I_{\eta\eta} = \nabla I - I_{\eta\eta}$, the energy function is defined as follow.

$$E = \int_c \langle \nabla I, \vec{n} \rangle ds - \iint_{\Omega_c} k_I |\nabla I| dxdy \tag{10}$$

In equation (10), k_I is curvature, $\vec{n}$ is unit direction vector, and Ω_c is area inside C.   Through the variation method for solving equation (10), when $I_{\varepsilon\varepsilon} \vec{n} = 0$, the Euler-Lagrange equation extremum is computed.

3.3   Neumann Boundary Condition

Consider the initial contour of ACM, the homogeneous Neumann boundary condition is adopted.

$$\partial u / \partial t = f(u) + \partial^2 u / \partial x^2 + \partial^2 u / \partial y^2 \tag{11}$$

Where $u = u(t,x) : [0,T] \times D \rightarrow R, x = (x, y)$ is the dependent variable , $f=f(u)$ is a nonlinear funtion of u, and where $D \subseteq R^2$ of the space variable has a boundary ∂D . Then for any $t \in [0,T]$,

$$\forall x \in [0,1]: \partial u / \partial y(t; x, 0) = \partial u / \partial y(t; x, 1) = 0$$
$$\forall y \in [0,1]: \partial u / \partial x(t; 0, y) = \partial u / \partial x(t; 1, y) = 0$$

$$(12)$$

By the Neumann boundary condition, we can make the curve rapid convergence and make it stop at the boundary of object.

4   Experiments Results and Analysis

The proposed scheme has been trained and validated on a clinical dataset of MR scans. In Fig.2, (a) is the select point process, (b) is the initial contour display result. In Fig.3, (a) is the original MR image, (b) is the final CC segmentation result. In order to effectively analysis the segmentation results, GVF snake model   has been tested by us. The test results are shown in Fig.4. Based on comparison, there are better algorithm speed and segmentation accuracy in our scheme.

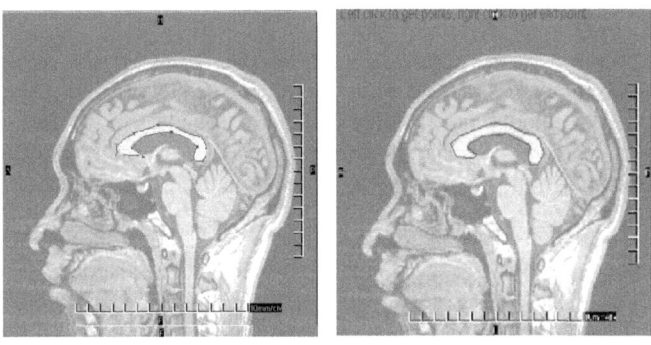

(a)   Select point process   (b) Initial contour compute result

Fig. 2. Initial contour compute result

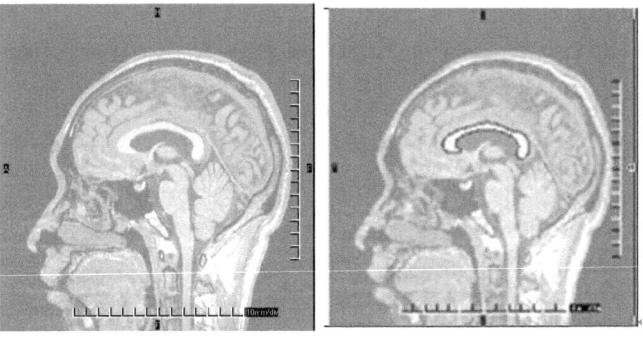

(a)   The original image   (b) CC segmentation   result

Fig. 3. CC segmentation result based on our scheme

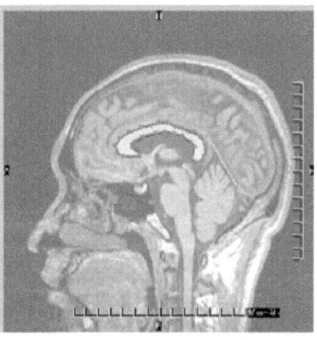 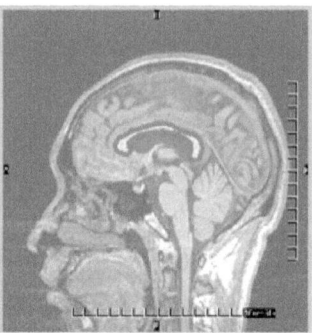

(a) Select point process   (b) CC segmentation result

Fig. 4. CC segmentation result based on snake model

5   Conclusion

This paper proposed a new approach to CC segmentation. Firstly, we adopt live-wire algorithm to compute the initial contour of ACM. Secondly, we segment CC based on an improved GAC model and take the Neumann boundary as the termination condition of the curve evolution. Finally, through test, the scheme's high performance for the CC segmentation shows much promise for further clinical applications.

References

1. Manuel, F.C., Andrew, E.S.: Dyslexia Diagnostics by Centerline-Based Shape Analysis of the Corpus Callosum. In: International Conference on Pattern Recognition, pp. 261–264 (2010)
2. Dong, H.-h., Guo, M.-x., Zhang, Y.-t., Fu, Y., Shi, H.-l.: Sex Differences in Brain Gray and White Matter in Healthy Young Adults: Correlations with Resting State ALFF. In: Proceedings of the 2010 3rd International Conference on Biomedical Engineering and Informatics (BMEI 2010), pp. 560–563 (2010)
3. Vatta, F., Mininel, S., Colafati, G.S., D'Errico, L.: A Novel Tool for the Morphometric Analysis of Corpus Callosum: Applications to the Diagnosis of Autism. Biomedical Sciences Instrumentation 45, 442–448 (2009)
4. Elnakib, A., El-Baz, A., Casanova, M.F., Farb, G.G.: Image Based Detection of Corpus Callosum Variability for More Accurate Discrimination Between Autistic and Normal Brains. In: Proceedings of 2010 IEEE 17th International Conference on Image Processing, pp. 4337–4340 (2010)
5. Hardan, A.Y., Minshew, N.J., Keshavan, M.S.: Corpus callosum size in autism. Neurology 55, 1033–1036 (2000)
6. Kontos, D., Megalooikonomou, V., Ge, J.C.: Morphometric Analysis of Brain Images with Reduced Number of Statistical Tests: A Study on the Gender-Related Differentiation of the Corpus Callosum. Artificial Intelligence in Medicine 47, 75–86 (2009)
7. Chung, M.K., Dalton, K.M., Alexander, A.L., Davidson, R.J.: Less White Matter Concentration in Autism: 2D Voxel-based Morphometry. Neuroimage 23, 242–251 (2004)

8. von Plessen, K., Lundervold, A., Duta, N., Heiervang, E., Klauschen, F., Smievoll, A.I., Ersland, L., Hugdahl, K.: Less Developed Corpus Callosum in Dyslexic Subjects – A Structural MRI study. Neuropsychologia 40, 1035–1044 (2002)
9. Kass, M., Witkin, A., Terzopoulos, D.: Snakes: Active Contour Models. Int. J. Comput. Vis 1, 321–331 (1988)
10. Caselles, V., Catte, F., Coll, T., Dibos, F.: A Geometric Model for Active Contours. Numer. Math. 66, 1–31 (1993)
11. Mortensen, E.N., Barrett, W.A.: Intelligent Scissors for Image Composition. In: Computer Graphics (SIGGRAPH 1995), pp. 191–198 (1995)

3D Simulation System for Subway Construction Site

Zheng Zheng and Jing Ji

Computer Science and Technology Department,
Tangshan Teachers' College Tang Shan, China

Abstract. With the development of computer technology, VRT has been commenced to apply to system simulation field, and extensively used in the aspects
including entertainment, medical treatment, education and training, engineering
and construction etc. The paper introduces the VRT and some related platform
software, discuss the train of thought and process for system exploitation, and
especially go into the details of the related technology application. On the
grounds of VRPplatform the system integrates the VRT and multimedia technology and ultimately accomplishes the full 3D visualization simulation for
subway construction process and HCI, allowing the visual and comprehensive
display of simulation menu of subway construction site and complete construction process via computer screen, which is inconvenient for outsiders to observe.

Keywords: Subway construction, VRT, Simulation.

1   Introduction

The respectively used construction method varies from each other since the construction method is greatly influenced by the factors including surface installation, road,
urban traffic, hydrogeology, environmental protection, construction equipment, financial condition etc when building subway in cities. However the inconveniences for
strangers to observe owing to the fact that construction site exists in danger and construction method is confidential bring about great obstacles for the mass who are
engaged in civil construction study and research, and also are unfavorable for the
improvement and growth for subway construction method.

The system development is designated to facilitate the guest to visit and study
around the subway construction site. The system applies to the instructional activities
and attains modernized instruction and demonstration effect, available for pupils to
watch panorama for ground surface and deep stratum construction through simulation
software and simulation model equipment and without travel instead. The guest is not
only capable of viewing outwards, but observes the details from any viewpoint and
location, clearing away the inconvenience of field teaching and limitation on touring
and faithfully realizing presence sensation interactive teaching of HC real-time interactive type. In addition if it is used by construction organization or activities including
technical seminars the traditional instrument such as paper characteristic tender documents can be replaced by 3D visualization scheme and in this way facilitating the
exploitation and growth of subway construction technology fundamentally.

B. Liu and C. Chai (Eds.): ICICA 2011, LNCS 7030, pp. 726–733, 2011.
© Springer-Verlag Berlin Heidelberg 2011

2   The Application of VRT in Simulation System

VRT is a kind of high-tech technologies developed over the years. It combines computer 3D graphic technology, computer simulation, multi-sensor interactive technology, display technology, highly parallel real-time computing technology and human behavior research etc. and it represents multi-discipline interleaving technology [1]. It has three basic features---immersion, interaction, imagination etc. namely, 3I[2]. Presently VRT performs very positively in many engineering realms and fundamental researches.

2.1   Current Research Situation for Visual Reality System

The growth for computer provides a kind of computation instrument and analysis instrument and accordingly results in the generation of many new methods solving problems. The generation and growth of VRT follows the same way. As far as visual reality itself is concerned it is primarily related to three research realm:

Establish real-time 3D visual effects by way of computer graphics;
Establish observation interface on visual world;
Utilize VRT to intensify the application on the aspects such as scientific computation technology.

2.2   Relevant Technical Characteristics and Composition for Visual Reality

The definition for visual reality can be generalized as:the technology which creates a simulation environment (such as aircraft flight deck, operation site etc) through computer and enables the user to penetrate into the environment by help of diversified sensor equipments, accomplishing the directly natural interaction between users and the environment. So VRT has the following 4 significant features[3]:

Perception. The so-called multiple perception implies that it is inclusive of aural perception, force perception, tactile perception, motion perception, even taste perception, olfactory perception apart from visual perception that is provided by common computer. The desired visual reality should possess all the human perception function.

Existence. Also known as present sensation, it refers to the authenticity extent that users comprehend when they are present in simulation environment as a protagonist. The desired simulation environment is that one in which users can't distinguish true from falsehood.

Interaction. Interaction indicates the operational extent of the object in simulation environment by users, and natural extent (real-time inclusive) of feedback acquisition from environment. For example, people could directly grab the object in the environment with hands, at this time hands have the holding feeling and can be aware of the weight of the object. The object in the field of view also can move with the hands motion.

Autonomous. It indicates the motion extent of the object in visual environment in compliance with physics law. For example, the object will move along the force direction, or roll over, or fall onto the floor from table when driven by force.

The model for visual reality system is shown as figure 1. Users perform operation on visual environment through sensor equipments and obtain real-time 3D display and other feedback information. Where the system constitutes feedback closed circuit with external world through sensor equipment the interaction between the user and visual environment could have an effect on the outsides under the users control.

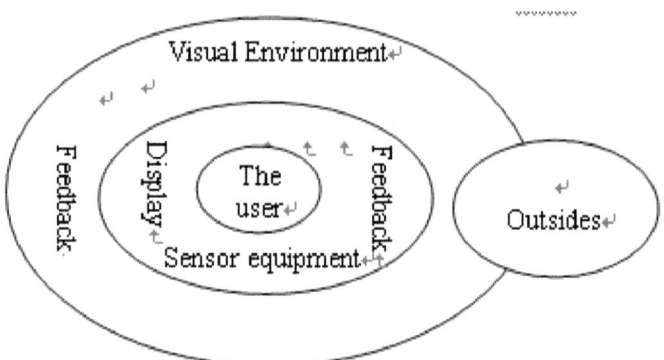

Fig. 1. The model for visual reality system

Visual reality system is primarily composed of the following five modules (such as figure 2):

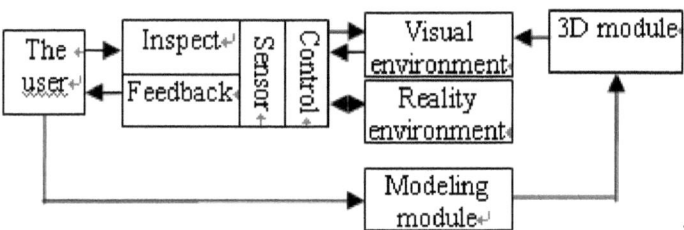

Fig. 2. The composition of visual reality system

Inspect module: Check up on the operational command of the user and allow it act on visual environment through sensor module.

Feedback module: Receive the information from sensor module, which offers real-time feedback for the user.

Sensor module: On the one hand receive the operational command from the user and allow it to act on visual environment; On the other hand offer the operation results to the user in diversified feedback forms.

Control module: Control the sensor, allowing it to act on the user, visual environment and reality world..

Modeling module: Acquire 3D manifestation of components of the reality world and accordingly establish corresponding visual environment.

2.3   The Application of VRT in the Simulation System

Visual reality system and system simulation or visualization simulation are both complementary and distinguish from each other in some practical functions. Simple visual reality system has incomparable advantages in aural sense, visual sense and tactile sense in comparison to generalized visualization simulation, namely, immersion, the people feel as if they were there, but the system is unsatisfactory in management of simulation process and decision assistance. By incorporate of visual reality with system simulation in an organic way, visual reality simulation system is created and the organism generated by combination of the advantages of the two systems allow the user to infuse the personal priorities and independent behavior into visual reality simulation process within a visual space, achieving the personal simulation in the circuit.. Visual reality simulation system is an upgrade for generalized visualization simulation and brings about qualitative leap, the application potential and developmental prospect is immense. Visual reality simulation is a progressively developed interdisciplinary technology which incorporates the technical characteristics of each application field with the application requirement on the basis of system science, computer science, probability theory and mathematical statistics. In the meantime it is an experimental science. With the development of each discipline the visual reality simulation is changing for the better with each passing by and has already evolve as an active emerging technology over the years.

Subway construction site simulation system is an application case of visual reality technology. It is a 3D space developed by use of diversified VR software platform and hardware unit, which embraces all sorts of site entity and manual entity. The user can interact with diversified visual entities in an natural manner and learn about the each link of the construction site by walking and roaming around the site.

3   Overall Design of 3D Simulation System for Subway Construction Site

Simulation system is a package of visual software platform including visual site scene roaming and construction animation. It is a recurrence of genuine scene for subway construction site and the user can learn about the integral figure and appearance of the site visually and omni-directionally and all manner of relevant information. According to design requirement the system is required to produce specific animation for specific scene to accomplish demonstration of construction animation apart from fundamental function of visual site independent roaming and regional routine path roaming.

On the basis of VRP and 3Dmax the subway construction site demonstration system again incorporates the graphic and text modification tools in Photoshop to produce lively demo effect. The user can appreciate the construction atmosphere of the site and learn about and experience the complete construction process just in front of the computer

3.1   Preliminary Preparation of Data and Information

The establishment of simulation system is orientated at the entire subway construction site and it is characteristic by broad coverage and massive amount. The required data not only embraces the extensive topography data, enormous surface object data, but the subterranean tunnel layout structure data, excavation and stratum data etc. Prior to establishment of simulation system for the complete subway construction site, it is required to perform collection and preliminary processing of local information which consists of image information, video information, drivage data etc. to prepare for deployment by construction site system in the future.

For the purpose of authentic effects the system requires the actual photos as elements, which entails substantial data acquisition tasks. The photo elements is obtained through the manners of photographing, scanning, shoot and processed with the software including Photoshop to achieve the desired effects. The details for construction simulation scene can be interpolated into the system by video complement for better impression.

With the aim of visit and enlightenment for the user, the system must be entered with data information as much as possible to facilitate the user to acquire the knowledge by emulating system from computer screen. Relevant data information is converted into textual information or multimedia information and insert into the system, in such way people can acquire the necessary facts by the data information which influx into the system while experiencing the lifelike construction site scene.

The acquisition of geography data primarily involves collection of CAD graphic of construction site topography, then simplifies and modifies the planning CAD documents and loads them into 3DSMAX, detaches the required contents, establishes the topography contour in expectation. Then perform modeling of construction site topography by employment of 3DSMAX topography generation plug-ins.

3.2   Construction Site Modelling

The construction of construction site visual scene is essential for digitalized simulation system. Vivid local model and appropriate scene driving is a necessary condition to ensure digitalized simulation system construction to be well afoot.

Construct the complete model and relevant animation with the help of 3D modeling software, in the meanwhile bake and export the entire scene so as to create visual entity for simulation system.

The primary tasks for simulation system concentrates on the establishment of 3D model, the fairness of the model has direct effect on the performance of the entire simulation system. As a result the establishment of 3D model in virtual reality system is the foundation as far as the entire system is concerned. But confined by the software computation character of the common computer and the load range bearable by the system itself the 3D model fails to attain delicate, or otherwise shutdown will occur in operation due to system erratic. Adequately accounting for the specific requirements of this system and want of model details grade, the system uses 3DSMAX to accomplish modeling of construction site realistic scene.

The scene model format mostly uses Polygon since the created model by Polygon is convenient for future arrangement of UVs; Topography model uses(non uniform rational B-spline curve) NUBRS modeling due to the fact that it can construct more vivid and picturesque topography structure; underworkings normally apply lofting skills to accomplish the establishment of intricate laneway structure.

Building plays an important role in simulation system. Building modeling normally applies polygon. Render relevant building extrusion polygon based on the simplified CAD original drawing, then fabricate texture mapping by use of processed photograph so as to reduce the complexity of the building model.

The tunnel constitutes the most important part in the system. Normally apply lofting skills to accomplish modeling, construct intricate laneway structure by creation of laneway section and tunnel layout lines. At length perform material mapping process on the laneway.

Diversified equipments are also indispensable component parts in the system. Distinguished modeling method can be applied in case of different equipments. At length convert the model into MESH model.

Where perform modeling on visual reality, the model should be tried to simply and details on model should be exhibited to the utmost by use of mapping skills. Finally create the scene.

The operation condition in the system requires to set up rigid body animation so as to enhance the authenticity and vitality of the system, avoiding just dull static vision in the construction site. VRPlatform is available for implanting diversified animation of 3 DS Ma x. We only need to produce relevant key frame animation orientated at the equipments including entry driving machine, belt conveyer, hauling unit in 3 DS Max.

Camera view is that is found by end user. In VRPlatform and 3DSMax camera is general-purpose. To make it convenient to adjust the camera the location and camera animation route can be preset in 3DSMax so as it available for view from designated visual angle and roaming in the local scene.

For increasing the authenticity of the visual scene and meanwhile accelerating the computation speed, render to textures skill is required. It renders the illumination information of 3DS MAX into texture mapping and re-maps the baked texture to the scene. Generally the scene is baked into two formats of Completemap and Lightingmap, the former is strong in lightening and supports most materials of 3DS MAX such as composite material and multidimension material, but the mapping is obscure and consumes much memory; for the latter the texture mapping is sharply but feeble in lightening, only supports the material defaulted by 3DS MAX. In view of these appropriate baking format should be selected as requirements in terms of the features of different scenes. The determined local model established in 3DS MAX is shown as figure 3:

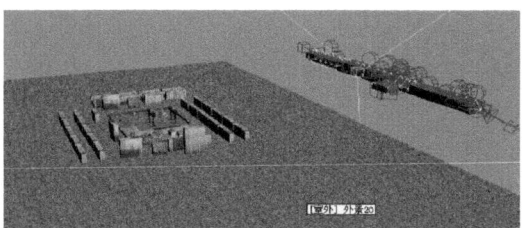

Fig. 3. 3DS MAX model

3.3   Driving of Visual Scene

Upon completing the establishment of the scene we should proceed interactive development through VR software platform. The established model is relatively independent and has not realized the liaison in the very meaning yet. It still requires development by real-time scene driving tools so as that the user can observe the object in the visual scene from any angle and feel as if they were there.

Presently the professional real-time scene driving software includes Vega•Virtools•Quest3d•VRP etc. As design requirements the real-time driving of animation driving for visual scene roaming of subway construction site simulation system is accomplished under the environment of VRP 3D interactive simulation platform which is the most representative product among home made VR software and is 3D visual reality platform software with complete independent intellectual property right. The software is developed by domestic Vistand Digital Technology Co.,Ltd. alone and can be integrated with 3DS MAX seamlessly. It has the real-time rendering image with the top quality among the industry and supports independent roaming, manual roaming. The user can customize roaming track and meanwhile possesses collision detection algorithm with high efficiency and high accuracy. It is accessible for SDK second development and has good expandability performance.

Upon loading the model into VRPlatform we should perform processing on the model, including establishment of air environment, adding in special object, compiling animation camera trigging and setting up the back parameters such as model location, collision test, the motion ways for the roamer etc.

The environment, namely skybox, is a panorama created by 6 photos and a simulation on visual air environment. The skybox can be made through some panorama software. Some special object, such as character, grass and flower can be produced through texture mapping by use of Billboard skill thanks to that its model is intricate. The object in Billboard can rotate omni-directionally and the rotation method should meet the following requirements: Billboard rotates around Z-axis with the variation of viewpoint location on xoy plane, quadrangle is always parallel with Z-axis and orients at the viewpoint all the time. In such way all the Billboard is directed towards viewpoint location[6] when the user proceeds visual roaming in the scene. The interactive development of VRPlatform primarily involves some logic triggering, such as the transformation of camera, the motion of rigid body animation, simultaneous occurrences etc. These logic triggering is compiled through VRP event compiler and script complier. Diversified commands have been offered in event complier and script complier and logic control of visual object in the system can be accomplished through script compiling.

4   Conclusion

With the grandness of VRT the development of simulation system has been broad used in many areas, such as offering practice for major pupils, performing prepost education and training for works or technical personnel, assisting in 3D spacious relation communication. The application prospect for VRT is overwhelming from a long-term point of view. The panoramic presentation process in the simulation system for subway construction site can basically represent the complete construction procedure and anticipated aim can be achieved in the application field.

References

1. Wang, C., Zhou, J., Li, L.: Creator visualization simulation modeling technology IM. Hua-zhong University of Science Press, Wuhan (2005)
2. Hong, B., Chai, Z.: Visual reality and application. National defence industrial press, Beijing (2005)
3. Burdea, G., Coifet, P.: VirtualReality Technology. John Wiley & Sons, Chichester (1994)
4. Weik, Y., Yang, X., Wang, F.: Visual reality and system simulation. National defence industrial press, Beijing (2004)
5. Liu, X., Wu, X.: Modern System Modeling and Simulation Technology. Northwestern Polytechnical University Press, Xi'an
6. Li, Y., Zhu, Q.: Construction of 3D visual simulation scene based on VRML. Computer and Applicable Chemistry 21(3) (2004)
7. Wang, D., Zhang, J.: Architecture design based on VRT. Structural Engineer (10) (2004)
8. ISO/IEC 14772-1:1997/Amd. 1:2002(E), Information Processing Systems - Computer Graphics The Virtual Reality Modeling LanguagePatr1-Functional specification and UTF-8 encoding Amendment 1-Enhancedinteroperability (April 12, 2002)
9. Koop, T., Broll, W.: VRML:Today and Tomorrow Computers and Graphics. An International Journal 3(20) (1996): Kromker, D., Encarnacao, J. (eds.), pp. 427–434. Pergamon Press, Oxford (1996)

HLA Based Collaborative Simulation
of Civil Aircraft Ground Services

Wu Bo[1], Hu Liu[1], Zhang Yibo[2], and Sun Yijie[3]

[1] School of Aeronautics Science and Engineering,
Beijing University of Aeronautics and Astronautics,
Beijing, China
[2] School of Astronautics,
Beijing University of Aeronautics and Astronautics,
Beijing, China
[3] China Academy of Civil Aviation Science and Technology,
Beijing, China
hiwubo@gmail.com

Abstract. Civil airport ground service is one of the important factors in airport adaptability evaluation. Through the civil aircraft ground service process analysis and HLA (High Level Architecture) distributed collaborative simulation technology applications, HLA simulation framework was established and airport ground service visualization collaborative simulation program was developed. By multiple terminals collaborative simulation, program intuitively displays aircraft ground service process for airport adaptability evaluation.

Keywords: High Level Architecture, airport ground service, visualization collaborative simulation.

1   Introduction

Airport adaptability is a very important factor in civil aircraft design. If the design of the aircraft cannot meet civil aviation regulations of the airworthiness requirements, and cannot be compatible with the airport, it will bring great loss for airport and aircraft manufacturers. Current research on the airport ground services mainly focus on arranging multi flights scheduled, through the establishment of mathematical models, application of various algorithms, optimization scheduling process [1] and [2]. Before the appearance of virtual reality technology, the simulation of a single aircraft ground services process usually use sand table model. The simulation need to move the sand table model to simulate the process of ground service. This method was not a quantitative analysis and was less convincing. And there are many details are difficult to simulate, such as ground vehicle service process. When using the geometric model or physical prototype of aircraft to simulate ground services process, many parameters have been determined. if it is found design problems, remediation is difficult and complex.

B. Liu and C. Chai (Eds.): ICICA 2011, LNCS 7030, pp. 734–741, 2011.

On the other hand, many virtual reality simulations has played an important role in engineering. Such as the use of high-level architecture (High Level Architecture, HLA) simulation environment has been related to the carrier aircraft flight simulator, launch vehicle simulation, multi-missile simulation, integrated environmental management [3-6]. Therefore, this study will explore to use virtual simulation technology to solve the above details and time period problems. Appling three-dimensional virtual simulation technology to simulate civil aircraft ground service process can show more details and accomplish quantitative analysis through the computer method of calculation. In the initial stages of aircraft design, the simulation can begin and timely feedback design.

Based on a typical civil aircraft airport ground service process, a airport ground service visualization collaborative simulation program was developed, which show the service process with interactive three-dimensional scenes. Through multi computer terminals simulate discharge of ground services in co-operation, the program complete the discharge process simulation.

2   HLA-Based Co-simulation Design

Collaborative simulation of civil aircraft ground services use three-dimensional virtual reality editing software VirTools [7] to develop simulation program and select the High Level Architecture (HLA) as the distributed interactive simulation architecture. Program's data transferring used HLA architecture to complete the co-simulation.

2.1   Program Design

Civil airport services include many kinds of services (Fig. 1), divided into the main cabin services, storage services, aircraft services [8].

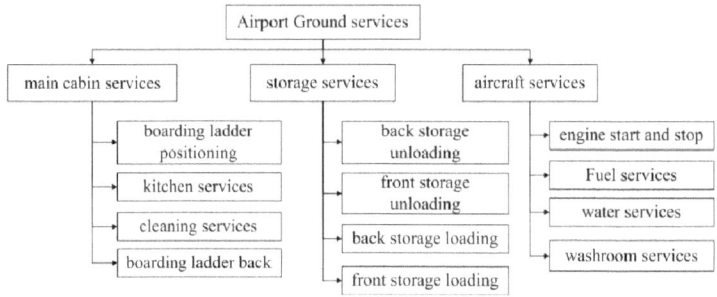

Fig. 1. Classification of civil airport services

Simulation program will be based around the cabin service as the main simulation object, through different computer network terminals operating to complete the unloading process together. Another airport ground services are not as the focus of simulation. The program will automatically simulate without terminal control. Un-loading process of civil aircraft requires baggage car and lift car to complete, so the two terminals operate baggage car and lift car. In order to assess and study the entire service process, a roaming terminal was arranged, which can observe the service process from various perspectives.

2.2   HLA Design

The data transmission technology of this simulation program use HLA architecture framework. There are three terminals in this programs baggage car, lift car and roam-ing. So the HLA federation defined three federation members. The federation members release and order data by RTI (Run Time Infrastructure) system. System architecture was shown in Fig. 2 [9] [10].

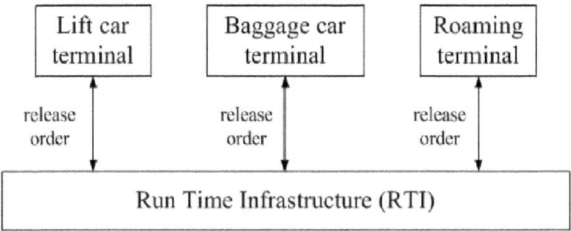

Fig. 2. System architecture

The function design of the federation members as follows:

(1) **Lift car terminal.** This terminal can control the lift car's movement, release car's movement data, order the other terminal's data and display the three-dimensional sense using the data. The class of lift car has the following functions: release the axis coordinates and Euler angles of all parts of the lift car; control the long board and short board up and down; order other federation members' data; set time management and container exchange.

(2) **Baggage car terminal.** This terminal can control the baggage car's movement, release car's movement data, order the other terminal's data and display the three-dimensional sense using the data. The class of baggage car has the following func-tions: release the axis coordinates and Euler angles of all parts of the baggage car; order other federation members' data; set time management and container exchange.

(3) **Roaming terminal.** This terminal can control aircrafts and other ground service vehicles, release vehicles and aircrafts movement data, orders the other terminal's data and display the three-dimensional sense using the data. The class of roaming has the following functions: release the axis coordinates Euler angles and other attributes of aircrafts and all other service cars; order other federation members' data; set time management and container exchange.

System object class and interaction class design as shown in Table 1 and Table 2.

Table 1. Object Class

Object class	Attribute/Parameter	Data type	Amount
Lift car class	axis coordinates(X/Y/Z)	float	3
	Euler angles(X/Y/Z)	float	3
	Long board and short board position	float	2
Baggage car class	Tractor axis coordinates(X/Y/Z)	float	3
	Tractor Euler angles(X/Y/Z)	float	3
	Trailer axis coordinates(X/Y/Z)	float	15
	Trailer Euler angles(X/Y/Z)	float	15
Container class	axis coordinates(X/Y/Z)	float	3
	Euler angles(X/Y/Z)	float	3
Aircraft class	axis coordinates(X/Y/Z)	float	3
	Euler angles(X/Y/Z)	float	3
	Front / rear door status	int	2
	Landing Gear opening state	int	1

Table 2. Interaction Class

Interactive class	Attribute/Parameter	Data type
Federal class	Join the federal	int
	Quit the federal	int
Message class	aircraft docked and front door opened	int
	Lift car is ready	int
	Baggage car is ready	int
	Container in place	int
	Front/rear unload completed	int
	Unload completed	int
Position check class	Long/short board height is correct	int
	Tractor is collided	int
	Trailer is collided	int

3   Development of Collaborative Simulation

The design of visualization collaborative simulation program is divided into modeling and VirTools programming.

3.1   Modeling

Modeling use three-dimensional modeling software Maya. To build the ground services environment, this paper built the T3 terminal building and the corresponding floor, runway and Parking apron (Fig. 3). Aircrafts and services vehicles were built. This paper established Airbus A320 aircraft, lift car, baggage car, garbage truck, water truck, tractor, fuel truck and compressed air truck as shown in Fig. 4. Lift car, which is responsible for unloading the container in the process of moving transport, is a important role in this simulate. The car has a long board and short board and can moved cargo containers from aircraft to the baggage car through caterpillar track.

Fig. 3. Airport environment models

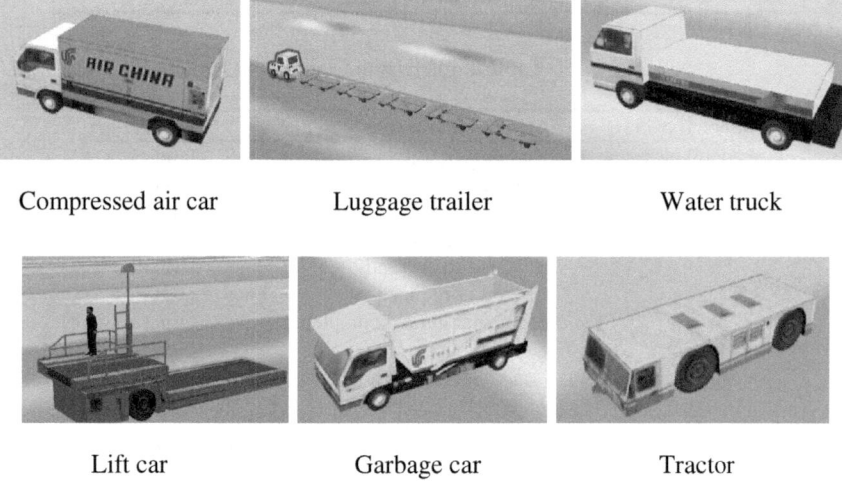

Compressed air car   Luggage trailer   Water truck

Lift car   Garbage car   Tractor

Fig. 4. Vehicle models

3.2   HLA in VirTools

Because VirTools doesn't have HLA module, it cannot be called directly. This paper used VirTools SDK [11] to program a HLA network transmission module for this simulation. VirTools SDK allow using C++ language. HLA design has been described previously, simulation program can use HLA module to transmit data. This program applied the array type in VirTools as data structure for HLA data transmit.

3.3   VirTools Visual Simulation Process

The modeling and HLA network transmission module need to import into VirTools simulation system. The entire simulation process is shown in Fig. 5. In this task baggage car and lift car need to work together. When one does not complete the assigned step, the other cannot be the next step. Program need to call messaging function in VirTools and send and receive messages to other terminals applying HLA module.

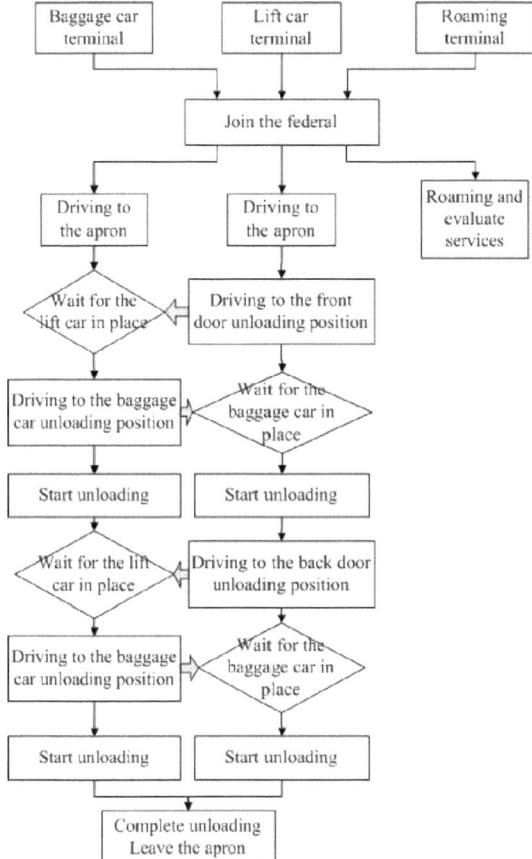

Fig. 5. Simulation flow chart

Because simulation also involves the assessment of the service content, the process of unloading services will record every step time, such as the step of cargo containers from the aircraft cabin door moved to the baggage car's short broad. Through record every step time, the program will evaluate the service process. At the same time, some errors in the simulation, such as vehicle collision, the trunk is not aligned and other issues, will be recorded and analysis to provide a reference for evaluation.

For the simulation of other ground services, such as cleaning services, water services, program will automatic simulate without terminal controlling. When aircraft parked, the service vehicle will automatically enter the corresponding service aircraft position and complete the task. In VirTools, these automatic simulations was programmed in roaming terminal (Fig. 6) and send these service vehicles' movement data to other terminals by HLA module.

After the whole process simulation, the program will output the every step time and error operation messages for the user to assess reference. Programming of the simulation system and each terminal is complex, so this paper does not make a detailed description.

4   Example

Applying the collaborative simulation program of civil aircraft ground services can visually demonstrate the entire process of airport ground services. To complete the ground service, the baggage car terminal and the lift car (Fig. 6) terminal need be controlled together. In the actual discharge process, people also need collaborate complete the service task as the program flow chart (Fig. 5). Users can assess the ground service through the program's outputs, which include movement of service vehicles, service order, every step time, frequency of operational errors in the process.

Fig. 6. Screenshots in roaming terminal and lift car terminal

In the process of simulation, if the lift car's broad lifting spend too many time and the process of passing the container is too slow, the aircraft designers should consider whether the door of aircraft body is too high and should consider to improve the aircraft fuselage design. If the baggage car always collides with aircraft wings or wing engine crane, the aircraft designer should consider whether the door position is too close the wings and need to adjust the door or engine position. In aircraft design, if designers adjust the position of fuselage openings or engines, it will need check the structural strength requirement. So if there are adjustments, the designers must recalculate which will waste a lot of time. The simulation program can simulate the process of ground services early in the design cycle of aircraft design, so it can detect these problems and the designers can assess and correct the design timely.

5   Conclusion

According to the demand of the simulation of civil aircraft ground services, this paper combined HLA and visual simulation technology and developed interactive visualization collaborative simulation program. Through the visual simulation of civil aircraft ground services, the Example program shows that the simulation is intuitive and interactive. Program can provide a more intuitive simulation tool to evaluate the aircraft design and the process of airport ground services for aircraft designers.

References

1. Wang, L., Liu, C., Tu, F.: Optimized Assignment of Civil Airport Gate. Journal of Nanjing University of Aeronautics and Astronautic 38(4), 433–437 (2006) (in Chinese)
2. Wei, L.: Investigation of Airport Ground Services and Aircraft Departures process optimization. Civil Aviation University of China, Beijing (2006) (in Chinese)
3. Li, J., Bian, X., Xia, G., et al.: Research on Visualization of Fly Simulation for Carrier Aircraft Based on HLA. Journal of System Simulation 20(9), 2352–2356 (2008) (in Chinese)
4. Li, H., Mao, W., Jing, J.: Research on Integration Method of Scene Simulation and Numerical Simulation Based on HLA. Journal of System Simulation 20(7), 1749–1753 (2008) (in Chinese)
5. Li, X., Zhang, Z., Gao, Q., et al.: Application of Data Distribution Management in Multimissile Launching Units Simulation System Based on HLA. Journal of System Simulation 20(18), 5044–5048 (2008) (in Chinese)
6. Wang, J., Wang, X.: Integrated Environment Based on HLA for Collaborative Simulation Run-Time Management. Journal of Beijing University of Aeronautics and Astronautics 29(3), 273–277 (2003) (in Chinese)
7. Dassault Systems: VirTools 5.0 Online Reference (2009)
8. Li, Y., Zhu, T.: Introduction to Civil Aviation Airport Ground Services. Civil Aviation Press of China, Beijing (2006) (in Chinese)
9. IEEE Std 1516.3-2003.IEEE Recommended Practice for High Level Architecture (HLA) Federation Development and Execution Process, FEDEP (2003)
10. Huang, J., Hao, J.: HLA Simulation Program Design. Publishing House of National University of Defense Technology, Hunan (2008) (in Chinese)
11. Dassault Systems. VirTools SDK Documentation (2009)

Solving Singular Perturbation Problems by B-Spline and Artificial Viscosity Method

Jincai Chang[*], Qianli Yang, and Long Zhao

College of Science,
Hebei United University Tangshan Hebei 063009, China
jincai@heut.edu.cn,
yangqianli2010@yahoo.cn

Abstract. In this paper, we propose a B-spline collocation method using artificial viscosity for solving singularly perturbed two-point boundary-value problems (BVPs). The artificial viscosity has been introduced to capture the exponential features of the exact solution on a uniform mesh and the scheme comprises a B-spline collocation method, which leads to a tri-diagonal linear system. The design of artificial viscosity parameter is confirmed to be a crucial ingredient for simulating the solution of the problem. A relevant numerical example is also illustrated to demonstrate the accuracy of the method and to verify computationally the theoretical aspects. The result shows that the B-spline method is feasible and efficient and is found to be in good agreement with the exact solution.

Keywords: B-spline method; Singular perturbation; Boundary layers; Artificial viscosity.

1 Introduction

In the field of singularly perturbed differential equation, the computation of its solution has been a great challenge and is of great importance due to the versatility of such equations in the mathematical modeling of processes in various application fields. They arise in several branches of engineering and applied mathematics. For example, fluid mechanics, quantum mechanics, optimal control, chemical-reactor theory, aerodynamics, reaction-diffusion process, geophysics etc. They provide the best simulation of observed phenomena and hence the numerical approximation of such equations has been of growing interest.

This class of problems possess boundary layers. That is, there are thin layers where the solution varies very rapidly, while away from the layers the solution behaves regularly and varies slowly. The presence of small parameter(ε) in these problems prevents us from obtaining satisfactory numerical solutions. In recent years spline methods for solution of singularly perturbed boundary value problems are given in [1-6] [9-13]to provide accurate numerical solutions. In the paper, we have used the B-spline

---

[*] Project supported by National Natural Science Foundation of China (No.61170317) and Natural Science Foundation of Hebei Province of China (No. A2009000735, A2010000908).

B. Liu and C. Chai (Eds.): ICICA 2011, LNCS 7030, pp. 742–749, 2011.

method and artificial viscosity. In this method, we replace the perturbation parameter affecting the highest derivative by artificial viscosity. The artificial viscosity is determined using the two-variable and perturbation parameter.

In the present paper, a cubic B-spline is used to solve singular perturbation boundary value problems of linear ODEs as the following systems which are assumed to have a unique solution in the interval [0,1].

$$\begin{cases} -\varepsilon^2 y'' + m(x)y' + n(x)y = f(x), & x \in [0,1] \\ \quad y(0) = \alpha, & y(1) = \beta \end{cases} \tag{1}$$

Where, ε is a small positive parameter($0 < \varepsilon << 1$), $m(x)$, $n(x)$ and $f(x)$ are sufficiently function, and $m(x) \geq a > 0, n(x) \geq b > 0, a,b$ are constant.

In section 2 we have given the definition of the B-spline method. Design of the artificial viscosity in section 3. In section 4, The spline technique presents to approximate the solution of singular perturbation two-point boundary value problems. we have solved there problems using the method and the max-absolute errors and graphs have also been shown in section 5, In section 6,reports the major conclusion and further developments.

Table 1. The case of boundary layer

$m(x)$	Boundary Layer
$m(x) \neq 0, 0 \leq x \leq 1$	
$m(x) > 0$	Boundary layer at $x = 0$
$m(x) < 0$	Boundary layer at $x = 1$
$m(x) = 0$	
$n(x) < 0$	Boundary Layer at $x = 0$ and $x = 1$
$n(x) > 0$	Rapidly oscillating solution
$n(x)_{\text{change sign}}$	Classical turning point
$m'(x) \neq n(x), m(0) = 0$	
$m'(x) > 0$	No boundary layer, interior layer at $x = 0$
$m'(x) < 0$	Boundary layer at $x = 0$ and $x = 1$, No interior layer at $x = 0$

2   The cubic B-spline

Let $\Omega = \{x_0, x_1, \cdots, x_n\}$ be a set of partition of $[0,1]$, The zero degree B-spline is defined as follows [7]:

$$B_{i,0}(x) = \begin{cases} 1, & x \in [x_i, x_{i+1}) \\ 0, & otherwise \end{cases}$$

and for positive p, it is defined in the following recursive form:

$$B_{i,p}(x) = \frac{x - x_i}{x_{i+p} - x_i} B_{i,p-1}(x) + \frac{x_{i+p+1} - x}{x_{i+p+1} - x_{i+1}} B_{i+1,p-1}(x), p \geq 2$$

We apply this recursion to get the cubic B-spline , it is defined as follows:

$$B_{i,3}(x) = \begin{cases} \dfrac{(x - x_i)^3}{(x_{i+1} - x_i)(x_{i+2} - x_i)(x_{i+3} - x_i)}, x \in [x_i, x_{i+1}), \\[2ex] \dfrac{(x - x_i)^2(x_{i+2} - x)}{(x_{i+2} - x_i)(x_{i+2} - x_{i+1})(x_{i+3} - x_i)} + \\[2ex] \dfrac{(x - x_i)(x_{i+3} - x)(x - x_{i+1})}{(x_{i+2} - x_{i+1})(x_{i+3} - x_{i+1})(x_{i+3} - x_i)} + \\[2ex] \dfrac{(x_{i+4} - x)(x - x_{i+1})^2}{(x_{i+2} - x_{i+1})(x_{i+3} - x_{i+1})(x_{i+4} - x_{i+1})}, \\ x \in [x_{i+1}, x_{i+2}) \\[2ex] \dfrac{(x - x_i)(x_{i+3} - x)^2}{(x_{i+3} - x_i)(x_{i+3} - x_{i+1})(x_{i+3} - x_{i+2})} + \\[2ex] \dfrac{(x - x_{i+1})(x_{i+3} - x)(x_{i+4} - x)}{(x_{i+3} - x_{i+2})(x_{i+3} - x_{i+1})(x_{i+4} - x_{i+1})} + \\[2ex] \dfrac{(x_{i+4} - x)^2(x - x_{i+2})}{(x_{i+4} - x_{i+1})(x_{i+4} - x_{i+2})(x_{i+3} - x_{i+2})}, \\ x \in [x_{i+2}, x_{i+3}) \\[2ex] \dfrac{(x_{i+4} - x)^3}{(x_{i+4} - x_{i+1})(x_{i+4} - x_{i+2})(x_{i+4} - x_{i+3})}, x \in [x_{i+3}, x_{i+4}) \\[2ex] 0, otherwise \end{cases}$$

The cubic B-splines are defined by the de Boor-Cox

$$B_{i,3}(x) = \frac{1}{6h^3} \begin{cases} (x-x_i)^3, x \in [x_i, x_{i+1}), \\ (x-x_i)^2(x_{i+2}-x)+(x-x_i)(x_{i+3}-x)(x-x_{i+1})+(x_{i+4}-x)(x-x_{i+1})^2, \\ \qquad x \in [x_{i+1}, x_{i+2}) \\ (x-x_i)(x_{i+3}-x)^2+(x-x_{i+1})(x_{i+3}-x)(x_{i+4}-x)+(x_{i+4}-x)^2(x-x_{i+2}), \\ \qquad x \in [x_{i+2}, x_{i+3}) \\ (x_{i+4}-x)^3, x \in [x_{i+3}, x_{i+4}) \\ 0, otherwise \end{cases}$$

Then,

$$B_{i,3}(x) = \frac{1}{6h^3} \begin{cases} (x-x_{i-2})^3, x \in [x_{i-2}, x_{i-1}), \\ h^3 + 3h^2(x-x_{i-1}) + 3h(x-x_{i-1})^2 - 3(x-x_{i-1})^3, \\ \qquad x \in [x_{i-1}, x_i) \\ h^3 + 3h^2(x_{i+1}-x) + 3h(x_{i+1}-x)^2 - 3(x_{i+1}-x)^3, \\ \qquad x \in [x_i, x_{i+1}) \\ (x_{i+2}-x)^3, x \in [x_{i+1}, x_{i+2}) \\ 0, otherwise \end{cases}$$

and

$$B_i(x_j) = \begin{cases} 4/6, & i = j \\ 1/6, & i - j = \pm 1 \\ 0, & i - j = \pm 2 \end{cases}$$

$$B_i'(x_j) = \begin{cases} 0, & i = j \\ \pm \dfrac{1}{2h}, & i - j = \pm 1 \\ 0, & i - j = \pm 2 \end{cases}$$

$$B_i''(x_j) = \begin{cases} -2/h^2, & i = j \\ 1/h^2, & i - j = \pm 1 \\ 0 & i - j = \pm 2 \end{cases}$$

3   Design of the Artificial Viscosity

When $m(x) \equiv 0$, therefore, we define the artificial viscosity as

$$\eta_i = \frac{h^2 n_i}{6\pi(p_i)}$$

Where $\pi(p_i) = \dfrac{1 - \cos(p_i h)}{2 + \cos(p_i h)}$, $p_i = \sqrt{n_i / \varepsilon^2}$, $n_i = n(x_i)$

When $n(x) \equiv 0$, therefore, we define the artificial viscosity as

$$\eta_i = \frac{hm_i}{2} \coth\left(\frac{hm_i}{2\varepsilon^2}\right)$$

Where $\coth\left(\dfrac{hm_i}{2\varepsilon^2}\right) = \dfrac{\exp(hm_i / 2\varepsilon^2) + \exp(-hm_i / 2\varepsilon^2)}{\exp(hm_i / 2\varepsilon^2) - \exp(-hm_i / 2\varepsilon^2)}$, $m_i = m(x_i)$

4   B-Spline Solution for Singular Perturbation

We redefine the problem by using the artificial viscosity, which be determined above, we rewrite the problem(1) as

$$\begin{cases} -\eta(x, \varepsilon^2) y'' + m(x) y' + n(x) y = f(x), & x \in [0,1] \\ y(0) = \alpha, & y(1) = \beta \end{cases}$$

Let

$$y(x) = \sum_{i=-3}^{n-1} c_i B_{i,3}(x) \tag{2}$$

be an approximate solution of Eq.(1),where c_i is unknown real coefficient and $B_{i,3}(x)$ are cubic B-spline functions. Let $x_0, x_1, \cdots, x_n$ are $n+1$ grid points in the interval $[a,b]$,so that $x_i = a + ih$, $i = 0,1,\cdots,n$, $x_0 = a, x_n = b$, $h = (b-a)/n$.It is required that the approximate solution Eq.(2)satisfies the differential equation at the points $x = x_i$. Putting Eq.(3) in (1),it follow that

$$-\eta_i y''(x_i) + m(x_i) y'(x_i) + n(x_i) y(x_i) = f(x_i), i = 0,1,\cdots,n \tag{3}$$

and boundary condition can be written as

$$y(0) = \alpha \tag{4}$$

$$y(1) = \beta \tag{5}$$

The spline solution of Eq.(1) is obtained by solving the following matrix equation. Then, a systems of $n + 3$ linear equations in the $n + 3$ unknowns $C_{-3}, C_{-2}, \cdots, C_{n-1}$ are obtained, using Eq.(2)can obtain the numerical solution. This systems can be written in the matrix-vector form as follows:

$$AC = F \tag{6}$$

where

$$C = [c_{-3}, c_{-2}, \cdots, c_{n-1}]^T, F = [6\alpha, 6h^2 f(x_0), 6h^2 f(x_1), \cdots, 6h^2 f(x_n), 6\beta]^T,$$

and A is an $(n+3) \times (n+3)$-dimensional tri-diagonal matrix given by

$$A = \begin{bmatrix} 1 & 4 & 1 & 0 & 0 & \cdots & 0 & 0 & 0 \\ p_0(x_0) & q_0(x_0) & r_0(x_0) & 0 & 0 & \cdots & 0 & 0 & 0 \\ 0 & p_1(x_1) & q_1(x_1) & r_1(x_1) & & & & & \\ \vdots & \vdots & \vdots & \vdots & \vdots & \vdots & \vdots & \vdots & \vdots \\ 0 & 0 & 0 & 0 & 0 & \cdots & p_n(x_n) & q_n(x_n) & r_n(x_n) \\ 0 & 0 & 0 & 0 & 0 & \cdots & 1 & 4 & 1 \end{bmatrix} \tag{7}$$

also the coefficients in the matrix A have the following form

$$p_i(x_i) = -6\eta_i - 3m_i h + n_i h^2, (i = 0, 1, \cdots, n)$$
$$q_i(x_i) = 12\eta_i + 4n_i h^2, (i = 0, 1, \cdots, n)$$
$$r_i(x_i) = -6\eta_i + 3m_i h + n_i h^2, (i = 0, 1, \cdots, n)$$

Then, we get a systems of $n + 3$ linear equations

$$p_i(x_i)c_{i-3} + q_i(x_i)c_{i-2} + r_i(x_i)c_{i-1} = 6f_i h^2, (i = 0, 1, \cdots, n)$$

5   Numerical Results

In this section, one numerical example is studied by B-spline. The results obtained by the method are compared with the analytical solution, so that we get the maximum absolute errors, then demonstrate the accuracy of the B-spline method. We can find that our method is much better with a view to accuracy and utilization. Moreover, the numerical results are illustrated in Figs.1.

Example: Consider the free motion of the undamped linear spring mass system with a very resistant spring. Let the prescribed specific displacement be at times $t = 0$ and $t = 1$. Then one can obtain the two point problem

$$\begin{cases} -\varepsilon^2 y''(t) + y(t) = 0, & t \in [0,1] \\ y(0) = 1, & y(1) = 1 \end{cases} \tag{8}$$

Where ε^2 (the ratio of the mass to the spring constant) is small. For non-exceptional small positive values of ε the exact solution oscillates rapidly, so no-pointwise limit exists as $\varepsilon \to 0$.

Then, Solve the above singularly perturbed boundary value problem, and

$$N = 16, \varepsilon = 2^{-8}, h = \frac{1}{16}$$

The analytical solution:

$$y(t) = \frac{(1 - \exp(-1/\varepsilon))(\exp(-t/\varepsilon) + \exp(-(1-t)/\varepsilon))}{(1 - \exp(-2/\varepsilon))}.$$

Thus, the max-absolute error is given by

$$\delta = -0.0916$$

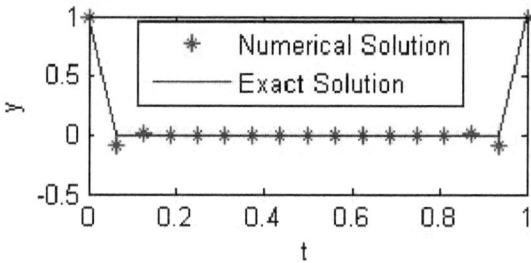

Fig. 1. Graph obtained by B-splines

6   Conclusion and Outlook

A family of B-spline method has been considered for the numerical solution of singularly-perturbed boundary value problems of linear ordinary differential equations. The cubic B-spline has been tested on a problem. The numerical results showed that the present method is an applicable technique and approximates the solution very well. The implementation of the present method is a very easy, acceptable, and valid scheme. This method gives comparable results and is easy to compute. Also this method produces a spline function which may be used to obtain the solution at any point in the range, whereas the finite difference method gives the solution only at the chosen knots. This method is easily tractable and can readily be applied to other problems of differential equations. This matter will be one of the future research targets.

References

1. Lin, B., Li, K., Cheng, Z.: B-spline solution of a singularly perturbed boundary value problem arising in biology. Chaos, Solitons and Fractals 42, 2934–2948 (2009)
2. Kadalbajoo, M.K., Gupta, V., Awasthi, A.: A uniformly convergent B-spline collocation method on a no-uniform mesh for singularly perturbed one-dimensional time-dependent linear convection-diffusion problem. Journal of Computational and Applied Mathematics 220, 271–289 (2008)
3. Kadalbajoo, M.K., Arorar, P.: B-spline collocation method for the singular perturbation problem using artificial viscosity. Computers and Mathematics with Applications 57, 650–663 (2009)
4. Kadalbajoo, M.K., Kumar, V.: B-spline solution of singular boundary value problems. Applied Mathematics and Computation 182, 1509–1513 (2006)
5. Kadalbajoo, M.K., Arorar, P.: B-splines with artificial viscosity for solving singularly perturbed boundary value problems. Mathematical and Computer Modelling 52, 654–666 (2010)
6. Kadalbajoo, M.K., Arorar, P., Gupta, V.: Collocation method using artificial viscosity for solving stiff singularly perturbed turning point problem having twin boundary layers. Computers and Mathematics with Applications 61, 1595–1607 (2011)
7. Wang, R.-h., Li, C.-j., Zhu, C.-g.: Computational Geometry. Science Press, Beijing (2008)
8. Ren, Y.-j.: Numerical Analysis and MATLAB Implementation. Higher Education Press (2008)
9. Kadalbajoo, M.K., Yadaw, A.S., Kumar, D., Gupta, V.: Comparative study of singularly perturbed two-point BVPs via:Fitted-mesh finite difference method, B-spline collocation method and finite element method. Applied Mathematics and Computation 204, 713–725 (2008)
10. Bawa, R.K., Natesan, S.: A Computational Method for Self-Adjoint Singular Perturbation Problems Using Quintic Spline 50, 1371–1382 (2005)
11. Kadalbajoo, M.K., Patidar, K.C.: A survey of numerical techniques for solving singularly perturbed ordinary differential equations. Applied Mathematics and Computation 204, 713–725 (2008)
12. Rao, S.C.S., Kumar, M.: Optimal B-spline collocation method for self-adjoint singularly perturbed boundary value problems. Applied Mathematics and Computation 188, 749–761 (2007)
13. Jayakumar, J.: Improvement of numerical solution by boundary value technique for singularly perturbed one dimensional reaction diffusion problem. Applied Mathematics and Computation 142, 417–447 (2003)

Computer Simulation and Calculation for Luminance Uniformity of Straight Line Radiation Sources

Jia Chen[1], Jia-wei Tan[1], Qing-huai Liu[1], and Fan-rong Zhang[2]

[1] School of Basic Science, Changchun University of Technology,
130012, Changchun, China
[2] College of Information Technology, Luoyang Normal University,
471022, Luoyang, China
{Jia Chen,Jia-wei Tan,Qing-huai Liu,
Fan-rong Zhang}singer310@126.com

Abstract. In order to avoid the macroscopically inhomogeneous characteristics of large area films which prepared by Photo-assisted MOCVD (Metal Organic Chemical Vapor Deposition), processed by RIP (Rapid isothermal processing) technique, etc., the calculation of luminance uniformity is necessary. This thesis gives a basic calculation method about the radiation intensity of illumination caused by halogen-tungsten lamp like line radiation sources. A concept of average radiation intensity (ARI) of illumination was proposed to give quantification when radiation intensity is Rapidly Changed. The distribution of radiation sources influence on ARII is discussed. The numerical solution carried out by computer simulation directly confirms that the layout of radiation sources in this paper enhances the luminance uniformity. Consequently, the macroscopically inhomogeneous of the film is improved.

Keywords: radiation intensity of illumination; computer simulation; Photo-assisted.

1 Introduction

In the area of electronic film materials, the film macroscopically inhomogeneous characteristics should be usually avoided [1-4]. One of the radiation heating methods is light form halogen-tungsten lamps, such as Photo-assisted Metal Organic Chemical Vapor Deposition (MOCVD), Rapid isothermal processing (RIP), and so on [5-8]. The distribution of radiation intensity of illumination on substrate surface mainly depends on the Halogen tungsten lamp arrangement. That is one of the most important factors which affect substrate surface temperature uniformity, and then result in macroscopically inhomogeneous characteristics [9-11]. Therefore, the film uniform certainly depends on Halogen tungsten lamp arrangement.

In order to calculate Light uniformity, the concept of average radiation intensity (ARI) of illumination is proposed, and then the distribution of the radiation intensity of illumination in theory is deduced to attain the optimized arrangement. In addition, the computer simulation results are given, by which the problems that the epitaxial film is of less uniformity can be analyzed and explained, According to simulation results, the method to improve the Light uniformity can be offered.

B. Liu and C. Chai (Eds.): ICICA 2011, LNCS 7030, pp. 750–757, 2011.
© Springer-Verlag Berlin Heidelberg 2011

2   Radiation Intensity Produced by Point Source

It is supposed that radiant power is Ψ in point source Q, and then radiant intensity is [11]:

$$I = \frac{\Psi}{4\pi} \tag{1}$$

As shown in figure 1,take any point A in plane Π, dS is plane of illumination, and $d\Omega$ is solid angle,r is the distance from A to Q, then

$$d\Omega = \frac{dS \cos\theta}{r^2} \tag{2}$$

θ is the angle between normal vector n and AQ,for the radiant energy flux

$$d\Psi = Id\Omega \tag{3}$$

So the radiation intensity produced by point source Q is

$$E = \frac{d\Psi}{dS} = \frac{\Psi}{4\pi} \cdot \frac{\cos\theta}{r^2} \tag{4}$$

3   Radiation Intensity Produced by Line Source

As is shown in figure 2, length of homogeneous line source is $2l$, and Q is its central point.Set Cartesian coordinate system to satisfy that x axis is parallel to line source, coordinate of Q is $(0, y_0, z_0)$. dx is line element at any point (x, y_0, z_0) in the line source.

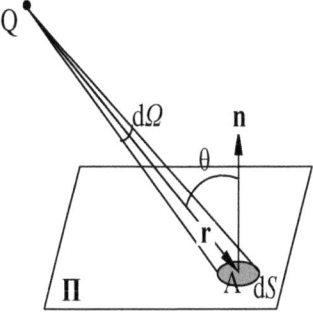

Fig. 1. Diagram of point source Q and plane Π

On the basis of result of radiation intensity produced by point source, we get that the radiation intensity in $(x_1, y_1, 0)$ produced by line element dx is

$$dE = \frac{\Psi z_0}{4\pi L} \cdot \frac{dx}{\left[(x_1 - x)^2 + (y_1 - y_0)^2 + z_0^2 \right]^{3/2}} \tag{5}$$

then it is derived that the radiation intensity in $(x_1, y_1, 0)$ on the plane produced by line source with length being 2l is

$$E = \int dE = \int_{-l}^{l} \frac{\Psi z_0}{8\pi L} \cdot \frac{dx}{\left[(x_1 - x)^2 + (y_1 - y_0)^2 + z_0^2 \right]^{3/2}}$$

$$= \frac{\Psi z_0}{8\pi l (y_1 - y_0)^2 + 8\pi L z_0^2} \left[\frac{l - x_1}{\sqrt{(l - x_1)^2 + (y_0 - y_1)^2 + z_0^2}} + \frac{l + x_1}{\sqrt{(l + x_1)^2 + (y_0 - y_1)^2 + z_0^2}} \right] \tag{6}$$

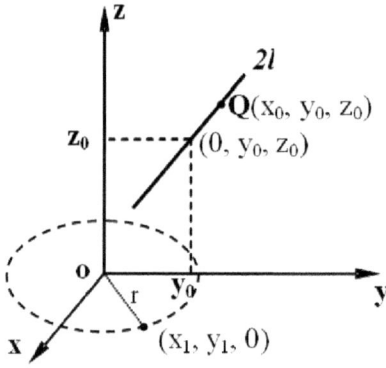

Fig. 2. Diagram of line source Q and plane Π

4   Average Radiation Intensity

In fact, substrate is not always fixed. For example, substrate is doing uniform circular motions when 1″ and 2″ epitaxial film are made. So the radiation intensity of every point in the substrate changes periodically, and we can sign it with periodic function E(t), the period of which is T. i.e.

$$E(t) = E(t+T) \tag{7}$$

In order to characterize the effect of radiation intensity in Rotating substrate, we give a new concept : average radiation intensity (ARI), which is expressed as the quotient of the sum of radiation intensity received in unit area within the time t divided by the time t. we denote it with $\overline{E}_t$, then we have

$$\overline{E}_t = \frac{1}{t} \int_0^t E(t)dt \tag{8}$$

Suppose Rotation of the substrate is at a constant speed n .let r be the distance from some point in the substrate to the center of the substrate. When t = 0,the point is designated by (r,0) in polar coordinates, and when $t = t_0$ the point is designated by $(r\cos 2n\pi, r\sin 2n\pi)$ in Cartesian coordinate system. According to 6,it can be derived that the radiation intensity is $E(r\cos 2n\pi, r\sin 2n\pi)$ at the point when the time is t. So the (ARI) at arbitrarily Periods of time $t_1 \sim t_2$ is shown as following formula:

$$\overline{E}_t = \frac{1}{t_2 - t_1} \int_{t_1}^{t_2} E(r\cos 2n\pi, r\sin 2n\pi)dt$$

Set $t_2 - t_1 = \dfrac{k}{n} + t_0 \ (0 \le t_0 < \dfrac{1}{n}, k \in Z^+)$, then:

$$\overline{E}_t = \frac{1}{\dfrac{k}{n}+t_0} \int_{t_1}^{t_1+k/n} E(r\cos 2n\pi, r\sin 2n\pi)dt + \frac{1}{\dfrac{k}{n}+t_0} \int_{t_1+k/n}^{t_1+k/n+t_0} E(r\cos 2n\pi, r\sin 2n\pi)dt$$

$$= \frac{n}{1+\dfrac{nt_0}{k}} \int_0^{1/n} E(r\cos 2n\pi, r\sin 2n\pi)dt + \frac{1}{\dfrac{k}{n}+t_0} \int_{t_1}^{t_1+t_0} E(r\cos 2n\pi, r\sin 2n\pi)dt$$

Supposed k>>1, then

$$\overline{E}_t = n \int_0^{1/n} E(r\cos 2n\pi, r\sin 2n\pi)dt \tag{9}$$

When growth time of epitaxial film $t_2 - t_1$ is much larger than period of rotation of the substrate $1/n$, it is clear that the ARI of a point in the substrate is only relevant to the distance between the point and the center of the substrate.

5   Simulation of Light Uniformity

Although the diversity of small area film is hardly considered, considering the large area film preparation, such as 2″ film, the problem caused by the diversity of radiation uniformity is serious. Radiation uniformity is indispensable for preparation of larger diameter circle, more pieces of film and superconductive strap. 5 lamps tiled are designed on the quartz clapboard. As is shown in figure 4.The design is simple arrangement, and in favor of shorting the distance between the lamp and the substrate, besides it can increase the lighting efficiency, however it can cause light no uniformity. In order to ensure radiation intensity uniformity, the rotation substrate is designed. Analog computation is given as the following:

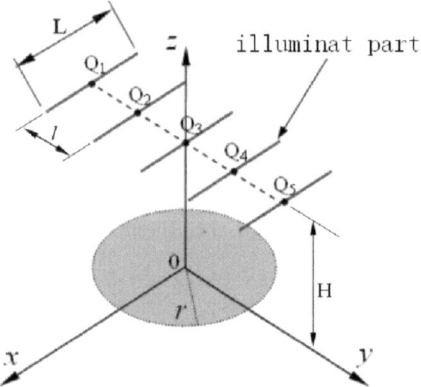

Fig. 3. Simplified model of 5 Halogen tungsten lamps tiled in system of coordinates

Firstly, we set up the assumptions model.

1, Lamps are simple regarded as line source with favorable luminance uniformity, Length of which is L.

2, influence of the reflected light from Reaction chamber wall and lampshade to radiation intensity in the substrate is neglected.

3, influence of light refracting from Quartz baffle to radiation intensity in the substrate is neglected.

The symbol of the model is shown in figure 4, the center of the five segments are (0,-2l,H), (0,-l,H), (0,0,H), (0,l,H), (0,2l,H) respectively. Make use of formula (6) we can calculate the radiation intensity at the same point (x, y) produced by every lamp. Use $E_1(x, y), E_2(x, y)$ ... $E_5(x, y)$ to show the radiation intensity in the point (x, y) in the substrate which is produced by the lamp accordingly, $E(x, y)$ signify total radiation intensity in the point (x, y) produced by 5 lamps. Then we have

$$E_i(x,y) = \frac{\Psi H}{4\pi L\left(y+(3-i)l\right)^2 + 2\pi L z_0^2} \left[\frac{\frac{L}{2}-x}{\sqrt{\left(\frac{L}{2}-x\right)^2 + \left((3-i)l+y\right)^2 + H^2}} + \frac{\frac{L}{2}+x}{\sqrt{\left(\frac{L}{2}+x\right)^2 + \left((3-i)l+y\right)^2 + H^2}} \right]$$

$$i=12345 \quad E(x, y) = E_1(x, y) + E_2(x, y) + E_3(x, y) + E_4(x, y) + E_5(x, y) \tag{10}$$

Suppose Rotation of the substrate is at a constant speed n, use the result of part 4, then $\overline{E}_t = n \int_0^{1/n} E\left(r \cos 2n\pi t, r \sin 2n\pi t\right) dt$, and $E(x, y)$ is given in (10)

Based on the actual data, for example L = 63 mm, $\Psi_0 = 1.5kw$, l = 25mm, H = 70 mm, numerical result of ARI produced from 5 lamps tiled is given by computer program Matlab [12], and the figure of ARI, radiation intensity contour line are given as the following:

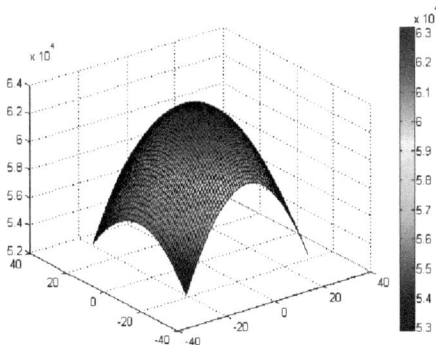

Fig. 4. Three-dimensional diagram ARI produced from 5 lamps tiled

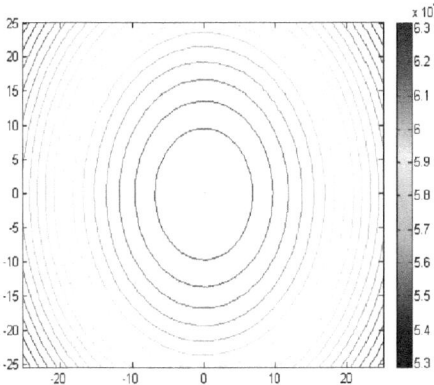

Fig. 5. Radiation intensity contour line produced from 5 lamps tiled

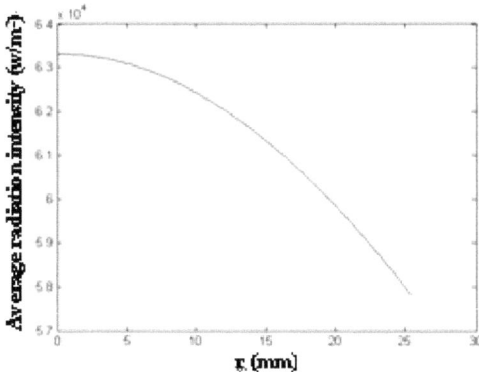

Fig. 6. The relationship between the ARI and radius of substrate

According to numerical calculation, we derive 3 conclusions:

(1) Tendency of radiation intensity distribution in the substrate is strong in the center and weak in the fringe. (2) In the substrate with high speed rotation, the gradient of ARI is high near the center and low near the fringe. And luminance uniformity increase as proximity to the center. (3) In the substrate with high speed rotation, the diversity between the center and the fringe increase with the increase of film area. For example, the diversity of 2″ sample reaches 8.7%, while the diversity of 1″ sample only reaches 2.3%.

Based on the basic principle of light, using software MATLAB, a method calculating ARI produced by several light sources and the computer simulation results about 5 lamps tiled are given. According to contrast of luminance uniformity about the 2″ sample and 1″ sample, we analyze and explained the problems that the epitaxial film is of less uniformity.

Establishment of this model facilitate the research about light uniformity photo assisted MOCVD system and another photo assisted equipment

References

1. Matsumura, T., Sato, Y.: A Theoretical Study on Van Der Pauw Measurement Values of Inhomogeneous Compound Semiconductor Thin Films. J. Mod. Phys. 1, 340–347 (2010)
2. Yom-Tov, N., Saguy, C., Bolker, A., Kalish, R., Yaish, Y.E.: Accurate Carrier-Type Determination of Nonhomogeneously Doped Diamond. Journal of Applied Physics 108, 043711 (2010)
3. Bierwagen, O., Ive, T., Van de Walle, C.G., Speck, J.S.: Causes of incorrect carrier-type identification in van der Pauw–Hall measurements. Applied Physics Letters 93, 242108 (2008)
4. Ikegami, K., Yoshiyama, T., Maejima, K., Shibata, H., Tampo, H., Niki, S.: Optical dielectric constant inhomogeneity along the growth axis in ZnO-based transparent electrodes deposited on glass substrates. Journal of Applied Physics 105, 093713 (2009)
5. Li, G., Fang, X., Zhao, L., Shanwen, Gao, Z., Li, W., Yin, J., Zhang, B., Du, G., Chou, P., He, L.: Chinping Chen. Physica C 468, 2213–2218 (2008)
6. Singh, R., Radpour, F., Chou, P.: Comparative study of dielectric formation by furnace and rapid isothermal processing. J. Vac. Sci. Technol. A7, 1456 (1989)
7. Michalowski, P., Schmidt, M., Schmidl, F., Grosse, V., Kuhwald, D., Katzer, C., Hübner, U., Seidel, P.: Engineering of YBa2Cu3O7−δ grain boundary Josephson junctions by Au nanocrystals. Physica Status Solidi - Rapid Research Letters 5, 268–270 (2011)
8. Ignatiev, A., Chou, P.C., Chen, Y., Zhang, X., Tang, Z.: Coated conductor development by photo-assisted MOCVD growth of YBCO thick films and buffer layers. Physica C 341–348, 2309–2313 (2000)
9. Zeng, J., Chou, P., Zhang, X., Tang, Z., Ignatiev, A.: Single liquid precursor delivery photo-assisted metalorganic chemical vapor deposition of high quality YBa2Cu3O7-x thin/thick films. Physica C 377, 235–238 (2002)

10. Zhong, Q., Chou, P.C., Li, Q.L., Taraldsen, G.S., Ignatiev, A.: High-rate growth of purely a-axis oriented YBCO high-Tc thin films by photo-assisted MOCVD. Physica C 246, 288–296 (1995)
11. Zhao, K., Zhong, X.: Optics, 1st edn., page 124. Beijin University Publishing Company (1984)
12. He, Z., Zheng, f.: Matlab, Numerical Analysis. Machine Press, Beijing (2009)

The Infeasible Combined Homotopy Method for a Class of Programming Problems

Jia-wei Tani, Jia Chen, and Qing-huai Liu[*]

School of Basic Science, Changchun University of Technology,
130012, Changchun, China
tanjiawei@mail.ccut.edu.cn,
liuqh6195@126.com

Abstract. In this paper, a class of nonconvex programming problems with inequality constraint functions was studied, under the conditions that the boundary is regular and the feasible set is connected and bounded. At first, the infeasible constraint functions were constructed. Then, the infeasible combined homotopy equation was constructed to solve this problem. At last the existence of a smooth homotopy path from any initial point to the solution of the problem was established and the convergence of the method is proved.

Keywords: Infeasible point, Combined Homotopy, Nonlinear Programming.

1   Introduction

In 1976, the homotopy method was introduced and studied by Kellogg [1], Smale [2], and Chow [3]. From then on, this method has become a powerful toll in finding solutions of algebra equations, fixed-point, mathematical programming et. Many researchers studied and modified this method. Reference [4-14] presented a series new interior combined homotopy method. In 1996, the well known combined homotopy method was established by Feng, Yu, Liu and Xu for solving nonlinear programming problems and the algorithm generated by this method has been proved to be globally convergent for a wide range of nonlinear programming problems. In 2006, boundary moving combined homotopy method was established by Yu et al [13] for solving nonlinear programming problems, which need not interior initial point. In 2010, Wang and Liu [14] improved the constraint shifting combined homotopy method established by Yu and his students for nonlinear programming general problems when the feasible set satisfies normal cone conditions. In this paper, the following nonconvex programming problem (P) was studied.

$$\begin{cases} \min f(x) \\ s.t.\ g_i(x) \le 0, i \in M \end{cases}$$

---

[*] Corresponding author.

B. Liu and C. Chai (Eds.): ICICA 2011, LNCS 7030, pp. 758–764, 2011.

Where $x \in R^n$, $M = \{1, 2, \cdots, m\}$, $f, g_i : R^n \to R$ are sufficiently smooth functions. But the interior initial point can not be find easily and the shifting constraint functions can not be established following the same formula, so that for solving this problem (P), under the conditions that the feasible set satisfied the normal cone condition, the infeasible constraint shifting functions are first constructed. Then the homotopy method for solving this problem is developed. We show that the algorithm generated by this method is feasible and convergent regardless the position of the initial point, a feasible interior point or an infeasible interior point.

2   Notation and Preliminaries

Throughout this paper the notation R^n was used to denote the n-dimensional real vector space and $\|\bullet\|$ to denote the Euclidean norm. For convenient, we provided that in the real vector space all the vectors were used as column vectors and we usually abbreviated the vector $(x^T, y^T) \in R^n \times R_+^m$ to (x, y). The following notations were also used to solve this problem (P).

$$\Omega = \{x \in R^n | g_i(x) \leq 0, i \in M\}$$
$$\Omega^0 = \{x \in R^n | g_i(x) < 0, i \in M\}$$
$$\partial\Omega = \Omega \setminus \Omega^0$$
$$g(x) = (g_1(x), g_2(x), \cdots, g_m(x))^T$$
$$I(x) = \{i | g_i(x) = 0, i \in M\}$$

For solving problem (P), we made a series of assumptions:

H1 $f(x), g_i(x) (i \in M)$ are all at least twice continuously differentiable functions.

H2 the feasible set Ω was connected and bounded and the interior of Ω are not empty.

H3 $\forall x \in \partial\Omega(t), \{\nabla_x g_i(x,t) | i \in I(x)\}$ is linearly positive independent.

H4 $\Omega(1)$ satisfies   the   normal   cone   conditions.
$\forall x \in \partial\Omega(1), \{x + \sum_{i \in I(x)} y_i \nabla_x g_i(x, 1) | y_i \geq 0\} \cap \Omega(1) = \{x\}$

Let $U \in R^n$ be an open set and let $\varphi : U \to R^p$ be a $C^\alpha (\alpha > \max\{0, n - p\})$ mapping.

We say that $y \in R^p$ is a regular value for φ, if

$$Range\left[\frac{\partial\varphi(x)}{\partial x}\right] = R^p, \forall x \in \varphi^{-1}(y).$$

In this part, some necessary lemmas were given.

Lemma 1. Let $V \subset R^n, U \subset R^m$ are open sets, and let $\varphi : V \times U \to R^k$ be a C^α mapping, where $\alpha > \max\{0, m-k\}$. If $0 \in R^k$ is a regular value of φ, then for almost all $a \in V$, 0 is a regular value of $\varphi_a = \varphi(a, \bullet)$. (Parameterized Sard Theorem on smooth manifold)[11]

Lemma 2. let $\varphi : U \subset R^n \to R^P$ be a $C^\alpha (\alpha > \max\{0, n-p\})$ mapping. If $0 \in R^P$ is a regular value of φ, then $\varphi^{-1}(0)$ consists of some $(n-p)$ – dimensional C^α manifolds. (The inverse image theorem)[11]

Lemma 3. One-dimensional smooth manifold is diffeomorphic to a unit circle or a unit interval. (Classification theorem of one-dimensional smooth manifold)[12]

3   Main Results

For a given $x^{(0)} \in R^n$, we supposed it was not a feasible point of problem (P), let $C = \max\{g_i(x^{(0)}), i \in M\} + \varepsilon$ $(\varepsilon > 0)$.

The infeasible constraint functions were constructed as followed:

$$g_i(x, t) = g_i(x) + tC, (t \in [0,1]), i \in M .$$

Let

$$\Omega(t) = \{x \in R^n | g_i(x, t) \leq 0, i \in M\}$$
$$\Omega^0(t) = \{x \in R^n | g_i(x, t) < 0, i \in M\}$$
$$I(x, t) = \{i | g_i(x, t) = 0, i \in M\}$$
$$g(x, t) = (g_1(x, t), g_2(x, t), \cdots, g_m(x, t))^T$$

It was easy to see that $\Omega^0(0) = \Omega^0$.

In this paper, we always suppose that H1, H2, H3 and H4 hold. We construct an infeasible combined homotopy equation as follows

$$H(\omega, \omega^{(0)}, t) = \begin{pmatrix} (1-t)(\nabla f(x) + \sum_{i=1}^{m} y_i \nabla_x g_i(x, t)) + t(x - x^{(0)}) \\ Yg(x, t) - tY^{(0)} g(x^{(0)}, 1) \end{pmatrix} = 0 \quad (1)$$

Where $x^{(0)} \in \Omega(1)$, $y^{(0)} = (y_1^{(0)}, y_2^{(0)}, \cdots, y_m^{(0)}) \in R_{++}^m$,

$$y = (y_1, y_2, \cdots, y_m)^T , \quad \omega^{(0)} = (x^{(0)}, y^{(0)}), \omega = (x, y),$$
$$Y = diag(y_1, y_2, \cdots, y_m).$$

If $t = 1$, the infeasible homotopy (1) was equivalent to

$$H(\omega, \omega^{(0)}, 1) = \begin{pmatrix} x - x^{(0)} \\ Yg(x,1) - Y^{(0)}g(x^{(0)},1) \end{pmatrix} = 0$$

Clearly this equation had a unique solution $\omega^{(0)} = (x^{(0)}, y^{(0)})$.

If $t = 0$, the infeasible homotopy (1) was equivalent to

$$H(\omega, \omega^{(0)}, 0) = \begin{pmatrix} \nabla f(x) + \sum_{i=1}^{m} y_i \nabla_x g_i(x,0) \\ Yg(x,0) \end{pmatrix} = \begin{pmatrix} \nabla f(x) + \sum_{i=1}^{m} y_i \nabla g_i(x) \\ Yg(x) \end{pmatrix} = 0$$

It is the K-K-T system of problem (P). Because

$$\lim_{t \to 0^+} \nabla_x g_i(x,t) = \nabla g_i(x).$$

Obviously, $H(\omega^{(0)}, \omega^{(0)}, 1) = 0$.

Let $H^{-1}(0) = \{(\omega, t) \in \Omega(t) \times R_+^m \times (0,1] | H(\omega, \omega^{(0)}, t) = 0\}$

Lemma 4. The homotopy equation (1) generates a smooth curve $\Gamma_{\omega^{(0)}}$ starting from $(x^{(0)}, y^{(0)}, 1)$, if 0 is a regular value of H for almost all $\omega^{(0)} \in \Omega(1) \times R_+^m$.

Proof. This $H'(\omega, \omega^{(0)}, t)$ was used to represent the Jacobi matrix of H. So that

$$H'(\omega, \omega^{(0)}, t) = \left(\frac{\partial H}{\partial \omega}, \frac{\partial H}{\partial \omega^{(0)}}, \frac{\partial H}{\partial t} \right).$$

If $\omega^{(0)} = (x^{(0)}, y^{(0)}) \in \Omega(1) \times R^m$, we knew that

$$\frac{\partial H}{\partial \omega^{(0)}} = \begin{pmatrix} -tE & 0 \\ -tY^{(0)}\nabla_x g(x^{(0)},1) & -tG(x^{(0)},1) \end{pmatrix},$$

This notation E was used to represent the identity matrix.

$$G(x^{(0)},1) = diag(g_1(x^{(0)},1), g_2(x^{(0)},1), \cdots, g_m(x^{(0)},1)).$$

Clearly, $\left| \frac{\partial H}{\partial \omega^{(0)}} \right| = (-t)^{m+n} \prod_{i=1}^{m} g_i(x^{(0)},1) \neq 0$. So that, $H'(\omega, \omega^{(0)}, t)$ is a full row rank matrix. By lemma 1, we know that 0 is a regular value of H and by Lemma 2, $H^{-1}(0)$ consists of some smooth curves. The equation $H(\omega^{(0)}, \omega^{(0)}, 1) = 0$ implies that there exists a smooth curve $\Gamma_{\omega^{(0)}}$ starting from $(\omega^{(0)}, 1)$.

Lemma 5. $\Gamma_{\omega^{(0)}}$ is a bounded curve in $\Omega(t) \times R_+^m \times (0,1]$, if 0 is a regular value of H, For a given $\omega^{(0)} = (x^{(0)}, y^{(0)}) \in \Omega(1) \times R_{++}^m$.

Proof. By equation (1), it is easy to see that $\Gamma_{\omega^{(0)}} \subset \Omega^*(t) \times R_+^m \times (0,1]$. If the smooth curve $\Gamma_{\omega^{(0)}}$ generated by the homotopy method was an unbounded curve, then there must exist a sequence of points $(x^{(k)}, y^{(k)}, t_k) \in \Gamma_{\omega^{(0)}}$, such that $\|(x^{(k)}, y^{(k)}, t_k)\| \to \infty$ as $k \to \infty$. By the second part of equation (1), we knew

$$y_i^{(k)} g_i(x^{(k)}, t_k) - t_k y_i^{(0)} g_i(x^{(0)}, 1) = 0, i \in M \tag{2}$$

It follows immediately that $g_i(x^{(k)}, t_k) \le 0$, $x^{(k)} \in \Omega(t_k)$. $\Omega(t_k)$ is a bounded set and thus $\{x^{(k)}\}$ is a bounded sequence. Such that, there exists a convergent subsequence, for convenient also denoted by $\{x^{(k)}\}$. Such that

$$x^{(k)} \to x^*, t_k \to t_*, \|y^{(k)}\| \to \infty$$

Let $I = \{i \in M \mid y_i^{(k)} \to \infty\}$, if $y_i^{(k)} \to \infty$ as $k \to \infty$, we have, from (2),

$$g_i(x^{(k)}, t_k) = t_k y_i^{(0)} (y_i^{(k)})^{-1} g_i(x^{(0)}, 1)$$

And hence

$$g_i(x^*, t_*) = \lim_{k \to \infty} g_i(x^{(k)}, t_k) = 0$$

Which implies $x^* \in \Omega(t_*)$, and obviously, $I \subseteq I(x^*, t_*)$.

If $t_* = 1$, the first equation of (1) was rewrote as

$$(1-t_k) \sum_{i \in I(x^*, t_*)} y_i \nabla_x g_i(x^{(k)}, t_k) + t_k(x^{(k)} - x^{(0)}) = -(1-t_k)(\nabla f(x^{(k)}) + \sum_{i \notin I(x^*, t_*)} y_i \nabla_x g_i(x^{(k)}, t_k))$$

$\{x^{(k)}\}$ and $\{y_i^{(k)}\}$ were bounded sets when $i \in (x^*, t^*)$. Such that

$$x^* - x^{(0)} + \sum_{i \in I(x^*, t^*)} \alpha_i \nabla_x g_i(x^*, t^*) = 0$$

As $k \to \infty$. This was impossible by H4.

If $t_* \in [0,1)$, the first equation of (1) was rewrote as,

$$(1-t_k)(\nabla f(x^{(k)}) + \sum_{i \in I(x^*, t_*)} y_i \nabla_x g_i(x^{(k)}, t_k)) + t_k(x^{(k)} - x^{(0)}) = -(1-t_k) \sum_{i \in I(x^*, t_*)} y_i \nabla_x g_i(x^{(k)}, t_k) \tag{3}$$

Let

$$\left\| \sum_{i \in I(x^*, t_*)} y_i^{(k)} \nabla_x g_i(x^{(k)}, t_k) / \max_{i \in II(x^*, t_*)} y_i^{(k)} \right\| = \alpha$$

It was easy to see that

$$\alpha > 0 \left\| \sum_{i \in I(x^*, t_*)} y_i^{(k)} \nabla_x g_i(x^{(k)}, t_k) \right\| = \max_{i \in II(x^*, t_*)} y_i^{(k)} \left\| \sum_{i \in I(x^*, t_*)} y_i^{(k)} \nabla_x g_i(x^{(k)}, t_k) / \max_{i \in II(x^*, t_*)} y_i^{(k)} \right\| \ge \alpha \max_{i \in II(x^*, t_*)} y_i^{(k)}$$

One could see that the left hand of (3) is bounded, while the right hand of (3) is unbounded, clearly, this is impossible.

Theorem 1. Let H be defined by (1), then for almost all $\omega^{(0)} = (x^{(0)}, y^{(0)}) \in \Omega(1) \times R_{++}^m$, the zero-point set $H^{-1}(0)$ of homotopy map (1) contains a smooth curve $\Gamma_{\omega^{(0)}}$, which starts from $(\omega^{(0)},1)$. As $t \to 0$, the limit point of $\Gamma_{\omega^{(0)}}$ is a K-K-T point of the problem (P).

Proof by Lemma 6 and Lemma 7, we knew that theorem 1 hold.

4   Tracing the Homotopy Path

We studied how to trace the homotopy path $\Gamma_{\omega^{(0)}}$ numerically in this section.

A classical program was the 'predictor-corrector' method. This method used an explicit difference scheme for solving numerically to give a predictor point and then uses a global convergent iterative method for solving the nonlinear system of equation to give a corrector point. We formulate a simple predictor-corrector procedure as following.

Algorithm 4.1 (NLP's Euler-Newton method) (see [15])

Step 0: Selected a initial point $\omega^{(0)} = (x^{(0)}, y^{(0)}) \in \Omega(1) \times R_{++}^m$ randomly, an suitable initial step-length $h_0 > 0$ and three small positive number $\varepsilon_1 > 0, \varepsilon_2 > 0, \varepsilon_3 > 0$ and note $k = 1$;

Step 1: (a) Solve an unite tangent vector $\xi^{(k)} \in R^{n+m+1}$;

(b) Choose the direction $\eta^{(k)}$ of the predictor step. If the sign of the determinant

$$\left| \begin{pmatrix} H'_{\omega^{(0)}}(\omega^{(k)}, t_k) \\ \xi^{(k)} \end{pmatrix} \right|$$ is $(-1)^{m+1}$, let $\eta^{(k)} = \xi^{(k)}$. If the sign of the determinant

$$\left| \begin{pmatrix} H'_{\omega^{(0)}}(\omega^{(k)}, t_k) \\ \xi^{(k)} \end{pmatrix} \right|$$ is $(-1)^m$, let $\eta^{(k)} = -\xi^{(k)}$;

Step 2: Solve a corrector point $(\omega^{(k+1)}, t_{k+1})$:

$$(\bar{\omega}^{(k)}, \bar{t}_k) = (\omega^{(k)}, t_k) + h_k \eta^{(k)}$$
$$(\omega^{(k+1)}, t_{k+1}) = (\bar{\omega}^{(k)}, \bar{t}_k) - H'_{\omega^{(0)}}(\bar{\omega}^{(k)}, \bar{t}_k)^+ H_{\omega^{(0)}}(\bar{\omega}^{(k)}, \bar{t}_k)$$

Case (a) $\left\| H_{\omega^{(0)}}(\omega^{(k+1)}, t_{k+1}) \right\| \le \varepsilon_1, h_{k+1} = \min\{h_0, 2h_k\}$, go to step 3;

Case (b) $\left\| H_{\omega^{(0)}}(\omega^{(k+1)}, t_{k+1}) \right\| \in (\varepsilon_1, \varepsilon_2), h_{k+1} = h_k$, go to step 3;

Case (c) $\left\| H_{\omega^{(0)}}(\omega^{(k+1)}, t_{k+1}) \right\| \ge \varepsilon_2, h_{k+1} = \max\{2^{-25} h_0, 0.5 h_k\}, k = k+1$, go to step 2;

Step 3: If $t_{k+1} \le \varepsilon_3$, then stop, else $k = k+1$, and go to Step 1.

In Algorithm 4.1, $H'_{\omega^{(0)}}(\omega, t)^+ = H'_{\omega^{(0)}}(\omega, t)^T (H'_{\omega^{(0)}}(\omega, t) H'_{\omega^{(0)}}(\omega, t)^T)^{-1}$ is the Moore-Penrose inverse of $H'_{\omega^{(0)}}(\omega, t)$.

In Algorithm 4.1, we could not compute the arclength parameter s explicitly. At any random point on $\Gamma_{\omega^{(0)}}$, the tangent vector had two opposite directors. When s moved following the positive direction, its value would increase. When s moved following the negative direction, its value would decrease. If we chose the negative direction, we would go back to the initial point. Therefore we should follow the positive directions. The formula in step 1(b) of Algorithm 4.1 that determines the positive direction is based on a basic theory of homotopy method [12], that is, the position η at any point (ω, t) on $\Gamma_{\omega^{(0)}}$ keeps the sign of the determinant $\left| \begin{pmatrix} H'_{\omega^{(0)}}(\omega^{(k)}, t_k) \\ \eta^T \end{pmatrix} \right|$ invariant. We have the following proposition. [16]

Proposition If $\Gamma_{\omega^{(0)}}$ is smooth, then the positive direction $\eta^{(0)}$ at the initial point $\omega^{(0)}$ satisfies sign $\left| \begin{pmatrix} H'_{\omega^{(0)}}(\omega^{(k)}, t_k) \\ \eta^{(0)T} \end{pmatrix} \right| = (-1)^{m+1}$

References

1. Smale, S.: A convergent process of price adjustment and global Newton method. J.Math. Econom. 3, 1–14 (1976)
2. Smale, S.: A convergent process of price adjustment and global Newton method. J.Math. Econom. 3, 1–14 (1976)
3. Chow, S.N., Mallet-Paret, J., Yorke, J.A.: Finding zeros of maps: homotopy methods that are constructive with probability one. Math. Comput. 32, 887–899 (1978)
4. Lin, Z.H., Li, Y., Feng, G.C.: A combined homotopy interior method for general nonlinear programming problem. Appl. Math. Comput. 80, 209–226 (1996)
5. Lin, Z.H., Zhu, D.L., Sheng, Z.P.: Finding a Minimal Efficient Solution of a Convex Multiobjective Program. Journal of oprimization Theory an Applications 118, 587–600 (2003)
6. Feng, G.C., Lin, Z.H., Bo, Y.: Existence of Interior Pathway to a Karush-Kuhn-Tucker Point of a Nonconvex Programming Problem. Nonlinear Analysis, Theory, Methods Applications 32(6), 761–768 (1998)
7. He, L., Dong, X.G., Tan, J.W., Liu, Q.H.: The Aggregate Homotopy Method for Multiobjective Max-min Problems. I.J. Image, Graphics and Signal Processing 2, 30–36 (2011)
8. Yu, B., Liu, Q.H., Feng, G.C.: A combined homotopy interior point method for nonconvex programming with pseudo cone condition. Northeast. Math. J. 16, 587–600 (2000)
9. Liu, Q.H., Yu, B., Feng, G.C.: A combined homotopy interior point method for nonconvex nonlinear programming problems under quasi normal cone condition, Acta. Math. Appl. Sinica. 26, 372–377 (2003)
10. Yang, Y., Lu, X., Liu, Q.: A Combined homotopy infeasible interior-point for convex nonlinear programming. Northeast Math. J. 22(2), 188–192 (2005)
11. Zhang, Z.S.: Introduction to Differential Topology. Bejing University Press, Beijing (2002)
12. Naber, G.L.: Topological Methods in Euclidean Spaces. Cambridge Unversity Press, London (1980)
13. Yu., B., Shang, Y.F.: Boundary Moving Combined Homotopy Method for Nonconvex Nonlinear Programming. Journal of Mathematical Research and Exposition (4), 831–834 (2006)
14. Wang, X.Y., Jiang, X.W., Liu, Q.H.: The Constraint Shifting Combined Homotopy Method For Bounded General Programming Problems. Journal of Information & Computational Science 7, 1–10 (2010)
15. Allgower, E.L., Georg, K.: Numerical Continuation Methods: An Introduction. Springer, Berlin (1990)

Generating Query Plans for Distributed Query Processing Using Genetic Algorithm

T.V. Vijay Kumar and Shina Panicker

School of Computer and Systems Sciences,
Jawaharlal Nehru University,
New Delhi-110067, India
tvvijaykumar@hotmail.com,
shinapanicker@gmail.com

Abstract. Query Processing is a key determinant in the overall performance of distributed databases. It requires processing of data at their respective sites and transmission of the same between them. These together constitute a distributed query processing strategy (DQP). DQP aims to arrive at an efficient query processing strategy for a given query. This strategy involves generation of efficient query plans for a distributed query. In case of distributed relational queries, the number of possible query plans grows exponentially with an increase in the number of relations accessed by the query. This number increases further when the relations, accessed by the query, have replicas at different sites. Such a large search space renders it infeasible to find optimal query plans. This paper presents a query plan generation algorithm that attempts to generate optimal query plans, for a given query, using genetic algorithm. The query plans so generated involve fewer sites, thus leading to efficient query processing. Further, experimental results show that the proposed algorithm converges quickly towards optimal query plans for an observed crossover and mutation probability.

Keywords: Distributed Query Processing, Genetic Algorithm.

1 Introduction

The enormous requirement of timely and accurate information in the current competitive era has led to databases exceeding the physical limitations of a centralized system and necessitates integration of the already existing sources of data, which may be geographically dispersed, to form a distributed database system[16]. The performance of a distributed database system is ascertained by its ability to process queries in an efficient manner [17][19]. Generally, query processing in distributed databases is carried out by decomposing the query into sub-queries which, thereafter, are processed at the respective sites. Their results are thereafter integrated and presented to the user. Thus, query processing in distributed database systems requires transmission of data among sites distributed over a network[11][14][15]. The data transmission between sites, along with its processing at the local site, is used to determine a distribution strategy for a query. This strategy, referred to as Distributed

B. Liu and C. Chai (Eds.): ICICA 2011, LNCS 7030, pp. 765–772, 2011.
© Springer-Verlag Berlin Heidelberg 2011

Query Processing (DQP), aims to determine the sites and the order in which they would be used for processing the query so that the overall cost (CPU, I/O, local processing, communication) is minimized [2][3][5] [16][20][21][22].

Distributed relational queries require access to relations, replicated and fragmented across various sites of a global network, for their processing. This involves costs due to joins between the relations and due to communication between various sites involved in processing. Thus, the cost incurred in processing a distributed query can be envisioned as the sum of the communication cost incurred and the local processing cost, or the join-cost, of the participating relations. This cost needs to be minimized in order to ensure efficient query processing. This problem can be addressed by devising a strategy that generates query plans involving fewer sites with low communication cost for processing the query[6][7][10].

A query plan for relational queries involves relations and the sites used for processing the query. The number of possible query plans grows, atleast, exponentially with an increase in the number of relations accessed by the query[1][12]. For large numbers of relations, it becomes infeasible to find optimal query plans. This problem intensifies when relations, accessed by the query, have replicas at different sites. Such a large search space renders it impossible to perform an exhaustive search for optimal query plans for a given query. This problem in distributed databases is referred to as a combinatorial optimization problem[10].

Simulated annealing, random search algorithms, dynamic programming etc. [12][13][16] have been used to generate such query plans by reducing the search space based on a cost heuristic. Their efficiency is however affected by the unconventional behavior in case of some specific problems, where they may get stuck at local minima[8]. An alternate way to address this problem is to explore and exploit the search space to generate good sets of query plans using the Genetic Algorithm (GA)[18]. This paper proposes a GA based algorithm to generate top query plans for processing a distributed query. As per the algorithm, a query plan involving fewer sites would engender a lower cost of processing as compared to query plans involving a large number of sites. Further, amongst query plans with fewer sites, query plans having lower communication costs are considered more appropriate for query processing. The query plans so generated lead to efficient query processing.

The paper is organized as follows: Section 2 discusses the proposed GA based query plan generation algorithm and an example based on it is given in section 3. The experimental results are given in section 4. Section 5 is the conclusion.

2   Proposed Algorithm

As discussed above, determining a good set of query plans, from among a large number of possible query plans, is a complex problem. The proposed algorithm attempts to generate such query plans using GA by considering their local processing cost and communication cost. These costs are used to define the total cost of processing a query i.e. TC, which needs to be minimized. This total cost TC is defined as

$$TC = \sum_{i=1}^{s-1} LPC_i \times a_i + \sum_{i=1, j=i+1}^{i=s-1, j=s} CC_{ij} \times b_i$$

where LPC_i is the local processing cost per byte at site i, CC_{ij} is the communication cost per byte between sites i and j, a_i is the bytes to be processed at site i, b_i is the bytes to be communicated from site i and s is the number of sites accessed in the query plan. The approach considers the cardinality and size of single tuple in bytes of each relation in the set of relations R accessed by the query. At each site, the relations are integrated, using equi-joins, on the common attributes, which are uniformly distributed in their respective relations, to arrive at a single relation.

The cardinality of the relation arrived at after integrating relations R_p and R_q in R at site i is given by:

$$Card_i = \frac{Card(R_p) \times Card(R_q)}{Dist_{pq} \times \min(Card(R_p), Card(R_q))}$$

where $Card(R_p)$ and $Card(R_q)$ are the cardinality of relations R_p and R_q respectively and $Dist_{pq}$ is the number of distinct tuples in the smaller relation among R_p and R_q.

The size of the relation arrived at after integrating relations R_p and R_q in R at site i, determined using the formula given in [4], is

$$Size_i = Size(R_p) + Size(R_q)$$

where $Size(R_p)$ and $Size(R_q)$ are size of single tuple in bytes in relations R_p and R_q respectively

The communication between sites occur in the order from site having relation with lower cardinality to site having relation with higher cardinality for a given query plan. The communication cost CC_{ij} and local processing cost LPC_i are known. The number of bytes to be processed i.e. a_i and communicated i.e. b_i for a site i is given by $Card_i \times Size_i$.

The proposed GA based algorithm for query plan generation is given in Fig. 1.

Input: R : Relations accessed by the query, P_c : Probability of crossover,
      P_m : Probability of mutation, G : Pre-defined number of generations
Output: Top-n query plans.
Method:
    Generate a random population of query plans P_{QP}
    WHILE generation $\leq G$ DO
    Use the following fitness function TC to evaluate each individual in P_{QP}

$$TC = \sum_{i=1}^{s-1} LPC_i \times a_i + \sum_{i=1, j=i+1}^{i=s-1, j=s} CC_{ij} \times b_i$$

      where s is the number of sites accessed in the query plan
          LPC_i is the local processing cost per byte at site i
          CC_{ij} is the communication cost per byte between sites i and j
          a_i is the bytes to be processed at site i
          b_i is the bytes to be communicated from site i
    Select the query plans from P_{QP} using binary tournament selection technique
    Apply crossover on selected chromosomes with crossover probability P_c
    Apply mutation with mutation probability P_m
    Place the new population into P_{QP}
    Increment Generation by 1
    END DO
    Return Top-n Query Plans

Fig. 1. GA Based Query Plan Generation Algorithm

The algorithm takes relations in the FROM clause of the query, probability of crossover and mutation and the number of generations the algorithm would repeatedly run, as input, and produces the Top-n query plans as output. The algorithm first generates a population of valid query plans(chromosomes), each having size equal to number of relations accessed by the query. Each gene in a chromosome represents a relation and is arranged in increasing order of the corresponding relation's cardinality. The value of gene is one of the site numbers in which the corresponding relation resides. As an example, for a query accessing four relations (R1, R2, R3, R4), one of the encoding scheme for the chromosome representation can be (1, 3, 1, 2) implying that R1 and R3 are in site 1, R2 is in site 3 and R4 is in site 2. The individuals in the population are then evaluated based on the fitness function *TC*. The individuals are then selected using binary tournament selection technique[9] wherein two individuals are randomly picked from the population and the one having lower TC is selected. The selected individuals undergo random single-point crossover, with probability P_c, and mutation, with probability P_m. The resultant new population replaces the old population and the above process is repeated until it has run for G generations whereafter top-n query plans are produced as output. The algorithm can be further illustrated with help of an example given next.

3   An Example

Consider a query that accesses four relations (R1, R2, R3, R4). Suppose there are five sites (S1, S2, S3, S4, S5) available to answer the query. The relation-site matrix is shown in Fig. 2(a). Let the population of query plans P_{QP} be as shown in Fig. 2(b).

	R1	R2	R3	R4
S1	0	1	1	1
S2	1	0	0	0
S3	1	1	1	1
S4	0	1	1	1
S5	0	0	1	1

(a)

2	1	5	1	R1 in site 2, R2 in site 1, R3 in site 5, R4 in site 1
3	3	1	1	R1 in site 3, R2 in site 3, R3 in site 1, R4 in site 1
3	3	3	3	R1 in site 3, R2 in site 3, R3 in site 3, R4 in site 3
2	4	1	5	R1 in site 2, R2 in site 4, R3 in site 1, R4 in site 5

(b)

Fig. 2. Site-Relation Matrix and Population of query plans P_{QP}

The communication-cost matrix, local processing cost matrix, distinct-tuple matrix and size matrix used to computing the fitness of individual query plans are shown in Fig. 3(a). The fitness of each individual query plan in P_{QP} is computed using fitness function TC given Fig. 1. These TC values are shown in Fig. 3(b).

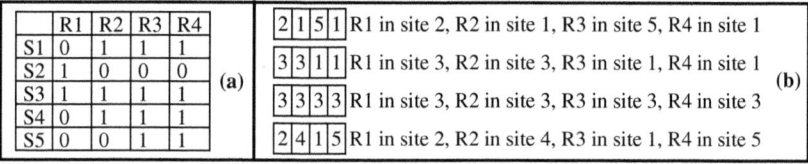

	S1	S2	S3	S4	S5
S1	-	150	180	150	160
S2	165	-	185	170	170
S3	160	160	-	170	150
S4	150	150	180	-	160
S5	170	160	190	150	-

Communication-Cost Matrix

	R1	R2	R3	R4
R1	0.4	0.5	0.4	0.4
R2	0.5	0.2	0.2	0.3
R3	0.4	0.2	0.2	0.1
R4	0.4	0.3	0.1	0.3

Distinct-Tuple-matrix for join
(as the % of smaller relation)

S1	S2	S3	S4	S5
2	2	3	3	2

Local Processing Cost Matrix

R1	R2	R3	R4
30	40	50	20

Size Matrix (a)

SNo	Query Plans	$TC = \sum\limits_{i=1}^{m-1} LPC_i a_i + \sum\limits_{i=1, j=i+1}^{i=m-1, j=m} CC_{ij} b_{ij}$
1	[2, 1, 5, 1]	21766700
2	[3, 3, 1, 1]	6006000
3	[3, 3, 3, 3]	4200000
4	[2, 4, 1, 5]	50519333

Fitness of query plans in P_{QP} (b)

Fig. 3. Matrix used for Fitness computations and TC of each query plan in P_{QP}

The binary tournament selection technique is then applied as shown in Fig. 4.

Randomly generated indexes [i] and [j]	Tournament between individual query plans [P(i)] and [P(j)]	Fitness [TC(P(i)] and [TC(P(j)]	Query Plan Selected (Lesser TC)
[1] and [4]	[2,1,5,1] and [2,4,1,5]	[21766700] and [50519333]	[2,1,5,1]
[2] and [3]	[3,3,1,1] and [3,3,3,3]	[6006000] and [4200000]	[3,3,3,3]
[1] and [3]	[2,1,5,1] and [3,3,3,3]	[21766700] and [4200000]	[3,3,3,3]
[2] and [4]	[3,3,1,1] and [2,4,1,5]	[6006000] and [50519333]	[3,3,1,1]

Fig. 4. Selection of query plans using binary tournament selection technique

The selected individuals undergo crossover, with crossover probability P_c=0.5, implying that two individuals would be selected randomly for crossover, and mutation, with mutation probability P_m=0.05, implying that atmost one bit in the population is changed. The crossover and mutation are shown in Fig. 5(a) and 5(b) respectively. The resultant new population is the population P_{QP} after mutation.

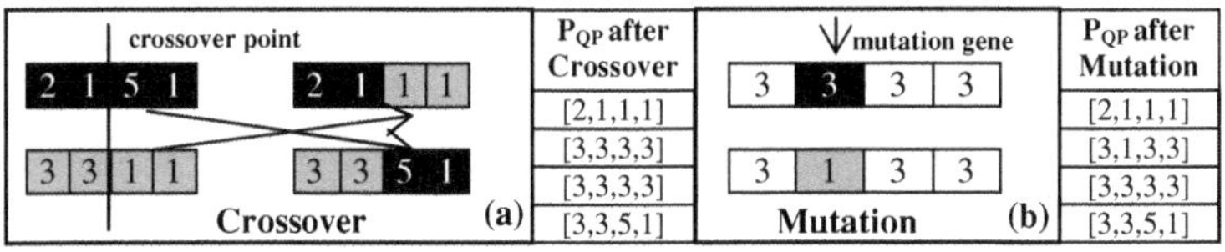

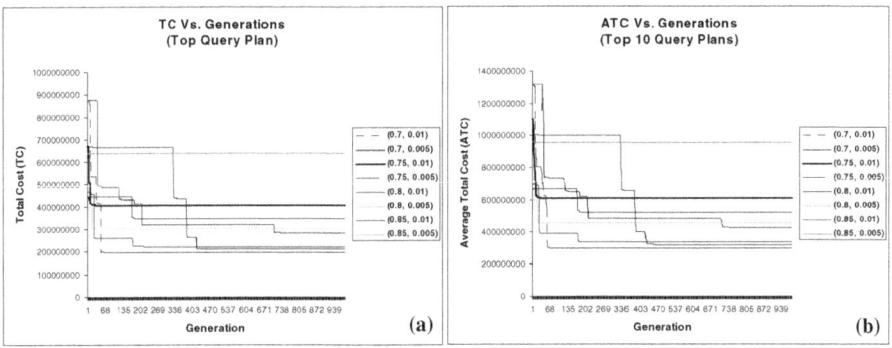

Fig. 5. Crossover and Mutation of query plans in P_{QP}

The process discussed above is repeated until it has run for G generations.

4   Experimental Results

The proposed GA based algorithm is implemented in MATLAB 7.7 in Windows 7 professional 64 bit OS, with Intel core i3 CPU at 2.13 GHz having 4 GB RAM. Experiments were carried out to ascertain minimum TC of 100 query plans with each query plan involving 10 relations distributed over 50 sites.

First, a graph is plotted to determine TC of the top query plan for a given query over 1000 generations for distinct pair of crossover and mutation probability {P_c, P_m} = {0.7, 0.01}, {0.7, 0.005}, {0.75, 0.01}, {0.75, 0.005}, {0.8, 0.01}, {0.8, 0.005}, {0.85, 0.01}, {0.85, 0.005}. This graph is shown in Fig. 6(a).

Fig. 6. TC Vs. Generation and ATC Vs. Generation for distinct values of (P_c, P_m)

It is observed from the graph that the top query plan with minimum TC was attained for P_c=0.7 and P_m=0.01. The next best result was obtained for P_c= 0.8 and P_m= 0.01. These crossover and mutation probabilities are even able to achieve minimum Average Total Cost (ATC) for generating top-10 query plans as can be observed from the graph shown in Fig. 6(b).

Next, the effectiveness of the above identified crossover and mutation probability is observed on five different data sets for generating the top query plan. The graphs are shown in Fig. 7.

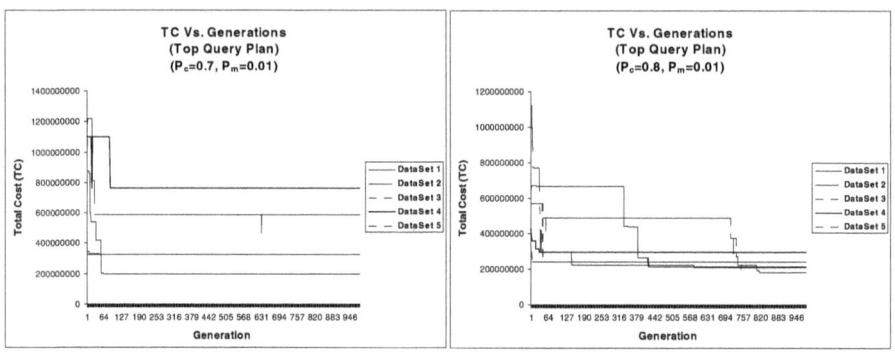

Fig. 7. TC Vs. Generation for P_c=0.7 and P_m=0.01 and P_c= 0.8 and P_m= 0.01

It is observed from the graph that for P_c= 0.8 and P_m= 0.01, in comparison to P_c=0.7 and P_m=0.01, convergence is faster to better ATC. A similar trend is also observed in the graph, shown in Fig. 8, for ATC value for top-10 query plans. This shows that P_c= 0.8 and P_m= 0.01 is able to give better results in terms of generating query plans with lesser total cost of processing a query.

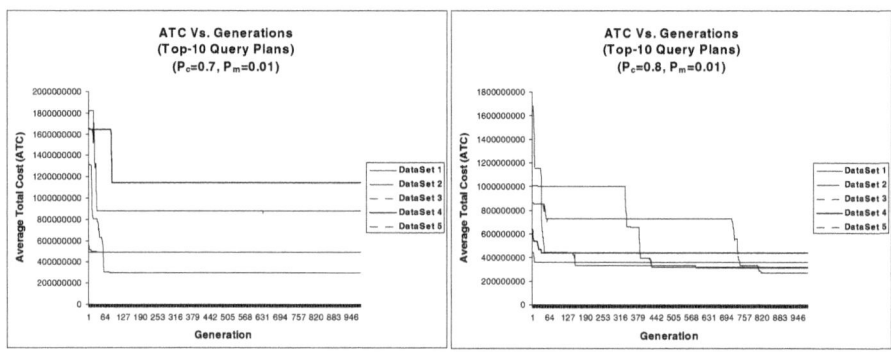

Fig. 8. ATC Vs. Generation for P_c=0.7 and P_m=0.01 and P_c= 0.8 and P_m= 0.01

Furthermore, in order to ascertain the suitability of P_c=0.8 and P_m=0.01, graphs showing ATC Vs. Generation for top-5, top-10, top-15 and top-20 query plans for P_c= 0.8 and P_m= 0.01 were plotted. These graphs, shown in Fig. 9, were plotted for 500 and 1000 generations.

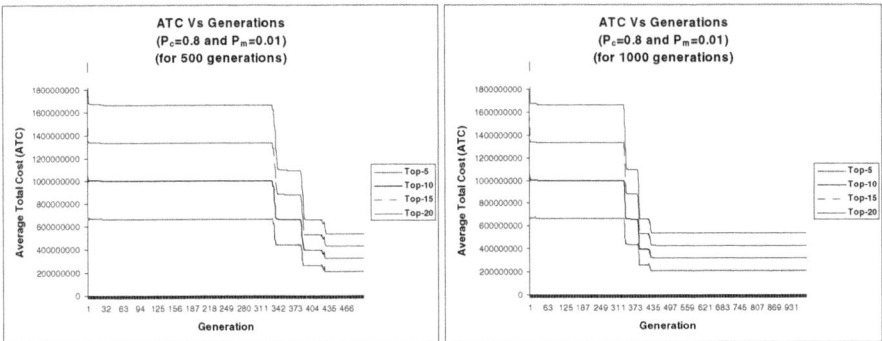

Fig. 9. ATC Vs. Generation for Top-{5, 10, 15, 20} query plans

The above graph shows that top-5, top-10, top-15 and top-20 query plans converge to their optimal ATC in less than 500 generations whereafter no further improvement was observed. This further validates the claim that $P_c = 0.8$ and $P_m = 0.01$ are the most fitting values with regard to generating query plans with minimum total cost of processing a query using the proposed algorithm.

5   Conclusion

This paper proposes an algorithm that generates top query plans for distributed queries using a simple genetic algorithm. The generation of query plans is based on the heuristic that query plans involving fewer sites are preferred over other query plans as they would incur lesser local processing and communication costs. These costs are used to define the total cost of processing a query. The proposed algorithm generates query plans that have minimum total cost of processing a query. Such query plans are likely to improve the efficiency of query processing.

Further, experiments were carried out to determine the TC value attained by the algorithm for different crossover and mutation probabilities. The results showed that the proposed GA based algorithm was able to converge quickly to a minimum TC value at an observed crossover and mutation probability. These observations may be helpful while generating optimal query plans, for a given query, using the proposed GA based algorithm.

References

1. Bennett, K., Ferris, M.C., Ioannidis, Y.E.: A Genetic Algorithm for Database Query Optimization. In: Proceedings of The Fourth International Conference on Genetic Algorithms, pp. 400–407 (1991)
2. Black, P., Luk, W.: A new heuristic for generating semi-join programs for distributed query processing. In: Proceedings of the IEEE 6th International Computer Software and Application Conference, Chicago, November 8-12, vol. III, pp. 581–588. IEEE, New York (1982)

3. Bodorik, P., Riordon, J.S.: A threshold mechanism for distributed query processing. In: Proceedings of The Sixteenth Annual Conference on Computer Science Table of Contents, Georgia, United States, pp. 616–625 (1988)
4. Ceri, S., Pelgatti, G.: Distributed Databases Principles & Systems, McGraw-Hill international edn., Computer Science Series (1985)
5. Chang, J.: A heuristic approach to distributed query processing. In: Proceedings of the 8th Internatmnal Conference on Very Large Data Bases, VLDB Endowment, Saratoga, California, pp. 54–61 (1982)
6. Chen, A.L.P., Brill, D., Templeton, M., Yu, C.T.: Distributed Query Processing in a Multiple Database System. IEEE Journal on Selected Areas in Communications I(3) (April 1989)
7. Chu, W.W., Hurley, P.: Optimal Query Processing for Distributed Database Systems. IEEE Transactions on Computers C-31(9) (september 1982)
8. Dong, H., Liang, Y.: Genetic Algorithms for Large Join Query Optimization. In: The Proceedings of GECCO 2007, London, UK, pp. 1211–1218 (2007)
9. Goldberg, D.E., Deb, K.: A comparative analysis of selection schemes used in Genetic Algorithms. In: Foundations of Genetic Algorithms, pp. 69–93. Morgan Kaufman (1991)
10. Gregory, M.: Genetic Algorithm Optimization of Distributed Database Queries. In: Proceedings of International Conference on Evolutionary Computation, pp. 71–276 (1998)
11. Hevner, A.R., Yao, S.B.: Query processing in distributed database systems. IEEE Transactions on Software Engineering SE-5, 177–187 (1979)
12. Ioannidis, Y.E., Kang, Y.C.: Randomized Algorithms for Optimizing Large Join Queries. In: Proceedings of ACM-SIGMOD Conference on Management of Data, Atlantic City, NJ, pp. 312–321 (May 1990)
13. Ioannidis, Y.E., Kang, Y.C.: Query Optimization by Simulated Annealing. In: Proceedings of the 1987 ACM-SIGMOD Conference, San Franscisco, CA, pp. 9–22 (1987)
14. Kambayashi, Y., Yoshikawa, M.: Query processing utilizing dependencies and horizontal decomposition. In: Proceedings of the ACM-SIGMOD International Conference on Management of Data, San Jose, Calif., May 23-26, pp. 55–67. ACM, New York (1983)
15. Kambayashi, Y., Yoshikawa, M., Yajima, S.: Query processing for distributed databases using generalized semijoins. In: Proceedings of the ACM-SIGMOD International Conference on Management of Data, Orlando, Fla., June 2-4, pp. 151–160. ACM, New York (1982)
16. Kossmann, D.: The State of the Art in Distributed Query Processing. ACM Computing Surveys 32(4), 422–469 (2000)
17. Liu, C., Yu, C.: Performance issues in distributed query processing. IEEE Transactions on Parallel and Distributed System 4(8), 889–905 (1993)
18. Mitchell, M.: An Introduction to Genetic Algorithm. Prentice Hall of India (1998)
19. Rho, S., March, S.T.: Optimizing distributed join queries: A genetic algorithmic approach. Annals of Operations Research 71, 199–228 (1997)
20. Stiphane, L., Eugene, W.: A State Transition Model for Distributed Query Processing. ACM Transactions on Database Systems 11(3), Pages 2 (1986)
21. Vijay Kumar, T.V., Singh, V., Verma, A.K.: Distributed Query Processing Plans Generation using Genetic Algorithm. International Journal of Computer Theory and Engineering 3(1), 38–45 (2011)
22. Yu, C.T., Chang, C.C.: Distributed Query Processing. ACM Computing Surveys 16(4), 399–433 (1984)

Author Index